D1064889

Knots 90

Knots 90

Proceedings of the International Conference
on Knot Theory and Related Topics
held in Osaka (Japan), August 15–19, 1990

Edited by

Akio Kawauchi

Walter de Gruyter · Berlin · New York 1992

Editor

Akio Kawauchi
Department of Mathematics
Osaka City University
Sugimoto, Sumiyoshi-ku
Osaka, 558, Japan

1991 Mathematics Subject Classification: 57-06; 57M25,57Q45; 81T10

⊗ Printed on acid-free paper which falls within the guidelines of the ANSI
to ensure permanence and durability.

Library of Congress Cataloging-in-Publication Data

International Conference on Knot Theory and Related Topics (1990 : Osaka, Japan)
Proceedings of the International Conference on Knot Theory and Related Topics ; held in Osaka (Japan), August 15–19, 1990 / edited by Akio Kawauchi.
p. cm.
Includes bibliographical references.
ISBN 3-11-012623-0
1. Knot theory–Congresses. I. Kawauchi, Akio, 1948–
II. Title.
QA612.2.I57 1990
514′.224–dc20 92-12201
CIP

Die Deutsche Bibliothek – Cataloging-in-Publication Data

Knots 90 : proceedings of the International Conference on Knot Theory and Related Topics held in Osaka (Japan), August 15–19, 1990 / ed. by Akio Kawauchi. – Berlin ; New York : de Gruyter, 1992
ISBN 3-11-012623-0
NE: Kawauchi, Akio [Hrsg.]; International Conference on Knot Theory and Related Topics <1990, Osaka>

Printed in Germany
Printing: Gerike GmbH, Berlin. Binding: Lüderitz & Bauer GmbH, Berlin.
Cover design: Thomas Bonnie, Hamburg

Acknowledgements

The conference was made possible by generous grants and gifts from the following foundations, corporations, and individuals. I would like to express my deepest appreciation to all of them.

Foundations. Japan World Exposition Commemorative Fund, Taniguchi Foundation, Society for the Support of Osaka City University

Corporations. Teijin, Chizaki Kogyo Co., Ltd., Citibank, N.A., Marion, Hitachi Information Systems, Ltd., NTT, NEC, Toho Mutual Life Insurance Co., Sumitomo Life Insurance Mutual Co., Mitsubishi Kasei Co., Ltd., Mitsubishi Electric Co., Ltd., Suntory, Mita Kogyo Co., Ltd.

Individuals. Hideaki Sudo, Seinosuke Yoshida, Noritaka Sasaki, Yukichi Kubo, Eiichi Kaneda, Yoshinori Shinohara, Hironobu Akagawa, Yōichi Noguchi

I would like to thank Ichiro Ikeda and Shiro Amano and the mathematicians Fujitsugu Hosokawa, Yukio Matsumoto, Takao Matumoto, Shingo Murakami, Shin'ichi Suzuki and Junzo Tao for help in negotiating financial support.

Some Japanese participants have been supported by a program represented by Mitsuyoshi Kato, of Grant-in-Aid for Co-operative Research (A), the Ministry of Education, Science and Culture. I would like to thank them for their cooperation.

Yuzo Hara, the administrative official of Mathematics Department of Osaka City University and Kook Seminar members helped me very much in preparing the conference. I would like to thank all of them, especially, Tsuyoshi Kobayashi, Yoshihiko Marumoto, Hitoshi Murakami, Makoto Sakuma, Seiichi Kamada, Masae Okada, Tatsuyuki Shiomi and Masako Kobayashi. A special thanks goes to Seiichi Kamada for help in preparing these Proceedings.

Finally, I wish to thank the organizers: Cameron M. Gordon, Jerome P. Levine, W.B. Raymond Lickorish, Kunio Murasugi and Laurence C. Siebenmann for having performed important roles, and the other participants for their contributions (in particular, to Ruth and Harry Lawrence concerning Mainichi TV Photo), without which the conference would not have succeeded so great.

January 1992 *A. Kawauchi*

Preface

The International Conference on Knot Theory and Related Topics (briefly, KNOTS 90) was held at Naniwa-Kaikan Hotel in Osaka City, Osaka, Japan from August 15 to August 19, 1990. There were 174 participants from 14 countries. The participant number by country (on the research institution basis) was as follows: Australia 3, Canada 3, France 3, Germany 1, Israel 3, Italy 1, Japan 114, Korea 2, Mexico 2, Spain 1, U.K. 3, U.S.A. 31, U.S.S.R. 6, and Yugoslavia 1.

In the organized session, three special invited speakers have delivered 20 minute lectures, and 19 invited speakers and four organizers have delivered 45 minute lectures and in the three parallel sessions, 67 speakers and one organizer have delivered 20 minute lectures. S. CAPPELL, A. FLOER and V. JONES were the special invited speakers.

A. FLOER talked about the calculation of Floer homology which is reduced to the calculation of certain groups associated to links in the 3-sphere satisfying the skein relation. Sadly, Professor Floer, who was doing important work, passed away on May 15, 1991.

The following individuals were invited speakers:

M. BOILEAU, T. COCHRAN, D. COOPER, M. FARBER, J. HILLMAN, G. JIN, L. KAUFFMAN, P. KIRK, T. KOBAYASHI, T. KOHNO, J. LUECKE, Y. NAKANISHI, K. ORR, M. SCHARLEMANN, M. THISTLETHWAITE, V. TURAEV, P. VOGEL, K. WALKER, and S. YAMADA.

The organizers of the conference were

C. GORDON, J. LEVINE, W. LICKORISH, K. MURASUGI, L. SIEBENMANN and A. KAWAUCHI, though L. SIEBENMANN did not attend the conference.

This volume is a collection of 42 (refereed) research or expository papers on Knot Theory and Related Topics, which are based on lectures delivered at the conference. A few papers by individuals associated with the conference were also included. The order in which the papers appear is the order in which they were received. Abstracts of lectures that were not published can be found in the Conference Report of the conference.

Osaka, January 1992 *Akio Kawauchi*

Table of contents

Variants on 3-manifold invariants

W.B.R. Lickorish

It was shown in [**7**] and [**8**] that some of the 3-manifold invariants introduced by Witten [**14**] (see also [**10**] and [**4**]) can be expressed in terms of the bracket polynomial of Kauffman [**2**], [**6**], and substantiated by means of the intricacies of the Temperley-Lieb algebra [**1**], [**3**]. The method was as follows. First consider an oriented 3-manifold as being defined by surgery on a framed link (L, f) in S^3. Then represent (L, f) by a planar diagram in which the writhe of each link component (that is the sum of the signs of its crossings) is the framing of that component. Evaluate, at a certain root of unity, a weighted sum of the bracket polynomials of various planar parallels of that diagram; then multiply by a 'correction' term. More details appear below. The first result of this paper is that if the diagram is instead chosen so that the writhe of the diagram of each component is the framing of that component *plus* a constant k, then essentially the same procedure also gives a 3-manifold invariant. The second result is that this invariant is independent of k so that it is in fact the same invariant as before. At least in theory, this anticlimax could be advantageous, for adding a constant to the writhe of each component can simplify considerably the parallels of certain diagrams. However the real interest in this modification lies in understanding the nature of the weighting coefficients that occur. Calculation of these coefficients for $k = 0$ is not easy. However it will be seen that, when $k = -1$, the coefficients are in principle accessible, being determined as the coefficients of a specific polynomial.

The linear skein theory associated to the bracket polynomial has been described several times [**5**] [**8**] [**9**] [**11**]; it is necessary to do so briefly again to establish notation. Let A be a non-zero complex number, fixed at least for the time being. Consider a square with i specified points on its left-hand edge and i on its right-hand edge. Let V_i be the complex vector space of formal linear sums of (planar isotopy classes of) link diagrams D, of arcs and closed curves in the square with the specified points as boundary, quotiented by

(i) $\qquad D \cup \bigcirc = \delta D \quad \text{where} \quad \delta = -A^{-2} - A^2,$

(ii) $\qquad \times = A \asymp + A^{-1} \supset\subset .$

Knots 90

As usual (i) refers to the removal of a closed curve containing no crossing at all, and (ii) refers to three diagrams identical except where shown. When a product is defined by means of juxtaposition of diagrams V_i is the ith Temperley-Lieb algebra. Not to be despised is V_0, a copy of $\mathbb{C}$ generated by the empty diagram; the element of V_0 represented by a diagram D is its Kauffman bracket $< D > \in \mathbb{C}$. More details are in [**8**]. Similarly, let $\mathcal{A}$ denote the space of link diagrams in an annulus quotiented by the same relations. In both V_i and $\mathcal{A}$, regularly isotopic diagrams represent the same vector space elements.

In the diagrams of this paper, a non-negative integer m beside a diagrammatic component will signify the presence of m copies of that component all parallel in the plane.

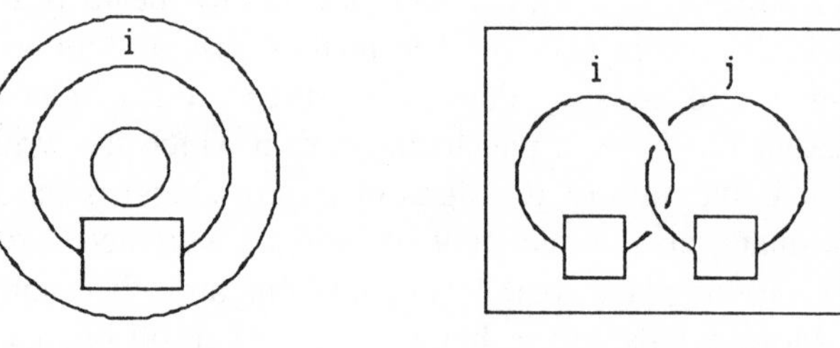

Figure 1 **Figure 2**

Consider the insertion, into the small square of Figure 1, of the diagrams in a square as defined above. That insertion induces at once a linear map $\Theta_i : V_i \to \mathcal{A}$. Insertion of diagrams into the two squares of Figure 2 similarly induces a bilinear map

$$V_i \times V_j \xrightarrow{\Phi_{i,j}} V_0 = \mathbb{C}.$$

Elements of specific interest in V_j are $1_j, z_j$, and its inverse z_j^{-1}, that are represented by the diagrams of Figure 3, (1_0 is the empty diagram).

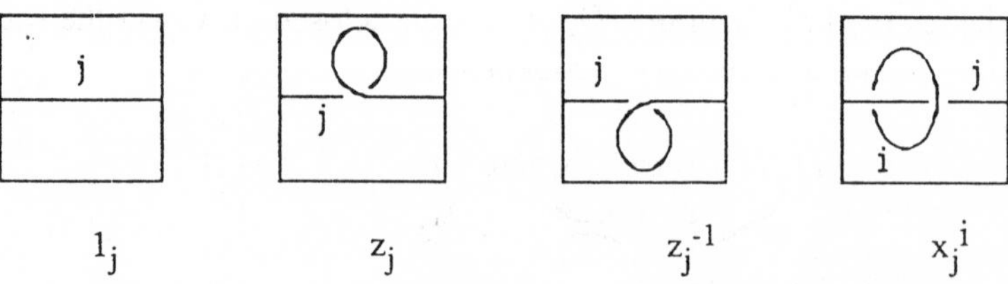

Figure 3

The element of $\mathcal{A}$ that is represented by i parallel curves encircling the annulus without a crossing will be denoted α_i. Note that $\Theta_i(1_i) = \alpha_i$. In this notation the trace of $x \in V_i$ is given by $tr(x) = \Phi_{i,0}(x, 1_0)$.

Let k be a fixed integer. A link diagram D in $\mathbb{R}^2 \cup \infty$ will be said to k–*represent* the framed link (L, f) if D is a diagram for L in the usual sense, and, denoting by $w(D_s)$ the writhe of the part of D corresponding to the component L_s, $w(D_s) = f(s) + k$ for each s. The Trace result of [**11**] implies that if framed links (L, f) and (L', f') are k-represented by diagrams D and D', then (L, f) and (L', f') are ambient isotopic in S^3 if and only if D and D' are regularly isotopic in $\mathbb{R}^2 \cup \infty$. If D is a link diagram with $D_1, D_2, \ldots, D_n$ corresponding to the link's components, and c is a function, $c : \{1, 2, \ldots, n\} \to \mathbb{Z}_+$, let $c * D$ be the diagram in which each D_s has been replaced by $c(s)$ copies all parallel *in the plane* to D_s. Let $C(n, r)$ be all functions $c : \{1, 2, \ldots, n\} \to \{0, 1, \ldots, r - 2\}$.

Theorem 1. *Let A be a primitive $4r^{\text{th}}$ root of unity for some (fixed) $r \geq 3$ and let $\delta = -A^{-2} - A^2$. Let k also be a fixed integer.*

(i) There is a unique solution $\mu_0, \mu_1, \ldots, \mu_{r-2}$ in the complex numbers to the set of linear equations

$$\sum_{i=0}^{r-2} \mu_i \Phi_{i,j}({z_i}^{k+1}, z_j) = \delta^j, j = 0, 1, \ldots, r - 2$$

(ii) Suppose that M^3 is obtained from S^3 by surgery on an n-component framed link (L, f), for which σ and ν are the signature and nullity of the linking matrix; suppose that (L, f) is k-represented by the diagram D. Then the expression

$$\kappa^{\frac{\sigma+\nu-n}{2}} \sum_{c \in C(n,r)} \mu_{c(1)} \mu_{c(2)} \cdots \mu_{c(n)} < c * D >$$

where $\kappa = \sum_{i=0}^{r-2} \mu_i tr(z_i^{k-1})$, is an invariant of the 3-manifold, a complex number independent of the choice of (L, f) or of D.

The proof of this theorem is the same as the proof for $k = 0$ given in [**8**]. It will not be repeated here. (Alternatively, from the result for $k = 0$, the theorem follows from the proof of Theorem 2.)

Theorem 2. *The above invariant does not depend upon k.*

Proof. Fix k, and let $\mu_0, \mu_1, \ldots, \mu_{r-2}$ be the coefficients defined in the theorem. Let $\lambda_0, \lambda_1, \ldots, \lambda_{r-2}$ be the similar coefficients when $k = 0$. It is clear that for some $\nu_{i,a} \in \mathbb{C}$

$$\Theta_i({z_i}^k) = \sum_{a \geq 0} \nu_{i,a} \alpha_a,$$

where $\nu_{i,a} = 0$ when $a > i$. Immersing the annulus around the circle obtained by inserting z_i into the left-hand small square of Figure 2 (and inserting z_j in the other square) shows that

$$\Phi_{i,j}({z_i}^{k+1}, z_j) = \sum_{a \geq 0} \nu_{i,a} \Phi_{a,j}(z_a, z_j).$$

This is essentially the annulus trick used in [**8**]. Thus

$$\sum_{i=0}^{r-2} \mu_i \sum_{a \geq 0} \nu_{i,a} \Phi_{a,j}(z_a, z_j) = \delta^j.$$

Hence, using the uniqueness of solution for the λ's, $\sum_{i=0}^{r-2} \mu_i \nu_{i,a} = \lambda_a$. Suppose that D is a diagram that represents the framed n-component link (L, f) and $D^{(k)}$ is D with k kinks inserted in each component, so that $D^{(k)}$ k-represents (L, f). Suppose that $c \in C(n, r)$, then, inserting an annulus around each component of D shows that

$$< c * D^{(k)} >= \sum_{a \in C(n,r)} \nu_{c(1),a(1)} \nu_{c(2),a(2)} \dots \nu_{c(n),a(n)} < a * D > .$$

So that

$$\sum_{c \in C(n,r)} \mu_{c(1)} \mu_{c(2)} \dots \mu_{c(n)} < c * D^{(k)} >= \sum_{a \in C(n,r)} \lambda_{a(1)} \lambda_{a(2)} \dots \lambda_{a(n)} < a * D > .$$

When D is the diagram of the unknot with one negative crossing this implies that

$$\sum_{i=0}^{r-2} \mu_i tr({z_i}^{k-1}) = \sum_{i=0}^{r-2} \lambda_i tr({z_i}^{-1})$$

(so that κ is independent of k) and these two equalities imply that the invariant when $k = 0$ is the same as that for all other values of k.

From now on k will be taken to be -1, and $\mu_0, \mu_1, \dots, \mu_{r-2}$ wil be the coefficients for that value of k.

Theorem 3. *Let A be a primitive $4r^{\text{th}}$ root of unity and $A^2 = e^{i\theta}$. Then, with $k = -1$ and $0 \leq s \leq r-2$, μ_s is the coefficient of t^s in the polynomial $F(t)$ where*

$$F(t) = \sum_{j=0}^{r-2} (-1)^j A^{-j^2-2j} \prod_{\substack{i=0 \\ i \neq j}}^{r-2} \frac{(t + 2\cos(i+1)\theta)}{2\cos(i+1)\theta - 2\cos(j+1)\theta}.$$

A preliminary lemma precedes the proof of this theorem. The standard base for (the vector space) of the Temperley-Lieb algebra V_i consists of all diagrams with no crossing and no closed curve. Of course, 1_i is an element of this base; $1_i{}^*$ will denote the corresponding element of the dual base, so that for any $x \in V_i$, $1_i{}^*(x)$ is the coefficient of 1_i in the expansion of x as a linear sum of the base elements.

Lemma 4. $1_i{}^*(z_i{}^{-1}) = (-1)^i A^{-i^2-2i}$.

Proof. Recall that  $= A^{-2}$  $+(1-A^4)$. Thus, as indicated in Figure 4,

$$1_i{}^*(z_i{}^{-1}) = (-A)^{-3} A^{-2(i-1)} 1_{i-1}{}^*(z_{i-1}{}^{-1}),$$

because any diagram without i arcs connecting the left edge to the right edge of the square can be disregarded. This simple recurrence, beginning with $1_0{}^*(z_0{}^{-1}) = 1$, gives $1_i{}^*(z_i{}^{-1}) = (-A)^{-3i} A^{-2\{(i-1)+(i-2)+\cdots+1\}}$.

$$1_i^*(z_i^{-1}) = A^{-2(i-1)}\, 1_i^*$$

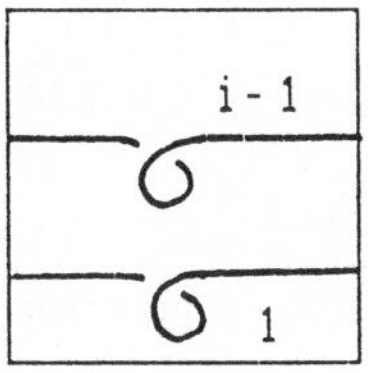

i - 1

1

Figure 4

Proof of Theorem 3. As in [**8**], let $<\ ,\ >: V_j \times V_j \to \mathbb{C}$ be the bilinear form defined by $< x, y >= tr(xy)$. When $k = -1$, the definition of the μ_i implies that

$$\sum_{i=0}^{r-2} \mu_i < 1_j, z_j x_j{}^i >=< 1_j, 1_j >, \qquad 0 \le j \le r-2.$$

The diagram of $x_j{}^i$ is included in Figure 3; note that, in the Temperley-Lieb algebra, $x_j{}^i$ is indeed the ith power of $x_j{}^1$ since regularly isotopic diagrams represent the same element of the algebra. The annulus trick of [**8**] shows that $\sum_{i=0}^{r-2} \mu_i <\ , z_j x_j{}^i >$ and $<\ , 1_j >$ are equal as maps of V_j to $\mathbb{C}$ for $0 \le j \le r-2$. But in [**8**] it was shown, using [**13**], that for $0 \le j \le r-2$ there exists an element $\mathbf{p}(j) \in V_j$ with the property that $< \mathbf{p}(j),\ >= 1_j{}^*$. Thus the equality

$$\sum_{i=0}^{r-2} \mu_i < \mathbf{p}(j) z_j{}^{-1}, z_j x_j{}^i >=< \mathbf{p}(j) z_j{}^{-1}, 1_j >$$

implies that $\sum_{i=0}^{r-2} \mu_i(1_j{}^* x_j{}^i) = 1_j{}^* z_j{}^{-1}$. But it was shown (as Corollary 7) in [**8**] that $1_j{}^* x_j{}^i = (-2\cos(j+1)\theta)^i$ and shown in Lemma 4 above that

${1_j}^* {z_j}^{-1} = (-1)^j A^{-j^2-2j}$. Thus, if $F(t) = \sum_{s=0}^{r-2} \mu_s t^s$, then $F(-2\cos(j+1)\theta) = (-1)^j A^{-j^2-2j}$ for $j = 0, 1, \ldots, r-2$. The result follows at once.

If D is a link diagram, with components $D_1, D_2, \ldots, D_n$, (-1)-representing M^3 and $c \in C(n,r)$, let $< c * D >$ be denoted by the monomial ${D_1}^{c(1)} {D_2}^{c(2)} \ldots {D_n}^{c(n)}$. Then the invariant of M^3 can be written $\kappa^{\frac{\sigma+\nu-n}{2}} F(D_1)F(D_2)\ldots F(D_n)$.

Theorems 1 and 2 can be interpreted somewhat differently in the following way. Suppose the n-component link diagram D represents a framed link (L, f), and consider the n-fold product $\mathcal{A} \times \mathcal{A} \times \cdots \times \mathcal{A}$. Let $\Phi_D : \mathcal{A} \times \mathcal{A} \times \cdots \times \mathcal{A} \to V_0 \equiv \mathbb{C}$ be induced by taking an n-tple of diagrams in the annulus and inserting them around the n annular neighbourhoods of the components of D. This gives a well-defined multi-linear map. Then, if $\Delta : \mathcal{A} \to \mathcal{A} \times \mathcal{A} \times \cdots \times \mathcal{A}$ is the diagonal map,

$$\Phi_D \Delta(\sum_{i=0}^{r-2} \lambda_i \alpha_i) = \sum_{a \in C(n,r)} \lambda_{a(1)} \lambda_{a(2)} \ldots \lambda_{a(n)} < a * D > .$$

Thus the element $(\sum_{i=0}^{r-2} \lambda_i \alpha_i)$ of $\mathcal{A}$ is the essence of the 3-manifold invariant. Now $\{\alpha_i : i = 0, 1, 2, \ldots\}$ and, for fixed k, $\{\Theta_i({z_i}^k) : i = 0, 1, 2, \ldots\}$ are two bases of $\mathcal{A}$; they are related, as already described by the matrix $\{\nu_{i,j}\}$. Then, simply changing coordinates one sees that $\sum_{i=0}^{r-2} \lambda_i \alpha_i = \sum_{i=0}^{r-2} \mu_i \Theta_i({z_i}^k)$ and clearly

$$\Phi_D \Delta(\sum_{i=0}^{r-2} \mu_i \Theta_i({z_i}^k)) = \sum_{c \in C(n,r)} \mu_{c(1)} \mu_{c(2)} \ldots \mu_{c(n)} < c * D^{(k)} > .$$

Of course, any base of $\mathcal{A}$ can be used in this way. In **[8]** there is described a sequence of elements $f_0, f_1, \ldots, f_{i-1}$ (due to Jones and Wenzl) in the Temperley-Lieb algebra V_i that have seemingly magical properties! Let this f_{i-1} be renamed $f^{(i)}$, so that $f^{(i)}$ is the final element in the sequence of f_j's in V_i. Then $\{\Theta_i(f^{(i)}) : i = 0, 1, 2, \ldots\}$ is another base of A. With respect to this base it is *comparatively* easy to calculate the coordinates of $\sum_{i=0}^{r-2} \lambda_i \alpha_i$ thus giving a tangible expression for the 3-manifold invariant. That expression agrees (with the possible exception of some sign conventions) with the expression in **[4]** derived from the quantum group approach of **[10]**. In particular, allocation of the $\Theta_i(f^{(i)})$ to link components followed by evaluation of the Kauffman bracket gives the same framed link invariant as allocation of irreducible representations of $SU(2)_q$ to those components followed by evaluation of the Reshetikhin-Turaev invariant. Details of this will appear later. It is interesting to note that Yamada **[15]** has used these same $\{f^{(i)}\}$ in an analogous way in the construction of invariants for graphs in S^3.

References

[1] R.J. Baxter, "Exactly solved models in statistical mechanics," (chapter 12), Academic Press, 1982.

[2] L.H. Kauffman, *State models and the Jones polynomial,* Topology **26** (1987), pp. 395–407.

[3] L.H. Kauffman, *Statistical mechanics and the Jones polynomial,* in "Braids," Ed. J.S. Birman and A. Libgober, Contemp. Math. **78** (1988), pp. 263–297.

[4] R. C. Kirby and P. Melvin, *Evaluations of the 3-manifold invariants of Witten and Reshetikhin-Turaev for* $\mathbf{sl}(2, \mathbf{C})$, Invent. Math. **105** (1991), pp. 473–545.

[5] W.B.R. Lickorish, *Linear skein theory and link polynomials,* Topology and its Applications **27** (1987), pp. 265–274.

[6] W.B.R. Lickorish, *Polynomials for links,* Bull. London Math. Soc. **20** (1988), pp. 558–588.

[7] W.B.R. Lickorish, *Three-manifold invariants from the combinatorics of the Jones polynomial,* Pacific J. Math. **149** (1991), pp. 337–347.

[8] W.B.R. Lickorish, *Three manifolds and the Temperley-Lieb algebra,* Math. Ann. (to appear).

[9] J.H. Przytycki, *Skein modules of 3-manifolds,* Bull. Pol. Acad. Sci. Math. (to appear).

[10] N.Y. Reshetikhin and V.G. Turaev, *Invariants of 3-manifolds via link polynomials and quantum groups,* Invent. Math. **103** (1991), pp. 547–597.

[11] B. Trace, *On the Reidemeister moves of a classical knot,* Proc. Amer. Math. Soc. **89** (1983), pp. 722–724.

[12] V.G. Turaev, *The Conway and Kauffman modules of the solid torus,* LOMI preprint, Leningrad (1988).

[13] H. Wenzl, *On sequences of projections,* C.R. Math. Rep. Acad. Sci. Canada **IX** (1987), pp. 5–9.

[14] E. Witten, *Quantum field theory and Jones' polynomial,* Comm. Math. Phys. **121** (1989), pp. 351–399.

[15] S. Yamada, *A topological invariant of spatial graphs,* these proceedings.

Fibered links which are band connected sum of two links

*Tsuyoshi Kobayashi**

Dedicated to Professor Junzo Tao on his sixtieth birthday

1. Introduction

Let L be an oriented link in the 3-sphere S^3. Let I be the unit interval $[0,1]$, and $b : I \times I \to S^3$ an embedding such that $b^{-1}(L) = \partial I \times I$, where an orientation on $b(\partial(I \times I))$ is coherent to that of L. Then let $L_b = ((L_1 \cup L_2) - b(\partial I \times I)) \cup b(I \times \partial I)$. L_b is a link and has the orientation coherent to that of L. We say that the oriented link L_b is obtained from L by a *band operation* (along the band b). We say that L is a *split union* of two links L_1 and L_2 if $L = L_1 \cup L_2$ and there exists a 2-sphere in S^3 which separates L_1 and L_2. If L is a split union of L_1 and L_2, then we say that L is a *split link* and we denote the link L by $L_1 \circ L_2$. A link L' is called a *band connected sum* of L_1 and L_2 if L' is obtained from a split link $L_1 \circ L_2$ by a band operation along a band b such that $b(I \times I) \cap L_1 \neq \emptyset$, and $b(I \times I) \cap L_2 \neq \emptyset$. The link L' is a connected sum of L_1 and L_2 if there is a 2-sphere Q which separates L_1 and L_2 and $Q \cap b$ an arc of the form $b(\{*\} \times I)$.

Let L be an oriented link in a 3-manifold M. A *Seifert surface* S for L is a compact oriented surface, without closed components, such that $\partial S = L$. Then $\chi(L)$ denotes the maximal Euler characteristic of all Seifert surfaces for L. We say that a Seifert surface S for L is of *minimal genus* if $\chi(S) = \chi(L)$. L is a *fibered link* if the exterior $E(L)$ of L is a surface bundle over S^1 such that a Seifert surface represents a fiber. An oriented surface F in M is a *fiber surface* if ∂F is a fibered link and $F \cap E(\partial F)$ is a fiber of a fibration of $E(\partial F)$.

* This work was supported by Grant-in Aid for Scientific Research, The ministry of Education, Science and Culture

Knots 90

Then the first result of this paper is the following theorem which is essentially due to Scharlemann-Thompson [**10**].

Theorem 1. *Let L be a link in S^3, which is a band connected sum of two links L_1 and L_2 along a band b. Then we have $\chi(L) \leq \chi(L_1) + \chi(L_2) - 1$. Moreover the equality holds if and only if there are mutually disjoint minimal genus Seifert surfaces S_1, S_2 for L_1, L_2 respectively such that $S_i \cap b(I \times I) = L_i \cap b(I \times I)$ $(i = 1, 2)$.*

Theorem 1 is a link version of the superadditivity of genera of knots under band connected sum, which was proved by D. Gabai [**3**], and M. Scharlemann [**9**]. We prove Theorem 1 in section 3. In sections 4 and 6, we investigate fibered links which are band connected sum of two links. In particular, we show:

Theorem 2. *Let L be a fibered link in S^3, which is a band connected sum of two links L_1 and L_2. Then we have:*

(1) L_1 and L_2 are fibered links,

(2) $\chi(L) = \chi(L_1) + \chi(L_2) - 1$ holds if and only if the band connected sum is a connected sum, and

(3) if $\chi(L) < \chi(L_1) + \chi(L_2) - 1$, then $\chi(L) \leq \chi(L_1) + \chi(L_2) - 5$.

In [**7**] we introduced the notion of pre-fiber surfaces. In section 5, we give a characterization of pre-fiber surfaces bounding a split link $L_1 \circ L_2$ (Theorem 3). By using this together with a result in [**7**], we give a characterization of bands for $L_1 \circ L_2$ which produce fibered links in section 6. Let K be a knot in S^3 bounding a 4-ball D^4. The *genus* of K, $g(K)$, is the minimal genus of all Seifert surfaces of K. The *4-genus* of K, $g^*(K)$, is the minimal genus of all oriented, connected, compact surfaces G in D^4 such that $\partial G = K$. As a consequence of Theorem 2 (3), we immediately have:

Corollary. *Let K be a fibered knot which is a band connected sum of two knots K_1 and K_2. Suppose that K is prime, and the band connected sum is non-trivial, i. e. $K \neq K_1, K_2$. Then we have $g(K_1) + g(K_2) \leq g(K) - 2$. In particular, $g^*(K) \leq g(K) - 2$ holds.*

2. Preliminaries

Throughout this paper, we work in the piecewise linear category, all manifolds, including knots, links, Seifert surfaces, are oriented, and all submanifolds are in general position unless otherwise specified. For the definitions of standard terms of 3-dimensional topology, knot and link theory, we refer to [**5**], [**6**], and [**8**]. For a topological space B, $\sharp B$ denotes the number of the components of B. Let H be a subcomplex of a complex K. Then $N(H; K)$ denotes a regular neighborhood of H in K.

For the definitions of a *sutured manifold*, (M, γ), *product sutured manifold*, *product disk*, and the notation $R_{\pm}(\gamma)$ in this context, we refer to [**7**]. Let L be a link in a 3-manifold M. The *exterior* $E(L)$ of L is the closure of the complement of $N(L; M)$. A *meridian loop* for L is a non-contractible simple loop in $\partial E(L)$, which bounds a disk in $N(L; M)$. Let S be a Seifert surface for L. Then we often abbreviate $S \cap E(L)$ to S. For the definitions of a *sutured manifold obtained from* S, and a *complementary sutured manifold for* S, we refer to [**7**]. Then we recall some fundamental results.

Proposition 2.1. *A surface $S(\subset S^3)$ is a fiber surface if and only if the complementary sutured manifold is a product sutured manifold.*

Proposition 2.2 [7]. *Let $S(\subset S^3)$ be a Seifert surface for a fibered link L. Then the following three conditions are equivalent.*
(1) S is a fiber surface.
(2) S is a minimal genus Seifert surface for L.
(3) S is incompressible in $E(L)$.

We recall some definitions from [**7**]. Let S be a fiber surface. There is an orientation preserving self-homeomorphism φ of S such that $\varphi|_{\partial S} = \mathrm{id}_{\partial S}$, and $E(L)$ is homeomorphic to $S \times I/\sim$, where $(x, 1) \sim (\varphi(x), 0)$ $(x \in S)$. φ is called a *monodromy map*. Then $\partial S \times I$ has an I-bundle structure such that each fiber projects to a meridian loop of L. Let $p : S \times I \to E(L)$ be a natural map, $D(\subset S \times I)$ a product disk for the product sutured manifold $(S \times I, \partial S \times I)$ such that $D \cap (\partial S \times I)$ is a union of two fibers of the above fibration of $\partial S \times I$. Then the 2-complex $\square = p(D)$ is called a *projected product disk* (or *pp disk* for short). For the pp disk $\square$, $\partial_-\square$, $\partial_+\square$ denote $p(D \cap (S \times \{0\}))$, $p(D \cap (S \times \{1\}))$ respectively (hence, $\partial_+\square$ corresponds to $\varphi(\partial_-\square)$). Suppose that there is an ambient isotopy f_t $(0 \leq t \leq 1)$ of $S \times I$ such that $f_0 = \mathrm{id}$, $f_t(D) \cap (\partial S \times I)$ is a union of two fibers. Then we say that the pp disk $\square' = p(f_1(D))$ is isotopic to $\square$ by an isotopy as a pp disk.

Let S be a *connected* surface in a 3-manifold M, and (N^c, δ^c) a complementary sutured manifold for S. S is a *pre-fiber surface* if there exist pairwise disjoint compressing disks D^+, D^- $(\subset N^c)$ for $R_+(\delta^c)$, $R_-(\delta^c)$ respectively such that $(\bar{N}, \delta^c)$ is homeomorphic to a (not necessarily connected) product sutured manifold, where $\bar{N}$ denotes the manifold obtained from N^c by cutting along $D^+ \cup D^-$. Then there is a pair of compressing disks $\bar{D}^+$, $\bar{D}^-$ for S such that $\bar{D}^\pm \cap N^c = D^\pm$. We say that $\bar{D}^+$, $\bar{D}^-$ is a *pair of canonical compressing disks* for S. We say that the pre-fiber surface S is of *type 1* (*type 2* resp.) if $\bar{N}$ is connected (disconnected resp.). For fundamental properties of a pre-fiber surface and a pair of canonical compressing disks, see [**7, Sect.4**].

3. Proof of Theorem 1

In this section, we prove Theorem 1. We assume that the reader is familiar with [**4**], [**9**], and [**10**]. Let D be a disk in S^3 such that $D \cap (b(I \times I) \cup (L_1 \circ L_2)) = b(\{1/2\} \times I)$, and $D \subset N(b(\{1/2\} \times I); S^3)$. Let S be a minimal genus Seifert surface for L in $E(\partial D)$.

By pushing out circle components from D if necessary, we may suppose that $S \cap D$ consists of an arc joining the two points $L \cap D$. By regarding $N(S \cap D; S)$ as $b(I \times I)$, we may suppose that $b(I \times I) \subset S$. Then the main result of this section is the following proposition which is essentially included in [**10**].

Proposition 3.1. *S is a minimal genus Seifert surface for L in S^3.*

Proof of Proposition 3.1.

Claim 3.1. *If there exists a 2-sphere Q in S^3, which separates L and ∂D, then S is a minimal genus Seifert surface for L in S^3.*

Proof. Let B^3 be the closure of the component of $S^3 - Q$ which contains ∂D. Since ∂D is a trivial knot, it is easy to see that there is a disk D_1 in B^3 such that $\partial D_1 = \partial D$. Then, by applying standard innermost disk argument, we may suppose that $S \cap D_1 = \emptyset$. Hence, it is clear that S is a minimal genus Seifert surface for L in S^3.

Claim 3.2. *If $E(L \cup \partial D)$ is irreducible, then S is a minimal genus Seifert surface for L in S^3*

Proof. We divide the proof of Claim 3.2 into two cases.

Case 1. *$E(L)$ is $S_{\partial D}$-atoroidal.*

In this case, by the argument of Case 2 of the proof of [**10,1.14**], we have either (1) S is a minimal genus Seifert surface for L in S^3, or (2) $E(L_1 \circ L_2)$ is irreducible and the surface obtained from S by cutting along $b(\{1/2\} \times I)$ is a minimal genus Seifert surface for $L_1 \circ L_2$. We note that the conclusion (2) cannot occur, because $E(L_1 \circ L_2)$ is reducible. Hence S is a minimal genus Seifert surface for L.

Case 2. *$E(L)$ is not $S_{\partial D}$-atoroidal.*

In this case there exists a non-boundary parallel torus T in $E(L \cup \partial D)$ such that T separates $E(\partial D)$ into V_1 and V_2 with $\partial E(\partial D) \subset V_1$, $S \subset V_2$, and $i_* : H_1(T) \to H_1(V_1)$ is injective. We note that since $E(L \cup \partial D)$ is irreducible, T is incompressible in $E(L \cup \partial D)$. Let $\bar{V}_1 = V_1 \cup N(\partial D; S^3)$, and (N^c, δ^c) a complementary sutured manifold for S in S^3.

Subclaim. *$\bar{V}_1$ is homeomorphic to the exterior of a non-trivial knot.*

Proof. Assume that $\bar{V}_1$ is not homeomorphic to the exterior of a non-trivial knot. Then, by **[8,107p.]**, we see that $\bar{V}_1$ is homeomorphic to a solid torus. Since $E(L \cup \partial D)$ is irreducible, we may suppose that $T \cap D$ consists of a number of parallel simple loops which are essential in the annulus $D \cap N^c$ and T. Let l be the component of $T \cap D$ which is innermost in D, and D' the closures of the components of $D - l$, which is an open disk. Since $T \cap D$ consists of mutually parallel simple loops in T, we see that there is an annulus A $(\subset \bar{V}_1)$ such that one boundary component of A is ∂D and $A \cap \partial\bar{V}_1 = \partial A \cap \bar{V}_1 = l$. On the other hand, we note that by attaching $N(D'; S^3 - \mathrm{Int}\bar{V}_1)$ to $\bar{V}_1$ along l, we get a 3-cell by Schoenflies Theorem **[8,34p.]**. This means that ∂D is a core curve of the solid torus $\bar{V}_1$. But this contradicts the fact that T is not boundary parallel in $E(L \cup \partial D)$.

By Subclaim, and the argument of the proof of Subclaim b in the proof of **[10,1.4]**, we see that S is a minimal genus Seifert surface for L in S^3.

This completes the proof of Claim 3.2.

Completion of the proof. By Claims 3.1 and 3.2, we may suppose that $E(L \cup \partial D)$ is reducible and there does not exists a 2-sphere which separates L and ∂D. Then, by using the unique factorization theorem **[5,3.21]**, we see that there is a connected sum decomposition $E(L \cup \partial D) = M_0 \sharp M_1$ along a 2-sphere P such that $\partial N(\partial D; S^3) \subset M_1$, and M_1 is irreducible. Let $\bar{M}_1$ be the closure of the component of $S^3 - P$ corresponding to M_1, and L' the sublink of L which are contained in $\bar{M}_1$. Then $E(L' \cup \partial D)$ is homeomorphic to M_1. Since S is a minimal genus Seifert surface, we may suppose that $S \cap P = \emptyset$ by applying standard innermost disk argument if necessary. Then let S' be the union of the components of S which are contained in $\bar{M}_1$. By Claim 3.2, we see that S' is a minimal genus Seifert surface for L' in S^3. Hence it is clear that S is a minimal genus Seifert surface for L in S^3.

This completes the proof of Proposition 3.1.

Proof of Theorem 1. Let S^* be the surface obtained from S by cutting along $b(\{1/2\} \times I)$ (hence we have $\chi(S^*) = \chi(S) + 1$). Since S^* is a Seifert surface for

$L_1 \circ L_2$ we see that $\chi(L_1)+\chi(L_2)$ $(=\chi(L_1 \circ L_2)) \geq \chi(S^*)$. By Proposition 3.1, we have $\chi(S) = \chi(L)$. The above equalities and inequality imply $\chi(L) \leq \chi(L_1) + \chi(L_2) - 1$.

Suppose that $\chi(L) = \chi(L_1) + \chi(L_2) - 1$ holds. Then S^* is a minimal genus Seifert surface for $L_1 \circ L_2$. It is easy to see that S^* is a union of minimal genus Seifert surfaces for L_1 and L_2.

This completes the proof of Theorem 1.

4. Case of $\chi(L) = \chi(L_1) + \chi(L_2) - 1$

In this section, we prove Theorem 2 (2), and Theorem 2 (1) in case of $\chi(L) = \chi(L_1) + \chi(L_2) - 1$. In fact, we show:

Proposition 4.1. *Let L, L_1, L_2 be as in Theorem 2. Then $\chi(L) = \chi(L_1)+\chi(L_2)-1$ if and only if both L_1 and L_2 are fibered links and the band connected sum is a connected sum.*

The 'if' part of Proposition 4.1 is well known. Hence we give a proof of 'only if' part in the rest of this section. Let D, S be as in section 3. By Proposition 3.1, S is a minimal genus Seifert surface for L in S^3, so that S is a fiber surface by Proposition 2.2. Hence S is connected (see **[7,Lemma 2.1]**). Let Q be a 2-sphere in S^3 which separates L_1 and L_2. Then, by using standard innermost circle and outermost arc arguments, we may suppose that each component of $Q \cap b(I \times I)$ is an arc of the form $b(\{*\} \times I)$. Let $a_1, \cdots, a_n$ be the components of $Q \cap b(I \times I)$, which are on $b(I \times I)$ in this order. Clearly n is an odd integer. We may suppose that $a_i = b(\{i/(n+1)\} \times I)$. Let β be a component of the frontier of $N(b(I \times I); S)$ in S, $p : S \times I \to E(L)$ a natural map, $p(D) = \square$ a pp disk for S such that $\partial_- \square = \beta$, and $\varphi : S \to S$ a monodromy map. By the assumption, $\chi(L) = \chi(L_1) + \chi(L_2) - 1$, we see that β separates S. Assume that, by deforming φ by an isotopy rel. ∂, we can make $\varphi(\beta) \cap \beta = \partial\beta$. Then, by **[7,4.6]**, we see that there is a 2-sphere S^2 in S^3 such that $S^2 \cap S = \beta$, so that it is easy to see that the band connected sum is a connected sum. Hence, hereafter, we assume:

4.1 $\varphi(\beta)$ $(\subset S)$ is not rel. ∂ isotopic to an arc which does not intersect with Intβ.

We may suppose that each component of $\partial_+ \square \cap b(I \times I)$ is an arc of the form $b(I \times \{*\})$. Since S is connected, we may suppose that no component of $\square \cap Q$ is a simple loop by using standard innermost circle arguments to $\square$ if necessary. Hence each component of $\square \cap Q$ is an arc whose endpoints are contained in $\partial_+ \square$.

We get a planar tree T by corresponding each component of $D-(D\cap p^{-1}(Q))$ to a vertex, and each component of $D\cap p^{-1}(Q)$ to an edge. We regard that T is embedded in D and each edge of T intersects $D\cap p^{-1}(Q)$ in one point which is contained in the corresponding component of $D\cap p^{-1}(Q)$. Let γ be a component of $D\cap p^{-1}(Q)$, and e_γ the edge of T corresponding to γ. Then $\gamma\cap e_\gamma$ consists of a point, which separates γ into two arcs γ_1 and γ_2. One endpoint of γ_i lies in $\cup_{j=1}^{n} a_j$. Labell the corresponding side of e_γ by k if the endpoint lies in a_k. Then we can labell each side of the edges of T by using $\{1,\cdots,n\}$. We say that the vertex of T which corresponds to the component of $D-p^{-1}(Q)$ containing $D\cap R_+(\delta^c)$ is the *root* of T. For the definitions of an outermost vertex, outermost edge, and an outermost fork in this context, we refer to **[7,721p.]**.

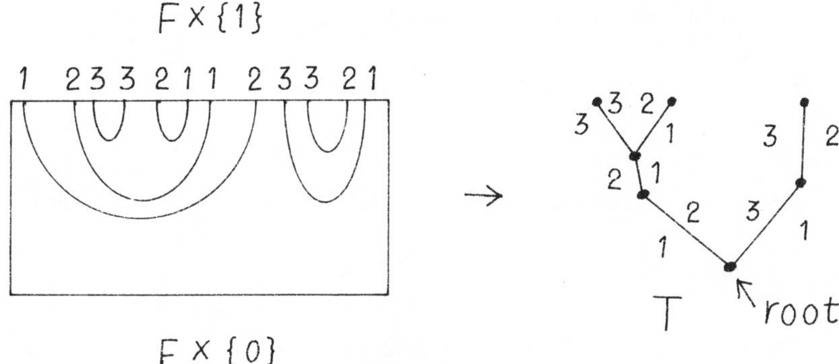

Figure 4.1

Claim 4.1. *Suppose that $n\geq 3$, and that there is an outermost edge e such that the root is not contained in ∂e, and e is labelled by i and $i+1$ for some $i\in\{1,2,\cdots,n-1\}$. Then there is a 2-sphere Q' in S^3 such that Q' separates L_1 and L_2, and $b(I\times I)\cap Q'$ consists of $n-2$ arcs of the form $b(\{*\}\times I)$.*

Proof. Let Δ be the closure of the component of $D-p^{-1}(Q)$ corresponding to the outermost vertex contained in ∂e, and $\alpha=b([i/(n+1),(i+1)/(n+1)]\times I)$. Then $N(Q\cup\alpha\cup\Delta;S^3)$ is homeomorphic to (2-sphere) $\times I$. It is clear that one boundary component of $N(Q\cup\alpha\cup\Delta;S^3)$ satisfies the conclusion.

Now we suppose the following minimality on b.

4.2 $\sharp\{Q\cap b(I\times I)\}$ is minimal among all bands rel.$L_1\circ L_2$ ambient isotopic to b.

Let $p(D)=\square$ be a pp disk as above. Then we suppose the following minimality on $\square$.

4.3 Each component of $\partial_+\Box \cap b(I \times I)$ is an arc of the form $b(I \times \{*\})$, and the number of the components is minimal among the rel.∂_- isotopy class of $\Box$ as a pp disk.

Recall that $n = \sharp\{Q \cap b(I \times I)\}$. If $n = 1$, then we have the conclusion of Proposition 4.1. Hence we assume $n \geq 3$. By the minimality (4.2) and Claim 4.1, we see that each outermost edge of T which does not contain the root is labelled by two 1's or n's. Then, by the argument of section 2 of [**11**], we see that for each a_i there is a component of $D \cap p^{-1}(Q)$, say α_i, such that $\partial p(\alpha_i) \subset a_i$. Let β_i be the subarc of a_i such that $\partial \beta_i = \partial p(\alpha_i)$. Then take α_j such that $\partial p(\alpha_j) \cup \beta_j$ is innermost, i. e. $p(\alpha_j) \cup \beta_j$ bounds a disk B in Q such that Int$B$ does not contain any a_i. Let Δ be the closure of the component of $D - \alpha_j$ such that $\Delta \cap (S \times \{0\}) = \emptyset$. Now we have the following two cases.

Case 1. *B does not contain ∂a_j.*

Let $G = B \cup p(\Delta)$. G is a disk in S^3 and $G \cap S = \partial G$. Moreover, by the minimality (4.3), we see that ∂G is an essential loop in S, a contradiction.

Case 2. *B contains ∂a_j.*

In this case, $\Box^* = B \cup p(\Delta)$ is a pp disk for S such that $\partial_-\Box^* = a_j$, and $\partial_+\Box^* = (a_j - \beta_j) \cup (p(\Delta) \cap S)$. By moving $\Box^*$ by a tiny isotopy as a pp disk rel. $\partial_-\Box^*$, we get a pp disk $\Box^{**}$ such that $\partial_-\Box^{**} \cap \partial_+\Box^{**} = a_j \cap \partial_+\Box^{**} = \partial a_j$. Since a_j is parallel to β in S, we see that $\varphi(\beta)$ $(\subset S)$ is rel. ∂ isotopic to an arc which does not intersect with Intβ, contradicting (4.1).

Hence we have $n = 1$ and this completes the proof of Proposition 4.1.

5. Pre-fiber surfaces for split links

For the statement of the result of this section, we prepare some notations. Let S be a 2-sided surface in a 3-manifold M, and $a(\subset M)$ an arc such that $a \cap S = \partial a(\subset$ Int $S)$, and the components of $N(\partial a; a)$ lie in one side of S. Then $\partial N(a; M) \cap S$ consists of two circles. Let A be the closure of the component of $\partial N(a; M) - S$, which is an open annulus. Then $S_a = (S -$ Int $N(a; M)) \cup A$ is a surface, and has the orientation coherent to that of S. We say that S_a is obtained from S by adding a *pipe* along a. Let $\mathcal{S} = S_1 \cup S_2$ be a split union of two surfaces S_1, S_2 in S^3, i. e. there is a 2-sphere in S^3 which separates S_1 and S_2. Let α be an arc in S^3 such that $\alpha \cap \mathcal{S} = \partial \alpha$ $(\subset$ Int $\mathcal{S})$, and the components of $N(\partial \alpha; \alpha)$ lie in one side of $\mathcal{S}$. We say that α is a *trivial arc* (with respect to S_1 and S_2) if there is a

2-sphere Q such that Q separates S_1 and S_2, and $Q \cap \alpha$ consists of one point. Let β be another arc which is trivial with respect to S_1, S_2. It is easy to see that $\mathcal{S}_\alpha$ and $\mathcal{S}_\beta$ are mutually isotopic.

Then the main result of this section is:

Theorem 3. *Let L be a link which is a split union of two links L_1, L_2, i. e. $L = L_1 \circ L_2$. Suppose that there is a pre-fiber surface S^* such that $\partial S^* = L$. Let $n = (\chi(L) - \chi(S^*))/2$. Then we have:*
(1) L_1 and L_2 are fibered links,
(2) S^* is of type 2 if and only if $n = 1$, and S^* is ambient isotopic to a surface obtained from $S_1 \cup S_2$ by adding a pipe along a trivial arc, where S_i is a fiber surface with $\partial S_i = L_i$ $(i = 1, 2)$. Moreover a pair of canonical compressing disks D^+, D^- for S^* looks as Figure 5.1,
(3) if S^* is of type 1, then S^* is ambient isotopic to a surface obtained from $S_1 \cup S_2$ by adding a pipe along a trivial arc, then adding $n - 1$ pipes successively as in Figure 5.1. Moreover a pair of canonical compressing disks D^+, D^- for S^* looks as Figure 5.1.

Conversely if S^* is obtained from a split union of two fiber surfaces by adding a number of pipes as above, then S^* is a pre-fiber surface.

Proof. Let D^+, D^- be a pair of canonical compressing disks for S^*. First, suppose that S^* is of type 2. Then ∂D^- separates S^*. Let S_1, S_2 be the components of the surface obtained from S^* by doing surgery along D^-. Then we have:

Claim 5.1. *S_1, S_2 are fiber surfaces, and by exchanging suffix if necessary, we have $\partial S_i = L_i$ $(i = 1, 2)$. Moreover S^* is ambient isotopic to a surface obtained from $S_1 \cup S_2$ by adding a pipe along a trivial arc, and a pair of canonical compressing disks for S^* looks as in Figure 5.1.*

Proof. Let $F = S_1 \cup S_2$, (M^c, γ^c) a complementary sutured manifold for S^*, and $B^\pm = D^\pm \cap M^c$. Namely $B^\pm$ are compressing disks for $R_\pm(\gamma^c)$, and $(\bar{M}^c, \gamma^c)$ is homeomorphic to $((S_1 \cup S_2) \times I, \partial(S_1 \cup S_2) \times I)$, where $\bar{M}^c = \mathrm{cl}(M^c - N(B^+ \cup B^-; M^c))$. Then we may identify S_1, S_2 to the surface corresponding to $S_1 \times \{1/2\}$, $S_2 \times \{1/2\}$, and we regard $((S_1 \cup S_2) \times [1/4, 3/4], \partial(S_1 \cup S_2) \times [1/4, 3/4])$ as the sutured manifold obtained from F, which is denoted by (N, δ). Let (N^c, δ^c) be the complementary sutured manifold for F, and C^+, C^- the closures of the components of $N^c - S^*$ with $(S_1 \cup S_2) \times \{3/4\} \subset C^+$, $(S_1 \cup S_2) \times \{1/4\} \subset C^-$. Since $C^\pm$ is homeomorphic to $((S_1 \cup S_2) \times I) \cup$ (1-handle), (C^+, C^-) gives a Heegaard splitting of (N^c, δ^c) in terms of Casson-Gordon **[C-G]**. Then hereafter

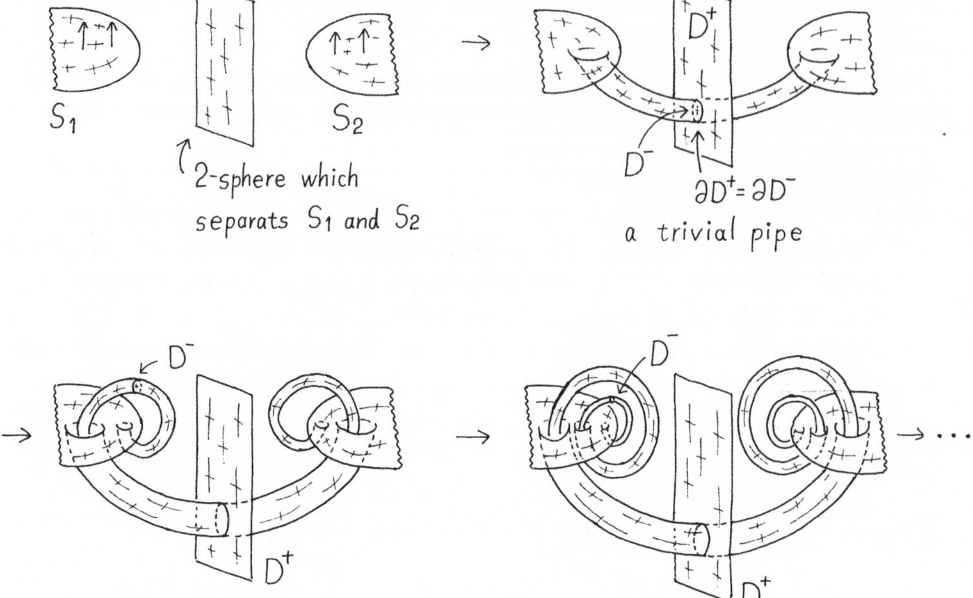

Figure 5.1

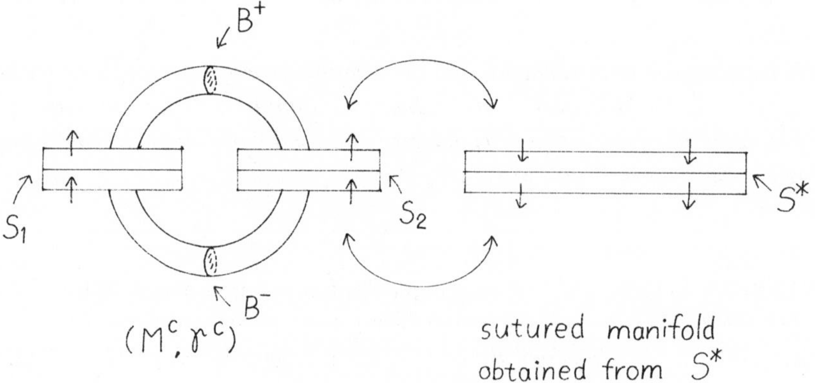

Figure 5.2

we identify the closures of the components of $(M^c, \gamma^c) - (N, \delta)$ to C^+, C^- for simplicity.

Subclaim. *F is an incompressible Seifert surface for L.*

Proof. Assume that F is compressible, i. e. there is a compressing disk D for $R_{\pm}(\delta^c)$ in N^c. Suppose that $\partial D \subset R_{-}(\delta^c)$. It is easy to see that the following arguments work for the case of $\partial D \subset R_{+}(\delta^c)$. Then, by **[1]**, we may suppose that $D \cap C^-$ consists of an essential disk in C^-, and $D \cap C^+$ consists of a product annulus. By **[7,Lemma 4.3]**, we see that $D \cap C^-$ is ambient isotopic to B^- in C^-. Hence, by cutting S^* along $\partial(D \cap C^-)$, we get a union of a punctured S_1 and a punctured S_2. On the other hand, by **[7,Lemma 5.3]**, we may suppose that $D \cap C^+$ is disjoint from B^+, so that $D \cap C^+$ is contained in $S_i \times I$ ($i = 1$ or 2) with the form (a simple loop)$\times[1/2, 1]$. Since $(D \cap C^+) \cap S^*$ is identified to $\partial(D \cap C^-)$, we see that $(D \cap C^+) \cap S^*$ is isotopic to ∂B^+ in S^* by Euler characteristic consideration. Hence $(D \cap C^+)$ is compressible in C^+, a contradiction.

This completes the proof of Subclaim.

We note that $E(L)$ is reducible. Then, by using standard innermost disk argument together with Subclaim, we get an essential 2-sphere Q in $E(L)$ such that $Q \cap S^* = \emptyset$, so that $Q \subset M^c$. Hence, by **[1]** and **[7,Lemma 4.3]**, we may suppose that $Q \cap C^{\pm} = B^{\pm}$. Hence the closures of the components of $N - Q$ is homeomorphic to $(S_i \times I) - \text{Int}\ B_i^3$ $(i = 1, 2)$, where B_i^3 is a 3-ball contained in Int $(S_i \times I)$. This means that the complementary sutured manifold for S_i is a product sutured manifod (hence S_i is a fiber surface by Proposition 2.1), and that S^* is obtained from $S_1 \cup S_2$ by adding a pipe along a trivial arc. Moreover the closures of the components of $Q - S^*$ give a pair of canonical compressing disks for S^*. Then it is easy to see that every essential 2-sphere in $E(L)$ is isotopic to the above Q. Hence Q is a 2-sphere which separates L_1 and L_2, and, by exchanging suffix if necessary, we have $\partial S_i = L_i$ $(i = 1, 2)$.

This completes the proof of Claim 5.1.

Then, by tracing the proof of Claim 5.1 conversely, we have:

Claim 5.2. *If S^* is obtained from a split union of two fiber surfaces by adding a pipe along a trivial arc, then S^* is a pre-fiber surface of type 2.*

For pre-fiber surface of type 1, we proved in **[7,Corollary 5.4]**:

Claim 5.3. *Suppose that S^* is of type 1. Let S_1^* be the surface obtained from S^* by doing a surgery along D^-. Then S_1^* is a pre-fiber surface.*

Completion of Proof. By Claim 5.1 we see that if $n = 1$ then Theorem 3 (1),(2) holds. Suppose that $n > 1$. Then we see that S^* is of type 1 by Claim 5.1. Let S_1^* be as in Claim 5.3. Then, by the induction on n and the argument of the proof of **[7,Theorem 3]**, we see that the conclusion (3) holds. Conversely if S^* is obtained from two fiber surfaces by attaching a number of pipes as in Figure 5.1, then by

Claim 5.2 and the argument of the proof of Claim of **[7,732p.]**, we see that S^* is a pre-fiber surface.

This completes the proof of Theorem 3.

6. Case of $\chi(L) < \chi(L_1) + \chi(L_2) - 1$

In this section, we prove the rest cases of Theorem 1. In fact, we show:

Proposition 6.1. *Let* L, L_1, L_2 *be as in Theorem 1. Suppose that* $\chi(L) < \chi(L_1) + \chi(L_2) - 1$. *Then we have:*
(1) L_1 *and* L_2 *are fibered links, and*
(1) $\chi(L) \leq \chi(L_1) + \chi(L_2) - 5$.

Then we give a characterization of the bands for $L_1 \circ L_2$ which produce fiber surfaces (see the remark of Proposition B below).

Proof of Proposition 6.1 (1). Let S be as in section 3. Then let S_0 be a surface obtained from S by adding a band as in Figure 6.1, and S_1 a surface obtained from S by cutting along $b(\{1/2\} \times I)$. Since S_0 is a plumbing of a fiber surface S and a Hopf band (cf. **[7,Figure 1.1 (ii)]**), we see that S_0 is a fiber surface **[12]**. Hence, by the assumption on $\chi(L)$ and **[7,Theorem 2(2)]**, we see that S_1 is a pre-fiber surface. Hence, by Theorem 3, we see that L_1 and L_2 are fibered links.

This completes the proof of Proposition 6.1(1).

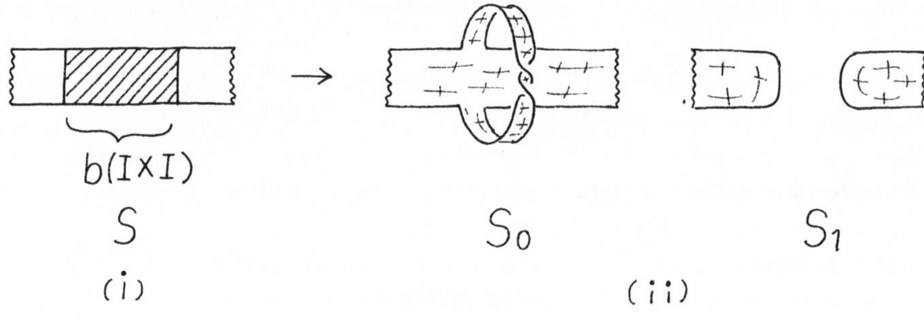

Figure 6.1

For the proof of Proposition 6.1(2), we recall a result in **[7]**. Let S_1 be a pre-fiber surface in a 3-manifold M, and $\beta : I \times I \to M$ a band for ∂S_1 with $\beta(I \times I) \cap \mathrm{Int} S_1 = \emptyset$. We call $\alpha = \beta(I \times \{1/2\})$ a *core arc* of β. Let (N^c, δ^c) be a complementary sutured manifold for S_1. Then we may suppose that $\alpha \cap N^c$ is an arc α' such that

$\partial\alpha' \subset \mathrm{Int}\, \delta^c$. We say that β is of *type F* (with respect to S_1) if there exist a pair of canonical compressing disks D^+, D^- with the following properties.

6.1 The core arc α intersects D^+, D^- in one points,

6.2 There is a disk Δ in N^c such that $\alpha' \subset \partial\Delta$, $\Delta \cap \partial N^c = \partial\Delta \cap \partial N^c = \mathrm{cl}(\partial\Delta - \alpha')$, and $\partial\Delta \cap R_+(\delta^c)$ ($\partial\Delta \cap R_-(\delta^c)$ resp.) consists of an arc.

Then, in [**7,8.2**] and its remark, we showed:

Proposition A. *Let S be a surface obtained from a pre-fiber surface S_1 in a rational homology 3-sphere by adding a band β. Then S is a fiber surface if and only if β is of type F.*

As a consequence of this result, we have:

Proposition B. *Let L be a link which is a band connected sum of two links L_1 and L_2 along a band b with $\chi(L) \leq \chi(L_1) + \chi(L_2) - 1$. Then L is a fibered link if and only if there exists a pre-fiber surface S^* such that $\partial S^* = L_1 \circ L_2$ and b is of type F with respect to S^**

Remark. In Theorem 3, we gave a complete characterization of pre-fiber surfaces bounding $L_1 \circ L_2$. Hence this result together with Proposition B gives a characterization of bands for $L_1 \circ L_2$ which produce fibered links.

Proof of Proposition B. Suppose that L is a fibered link. Then, by section 3, we see that there is a minimal genus Seifert surface S for L such that $b(I \times I) \subset S$. Let S^* be a surface obtained from S by cutting along $b(\{1/2\} \times I)$. By [**7,Theorem2 (2)**], we see that S^* is a pre-fiber surface. Then, by Proposition A, we see that S is obtained from a pre-fiber surface S^* by adding a band b of type F.

Suppose that there is a pre-fiber surface S^*, and a band b as above. Then, by Proposition A, we see that L is a fibered link.

This completes the proof of Proposition B.

Proof of Proposition 6.1(2). Let S_1 be a pre-fiber surface for $L_1 \circ L_2$ as above. Assume that $\chi(L) = \chi(L_1) + \chi(L_2) - 3$. Then, by the proof of Theorem 3, we see that S_1 is a pre-fiber surface of type 2, and there is a pair of canonical compressing disks D^+, D^- for S_1 such that $\partial D^+ = \partial D^-$, and $\partial D^\pm$ cuts S_1 into two components F_1, F_2 where F_i is a fiber surface for L_i with one hole. Then, by Proposition B, and the conditions (6.1),(6.2), we see that $b(I \times I) \cap L_1 = \emptyset$ or $b(I \times I) \cap L_2 = \emptyset$, a contradiction.

Example. Let S^* be a pre-fiber surface which bounds a two component trivial link T_2 as in Figure 6.2 (i). Then let b_1, b_2 be bands for T_2 as in Figure 6.2 (ii). We see that b_1, b_2 are of type F by Figure 6.2 (i) (note that the properly embedded arcs in S^* correspond to the arcs $\partial\Delta \cap R_\pm(\delta^c)$). Hence the knots obtained from T_2 by band moves along b_1, b_2 are fibered.

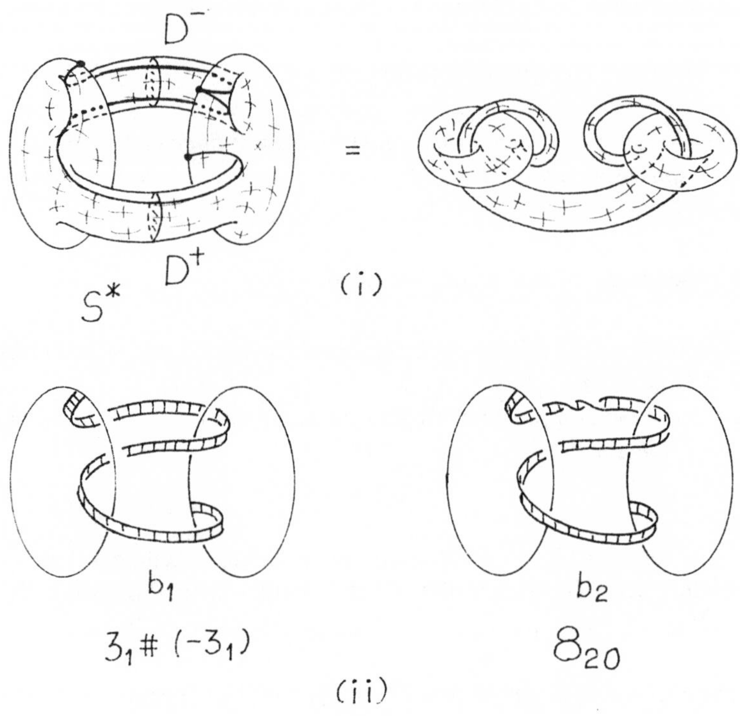

Figure 6.2

References

1. A. Casson and C. McA Gordon, *Reducing Heegaard splittings*, Topology and its Appl. **27** (1987), 275–283.
2. D. Gabai, *Detecting fibered links in* S^3, Comm. Math. Helv. **61** (1986), 519–555.
3. D. Gabai, *Genus is superadditive under band connected sum*, Topology **26** (1987), 209–210.
4. D. Gabai, *Foliations and the topology of 3-manifolds II*, J. Diff. Geom. **26** (1987), 461–478.

5. J. Hempel, *3-manifolds*, Ann. of Math Studies No. 86, Princeton University Press 1976, Princeton N. J..
6. W. Jaco, *Lectures on three manifold topology*, CBMS Regional Conference Series in Math. No. 43, 1983.
7. T. Kobayashi, *Fibered links and unknotting operations*, Osaka J. Math. **26** (1989), 699–742.
8. D. Rolfsen, *Knots and links*, Mathematical Lecture Series 7, Publish or Perish Inc., Berkeley Ca., 1976.
9. M. Scharlemann, *Sutured manifolds and generalized Thurston norms*, J. Diff. Geom. **29** (1989), 557–614.
10. M. Scharlemann and A. Thompson, *Link genus and the Conway moves*, Comm. Math. Helv. **64** (1989), 527–535.
11. M. Scharlemann and A. Thompson, *Detecting unknotted graphs in 3-space*, preprint.
12. J. Stallings, *Construction of fibered knots and links*, Proc. Symp. Pure Math. AMS **32** (1978), 55–60.

Geometries on 4-manifolds, Euler characteristic and elementary amenable groups

Jonathan A. Hillman

Abstract. We give criteria for a closed topological 4-manifold to be homeomorphic to an infrasolvmanifold or to have a finite covering space with a geometry of type $S^2 \times E^2$ or $S^3 \times E^1$ or which is s-cobordant to a manifold with a geometry of type $H^2 \times E^2, \tilde{SL} \times E^1, H^3 \times E^1$ or $S^2 \times H^2$. We determine the 2-knots K such that the manifold $M(K)$ obtained by surgery on K in S^4 is an infrasolvmanifold, and show that the only other geometries realizable in this way are $\tilde{SL} \times E^1, H^3 \times E^1, S^3 \times E^1$ and (possibly?) $H^2 \times E^2$. We also consider for which knots K the manifold $M(K)$ is homeomorphic to a complex surface.

AMS Subject classification: Primary 57N13, Secondary 57R19, 57R22.
Key words and phrases: elementary amenable, geometry, 2-knot, 4-manifold, topological surgery.

Introduction

A 4-dimensional geometry in the sense of Thurston is a pair (X, G_X) where X is a complete 1-connected 4-dimensional Riemannian manifold and G_X is a group of isometries which acts transitively on X and has discrete subgroups Γ such that $\Gamma\backslash X$ has finite volume [S1,Th]. Using an equivalent formulation, we shall say that a closed 4-manifold M is an X-manifold, or admits a geometry of type X, if it is homeomorphic to a quotient $\Gamma\backslash X$ for some torsion free discrete group Γ of isometries of a 1-connected homogeneous space $X = G/K$, where G is a connected Lie group and K is a compact subgroup of G such that the intersection of the conjugates of K is trivial, and X has a G-invariant metric. If G is solvable we shall say that the geometry is of solvable Lie type. We shall consider the problem of characterizing algebraically the closed topological 4-manifolds which admit such geometries.

Knots 90

Filipkiewicz has shown that there are 19 possible maximal 4-dimensional geometries [F,W1,W2]; one of these $(Sol^4_{m,n})$ is in fact a countably infinite family of closely related geometries, and one (F^4) is not realizable by any closed 4-manifold. Ue has characterized the closed orientable 4-manifolds admitting geometries of type $S^2 \times E^2, S^3 \times E^1, E^4, Nil^3 \times E^1, Nil^4, Sol^3 \times E^1, H^2 \times E^2$ or $\tilde{SL} \times E^1$ as fibred orbifolds over 2-orbifolds with general fibre $S^1 \times S^1$ and nonsingular total space [U1,2]. (There is one closed orientable E^4-manifold which is not such a "Seifert 4-manifold", and there are such Seifert manifolds which fibre over a hyperbolic base and which admit no geometry.)

There are only a handful of manifolds admitting one of the three geometries based on compact models $(S^4, P^2(C)$ and $S^2 \times S^2)$. Apart from these the most well understood geometry is perhaps E^4, the geometry of flat 4-manifolds. A closed Riemannian 4-manifold is flat if and only if it is finitely covered by R^4/Z^4. Thus it is aspherical and its fundamental group is torsion free and virtually abelian, i.e. has a subgroup of finite index isomorphic to Z^4. There are 75 such groups and each is the fundamental group of some flat 4-manifold, which is determined up to diffeomorphism by its group (cf. Chapter 3 of [Wo]). It is not hard to show that a closed 4-manifold M is homotopy equivalent to a flat 4-manifold if and only if $\chi(M) = 0$ and $\pi_1(M)$ is virtually Z^4. Using topological surgery it can then be shown that these conditions imply that M admits the geometry E^4.

In Theorem 1 we shall give a similar characterization of 4-dimensional infrasolvmanifolds, i.e. closed 4-manifolds admitting a geometry of solvable Lie type $E^4, Nil^3 \times E^1, Nil^4, Sol^4_0, Sol^4_1$ or $Sol^4_{m,n}$, as closed 4-manifolds with Euler characteristic 0 and having torsion free elementary amenable normal subgroups of Hirsch length at least 3 in the fundamental group. Properties of the fundamental group may be used to distinguish between these geometries. In particular, the mapping torus of a self homeomorphism of a 3-manifold with a geometry of type E^3, Nil^3 or Sol^3 has the corresponding product geometry if and only if the induced outer automorphism of the fundamental group has finite order (Theorem 2). (Note that the product geometry $Sol^3 \times E^1$ is equivalent to $Sol^4_{m,m}$ for any m.) Theorem 3 provides an algebraic complement: a torsion free elementary amenable normal subgroup of Hirsch length 2 and infinite index in the fundamental group of a closed 4-manifold with Euler characteristic 0 must be virtually abelian. In Theorems 4 and 5 we show that a closed 4-manifold M has a finite covering space homeomorphic to $S^2 \times S^1 \times S^1$ or $S^3 \times S^1$ if and only if $\chi(M) = 0$ and $\pi_1(M)$ is virtually Z^2 or Z. (Not all such manifolds M admit geometric structures.) In Theorem 6 we show that the above mentioned criterion for mapping tori holds also for the other 3-dimensional geometries of aspherical type.

The 5-dimensional s-cobordism theorem has been established so far only over elementary amenable fundamental groups [Fr,FQ]. (This is the smallest class of groups containing all finite groups and Z and which is closed under extension and increasing union.) Thus it is not yet known to hold over the fundamental groups of closed 4-manifolds admitting one of the remaining geometries $\tilde{SL} \times E^1, H^3 \times E^1, H^2 \times E^2, H^2 \times H^2, S^2 \times H^2, H^4$ or $H^2(C)$. In principle, the Euler characteristic and fundamental group should suffice to characterize 4-manifolds with a finite covering space s-cobordant to a 4-manifold admitting one of these geometries, but we only have intrinsic characterizations of the groups arising in the cases $\tilde{SL} \times E^1, H^2 \times E^2$ and $S^2 \times H^2$. These cases are considered in Theorems 7 and 8. In the $H^2 \times H^2$ case we can only treat manifolds finitely covered by the cartesian product of two hyperbolic surfaces (Theorem 9).

In Theorem 10 we restrict our attention to the 4-manifolds $M(K)$ obtained by surgery on 2-knots $K : S^2 \to S^4$. Here we show that $M(K)$ is an infrasolvmanifold if and only if K is a Cappell-Shaneson 2-knot or πK is one of two virtually abelian groups $G(+)$ or $G(-)$ or is the group $\pi(b, \epsilon)$ of the 2-twist spin of a Montesinos knot $K(o|b; (3,1), (3,1), (3,\epsilon))$ for some even b and $\epsilon = \pm 1$ or if the commutator subgroup $\pi K'$ is a torsion free nilpotent group of class 2 and Hirsch length 3. The corresponding knot groups are well understood (c.f. Chapter 6 of [H]). The geometries realized are $E^4, Nil^3 \times E^1, Sol_0^4, Sol_1^4$ and certain of the $Sol_{m,n}^4$ geometries. Branched r-twist spins of torus knots and simple (non-torus) knots with $r > 2$ lead to manifolds with geometries of type $\tilde{SL} \times E^1$ and $H^3 \times E^1$ respectively, except for the 3-, 4- and 5-twist spins of the trefoil knot and the 3-twist spin of the (2,5)-torus knot, the 6-twist spin of the trefoil knot and the 3-twist spin of the figure eight knot (and their cyclic branched covers), which lead to geometries of type $S^3 \times E^1, Nil^3 \times E^1$ and E^4, respectively. Other examples with geometry $S^3 \times E^1$ arise from 2-twist spins of 2-bridge knots and certain other "small" simple knots. None of the other geometries (excepting perhaps(?) $H^2 \times E^2$) are realizable by such knot manifolds.

We conclude with some remarks on such knot manifolds which are homeomorphic to complex analytic surfaces. In all cases $M(K)$ admits a geometry compatible with the complex structure and is either an Inoue surface with geometry of type Sol_0^4 or Sol_1^4 or an elliptic surface with a geometry of type $S^3 \times E^1$, $Nil^3 \times E^1$, or $\tilde{SL} \times E^1$. The first four of these cases are well understood, while every known example of the fifth case is homeomorphic to $M(K_1)$ where K_1 is some branched twist spin of a torus knot. In particular, in Theorem 11 we show that $M(K)$ is homeomorphic to a Hopf surface if and only if K is a branched r-twist spin of the (p,q)-torus knot for some p, q and r such that $p^{-1} + q^{-1} + r^{-1} > 1$. In Theorem 12 we characterize the knot manifolds which are s-cobordant to elliptic surfaces of Kodaira dimension 1 in terms of properties of the knot group.

Notation and terminology

A group virtually has some property if it has a normal subgroup of finite index with that property. Thus a virtually poly-Z group is one which has a subgroup of finite index with a compositon series whose factors are all infinite cyclic. The number of infinite cyclic factors is independent of the composition series used, and is called the Hirsch length of the group. This notation has been extended to the class of elementary amenable groups in [H2], and $h(G)$ shall denote the Hirsch length of such a group G. A $PD_n^{(+)}$-group is an (orientable) Poincaré duality group of formal dimension n.

Let $\pi', \zeta\pi$ and $\sqrt{\pi}$ denote respectively the commutator group, centre and Hirsch-Plotkin radical of the group π. The Hirsch-Plotkin radical is the maximal locally nilpotent normal subgroup; in a virtually poly-Z group π every subgroup is finitely generated, and so $\sqrt{\pi}$ is then the maximal nilpotent normal subgroup. (In [Ra,W2] $\sqrt{\pi}$ is called the nilradical.) For each natural number q let Γ_q denote the group with presentation $< x, y, z | xz = zx, yz = zy, [x, y] = z^q >$; Γ_q is torsion free nilpotent and has Hirsch length 3. Conversely any finitely generated torsion free nilpotent group of Hirsch length 3 is either free abelian of rank 3 or is isomorphic to Γ_q for some $q \geq 1$.

Motivated by the group theoretic terminology we shall say that M is virtually (s-cobordant to) an X-manifold if it has a finite covering space which is (s-cobordant to) an X-manifold.

The solvable Lie cases

An infrasolvmanifold is a quotient $\pi \backslash G/K$ where G is a Lie group whose maximal connected subgroup G_o is solvable and of finite index in G, K is a maximal compact subgroup of G which meets each component of G and π is a torsion free discrete cocompact subgroup of G. An infrasolvmanifold M is aspherical since $G/K = G_o/G_o \cap K$ is homeomorphic to R^m [Iw] and so its fundamental group π is a PD_m-group, where m is the dimension of M; since π is also virtually solvable it is thus virtually poly-Z of Hirsch length m by Theorem 9.23 of [B]; moreover $\chi(M) = 0$. As K is compact we may equip $X = G/K$ with a G-invariant metric, and then π acts isometrically, so infrasolvmanifolds have geometric structures of solvable Lie type. Conversely, a closed manifold with a geometry of solvable Lie type is an infrasolvmanifold. In general we cannot assume that π is a subgroup of G_o, even if π is solvable (cf. Section 4.29 of [Ra]). Every torsion free virtually poly-Z group of Hirsch length m admits a representation as a discrete subgroup of $Aff(R^m)$, the group of affine motions of R^m, and so infrasolvmanifolds have complete affinely flat structures [Ni]. If π is virtually nilpotent then we may take

G to be a subgroup of the affine group $Aff(N)$ of some 1-connected nilpotent Lie group N such that $\pi \cap N$ is a cocompact subgroup of N and $[\pi : \pi \cap N]$ is finite [Ru].

Theorem 1. *Let M be a closed 4-manifold with fundamental group π. The following conditions are equivalent:*
(i) M is homeomorphic to an infrasolvmanifold;
(ii) $\chi(M) = 0$, π is torsion free and virtually poly-Z and $3 \leq h(\sqrt{\pi}) \leq h(\pi) = 4$;
(iii) $\chi(M) = 0$ and π has a nilpotent normal subgroup ν such that $h(\nu) \geq 3$; and
(iv) $\chi(M) = 0$ and π has an elementary amenable normal subgroup ρ whose finite subgroups have bounded order and which has no nontrivial finite normal subgroup and such that $h(\rho) \geq 3$.
Moreover any torsion free virtually poly-Z group of Hirsch length 4 is the fundamental group of such a manifold, which is determined up to homeomorphism by the group.

Proof. If M is homeomorphic to an infrasolvmanifold then $\chi(M) = 0$ and π is torsion free and virtually poly-Z and $h(\pi) = 4$, by the remarks above. Let G be a normal poly-Z subgroup of finite index in π. Then $1 \neq \sqrt{G} \leq \sqrt{\pi}$. If $h(\sqrt{\pi}) \leq 2$ then $\sqrt{G}$ would be Z or Z^2. It is easy to see that if G is solvable and $\sqrt{G}$ is abelian the natural map from $G/\sqrt{G}$ to $Aut(\sqrt{G})$ induced by conjugation in G is 1-1. Since $Aut(Z)$ is finite and $Aut(Z^2) = GL(2, Z)$ has no noncyclic poly-Z subgroup we would have $h(G/\sqrt{G}) \leq 1$. But then $h(\pi) = h(G) \leq 3$. Therefore we must have $h(\sqrt{\pi}) \geq 3$ and so (i) implies (ii). It is clear that (ii) implies (iii) and (iv).

Suppose that (iii) holds. The subset T of elements of finite order in ν is a characteristic subgroup [Ro:5.2.7] and hence is normal in π. Moreover ν/T is torsion free and $h(\nu/T) = h(\nu) \geq 3$. Therefore π/T is a PD_4-group over Q, by Theorem 3 of Chapter 3 of [H], and so $3 \leq c.d._Q\nu/T \leq 4$ (as in any case $h(\nu/T) \leq c.d._Q\nu/T$). If $c.d._Q\nu/T = 3$ then $h(\nu/T) = c.d._Q\nu/T$ and there is a free $Q[\nu/T]$-module W such that $H^3(\nu/T; W) \neq 0$. It is not hard to see that ν/T has a finitely generated subgroup F with $h(F) = 3$. Any such subgroup lies in a subnormal sequence $F \leq F\zeta(\nu/T) \leq \nu/T$. Two applications of an LHS spectral sequence argument as in Lemma 1 of Chapter 3 of [H] show that if F has infinite index in ν/T then $H^s(\nu/T; W) = 0$ for $s \leq 3$. Therefore this index must be finite and so ν/T is finitely generated. Since it is torsion free and nilpotent of Hirsch length 3 it is a PD_3-group. The LHS spectral sequence for π/T as an extension of π/ν by ν/T (with coefficients $Q[\pi/T]$) then implies that $H^4(\pi/T; Q[\pi/T]) \cong H^1(\pi/\nu; H^3(\nu/T; Q[\pi/T])) \cong H^1(\pi/\nu; Q[\pi/\nu])$. Since $H^4(\pi/T; Q[\pi/T]) \cong Q$ it follows that π/ν has two ends. Thus π has a subgroup σ of finite index such that $\nu \leq \sigma$ and $\sigma/\nu \cong Z$. By Theorem A of [BS] σ is an ascending HNN-group with base a finitely generated subgroup of ν. Therefore σ and hence π are constructable and so are virtually torsion free [BB]. If $c.d._Q\nu/T = 4$ then $[\pi:\nu] = [\pi/T:\nu/T]$ is finite [Sl], so ν is finitely generated, and therefore virtually poly-Z. Thus in either case π has a torsion free solvable subgroup of finite index and Hirsch length 4, and so (iii) implies (iv).

If (iv) holds then M is aspherical by Theorems 3 and 6 of [H2], so π is a PD_4-group and $3 \leq h(\rho) \leq c.d.\rho \leq 4$. Therefore ρ is virtually solvable [HL]. If $c.d.\rho = 4$ then $[\pi:\rho]$ is finite [Sl] and so π is virtually solvable also. If $c.d.\rho = 3$ then $c.d.\rho = h(\rho)$ and so ρ is a duality group and is FP [Kr]. Therefore $H^q(\rho; Q[\pi]) \cong H^q(\rho; Q[\rho]) \otimes Q[\pi/\rho]$ and is 0 unless $q = 3$. It then follows from the LHS spectral sequence for π as an extension of π/ρ by ρ (with coefficients $Q[\pi]$) that $H^4(\pi; Q[\pi]) \cong H^1(\pi/\rho; Q[\pi/\rho]) \otimes H^3(\rho; Q[\rho])$. Therefore $H^1(\pi; Q[\pi]) \cong Q$, so π/ρ has two ends and we again find that π is virtually solvable. In all cases π is torsion free and virtually poly-Z, by Theorem 9.23 of [B], and $h(\pi) = 4$. It is shown in [AJ] that any such group is the fundamental group of an infrasolvmanifold. Thus (iv) implies (i). The uniqueness up to homeomorphism follows from Theorem 2.16 of [FJ] together with the fact that the topological surgery arguments remain valid for 4-manifolds with such fundamental groups [Fr,FQ]. □

Is the hypothesis on torsion needed in condition (iv)?

There is an analogous characterization of higher dimensional infrasolvmanifolds. The conditions (ii),(iii) and (iv) must be strengthened to require also that $\pi_j(M) = 0$ for $1 < j < n/2$ (or equivalently the homology of the universal covering space of M should vanish in this range) and the normal subgroup ($\sqrt{\pi}$, ν, or ρ respectively) should have Hirsch length at least $n-1$. The example of $M = S^2 \times (S^1)^{n-2}$ shows that this bound is best possible in general. (Cf. also Theorem 3 below.)

Which geometry is applicable can be seen as follows. (Cf. also Proposition 10.4 of [W2].) Let π be a torsion free virtually poly-Z group with $h(\pi) = 4$ and let A be an abelian normal subgroup of π which is of maximal rank. If A has rank 4 then π is virtually Z^4 and M is an E^4-manifold. If A has rank 3 then $h(\pi/A) = 1$ so π/A has an infinite cyclic normal subgroup of finite index. Thus π has a normal subgroup σ of finite index which is a semidirect product $A \times_\theta Z$ where the action of a generator t of Z by conjugation on the normal subgroup A is given by a matrix θ in $GL(3, Z)$. On replacing σ by a subgroup of index 2 if necessary we may assume that θ has no negative eigenvalues and so is in $SL(3, Z)$. This matrix must have infinite order for otherwise the subgroup generated by A and a suitable power of t would be abelian of rank 4. If the eigenvalues of θ are distinct and positive then σ is a discrete cocompact subgroup of the group $Sol^4_{m,n}$ where $m = trace(\theta)$ and $n = trace(\theta^{-1})$ (i.e., the characteristic polynomial of θ is $X^3 - mX^2 + nX - 1$). If two of the eigenvalues are not real then they are complex conjugates and σ is a discrete cocompact subgroup of $Isom(Sol^4_0)$. If two of the eigenvalues are equal then they are all 1 and so $\theta - I$ is nilpotent, i.e. $(\theta - I)^3 = 0$. If $(\theta - I)^2 = 0$ then σ is isomorphic to $\Gamma_q \times Z$ for some $q \geq 1$ and so is a discrete cocompact subgroup of the group $Nil^3 \times R$; otherwise σ is a discrete cocompact subgroup of Nil^4.

If π has an abelian normal subgroup B of rank 2 then the action of π on B by conjugation induces a homomorphism from π to $Aut(B) \cong GL(2, Z)$ with kernel $C_\pi(B)$, the centralizer of B in π. Since $GL(2, Z)$ is virtually free it does not have poly-Z subgroups of Hirsch length greater than 1 and so $h(C_\pi(B)) \geq 3$. If $h(C_\pi(B)) = 3$ then $\sqrt{C_\pi(B)} \cong Z^3$; otherwise B is central in a subgroup of finite index in π. In either case after replacing π by a subgroup σ of finite index if necessary we may assume that it is isomorphic to a product $\mu \times Z$ where B is a normal subgroup of μ. As σ then has an abelian normal subgroup of rank at least 3 the discussion of the above paragraph applies.

As every poly-Z group clearly has nontrivial abelian normal subgroup it only remains for us to consider the case when A is infinite cyclic. On replacing π by a subgroup σ of finite index we may assume that A is central and that σ/A is poly-Z. Let B be the preimage in σ of a nontrivial abelian normal subgroup of σ/A. Then B is nilpotent (since A is central and B/A is abelian) and $h(B) > 1$, so B is nonabelian (by maximality of the rank of A). Therefore $h(B) \geq 3$. If $h(B) = 3$ then $B \cong \Gamma_q$ for some $q \geq 1$ and then π contains a normal subgroup of finite index which is a semidirect product $\Gamma_q \times_\theta Z$. Such a semidirect product is either a discrete cocompact subgroup of Sol_1^4 or is virtually nilpotent of Hirsch length 4. But in the latter case it has abelian normal subgroups of rank at least 2 and so our earlier discussion applies.

In all cases we see that $h(\sqrt{\pi}) \geq 3$ and $\sqrt{\pi}$ is $Z^3, \Gamma_q, Z^4, \Gamma_q \times Z$ or nilpotent of class 3. If $h(\sqrt{\pi}) = 3$ then $h(\pi/\sqrt{\pi}) = 1$, so $\pi/\sqrt{\pi}$ has a finite normal subgroup with quotient Z or $Z/2Z * Z/2Z$. Thus π has a subgroup of index at most 2 which maps onto Z with kernel containing $\sqrt{\pi}$ as a subgroup of finite index. The corresponding covering space of M is the mapping torus of a self homeomorphism of an E^3- or Nil^3-manifold. (When $\sqrt{\pi} \cong Z^3$ it can be shown that $\pi/\sqrt{\pi}$ itself maps onto Z except perhaps when the matrix θ invoked above is conjugate to its inverse, which can happen only for the geometry $Sol^3 \times E^1 = Sol_{m,m}^4$. Thus all Sol_0^4-manifolds and most $Sol_{m,n}^4$-manifolds are mapping tori of self homeomorphisms of flat 3-manifolds.)

It follows from the above that a closed 4-manifold M has a product geometry of type $E^4, Nil^3 \times E^1$ or $Sol^3 \times E^1$ if and only if $\pi_1(M)$ is virtually poly-Z and has a subgroup of finite index of the form $\nu \times Z$.

Theorem 2. *Let ϕ be a self homeomorphism of a 3-manifold N with a geometry of type E^3 or Nil^3. Then the mapping torus $M(\phi) = N \times_\phi S^1$ admits the corresponding product geometry if and only if the outer automorphism $[\phi_*]$ induced by*

ϕ has finite order in $Out(\pi_1(N))$. The mapping torus of a self homeomorphism of a Sol^3-manifold admits the geometry $Sol^3 \times E^1$.

Proof. Let $\nu = \pi_1(N)$ and let t be an element of $\pi = \pi_1(M(\phi))$ which projects to a generator of $\pi_1(S^1)$. (Thus t determines a splitting of the extension $1 \to \nu \to \pi \to Z \to 1$ and the outer automorphism of ν induced by conjugation by t in π is $[\phi_*]$.)

Suppose first that $M(\phi)$ is flat and let A be the maximal abelian normal subgroup of π. Then $[\pi : A]$ and $[\nu : \nu \cap A]$ are finite, so N is flat and the automorphism of ν induced by conjugation by $t^{[\pi:A]}$ is the identity on $\nu \cap A$. It follows from the "Second Theorem" of Bieberbach (automorphisms of a crystallographic group are induced by conjugations in $Aff(R^3)$, cf. Theorem 3.2.2 of [Wo]) that the image of such an automorphism in $Out(\nu)$ has finite order. Therefore $[\phi_*]$ has finite order also. If N is a Nil^3-manifold and $M(\phi)$ is a $Nil^3 \times E^1$-manifold then $\zeta\nu$ is characteristic in ν and hence normal in π, and $\nu/\zeta\nu$ and $\pi/\zeta\nu$ are virtually abelian. A similar argument then shows that the image of $\phi*$ in $Out(\nu/\zeta\nu)$ has finite order. The kernel of the natural map from $Aut(\nu)$ to $Aut(\nu/\zeta\nu)$ is isomorphic to $Hom(\nu/\nu', \zeta\nu)$. If ν/ν' is finite this kernel is trivial; otherwise ν is isomorphic to Γ_q for some q. In either case the kernel of the natural map from $Out(\nu)$ to $Out(\nu/\zeta\nu)$ is finite and so $[\phi_*]$ has finite order also. Conversely, if $[\phi_*]$ has finite order in $Out(\nu)$ then π has a subgroup of finite index isomorphic to $\nu \times Z$. Hence if moreover N has a geometry of solvable Lie type the mapping torus has the product geometry, by the discussion above.

If N is a Sol^3-manifold then $\sqrt{\nu} \cong Z^2$ and $h(\nu/\sqrt{\nu}) = 1$ so ν has a characteristic subgroup ν_1 of finite index which is an extension of Z by $\sqrt{\nu}$; the action of Z on $\sqrt{\nu}$ is determined by a matrix of infinite order in $GL(2, Z)$. It is easily verified that $Out(\nu_1)$ is finite; let e be its order. The subgroup of π generated by ν_1 and t^e has finite index in π and is isomorphic to $\nu_1 \times Z$. Thus $M(\phi)$ has a finite covering space with geometry $Sol^3 \times E^1$ and so also admits this geometry. □

Theorem 4.2 of [KLR] (which extends Bieberbach's theorem to the virtually nilpotent case) may be used to show directly that every outer automorphism class of finite order of the fundamental group of an E^3- or Nil^3-manifold is realizable by an isometry of an affinely equivalent manifold. If N is flat and $[\phi_*]$ has infinite order then $M(\phi)$ may admit one of the other product geometries $Sol^3 \times E^1$ or $Nil^3 \times E^1$. We shall obtain similar results for the other 3-dimensional geometries of aspherical type in the next section.

Theorem 3. *Let M be a closed 4-manifold with $\chi(M) = 0$ and fundamental group π, and suppose that π has a torsion free elementary amenable normal subgroup ρ*

with $h(\rho) = 2$ and which has infinite index in π. Then M is aspherical and ρ is virtually abelian.

Proof. Since ρ is torsion free elementary amenable and $h(\rho) = 2$ there is an exact sequence $1 \to A \to \rho \to B \to 1$ where A and B are isomorphic to nontrivial subgroups of Q (cf. Theorem 2 of [H2]). In particular $Aut(A)$ is isomorphic to a subgroup of $Q^\times$. Suppose that ρ is not virtually abelian. Then the homomorphism from B to $Aut(A)$ induced by conjugation in ρ must map B onto an infinite subgroup of $Q^\times$. Therefore $B \cong Z$, while A cannot be finitely generated. Moreover $\zeta\rho = 1$ and A is the maximal abelian normal subgroup of ρ, and thus is a characteristic subgroup. Therefore every automorphism of ρ restricts to an automorphism of A. This restriction from $Aut(\rho)$ to $Aut(A)$ is an epimorphism, with kernel isomorphic to A, and so $Aut(\rho)$ is solvable.

Since A is characteristic in ρ it is normal in π. The quotient group π/A has one end as the image of ρ is an infinite cyclic normal subgroup of infinite index. Therefore π is 1-connected at ∞ by Theorem 1 of [Mh], and so $H^s(\pi; Z[\pi]) = 0$ for $s \leq 2$ [GM]. Therefore M is aspherical and π is a PD_4-group by Theorem 6 of [H2]. Let C be the centralizer of ρ in π. Then C in nontrivial, for otherwise π would be isomorphic to a subgroup of $Aut(\rho)$, which is solvable, and so π would be virtually poly-Z. But then A would be finitely generated, and ρ would be virtually abelian. Moreover $C \cap \rho = \zeta\rho = 1$, so $C\rho \cong C \times \rho$ and $c.d.C + c.d.\rho = c.d.C\rho \leq c.d.\pi = 4$. The quotient group $\pi/C\rho$ is isomorphic to a subgroup of $Out(\rho)$.

If $c.d.C\rho \leq 3$ then as C is nontrivial and $h(\rho) = 2$ we must have $c.d.C = 1$ and $c.d.\rho = h(\rho) = 2$. Therefore C is free and ρ is of type FP [Kr]. By Theorem A of [BS] ρ is an ascending HNN-group with base a finitely generated subgroup of A and so has a presentation of the form $< a, t | tat^{-1} = a^n >$ for some nonzero integer n. The subgroup of $Aut(\rho)$ represented by $(n-1)A$ consists of inner automorphisms. Since $n \neq 1$ the quotient $A/(n-1)A \cong Z/(n-1)Z$ is finite, and as $Aut(A) \cong Z[1/n]^\times$ it follows that $Out(\rho)$ is virtually abelian. Therefore π has a subgroup σ of finite index which contains $C\rho$ and such that $\sigma/C\rho$ is a finitely generated free abelian group, and in particular $c.d.\sigma/C\rho$ is finite. As σ is a PD_4-group it follows from Theorem 9.11 of [B] that $C\rho$ is a PD_3-group and hence that ρ is a PD_2-group. If $c.d.C\rho = 4$ then $[\pi : C\rho]$ would be finite [Sl], so $C \times \rho$ would be a PD_4-group. But then ρ would again be a PD_2-group, by Theorem 9.11 of [B]. As a solvable PD_2-group is virtually abelian our original assumption must have been wrong. This proves the theorem. □

Every group with a presentation of the form $< a, t | tat^{-1} = a^n >$ is torsion free and solvable, and is the fundamental group of some orientable closed 4-manifold M with $\chi(M) = 0$. Thus the hypothesis that the subgroup ρ have infinite index in π is necessary for the above theorem. Must ρ be finitely generated? If so $\sqrt{\rho}$

is a normal subgroup of π isomorphic to Z^2. We shall consider 4-manifolds with Euler characteristic 0 and whose fundamental groups have free abelian normal subgroups of rank 2 and of infinite index again in Theorem 7.

The spherical solvable cases

When the model space is not contractible the fundamental group may have nontrivial torsion and there may be nongeometric manifolds which are finitely covered by geometric manifolds. (An example of this kind is given after Theorem 5 below.)

Theorem 4. *Let M be a closed 4-manifold with fundamental group π. The following conditions are equivalent:*
(i) M has $S^2 \times S^1 \times S^1$ as a covering space of degree dividing 8;
(ii) M is virtually an $S^2 \times E^2$-manifold; and
(iii) $\chi(M) = 0$ and π is virtually Z^2.

Proof. It is clear that (i) implies (ii) and (iii). An orientable manifold is an $S^2 \times E^2$-manifold if and only if it is homeomorphic to the mapping torus of an orientation preserving self homeomorphism of $S^2 \times S^1$ or $RP^3 \sharp RP^3$, by Theorem 12 of [U2]. Since any such mapping torus has $S^2 \times S^1 \times S^1$ as a 4-fold covering space (ii) implies (i).

Suppose that (iii) holds. Let A be a torsion free subgroup of finite index in π. By Theoem 2 of [H1] the corresponding covering space M_A of M is homotopy equivalent to the total space of an S^2-bundle over the torus or the Klein bottle, and so the universal covering space $\tilde{M}$ is homotopy equivalent to S^2. A Gysin sequence argument then shows that if g is a nontrivial element of finite order in π then it acts nontrivially on $H_2(\tilde{M}; Z) \cong Z$. In particular, the only finite subgroups of π are of order at most 2.

Let T be the maximal finite normal subgroup of π and let U be the maximal normal subgroup of π containing T and such that U/T is abelian. Then $U/T \cong Z^2$ and π/U is a finite group which acts effectively on U/T. (Thus π/T is a 2-dimensional crystallographic group, i.e., is the orbifold group of a 2-dimensional euclidean orbifold.) Since π/T has no odd torsion $[\pi : U]$ divides 4. If T is nontrivial then it is central and π/T is torsion free, so $[\pi : U] \leq 2$. Moreover M then has a finite covering space with fundamental group isomorphic to $Z^2 \oplus T \cong Z^2 \oplus (Z/2Z)$ which is therefore homotopy equivalent to an RP^2-bundle over the torus, by Theorem 3 of [H1]. Therefore M is nonorientable and the restriction of $w_1(M) : \pi \to Z/2Z$ to T is an isomorphism, so $\pi \cong (\pi/T) \times T$.

In all cases π has a torsion free subgroup of index at most 2 which we may take as A. On passing to a subgroup of index at most 4 in A we may assume that the corresponding covering space M_A is homotopy equivalent to $S^2 \times S^1 \times S^1$. It is then homeomorphic to $S^2 \times S^1 \times S^1$ by [Kw], and so (iii) implies (i). □

Is the 2-fold covering space of M corresponding to a torsion free subgroup of π always geometric?

Theorem 5. *Let M be a closed 4-manifold with fundamental group π. Then M is virtually an $S^3 \times E^1$-manifold if and only if $\chi(M) = 0$ and π is virtually Z. If these conditions hold and F is the maximal finite normal subgroup of π then F has cohomological period dividing 4, the corresponding covering space of M is a finite PD_3-complex and $\pi/F \cong Z$ or $(Z/2Z) * (Z/2Z)$.*

Proof. An orientable manifold is an $S^3 \times E^1$-manifold if and only if it is homeomorphic to the mapping torus of an orientation preserving self homeomorphism of an S^3-manifold, by Theorem A(1) of [U2]. Thus the conditions are clearly necessary.

Conversely, suppose that they hold. Let $\hat{M}$ be a finite covering space of M which is orientable and has fundamental group Z. As in Theorem 1 of Chapter 4 of [H] the universal cover $\tilde{M}$ is homotopy equivalent to S^3, by equivariant Poincaré duality. Since $\hat{M}$ is orientable the generator of the group of covering automorphisms $Aut(\tilde{M}/\hat{M})$ is homotopic to the identity. Therefore $\hat{M} \simeq \tilde{M} \times S^1 \simeq S^3 \times S^1$ and so is homeomorphic to $S^3 \times S^1$, by surgery over Z (cf. [FQ]).

Since F acts freely on $\tilde{M} \simeq S^3$ it has cohomological period dividing 4 and $M_F = \tilde{M}/F$ is a finitely dominated PD_3-complex. The image of the finiteness obstruction for M_F under the "geometrically significant injection" of $\tilde{K}_0(Z[F])$ into $Wh(F \times Z)$ of [Rn] is the obstruction to $M_F \times S^1$ being a simple PD-complex. Since the group of self homotopy equivalences of M_F is finite [P1] M has a finite covering space which is homotopy equivalent to $M_F \times S^1$ and so this PD_4-complex is simple. Therefore M_F is finite. The final assertion is well known. □

It can be shown that if $\chi(M) = 0$ and π is virtually Z then the maximal finite normal subgroup F of π is either the group of an S^3-manifold or is one of the groups $Q(8a, b, c) \times Z/dZ$. If $\pi/F \cong Z$ then M is orientable unless $\pi \cong\ < a, t | tat^{-1} = a^r, a^s = 1 >$ where $r^2 \equiv -1$ modulo (s); otherwise $\pi/F \cong (Z/2Z) * (Z/2Z)$ and M is nonorientable. The argument of Theorem 1 of Chapter 7 of [H] extends to show that the homotopy type of M is determined by π and the class of the first nontrivial k-invariant in $H^4(\pi; \tilde{Z})$ (where $\tilde{Z} \cong \pi_3(M)$ as a $Z[\pi]$-module), modulo the action of $Aut(\pi)$ and $Aut(Z) = \{\pm 1\}$.

It is not true in general that a manifold M finitely covered by an $S^3 \times E^1$-manifold need admit a geometry. For instance, as $Wh(Z \oplus (Z/2Z)) = 0$ [Kw] and

$L_5(Z \oplus (Z/2Z), +) \cong Z^2$ by Theorem 13A.8 of [W] the set $S_{TOP}(RP^3 \times S^1)$ is infinite. As the group of self homotopy equivalences of $RP^3 \times S^1$ is finite there must be infinitely many distinct manifolds of this (simple) homotopy type. Although they are all doubly covered by $S^3 \times S^1$ only $RP^3 \times S^1$ itself is geometric. Moreover the arguments of [HM] may be extended to the 4-dimensional case by means of [FQ] and then imply that there is a closed orientable 4-manifold M with $\pi_1(M) = Q(24, 13, 1) \times Z$ and $\chi(M) = 0$; as $Q(24, 13, 1)$ is not isomorphic to a subgroup of $SO(4)$ which acts freely on S^3 such a manifold is not even homotopy equivalent to an $S^3 \times E^1$-manifold.

The nonsolvable cases

For the remaining geometries (apart from those with compact models) it is not yet known whether surgery arguments are valid, for the fundamental groups arising all contain nonabelian free subgroups. In the next theorem we use 3-manifold theory instead, to extend the characterization of aspherical mapping tori with product geometries.

Theorem 6. *Let ϕ be a self homeomorphism of a 3-manifold N with a geometry of type $H^2 \times E^1$ or $\tilde{SL}$. Then the mapping torus $M(\phi) = N \times_\phi S^1$ admits the corresponding product geometry if and only if the outer automorphism $[\phi_*]$ induced by ϕ has finite order in $Out(\pi_1(N))$. The mapping torus of a self homeomorphism ϕ of a hyperbolic 3-manifold N which is also Haken admits the geometry $H^3 \times E^1$.*

Proof. Let $\nu = \pi_1(N)$ and let t be an element of $\pi = \pi_1(M(\phi))$ which projects to a generator of $\pi_1(S^1)$. If $M(\phi)$ has geometry $\tilde{SL} \times E^1$ then π is a discrete cocompact subgroup of $Isom(\tilde{SL}) \times Isom(E^1)$. On replacing $M(\phi)$ by the 2-fold covering space $M(\phi^2)$, if necessary, we may assume that π is a discrete cocompact subgroup of $Isom(\tilde{SL}) \times R$. The radical of this group is R^2 and is central, and π meets the radical in a lattice subgroup L by [Ra:8.27]. Since the centre of ν is Z the image of L in π/ν is nontrivial. It follows easily that π has a subgroup σ of finite index isomorphic to $\nu \times Z$, and in particular that conjugation by $t^{[\pi:\sigma]}$ induces an inner automorphism of ν.

If $M(\phi)$ has geometry $H^2 \times E^2$ then π is a discrete cocompact subgroup of $Isom(H^2) \times Isom(E^2)$. In this case π meets $Isom(E^2)$ in a (torsion free) discrete cocompact subgroup and thus has a subgroup σ of finite index isomorphic to $\rho \times Z^2$, where ρ is a discrete cocompact subgroup of $Isom(H^2)$, and is a subgroup of ν (cf. Section 2 of [W2]). It again follows that $t^{[\pi:\sigma]}$ induces an inner automorphism of ν.

Conversely, suppose that N has a geometry of type $H^2 \times E^1$ or $\tilde{SL}$ and that $[\phi_*]$ has finite order in $Out(\nu)$. Then ϕ is homotopic to a self homeomorphism of (perhaps larger) finite order [Zn] and is therefore isotopic to such a self homeomorphism [S2,BO], which may be assumed to preserve the geometric structure [MS]. Thus we may assume that ϕ is an isometry. The self homeomorphism of $N \times R$ sending (n, r) to $(\phi(n), r+1)$ is then an isometry for the product geometry and the mapping torus has the product geometry.

If N is hyperbolic then by Mostow rigidity ϕ is homotopic to an isometry of finite order [Mo], and if also N is Haken then ϕ is isotopic to such an isometry [Wd], and again the mapping torus has the product geometry. □

If $[\phi_*]$ has infinite order and N is an $H^2 \times E^1$-manifold then $M(\phi)$ may admit the geometry $H^3 \times E^1$. However if $[\phi_*]$ has infinite order and N is an $\tilde{SL}$-manifold then $M(\phi)$ admits no geometric structure.

Theorem 7. *Let M be a closed 4-manifold with fundamental group π. Then M is virtually s-cobordant to an $H^2 \times E^2$- or $\tilde{SL} \times E^1$-manifold if and only if $\chi(M) = 0$, π has a free abelian normal subgroup A of rank 2 such that π/A is not virtually abelian and π has a normal subgroup of finite index isomorphic to $Z \times \rho$ where ρ/ρ' is infinite. The geometry is of type $H^2 \times E^2$ if and only if there is such a subgroup ρ which is itself isomorphic to a product $Z \times \sigma$.*

Proof. The conditions are necessary, for a manifold with such a geometry is finitely covered by the cartesian product of a surface of hyperbolic type with a torus or by the product of an $\tilde{SL}$-manifold with a circle.

Conversely, suppose that they hold. On replacing π by a subgroup of finite index if necessary we may assume that M is orientable and $\pi = Z \times \rho$ where $\rho \cap A = Z$ and is central in ρ, $\rho/\rho \cap A$ is infinite and ρ/ρ' is infinite. Since π/A is infinite $H^0(\pi/A; Z[\pi/A]) = 0$ by Lemma 8.1 of [B] and it then follows from the LHS spectral sequence for π as an extension of π/A by A (with coefficients $Z[\pi]$) that $H^s(\pi; Z[\pi]) = 0$ for $s \leq 2$. Therefore M is aspherical by Theorem 3 of Chapter 3 of [H] and so π is a PD_4^+-group. Hence ρ is a PD_3^+-group. By Theorem 6 of Chapter 5 of [H] ρ is the fundamental group of an orientable aspherical Seifert fibred 3-manifold N, which must be of $H^2 \times E^1$- or $\tilde{SL}$-type since $\rho/\rho \cap A$ is not virtually abelian. (Note that N is also Haken, since ρ/ρ' is infinite.) Since M is aspherical there is a homotopy equivalence $f : M \to N \times S^1$. Since $Wh(\pi) = 0$ by Lemma 1.1 of [Sk] this homotopy equivalence is simple, and since the surgery obstruction maps $\sigma_i : [S^i(N \times S^1), G/TOP] \to L_i(\pi)$ are isomorphisms by Proposition 2.6 of [Sk] M is in fact s-cobordant to $N \times S^1$, which admits a product geometry. The final assertion is clear. □

A similar argument shows that a 4-manifold M with fundamental group π is virtually s-cobordant to an $H^3 \times E^1$-manifold if and only if $\chi(M) = 0, \sqrt{\pi} = Z$ and π has a normal subgroup of finite index which is isomorphic to $Z \times \rho$ where ρ is a discrete cocompact subgroup of $PSL(2, C)$. If every PD_3-group is the fundamental group of an aspherical closed 3-manifold and if every atoroidal aspherical closed 3-manifold is hyperbolic we could replace the last assertion by the more intrinsic conditions that ρ have one end (which would suffice with the other conditions to imply that M aspherical and hence that ρ is a PD_3-group) and no noncyclic abelian subgroups (which would imply that any irreducible 3-manifold with fundamental group ρ is atoroidal). The surgery arguments may be justified by reference to the work of Farrell and Jones [FJ1,FJ2].

Theorem 8. *Let M be a closed 4-manifold with fundamental group π. Then M is virtually s-cobordant to an $S^2 \times H^2$-manifold if and only if π is virtually a PD_2-group, has no infinite abelian normal subgroup and $\chi(M) = 2\chi(\pi)$.*

Proof. Let Γ be a discrete cocompact subgroup of $Isom(S^2 \times H^2)$. On replacing Γ by a subgroup of finite index if necessary we may assume that it is a subgroup of $SO(3) \times PSL(2, R)$. Since $\Gamma \cap SO(3) \times 1$ acts freely and by orientation preserving maps on $S^2 \times H^2$ it is trivial. Therefore projection onto the second factor maps Γ monomorphically to a discrete cocompact subgroup of $PSL(2, R)$, and so it has a torsion free subgroup γ of finite index (cf. Theorem 4.10.1 of [ZVC]). The quotient space $\gamma\backslash S^2 \times H^2$ is the total space of an S^2-bundle over a closed orientable surface of hyperbolic type. Thus the conditions are necessary.

Conversely, suppose that they hold. On replacing π by a subgroup of finite index if necessary we may assume that M is orientable and π is a PD_2^+-group with no nontrivial abelian normal subgroup. By Theorem 2 of [H1] there is a simple homotopy equivalence f from M to the total space E of an S^2-bundle over a closed orientable surface of hyperbolic type. As we may assume that the structure group of the bundle is $SO(3)$ it follows easily that E is an $S^2 \times H^2$-manifold. Let $j : S^2 \to E$ be the inclusion of a fibre. Then $j_*[S^2]$ generates the kernel of the natural homomorphism from $H_2(E; Z/2Z) \cong (Z/2Z)^2$ to $H_2(\pi; Z/2Z) \cong Z/2Z$ and so is nontrivial, and $j^* w_2(E) = 0$. The argument of Theorem 5.1 of [CH] may be used to show that the normal invariant of the self homotopy equivalence f_α of E obtained by pinching off a 4-cell and mapping it in via $\alpha = j\eta S\eta$ (the generator of $\pi_4(E) = \pi_4(S^2)$) is nontrivial. In Section 2 of [H1] it is shown that there are at most 2 s-cobordism classes of (simple) homotopy equivalences with range E. Since id_E and f_α are distinguished by their normal invariants it follows that M is s-cobordant to E. □

Theorem 9. *Let M be a closed 4-manifold with fundamental group π. Then M is virtually s-cobordant to the cartesian product of two surfaces of hyperbolic type if* *and only if π is a PD_4-group, π has no nontrivial abelian normal subgroup, and π has a normal subgroup of finite index and of orientable type which is isomorphic to a nontrivial product $\sigma \times \tau$ where $\chi(M)[\pi : \sigma \times \tau] = (2 - \beta_1(\sigma))(2 - \beta_1(\tau))$.*

Proof. The conditions are clearly necessary. Conversely, suppose that they hold. On replacing π by a normal subgroup of finite index if necessary we may assume that M is orientable, π is orientable type and $\pi = \sigma \times \tau$. Since π has no nontrivial abelian normal subgroup neither factor can be infinite cyclic, and so σ and τ are each PD_2^+-groups, and therefore are fundamental groups of closed orientable surfaces, F_σ and F_τ say. Since $\sqrt{\sigma} = \sqrt{\tau} = 1$ these surfaces must be hyperbolic type. By Theorem 1 of [H1] the classifying map from M to $F_\sigma \times F_\tau = K(\pi, 1)$ is a simple homotopy equivalence, and by Theorem 5 of [H1] M is s-cobordant to $F_\sigma \times F_\tau$, which is clearly an $H^2 \times H^2$-manifold. □

Let $P = PSL(2, R)$ and let Γ be a dicrete cocompact subgroup of $P \times P$. Then $M = \Gamma \backslash H^2 \times H^2$ is a compact complex analytic surface. If $\Gamma \cap P \times 1$ or $\Gamma \cap 1 \times P$ is nontrivial then we may argue as in Theorem 6.3 of [W1] to conclude that M is virtually a cartesian product. Otherwise the natural foliations of $H^2 \times H^2$ descend to give a pair of transverse foliations of M by copies of H^2. If Γ is the group of units of a quaternion algebra over a real quadratic field F which splits at both infinite primes then it may be viewed as a discrete cocompact subgroup of $PSL(2, F_{\infty_1}) \times PSL(2, F_{\infty_2}) \cong P \times P$, and in [W2] it is observed that in such a case M is not virtually a cartesian product. Thus not every $H^2 \times H^2$-manifold satisfies the hypotheses of Theorem 7. (Cf. also page 177 of [BPV].)

For the other geometries of hyperbolic type H^4 and $H^2(C)$ necessary conditions for M to be s-cobordant to such a manifold are that π be isomorphic to a discrete torsion free cocompact subgroup of the isometry group and that $\chi(M) = \chi(\pi)$. These conditions are also sufficient if π is of orientable type (as always holds for $H^2(C)$) and the classifying map from M to $K(\pi, 1)$ has nonzero degree, by Theorem 7 of Chapter 3 of [H], which implies that the classifying map is then a homotopy equivalence, together with the recent work of Farrell and Jones [FJ3], which shows that $Wh(\pi) = 0$ and that the surgery obstruction maps are isomorphisms. At present there are not even conjectural intrinsic characterizations of the fundamental groups arising in these cases.

2-knots

A 2-knot is a locally flat (TOP) embedding $K : S^2 \to S^4$. If $M(K)$ is the closed orientable 4-manifold obtained from S^4 by surgery on K then $\pi K = \pi_1(M(K))$ has weight 1, $H_1(M(K); Z) = \pi K/\pi K' = Z$ and $\chi(M(K)) = 0$. Conversely, if M is a closed orientable 4-manifold with fundamental group π of weight 1 and $\chi(M) = 0$ then it may be obtained in this way, for surgery on a loop in M representing a normal generator of π gives a 1-connected 4-manifold Σ with $\chi(\Sigma) = 2$ which is thus homeomorphic to S^4 and which contains an embedded 2-sphere as the cocore of the surgery. (In particular, these conditions imply that $H_1(M; Z) = Z$.) If π is solvable then it has weight 1 if and only if $\pi/\pi' = H_1(M; Z)$ is cyclic. (Cf. [H] for more details.)

Theorem 10. *Let K be a 2-knot. Then $M(K)$ is homeomorphic to an infrasolvmanifold if and only if πK has a torsion free elementary amenable normal subgroup of Hirsch length greater than 2. No such knot manifold $M(K)$ admits a geometry of type $Nil^4, S^4, P^2(C), S^2 \times S^2, S^2 \times E^2, S^2 \times H^2, H^4, H^2(C), H^2 \times H^2$ or F^4.*

Proof. If $M(K)$ is an infrasolvmanifold then the above conditions on πK are certainly necessary. Conversely, if they hold πK is torsion free and virtually poly-Z of Hirsch length 4 and determines $M(K)$ up to homeomorphism, by Theorem 1. By Theorem 14 of Chapter 6 of [H] πK must be either $G(+)$ or $G(-), \pi(b, \epsilon)$ for some even b and $\epsilon = \pm 1$ or $\pi K' \cong Z^3$ or Γ_q for some odd q. If $\pi K = G(+)$ or $G(-)$ then it is virtually Z^4 and so $M(K)$ admits a geometry of type E^4. If $\pi K = \pi(b, \epsilon)$ then $M(K)$ is the mapping torus of the canonical involution of the 2-fold branched cover of S^3, branched over the Montesinos knot $K(o|b; (3,1), (3,1), (3,\epsilon))$, which is a Nil^3-manifold, and so $M(K)$ admits a geometry of type $Nil^3 \times E^1$. If $\pi K' \cong Z^3$ then K is a Cappell-Shaneson 2-knot, and up to change of orientations we may assume that the Alexander polynomial of K is $X^3 - mX^2 + (m-1)X - 1$ for some integer m. If $m \geq 6$ all the roots are positive and $M(K)$ admits a geometry of type $Sol^4_{m,m-1}$. If $0 \leq m \leq 5$ two of the roots are complex conjugates and $M(K)$ admits a geometry of type Sol^4_0. If $m < 0$ two of the roots are negative and πK has a subgroup of index 2 which is a discrete cocompact subgroup of the Lie group $Sol^4_{m',n'}$ where $m' = m^2 - 2m + 2$ and $n' = m^2 - 4m + 1$, so $M(K)$ admits a geometry of type $Sol^4_{m',n'}$.

If $\pi K' \cong \Gamma_q$ and the meridianal outer automorphism is of finite order then $q = 1$ and K is the 6-twist spin of the trefoil knot or its 5-fold branched cover (which is also its Gluck reconstruction). In this case $M(K)$ admits a geometry of type $Nil^3 \times E^1$. Otherwise if $\pi K' \cong \Gamma_q$ and the meridianal outer automorphism is of infinite order then $M(K)$ admits a geometry of type Sol^4_1.

It follows from the above discussion that there is no 2-knot K for which $M(K)$ admits a geometry of type Nil^4 and that many of the geometries of type $Sol^4_{m,n}$ do not occur. (In particular, $Sol^3 \times E^1$ does not occur.) The geometries $S^4, P^2(C)$ and $S^2 \times S^2$ cannot arise since πK must be infinite; the geometries $H^4, H^2(C), H^2 \times H^2$ and $S^2 \times H^2$ cannot arise since $\chi(M(K)) = 0$ (cf. Theorem 6.1 of [W2]); the geometry $S^2 \times E^2$ cannot arise since no group which is virtually Z^2 can have infinite cyclic abelianization and the geometry F^4 is not realized by any closed 4-manifold. □

If πK has a torsion free elementary amenable normal subgroup ρ with $h(\rho) = 2$ then by Theorem 3 either ρ has finite index in πK, in which case $\pi K \cong \Phi$, the group represented by $< a, t | tat^{-1} = a^2 >$ (by Theorem 6 of Chapter 4 of [H]), or ρ is virtually abelian. In the latter case πK has a torsion free abelian normal subgroup of rank 2 and $M(K)$ is aspherical. (See Chapter 5 of [H] for further constraints on such groups.) Less is known about the cases when $h(\rho) = 1$ (and so ρ is abelian of rank 1) and there is no such subgroup of greater length. (See Chapter 4 of [H] for results on such groups.)

We do not know whether there are any 2-knots K such that $M(K)$ admits a geometry of type $H^2 \times E^2$. If there is such a knot then $\sqrt{\pi K} \cong Z^2$ and $G = \pi K/\sqrt{\pi K}$ is virtually a surface group. If $\sqrt{\pi K} \leq \pi K'$ then G/G' is infinite cyclic and so G has a finite normal subgroup N such that G/N is a plane motion group [EM]. But examination of the presentations of such groups, as given in Theorem 4.5.6 of [Zg], shows that no such group has infinite cyclic abelianization. Therefore $\sqrt{\pi K}$ is not contained in $\pi K'$ and $M(K)$ must fibre over S^1 with fibre an $H^2 \times E^1$-manifold and monodromy of finite order.

The manifolds obtained from branched r-twist spins of (p, q)-torus knots with $p^{-1}+q^{-1}+r^{-1} < 1$ have geometry $\tilde{SL} \times E^1$, while those obtained from branched r-twist spins of simple (nontorus) knots with $r > 2$ have geometry $H^3 \times E^1$, excepting only the one from the 3-twist spin of the figure eight knot and its 2-fold branched cover (which is also its Gluck reconstruction), which has group $G(+)$ and geometry E^4. (The latter manifold is the only closed orientable E^4-manifold which is not "Seifert fibred" [U1].) Examples with geometry $S^3 \times E^1$ arise from 2-twist spins of 2-bridge knots and certain other "small" simple knots, and from branched r-twist spins of (p, q)-torus knots with $p^{-1} + q^{-1} + r^{-1} > 1$. (The case $p^{-1} + q^{-1} + r^{-1} = 1$ gives a $Nil^3 \times E^1$-manifold.)

If $M(K)$ is homeomorphic to a complex surface S then S is minimal, since $\beta_2(S) = 0$, and has Kodaira dimension $\kappa(S) = 1, 0$ or -1, since $\beta_1(S) = 1$ is odd. If $\kappa(S) = 1$ or 0 then S is elliptic and admits a compatible geometric structure, of type $\tilde{SL} \times E^1$ or $Nil^3 \times E^1$, respectively [U1,U2,W2]. The only complex surfaces with $\kappa(S) = -1$, $\beta_1(S) = 1$ and $\beta_2(S) = 0$ are Inoue surfaces, which are not elliptic, but admit compatible geometries of type Sol_0^4 or Sol_1^4, and Hopf surfaces[LYZ]. A Hopf surface is a complex surface whose universal covering space is homeomorphic to $S^3 \times R \cong C^2 \backslash \{0\}$. Some Hopf surfaces admit no compatible geometry [W1], and there are $S^3 \times E^1$-manifolds that admit no complex structure. The geometric Hopf surfaces are the elliptic surfaces of Kodaira dimension -1. By Theorem 4.5 of [W2], if $M(K)$ has a complex structure compatible with a geometry then the geometry is one of $Sol_0^4, Sol_1^4, Nil^3 \times E^1, S^3 \times E^1$ or $\tilde{SL} \times E^1$. Conversely, if $M(K)$ admits one of the first three of these geometries then it is homeomorphic to an Inoue surface of type S_M , an Inoue surface of type $S^{(+)}_{N,p,q,r;t}$ or $S^{(-)}_{N.p,q,r}$ or an elliptic surface of Kodaira dimension 0 respectively. (Cf. [In] and Chapter V of [BPV].)

Lemma. *Let K be a branched r-twist spin of the (p,q)-torus knot. Then $M(K)$ is homeomorphic to an elliptic surface.*

Proof. We shall adapt the argument of Lemma 1.1. of [Ml]. (Cf. also [Ne].) Let $V_0=\{(z_1+z_2+z_3)\in C^3\backslash\{0\}|z_1^p+z_2^q+z_3^r=0\}$, and define an action of $C^\times$ on V_0 by $u.v=(u^{qr}z_1,u^{pr}z_2,u^{pq}z_3)$ for all u in $C^\times$ and $v=(z_1,z_2,z_3)$ in V_0. Define fuctions $m:V_0\to R^+$ and $n:V_0\to m^{-1}(1)$ by $m(v)=(|z_1|^p+|z_2|^q+|z_3|^r)^{1/pqr}$ and $n(v)=m(v)^{-1}.v$ for all v in V_0. Then the map $(m,n):V_0\to m^{-1}(1)\times R^+$ is an R^+-equivariant homeomorphism, and so $m^{-1}(1)$ is homeomorphic to V_0/R^+. Therefore there is a homeomorphism from $m^{-1}(1)$ to the Brieskorn manifold $M(p,q,r)$, under which the action of the group of rth roots of unity on $m^{-1}(1)=V_0/R^+$ corresponds to the group of covering homeomorphisms of $M(p,q,r)$ as the branched cyclic cover of S^3, branched over the (p,q)-torus knot [Ml]. The manifold $M(K)$ is the mapping torus of some generator of this group of self homeomorphisms of $M(p,q,r)$. Let ω be the corresponding primitive r^{th} root of unity. If $t>1$ then $t\omega$ generates a subgroup Ω of $C^\times$ which acts freely and holomorphically on V_0, and the quotient V_0/Ω is an elliptic surface over the curve $V_0/C^\times$. Moreover V_0/Ω is homeomorphic to the mapping torus of the self homeomorphism of $m^{-1}(1)$ which maps v to $m(t\omega.v)^{-1}.t\omega.v=\omega m(t.v)^{-1}t.v$. Since this map is isotopic to the map sending v to $\omega.v$ this mapping torus is homeomorphic to $M(K)$. This proves the Lemma. □

The Kodaira dimension of the elliptic surface in the above lemma is 1,0 or -1 according as $p^{-1}+q^{-1}+r^{-1}$ is <1, 1 or >1. In the next theorem we shall settle the case of elliptic surfaces with $\kappa=-1$.

Theorem 11. *Let K be a 2-knot. Then $M(K)$ is homeomorphic to a Hopf surface if and only if K is a branched r-twist spin of the (p,q)-torus knot for some p,q and r such that $p^{-1}+q^{-1}+r^{-1}>1$.*

Proof. If K is such a branched twist spin then $M(K)$ is homeomorphic to an elliptic surface, by the Lemma, and the surface must be a Hopf surface if $p^{-1}+q^{-1}+r^{-1}>1$.

If $M(K)$ is homeomorphic to a Hopf surface then either K is trivial or πK is nonabelian; in either case πK is isomorphic to a subgroup of $GL(2,C)$ which contains a contraction (Kodaira - cf. [Ka]). Since the image of πK under the homomorphism $det:GL(2,C)\to C^\times$ is infinite and abelian $\pi K\cap SL(2,C)=\{g||det(g)|=1\}=\pi K'$. It then follows from Proposition 8 of [Ka] and Theorem 3 of Chapter 4 of [H] that $\pi K'$ is either cyclic of odd order n, the quaternion group (B_2 in the notation of [Ka]), binary tetrahedral or binary icosahedral. (There is an error in Lemma 4 of [Ka] - the normalizer $N_{SL(2,C)}(B_2)$ has an element of order 3.) Hence πK is isomorphic to the group of the 2-twist spin of the $(2,n)$-torus knot, or of the 3-,4- or 5-twist spin of the trefoil knot, respectively. Since such Hopf surfaces are determined up to diffeomorphism by their fundamental groups (by Theorems 10 and 12 of [Ka]) $M(K)$ is homeomorphic to the manifold of the corresponding torus knot. Since each weight class in one of these groups is realized by some branched twist spin of such a torus knot (cf. Table 1 of [PS]) and the Gluck reconstruction of a branched r-twist spin of a classical knot is another branched r-twist spin of that knot [Pl] the theorem follows. □

As observed above a knot manifold $M(K)$ is homeomorphic to an elliptic surface with $\kappa = 0$ if and only if it admits a geometry of type $Nil^3 \times E^1$. Such knots may be characterized algebraically by the conditions πK is virtually poly-Z and $\zeta\pi K \cong Z^2$. For the remaining class of elliptic surfaces (those with $\kappa = 1$) we must settle for a characterization up to s-cobordism.

Theorem 12. *Let K be a 2-knot. Then $M(K)$ is s-cobordant to an elliptic surface of Kodaira dimension 1 if and only if $\zeta\pi K \cong Z^2$ and $\pi K'$ has a subgroup of finite index with infinite abelianization but is neither virtually poly-Z nor virtually a product $Z \times \sigma$.*

Proof. If $M(K)$ is an elliptic surface of Kodaira dimension 1 then it admits a compatible geometry of type $\tilde{SL} \times E^1$ and πK is isomorphic to a discrete cocompact subgroup of $Isom_0(\tilde{SL}) \times R$, the maximal connected subgroup of $Isom(\tilde{SL} \times E^1)$, for the other components consist of orientation reversing or antiholomorphic isometries (cf. Theorem 3.3. of [W2]). Since πK meets $\zeta(Isom_0(\tilde{SL} \times R) \cong R^2$ in a lattice subgroup $\zeta\pi K \cong Z^2$ and since πK projects nontrivially onto the second factor $\pi K' = \pi K \cap Isom_0(\tilde{SL})$ and is the fundamental group of an $\tilde{SL}$-manifold. The necessity of the conditions is clear.

Conversely, suppose that they hold. Then $M(K)$ is aspherical, $\pi K'$ is the fundamental group of an aspherical Seifert fibred 3-manifold N and $\zeta\pi K$ is not contained in $\pi K'$, by Theorems 1,2 and 6 of Chapter 5 of [H]. Since $\pi K'$ is neither virtually poly-Z nor virtually a product N must be of type $\tilde{SL}$. Since $\zeta\pi K$ is not contained in $\pi K'$ the meridianal outer automorphism has finite order. This outer automorphism can be realized by a self homeomorphism Θ of N and the mapping torus $M(\Theta)$ has the product geometry by Theorem 6 above. As moreover the meridianal automorphism is orientation preserving and induces the identity on $\zeta\pi K' \cong Z$ the group πK is contained in $Isom_0(\tilde{SL}) \times R$, and so $M(\Theta)$ has a compatible structure as an elliptic surface of Kodaira dimension 1, by Theorem 3.3 of [W2]. Since $M(K)$ is aspherical it is homotopy equivalent to

$M(\Theta)$. The torus $\zeta(Isom_0(\tilde{SL}) \times R)/\zeta\pi K \cong R^2/Z^2$ acts effectively on $M(\Theta)$. Therefore $Wh(\pi K) = 0$ and the surgery obstruction maps are bijections [NS] and so any such homotopy equivalence is s-cobordant to a homeomorphism. □

An elliptic surface with Euler characteristic 0 is a Seifert fibred 4-manifold, and so is determined up to diffeomorphism by its fundamental group if the base orbifold is euclidean or hyperbolic [U1,U2]. Using this result (instead of [Ka]) together with Theorem 8 of Chapter 5 of [H] and the lemma above it may be shown that if $M(K)$ is homeomorphic to such an elliptic surface, $\zeta\pi K \cong Z^2$ and some power of a weight element is central in πK then $M(K)$ is homeomorphic to $M(K_1)$ where K_1 is some branched twist spin of a torus knot. However in general there may be infinitely many algebraically distinct weight classes in πK and we cannot conclude that K is itself such a branched twist spin.

References

[AJ] Auslander, L. and Johnson, F.E.A. *On a conjecture of C.T.C.Wall*, J.London Math. Soc.**14** (1976), 331-332.

[BPV] Barth, W.,Peters, C. and Van de Ven, A. Compact Complex Surfaces, Ergebnisse der Mathematik und ihrer Grenzgebiete 3 Folge, Bd 4, Springer-Verlag, Berlin-Heidelberg-New York-Tokyo (1984).

[BB] Baumslag, G. and Bieri, R. *Constructable solvable groups*, Math.Z.**151** (1976), 249-257.

[B] Bieri, R. Homological Dimension of Discrete Groups, Queen Mary College Mathematical Notes, London (1976).

[BS] Bieri, R. and Strebel, R. *Almost finitely presentable soluble groups*, Commentarii Math.Helvetici **53** (1978), 258-278.

[BO] Boileau, M. and Otal, J.-P. *Groupe des difféotopies de certaines variétés de Seifert*, C.R.Acad.Sci.Paris **303** (1986), 19-22.

[CH] Cochran, T.D. and Habegger, N. *On the homotopy theory of simply-connected four-manifolds*, Topology **29** (1990), 293-310.

[EM] Eckmann, B. and Müller, H. *Plane motion groups and virtual Poincaré duality groups*, Inventiones Math. **69** (1982), 293-310.

[FJ] Farrell, F.T. and Jones, L.E. Classical Aspherical Manifolds, CBMS Regional Conference series **75**, American Mathematical Society, Providence (1990).

[FJ1] Farrell, F.T. and Jones, L.E. *K-Theory and dynamics*, Ann.Math. **124** (1986), 531-569.

[FJ2] Farrell, F.T. and Jones, L.E. *A topological analogue of Mostow's rigidity theorem*, J.Amer.Math.Soc.**2** (1989), 257-370.

[FJ3] Farrell, F.T. and Jones, L.E. *Rigidity and other topological aspects of compact nonpositively curved manifolds*, Bull.Amer. Math.Soc. **22** (1990), 59-64.

[F] Filipkiewicz, R.O. Four-Dimensional Geometries, Ph.D thesis, University of Warwick (1984).

[Fr] Freedman, M.H. *The disk theorem for four-dimensional manifolds*, in Proceedings of the International Congress of Mathematicians, Warsaw (1983), 647-663.

[FQ] Freedman,. M.H. and Quinn, F. The Topology of Four-Manifolds, Princeton University Press, Princeton (1990).

[GM] Geoghegan, R. and Mihalik, M.L. *A note on the vanishing of* $H^n(G; Z[G])$, J.Pure Appl.Alg.**39** (1986), 301-304.

[HM] Hambleton, I. and Madsen, I. *Actions of finite groups on* R^{n+k} *with fixed point set* R^k, Canadian Math. J. **38** (1986), 781-860.

[H] Hillman, J.A. 2-Knots and their Groups, Australian Mathematical Society Lecture Series **5**, Cambridge University Press, London-New York-Melbourne (1989).

[H1] Hillman, J.A. *On 4-manifolds homotopy equivalent to surface bundles over surfaces*, Topology Appl., in press.

[H2] Hillman, J.A. *Elementary amenable groups and 4-manifolds with Euler characteristic 0*, J.Austral. Math. Soc. **50** (1991), 160-170.

[HL] Hillman, J.A. and Linell, P.A. *Elementary amenable groups of finite Hirsch length are locally finite by virtually solvable*, J.Austral. Math. Soc. (to appear).

[Iw] Iwasawa, K. *On some types of topological groups*, Ann. Math. **50** (1949), 507-558.

[In] Inoue, M. *On surfaces of type of class* VII_0, Inventiones Math. **24** (1974), 269-310.

[KLR] Kamishima, Y., Lee, K.B. and Raymond, F. *The Seifert construction and its applications to infranilmanifolds*, Quarterly J.Math. Oxford **34** (1983), 433-452.

[Ka] Kato, M. *Topology of Hopf surfaces*, J.Math.Soc.Japan **27** (1975), 222-238. *Erratum*, ibid.**41** (1989), 173.

[Kr] Kropholler, P.H. *Cohomological dimensions of soluble groups*, J.Pure Appl.Alg. **43** (1986), 281-287.

[Kw] Kwasik, S. *On low-dimensional s-cobordisms*, Commentarii Math. Helvetici **61** (1986), 415-428.

[LYZ] Li, J., Yau, S.-T. and Zheng, F. *A simple proof of Bogomolov's theorem on Class* VII_0 *surfaces with* $b_0 = 0$, Illinois J. Math. **34** (1990), 217-220.

[MS] Meeks, W.H.,III and Scott, G.P. *Finite group actions on 3-manifolds*, Inventiones Math. **86** (1986), 287-346.

[Mh] Mihalik, M.L. *Solvable groups that are simply connected at* ∞, Math.Z. **195** (1987), 79-87.

[Ml] Milnor, J.W. *On the 3-dimensional Brieskorn manifolds* $M(p, q, r)$, in Knots, Groups and 3-Manifolds (edited by L.P.Neuwirth), Princeton University Press, Princeton (1975), 175-225.

[Mo] Mostow, G.D. *Quasi-conformal mappings in* n*-space and the rigidity of hyperbolic space forms*, Publ.Math.I.H.E.S.**34** (1986), 53-104.

[Ne] Neumann, W.D. *Geometry of quasihomogeneous surface singularities*, in Singularities (edited by P.Orlik), PSPM **40**, American Mathematical Society, Providence (1983), vol. 2, 245-258.

[NS] Nicas, A. and Stark, C.W. K*-theory and surgery of codimension-two torus actions on aspherical manifolds*, J.London Math.Soc. **31** (1985), 173-183.

[Ni] Nisse, M. *Structure affine des infranilvariétés et des infrasolvariétés*, C.R.Acad. Sci. Paris **310** (1990), 667-670.

[P1] Plotnick, S.P. *Homotopy equivalences and free modules*, Topology **21** (1982), 91-99.

[P2] Plotnick, S.P. *Fibered knots in S^4 - twisting, spining, rolling, surgery, and branching*, in Four-Manifold Theory (edited by C.McA.Gordon and R.C.Kirby), CONM **35**, American Mathematical Society, Providence (1984), 437-459.

[PS] Plotnick, S.P. and Suciu, A.I. *Fibered knots and spherical space forms*, J.London Math.Soc. **35** (1987), 514-526.

[Ra] Ragunathan, M.S. Discrete Subgroups of Lie Groups, Ergebnisse der Mathematik **68**, Springer-Verlag, Berlin-Heidelberg-New York (1972).

[Rn] Ranicki, A. *Algebraic and geometric splittings of the K- and L-groups of polynomial extensions* in "Transformation Groups", Poznan 1985 (edited by S. Jackowski and K. Pawalowski), Lecture Notes in Mathematics 1217, Springer-Verlag, Berlin-Heidelberg-New York (1986), 321-363.

[Ro] Robinson, D.J.S. A Course in the Theory of Groups, Graduate Texts in Mathematics **80**, Springer-Verlag, Berlin-Heidelberg-New York (1982).

[Ru] Ruh, E.A. *Almost flat manifolds*, J.Diff.Geom. **17** (1982), 1-14.

[S1] Scott, G.P. *The geometries of 3-manifolds*, Bull.London. Math. Soc. **15** (1983), 401-487.

[S2] Scott, G.P. *Homotopy implies isotopy for some Seifert fibre spaces*, Topology **24** (1985), 341-351.

[Sk] Stark, C.W. *Structure sets vanish for certain bundles over Seifert manifolds*, Trans. Amer.Math.Soc.**285** (1984), 603-615.

[Sl] Strebel, R. *A remark on subgroups of infinite index in Poincaré duality groups*, Commentarii Math.Helvetici **52** (1977), 317-324.

[Th] Thurston, W. *Three-dimensional manifolds, Kleinian groups and hyperbolic geometry*, Bull.Amer.Math.Soc. **6** (1982), 357-381.

[U1] Ue, M. *Geometric structures in the sense of Thurston and Seifert 4-manifolds I*, J.Math.Soc.Japan **42** (1990), 511-540.

[U2] Ue, M. *Geometric structures in the sense of Thurston and Seifert 4-manifolds II*, J.Math.Soc.Japan **43** (1991), 149-183.

[Wd] Waldhausen, F. *On irreducible 3-manifolds which are sufficiently large*, Ann. Math. **87** (1968), 56-88.

[W] Wall, C.T.C. Surgery on Compact Manifolds, Academic Press, London-New York (1970).

[W1] Wall, C.T.C. *Geometries and geometric structures in real dimension 4 and complex dimension 2*, in Geometry and Topology (edited by J.Alexander and J.Harer), Lecture Notes in Mathematics **1167**, Springer-Verlag, Berlin-Heidelberg-New York (1985), 268-292.

[W2] Wall, C.T.C. *Geometric structures on compact analytic surfaces*, Topology **25** (1986), 119-153.

[Wo] Wolf, J.A. Spaces of Constant Curvature, fifth edition, Publish or Perish Inc., Wilmington (1984).

[ZVC] Zieschang, H., Vogt, E. and Coldewey, H.D. Surfaces and Planar Discontinuous Groups, Lecture Notes in Mathematics **835**, Springer-Verlag, Berlin-Heidelberg-New York (1980).

[Zg] Zieschang, H. Finite Groups of Mapping Classes of Surfaces, Lecture Notes in Mathematics **875**, Springer-Verlag, Berlin-Heidelberg-New York (1981).

[Zn] Zimmermann, B. *Uber Gruppen von Homoömorphismen Seifertscher Faserräume und flacher Mannigfaltigkeiten*, Manuscripta Math. **30** (1980), 361-373.

Satellites and surgery invariants

H.R. Morton and P.M. Strickland

Abstract. A satellite formula relating the quantum invariants of a satellite knot and those of its companion and pattern links will be described briefly. The $SU(2)_q$ invariants of a framed k-component link L, when the variable q is replaced by an rth root of unity yield a natural map

$$J_r(L) : \mathcal{R}_r^{\otimes k} \to \Lambda_r,$$

where $\mathcal{R}_r$ is a finite-dimensional truncation of the representation ring of $SU(2)$, and $\Lambda_r = \mathbf{Z}[e^{\pi i/2r}]$. The effect on J_r of a framing change on L is given by applying a suitable power of an automorphism F_r of $\mathcal{R}_r$ to each factor.

Use of the satellite formula for a simple choice of companion exhibits Reshetikhin and Turaev's invariant of the 3-manifold given by surgery on L as the evaluation of $J_r(L)$ on a fixed element M_0 in each $\mathcal{R}_r$, after slight normalisation. Explicit calculation of M_0 can be made easily because of a beautiful relation between F_r and the invariants $J_r(H)$ of the Hopf link. This relation can be viewed in terms of an action of $\mathrm{PSL}(2, \mathbf{Z})$ on $\mathcal{R}_r$, at least up to scalar multiples by roots of unity, and shows how the invariant of the manifold given by Dehn surgery with coefficients a_i/b_i on a link L can be found by evaluating $J_r(L)$ on suitable elements $M_{a_i/b_i} \in \mathcal{R}_r$. An indication is also given of how these results extend to other quantum groups.

In the final section we give an explicit formula for the invariant when any Dehn surgery is used, confirming its correctness via the Rolfsen moves.

Introduction

This is an account of a 3-manifold invariant for $SU(2)_q$ which was conceived, following Reshetikhin and Turaev's original description, as a direct approach with the emphasis on using the multilinearity and the explicit formula of the satellite calculations in [MS], avoiding specialisation of link invariants to a root of unity until as late as possible. Much of the paper is an expansion of a talk presented in Oberwolfach in September 1989. Its eventual form followed the unexpected discovery, prompted by explicit calculations, that apart from scalar factors, as

Knots 90

detailed later, the invariants of the Hopf link, together with the factors associated with a change of framing, can be organised to represent the modular group $\mathrm{PSL}(2, \mathbf{Z})$, once the variable is specialised to a root of unity. This suggests a way to calculate the invariant for manifolds given by general Dehn surgery from a link, in terms of the link invariants. The conformal field theory approaches outlined by Segal as a concrete means of handling Witten's ideas make this very plausible, although suitable normalisations to deal with the scalar factors have been surprisingly elusive.

Independent work, both by Kirby and Melvin [KM], and very elegantly by Lickorish [L], has provided for different aspects of the invariants to be explored. Our approach is somewhere between these two. We use the quantum group $SU(2)_q$ as in our development [MS] of Kirillov and Reshetikhin [KR], and draw on the explicit form of multilinearity of parallels and satellites given there. We do not use the finite-dimensional Hopf algebras, where a root of unity has been introduced. There is then no need to take account of the more complicated representation theory which arises in that case, as is done in [RT], and avoided by the use of clever arguments in [KM].

Roots of unity will appear here only in specialisations of existing invariants, and we develop enough background to ensure that such moves are completely legitimate where we need them.

1. Link invariants

The $SU(2)_q$ invariants of a framed link L with k components are described in [KR]. Assign irreducible $SU(2)_q$ modules W_{i_j} to the jth component of L, and there is an invariant $J(L; W_{i_1}, \ldots, W_{i_k}) \in \Lambda = \mathbf{Z}[q^{\pm 1/4}]$. The definition can be extended multilinearly to allow the use of a Λ-linear combination of modules on each component. The invariants for L can then be viewed as a single Λ-linear map

$$J(L) : \mathcal{R}^{\otimes k} \to \Lambda,$$

where $\mathcal{R}$ is the representation ring of $SU(2)_q$. It can be shown, for example in [MS], that the product in the ring $\mathcal{R}$ has a nice interpretation in terms of invariants of parallel links. The result can be summarised as follows.

Let $m : \mathcal{R} \otimes \mathcal{R} \to \mathcal{R}$ be the product in the ring, and let L be a framed knot. Then

$$J(L) \circ m = J(L^{(2)}),$$

where $L^{(2)}$ is the framed 2-parallel of L.

This result extends naturally to multiparallels of links with more than one component, as in [MS]. It allows for an alternative description of $J(L)$ to that in terms of the irreducible module assignments.

The ring $\mathcal{R}$ is spanned by the irreducible representations W_i, one of each dimension i, or equally well by the powers $(W_2)^j$, since $\mathcal{R} \cong \Lambda[W_2]$ as a ring. Since the evaluation of $J(L)$ on the element W_2 for each component is essentially the bracket polynomial version of the Jones polynomial of L it is then possible to calculate $J(L)$ in terms of the Jones polynomials for multiparallels of L. This corresponds to the use of powers of W_2 as the Λ-basis for $\mathcal{R}$ rather than the irreducibles. The change of basis information needed to pass from one basis to the other is noted in [MS], and also in a nice form in [KM]. The basis of powers has been used by Lickorish [L] in his approach to the 3-manifold invariants, allowing him to avoid any use of the other irreducibles $W_i, i > 2$, and so present the manifold invariants without having to consider the quantum group $SU(2)_q$ at all.

Satellites. Where a link P is given which has one distinguished unknotted component, the remaining components form a closed tangle relative to the distinguished component as axis. The tangle can then be used as a pattern to form satellites of given companion knots or links, based on the pattern P. The relation between the invariants of the companion, satellite and pattern is summarised in the next theorem, from [MS].

Theorem 1.1. *Let P be a pattern link with k components, with one distinguished unknotted component. There is a Λ-linear map $G : \mathcal{R}^{\otimes(k-1)} \to \mathcal{R}$ such that any satellite K formed from a companion C using P as pattern has invariant*

$$J(K) = J(C) \circ G.$$

Proof. This is given in [MS] by constructing G in terms of the basis of irreducibles for $\mathcal{R}$. If C has more than one component then G is used on that component which is to be embellished by the pattern. □

Remark. Although it is not initially clear that the ring Λ can be used without extension to permit some denominators, this follows by working with the basis of powers of W_2 and using the skein relations for the bracket polynomials on the resulting tangles in the construction of G.

The pattern link P itself can be considered as a satellite of the Hopf link H using P as pattern, so that

$$J(P) = J(H) \circ (G \otimes \mathrm{id}_{\mathcal{R}}).$$

The map G can then be recovered from the invariants $J(H)$ and $J(P)$.

We always assume that we are considering links with a given choice of framing, and that when satellite and parallel constructions are made they respect the framing. It is easy to calculate the change which takes place in the invariant when the same underlying link is used, but the framing on one or more components is altered.

Theorem 1.2. *There is a linear isomorphism $F : \mathcal{R} \to \mathcal{R}$ which can be used on the copy of $\mathcal{R}$ corresponding to one component, L_i say, of L before applying $J(L)$, and will then give the invariant J for the link whose framing on L_i has been increased by one.*

Proof. It is known [KR] how the framing change on a component affects the invariant when an irreducible is selected in $\mathcal{R}$ for that component. This determines the map F explicitly by $F(W_i) = f_i W_i$ where the 'framing factor' f_i is given by $f_i = (-1)^{i-1} a^{i^2-1}$. □

Remark. We retain the sign in f_i as in [KR] and [MS]. Kirby and Melvin use a variant where the sign does not appear, but this needs a little caution in interpreting the relation with the bracket polynomial. One source of signs can be accounted for by considering $-W_2$ in place of W_2 as the polynomial generator for $\mathcal{R}$.

2. Roots of unity

We now consider the behaviour of J when the variable q in Λ is specialised to be an rth root of unity. We shall suppose that $q^{1/4}$ is a primitive $4r$th root of unity, and we consider the ring Λ_r given by factoring out the cyclotomic polynomial φ_{4r} generated by the $4r$-th root a in $\Lambda = \mathbf{Z}[a^{\pm 1}]$. We then have a specialisation homomorphism $e_r : \Lambda \to \Lambda_r = \Lambda / < \varphi_{4r}(a) >$.

Proposition 2.1. *For any link L the evaluation $e_r(J(L; W_{i_1}, \ldots, W_{i_k})) = 0$ if $W_{i_j} = W_r$ for any j.*

Proof. For each j we can find $\lambda_j \in \Lambda$ such that $J(L; W_{i_1}, \ldots, W_{i_k}) = \lambda_j \delta_{i_j}$, where $\delta_i = (-1)^{i-1} \dfrac{a^{2i} - a^{-2i}}{a^2 - a^{-2}}$, as in [MS]. Now $e_r(\delta_r) = 0$. □

Corollary 2.2. *If $V_j \in \mathcal{R}$ lies in the ideal generated by W_r, for some j, then $e_r(J(L; V_1, \ldots, V_k)) = 0$.*

Proof. Without loss of generality we may consider the case where $k = 1$, and $V = W_r V'$ with $V' \in \mathcal{R}$. Then

$$e_r(J(L;V)) = e_r(J(L^{(2)};W_r,V')) = 0.\square$$

Notation. Write $W_r = \Delta_r(W_2)$ as a polynomial in W_2. The polynomial Δ_r satisfies $\Delta_r(2\cos\theta) = \sin r\theta/\sin\theta$, and is closely related to the Tchebychev polynomial T_r with $T_r(2\cos\theta) = \cos r\theta$.

Write $\mathcal{R}_r = (\mathcal{R} \otimes \Lambda_r)/< W_r > \cong \Lambda_r[W_2]/< \Delta_r(W_2) >$, and write $p_r : \mathcal{R} \to \mathcal{R}_r$ for the projection. The ring $\mathcal{R}_r$ is known as a *Verlinde algebra*; similar algebras may be defined for representation rings of other simple Lie groups.

We shall extend the definition in $\mathcal{R}$ to allow W_i with $i \leq 0$ by setting $W_{-k} = -W_k$ for $k > 0$. Multiplication in $\mathcal{R}$ can then be described simply, for $i,j > 0$ as

$$W_i W_j = \sum_{k=i-j+1}^{i+j-1} W_k = \sum_{k=j-i+1}^{j+i-1} W_k,$$

where the sum is in steps of 2, [MS], since the excess terms in one of these sums will cancel. (Under a suitable convention for sums, the same result holds for all i,j.) It then follows that

$$(W_{j+1} - W_{j-1})W_r = W_{r+j} + W_{r-j}, \text{ for all } r,j.$$

Proposition 2.3. *The invariant* $J(L) : \mathcal{R}^{\otimes k} \to \Lambda$ *induces a* Λ_r*-linear map* $J_r(L) : \mathcal{R}_r^{\otimes k} \to \Lambda_r$ *with* $e_r \circ J(L) = J_r(L) \circ p_r^{\otimes k}$.

Proof. The map $e_r \circ J(L)$ is zero on the kernel of $p_r^{\otimes k}$, by corollary 2.2. □

Proposition 2.4. *The isomorphism* $F : \mathcal{R} \to \mathcal{R}$ *induces* $F_r : \mathcal{R}_r \to \mathcal{R}_r$ *such that* $F_r \circ p_r = p_r \circ F$.

Proof. We must show that $p_r \circ F$ is zero on the ideal generated by W_r. This ideal is spanned by the elements W_r and

$$(W_{j+1} - W_{j-1})W_r = W_{r-j} + W_{r+j} \quad \text{for } j \in \mathbf{N}.$$

Now

$$p_r \circ F(W_r) = f_r p_r(W_r) = 0$$

and

$$\begin{aligned}
&p_r \circ F(W_{r-j} + W_{r+j}) \\
&\qquad = p_r\left((-1)^{r-j-1}a^{(r-j)^2-1}W_{r-j} + (-1)^{r+j-1}a^{(r+j)^2-1}W_{r+j}\right) \\
&\qquad = (-1)^{r-j-1}a^{r^2+j^2-1}\left(a^{-2rj}p_r(W_{r-j}) + a^{2rj}p_r(W_{r+j})\right) = 0
\end{aligned}$$

as $p_r(W_{r-j}) = -p_r(W_{r+j})$ and $e_r(a^{4r}) = 1$. □

Notation. We abuse notation by writing W_i for $p_r(W_i) \in \mathcal{R}_r$. We shall write $i \in j \otimes k$ for i, j, k between 1 and $r-1$ to mean that W_i has non-zero coefficient in the product $W_j \otimes W_k$ in $\mathcal{R}_r$.

Then we have

Lemma 2.5. $j \in i \otimes k \Leftrightarrow k \in i \otimes j$.

Proof. We have $W_{r+n} = -W_{r-n}$ in $\mathcal{R}_r$. Thus

$$W_i \otimes W_k = \sum_{|i-k|+1}^{m} W_j$$

where $m = \min(i+k-1, 2r-i-k-1)$, and the sum runs in steps of two. If we let $\alpha = (i-1)/2$, $\beta = (j-1)/2$ and $\gamma = (k-1)/2$ then the condition for W_i to be a summand of $W_j \otimes W_k$ is that α, β and γ form the sides of a triangle, with perimeter an integer less that $r-1$. Since this is clearly a symmetric condition the lemma is proved. □

Note: In the case of a deformation of a general simple Lie algebra, the statement would need to be altered to allow for multiplicities, and for the fact that conjugate representations enter in.

We may now use the reduced invariants $J_r(L)$ of framed links L, which can be calculated from their standard $SU(2)_q$ invariant $J(L)$, to determine the Reshetikhin-Turaev invariant of the manifold given by surgery on L. We construct $U \in \mathcal{R}_r$ so that the invariant in Λ_r is a simple multiple of $J_r(L; U, \ldots, U)$.

Write $M^3(L)$ for the manifold constructed from a framed link L by surgery.

Theorem (Kirby, Fenn-Rourke).

1. *Every closed oriented M^3 arises in this way.*
2. *There is an orientation preserving homeomorphism $M^3(L) \cong M^3(L')$ if and only if L, L' are related by a sequence of Kirby moves.*

Kirby moves are of two types, shown in figure 1.

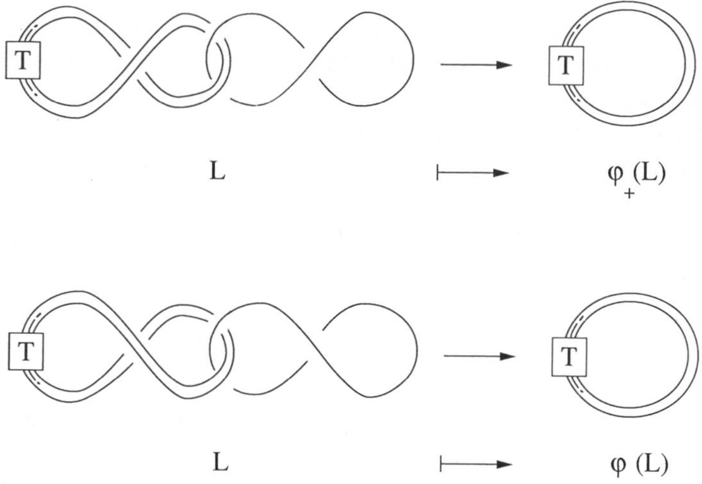

Figure 1

Remark. Fenn and Rourke [FR] showed that Kirby's original moves could be reduced to these.

In all diagrams we shall use the convention that the link is framed with the *planar* framing, i.e. the chosen parallel is given by the edge of a ribbon following one side of the curve in the plane of the diagram.

To a framed *oriented* link $L = L_1 \cup L_2 \cup \ldots \cup L_k$ we can associate a quadratic form with $k \times k$ matrix (ℓ_{ij}) where

$$\ell_{ij} = \mathrm{lk}(L_i, L_j), \ i \neq j,$$

$$\ell_{ii} = \text{framing on } L_i.$$

Write $\mathrm{sig}(L)$ for the signature of this form. (This is not generally the signature of the link L in the usual sense.)

Then $\mathrm{sig}(L)$ is independent of the choice of orientation of L, and

$$\mathrm{sig}\,\varphi_\pm(L) = \mathrm{sig}\,L \mp 1.$$

Invariants of $M^3(L)$

For each root of unity a, find $c(a), T_L(a) \in \Lambda_r$, for each L, such that

$$\begin{aligned} T_{\varphi_+(L)} &= c(a) T_L \\ T_{\varphi_-(L)} &= (c(a))^{-1} T_L \end{aligned}.$$

Then $\mathcal{I}(M) = c(a)^{\mathrm{sig}\, L} T_L$ depends only on $M(L)$.

To demonstrate independence of the invariant on the choice of framed link it is sufficient to ensure that it is unaltered by the Kirby moves. We must then compare the invariants of the two links L and $\varphi_+(L)$.

The pattern determined by the tangle T defines $G : \mathcal{R}^{\otimes k-1} \to \mathcal{R}$. For a fixed assignment of elements $V_1, \ldots, V_{k-1}$ of $\mathcal{R}$ to the strings of T write $X = G(V_1, \ldots, V_{k-1})$. The satellite formula then shows that, for a choice of element $Y \in \mathcal{R}$ on the unknotted component of the first link, this link has invariant $J(H; F(X), F(Y))$. This must be compared with the invariant of the second link, which is $J(O; X) = J(H; X, W_1)$, where O is the unknot, as indicated in figure 2.

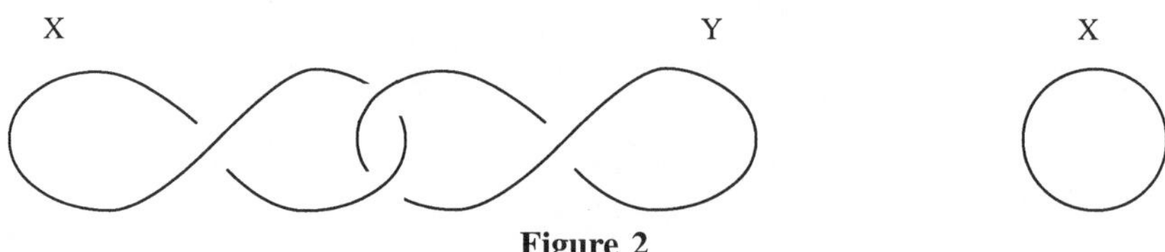

Figure 2

The reduced invariants of the two links will then be $J_r(H; F_r(X), F_r(Y))$ and $J_r(H; X, W_1)$.

Notation. Write $< \ , \ >_r$ for the bilinear form on $\mathcal{R}_r$ determined by $J_r(H)$ as

$$< V, W >_r = J_r(H; V, W).$$

Theorem 2.6. *The element $U = \sum_{j=1}^{r-1} \delta_j W_j \in \mathcal{R}_r$ satisfies*

$$< F_r(X), F_r(U) >_r = c_+ < X, W_1 >_r$$
$$< F_r^{-1}(X), F_r^{-1}(U) >_r = c_- < X, W_1 >_r$$

for every $X \in \mathcal{R}_r$, where $c_+ = \sum f_j \delta_j^2$, and $c_- = \bar{c}_+ \in \Lambda_r$.

Notation. Write $c_+ = \rho(a) c(a)$ with $\rho(a) > 0$ and $|c(a)| = 1$.

Remark. The element $c(a)^2$ turns out to be the power a^{2r^2-r-6} of a.

Corollary 2.7. *The element $\rho(a)^{-k} c(a)^{\mathrm{sig}\,(L)} J_r(L; U, \ldots, U)$ depends only on the manifold given from the k component framed link L by surgery, where* sig (L) *is the signature of the quadratic form determined by the linking numbers and framings of L.*

Proof. Take $T_L = J_r(L; \rho(a)^{-1} U, \ldots, \rho(a)^{-1} U)$ and use theorem 2.6 to compare the invariants arising from L and $\varphi_\pm(L)$. □

Remark. Apart from a factor of $c(a)$ this gives the invariant of Reshetikhin and Turaev. Kirby and Melvin use exactly this normalisation, which ensures that oppositely oriented manifolds have conjugate invariants.

We shall give the proof of theorem 2.6 shortly, in the context of further properties of F_r and $< \ , \ >_r$, leading to a means for finding the invariant for the manifold given by general Dehn surgery on a link in terms of the invariant of the link.

It is helpful to view each copy of $\mathcal{R}_r$ associated to a link component as depending on a choice of parallel and meridian for the peripheral torus. The automorphism F_r corresponds to altering the choice of parallel, by Dehn twists about the meridian, to allow for integer framing change when calculating the invariant of the link exterior. We use the bilinear form $< \ , \ >_r$ to construct another automorphism which will correspond to a Dehn twist about the parallel.

We shall prove

Theorem 2.8. *For any $X \in \mathcal{R}_r$ we have*

$$< X, U >_r = < F_r(X), U >_r.$$

Theorem 2.9. *The symmetric bilinear form $< \ , \ >_r$ is non-degenerate. Its matrix H_r relative to the basis of irreducibles $W_1, \ldots, W_{r-1}$ satisfies*

$$H_r^2 = \rho(a)^2 I.$$

Definition. We may then define $\Phi_r : \mathcal{R}_r \to \mathcal{R}_r$ to be the adjoint of F_r, that is $< F_r(X), Y >_r = < X, \Phi_r(Y) >_r$ for all X, Y.

The matrices of F_r and Φ_r in the basis of irreducibles are then related by $\Phi_r = H_r^{-1} F_r H_r$. This necessitates extending the coefficient ring Λ_r to include an inverse for $\det(H)$; making $2r$ invertible will be sufficient, as we shall see in the proof of theorem 2.9. In order to include $\rho(a)$ and $c(a)$ we also need the square root of two, or equivalently of i; one possibility would be to take Λ_r to be the cyclotomic field generated by the $8r$th roots of unity.

Corollary 2.10. *The element U is an eigenvector of Φ_r with eigenvalue* 1.

Proof. We have $< X, U >_r = < F_r(X), U >_r = < X, \Phi_r(U) >_r$ for all X, so $U = \Phi_r(U)$. □

Remark. From theorem 2.6 we have

$$c_+ < X, W_1 >_r = < F_r(X), F_r(U) >_r = < X, \Phi_r F_r(U) >_r$$

so that $\Phi_r F_r(U) = c_+ W_1$. This could be used as a definition of U.

Proof of theorem 2.6. To give a self-contained argument we shall work in coordinates relative to the basis of irreducibles in $\mathcal{R}_r$. Let X have coordinate vector $\mathbf{x} = (x_1, \ldots, x_{r-1})$. Now U has coordinate vector $\delta = (\delta_1, \ldots, \delta_{r-1})$ which is also the first column of the matrix H_r so we must show that

$$\mathbf{x}^T F_r H_r F_r \delta = c_+ \mathbf{x}^\mathbf{T} \delta,$$

for all $\mathbf{x}$.

It is then enough to show that the vector $\delta = (\delta_1, \delta_2, \ldots \delta_{r-1})$ is an eigenvector of the matrix $F_r H_r F_r$ with eigenvalue $c_+ = \sum_1^{r-1} f_k \delta_k^2$.

The ijth entry of H_r is given by $\sum\limits_{k \in i \otimes j} f_k f_j^{-1} f_i^{-1} \delta_k$.

So the ith entry of $F_r H_r F_r \delta$ is $\sum\limits_{j=1}^{r-1} \sum\limits_{k \in i \otimes j} f_k \delta_k \delta_j$. By lemma 2.5, we can rearrange this sum as

$$\sum_{k=1}^{r-1} f_k \delta_k \sum_{j \in i \otimes k} \delta_j = \sum_{k=1}^{r-1} f_k \delta_k \delta_k \delta_i = c_+ \delta_i,$$

and the result is proved.

Conjugation gives the other half of the result, since $\bar{\delta} = \delta$ and $\overline{H}_r = H_r$ while $\overline{F}_r = F_r^{-1}$ and $\bar{c}_+ = c_-$. □

We can represent theorem 2.6 diagrammatically by figure 3.

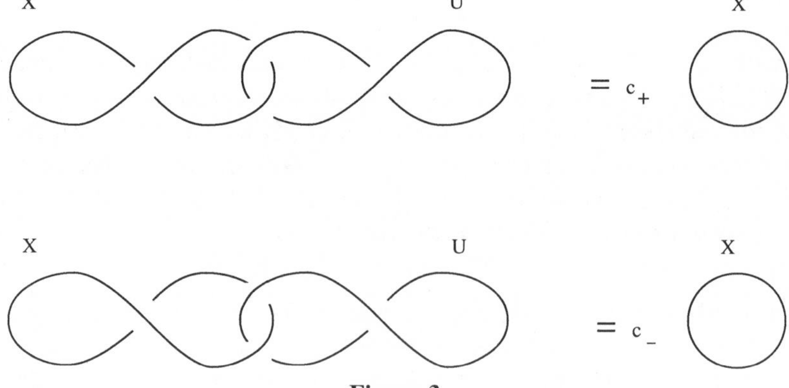

Figure 3

As noted before theorem 2.6, it then follows from the satellite formula that the invariants of the two links L and $\varphi_+(L)$ are the same, up to the factor c_+, when U is used on the unknotted string, with the same assignments made to the strings of the tangle T in each case.

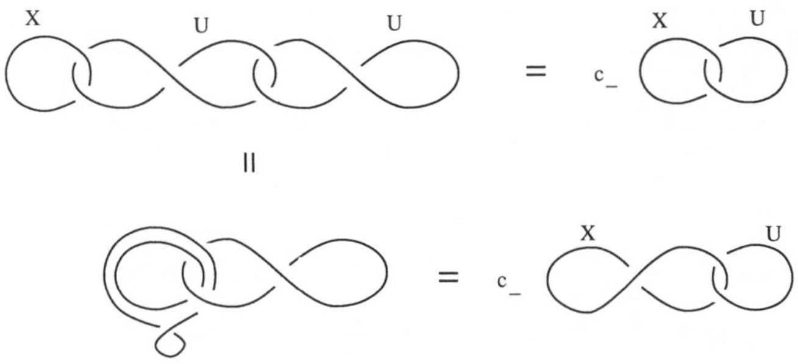

Figure 4

Proof of theorem 2.8. Applying this result to the link shown in figure 4 in which each component labelled with the element U plays the role of the unknotted curve in turn, for a suitable choice of T, gives two links with the same invariant, one being $< X, U >_r$ and the other $< F_r(X), U >_r$. □

The original proof of theorem 2.9, which we give here, relies on explicit knowledge of the entries in H_r, and follows the details in [S]. A similar technique has also been used to show that $(H_r F_r)^3$ is a scalar matrix. In the next section we give a more diagrammatic argument for these results, using connected sums of links, and the other results of this section.

Proof of theorem 2.9. As shown in [MS], the invariant $J(H; W_i, W_j)$ of a Hopf link labelled by irreducibles W_i and W_j is

$$H_{ij} = (-1)^{i+j} \frac{s^{ij} - s^{-ij}}{s - s^{-1}},$$

where $s = a^2$. Let c_{ik} be the general entry in H_r^2, then

$$\begin{aligned}(-1)^{i+k}(s - s^{-1})^2 c_{ik} &= \sum_{j=1}^{r-1} (s^{ij} - s^{-ij})(s^{jk} - s^{-jk}) \\ &= \sum_{j=1}^{r-1} (s^{i+k})^j + (s^{-(i+k)})^j - \sum_{j=1}^{r-1} (s^{i-k})^j + (s^{-(i-k)})^j.\end{aligned}$$

Writing $z_1 = s^{i+k}$, $z_2 = s^{i-k}$ we have

$$\begin{aligned}\sum_{j=-(r-1)}^{r-1} (z_1^j - z_2^j) &= \sum_{j=1}^{r-1} (z_1^j + z_1^{-j}) - \sum_{j=1}^{r-1} (z_2^j + z_2^{-j}) + z_1^r - z_2^r + z_1^0 - z_2^0 \\ &= (s - s^{-1})^2 c_{ik} + (-1)^{i+k} - (-1)^{i-k} \\ &= (s - s^{-1})^2 c_{ik}.\end{aligned}$$

Now the sum of any $2r$ consecutive powers of a $2r$th root of unity, other than 1, is zero; and $z_2 = 1$ exactly when $i = j$, whereas $z_1 \neq 1$. Then

$$(-1)^{i+j}(s - s^{-1})^2 c_{ij} = \begin{cases} 0, & i \neq j \\ -2r, & i = j. \end{cases}$$

This shows that $H_r^2 = \dfrac{-2r}{(s - s^{-1})^2} I$. We can now use the proof of theorem 2.6 to identify the scalar with $\rho(a)^2$. For we have $F_r H_r F_r \delta = c_+ \delta$ and $F_r^{-1} H_r F_r^{-1} \delta = c_- \delta$. It follows that

$$F_r H_r^2 F_r^{-1} \delta = c_+ c_- \delta = \rho(\alpha)^2 \delta.$$

On the other hand, $F_r H_r^2 F_r^{-1} = \dfrac{-2r}{(s - s^{-1})^2} I$, and so $\rho(a)^2 = \dfrac{-2r}{(s - s^{-1})^2}$, completing the proof. □

3. Modular group

In this section we shall show that F_r and Φ_r, as automorphisms of $\mathcal{R}_r$, obey the same relations as the generators $\begin{pmatrix} 1 & 1 \\ 0 & 1 \end{pmatrix}$ and $\begin{pmatrix} 1 & 0 \\ -1 & 1 \end{pmatrix}$ of $\mathrm{SL}(2, \mathbf{Z})$, up to multiplication by powers of the scalar $c(a)^2$, giving us a 'projective' representation for the modular group $\mathrm{PSL}(2, \mathbf{Z})$ on $\mathcal{R}_r$. As John Humphreys has pointed out, one could easily make this a genuine representation of a central extension of PSL(2,**Z**) by an element whose $4r$th power was the identity.

Theorem 3.1. *The automorphisms F_r and Φ_r of $\mathcal{R}_r$ satisfy*

$$F_r \Phi_r F_r = \Phi_r F_r \Phi_r.$$

Theorem 3.2. *The automorphism $(F_r \Phi_r F_r)^2$ is scalar multiplication by $c(a)^2$.*

These two results will follow by establishing

Proposition 3.3. *The matrices H_r and F_r satisfy*

$$(H_r F_r)^3 = c_+ \rho(a)^2 I = \rho(a)^3 c(a) I.$$

Proof of theorem 3.1. Using the matrices in the basis of irreducibles, we have

$$\begin{aligned} F_r\Phi_r F_r &= F_r H_r^{-1} F_r H_r F_r \\ &= \rho(a)^{-2} F_r H_r F_r H_r F_r \text{ by theorem 2.9} \\ &= c_+ H_r^{-1} \text{ by theorem 3.3} \end{aligned}$$

and

$$\begin{aligned} \Phi_r F_r \Phi_r &= H_r^{-1} F_r H_r F_r H_r^{-1} F_r H_r \\ &= \rho(a)^{-2} H_r^{-1} F_r H_r F_r H_r F_r H_r \\ &= c_+ H_r^{-1}. \end{aligned}$$

□

Proof of theorem 3.2. Again using the matrices we have

$$\begin{aligned} (F_r\Phi_r F_r)^2 &= F_r\Phi_r F_r\Phi_r F_r\Phi_r \\ &= \rho(a)^{-6}(F_r H_r)^6 \\ &= \rho(a)^{-6}(\rho(a)^3 c(a))^2 I \\ &= c(a)^2 I. \end{aligned}$$

□

In order to prove proposition 3.3, we will use the work of the previous section, together with the following result on the invariant of a connected sum of two links;

Lemma 3.4. *Let $K = K_1 \cup \ldots \cup K_m$ and let $L = L_1 \cup \ldots \cup L_n$ be two framed links. Let W be any irreducible representation of $SU(2)_q$, and let $L\#K$ denote the connected sum of L and K along the first components of L and K. Then*

$$\begin{aligned} &J(L; W, X_2, \ldots, X_m) J(K; W, Y_2, \ldots, Y_n) \\ &= \delta_W J(L\#K; W, X_2, \ldots, X_m, Y_2, \ldots, Y_n) \end{aligned}$$

for any X_i and Y_j in $\mathcal{R}$.

Proof. Present K and L as the closures of 1-1 tangles S and T on the first strings; then the connected sum will be the closure of the tangle ST. By Schur's lemma, the invariants of S and T are scalars σ and τ say times the identity map on W. The invariants of K, L and $K\#L$ are then $\sigma\delta_W$, $\tau\delta_W$ and $\sigma\tau\delta_W$ respectively, proving the lemma. □

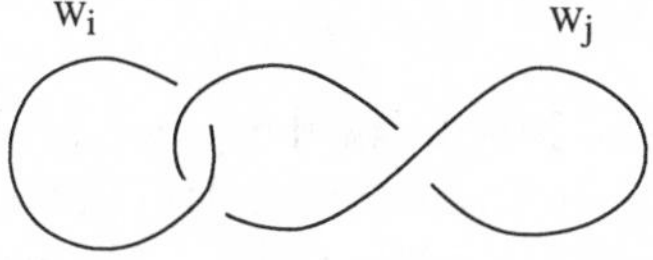

Figure 5

Proof of proposition 3.3. The ijth entry, b_{ij} say, of the matrix H_rF_r is $J_r(H; W_i, F_r(W_j))$. This is the $SU(2)_q$ invariant of a Hopf link with a positive twist on the second component, and labelled by representations W_i and W_j, as shown in figure 5. Let T_{ijk} be the invariant of the link shown in figure 6, with elements W_i, W_j and W_k assigned to the components as shown. Regarding the link as the connected sum of two links shows, by lemma 3.4, that $\delta_j T_{ijk} = b_{ij}b_{jk}$. The ikth entry, c_{ik} say, of $(H_rF_r)^2$ is then $\sum_{j=1}^{r-1} \delta_j T_{ijk}$.

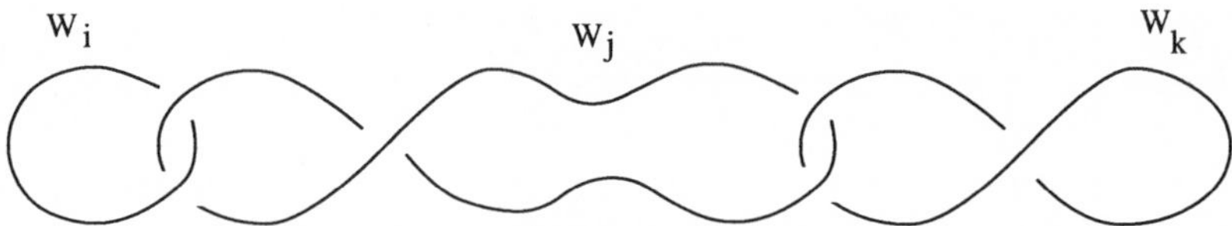

Figure 6

We may rewrite this sum as the invariant of the same 3-component link, in which the central component has the element $\sum_{j=1}^{r-1} \delta_j W_j = U$ attached, while the other two strings have W_i and W_k respectively. Make a positive Kirby move on this central string, to get a new framed link L, as in figure 7, and then $c_{ik} = c_+ J_r(L; W_i, W_k)$, by theorem 2.6. Now L is a Hopf link, with altered framing, and its invariant $J_r(L; W_i, W_k) = J_r(H; F_r^{-1}(W_i), W_k)$ is the ikth entry of $F_r^{-1}H_r$.

Then $(H_rF_r)^2 = c_+F_r^{-1}H_r$, giving $(H_rF_r)^3 = c_+\rho(a)^2 I$. □

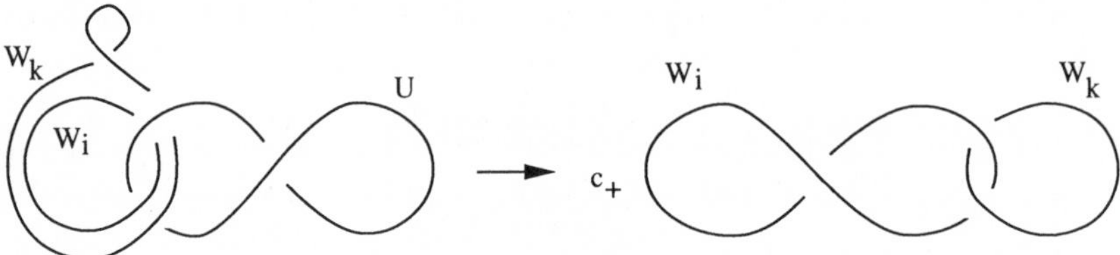

Figure 7

We now give an alternative proof of theorem 2.9, using the same methods. This proof does not use explicit knowledge of the Hopf link invariants, and can be modified to give a similar result for other quantum groups, with a permutation matrix (of order 2) in place of the identity.

Alternative proof of theorem 2.9. As in the proof of 3.3, we can write the ikth element, c_{ik}, of H_r^2 as the invariant $J_r(L; W_i, U, W_k)$ of a 3-component link, which is the connected sum of two Hopf links, this time with zero framing on all components, as shown in figure 8.

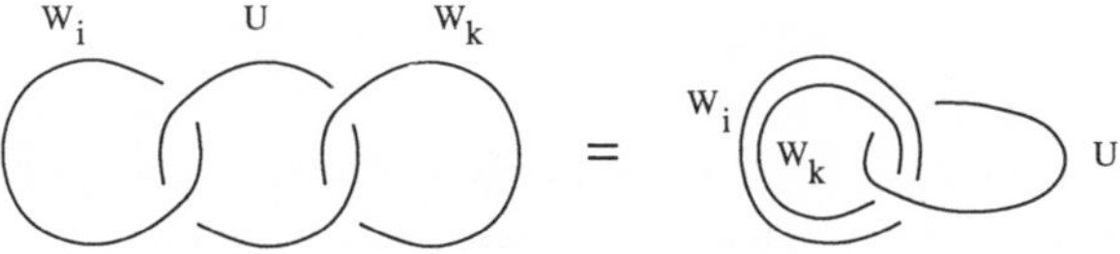

Figure 8

Now this 3-component link is the 2-parallel of a Hopf link, in which one of the components is replaced by two, carrying the elements W_i and W_k. By the result for invariants of parallels, we can then write

$$c_{ik} = J_r(L; W_i, U, W_k) = J_r(H; W_i W_k, U) = \sum_{j \in i \otimes k} J_r(H; W_j, U).$$

Since $1 \in i \otimes k$ if and only if $i = k$, for $1 \leq i, k \leq r-1$, it is enough to show that $J_r(H; W_j, U) = 0$ when $1 < j \leq r-1$, giving $c_{ik} = 0$ for $i \neq k$ and $H_r^2 = J_r(H; W_1, U)I$. Now $J_r(H; W_j, U) = J_r(H; F_r(W_j), U)$ $= e_r(f_j) J_r(H; W_j, U)$, using theorem 2.8, and $e_r(f_j) = a^{j^2-1} = 1$ in Λ_r for $j \leq r-1$ if and only if $j = 1$, at least for r prime. An explicit proof that $J_r(H; W_j, U) = 0$ for $j \neq 1$ is needed to complete the proof in general by this method.

As for the earlier proof of theorem 2.9, it now follows, knowing that H_r^2 is a scalar matrix, that the scalar is $\rho(a)^2$. This gives another calculation for $\rho(a)^2$ since $J_r(H; W_1, U) = \sum_{j=1}^{r-1} \delta_j^2$, which is clearly the product of the first row and column of H_r. □

For general simple Lie algebras, and hence for their quantum groups by [Ro], it can be shown that if U, V and W are three irreducible representations, then U is a summand of $V \otimes W$ if and only if $V^\dagger$ is a summand of $W \otimes U^\dagger$, where $\dagger$ denotes the conjugate representation. Thus the one dimensional representation will be contained as a summand of $V \otimes W$ exactly when V and W are conjugate. In the case of quantum groups having representations which are not self conjugate (so that the invariants are orientation dependent), this has the consequence that the matrices H_r do not square to a scalar, but to a multiple of the permutation matrix P_r which interchanges conjugate modules. On the other hand, it is still true that the universal module is a scalar times the sum of $\delta_W W$ as W runs over the relevant irreducibles; this follows because the δ for W is identical to that for $W^\dagger$, as changing the orientation of an unknotted component does not alter a link. In particular, following the proof of theorem 3.3 will show that $(H_r F_r)^3$ is now a multiple of P_r. It can be seen that the theorems 3.1 and 3.2 will also go through for other quantum groups, as P_r commutes with H_r and F_r, and hence does not affect the proof in any essential way.

4. General Dehn surgery

In the final section we indicate how the modular group action allows us to handle the manifold invariant for a manifold given from a framed link L by general Dehn surgery. The principle adopted is to regard the invariant $J_r(L)$ of a link as an invariant of a 3-manifold with boundary components which carry a choice of parallel and meridian. In the spirit of Segal's views of Witten's work it is appropriate to think of the map $J_r(L) : \mathcal{R}_r^{\otimes k} \to \Lambda_r$ as determined by the link exterior; each boundary component, with chosen parallel and meridian coordinates, having associated with it a copy of $\mathcal{R}_r$. Evaluation of $J_r(L)$ at $W_1 \in \mathcal{R}_r$ for a given boundary component gives the invariant of the link with that boundary component removed, and can be thought of as the manifold given by attaching a solid torus whose meridian disc spans the meridian of the boundary torus.

To perform any other surgery, say on a (p, q) curve (relative to the meridian and parallel coordinates on L), we apply an automorphism of the modular group, as represented by F_r and Φ_r on $\mathcal{R}_r$, to the appropriate copy of $\mathcal{R}_r$ which will carry the meridian $(1, 0)$ to the (p, q) curve before evaluating $J_r(L)$ at W_1. Thus the invariant of the new manifold, with one fewer boundary component, might be expected to be given by evaluating $J_r(L)$ on an element $M_{p/q} = \theta_{p/q}(W_1)$. In this notation we should write $W_1 = M_\infty$.

While there is some choice of automorphism to carry the meridian to the (p, q) curve there will be no ambiguity, apart from powers of $c(a)$, in the choice of $M_{p/q}$ because the automorphisms will differ on W_1 by an automorphism which carries the meridian to itself. These automorphisms are represented by powers of F_r, and $F_r(W_1) = W_1$. The precise choice will be governed by the fact that the signature of the generalised linking matrix of L, whose diagonal entries are now possibly fractional framings, may be changed under the third Rolfsen move described below; and this will affect the calculation of corollary 2.7.

We take the following to constitute the exact definition of M_α for each α. First, define $M_\infty = W_1$, and then let

$$\begin{aligned} F_r(M_\alpha) &= M_{\alpha+1}, \\ \Phi_r(M_\alpha) &= \begin{cases} M_{\alpha'} & \text{if } \alpha < 1 \\ c(a)^2 M_{\alpha'} & \text{if } \alpha > 1 \end{cases}, \quad \text{with } \alpha' = \frac{\alpha}{1-\alpha}. \\ \Phi_r(M_1) &= c(a) M_\infty \\ \Phi_r(M_\infty) &= c(a) M_{-1} \end{aligned}$$

To show that this leads to a well-defined choice, we must prove that

$$\begin{aligned} \Phi_r F_r \Phi_r(M_\alpha) &= F_r \Phi_r F_r(M_\alpha) \qquad \text{and} \\ (F_r \Phi_r F_r)^2(M_\alpha) &= c(a)^2 M_\alpha \end{aligned}$$

for all α. To do this we shall temporarily allow F and Φ to stand for their counterparts in $\mathrm{PSL}(2,\mathbf{Z})$ acting on $\mathbf{QP}^1$, so that we can represent the two products above by

$$\alpha \overset{\Phi}{\longmapsto} \frac{\alpha}{1-\alpha} \overset{F}{\longmapsto} \frac{1}{1-\alpha} \overset{\Phi}{\longmapsto} -\frac{1}{\alpha}$$

$$\alpha \overset{F}{\longmapsto} \alpha+1 \overset{\Phi}{\longmapsto} \frac{\alpha+1}{-\alpha} \overset{F}{\longmapsto} -\frac{1}{\alpha}$$

The second route will give rise to a factor of $c(a)^2$ exactly when $\alpha > 0$ (∞ is neither positive nor negative).The first route will do the same if $\alpha > 1$ (on the first application of Φ) or $1 > \alpha > 0$ (on the second); if $\alpha = 1$, the factor will also arise in two halves. For $\alpha = \infty$ or 0, either route introduces a factor of $c(a)$.

For the second identity above, we follow the second route throughout. If α is finite and non-zero, then exactly one of α and $-\frac{1}{\alpha}$ is positive, so the factor $c(a)^2$ comes in; for $\alpha = 0$ or ∞ we get two separate factors of $c(a)$.

Definition. A 3-manifold is said to be given Dehn surgery on a *framed* k-component link L, with surgery coefficients $(\alpha_1, \ldots, \alpha_k), \alpha_i \in \mathbf{Q} \cup \{\infty\}$, when it is constructed by gluing a solid torus to each boundary component of the link exterior along a curve of slope α_i relative to the meridian and chosen parallel for the ith component.

Thus slope ∞ will always refer to the meridian, while slope 0 will give the parallel chosen by the framing.

Rolfsen [R] shows that if two oriented manifolds given by Dehn surgeries on links L, L' are homeomorphic then L and L' are related by a sequence of moves of three types. (Rolfsen only used framing zero, but the modifications for arbitrary framing are straightforward.)

I. Change the framing, and the surgery coefficients so that the underlying link and surgery curves are unchanged. This has the effect of adding or subtracting an integer to the surgery coefficient when the framing is changed on a component.

II. Add or remove a component with surgery coefficient ∞.

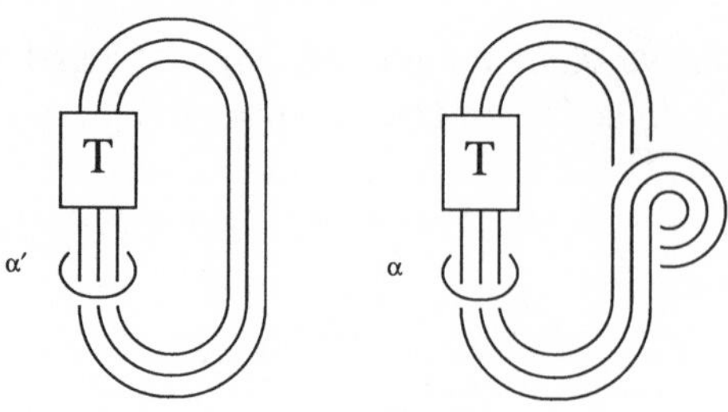

Figure 9

III. Replace a link L which has one distinguished unknotted component of framing 0 by L' as shown in figure 9. The framing in each diagram is planar (so that the framing on components has been changed depending on their linking number with the unknotted curve) and the surgery coefficients are unchanged on all except the unknotted component, where the coefficient α' becomes $\alpha = \dfrac{\alpha'}{1+\alpha'}$.

Theorem 4.1. *Let a closed 3-manifold be given by Dehn surgery on a framed k-component link L with surgery coefficients $\alpha_1, \ldots, \alpha_k$ relative to the framing coordinates on L. Then the invariant of the manifold can be calculated as*

$$c(a)^{\mathrm{sig}(L)} J_r(L; M_{\alpha_1}, \ldots, M_{\alpha_k}),$$

where $sig(L)$ is calculated as the signature of the linking matrix with the absolute surgery coefficients (i.e. relative to the topological framings) down the diagonal, omitting any components with infinite surgery coefficients.

In order to prove this result, we will need the following lemma

Lemma 4.2. *Let A and A' be the $k \times k$ matrices*

$$\begin{pmatrix} \alpha & \ell^T \\ \ell & B \end{pmatrix} \quad \text{and} \quad \begin{pmatrix} \alpha' & \ell^T \\ \ell & B - \ell\ell^T \end{pmatrix}$$

where $\alpha' = \frac{\alpha}{1-\alpha}$, $\alpha \neq 1$ and ℓ and B are $(k-1) \times 1$ and $(k-1) \times (k-1)$ matrices respectively. Then the signatures of these matrices are related by

$$\mathrm{sig}(A) = \mathrm{sig}(A') + \mathrm{sign}(\alpha) - \mathrm{sign}(\alpha').$$

Proof. Let P be the matrix $\begin{pmatrix} 1 & 0 \\ -\alpha^{-1}\ell & I \end{pmatrix}$; then, for $\alpha \neq 0$,

$$PAP^T = \begin{pmatrix} \alpha & 0 \\ 0 & B - \alpha^{-1}\ell\ell^T \end{pmatrix}.$$

So $\text{sig}(A) = \text{sign}(\alpha) + \text{sig}(B - \alpha^{-1}\ell\ell^T)$. Similarly,

$$\text{sig}(A') = \text{sign}(\alpha') + \text{sig}(B - \ell\ell^T - \alpha'^{-1}\ell\ell^T).$$

But $1 + \alpha'^{-1} = \alpha^{-1}$, so the last term in each of these expressions is the same, proving the lemma for $\alpha \neq 0$.

When $\alpha = 0$, we take $P = \begin{pmatrix} 1 & 0 \\ -\frac{1}{2}\ell & I \end{pmatrix}$, so that $PAP^T = A'$, completing the proof. □

Proof of theorem 4.1. We use Rolfsen's moves on framed links to pass between surgery descriptions of a manifold.

It is readily seen that the first two moves leave the invariant unaltered, since with the definitions chosen $M_\infty = W_1$ has the effect of ignoring a component, while the reframing works because $F_r(M_\alpha) = M_{\alpha+1}$. Neither of these affect the generalised linking matrix.

When move III is applied the two links L and L' will yield invariants $< X, M_{\alpha'} >_r$ and $< F_r(X), M_\alpha >_r$, for some X, by the satellite theorem. Now

$$< F_r(X), M_\alpha >_r = < X, \Phi_r(M_\alpha) >_r$$

by the adjoint property of F_r and Φ_r, and we have $\Phi_r(M_\alpha) = M_{\alpha'}$ (up to powers of $c(a)$). Any discrepancy in the power of $c(a)$ is compensated for by a change in the signature of the generalised linking matrix, as follows. Let A and A' be the matrices for the two links in figure 9; then these are related as in lemma 4.2. For $\alpha < 1$ we have $\text{sign}(\alpha) = \text{sign}(\alpha')$ and $\Phi_r(M_\alpha) = M_{\alpha'}$. For $\alpha > 1$ we have $\text{sign}(\alpha) = \text{sign}(\alpha') + 2$, compensating for the fact that $\Phi_r(M_\alpha) = c(a)^2 M_{\alpha'}$. Finally, for $\alpha = 1, \alpha' = \infty$ the method of the lemma shows that $\text{sig}(A) = 1 + \text{sig}(B - \ell\ell^T)$; and for $\alpha = \infty, \alpha' = -1$ we have $\text{sig}(A') = -1 + \text{sig(B)}$, which deals with the special cases. □

We have been able to make some calculations for lens spaces given both by framed surgery on torus knots, and also by Dehn surgery on the unknot, which confirm the above result.

Acknowledgment The second author was supported during this work by SERC grant no. GR/D/98662.

References

[FR] Fenn, R.A. and Rourke, C.P. *On Kirby's calculus of links*. Topology 18 (1979), 1-15.

[K] Kirby, R.C. *A calculus for framed links in* S^3. Invent. Math. 45 (1978), 35-56.

[KM] Kirby, R.C. and Melvin, P.M. *On the 3-manifold invariants of Witten and Reshetikhin-Turaev for* $sl(2, \mathbf{C})$. Preprint, Berkeley 1990.

[KR] Kirillov, A.N. and Reshetikhin, N. Yu. *Representations of the algebra* $U_q(sl_2)$, *q-orthogonal polynomials and invariants of links*. LOMI preprint E-9-88, Leningrad 1988.

[L] Lickorish, W.B.R. *3-manifolds and the Temperley-Lieb algebra*. Preprint UCLA 1990.

[MS] Morton, H.R. and Strickland, P.M. *Jones polynomial invariants for knots and satellites*. Math. Proc. Camb. Phil. Soc. **109** (1991), 83-103.

[RT] Reshetikhin, N.Yu. and Turaev, V.G. *Invariants of 3-manifolds via link polynomials and quantum groups*. To appear in Inv. Math.

[R] Rolfsen, D. *Rational surgery calculus: extension of Kirby's theorem*. Pacific Journal of Math. 110 (1984), 377-386.

[Ro] Rosso, M. *Finite dimensional representations of the quantum analog of the enveloping algebra of a complex simple Lie algebra*. Commun. Math. Phys. **117** (1988), 581-593.

[S] Strickland P.M. $SU(2)_q$*-invariants of manifolds and links at roots of unity*. Preprint, Liverpool University 1990.

[W] Witten, E. *Quantum field theory and the Jones polynomial*. Commun. Math. Phys. **121** (1989), 351-399.

Hermitian forms on periodic modules and even-dimensional links

*M. Farber**

In this paper we will study a class of Hermitian forms on $\mathbb{Z}$-torsion modules over the group ring of a free group F_μ. These forms emerge geometrically as higher Blanchfield forms on homology modules associated to even-dimensional links.

The first part of the paper (sections 1-3) is devoted to an algebraic study of these forms, using the general treatment of link modules developed in [F]. It is shown that the structure of the form is determined completely by a finite algebraic object consisting of a minimal lattice plus a scalar form on that lattice. In the case of knots ($\mu = 1$), but only in this case, the minimal lattice coincides with the whole module and the scalar form constructed here reduces to the linking form found by J. Levine [L1] and the author [F1]. This shows a sense in which the properties of knots ($\mu = 1$) and of links ($\mu > 1$) are quite different.

In the second part of the paper these algebraic results are applied to study torsion links. These are simple even-dimensional links with the middle-dimensional Alexander module being odd-$\mathbb{Z}$-torsion (the precise definition is given in §4). Any torsion link determines a form (a "secondary" Blanchfield form) which is precisely of the type studied in the first (algebraic) part of the paper. It is proved here that this gives a one-to-one correspondence between the isotopy classes of torsion links and the isomorphism classes of Hermitian forms. This result is in the spirit of the well-known Trotter-Kearton theorem [T],[K1] about simple odd-dimensional knots; formally it is a generalization of [Ko]; cf. also [F2].

The results of Browder-Levine [BL] and Browder [Br] show that any torsion knot (= torsion link of one component) admits a unique minimal Seifert manifold which is the fiber of the fibering of the knot complement over the circle. We will prove here that torsion links share this uniqueness property: the minimal Seifert manifold is unique up to ambient isotopy. It seems to be an interesting

* The research was supported by grant No. 88-00114 from the United States-Israel Binational Science Foundation (BSF), Jerusalem, Israel

Knots 90

open question to understand the structure of the link complement in terms of this minimal Seifert manifold.

We show also that the homology of the minimal Seifert manifold provides a complete finite algebraic invariant of a torsion link.

I would like to thank J. Levine and J. Hillman for useful discussions. I am also grateful to the Department of Mathematics of Sydney University, for the facilities they provided for my visit there.

§1. Lattices in link modules

This section gives a brief review of some definitions and results of [F] that will be used in the present paper.

Fix an integer $\mu > 0$ and a subring $k \subset \mathbb{Q}$. Let F_μ denote the free group on μ generators $t_1, \dots, t_\mu$ and let $\Lambda = k[F_\mu]$ be the group ring.

1.1 A left Λ-module M has the *Sato property* if $\mathrm{Tors}_q^\Lambda(k, M) = 0$ for all q, where k is regarded as a right Λ-module with trivial action via the augmentation map. As was shown by Sato [S], this condition is equivalent to the following: the map

$$M^\mu = \underbrace{M \times \cdots \times M}_{\mu \text{ times}} \to M \ ,$$

given by $(m_1, \dots, m_\mu) \mapsto \sum_{i=1}^{\mu}(t_i - 1)m_i$ is a bijection. In other words, each $m \in M$ has unique representation in the form

$$m = \sum_{i=1}^{\mu}(t_i - 1)m_i \ .$$

Let us define *"derivations"* $\partial_i : M \to M$, $i = 1, \dots, \mu$, by

$$\partial_i(m) = m_i \ ,$$

where $m_i \in M$ is the element appearing in the above decomposition. Thus,

$$m = \sum_{i=1}^{\mu}(t_i - 1)\partial_i(m) \ , \qquad m \in M \ .$$

If $\lambda \in \Lambda$, then

$$\partial_i(\lambda m) = \partial_i(\lambda)m + \varepsilon(\lambda)\partial_i(m) \ ,$$

where $\partial_i(\lambda) \in \Lambda$ is the Fox derivative with respect to t_i [CF], and $\varepsilon(\lambda) \in k$ is the augmentation.

We can think of M as also having a left module structure over the ring $D = k[\partial_1, \dots, \partial_\mu]$ of polynomials in the non-commuting variables $\partial_1, \dots, \partial_\mu$. Any Λ-

homomorphism $f : M_1 \to M_2$ between modules having the Sato property is also a D-homomorphism. The converse is also true. Thus

$$\mathrm{Hom}_\Lambda(M_1, M_2) = \mathrm{Hom}_D(M_1, M_2) \ .$$

1.2 The most important example of a module with the Sato property is the following:

Let $\Gamma = k\,[[x_1, \ldots, x_\mu]]$ be the ring of formal power series of non-commuting variables $x_1, \ldots, x_\mu$. The ring Λ may be embedded in Γ via the Magnus embedding $t_i \mapsto 1 + x_i,\ \ t_i^{-1} \mapsto 1 - x_i + x_i^2 - x_i^3 + \cdots$. Then Γ/Λ is a left Λ-module with the Sato property. The derivation $\partial_i : \Gamma/\Lambda \to \Gamma/\Lambda$ acts as cancellation of x_i from the left on monomials containing x_i on the leftmost position, and sends all other monomials to zero.

In fact, the above-mentioned rule defines an additive map $\partial_i : \Gamma \to \Gamma$ with the property

$$\gamma = \varepsilon(\gamma) + \sum_{i=1}^{\mu} x_i \partial_i(\gamma) \ ,$$

where $\varepsilon(\gamma) \in k$ is the augmentation. ∂_i maps Λ into itself and the restriction $\partial_i|_\Lambda$ coincides with the Fox derivative $\partial/\partial t_i$ [CF].

These remarks allow us to introduce a left D-module structure on Γ and Λ, which will be used later.

1.3 A *module of type* L is a left finitely generated Λ-module with the Sato property.

Γ/Λ is *not* a module of type L.

Modules of type L appear as homology modules of free coverings of boundary links [S], cf. also §4.

We shall now introduce some more operations in modules M having the Sato property. If $m \in M$ then the equation

$$m = \sum_{i=1}^{\mu} (t_i - 1)\partial_i(m)$$

is equivalent to

$$m = \sum_{i-1}^{\mu} \left(t_i^{-1} - 1\right) \overline{\partial}_i(m) \ ,$$

where $\overline{\partial}_i : M \to M,\ i = 1, \ldots, \mu$ is defined by $\overline{\partial}_i(m) = -t_i \partial_i(m)$. Define

$$\pi_i(m) = -\partial_i(m) - \overline{\partial}_i(m) = (t_i - 1)\partial_i(m) \ ,$$

which will be called the i-th *component* of m. Then

$$m = \pi_1(m) + \cdots + \pi_\mu(m), \quad m \in M,$$
$$\pi_i \circ \pi_i = \pi_i,$$
$$\pi_i \circ \pi_j = 0 \quad \text{for} \quad i \neq j,$$
$$\partial_i = \partial_i \circ \pi_i,$$
$$\overline{\partial}_i = \overline{\partial}_i \circ \pi_i \ .$$

Let us also introduce an operator $z : M \to M$ by

$$z = -\partial_1 - \cdots - \partial_\mu \ .$$

One can express ∂_i and $\overline{\partial}_i$ in terms of z and π_i:

$$\partial_i = -z\pi_i \ ,$$
$$\overline{\partial}_i = -\overline{z}\pi_i \ ,$$

where

$$\overline{z} = 1 - z : M \to M \ .$$

Thus, the whole structure is given by a decomposition of unity $\{\pi_i\}_{i=\overline{1,\mu}}$, which gives a splitting of M into a direct sum (over k)

$$M \approx X_1 \oplus \cdots \oplus X_\mu \ ,$$

and an endomorphism

$$z : M \to M \ .$$

1.4 Let M be a Λ-module of type L. A *lattice* in M is a k-submodule $A \subset M$ which:
(a) is invariant under $\partial_i, \ \overline{\partial}_i, \ i = 1, \ldots, \mu$;
(b) generates M over Λ;
(c) is finitely generated over k.

Condition (a) is equivalent to each of the following conditions (a′),(a″),(a‴):
(a′) A is invariant under z and $\pi_i, \quad i = 1, \ldots, \mu$;
(a″) A is invariant under ∂_i and $\pi_i, i = 1, \ldots, \mu$;
(a‴) A is invariant under $\overline{\partial}_i$ and $\pi_i, \ i = 1, \ldots \mu$.

1.5 Lemma. *(1) Each Λ-module M of type L contains a lattice; (2) If A_1 and A_2 are two lattices in M then $A_1 + A_2$ and $A_1 \cap A_2$ are also lattices; (3) Let $A \subset M$ be a lattice and B be a Λ-module with the Sato property. Then any D-homomorphism $A \to B$ can be uniquely extended to a Λ-homomorphism $M \to B$. Thus,* $\mathrm{Hom}_\Lambda(M, B) = \mathrm{Hom}_D(A; B)$. *In particular, two modules of type L are isomorphic if and only if they admit lattices which are isomorphic as D-modules.*

For the proof cf. [F], Lemmas 1.5 and 2.6.

1.6 Assume that $k = \mathbb{Z}$ and M is a module of type L. We will say that M is *periodic* if there is an integer $N \in \mathbb{Z}$, $N \neq 0$ with $NM = 0$. As follows, from Lemma 1.5.(1), this is equivalent to $M = \mathrm{Tors}_{\mathbb{Z}} M$.

Any lattice of a periodic module of type L is finite and conversely, any module of type L admitting a finite lattice is periodic.

1.7 Theorem. *Let M be a module of type L. Assume that either (i) $k = \mathbb{Q}$ or (ii) $k = \mathbb{Z}$ and M is periodic. Then M contains a minimal lattice $A \subset M$, which is the intersection of all lattices in M. A lattice $A \subset M$ is the minimal lattice if and only if for any $k = 1, \ldots, \mu$*

$$\pi_k z A = \pi_k A \quad \text{and} \quad \pi_k \overline{z} A = \pi_k A$$

(where $\overline{z} = 1 - z$).

Proof. Cf. [F], §1.

§2. The dual of a periodic module

2.1 Let S denote $\mathbb{Q}\Gamma/(\Gamma + \mathbb{Q}\Lambda)$; it is Λ-Λ-bimodule. If M is a periodic Λ-module of type L define the dual module $\widehat{M}$ as the set of all left Λ-homomorphisms $M \to S$:

$$\widehat{M} = \mathrm{Hom}_\Lambda(M; S) \ .$$

$\widehat{M}$ has a natural right Λ-module structure. We shall transform it into a left Λ-module structure by using the standard involution $t_i \to t_i^{-1}$ of Λ. In other words, we consider $\widehat{M}$ with the following left Λ-module structure

$$(t_i f)(m) = f(m) t_i^{-1}$$

for

$$f \in \widehat{M} \ , \quad m \in M \ , \quad i = 1, \ldots, \mu \ .$$

2.2 Proposition. *$\widehat{M}$ is a periodic Λ-module of type L.*

Proof. First of all, arguments similar to those of 2.3, 2.4, 2.5 of [F] show that $\widehat{M}$ has the Sato property.

Secondly, $\widehat{M}$ is periodic. Indeed, let A be the minimal lattice in M. There exists an integer $N \neq 0$ with $NA = 0$. Thus, for any $f \in \widehat{M}$, Nf vanishes on A and thus $Nf = 0$.

We only have left to show that $\widehat{M}$ is finitely generated over Λ. To do this we consider the following map

$$\psi : \mathrm{Hom}_D(A; \mathbb{Q}\Gamma/\Gamma) \to \mathrm{Hom}_D(A; S) .$$

We will establish the following facts:

(1) $\mathrm{Hom}_D(A; S)$ is naturally isomorphic to $\widehat{M}$;
(2) the image of ψ generates $\mathrm{Hom}_D(A; S)$ over Λ;
(3) the module $\mathrm{Hom}_D(A; \mathbb{Q}\Gamma/\Gamma)$ is isomorphic to $A^* = \mathrm{Hom}_{\mathbb{Z}}(A; \mathbb{Q}/\mathbb{Z})$ and so it is finite.

(1) follows directly from Lemma 1.5.

Let us prove (3). If $F : A \to \mathbb{Q}\Gamma/\Gamma$ is a D-homomorphism then

$$F(a) = \sum_\alpha x^\alpha f_\alpha(a) \quad \text{for} \quad a \in A ,$$

where α runs over all tuples $(i_1, \ldots, i_s)$ with $i_1, \ldots, i_s \in \{1, \ldots, \mu\}$ and x^α denotes the monomial

$$x_{i_1} x_{i_2} \ldots x_{i_s} ,$$

with the convention

$$x^\phi = 1 .$$

For each multi-index α, $f_\alpha : A \to \mathbb{Q}/\mathbb{Z}$ is a $\mathbb{Z}$-homomorphism. Since F is a D-homomorphism, we get

$$f_\alpha(\partial_i a) = f_{i\alpha}(a) ,$$

and so

$$f_\alpha(a) = f_\phi(\partial_{i_s} \ldots \partial_{i_1} a) , \qquad a \in A ,$$

for

$$\alpha = (i_1, \ldots, i_s) .$$

Thus, the whole D-homomorphism F is determined by $f_\phi : A \to \mathbb{Q}/\mathbb{Z}$. Conversely, given a $\mathbb{Z}$-homomorphism $f_\phi : A \to \mathbb{Q}/\mathbb{Z}$ one can define a map $F : A \to \mathbb{Q}\Gamma/\Gamma$ by

$$F(a) = \sum_\alpha x^\alpha f_\phi(\partial^\alpha a) ,$$

where

$$\partial^\alpha = \partial_{i_s} \ldots \partial_{i_1} ,$$

for

$$\alpha = (i_1, \ldots, i_s) .$$

It is clear that F is a D-homomorphism. This proves (3).

To prove (2), consider a D-homomorphism $f : A \to S$. It is clear that f admits a $\mathbb{Z}$-lifting $\widehat{f} : A \to \mathbb{Q}\Gamma/\Gamma$

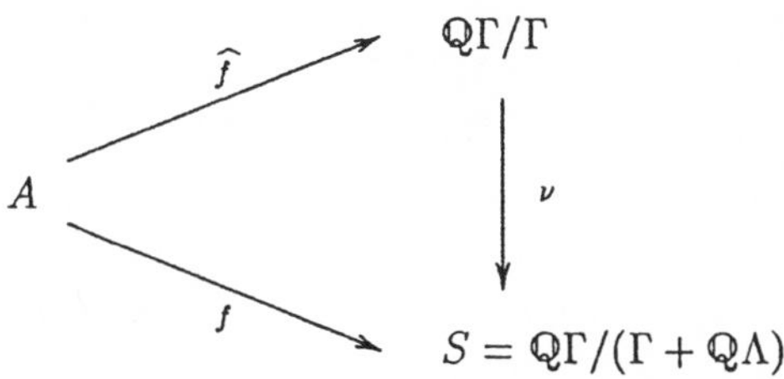

For $i = 1, \ldots, \mu$ consider the map $g_i : A \to \mathbb{Q}\Lambda/\Lambda$,

$$g_i(a) = \partial_i \widehat{f}(a) - \widehat{f}(\partial_i a) , \qquad a \in A .$$

They measure the obstructions for $\widehat{f}$ to be a D-homomororphism. Let us write

$$g_i(a) = \sum_{\pi \in F_\mu} \pi \cdot g^i_\pi(a) ,$$

where $g^i_\pi : A \to \mathbb{Q}/\mathbb{Z}$ is a $\mathbb{Z}$-homomorphism and g^i_π is nonzero only for a finite number of pairs (i, π), $i = 1, \ldots, \mu$, $\pi \in F_\mu$.

Let $\widehat{g}^i_\pi : A \to \mathbb{Q}\Gamma/\Gamma$ be the D-homomorphism

$$\widehat{g}^i_\pi(a) = \sum_\alpha x^\alpha g^i_\pi(\partial^\alpha a) .$$

Define

$$\widehat{g} : A \to \mathbb{Q}\Gamma/\Gamma$$

by

$$\widehat{g}(a) = \sum_{i,\pi} \widehat{g}^i_\pi(a) x_i \pi ,$$

the sum being in fact finite.

Let us show that $\widehat{h} = \widehat{g} - \widehat{f} : A \to \mathbb{Q}\Gamma/\Gamma$ is a D-homomorphism:

$$\begin{aligned} \partial_j \widehat{g}(a) - \widehat{g}(\partial_j a) &= \sum_{i,\pi,\alpha} x^\alpha g^i_\pi(\partial^\alpha \partial_j a) x_i \pi + \sum_\pi g^j_\pi(a)\pi - \sum_{i,\pi,\alpha} x^\alpha g^i_\pi(\partial^\alpha \partial_j a) x_i \pi \\ &= \sum_\pi g^j_\pi(a)\pi = g_j(a) \\ &= \partial_j \widehat{f}(a) - \widehat{f}(\partial_j a) . \end{aligned}$$

Denote, $h = \nu \circ \widehat{h}$, $h : A \to S$. It is a D-homomorphism. Obviously,

$$h \in \operatorname{im}(\psi) .$$

Denote, $g = \nu \circ \widehat{g}$, $g : A \to S$. This is also a D-homomorphism. The formula above for $\widehat{g}$ shows that

$$g \in \Lambda(\operatorname{im}(\psi)) .$$

Thus

$$f = g - h \in \Lambda(\ker(\psi)) .$$

This proves Proposition 2.2.

2.3 Theorem. *Let M be a periodic Λ-module of type L and let $A \subset M$ be its minimal lattice. Consider the following homomorphism*

$$\varphi : A^* = \operatorname{Hom}_{\mathbb{Z}}(A; \mathbb{Q}/\mathbb{Z}) \to \operatorname{Hom}_D(A; S) = \widehat{M} ,$$

$$\varphi(f)(a) = \sum_{i=1}^{\mu} \sum_{\alpha} x^{\alpha} f(\pi_i \partial^{\alpha} a) x_i \in S ,$$

where $f \in A^*$, $a \in A$ and α runs over all multi-indices $\alpha = (i_1, \ldots, i_s)$ with $i_1, \ldots, i_s \in \{1, \ldots, \mu\}$ and x^{α} denotes $x_{i_1} x_{i_2} \ldots x_{i_s}$ while ∂^{α} denotes $\partial_{i_s} \partial_{i_{s-1}} \cdots \partial_{i_1}$. Then φ is a monomorphism and its image coincides with the minimal lattice of $\widehat{M}$.

The proof makes use of Lemma 2.5 below and will therefore be postponed until after that lemma has been presented..

2.4 Let C be a D-module. We will say that C is a *D-module of type* 0 if $\partial_{\kappa} C = 0$ for all $\kappa = 1, \ldots, \mu$. We will say that C is a *D-module of type* i (where $i \in \{1, 2, \ldots, \mu\}$) if $\partial_{\kappa} C = 0$ for $\kappa \neq i$, $\kappa \in \{1, \ldots, \mu\}$ and $(1 + \partial_i) C = 0$.

A D-module Y will be called *primitive* if it has a filtration

$$0 = Y_0 \subset Y_1 \subset Y_2 \subset \ldots , \qquad \bigcup Y_i = Y$$

with a property that for each $j = 1, 2, \ldots$ there exists a number $i = i(j) \in \{0, 1, \ldots, \mu\}$ such that Y_j / Y_{j-1} is a D-module of type i.

Any submodule and any factor-module of a primitive D-module is also primitive. The direct sum of two primitive modules is also primitive.

A basic example of a primitive D-module is provided by $Y = \mathbb{Q}\Lambda$ with the D-module structure given by Fox derivatives. To show this, one can proceed as follows. Let

$$e_0, e_1, e_2, \ldots$$

be all elements of the free group F_{μ} in a linear ordering such that each e_n is a reduced word and can be written as

$$e_n = x_i^{\varepsilon} e_{\kappa} ,$$

where $\kappa < n, i \in \{1, \dots, \mu\}$ and $\varepsilon = \pm 1$. Let Y_n be the linear hull of $e_0, e_1, \dots, e_n$. Then Y_n/Y_{n-1} is a D-module of type 0 (if $\varepsilon = 1$) or of type i, $i \in \{1, \dots, \mu\}$ (if $\varepsilon = -1$).

2.5 Lemma. *Let M be a Λ-module of type L and let $X \subset M$ be a D-submodule which is finitely generated over $\mathbb{Z}$ and generates M over Λ. Let Y be a primitive D-module. Then the kernel of any D-homomorphism*

$$g : X \to Y$$

generates M over Λ.

Proof. See the proof of Lemma 4.4 in [F].

2.6 Proof of Theorem 2.3. Let us show that φ is a monomorphism. Assume that $f \in A^*$ and $\varphi(f) = 0$. This means that for each $a \in A$ the power series

$$\sum_{i=1}^{\mu} \sum_{\alpha} x^{\alpha} f(\pi_i \partial^{\alpha} a) x_i$$

represents an element of $\mathbb{Q}\Lambda/\Lambda$. For each $i = 1, \dots, \mu$ define

$$F_i : A \to \mathbb{Q}\Lambda/\Lambda$$

by

$$F_i(a) = \sum_{\alpha} x^{\alpha} f(\pi_i \partial^{\alpha} a) \ .$$

F_i is a D-homomorphism and, on the other hand, $\mathbb{Q}\Lambda/\Lambda$ is a primitive D-module. From Lemma 2.5 it follows that $K_i = \ker F_i$ generates M over Λ. If $a \in K_i$ then $\partial_{\kappa} a \in K_i$ and $\pi_{\kappa} a \in K_i$ for each $\kappa \in \{1, \dots, \mu\}$. Thus K_i is a lattice and $K_i \subset A$, which implies $K_i = A$. Because this is true for any i it follows that $f = 0$.

Our next step will be to show that $\operatorname{im}(\varphi)$ is invariant under $\overline{z}$ and $\pi_1, \dots, \pi_{\mu}$. To do this we will introduce operations

$$\overline{z}, \pi_1, \dots, \pi_{\mu} : A^* \to A^*$$

by

$$(\overline{z} f)(a) = f(za)$$
$$(\pi_i f)(a) = f(\pi_i a)$$

for

$$f \in A^* \ , \quad a \in A \ , \quad i = 1, \dots, \mu \ .$$

To show that φ commutes with $\overline{z}, \pi_1, \ldots, \pi_\mu$ we compute:

$$\begin{aligned}\varphi(\overline{z}f)(a) &= \sum_{i=1}^{\mu}\sum_{\alpha} x^\alpha f(z\pi_i\partial^\alpha a)x_i \\ &= -\sum_{i=1}^{\mu}\sum_{\alpha} x^\alpha f(\partial_i\partial^\alpha a)x_i \\ &= -\sum_{|\alpha|\geq 1} x^\alpha f(\partial^\alpha a) \\ &= (\overline{z}\varphi(f))(a) \qquad (\mathrm{mod}\, \mathbb{Q}\Lambda)\end{aligned}$$

and

$$\begin{aligned}\varphi(\pi_j f)(a) &= \sum_{i=1}^{\mu}\sum_{\alpha} x^\alpha f(\pi_j\pi_i\partial^\alpha a)x_i \\ &= \sum_{\alpha} x^\alpha f(\pi_j\partial^\alpha a)x_j \\ &= (\pi_j\varphi(f))(a)\ .\end{aligned}$$

To prove that $\mathrm{im}(\varphi)$ is a lattice there remains to show that $\mathrm{im}(\varphi)$ generates $\mathrm{Hom}_D(A; S) = \widehat{M}$ over Λ.

Consider the homomorphism

$$\psi : A^* = \mathrm{Hom}_{\mathbb{Z}}(A; \mathbb{Q}/\mathbb{Z}) \to \mathrm{Hom}_D(A, S) = \widehat{M}\ ,$$

$$\psi(f)(a) = \sum_{\alpha} x^\alpha f(\partial^\alpha a)$$

for $f \in A^*$ and $a \in A$, where α runs over all multi-indices. Note that ψ is essentially the homomorphism which appeared in the proof of Proposition 2.2; it was proved there that $\mathrm{im}(\psi)$ generates $\widehat{M}$. Now, it is easy to check that

$$\psi(f) = \varphi(g)\ ,$$

where $g \in A^*$, $g(a) = -f(za)$ for $a \in A$. Thus, $\mathrm{im}(\varphi) \supset \mathrm{im}(\psi)$ and so $\mathrm{im}(\varphi)$ generates $\widehat{M}$ over Λ.

In order to show that $\mathrm{im}(\varphi)$ is the *minimal* lattice we can check the condition of Theorem 1.7 for A^*. Since

$$(\pi_\kappa z f)(a) = f(\overline{z}\pi_\kappa a)$$

for $f \in A^*$, $a \in A$, the identity

$$\pi_\kappa z A^* = \pi_\kappa A^*$$

is equivalent to the following statement: for each $\mathbb{Z}$-homomorphism $g : \pi_\kappa A \to \mathbb{Q}/\mathbb{Z}$ there exists a $\mathbb{Z}$-homomorphism $h : A \to \mathbb{Q}/\mathbb{Z}$ such that the diagram

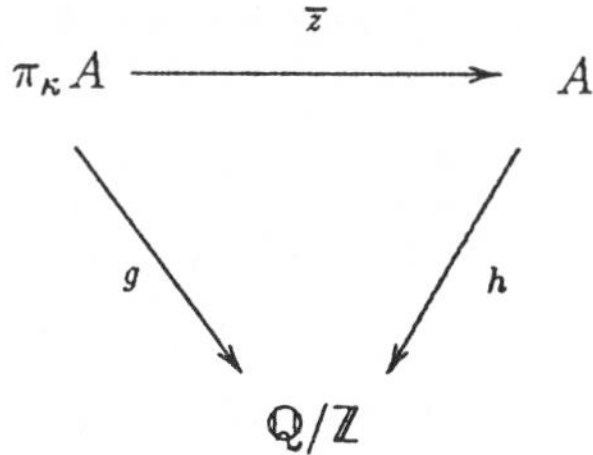

commutes. The last statement is equivalent to the fact that $\overline{z}|_{\pi_\kappa A}$ is a monomorphism, which is in fact true: if $a \in \pi_\kappa A$ and $\overline{z}a = 0$ then $\overline{\partial}_i a = 0$ for all $i = 1, \ldots, \mu$ and $a = \sum_{i=1}^{\mu} \overline{x}_i \overline{\partial}_i(a) = 0$. The identity

$$\pi_\kappa \overline{z} A^* = \pi_\kappa A^*$$

follows similarly.

This proves the theorem.

§3. Hermitian forms on periodic modules

3.1 Let M_1, M_2 be two periodic Λ-modules of type L. Consider a $\mathbb{Z}$-bilinear pairing

$$[\ ,\ ] : M_1 \times M_2 \to S = \mathbb{Q}\Gamma/(\Gamma + \mathbb{Q}\Lambda)$$

with the properties:

(a) $[\lambda a, b] = \lambda[a, b]$ for $\lambda \in \Lambda$, $a \in M_1$, $b \in M_2$;

(b) $[a, \lambda b] = [a, b]\overline{\lambda}^{-}$;

(c) $[\ ,\ ]$ is non-degenerate in the following sense: for $b \in M_2$ let $\varphi_b : M_1 \to S$ be the Λ-homomorphism defined by $\varphi_b(a) = [a, b]$, then the map

$$M_2 \to \widehat{M_1} = \mathrm{Hom}_\Lambda(M_1; S)\ , \qquad b \mapsto \varphi_b\ ,$$

is an isomorphism.

In the case $M_1 = M_2 = M$ we will consider an additional property:

(d) $[a, b] = \varepsilon\overline{[b, a]}$ for $a, b \in M$, $\varepsilon = \pm 1$.

3.2 Theorem. *Let M_1, M_2 be two periodic modules of type L supplied with a pairing*

$$[\ ,\ ] : M_1 \times M_2 \to S$$

satisfying (a),(b),(c) *of subsection* 3.1. *Then there exists a unique $\mathbb{Z}$-bilinear form*

$$\langle\ ,\ \rangle : A_1 \times A_2 \to \mathbb{Q}/\mathbb{Z}$$

(the scalar form) defined on the minimal lattices $A_1 \subset M_1$, $A_2 \subset M_2$ such that
(1) *for $a \in A_1$, $b \in A_2$*

$$[a,b] = \sum_{i=1}^{\mu} \sum_{\alpha} x^{\alpha} \langle \partial^{\alpha} a, \pi_i b \rangle x_i \qquad (\operatorname{mod} \mathbb{Q}\Lambda) ,$$

where α runs over all multi-indices $\alpha = (i_1, \ldots, i_s)$ with $i_j \in \{1, \ldots, \mu\}$, $x^{\alpha} = x_{i_1} x_{i_2} \ldots x_{i_s}$, and $\partial^{\alpha} = \partial_{i_s} \partial_{i_{s-1}} \ldots \partial_{i_1}$;
(2) *$\langle \pi_i a, b \rangle = \langle a, \pi_i b \rangle$ for all $a \in A_1$, $b \in A_2$, $i = 1, \ldots, \mu$;*
(3) *$\langle za, b \rangle = \langle a, \bar{z} b \rangle$, where $\bar{z} = 1 - z$;*
(4) *$\langle\ ,\ \rangle$ is non-degenerate, i.e. the associated map $A_2 \to A_1^* = \operatorname{Hom}_{\mathbb{Z}}(A_1; \mathbb{Q}/\mathbb{Z})$ is an isomorphism.*

Conversely, given a scalar form $\langle\ ,\ \rangle$ with the above properties, the formula in (1) defines a pairing $A_1 \times A_2 \to S$ which can be uniquely extended to a pairing $M_1 \times M_2 \to S$ satisfying (a),(b),(c) *of 3.1.*

Proof. Defining a pairing $M_1 \times M_2 \to S$ satisfying (a),(b),(c) of 3.1 is equivalent to specifying a Λ-isomorphism

$$M_2 \to \operatorname{Hom}_{\Lambda}(M_1; S) = \widehat{M_1} ,$$

and by Lemma 2.6 of [F] and Theorem 2.3 this is equivalent to specifying a D-isomorphism

$$A_2 \to A_1^*$$

which is the restriction of the above homomorphism to the minimal lattices and which represents the scalar form.

3.3 Theorem. *Let M be a periodic module of type L and $[\ ,\ ] : M \times M \to S$ be a pairing satisfying* (a),(b),(c) *of* 3.1. *The pairing $[\ ,\ ]$ satisfies* (d) *of* 3.3 *if and only if the scalar form $\langle\ ,\ \rangle$ is $(-\varepsilon)$-symmetric: $\langle a, b \rangle = -\varepsilon \langle b, a \rangle$ for $a, b \in A$; here A is the minimal lattice of M.*

The proof is similar to that of Theorem 3.3 of [F].

§4. Blanchfield forms

4.1 $\mathcal{F}$-links and Seifert surfaces. An *n-dimensional μ-component* link is an oriented smooth submanifold Σ^n of S^{n+2}, where $\Sigma^n = \Sigma_1^n \cup \ldots \cup \Sigma_{\mu}^n$ is the ordered disjoint union of μ submanifolds of S^{n+2}, each homeomorphic to S^n. Let $X = S^{n+1} - T(\Sigma)$ be the complement of a tubular neighbourhood $T(\Sigma)$ of Σ in

S^{n+2}. Fix a base point $* \in X$; for each $i = 1, \ldots, \mu$ the meridian $m_i \in \pi_1(X, *)$ is defined up to conjugation.

A *splitting* [CS] for Σ^n is a homomorphism (which is defined up to conjugation) $s : \pi_1(X, *) \to F_\mu$ onto the free group with μ generators $t_1, \ldots, t_\mu$ having the following property: the image of the ith meridian m_i is conjugate to $t_i \in F_\mu$, $i = 1, \ldots, \mu$.

An $\mathcal{F}$*-link* [CS] is a pair (Σ^n, s), where Σ^n is a link and s is a splitting for Σ. Two $\mathcal{F}$-links (Σ_1, s_1) and (Σ_2, s_2) of the same dimension and multiplicity are *equivalent* if there exists a diffeomorphism $h : S^{n+2} \to S^{n+2}$, taking Σ_1^n onto Σ_2^n, preserving the orientations of S^{n+2} and Σ_ν, $\nu = 1, 2$ and mapping s_2 onto s_1.

A link Σ^n is a *boundary link* if there is an oriented smooth submanifold V^{n+1} of S^{n+2} such that $V^{n+1} = V_1^{n+1} \cup \ldots \cup V_\mu^{n+1}$ is the disjoint union of submanifolds V_i^{n+1} satisfying $\partial V_i = \Sigma_i$ $(i = 1, \ldots, \mu)$. If each V_i is connected, we say that V is a Seifert manifold for Σ.

Any Seifert manifold V of a boundary link Σ defines an obvious splitting s_V: if α is a loop in X, then $s_V([\alpha])$ is the word in $t_1, \ldots, t_\mu$ obtained by writing down $t_i^{\varepsilon_i}$ $(\varepsilon_i = \pm 1)$ for each intersection point p of α and V_i (where ε_i is the local intersection number) and then multiplying these words in order of the appearance of the corresponding intersection points in α.

Conversely, any link admitting a splitting is a boundary link, cf. [G]. Choosing a splitting is equivalent to a choice of a Seifert surface up to embedded concordance, c.f. [CS],[F5]. All possible transformations of the Seifert surface which preserve the $\mathcal{F}$-structure were described in [F5].

4.2 Torsion links. In this paper we will consider a particular class of even-dimensional $\mathcal{F}$-links. An $\mathcal{F}$-link (Σ^{2q}, s) is called *torsion* if

(a) the splitting $s : \pi_1(S^{2q+2} - \Sigma^{2q}) \to F_\mu$ is an isomorphism;

(b) $\pi_i(S^{2q+2} - \Sigma) = 0$ for $1 < i < q$;

(c) the group $\pi_q(S^{2q+2} - \Sigma)$ is $\mathbb{Z}$-torsion and has no elements of order 2.

It is the goal of this paper to classify torsion links in purely algebraic terms for the case $q \geq 4$.

Let V be a Seifert manifold for Σ with $s_V = s$. We will say that V is *minimal* if each component V_j of V is $(q-1)$-connected and the maps $i_+, i_- : V_j \to S^{n+2} - V$ (defined as small shifts in directions of the positive and negative normal to V_j, respectively) induce monomorphisms in q-dimensional homology. Gutierrez [G] has shown that such minimal Seifert manifolds always exist.

4.3 Poincaré duality. Let (Σ^n, s) be an $\mathcal{F}$-link in S^{n+2} and let $X = S^{n+2} - T(\Sigma^n)$ be the complement of a tubular neighbourhood $T(\Sigma)$ of Σ in S^{n+2}. Fix a particular epimorphism $s_0 : \pi_1(X, *) \to F_\mu$ onto the free group F_μ in $t_1, \ldots, t_\mu$.

Consider the covering

$$\widetilde{X} \to X$$

corresponding to the kernel of s_0; the group F_μ acts on $\widetilde{X}$ as the group of covering transformations. The homology groups

$$H_\kappa(\widetilde{X};\mathbb{Z}) \ , \qquad \kappa = 1, 2, \ldots$$

are modules over $\Lambda = \mathbb{Z}[F_\mu]$. Sato [S] has shown that *these modules are of type L.*

Fix a triangulation of X and consider the corresponding equivariant triangulation of $\widetilde{X}$ and the simplicial chain complex $C_*(\widetilde{X})$. Let X^1 denote the dual triangulation of X and let $C_*(\widetilde{X}^1)$ denote the similar chain complex. $C_*(\widetilde{X})$ and $C_*(\widetilde{X}^1)$ are complexes of free finitely generated left Λ-modules. There is an intersection pairing (cf. Milnor [M1])

$$C_i(\widetilde{X}^1) \times C_j(\widetilde{X}, \partial\widetilde{X}) \to \Lambda \ , \qquad (\alpha, \beta) \mapsto \alpha \cdot \beta$$

for $i + j = n + 2$ with the properties:
(i) it is bilinear over $\mathbb{Z}$;
(ii) $(g\alpha) \cdot \beta = g(\alpha \cdot \beta)$, $\alpha \cdot (g\beta) = (\alpha \cdot \beta)g^{-1}$ for $g \in F_\mu$, $\alpha \in C_i(\widetilde{X}^1)$, $\beta \in C_j(\widetilde{X}, \partial\widetilde{X})$.
This pairing defines a chain map

$$C_j(\widetilde{X}, \partial\widetilde{X}) \to \overline{\mathrm{Hom}_\Lambda\,(C_i(\widetilde{X}^1); \Lambda)}$$

inducing the Poincaré duality isomorphism

$$H_j(\widetilde{X}, \partial\widetilde{X}) \to \overline{H^i(C(\widetilde{X}^1); \Lambda)} \ ,$$

where the bar means that the module under it, which is naturally a right Λ-module, should be converted into a left Λ-module using the standard involution $t_i \mapsto t_i^{-1}$ of Λ.

4.4 Ext-functors. The cohomology module $H^i(C(\widetilde{X}^1); \Lambda)$ could be computed using the universal coefficient spectral sequence [EC]. As shown in §5 of [F5], there is an exact sequence

$$0 \to e^2(H_{i-2}(\widetilde{X}^1)) \to H^i(C(\widetilde{X}^1); \Lambda) \to e^1(H_{i-1}(\widetilde{X}^1)) \to 0$$

and the image of $e^2(H_{i-2}(\widetilde{X}^1))$ coincides with the $\mathbb{Z}$-torsion part of $H^i(C; \Lambda)$. Here $e^\nu(M)$ denotes

$$\mathrm{Ext}^\nu_\Lambda(M; \Lambda) \ .$$

It was also shown in [F5] that for any module M of type L there are natural isomorphisms

$$e^1(M) \approx \mathrm{Hom}_\Lambda(M;\Gamma/\Lambda)\ ,$$
$$e^2(M) \approx \mathrm{Hom}_\Lambda(\tau M;\mathbb{Q}\Gamma/(\Gamma+\mathbb{Q}\Lambda))\ ,$$

where Γ denotes the ring of formal power series in non-commuting variables $x_1,\dots,x_\mu$ with integral coefficients, $\mathbb{Q}\Gamma$ is a similar ring with rational coefficients, Λ is embedded in Γ and $\mathbb{Q}\Lambda$ is embedded in $\mathbb{Q}\Gamma$ via the Magnus embedding.

4.5 The Poincaré duality isomorphism together with the above mentioned formulas for e^1 and e^2 produce two families of non-degenerate Hermitian forms:

$$B_{i-1}(\widetilde{X}) \times B_j(\widetilde{X}) \to \Gamma/\Lambda\ ,$$
$$T_{i-2}(\widetilde{X}) \times T_j(\widetilde{X}) \to \mathbb{Q}\Gamma/(\Gamma+\mathbb{Q}\Lambda)\ ,$$

where $i+j=n+2$, and $T_j(\widetilde{X}) = \mathrm{Tors}_{\mathbb{Z}}\, H_j(\widetilde{X})$, $B_j(\widetilde{X}) = H_j(\widetilde{X})/T_j(\widetilde{X})$. We refer to [F5] for a more detailed description of these forms.

Note that from the existence of these forms it follows that all Alexander modules $H_\kappa(\widetilde{X})$, $1<\kappa\le 2q$ of a torsion link are zero, except for the module $H_q(\widetilde{X})$.

4.6 Consider also the completed chain complex

$$C'_*(\widetilde{X}^1) = \Gamma \otimes_\Lambda C_*(\widetilde{X}^1)$$

(the completion of $C_*(\widetilde{X}^1)$ with respect to powers of the augmentation ideal of Λ). There is obviously an intersection pairing

$$C'_i(\widetilde{X}^1) \times C_j(\widetilde{X},\partial\widetilde{X}) \to \Gamma$$

with properties similar to those of 4.3. The completed complex is acyclic [F5]. Using this fact one can find an explicit description of the Hermitian forms constructed in subsection 4.5. We will do this now for the case of the pairing

$$[\ ,\ ] : T_q(\widetilde{X}) \times T_q(\widetilde{X}) \to \mathbb{Q}\Gamma/(\Gamma+\mathbb{Q}\Lambda)\ ,$$

under the assumption that $n=2q$.

Let the cycles $\alpha \in C_q(\widetilde{X}^1)$ and $\beta \in C_q(\widetilde{X})$ represent classes $[\alpha],[\beta] \in T_q(\widetilde{X})$. Then there is an "infinite" chain c in $C'_{q+1}(\widetilde{X}^1)$ with $\partial c = \alpha$. On the other hand there is an integer $N \neq 0$ and a chain $d \in C_{q+1}(\widetilde{X})$ with $N\beta = \partial d$. Then,

$$[a,b] = \frac{1}{N} c\cdot d \in \mathbb{Q}\Gamma/(\Gamma+\mathbb{Q}\Lambda)\ ,$$

where $a=[\alpha]$, $b=[\beta] \in T_q(\widetilde{X})$. The last formula is essentially contained in the arguments of [F5], §5.

§5. Seifert manifold and Alexander modules

5.1 Let us first introduce an algebraic notion which describes the homology structure of a Seifert manifold of an even-dimensional link.

An *ε-symmetric torsion isometry structure of multiplicity* μ is a tuple

$$(A, \langle\ ,\ \rangle, z, \pi_1, \ldots, \pi_\mu)$$

where A is a finite abelian group, $\langle\ ,\ \rangle : A \otimes A \to \mathbb{Q}/\mathbb{Z}$ is an ε-symmetric non-degenerate bilinear form and $z, \pi_1, \ldots, \pi_\mu : A \to A$ are endomorphisms satisfying
(i) $\langle za, b\rangle = \langle a, \bar{z}b\rangle$, where $a, b \in A$ and $\bar{z}$ denotes $1 - z : A \to A$'
(ii) $\langle \pi_i a, b\rangle = \langle a, \pi_i b\rangle$;
(iii) $\pi_1 + \pi_2 + \ldots + \pi_\mu = 1_A$;
(iv) $\pi_i \cdot \pi_j = \delta_{ij}\pi_j$.

Two torsion isometry structures are *isomorphic* iff there exists an isomorphism between the corresponding groups, commuting with $z, \pi_1, \ldots, \pi_\mu$, and bringing one form $\langle\ ,\ \rangle$ onto the other.

5.2 Non-degenerate Hermitian forms

$$[\ ,\ ] : M \times M \to \mathbb{Q}\Gamma/(\Gamma + \mathbb{Q}\Lambda)$$

on periodic modules, which were studied in §3, provide an algebraic source of examples of torsion isometry structures: if $A \subset M$ is the minimal lattice of M and $\langle\ ,\ \rangle : A \times A \to \mathbb{Q}/\mathbb{Z}$ is the scalar form, then the tuple

$$(A, \langle\ ,\ \rangle, z, \pi_1, \ldots, \pi_\mu)$$

is a torsion isometry structure (here $z, \pi_1, \ldots, \pi_\mu$ denote the restrictions to A of the corresponding operators on M). If the original form $[\ ,\ ]$ is ε-Hermitian, then $\langle\ ,\ \rangle$ is $(-\varepsilon)$-symmetric, cf. 3.3.

We will say that an abstract torsion isometry structure A *admits an embedding* in a periodic module M of type L supplied with a non-degenerate Hermitian form $[\ ,\ ] : M \times M \to \mathbb{Q}\Gamma/(\Gamma + \mathbb{Q}\Lambda)$ if A is isomorphic (as a torsion isometry structure) to the minimal lattice of M.

If a torsion isometry structure admits embeddings in two periodic Λ-modules $(M_\nu, [\ ,\ ]_\nu)$, $\nu = 1, 2$, then there exists an isomorphism $f : M_1 \to M_2$ preserving the forms; this follows from Lemma 1.5 and Theorem 3.2.

A torsion isometry structure A will be called *minimal* if for every $a \in A$, $\kappa = 1, \ldots, \mu$, either of the conditions $z\pi_\kappa a = 0$ or $\bar{z}\pi_\kappa a = 0$ implies $\pi_\kappa a = 0$

Every torsion isometry structure, admitting an embedding in the periodic Λ-module, is minimal.

5.3 Geometrically, torsion isometry structures appear as the middle dimensional homology of Seifert manifolds of even-dimensional links. Namely, let (Σ^{2q}, s) be a μ-component $\mathcal{F}$-link and $V^{2q+1} \subset S^{2q+2}$ be any Seifert surface for (Σ^{2q}, s). Denote $A = T_q(V) = \mathrm{Tors}_{\mathbb{Z}}\, H_q(V)$ and let

$$\langle\ ,\ \rangle : A \times A \to \mathbb{Q}/\mathbb{Z}$$

be the classical linking form (note, that orientation of V is specified by the orientation of Σ). Let $\pi_i : A \to A$, $i = 1, \dots, \mu$, be the restriction on the torsion subgroup of the composition $H_q(V) \to H_q(V_i) \to H_q(V)$. An operation $z : A \to A$ is defined as follows. Let Y denote the result of cutting the sphere S^{2q+2} along V. Let $i_+, i_- : V \to Y$ be small shifts of V in the directions of the positive and negative normals to V, respectively. Then the map $i_{+*} - i_{-*} : H_*(V) \to H_*(Y)$ is an isomorphism (cf. [F4], §1.1) and we define $z(v) \in H_q(V)$ for $v \in H_q(V)$ by

$$(i_{+*} - i_{-*})(z(v)) = i_{+*}(v)\ .$$

It is easy to see that the tuple $(A, \langle\ ,\ \rangle, z, \pi_1, \dots, \pi_\mu)$ is a $(-1)^{q+1}$-symmetric torsion isometry structure.

5.4 From now on we will be dealing with torsion links, cf. 4.2. A Seifert manifold V^{2q+1} of a torsion link is *minimal* iff each component is $(q-1)$-connected and the corresponding torsion isometry structure (described in 5.3) is minimal. In this and the subsequent sections we will show that the torsion isometry structure of the minimal Seifert manifold admits an embedding in the Alexander module supplied with the Blanchfield form. This result combined with the algebraic considerations of §3 will give a proof of the fact that the isomorphism type of the torsion isometry structure of a minimal Seifert manifold is uniquely determined by the link.

5.5 Let (Σ^{2q}, s) be a torsion link and V^{2q+1} be its minimal Seifert manifold. Consider the $(2q+2)$-dimensional manifold Y obtained by cutting $S^{2q+2} - \Sigma$ along V. The boundary of Y is the disjoint union of

$$\partial_1^+ Y \cup \partial_1^- Y \cap \partial_2^+ Y \cup \partial_2^- Y \cup \ldots \cup \partial_\mu^+ Y \cup \partial_\mu^- Y\ ,$$

where each $\partial_i^\varepsilon Y$, $\varepsilon = \pm$, is homeomorphic to V_i. There is an identification map

$$\psi : Y \to S^{2q+2} - \Sigma$$

which is a homeomorphism on $\operatorname{int} Y$ and maps $\partial_i^+ Y$ and $\partial_i^- Y$ onto V_i. The internal normal on $\partial_i^+ Y$ corresponds under ψ to the positive normal on V_i.

The map $\psi : Y \to S^{2q+2} - \Sigma = X$ can be lifted into $\widetilde{X}$, where $\widetilde{X} \to X$ is the universal covering. One can find a lifting $\widetilde{\psi} : Y \to \widetilde{X}$ such that

$$\widetilde{\psi}(Y) \cap g\widetilde{\psi}(Y) = \begin{cases} \partial_i^- Y & \text{for } g = t_i\ , \\ \partial_i^+ Y & \text{for } g = t_i^{-1}\ , \\ \emptyset & \text{for other } g \in F_\mu,\ g \neq 1. \end{cases}$$

Identify Y with its image in $\widetilde{X}$ under $\widetilde{\psi}$. It is clear that

$$\bigcup_{g\in F_\mu} gY = \widetilde{X} .$$

5.6 Let $f : H_q(V) \to H_q(\widetilde{X})$ be the composition of $i_{+*} - i_{-*} : H_q(V) \to H_q(Y)$ and $i_* : H_q(Y) \to H_q\widetilde{X})$, where $i : Y \to \widetilde{X}$ is the inclusion.

We claim that the map

$$f : H_q(V) \to H_q(\widetilde{X})$$

has the following properties:
(i) it is a monomorphism;
(ii) $f(\pi_i a) = \pi_i f(a)$ for $a \in H_q(V)$, $i = 1, \ldots, \mu$;
(iii) $f(za) = zf(a)$;
(iv) the image of f generates $H_q(\widetilde{X})$ over Λ.
The proof is almost identical to those of [F], 6.18-6.25, where odd-dimensional links where considered, and will therefore, be omitted.

Note that from (i) it follows that the group $H_q(V)$ is in fact finite.

5.7 Corollary. $\operatorname{im}(f)$ *is the minimal lattice in* $H_q(\widetilde{X})$.

Proof. Conditions (ii),(iii),(iv) mean that $A = \operatorname{im}(f)$ is a lattice. Let us show that it is minimal. Assuming the contrary, one may conclude that by Theorem 1.7 there exists an integer $\kappa \in \{1, \ldots, \mu\}$ with

$$\pi_\kappa z A \subsetneqq \pi_\kappa A$$

or

$$\pi_\kappa \bar{z} A \subsetneqq \pi_\kappa A .$$

If $\pi_\kappa z A \subsetneqq \pi_\kappa A$ then there is $x \in \pi_\kappa A$ with $x \neq 0$ and

$$\langle \pi_\kappa z A, x\rangle = 0 .$$

From this it follows that

$$\langle A, \bar{z}x\rangle = 0$$

and $\bar{z}x = 0$. Now $x = 0$ follows from the minimality of A; this gives a contradiction. The assumption

$$\pi_\kappa \bar{z} A \subsetneqq \pi_\kappa A$$

might be considered similarly.

§6. Computation of the Blanchfield form

6.1 Let (Σ^{2q}, s) be a torsion link, $X = S^{2q+2} - T(Z)$ be the complement of a tubular neighbourhood of Σ, and $\widetilde{X} \to X$ be the free covering. Further, let V be a minimal Seifert manifold for (Σ, s), and let $f : H_q(V) \to H_q(\widetilde{X})$ be the map constructed in 5.6. From §4 we know that there is a Blanchfield form

$$[\ ,\ ] : H_q(\widetilde{X}) \times H_q(\widetilde{X}) \to \mathbb{Q}\Gamma(\Gamma + \mathbb{Q}\Lambda)$$

and our aim is to compute this form on the image of f. We want to express the result in terms of the torsion isometry structure of the Seifert surface. The answer is as follows.

6.2 Theorem. *For $a, b \in H_q(V)$*

$$[f(a), f(b)] = \sum_{i=1}^{\mu} \sum_{\alpha} x^{\alpha} \langle \partial^{\alpha} a, \pi_i b\rangle\, x_i \quad \in \mathbb{Q}\Gamma/(\Gamma + \mathbb{Q}\Lambda)\ ,$$

where α runs over all multi-indices $\alpha = (i_1, \dots, i_s)$, with $i_1, i_2, \dots, i_s \in \{1, \dots, \mu\}$, x^{α} denotes the monomial

$$x_{i_1} x_{i_2} \dots x_{i_s}$$

and ∂^{α} denotes

$$\partial_{i_s} \partial_{i_{s-1}} \dots \partial_{i_1}$$

with $\partial_i = -z\pi_i : H_q(V) \to H_q(V)$. The brackets $\langle\ ,\ \rangle$ denote the linking form on V.

Comparing the formula of Theorem 6.2 with Theorem 3.2 and Corollary 5.7 we obtain

6.3 Corollary. *The torsion isometry structure of any minimal Seifert manifold of a torsion link admits an embedding (cf. 5.2) in the Alexander module of the link supplied with the Blanchfield form.*

In fact the map $f : H_q(V) \to H_q(\widetilde{X})$ provides such an embedding.

6.4 Corollary. *The isomorphism type of torsion isometry structure of any minimal Seifert manifold of the torsion link is uniquely determined by the link.*

This follows from results of §3: a minimal lattice with its scalar form is uniquely determined by the Hermitian form.

6.5 *Proof of Theorem 6.2.* According to 4.6, in order to compute the value of the Blanchfield form $[f(a), f(b)]$ we have, to find an "infinite" chain $c \in C'_{q+1}(\widetilde{X}^1)$

in the completed complex such that ∂c is finite and represents $f(a)$, and also a chain $d \in C_{q+1}(\widetilde{X})$ with $\partial d = Nx$, where x is a cycle representing $f(b)$ and N is a nonzero integer. Then

$$[f(a), f(b)] = \frac{1}{N} c \cdot d \in \mathbb{Q}\Gamma/(\Gamma + \mathbb{Q}\Lambda) .$$

Thus, our first aim is to construct this infinite chain c.

But before doing this we must discuss some general constructions of $(q+1)$-dimensional chains in $\widetilde{X}$.

6.6 Let Y be the manifold obtained from $S^{2q+2} - \Sigma$ by cutting along V. We will identify Y with its embedding in $\widetilde{X}$ constructed in 5.5, and will identify V with $\partial^+ Y \subset Y \subset \widetilde{X}$.

Let v be a q-dimensional cycle in V and let $\pi_i zv$ be a cycle representing $\pi_i z[v] \in H_q(V)$, $i = 1, \ldots, \mu$. By the definition of z (cf. 5.3), there exists a $(q-1)$-dimensional chain c_v in Y with $\partial c_v \in \partial Y$ such that

$$\partial c_v = i_+(v) - (i_+ - i_-)(zv) .$$

Identifying $i_+(v)$ with v and $i_+(zv)$ with zv, we should identify $i_-(zv)$ with

$$\sum_{i=1}^{\mu} t_i(\pi_i zv)$$

and we have

$$\partial c_v = v + \sum_{i=1}^{\mu} x_i(\pi_i zv) ,$$

where $x_i = t_i - 1 \in \Lambda$.

6.7 Lemma. *Let $[c_v]$ be the homology class in $H_{q+1}(Y, \partial Y)$ represented by the cycle c_v constructed above. Let*

$$\ell^Y : H_{q+1}(Y, \partial Y) \times H_q(Y) \to \mathbb{Q}/\mathbb{Z}$$

denote the linking pairing in Y and

$$\langle\, , \rangle : H_q(V) \times H_q(V) \to \mathbb{Q}/\mathbb{Z}$$

denote the linking pairing in Y. Then for any class $x \in H_q(V)$ the following formula holds

$$\ell^Y([c_v], (i_{+*} - i_{-*})(x)) = -\langle [v], x \rangle ,$$

where $[v] \in H_q(V)$ denotes the homology class of v.

Note first that $H_{q+1}(Y, \partial Y)$ as well as $H_q(V)$ are in fact $\mathbb{Z}$-torsion and thus the linking pairings are correctly defined.

The proof of Lemma 6.7 is identical to the proof of Lemma 6.22 in [F] and is therefore omitted.

6.8 We now proceed with the construction of the "infinite" chain c such that $\partial c = f(a)$.

We are given a homology class $a \in H_q(V)$. For each multi-index $\alpha = (i_1, \ldots, i_s)$ and each number $i \in \{1, \ldots, \mu\}$ define

$$a^i_\alpha = \pi_{i_s} z \pi_{i_{s-1}} z \ldots \pi_{i_1} z \pi_i a \in H_q(V) .$$

Let a cycle v^i_α realize a^i_α. By 6.6, there is a $(q+1)$-dimensional chain c^i_α in Y with $\partial c^i_\alpha \subset \partial Y$ and

$$\partial c^i_\alpha = v^i_\alpha + \sum_{j=1}^{\mu} x_j v^i_{\alpha j} ,$$

where $\alpha j = (i_1, \ldots, i_s, j)$ for $\alpha = (i_1, \ldots, i_s)$. Put

$$c = \sum_\alpha \sum_{i=1}^{\mu} (-1)^{|\alpha|+1} x_i x^\alpha c^i_\alpha ,$$

where α runs over all multi-indices and $|\alpha|$ denotes s for $\alpha = (i_1, \ldots, i_s)$. This is a convergent power series in $C'_{q+1}(\widetilde{X}^1)$ and a short computation identical with that given in 6.25 of [F] shows that ∂c is finite and represents $f(a)$.

6.9 Going back to the proof of Theorem 6.2, assume that we have two homology classes $a, b \in H_q(V)$. In the previous subsection we have found an infinite chain c with ∂c representing $f(a)$. Let d be a chain in V with $\partial d = N\beta$, where β is a cycle representing b. Then

$$\begin{aligned}
[f(a), f(b)] &= \frac{1}{N} c \cdot (i_+ - i_-)(d) \\
&= \sum_\alpha \sum_{i=1}^{\mu} (-1)^{|\alpha|+1} x_i x^\alpha \ell^Y([c^i_\alpha], (i_+ - i_-)(b)) \\
&= \sum_\alpha \sum_{i=1}^{\mu} (-1)^{|\alpha|} x_i x^\alpha \langle a^i_\alpha, b \rangle
\end{aligned}$$

(by virtue of Lemma 6.7), and then computations identical with those of 6.25 in [F] complete the proof of Theorem 6.2.

Combining Theorem 6.2 with 5.7 we obtain:

6.10 Theorem. *Let V^{2q+1} be a minimal Seifert manifold of a torsion link (Σ^{2q}, s). Then the torsion isometry structure determined by V (cf. 5.3) is isomorphic to the torsion isometry structure of the minimal lattice in $H_q\widetilde{X}$ supplied with the*

Blanchfield form

$$H_q(\widetilde{X}) \times H_q(\widetilde{X}) \to \mathbb{Q}\Gamma/(\Gamma + \mathbb{Q}\Lambda)$$

(cf. 5.2). Thus, the isomorphism type of the torsion isometry structure of a minimal Seifert manifold is determined uniquely by the link.

6.11 Corollary. *The Blanchfield form*

$$H_q(\widetilde{X}) \times H_q(\widetilde{X}) \to \mathbb{Q}\Gamma/(\Gamma + \mathbb{Q}\Lambda)$$

of a torsion link of dimension $2q$ is $(-1)^q$-Hermitian.

Proof. The scalar form on the minimal lattice is isomorphic to the torsion isometry structure of any minimal Seifert manifold, which is obviously $(-1)^{q+1}$-symmetric. The result now follows from Theorem 3.3.

§7. The main results

7.1 Theorem. *Let (Σ^{2q}, s), $q \geq 4$, be a torsion link. Then any two minimal Seifert manifolds V and W of Σ (with $s_V = s = s_W$) are ambient isotopic (i.e. there exists a smooth isotopy $h_t : S^{2q+2} \to S^{2q+2}$, $t \in [0,1]$, $h_0 = id$, with $h_1(V_j) = W_j$ for each component V_j of V and the corresponding component W_j of W, $j = 1, \ldots, \mu$. Moreover, $h_1|_{V_j} : V_j \to W_j$ preserves the orientations).*

Proof. We will use the stable-homotopy reduction of the classification problem established in [F5]. Let $(\widehat{V}, u_V, z_V, \pi_1^V, \ldots, \pi_\mu^V)$ and $(\widehat{W}, u_W, z_W, \pi_1^W, \ldots, \pi_\mu^W)$ be stable isometry structures of the Seifert manifolds V and W, respectively ($\widehat{V}$ denotes the bouquet $\bigvee_{i=1}^{\mu} V_i$ and similarly for W). By Theorem 2.6 of [F5] in order to show that V and W are ambient isotopic it is enough to construct an S-equivalence

$$f : \widehat{V} \to \widehat{W}$$

with

$$\begin{aligned} &f \circ z_V = z_W \circ f \;, \\ &f \circ \pi_i^V = \pi_i^W \circ f \;, \quad i = 1, \ldots, \mu \\ &u^W \circ (f \wedge f) = u^V \end{aligned}$$

(the sign "=" means "stably homotopic" here). Now, both $\widehat{V}$ and $\widehat{W}$ are $(q-1)$-connected CW-complexes. By Poincaré duality, we may assume that $\dim \widehat{V} =$

$\dim \widehat{W} = q + 1$. We also know that $H_{q+1}(\widehat{V}) = H_{q+1}(\widehat{W}) = 0$ and that $H_q(\widehat{V})$ and $H_q(\widehat{W})$ are finite groups with no element of order two. We can apply (an easy part of) the classification Theorems 6.2 and 7.2 of [F3] to show that such an S-equivalence f exists if and only if the torsion isometry structures of V and W are isomorphic. But this last fact follows from Theorem 6.10.

This completes the proof.

7.2 Theorem. *Two torsion links (Σ_ν^{2q}, s_ν), $\nu = 1, 2$, with $q \geq 4$, are equivalent if and only if the corresponding Alexander modules $H_q\widetilde{X}_\nu$, $\nu = s$, together with their Blanchfield forms*

$$H_q(\widetilde{X}_\nu) \times H_q(\widetilde{X}_\nu) \to \mathbb{Q}\Gamma/(\Gamma + \mathbb{Q}\Lambda)$$

are isomorphic.

Proof. Assume that there is a Λ-isomorphism

$$f : H_q(\widetilde{X}_1) \to H_q(\widetilde{X}_2)$$

preserving the Blanchfield forms. Let V_ν, $\nu = 1, 2$, be a minimal Seifert manifold of Σ_ν. The restriction of f to a minimal lattice of $H_q(\widetilde{X}_1)$ is an isomorphism between minimal lattices of $H_q(\widetilde{X}_1)$ and $H_q(\widetilde{X}_2)$. This restriction preserves the scalar forms (this follows from Theorem 3.2). By Theorem 6.10 we obtain that the torsion isometry structures of V_1 and V_2 are isomorphic. Then, using the results of [F5] we obtain by arguments similar to those of the proof of Theorem 7.1 that V_1 and V_2 are ambient isotopic; in particular Σ_1 and Σ_2 are equivalent (as $\mathcal{F}$-links).

7.3 Theorem. *Two torsion links (Σ_ν^{2q}, s_ν), $\nu = 1, 2$, with $q \geq 4$, are equivalent if and only if the torsion isometry structures of any pair of their respective minimal Seifert manifolds are isomorphic.*

Proof. This follows from Theorem 7.2 plus the remark in 5.2.

7.4 Theorem. *Given a minimal ε-symmetric torsion isometry structure A of multiplicity μ with no 2-torsion and an integer $q \geq 4$ with $(-1)^{q+1} = \varepsilon$, there exists a torsion link (Σ^{2q}, s) of μ components and its minimal Seifert manifold V^{2q+1} such that the torsion isometry structure of V is isomorphic to A.*

Proof. This follows from Theorem 2.6 of [F5] plus the homotopy classification of maps between A_q^1-spaces given in [F3].

7.5 Theorem. *Given a periodic Λ-module M of type L with no elements of order 2 and an ε-Hermitian non-degenerate form*

$$[\,,\,] : M \times M \to \mathbb{Q}\Gamma/(\Gamma + \mathbb{Q}\Lambda)\,,$$

for each $q \geq 4$, *with* $(-1)^q = \varepsilon$, *there exists a torsion link* (Σ^{2q}, s) *in* S^{2q+2} *such that the Alexander module* $H_q(\widetilde{X})$ *of* (Σ^{2q}, s), *considered together with its Blanchfield form*

$$H_q(\widetilde{X}) \times H_q(\widetilde{x}) \to \mathbb{Q}\Gamma/(\Gamma + \mathbb{Q}\Lambda)$$

is isomorphic to $(M, [\ ,\ ])$.

Proof. Consider the minimal lattice of M together with its scalar form (cf. §3). By Theorem 7.4, one may realize the corresponding torsion isometry structure by a minimal Seifert manifold. The result now follows from 5.2.

References

[B] R.C. Blanchfield, Intersection theory of manifolds with operators with applications to knot theory. Ann. Math. 65 (1957), 340-356.

[Br] W. Browder, Diffeomorphisms of 1-connected manifolds, Trans. Amer. Math. Soc. 128 (1967), 155-163.

[BL] W. Browder, J. Levine, Fibering manifolds over S^1, Comment. Math. Helv. 40 (1966), 153-160.

[CF] R.H. Crowell, R.H. Fox, Introduction to knot theory, 1963, Ginn and Company.

[CS] S.E. Cappell, J.L. Shaneson, Link cobordism, Comment. Math. Helv. 55 (1980), 29-49.

[D] J. Duval, Forme de Blanchfield et cobordisme d'entrelacs bords, Comment, Math. Helv. 61 (1986), 617-635.

[EC] S. Eilenberg, H. Cartan, Homological Algebra. Princeton Univ. Press, 1956.

[F] M. Farber, Hermitian forms on link modules, Comm. Math. Helv. 66 (1991), 189-236.

[F1] M. Farber, Duality in an infinite cyclic covering and even-dimensional knots, Math. USSR-Izv. 11 (1977), 749-781.

[F2] M. Farber, An algebraic classification of some even-dimensional spherical knots, I, II, Trans. Amer. Math. Soc. 281 (1984), 507-570.

[F3] M. Farber, Classification of stable fibred knots, Math. USSR, Sbornik, 43 (1982), 199-234. Preprint.

[F4] M. Farber, Classification of simple knots. Russian Math. Surveys 38 (1983), 63-117.

[F5] M. Farber, Stable-homotopy and homology invariants of boundary links, Transactions of AMS, to appear.

[G] M.A. Gutierrez, Boundary links and an unlinking thorem. Trans. Amer. Math. Soc. 171 (1972), 491-499.

[K] M.A. Kervaire, Les noeuds de dimensions superieures, Bull. Soc. Math. France 93 (1965), 225-271.

[K1] C. Kearton, Blanchfield duality and simple knots, Trans. Amer. Math. Soc. 202 (1975), 141-160.

[Ko] S. Kojima, Classification of simple knots by Levine pairing, Comment. Math. Helv. 59 (1979), 356-367.

[L] J. Levine, An algebraic classification of some knots of codimension two, Comment. Math. Helv. 45 (1970), 185-198.

[L1] J. Levine, Knot modules, I. Trans. Amer. Math. Soc. 229 (1977), 1-50.

[M] J. Milnor, Infinite cyclic coverings, Conference on the topology of manifolds, Boston, Prindle, Weber & Schmidt, 1968, 115-133.

[M1] J. Milnor, A duality theorem for Reidemeister torsion, Ann. Math. 76 (1962), 137-147.

[S] N. Sato, Free coverings and modules of boundary links, Trans. Amer. Math. Soc. 264 (1981), 499-505.

[T] H.F. Trotter, On S-equivalence of Seifert matrices, Invent. Math. 20 (1973), 173-207.

Unknotting, knotting by twists on disks and property (P) for knots in S^3

Yves Mathieu

1 Introduction and results

After Gordon and Luecke [GoL], we now know that knots in S^3 are determined by their complements, or equivalently that non trivial Dehn surgery on a nontrivial knot cannot yield S^3. A stronger conjecture is: *can surgery on a knot yield an homotopy sphere? If not, the knot has property (P),* cf [BM]. A history of this long studied problem, which appears in a paper of Tietze in 1908 [T], is made by W. Whitten [**W**].

Let K be a knot in S^3. A disk D is essential for K if K intersects D transversely and if D bounds no disk in $S^3 - K$. This notion appears in [MD] and [ST] for the particular case of a crossing disk. See §1 for more precision. By twisting n times on D a knot K, we get a family of knots. If D is inessential, K is unchanged. If K is trivial (i.e. boundary of a disk), and if D is an essential disk for K, we can ask if the new knot in S^3 obtained from K by Dehn twists on D remains the boundary of a disk. We want also to relate these twisting operations to property (P): We consider the following questions:

Q1 If D is essential for a knot k, can ∂D bound a disk in a homology sphere obtained by surgery on k? Can the trivial knot remain trivial after twisting on a disk?

Q2 What about property (P) for knots obtained by twisting a given knot on disks ?

The notations and definitions are in §2. We answer in §3, Q1 by giving a complete classification by describing the only links $(K, \partial D)$, with K trivial, for which K remains trivial after twisting on D. This completes Gabai's partial results, [Ga 2], and generalizes D. Goldsmiths [Gs]. As an application we have a sufficient condition to get a prime knot by twisting K on ∂D, and a direct proof of property

Knots 90

(P) for knots respected by an axial rotation of finite order (cf. corollary 7 in [CGLS]). In §4 we prove property (P) for knots trivialised by n-twisting ($|n| > 1$) on exactly one essential disk and we study property (P) for knots constructed by twisting a given nontrivial knot on an essential disk. In paticular twisting Stallings braids produces knots with property (P).

Finally we relate twisting operations to the *general placement problem* of R. H. Fox in *Quick trip through Knot Theory* (p.120) : can we find two knots k and h in an orientable 3-manifold M with $M - k$ homeomorphic to $M - h$, but no homeomorphism from the pair (M, k) to (M, h) ?

I wish to thank Michel Domergue and Michel Boileau for many discussions, and Cameron McA. Gordon and Marty Scharlemann for helpful comments.

2 Notations and definitions

The manifolds are orientable.

2.1 Dehn surgery on links

Let $\mathcal{L} = (\ell_1, \cdots, \ell_p)$ be a link in a 3-manifold M and $\alpha = (\alpha_1, \cdots, \alpha_p)$ a system of essential simple closed curves, with $\alpha_i \subset \partial N(\ell_i)$. We note $M(\mathcal{L}; \alpha)$ the identification $(M - \cup_i \overset{\circ}{N}(\ell_i)) \cup (\cup_{\varphi_i} D^2 \times S^1)$ where φ_i is an attaching homeomorphism from the ith copy of $\partial D^2 \times S^1$ to the boundary $\partial N(\ell_i)$ with $\varphi_i(\partial D^2 \times \{pt\}_i) = \alpha_i$. The manifold $M(\mathcal{L}; \alpha)$ is also denoted by $M(\ell_1; \alpha_1) \cdots (\ell_p; \alpha_p)$. For a knot k and a curve α, $M(k; \alpha)$ is the standard Dehn filling of the knot space $M - \overset{\circ}{N}(k)$.

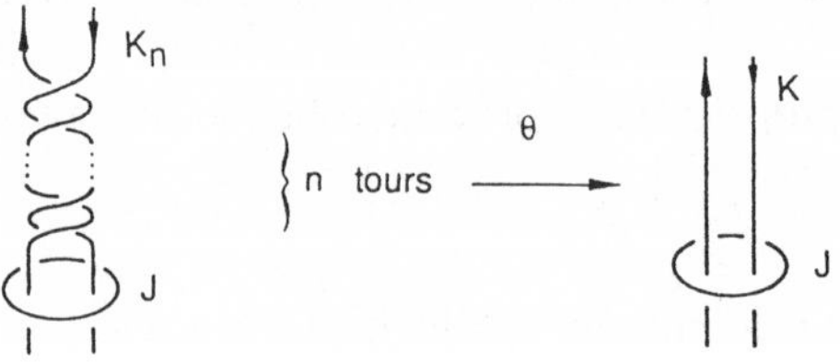

Figure 1

A curve γ in $M - k$ has a natural embedding in the manifold $M(k; \alpha)$, which we will call $\gamma(k; \alpha)$. The core of the surgery is the image of $\{0\} \times S^1$ in the quotient $M(k; \alpha)$. In homology spheres, we can parametrize the curve α by a rational number, by using a meridian-prefered longitude pair (μ, λ) of the knot k. If $\gamma \sim \mu^p \cdot \lambda^q$ we write $M(k; p/q)$ and M is the trivial surgery $M(k; \infty)$, with ∞ corresponding to the meridian μ of k.

2.2 Twisting on disks

Let K be a knot in S^3 cutting transversely a disk D, with $\partial D = J$. There is a Dehn twist homeomorphism θ_J (or simply θ) from $S^3 - J$ onto itself. Let (μ_J, λ_J) be a meridian-longitude pair with $\lambda_J \sim 0$ in $S^3 - J$; we have $\theta(\mu_J) = \mu_J \cdot \lambda_J$, for suitable orientations of the curves (μ_J, λ_J).

Definition. The knot $K_n = \theta_J^n(K)$ is said to be obtained by twisting n times K on D (or on J); n is a positive or negative integer, depending on the choice of orientation and on the sense of the twists. Further, since $\theta_J^n(\mu_J) = \mu_J \cdot \lambda_J^n$, there is a homeomorphism between S^3 and $S^3(J; 1/n)$, and K_n is K after surgery along the slope $1/n$ on J, i.e $K_n = K(J; 1/n)$.

Remarks.
(i) Consider now the case of linking number $lk(k, J) = 0$ and $D \cap K = 2pts$.
This is the classical notion of crossing link [ST], §1, and $K_{\pm 1}$ differs from K by a single crossing change. So if $K_{\pm 1}$ is trivial, K is an unknotting number one knot. If K_n is trivial ($|n| > 1$), then the knot K has property (P), from the proof given in [MD].
(ii) Twisting operations are available in arbitrary 3-manifolds.

3 Triviality before/after surgery. Question Q1

3.1 A criterion on knotting trivial curves in S^3

It is well known that we can find, for every knot k, a collection of disks cutting k in two points with opposite orientations and then give twists on them to trivialize the knot: a minimal such collection has exactly $u(k)$ disks, where $u(k)$ is the unknotting number of k. In [MD] is defined the generalized unknotting number $d(k)$, or simply d, for a knot k, with $d(k) \leq u(k)$, by using d disks D_i , with $J_i = \partial D_i$, $lk(k, J_i) = 0$ and d integers n_i such that k becomes trivial after n_i full twists on D_i, for $i = 1$ to d. Conversely, we have a generic construction for knots by starting from the trivial knot K and then by twisting n_i times on essential disks D_i. We obtain all the knots in a simply connected 3-manifold and homotopically trivial knots in arbitrary manifold. Therefore the natural question is to know what kind of disk D and then what surgery on ∂D yield the trivial knot from the trivial one in S^3 ?

Let us define 3 particular links with two trivial components (see Figure 2):

$\mathcal{L}_0$ is called the trivial link with 2 components and $\mathcal{L}_1$ the Hopf link. In $\mathcal{L}_2$, K is a (1,2)-cable of J.

We prove the following theorem of classification by characterizing links $\mathcal{L} = \{k, J\}$ which answer question Q1.

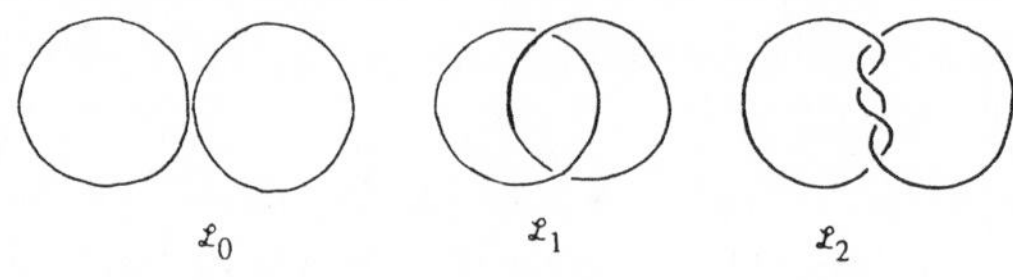

Figure 2

Theorem 3.1. *Let $\mathcal{L} = \{k, J\}$ be a link in S^3, k a knot, D an essential disk with $J = \partial D$. There exists an integer n such that the knot space of the knot $J(k; 1/n)$ remains boundary-compressible after $1/n$ Dehn surgery on k if and only if k is the trivial knot in S^3 and $\mathcal{L}$ is one of the two links $\mathcal{L}_1$ or $\mathcal{L}_2$. Moreover, for $\mathcal{L}_1$, $J(k; 1/n)$ remains trivial for every n, and for $\mathcal{L}_2$, only the knot $J(k; 1)$ is trivial.*

Proof. We have a Heegaard splitting of S^3 by $S^3 = N(J) \cup V$ where V is a solid torus containing k. The $1/n$-surgery on k yields a homology sphere $M = S^3(k; 1/n)$ with a new knot $j = J(k; 1/n)$, i.e. J after surgery, such that $M = N(j) \cup V(k; 1/n)$. Suppose the disk is essential for k. Then $M - int(N(j)) = V(k; 1/n)$ is boundary compressible and so by [Ga 1] is either a solid torus or $V(k; 1/n) \approx W_1 \# W_2$, where W_2 is closed and has $H_1(W_2)$ non-trivial : this case is here impossible since M is a homology sphere. (Remark : M is a union of two solid tori and is a lens space ; this lens space has trivial homology and is therefore the 3-sphere. The Gordon-Luecke theorem [GoL], $S^3(k; 1/n) \approx S^3$ and $n \neq 0$ imply that k is the trivial knot in S^3, but it is not necessary to use here the Gordon-Luecke theorem because it follows already from Gabai's results [Ga 1]). $\mathcal{L} = \{k, J\}$ is a link different from $\mathcal{L}_0$. Since $V(k; 1/n) \approx V$, the knot k is a 0- or a 1-bridge braid in the solid torus $S^3 - int(N(J)) = V$.

Case of 1-bridge braid. If k was a 1-bridge braid in V, with $S^3 = N(J) \cup V$ then $w(k) > 2$ and the manifold $V(k; 1/n)$ has incompressible boundary from Gabai's theorem : the case $V(J; \alpha) \approx V$ implies α is parametrized by $p \cdot \mu + \lambda$ where (μ, λ) is a prefered pair meridian-longitude when V is viewed with standard embedding in S^3 (λ is null homologous in $S^3 - k$) and $|p| = t \cdot d \cdot w > 1$, where t and d are nonnull integers depending on the braid k (see [Ga 2]). In this case the knot space of $J(k; 1/n)$ is never boundary-compressible.

Case of 0-bridge braid. The 0-bridge braid k is a (p, q) torus knot on ∂V. In fact the knot k is in $int(V)$ and is parallel by an annulus A to a (p, q) torus knot on ∂V. This annulus intersects the boundary of a regular neighborhood of k in V along a parallel ξ of k and the Dehn twist θ_A on A is a homeomorphism from $V - int(N(k))$ to itself which sends the curve μ on $\mu \cdot \xi$ and $\theta_A^n(\mu) = \mu \cdot \xi^n$; but ξ is homologous to $\mu^{pq} \cdot \lambda$ in $V - k$ so that $\theta_A^n(\mu) = \mu^{pqn+1} \cdot \lambda^n$ and by extending θ_A we have a homeomorphism from V to $V(k; \alpha)$ where α is the surgery curve $\mu^{pqn+1} \cdot \lambda^n$. Thus $pqn + 1 = \pm 1$, or $pqn = \pm 2$. This implies $\mathcal{L}$ is $\mathcal{L}_1$ if $(p, q) = (1, 1)$ or $(2, 1)$ and is $\mathcal{L}_2$ if $(p, q) = (1, 2)$. In the later case $n = 1$ or -1.

$\Leftarrow$ The only if part of the proof is trivial. $\heartsuit$.

As a direct consequence, we obtain a **knotting criterion** : consider $\mathcal{L} = \{K, J\}$ with two trivial components and apply our results on knots K_n obtained by twisting n times K on J.

Corollary 3.1. *Let $\mathcal{L} = \{K, J\}$ be a link in S^3 with two trivial components, where $\mathcal{L}$ is different from the trivial cases $\mathcal{L}_0$ and $\mathcal{L}_1$.*
(i) If $\mathcal{L} \neq \mathcal{L}_2$, all the K_n are non trivial knots for every $n \neq 0$
(ii) if $\mathcal{L} = \mathcal{L}_2$, K_n are non trivial knots for every $n \neq 0$ and $n \neq 1$.

As a consequence of the corollary, we can ask if this process of knotting the component K from any link $\mathcal{L} = \{K, J\}$ with two trivial components ($\mathcal{L} \neq \mathcal{L}_0$ and $\mathcal{L}_1$), by twisting on a disk bounded by J can yield any knot in S^3:

Questions. *If k is any knot in S^3, can we find a disk D such that Dehn twists on D trivialize k? In particular can we unknot non prime knots?* Such a disk is called an unknotting disk. The knot 9_{46} is a nice example, [NS], with one unknotting disk cutting the knot in 3 points.

If more than one disk is necessary to trivialize the knot, what about the minimal number of disks?

We don't know if any knot is obtained by twisting one component of a link $\mathcal{L} = \{K, J\}$ on the other one, but we have a condition to get a prime knot, with $w(K)$ being the winding number of K:

Proposition 3.1. *Let $\mathcal{L} = \{K, J\}$ be a link with 2 trivial components and n an integer such that $n = p \cdot q$, $p > 1$ and $gcd(p, w) = 1$.*

Then K_n is a prime knot in S^3. In particular if K is the closure $cl(\beta)$ of a braid β with w strings and axis J, and n is an integer prime to w, then $K_n = cl(\Delta^{2n}\beta)$ is a prime knot.

Proof. Take p a divisor of $n(n = p \cdot q)$ and $gcd(p, w) = 1$. The p-cyclic branched covering of S^3 with K as ramification is S^3 in which the lifting J' of J is a non trivial knot. If we perform downstairs $1/n$ Dehn surgery on J, we have a new knot $K_n = K(J; 1/n)$ and the p-cyclic branched covering of $S^3 = S^3(J; 1/n)$ with K_n as ramification is a manifold M_p which is also obtained by $1/q$ Dehn surgery on J' upstairs : $M_p = S^3(J'; 1/q)$. From theorem 3[GoL] the homology sphere M_p is irreducible and thus from [MY] the ramification K_n is a prime knot.

3.2 Application

As a consequence of theorem 3.1, we have a direct proof of corollary 7 in [CGLS] for property (P) of knots respected by an axial rotation of finite order, without using the "finite cyclic surgery theorem" of [CGLS]:

Corollary 3.2. *The homology spheres obtained by Dehn surgery on a knot in S^3 respected by a rotation of finite order with axis disjoint from the knot are not simply connected (property (P)).*

Proof. Let G be a finite cyclic group acting on S^3 and a knot k respected by the action. We set $J = Fix(G)$: by the Smith Conjecture J is the trivial knot in S^3, J disjoint from k. Necessarily the linking $lk(J, k) \neq 0$. Suppose now the manifold $S^3(k; +1/n) = \Sigma^3$ is a homotopy sphere. We can extend the action of G on Σ^3 : G_n is the action on Σ^3 with $Fix(G_n) = J(k; 1/n)$. This new knot is such that $\Sigma^3 - int(J(k; 1/n))$ is boundary-compressible by the generalized Smith Conjecture. From theorem 3.1, the knot k is the trivial one. ♡

Remark. In terms of branched covering, this generalizes Goldsmith's construction in [Gs].

4 Twisting and property (P)

We are now interested in property (P) on the knots K_n obtained by twisting n times K on J. The links $\mathcal{L} = \{K, J\}$ have two trivial components with arbitrary $lk(K, J)$. The disk D is such that $\partial D = J$. We first generalize results of [MD] in which $lk(K, J) = 0$ and $D \cap K = 2pts$.

4.1 Links with trivial components

Theorem 4.1. *Let $\mathcal{L} = \{K, J\}$ be a link in S^3 with two trivial components different from $\mathcal{L}_1$ and $\mathcal{L}_0$.*

Then the knot K_n , obtained by twisting K on D ($\partial D = J$) n times, is a non trivial knot with property (P), $\forall n$, $|n| > 1$. Moreover, one of the two knots K_1 or K_{-1} has property(P).

Case 1 $w(J) = 0$: $\mathcal{L} \neq \mathcal{L}_0$ and $\mathcal{L}_1$ implies that $K_n = K(J; 1/n)$ is always a nontrivial knot, from our theorem 3.1, for $\mathcal{L}$ cannot be $\mathcal{L}_2$. We now prove property (P). By the finite cyclic surgery theorem of [CGLS], only ± 1-surgery on K_n can yield a homotopy sphere. But $S^3(K_n; \pm 1) = S^3(K_n; \pm 1)(J; \infty)$. If $\theta_J : S^3 - J \to S^3 - J$ is the Dehn twist homeomorphism associated to a disk with boundary J, with $\theta_J^n(\mu_J) = \mu_J \cdot \lambda_J^n$, then θ_J^n trivializes $K_n : \theta_J^n(K_n) = K$, the trivial knot. Thus $S^3(K_n; \pm 1) \approx S^3(K; \pm 1)(J; 1/n)$. In the surgered model $S^3(K; \pm 1)$ of S^3, J becomes a knot $j = J(K; \pm 1)$ and $S^3(K_n; \pm 1) \approx S^3(j; 1/n)$. Since j is a knot and $|n| > 1$, the manifold $S^3(j; 1/n)$ is not simply connected from [CGLS], and K_n has property (P); so we generalize [MD].

Case 2 $w(j) \neq 0$: let $H = S^3(K_n; \pm 1)$ be the homology sphere. We note at first that H cannot be S^3 by the Gordon-Luecke theorem. Then we use the Kirby-Rolfsen calculation of surgery coefficients to get $H = S^3(K_n; \pm 1)(J; \infty) \approx S^3(K; -nw^2 \pm 1)(J; 1/n)$. The manifold $S^3(K; -nw^2 \pm 1)$ is a lens space and the curve J becomes a knot $j = J(K; -nw^2 \pm 1)$ in this space, with knot space $X = S^3(K; -nw^2 \pm 1) - int(N(j))$. We now identify the basis of $H_1(\partial X) = H_1(\partial N(j))$ with the previous one of $H_1(\partial N(J))$. The Dehn filling $X(1/n)$ is $H = X \cup V$, where V is a solid torus. We now prove H is not simply connected:

(i) If X is not a Seifert fiber space we can use [CGLS] : since $X(1/0)$ has a finite cyclic fundamental group, only $X(\pm 1)$ can be simply connected.

(ii) If X is a Seifert fiber space, either the meridian m_v of the filling torus is not homologous to a fiber, and $X(1/n) = S^3(K_n; \pm 1)$ is a non simply connected Seifert fiber space, since it is different from S^3 which is the only simply connected Seifert fiber space, or the meridian m_v is homologous to a fiber and by [H], the manifold $X(1/n)$ is S^3 (impossible), or contains a lens space as a factor, which is homologically impossible.

Applications. In the following, $\mathcal{L} = \{K, J\}$ is a link in S^3 with two trivial components and arbitrary linking number $lk(K, J)$. If we suppose J can be represented as a closed braid β with axis K (Stallings braids are a particular case), then we get nontrivial knots $\Delta^{2n}\beta$, by twisting n times β along the axis, in standard notation of braid theory.

Corollary 4.1. *If β is the unknotted closure of a braid, then all the knots $\Delta^{2n}\beta$ have property (P).*

4.2 Twists on knots

Let k be a knot in S^3 and D an essential disk, with $lk(k, \partial D) = 0$. Write $J = \partial D$ and $\theta_J : S^3 - J \to S^3 - J$ the above homeomorphism. We define the twist knots of a pair (k, J) by a family of knots $k_n = \theta_J^n(k)$, $\forall n \in \mathbb{Z}$ and write $C(k; J) = \{k_n\}_{n\in\mathbb{Z}}$ (or sometimes $C(k)$). A knot in the infinite family $C(k)$ is obtained by twisting k on J.

Theorem 4.2. *Among the family $C(k)$ of knots obtained by twisting a given knot k on a curve J, boundary of an essential disk, with $lk(k, J) = 0$, all knots have property (P) but at most two knots.*

If the trivial knot K is in the family $C(k)$, by theorem 4.1, at most two knots don't have property (P) corresponding to K_1 or K_{-1}, which are in $C(k)$, with other indices. If the trivial knot is not in $C(k)$, by [CGLS] it suffices to prove that the homology spheres $S^3(k_n; \pm 1)$ are not simply connected for all n, but for at most two integers. As above, $S^3(k_n; \pm 1) = S^3(k_n; \pm 1)(J; \infty)$ which is homeomorphic by the Dehn twist θ_J to a manifold $S^3(k_n; \pm 1)(J; \infty) = S^3(k; \pm 1)(J; -1/n)$. We call j the knot J after surgery in the manifold $S^3(k; \pm 1)$, i.e. $j = J(k; \pm 1)$. By setting $M = S^3(k; \pm 1)$ and $X = M - int(N(j))$ we have $S^3(k_n; \pm 1) = M(j; 1/n)$, and so two cases:

(i) X is not a Seifert fiber space : then we can apply the finite cyclic surgery theorem of [CGLS] and there exist at most two curves α and β, $|\alpha \cap \beta| = 1pt$, such that $X(\alpha)$ and $X(\beta)$ can be simply connected.

(ii) X is a Seifert fiber space : as in theorem 4.1, $X(1/n)$ cannot be simply connected.

4.3 Twists on knots and the generalized complement conjecture

The generalized complement conjecture asks the question : If M is an orientable 3-manifold, can we find two knots k and h such that $M - k$ and $M - h$ are homeomorphic but the knots non-equivalent by homeomorphism in M? Let $\mathcal{L} = \{J, K\}$ be a link in S^3 with two trivial components and the knots $K_n = K(J; 1/n)$ and $J_p = J(K; 1/p)$.

Question. Does there exist a link $\mathcal{L} = \{J, K\}$ in S^3 and integers $n \neq p$ such that K_n is equivalent by homeomorphism to J_p ? If such equivalent knots exist, then we produce a manifold $M = S^3(K, 1/p)(J; 1/n)$ with two distinct surgical presentations, $M \approx S^3(K_n; \alpha) \approx S^3(K_n; \beta)$ where α and β are two different slopes on the boundary of a regular neighborhood of the knots K_n, depending on n, p and $w = lk(J, K)$. In M the cores k and h of the two surgeries are knots having homeomorphic complements, but non-equivalent in M since by the Gordon-Luecke theorem there exists no homeomorphism from $(S^3 - K_n, \alpha)$ to $(S^3 - K_n, \beta)$. If $lk(K, J) = 0$, we get counter examples in a homology sphere.

We have a first answer in [M, chapter 3] from our link $\mathcal{L}_2$: the knots K_n are $(2, 2n + 1)$-torus knots and M is a Seifert manifold with 3 exceptional fibers. In this case, the cores of the two surgeries k and h in M have an orientation reversing homeomorphism of their complement, but are not equivalent by homeomorphism of M.

Addendum: In additional references, [MoS] the authors answer the question §3.1. See also [KMS].

References

[BM] R.H Bing, J. Martin, *Cubes with knotted holes*, Trans. A.M.S. **155** (1971), 217–231.

[CGLS] M. Culler, C. McA. Gordon, J. Luecke, P. Shalen, *Dehn surgery on knots*, Annals of Math. **125** (1987), 237–300.

[Ga 1] D. Gabai, *Surgery on knots in solid tori*, Topology **28** (1989), 1–6.

[Ga 2] D. Gabai, *1-Bridge braids in solid tori*, Topology and its Applications **37,** No. 3 (1990), 221–236.

[Go] C. McA. Gordon, *Dehn surgery and satellite knots*, Trans. A.M.S. **275** (1983), 687–708.

[GoL] C. McA. Gordon, J. Luecke, *Knots are determined by their complements*, Journal of A.M.S. **2-2** (1989), 374–415.

[Gs] D. Goldsmith, *Symmetric fibered links*, Knots, groups, and 3-manifolds, edited by L.P. Neuwirth,, Annals of Math. studies, Princeton University press, 1975, pp. 3–23.

[H] W. Heil, *Elementary surgery on Seifert fibered spaces*, Yokohama Math. J. **22** (1974), 135–142.

[M] Y. Mathieu, *Sur des Noeuds qui ne sont pas déterminés par leur complément et problèmes de chirurgie dans les variétés de dimension 3*, Thèse, Marseille (1990).

[MD] Y. Mathieu, M. Domergue, *Chirurgies de Dehn de pente* ± 1, Math. Annalen **280** (1988), 501–508.

[Mo] K. Motegi, *Homology 3-spheres which are obtained by Dehn surgery*, Math. Annalen **281** (1988), 483–493.

[MY] W. H. Meeks, S. T. Yau, *Topology of three dimensional manifolds*, Annals of Math. **112** (1980), 441–484.

[NS] Y. Nakanishi, S. Suzuki, *On Fox's congruence classes of knots*, Osaka J. Math. **24** (1987), 313–324.

[Se] S. Seifert, *Topologie dreidimensionaler gefaserter Räume*, Acta. Math. **60** (1933), 147–238.

[ST] M. Scharlemann, A. Thompson, *Unknotting number, genus and companion tori*, Math. Ann. **280** (1988), 191–205.

[T] H. Tietze, *Uber die Topologischen Invarianten*, Monatsh. Math. Phys. **19** (1908), 1–118.

[W] W. Whitten, *Knot complements and groups*, Topology **26** (1987), 41–44.

Additional references:

[MoS] K. Motegi, T. Shibuya, *Are knots obtained from a plain pattern prime?*, Preprint 1991.

[KMS] M. Kouno, K. Motegi, T. Shibuya, *Behavior of knots under twisting*, Advanced Studies in Pure Math. **20**, (to appear).

On the $Z_2 \oplus Z_2$ branched coverings of spatial K_4-graphs

Masahiro Nakao

Introduction

A K_4-*graph* G consists of four vertices and six edges, where a pair of two vertices are joined by one edge. Let $f : G \to S^3$ be an embedding of G into the 3-sphere S^3. Then $f(G)$ is called a *spatial* K_4-*graph*. We denote the six edges of G by x, x', y, y', z, and z' as shown in Fig.0.1.

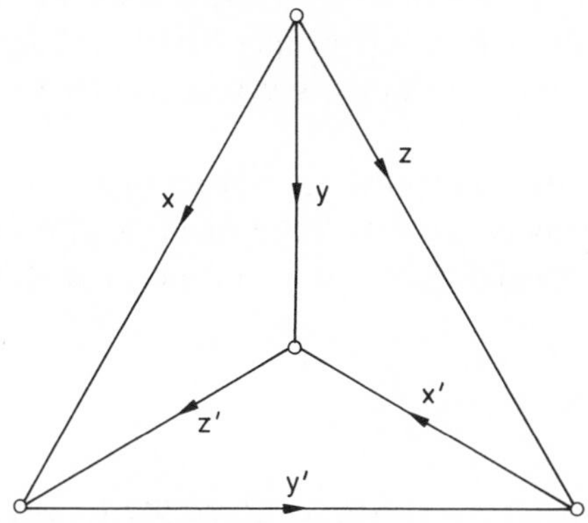

Figure 0.1

Then $f(x \cup y \cup z')$, $f(y \cup z \cup x')$, $f(z \cup x \cup y')$, $f(x' \cup y' \cup z')$, $f(x \cup y \cup x' \cup y')$, $f(y \cup z \cup y' \cup z')$, and $f(z \cup x \cup z' \cup x')$ are called *constituent knots* of G. From now on, we use G, instead of $f(G)$, as a spatial K_4-graph.

In this paper we consider the $Z_2 \oplus Z_2$ branched covering $\bar{M}_\phi(G)$ of a spatial K_4-graph and investigate relations between $\bar{M}_\phi(G)$ and the 2-fold branched coverings of three constituent knots of G each of which has four edges. The main theorem of the paper is as follows:

Knots 90

Theorem. *Let G be a spatial K_4-graph, and k_{xy}, k_{yz}, and k_{zx} three constituent knots of G each of which has four edges. Let $\bar{M}_2(k_{xy})$, $\bar{M}_2(k_{yz})$, and $\bar{M}_2(k_{zx})$ be the 2-fold branched coverings of k_{xy}, k_{yz}, and k_{zx}, respectively. Then we have*

$$H_1\left(\bar{M}_\phi(G), Z\right) \cong H_1\left(\bar{M}_2(k_{xy}), Z\right) \oplus H_1\left(\bar{M}_2(k_{yz}), Z\right) \oplus H_1\left(\bar{M}_2(k_{zx}), Z\right) \oplus Z_2.$$

Applying the main theorem to the spatial K_4-graph which is the vertex sum of a planar K_4-graph and a spatial θ_3-curve, we have the following corollary. (See also Nakao [3].)

Corollary. *Let $\bar{M}_\phi(\theta)$ be the $Z_2 \oplus Z_2$ branched covering of a spatial θ_3-curve θ. Let k_1, k_2, and k_3 be the three constituent knots of θ, and $\bar{M}_2(k_1)$, $\bar{M}_2(k_2)$, and $\bar{M}_2(k_3)$ the 2-fold branched coverings of k_1, k_2, and k_3, respectively. Then we have*

$$H_1\left(\bar{M}_\phi(\theta), Z\right) \cong H_1\left(\bar{M}_2(k_1), Z\right) \oplus H_1\left(\bar{M}_2(k_2), Z\right) \oplus H_1\left(\bar{M}_2(k_3), Z\right).$$

In section 1 we introduce the $Z_2 \oplus Z_2$ branched covering $\bar{M}_\phi(G)$ of a spatial K_4-graph G and show another way of constructing $\bar{M}_\phi(G)$. In section 2 we give some examples of spatial K_4-graphs and their $Z_2 \oplus Z_2$ branched coverings. In section 3, we give some lemmas which will be used in the main theorem. Though the proof of the main theorem in section 4 is mainly based on Tieze transformations, we also use some geometric consideration given in section 3.

The author would like to express his sincere gratitude to Professor Shin'ichi Kinoshita and Dr. Katsuyuki Yoshikawa for their helpful suggestion. In addition, he wishes to offer his special thanks to the members of KOOK Seminar for their encouragement.

1. Constructions of $Z_2 \oplus Z_2$ branched coverings of spatial K_4-graphs

Let G be a spatial K_4-graph. Suppose, for instance, that G has a diagram as shown in Fig.1.1.

Take a base point B in $S^3 - G$, and label each pass as $x_1,\dots,x_\ell$, $x_{\ell+1},\dots,x_L$, $y_1,\dots,y_m$, $y_{m+1},\dots,y_M$, $z_1,\dots,z_n$, $z_{n+1},\dots,z_N$, as shown in Fig.1.1. We also use each of x_i, y_j, or z_k, as a loop which goes around the corresponding pass once as shown in Fig.1.1. Further, we also use each as an element of $\pi_1(S^3 - G,\ B)$. So these notations should be adequately understood from the context.

Let us consider a monodromy map $\phi\ :\ \pi_1(S^3 - G,\ B)\ \rightarrow\ Z_2 \oplus Z_2 < S_4$ defined by $\phi(x_i) = (12)(34)$, $\phi(y_j) = (13)(24)$, and $\phi(z_k) = (14)(23)$, where $i = 1, \dots, L, j = 1, \dots, M$, and $k = 1, \dots, N$. Then there is a covering $M_\phi\ (G)$

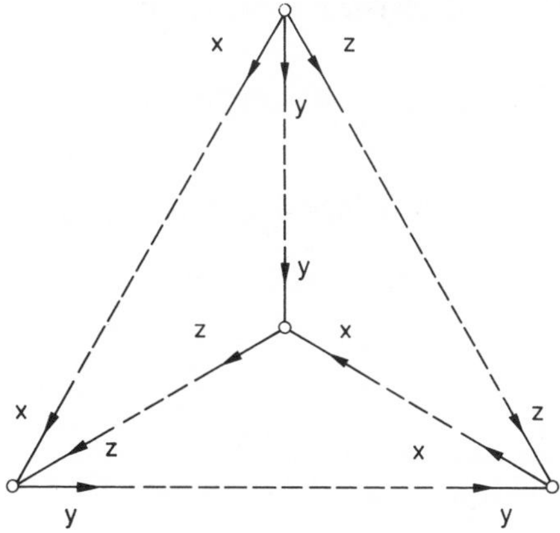

Figure 1.1

of $S^3 - G$ whose fundamental group is Ker ϕ. The $Z_2 \oplus Z_2$ branched covering $\bar{M}_\phi(G)$ of G is the completion of $M_\phi(G)$. (See Fox [1].)

Now we explain another way of constructing the $Z_2 \oplus Z_2$ branched covering of a spatial K_4-graph, which is as follows:
Let x, x', y, y', z, and z', be six edges of the spatial K_4-graph in Fig.0.1 and $k_{xy} = f(x \cup y \cup x' \cup y')$, $k_{yz} = f(y \cup z \cup y' \cup z')$, and $k_{zx} = f(z \cup x \cup z' \cup x')$ three constituent knots with four edges. First, construct the 2-fold covering of S^3 branched along k_{xy}. We denote it by $\bar{M}_2(k_{xy})$. Let $\tilde{z}$ be the lift of the edges $z \cup z'$ in $\bar{M}_2(k_{xy})$. The lift $\tilde{z}$ is a link which bounds an orientable surface in $\bar{M}_2(k_{xy})$. Hence we can construct the 2-fold branched covering of $\bar{M}_2(k_{xy})$ branched along $\tilde{z}$. It is easy to see that this covering coincides with the formerly defined $Z_2 \oplus Z_2$ branched covering $\bar{M}_\phi(G)$ of G. By the same way, we can also construct $\bar{M}_\phi(G)$ by choosing $\bar{M}_2(k_{yz})$ or $\bar{M}_2(k_{zx})$ first, instead of $\bar{M}_2(k_{xy})$. Hence, $\bar{M}_\phi$ (G) is independent of the choice of the constituent knots, when we first construct the 2-fold branched covering of one of its three constituent knots with four edges.

Proposition 1.1. *Let G be a planar K_4-graph, then we have*

$$\bar{M}_\phi(G) \approx P^3.$$

Proof. We use the second method to construct $\bar{M}_\phi(G)$. $\bar{M}_2(k_{xy})$ is S^3, and the lift $\tilde{z}$ in $\bar{M}_2(k_{xy})$ is a Hopf link. The 2-fold branched covering of S^3 branched along $\tilde{z}$ is P^3.

2. Some examples

Proposition 2.1. *Let G be a spatial K_4-graph which has a diagram as shown in Fig.2.1. Then we have*

$$\bar{M}_\phi(G) \approx \bar{M}_2(k) \sharp \bar{M}_2(k) \sharp P_3,$$

where three constituent knots with four edges are k, k, and a trivial knot 0_1.

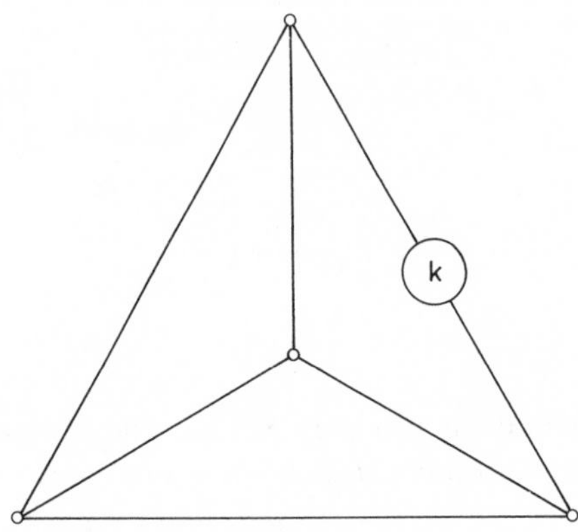

Figure 2.1

Proof. We give a brief proof by the second method of the construction. Without loss of generality we can suppose that $k_{yz} = k$, $k_{zx} = k$, and $k_{xy} = 0_1$. Then $\bar{M}_2\ (k_{xy})$ is S^3, and the lift $\tilde{z}$ is $(k \sharp k) \sharp L$, where L is a Hopf link. Next we construct the 2-fold branched covering of $\bar{M}_2(k_{xy})$ branched along $(k \sharp k) \sharp L$. Therefore we have

$$\bar{M}_\phi(G) \approx \bar{M}_2(k) \sharp \bar{M}_2(k) \sharp P^3,$$

which completes the proof.

Remark. The above example shows that the following formula (I) holds for some of the spatial K_4-graphs.

(I) $$\bar{M}_\phi(G) \approx \bar{M}_2(k_{xy}) \sharp \bar{M}_2(k_{yz}) \sharp \bar{M}_2(k_{zx}) \sharp P^3.$$

But formula (I) does not hold in general. Here is an example.

Example 2.2. Let G_1 be a spatial K_4-graph as shown in Fig.2.2. (Simon [4].) The three constituent knots with four edges of G_1 are one figure eight knot 4_1 and two trivial knots. It is easy to see that $\bar{M}_\phi(G_1) \approx L(5,\ 1) \sharp P^3$, and $\bar{M}_2(4_1) \sharp \bar{M}_2(0_1) \sharp \bar{M}_2(0_1) \sharp P^3 \approx L(5,\ 2) \sharp P^3$. Therefore we have $\bar{M}_\phi(G_1)$ is

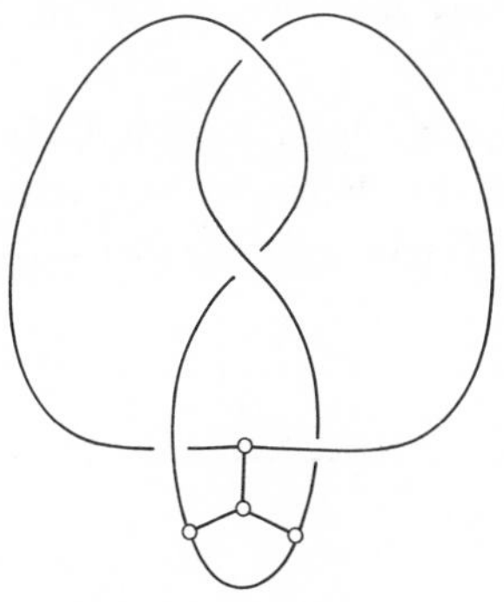

Figure 2.2

not homeomorphic to $\bar{M}_2(4_1) \sharp \bar{M}_2(0_1) \sharp \bar{M}_2(0_1) \sharp P^3$. We, however, obtain

$$\pi_1(\bar{M}_\phi(G_1)) \cong \pi_1(\bar{M}_2(4_0)) * \pi_1(\bar{M}_2(0_1)) * \pi_1(\bar{M}_2(0_1)) * Z_2.$$

Remark. The following formula (II) holds for some of the spatial K_4-graphs.

(II) $\pi_1(\bar{M}_\phi(G)) \cong \pi_1(\bar{M}_2(k_{xy})) * \pi_1(\bar{M}_2(k_{yz})) * \pi_1(\bar{M}_2(k_{zx})) * Z_2.$

But formula (II) does not holds in general. Here is an example.

Example 2.3. Let G_2 be a spatial K_4-graph as shown in Fig.2.3. ([4])

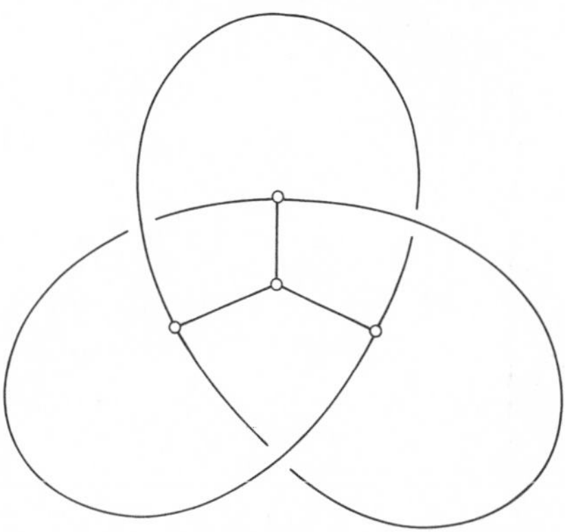

Figure 2.3

Each one of the three constituent knots of G_2 is trivial. However, we can show

$$\pi_1(\bar{M}_\phi(G_2)) \cong < X, Y;\ X^2 = Y^2 = (XY)^2 = 1 > .$$

Therefore we have $\pi_1(\bar{M}_\phi(G_2))$ is not isomorphic to $\pi_1(\bar{M}_2(0_1)) * \pi_1(\bar{M}_2(0_1)) * \pi_1(\bar{M}_2(0_1)) * Z_2$.

3. Preliminaries for the proof of the Theorem

We can put any K_4-graph into a "good" position. (Here a "good" position means that at least one pair of edges which are not adjacent each other have no crossings in the diagram of G.) So we may suppose that G has the diagram as shown in Fig.3.1.

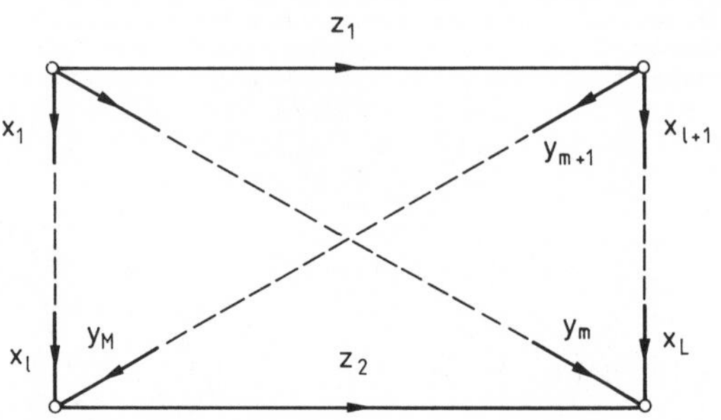

Figure 3.1

Classify the crossings to the following four types. (See Fig.3.2.)

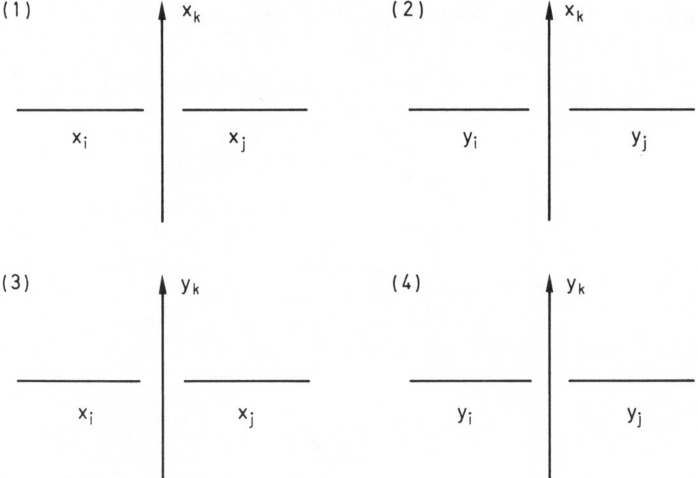

Figure 3.2

Then the fundamental group of $S^3 - G$ has the following presentation:

$$\pi_1(S^3 - G) \cong < x_1, ..., x_\ell,\ x_{\ell+1}, ..., x_L,\ y_1, ..., y_m,\ y_{m+1}, ..., y_M,\ z_1,\ z_2;$$
$$x_1 z_1 y_1 = 1,\ x_\ell z_2^{-1} y_M = 1,\ x_{\ell+1} y_{m+1} z_1^{-1} = 1,\ x_L y_m z_2 = 1,$$
$$(1)\ x_i x_k = x_k x_j, ...,$$

(2) $y_i x_k = x_k y_j, \ldots,$
(3) $x_i y_k = y_k x_j, \ldots,$
(4) $y_i y_k = y_k y_j, \ldots >.$

Similarly, the fundamental groups of the complementary domains of the three constituent knots with four edges have the following presentations:

$\pi_1(S^3 - k_{xy}) \cong < x_1, \ldots, x_\ell,\ x_{\ell+1}, \ldots, x_L,\ y_1, \ldots, y_m,\ y_{m+1}, \ldots, y_M;$
$x_1 y_1 = 1,\ x_\ell y_M = 1,\ x_{\ell+1} y_{m+1} = 1,\ x_L y_m = 1,$
(1) $x_i x_k = x_k x_j, \ldots,$
(2) $y_i x_k = x_k y_j, \ldots,$
(3) $x_i y_k = y_k x_j, \ldots,$
(4) $y_i y_k = y_k y_j, \ldots >,$
$\pi_1(S^3 - k_{yz}) \cong < y_1, \ldots, y_m,\ y_{m+1}, \ldots, y_M,\ z_1,\ z_2;$
$z_1 y_1 = 1,\ z_2^{-1} y_M = 1,\ y_{m+1} z_1^{-1} = 1,\ y_m z_2 = 1,$
(2) $y_i = y_j, \ldots,$
(4) $y_i y_k = y_k y_j, \ldots >$, and
$\pi_1\ (S^3 - k_{zx}) \cong < x_1, \ldots, x_\ell,\ x_{\ell+1}, \ldots, x_L,\ z_1,\ z_2;$
$x_1 z_1 = 1,\ x_\ell z_2^{-1} = 1,\ x_{\ell+1} z_1^{-1} = 1,\ x_L z_2 = 1,$
(1) $x_i x_k = x_k x_j, \ldots,$
(3) $x_i = x_j, \ldots >.$

Choosing $\{\ 1,\ x_1,\ y_1,\ x_1 y_1\ \}$ as a Schreier system, we calculate $\pi_1(\bar{M}_\phi(G))$ from $\pi_1\ (S^3 - G)$ by the Reidemeister Schreier Method. (See [1] and Nakao [3].) By abelianizing $\pi_1(\bar{M}_\phi(G))$, we have

$H_1(\bar{M}_\phi(G))$
$\cong < x_{11}, \ldots, x_{L1},\ x_{13}, \ldots, x_{L3},\ y_{11}, \ldots, y_{M1}, y_{12}, \ldots, y_{M2},\ z_{11},\ z_{21},\ z_{12},\ z_{22};$
$x_{11} = x_{13} = y_{11} = y_{12} = z_{11} = z_{12} = 0,$
$x_{\ell 1} - y_{M1} + z_{22} = 0,\ -x_{\ell 1} - y_{M2} + z_{21} = 0,$
$x_{\ell 3} + y_{M1} - z_{21} = 0,\ -x_{\ell 3} + y_{M2} - z_{22} = 0,$
$x_{\ell+1,1} + y_{m+1,2} - z_{11} = 0,\ -x_{\ell+1,1} + y_{m+1,1} - z_{12} = 0,$
$x_{\ell+1,3} - y_{m+1,2} + z_{12} = 0,\ -x_{\ell+1,3} - y_{m+1,1} + z_{11} = 0,$
$x_{L1} + y_{m2} - z_{21} = 0,\ -x_{L1} + y_{m1} - z_{22} = 0,$
$x_{L3} - y_{m2} + z_{22} = 0,\ -x_{L3} - y_{m1} + z_{21} = 0,$
(1) $2\ x_{k1} - x_{i1} - x_{j1} = 0,\ 2\ x_{k3} - x_{i3} - x_{j3} = 0, \ldots,$
(2) $x_{k1} - x_{k3} - y_{i1} + y_{j2} = 0,\ x_{k1} - x_{k3} - y_{j1} + y_{i2} = 0, \ldots,$
(3) $y_{k1} - y_{k2} - x_{i1} + x_{j3} = 0,\ y_{k1} - y_{k2} - x_{j1} + x_{i3} = 0, \ldots,$
(4) $2\ y_{k1} - y_{i1} - y_{j1} = 0,\ 2\ y_{k2} - y_{i2} - y_{j2} = 0, \ldots >.$

Similarly, we have

$$H_1(\bar{M}_2(k_{xy}))\cong< X_{11}, ..., X_{L1},\ Y_{11}, ..., Y_{M1};$$
$$X_{11}=Y_{11}=0,\ X_{\ell 1}-Y_{M1}=0,\ -X_{\ell+1,1}+Y_{m+1,1}=0,\ -X_{L1}+Y_{m1}=0,$$
$$(1)\ 2\ X_{k1}-X_{i1}-X_{j1}=0, ...,$$
$$(2)\ 2\ X_{k1}-Y_{i1}-Y_{j1}=0, ...,$$
$$(3)\ 2\ Y_{k1}-X_{i1}-X_{j1}=0, ...,$$
$$(4)\ 2\ Y_{k1}-Y_{i1}-Y_{j1}=0, ... >,$$
$$H_1(\bar{M}_2(k_{yz}))\cong\ < Y_{12}, ..., Y_{M2},\ Z_{12},\ Z_{22};$$
$$Y_{12}= Z_{12}=0,\ Y_{M2}-Z_{22}=0,\ Y_{m+1,2}=0,\ -Y_{m2}+Z_{22}=0,$$
$$(2)\ Y_{i2}-Y_{j2}=0, ...,$$
$$(4)\ 2\ Y_{k2}-Y_{i2}-Y_{j2}=0, ... >, \text{ and}$$
$$H_1(\bar{M}_2(k_{zx}))\cong\ < X_{13}, ..., X_{L3},\ Z_{11},\ Z_{21};$$
$$X_{13}= Z_{11}=0,\ X_{\ell 3}-Z_{21}=0,\ X_{\ell+1,3}=0,\ X_{L3}-Z_{21}=0,$$
$$(1)\ 2\ X_{k3}-X_{i3}-X_{j3}=0, ...,$$
$$(3)\ X_{i1}-X_{j3}=0, ... > .$$

Hence, we have

$$H_1(\bar{M}_2(k_{xy}))\oplus H_1(\bar{M}_2(k_{yz}))\oplus\ H_1(\bar{M}_2(k_{zx}))$$
$$\cong< X_{11}, ..., X_{L1},\ X_{13}, ..., X_{L3},\ Y_{11}, ..., Y_{M1},\ Y_{12}, ..., Y_{M2},\ Z_{11},\ Z_{21},\ Z_{12},\ Z_{22};$$
$$X_{11}= X_{13}= Y_{11}= Y_{12}= Z_{11}= Z_{12}=0,$$
$$X_{\ell 1}= Y_{M1},\ X_{\ell+1,1}= Y_{m+1,1},\ X_{L1}= Y_{m1},\ Y_{M2}= Z_{22},\ Y_{m+1,2}=0,$$
$$Y_{m2}= Z_{22},\ X_{\ell 3}= Z_{21},\ X_{\ell+1,3}=0,\ X_{L3}= Z_{21},$$
$$(1)\ 2\ X_{k1}-X_{i1}-X_{j1}=0,\ 2\ X_{k3}-X_{i3}-X_{j3}=0, ...,$$
$$(2)\ 2\ X_{k1}-Y_{i1}-Y_{j1}=0,\ Y_{i2}-Y_{j2}=0, ...,$$
$$(3)\ 2\ X_{k1}-Y_{i1}-Y_{j1}=0,\ X_{i3}-X_{j3}=0, ...,$$
$$(4)\ 2\ Y_{k1}-Y_{i1}-Y_{j1}=0,\ 2\ Y_{k2}-Y_{i2}-Y_{j2}=0, ... > .$$

Lemma 3.1. *Any spatial K_4-graph can be deformed to a planar K_4-graph by an ambient isotopy and the "operations", which change the crossings within the same pair edge group, where {x, x'}, {y, y'}, and {z, z'} are pair edge groups.*

Proof. Consider a diagram of G. Let x, x', y, y', z, and z', be six edges as shown in Fig.0.1. First put G into a "good" position. (Without loss of generality, we may assume that edges z and z' have no crossings.) Next, by an ambient isotopy and the "operations", we can deform G into such a position that one constituent knot with four edges is trivial. (Without loss of generality, we may assume that k_{yz} is trivial.) The knot k_{yz} bounds a two-disk D^2. Now let m be the sum of the number of intersection of x and D^2 and that of x' and D^2. Then the lemma can be proved by an induction on m. (See [3].)

Lemma 3.2. *Let G and G' be two spatial K_4-graphs. Suppose that G' is obtained from G by changing a crossing of edges in a pair edge group. If $H_1(\bar{M}_\phi(G))$ is a finite group of even order, so is $H_1(\bar{M}_\phi(G))$.*

Proof. This follows from lemma 3.1. (See lemma 3.2 in [3].)

Proposition 3.3. *Let G be a K_4-graph. Then, $H_1(\bar{M}_\phi(G))$ is a finite group of even order.*

Proof. If G is a planar K_4-graph, then $H_1(\bar{M}_\phi(G)) \cong Z_2$ by proposition 1.1. Then, apply lemmas 3.1 and 3.2. (See proposition 3.3 in [3].)

Proposition 3.4. *Let G be a K_4-graph. Then, we have*

$$H_1(\bar{M}_\phi(G)) \cong Z_{2(2K-1)},$$

where K is a natural number.

The proof of the above proposition will be given in Section 4.

By proposition 3.4, each element g of $H_1(\bar{M}_\phi(G))$ has either property 1 or property 2 defined in the below.
An element g has property 1 if there exists a natural number K such that $(2K-1)g = 0$.
An element g has property 2 if (1) $(2K-1)g \neq 0$ for any natural number K and (2) there exists a natural number K' such that $2(2K'-1)g = 0$.

Consider a diagram shown in Fig.3.1. If all of the elements of $H_1(\bar{M}_\phi(G))$ have property 1, $H_1(\bar{M}_\phi(G))$ is a finite group of odd order. Therefore there exists an element g of $H_1(\bar{M}_\phi(G))$ which has property 2. Moreover, if one element has property 2, all the other elements in the same edge have property 2. Note that there always exists $g \in H_1(\bar{M}_\phi(G))$ which has property 1. So we may assume that $x_{11}, \ldots, x_{\ell 1}, x_{13}, \ldots, x_{\ell 3}, y_{11}, \ldots, y_{m1}, y_{12}, \ldots, y_{m2}$, and z_{11}, z_{12}, have property 1 and that $x_{\ell+1,1}, \ldots, x_{L1}, x_{\ell+1,3}, \ldots, x_{L3}, y_{m+1,1}, \ldots, y_{M1}, y_{m+1,2}, \ldots, y_{M2}$, and z_{21}, z_{22}, have property 2.

4. The proof of the Theorem

By proposition 3.4, we suppose that the order of $H_1(\bar{M}_\phi(G))$ is equal to $2(2K'-1)$, where K' is a natural number. It is well-known that $H_1(\bar{M}_2(k_{xy}))$, $H_1(\bar{M}_2(k_{yz}))$, and $H_1(\bar{M}_2(k_{zx}))$ are finite groups of odd order. So we also suppose that the order of $H_1(\bar{M}_2(k_{xy}))\oplus H_1(\bar{M}_2(k_{yz}))\oplus H_1(\bar{M}_2(k_{zx}))$ is equal to $(2K''-1)$, where K'' is a natural number. Let $(2K-1)$ be the least common multiple of $(2K'-1)$ and $(2K''-1)$. To prove the main theorem we shall deform a presentation of $H_1(\bar{M}_\phi(G))$ to that of $H_1(\bar{M}_2(k_{xy}))\oplus H_1(\bar{M}_2(k_{yz}))\oplus H_1(\bar{M}_2(k_{zx}))\oplus Z_2$ by Tieze transformations. We show these steps in the below.

$$
\begin{aligned}
&H_1(\bar{M}_\phi(G)) \cong < A; R_1, R_2 >, \text{ where} \\
&\quad A = \{x_{11},\ldots,x_{L1},\ x_{13},\ldots,x_{L3},\ y_{11},\ldots,y_{M1},\ y_{12},\ldots,y_{M2},\ z_{11},\ z_{21},\ z_{12},\ z_{22}\}, \\
&\quad R_1 = \{x_{11}=x_{13}=y_{11}=y_{12}=z_{11}=z_{12}=0, \\
&\qquad x_{\ell 1}-y_{M1}+z_{22}=0,\ -x_{\ell 1}-y_{M2}+z_{21}=0, \\
&\qquad x_{\ell 3}+y_{M1}-z_{21}=0,\ -x_{\ell 3}+y_{M2}-z_{22}=0, \\
&\qquad x_{\ell+1,1}+y_{m+1,2}-z_{11}=0,\ -x_{\ell+1,1}+y_{m+1,1}-z_{12}=0, \\
&\qquad x_{\ell+1,3}-y_{m+1,2}+z_{12}=0,\ -x_{\ell+1,3}-y_{m+1,1}+z_{11}=0, \\
&\qquad x_{L1}+y_{m2}-z_{21}=0,\ -x_{L1}+y_{m1}-z_{22}=0, \\
&\qquad x_{L3}-y_{m2}+z_{22}=0,\ -x_{L3}-y_{m1}+z_{21}=0\}, \text{ and} \\
&\quad R_2 = \{(1)\ 2\,x_{k1}-x_{i1}-x_{j1}=0,\ 2\,x_{k3}-x_{i3}-x_{j3}=0,\ldots, \\
&\qquad (2)\ x_{k1}-x_{k3}-y_{i1}+y_{j2}=0,\ x_{k1}-x_{k3}-y_{j1}+y_{i2}=0,\ldots, \\
&\qquad (3)\ y_{k1}-y_{k2}-x_{i1}+x_{j3}=0,\ y_{k1}-y_{k2}-x_{j1}+x_{i3}=0,\ldots, \\
&\qquad (4)\ 2\,y_{k1}-y_{i1}-y_{j1}=0,\ 2\,y_{k2}-y_{i2}-y_{j2}=0,\ldots\}.
\end{aligned}
$$

We add new generators B and their defining relations R_3.

$$
\begin{aligned}
&H_1(\bar{M}_\phi(G)) \cong < A, B; R_1, R_2, R_3 >, \\
&\text{where } B = \{X_{11},\ldots,X_{L1},\ X_{13},\ldots,X_{L3},\ Y_{11},\ldots,Y_{M1},\ Y_{12},\ldots,Y_{M2},\ Z_{11},\ Z_{21}, \\
&\qquad Z_{12},\ Z_{22}\} \\
&\text{and} \\
&R_3 = \{X_{11}=x_{11}-x_{13},\ldots,X_{L1}=x_{L1}-x_{L3}, X_{13}=x_{11}+x_{13},\ldots,X_{13}=x_{L1}+x_{L3}, \\
&\qquad Y_{11}=y_{11}-y_{12},\ldots,Y_{M1}=y_{M1}-y_{M2}, Y_{12}=y_{11}+y_{12},\ldots,Y_{M2}=y_{M1}+y_{M2}, \\
&\qquad Z_{11}=z_{11}-z_{12},\ Z_{21}=z_{21}-z_{22},\ Z_{12}=z_{11}+z_{12},\ Z_{22}=z_{21}+z_{22}\ \}.
\end{aligned}
$$

We add relations R_4 and R_5 reducible from R_1 and R_2.

$$
\begin{aligned}
&H_1(\bar{M}_\phi(G)) \cong < A, B; R_1, R_2, R_3, R_4, R_5 >, \text{ where} \\
&R_4 = \{x_{11}-x_{13}=0,\ x_{11}+x_{13}=0,\ y_{11}-y_{12}=0, \\
&\qquad y_{11}+y_{12}=0,\ z_{11}-z_{12}=0,\ z_{11}+z_{12}=0, \\
&\qquad (x_{\ell 1}-y_{M1}+z_{22})+(-x_{\ell 3}+y_{M2}-z_{22})=0, \\
&\qquad (x_{\ell+1,1}+y_{m+1,2}-z_{11})+(-x_{\ell+1,3}-y_{m+1,1}+z_{11})=0, \\
&\qquad (x_{L1}+y_{m2}-z_{21})+(-x_{L3}-y_{m1}+z_{21})=0, \\
&\qquad (x_{\ell 1}-y_{M1}+z_{22})+(-x_{\ell 1}-y_{M2}+z_{21})=0, \\
&\qquad (x_{\ell+1,1}+y_{m+1,2}-z_{11})+(-x_{\ell+1,1}+y_{m+1,1}-z_{12})=0, \\
&\qquad (x_{L1}+y_{m2}-z_{21})+(-x_{L1}+y_{m1}-z_{22})=0, \\
&\qquad (x_{\ell 1}-y_{M1}+z_{22})+(x_{\ell 3}+y_{M1}-z_{21})=0,
\end{aligned}
$$

$$
\begin{aligned}
&(x_{\ell+1,1}+y_{m+1,2}-z_{11})+(x_{\ell+1,3}-y_{m+1,2}+z_{12}) = 0,\\
&(x_{L1}+y_{m2}-z_{21})+(x_{L3}-y_{m2}+z_{22}) = 0\}, \text{ and}\\
R_5 = \{&(1)\ (2\ x_{k1}-x_{i1}-x_{j1})-(2\ x_{k3}-x_{i3}-x_{j3}) = 0,\\
&\quad (2\ x_{k1}-x_{i1}-x_{j1})+(2\ x_{k3}-x_{i3}-x_{j3}) = 0, \ldots,\\
&(2)\ (x_{k1}-x_{k3}-y_{i1}+y_{j2})-(x_{k1}-x_{k3}-y_{j1}+y_{i2}) = 0,\\
&\quad (x_{k1}-x_{k3}-y_{i1}+y_{j2})+(x_{k1}-x_{k3}-y_{j1}+y_{i2}) = 0, \ldots,\\
&(3)\ (y_{k1}-y_{k2}-x_{i1}+x_{j3})-(y_{k1}-y_{k2}-x_{j1}+x_{i3}) = 0,\\
&\quad (y_{k1}-y_{k2}-x_{i1}+x_{j3})+(y_{k1}-y_{k2}-x_{j1}+x_{i3}) = 0, \ldots,\\
&(4)\ (2\ y_{k1}-y_{i1}-y_{j1})-(2\ y_{k2}-y_{i2}-y_{j2}) = 0,\\
&\quad (2\ y_{k1}-y_{i1}-y_{j1})+(2\ y_{k2}-y_{i2}-y_{j2}) = 0, \ldots\}.
\end{aligned}
$$

We add relations R_6 and R_7 reducible from R_3, R_4, and R_5.

$$
\begin{aligned}
&H_1(\bar{M}_\phi(G)) \cong < A, B; R_1, R_2, R_3, R_4, R_5, R_6, R_7 >, \text{ where}\\
&R_6 = \{X_{11}= X_{13}= Y_{11}= Y_{12}= Z_{11}= Z_{12} = 0,\\
&\quad X_{\ell 1}= Y_{M1},\ X_{\ell+1,1}= Y_{m+1,1},\ X_{L1}= Y_{m1},\ Y_{M2}= Z_{22},\ Y_{m+1,2} = 0,\\
&\quad Y_{m2}= Z_{22},\ X_{\ell 3}= Z_{21},\ X_{\ell+1,3} = 0,\ X_{L3}= Z_{21}\}, \text{ and}\\
&R_7 = \{(1)\ 2\ X_{k1}-X_{i1}-X_{j1} = 0,\ 2\ X_{k3}-X_{i3}-X_{j3} = 0, \ldots,\\
&\quad (2)\ 2\ X_{k1}-Y_{i1}-Y_{j1} = 0,\ Y_{i2}-Y_{j2} = 0, \ldots,\\
&\quad (3)\ 2\ X_{k1}-Y_{i1}-Y_{j1} = 0,\ X_{i3}-X_{j3} = 0, \ldots,\\
&\quad (4)\ 2\ Y_{k1}-Y_{i1}-Y_{j1} = 0,\ 2\ Y_{k2}-Y_{i2}-Y_{j2} = 0, \ldots\}.
\end{aligned}
$$

Since $< B; R_6, R_7 >$ is a presentation of $H_1(\bar{M}_2(k_{xy})) \oplus H_1(\bar{M}_2(k_{yz})) \oplus H_1(\bar{M}_2(k_{zx}))$, the following relations R_8 are reducible.

$$
\begin{aligned}
&H_1(\bar{M}_\phi(G)) \cong < A, B; R_1, R_2, R_3, R_4, R_5, R_6, R_7, R_8, R_9 >, \text{where}\\
&R_8= \{(2K-1)X_{11} = 0, \ldots, (2K-1)X_{L1} = 0,\\
&\quad (2K-1)X_{13} = 0, \ldots, (2K-1)X_{L3} = 0,\\
&\quad (2K-1)Y_{11} = 0, \ldots, (2K-1)Y_{M1} = 0,\\
&\quad (2K-1)Y_{12} = 0, \ldots, (2K-1)Y_{M2} = 0,\\
&\quad (2K-1)Z_{11} = 0,\ (2K-1)Z_{21} = 0,\\
&\quad (2K-1)Z_{12} = 0,\ (2K-1)Z_{22} = 0\}.
\end{aligned}
$$

The following relations R_9 are reducible from R_3.

$$
\begin{aligned}
R_9 = \{&2\ x_{11}= X_{11}+X_{13}, \ldots, 2\ x_{L1}= X_{L1}+X_{L3},\\
&2\ x_{13}= -X_{11}+X_{13}, \ldots, 2\ x_{L3}= -X_{L1}+X_{L3},\\
&2\ y_{11}= Y_{11}+Y_{12}, \ldots, 2\ y_{M1}= Y_{M1}+Y_{M2},\\
&2\ y_{12}= -Y_{11}+Y_{12}, \ldots, 2\ y_{M2}= -Y_{M1}+Y_{M2},\\
&2\ z_{11}= Z_{11}+Z_{12},\ 2\ z_{21}= Z_{21}+Z_{22},\\
&2\ z_{12}= -Z_{11}+Z_{12},\ 2\ z_{22}= -Z_{21}+Z_{22}\}.
\end{aligned}
$$

The proof of Proposition 3.4. By relations R_8 and R_9 the following relations are reducible.

$$2(2K-1)x_{11}=0,...,2(2K-1)x_{L1}=0,$$
$$2(2K-1)x_{13}=0,...,2(2K-1)x_{L3}=0,$$
$$2(2K-1)y_{11}=0,...,2(2K-1)y_{M1}=0,$$
$$2(2K-1)y_{12}=0,...,2(2K-1)y_{M2}=0,$$
$$2(2K-1)z_{11}=0,\ 2(2K-1)z_{21}=0,$$
$$2(2K-1)z_{12}=0, \text{ and } 2(2K-1)z_{22}=0.$$

Since $H_1(\bar{M}_\phi(G))$ is a finite group of even order and $2(2K-1)g=0$ for all of the generators, the proof is complete.

We add new generators C and their defining relations R_{10}, and other relations R_{11} reducible from R_8 and R_9.

$$H_1(\bar{M}_\phi(G)) \cong < A, B, C; R_1, R_2, R_3, R_4, R_5, R_6, R_7, R_8, R_9, R_{10}, R_{11} >, \text{ where}$$
$$C=\{\alpha_{11},...,\alpha_{L1},\alpha_{13},...,\alpha_{L3},\beta_{11},...,\beta_{M1},\beta_{12},...,\beta_{M2},\gamma_{11},\ \gamma_{21},\gamma_{12},\ \gamma_{22}\},$$
$$R_{10}=\{\alpha_{11}=K\ X_{11}+K\ X_{13}-x_{11},\ ...,\alpha_{L1}=K\ X_{L1}+K\ X_{L3}-x_{L1},$$
$$\alpha_{13}=-K\ X_{11}+K\ X_{13}-x_{13},\ ...,\alpha_{L3}=-K\ X_{L1}+K\ X_{L3}-x_{L3},$$
$$\beta_{11}=K\ Y_{11}+K\ Y_{12}-y_{11},\ ...,\beta_{M1}=K\ Y_{M1}+K\ Y_{M2}-y_{M1},$$
$$\beta_{12}=-K\ Y_{11}+K\ Y_{12}-y_{12},\ ...,\beta_{M2}=-K\ Y_{M1}+K\ Y_{M2}-y_{M2},$$
$$\gamma_{11}=K\ Z_{11}+K\ Z_{12}-z_{11},\ \gamma_{21}=K\ Z_{21}+K\ Z_{22}-z_{21},$$
$$\gamma_{12}=-K\ Z_{11}+K\ Z_{12}-z_{12},\ \gamma_{22}=-K\ Z_{21}+K\ Z_{22}-z_{22}\}, \text{ and}$$
$$R_{11}=\{2\ x_{11}=2K\ X_{11}+2K\ X_{13},\ ...,2\ x_{L1}=2K\ X_{L1}+2K\ X_{L3},$$
$$2\ x_{13}=-2K\ X_{11}+2K\ X_{13},...,2\ x_{L3}=-2K\ X_{L1}+2K\ X_{L3},$$
$$2\ y_{11}=2K\ Y_{11}+2K\ Y_{12},...,2\ y_{M1}=2K\ Y_{M1}+2K\ Y_{M2},$$
$$2\ y_{12}=-2K\ Y_{11}+2K\ Y_{12},...,2\ y_{M2}=-2K\ Y_{M1}+2K\ Y_{M2},$$
$$2\ z_{11}=2K\ Z_{11}+2K\ Z_{12},\ 2\ z_{21}=2K\ Z_{21}+2K\ Z_{22},$$
$$2\ z_{12}=-2K\ Z_{11}+2K\ Z_{12},\ 2\ z_{22}=-2K\ Z_{21}+2K\ Z_{22}\}.$$

By R_{10} and R_{11}, we obtain the following relations S_1.

$$S_1=\{2\ \alpha_{11}=\ ...\ =2\ \alpha_{L1}=0,\ 2\ \alpha_{13}=\ ...\ =2\ \alpha_{L3}=0,$$
$$2\ \beta_{11}=\ ...\ =2\ \beta_{M1}=0,\ 2\ \beta_{12}=\ ...\ =2\ \beta_{M2}=0,$$
$$2\ \gamma_{11}=2\ \gamma_{12}=0,\ 2\ \gamma_{21}=2\ \gamma_{22}=0\}.$$

Since $x_{11},...,x_{\ell 1},x_{13},...,x_{\ell 3},\ y_{11},...,y_{m1},y_{12},...,y_{m2},$ and $z_{11},z_{12},$ have property 1 as stated before, we obtain the following relations S_2.

$$S_2=\{(2K-1)x_{11}=0,...,(2K-1)x_{\ell 1}=0,$$
$$(2K-1)x_{13}=0,...,(2K-1)x_{\ell 3}=0,$$
$$(2K-1)y_{11}=0,...,(2K-1)y_{m1}=0,$$
$$(2K-1)y_{12}=0,...,(2K-1)y_{m2}=0,$$
$$(2K-1)z_{11}=0,\ (2K-1)z_{21}=0\}.$$

By R_8, R_{10} and S_2 we obtain the following relations S_3.

$$S_3 = \{(2K-1)\alpha_{11} = 0, \dots, (2K-1)\alpha_{\ell 1} = 0,\\ (2K-1)\alpha_{13} = 0, \dots, (2K-1)\alpha_{\ell 3} = 0,\\ (2K-1)\beta_{11} = 0, \dots, (2K-1)\beta_{m1} = 0,\\ (2K-1)\beta_{12} = 0, \dots, (2K-1)\beta_{m2} = 0,\\ (2K-1)\gamma_{11} = 0,\ (2K-1)\gamma_{21} = 0\}.$$

By S_1 and S_3 we obtain the following relations

$$\alpha_{11} = \dots = \alpha_{\ell 1} = 0,\ \alpha_{13} = \dots = \alpha_{\ell 3} = 0,\ \beta_{11} = \dots = \beta_{m1} = 0,\\ \beta_{12} = \dots = \beta_{m2} = 0,\ \gamma_{11} = \gamma_{12} = 0.$$

Now consider a crossing of type (1) in Fig.3.2. Then, substitute $x_{\lambda 1} = K\ X_{\lambda 1} + K\ X_{\lambda 3} - \alpha_{\lambda 1}$ and $x_{\lambda 3} = -K\ X_{\lambda 1} + K\ X_{\lambda 3} - \alpha_{\lambda 3}$, where $\lambda = k, i, or j$ in the relations $2\ x_{k1} - x_{i1} - x_{j1} = 0$ and $2\ x_{k3} - x_{i3} - x_{j3} = 0$. Then, by R_2 and R_7, we have $\alpha_{i1} = \alpha_{j1}$ and $\alpha_{i3} = \alpha_{j3}$. We consider crossings of type (2), (3), and (4) in a similar way. Then, we obtain

$$\alpha_{11} = \dots = \alpha_{\ell 1} = 0,\ \alpha_{13} = \dots = \alpha_{\ell 3} = 0,\ \beta_{11} = \dots = \beta_{m1} = 0,\\ \beta_{12} = \dots = \beta_{m2} = 0,\ \gamma_{11} = \gamma_{12} = 0, \text{ and}\\ \alpha_{\ell+1,1} = \dots = \alpha_{L1} = \alpha_{\ell+1,3} \dots = \alpha_{L3} = \beta_{m+1,1} = \dots = \beta_{M1}\\ = \beta_{m+1,2} = \dots = \beta_{M2} = \gamma_{21} = \gamma_{22}.$$

We add reducible relations stated in the above.

$$H_1(\bar{M}_\phi(G)) \cong < A, B, C; R_1, R_2, R_3, R_4, R_5, R_6, R_7, R_8, R_9, R_{10}, R_{11}, R_{12} >,$$
where

$$R_{12} = \{\alpha_{11} = \dots = \alpha_{\ell 1} = 0,\ \alpha_{13} = \dots = \alpha_{\ell 3} = 0,\ \beta_{11} = \dots = \beta_{m1} = 0,\\ \beta_{12} = \dots = \beta_{m2} = 0,\ \gamma_{11} = \gamma_{12} = 0,\\ \alpha_{\ell+1,1} = \dots = \alpha_{L1} = \alpha_{\ell+1,3} \dots = \alpha_{L3} = \beta_{m+1,1} = \dots = \beta_{M1}\\ = \beta_{m+1,2} = \dots = \beta_{M2} = \gamma_{21} = \gamma_{22},\\ 2\ \alpha_{\ell+1,1} = \dots = 2\ \alpha_{L1} = 2\ \alpha_{\ell+1,3} = \dots = 2\ \alpha_{L3}\\ = 2\ \beta_{m+1,1} = \dots = 2\ \beta_{M1} = 2\ \beta_{m+1,2} = \dots = 2\ \beta_{M2}\\ = 2\ \gamma_{21} = 2\ \gamma_{22} = 0\}.$$

This can be transformed to

$$H_1(\bar{M}_\phi(G)) \cong < B, D; R_6, R_7, R_8, R_{13} >,$$
where $D = \{\alpha\}$ and $R_{13} = \{2\ \alpha = 0\}$.

Thus the proof of the main theorem is complete.

References

1. R. H. Fox, *A quick trip throuth knot theory*, Topology of 3-manifolds and Related Topics, (edited by M. K. Fort, Jr.), Prentice-Hall, Englewood Cliffs, New Jersey, 1962, pp. 120–167.
2. M. Hall, Jr., *The Theory of Groups*, Macmillan Co., New York, 1959.
3. M. Nakao, *On the* $Z_2 \oplus Z_2$ *branched coverings of spatial* θ*-curves*, preprint.
4. J. Simon, *A topological approach to the stereochemistry of nonrigid molecules*, Graph Theory and Topology in Chemistry, (edited by R. B. King and D. Rouvray), Elsevier Science Publishers, New York, 1987, pp. 43–75.

Geometric link homotopy invariants

Ulrich Koschorke

In the last six years there has been tremendous progress in link homotopy theory. With one single exception (namely the work of N. Habegger and X.S. Lin, see example 2.2) all the big breakthroughs seem to have arisen directly from the discovery of new link homotopy invariants. In this paper we discuss some of the new results on spherical linking phenomena and the highly geometric invariants which lie at their heart.

§1. The central question of link homotopy theory

Throughout this paper we fix r nonnegative integers $p, q, \ldots, u$ as well as a further natural number m; to keep the exposition simple we assume that $p, q, \ldots, u \leq m-2$, although much of the time this condition is not really essential.

Definition 1.1. *A map*

$$f = f_1 \amalg \ldots \amalg f_r : S^p \amalg S^q \amalg \ldots \amalg S^u \longrightarrow \mathbf{R}^m$$

is called a link map if the images of the r component spheres S^p, S^q,... are pairwise disjoint in euclidean space $\mathbf{R}^m$.

Two such link maps are called link homotopic if there is a deformation from one to the other through link maps.

Let $LM^m_{p,q,\ldots,u}$ denote the resulting set – and often (semi) group – of all link homotopy classes of link maps f.

Problem. *What is the size and structure of $LM^m_{p,q,\ldots,u}$?*

Knots 90

§2. Global invariants: κ, μ and α

Definition 1.1 above allows the following equivalent formulation: f is a link map if and only if the image of the product map

$$\hat{f} = f_1 \times f_2 \times ... \times f_r : S^p \times S^q \times ... \times S^u \longrightarrow (\mathbf{R}^m)^r$$

lies in the configuration space

$$\tilde{C}_r(\mathbf{R}^m) := \{(y_1, ..., y_r) \in (\mathbf{R}^m)^r | y_i \neq y_j \text{ for } 1 \leq i \leq j \leq r\}.$$

Thus we obtain the link homotopy invariant

$$\kappa[f] := [\tilde{f}] \in [S^p \times S^q \times ... \times S^u, \tilde{C}_r(\mathbf{R}^m)] \qquad (2.1)$$

which measures to some degree how every full component map $f_1 = f|S^p, f_2 = f|S^q, ...$ links with every other full component.

In general the (standard) homotopy set to the right in (2.1) seems to be quite unmanageable, and $\kappa[f]$ contains many redundancies. Fortunately, however, it can be broken up into simpler pieces: we can use the natural cell decomposition of $S^p \times S^q \times ... \times S^u$ on the one hand, and (homotopy) fibrations like

$$\bigvee^{r-1} S^{m-1} \subset \tilde{C}_r(\mathbf{R}^m) \underset{\text{section}}{\overset{\text{(forget } y_r)}{\rightleftarrows}} \tilde{C}_{r-1}(\mathbf{R}^m)$$

on the other hand to extract successively defined families of π_*^S-valued μ-invariants for $[f]$ (for more details see [**Ko1**]).

Example 2.2 (Classical link homotopy). In the special case when $p = q = \cdots = u = 1$ and $m = 3$ our μ-invariants all lie in the 0-stem $\pi_0^S = \mathbf{Z}$, and they coincide with the ones which J. Milnor defined in his fundamental paper [**M**]. Hence his basic triviality criterion can be rephrased as follows.

Milnor's Theorem (1954). *For an arbitrary number r of components the map of pointed sets*

$$\kappa : LM^3_{1,...,1} \longrightarrow [(S^1)^r, \tilde{C}_r(\mathbf{R}^3)]$$

has a trivial kernel.

In the absence of group structures this was but the first (though very important) step towards the homotopy classification of classical links. A full description of $LM^3_{1,...,1}$ (in terms of a certain group action) was only given more than three decades later by N. Habegger and X. S. Lin who write, however, that the complicated nature of the relevant group action "conspires to make it difficult, if not impossible, to find a complete set of invariants" (see [**HL**], at the beginning of §3). In view of Milnor's theorem, why not consider κ ?

Open question. *Is* κ *injective on all of* $LM^3_{1,\dots,1}$*?* □

Example 2.3 (Higher dimensions). In the case of *two* components the κ-invariant coincides with the first (and only) μ-invariant via the bijection

$$[S^p \times S^q, \tilde{C}_2(\mathbf{R}^m)] \approx [S^p \times S^q, S^{m-1}] \xrightarrow[\approx]{degree} \pi^S_{p+q-m+1}$$

and is now commonly known as the α-invariant. (Here we use the homotopy equivalence

$$\begin{array}{ccccc} \tilde{C}_2(\mathbf{R}^m) & \longrightarrow \mathbf{R}^m \times (\mathbf{R}^m - \{0\}) & \xrightarrow{\sim} & S^{m-1} & ; \\ (y_1, y_2) & \longrightarrow (y_1, y_1 - y_2) & \longrightarrow & \dfrac{y_1 - y_2}{\|y_1 - y_2\|} & \end{array}$$

moreover, given a smooth map $g : S^p \times S^q \longrightarrow S^{m-1}$ with regular value z, we can define the degree of $[g]$ by the framed bordism class $[g^{-1}\{z\}] \in \Omega^{\text{fr}}_* \cong \pi^S_*$. E.g. if $p + q = m - 1$, we obtain the well-known integer linking number whose origins date back at least as far as C.F. Gauss. In more general higher dimensions the α-invariant was introduced by P. Scott and denoted by S (see [**S**], §2; compare also [**MR**] and [**Ko2**]); often it plays a significant auxiliary role for the calculation of $LM^m_{p,q}$ (supplementing the $\tilde{\beta}$-invariant of §3), see [**Ko3**] and e.g. [**HK**].

The real importance of the new κ- and μ- invariants, however, stems from the following fact: in a large metastable dimension range a link map f with an *arbitrary* number of components is entirely determined, up to link homotopy , by all the sub-link maps of f with two components and by the (first truly higher order) μ-invariant of each 3-component sub-link map $f_i \amalg f_j \amalg f_k$ (see [**Ko5**] for details). □

§3. Semiglobal invariants: σ, $\overset{(\sim)}{\beta}$, γ and $\overset{(\sim)}{\delta}$

In view of the last remark we can center our attention on a link map

$$f = f_1 \amalg f_2 : S^p \amalg S^q \longrightarrow \mathbf{R}^m$$

with two components. The pairing

$$\varphi(y)(x) = \frac{f_1(x) - y}{\|f_1(x) - y\|} \qquad \text{for } y \in \mathbf{R}^m - f_1(S^p),\ x \in S^p$$

gives rise to a canonical map

$$\varphi : \mathbf{R}^m - f_1(S^p) \longrightarrow \mathrm{Map}(S^p, S^{m-1}) \tag{3.1}$$

where the mapping space to the right is endowed with the compact-open topology. Thus we obtain a link homotopy invariant whenever we have a reasonably compatible procedure to construct maps which go into a neighborhood of $f_2(S^q)$ (and hence into $\mathbf{R}^m - f_1(S^p)$).

Procedure I: Selfintersections. Choose a framed zerobordism B of S^q and a framed selftransverse immersion

$$\tilde{f} : B \looparrowright \mathbf{R}^m$$

which, when restricted to $\partial B = S^q$, approximates f_2. Given any natural number r, consider the boundary N of the r-tuple point manifold of $\tilde{f}$ (in $\mathbf{R}^m$). After smoothing corners, N is a zerobordant $(m-q-1)\lambda_r$-manifold of dimension

$$t = m - r(m-q-1) - 1, \tag{3.2}$$

where λ_r denotes the canonical $(r-1)$-plane bundle over the classifying space $B\Sigma_r$ of the permutation group Σ_r. In other words, N is naturally equipped with

(i) a map

$$g : N \longrightarrow B\Sigma_r$$

(extracted from the action of Σ_r which permutes the branches of f at the r-tuple points), and

(ii) a stable trivialization

$$\bar{g} : TN \oplus g^*((m-q-1)\lambda_r) \cong N \times \mathbf{R}^q$$

(derived from the framing of the immersion f, i.e. from the trivialization of its normal bundle).

Moreover, we can compose the natural map j from N into a neighborhood of $f_2(S^q)$ with the map φ defined in (3.1). The resulting normal bordism class

$$\gamma_r(f) := [N, (g, \varphi \circ j), \bar{g}] \in \Omega_t(B\Sigma_r \times (\mathrm{Map}(S^p, S^{m-1}), *); (m-q-1)\lambda_r) \tag{3.3}$$

(cf. [**Ko3**]) depends only on the link homotopy class of f and measures to some extent the r-tuple point geometry of f_2 and how it links with all of f_1.

The normal bordism group on the right hand side in (3.3) has also a natural geometric interpretation: by the Thom construction we can identify it with the *joint bordism group* of $(m-p-1)$-codimensional framed smooth immersions

$$i : M \looparrowright N \quad (\text{where} N \text{is zerobordant})$$

between closed $(m-q-1)\lambda_r$-manifolds. Again we can analyse selfintersections of every order $s \geq 1$ and we obtain the "multiple point Hopf invariant homomorphism "

$$\begin{array}{c} \Omega_t(B\Sigma_r \times (\mathrm{Map}(S^p, S^{m-1}), *); (m-q-1)\lambda_r) \\ \Big\downarrow{\scriptstyle h=\oplus h_s} \\ \bigoplus_{s=1}^{\infty} \Omega_u(B\Sigma_r \times B\Sigma_s; (m-q-1)\lambda_r + (m-p-1)\lambda_s) \end{array} \tag{3.4}$$

where $u = m - r(m-q-1) - s(m-p-1) - 1$.

The following results were proved in [**Ko3**].

Isomorphism theorem. *h is bijective.*

Symmetry theorem. *For all $r, s \geq 1$*

$$h_r\gamma_s(f_2 \amalg f_1) \approx \pm h_s\gamma_r(f)$$

In particular, the complete invariants

$$\gamma(f) = \bigoplus_{r=1}^{\infty} \gamma_r(f), \qquad \bigoplus_{r,s\geq 1} h_s\gamma_r(f) \qquad \text{and} \qquad \gamma(f_2 \amalg f_1)$$

are all equally strong.

In the construction of $\gamma_r(f)$ above we may assume from the very beginning that f_2 itself is a framed selftransverse immersion whose r-tuple points form an $(r-1)$–codimensional submanifold of N. In the special case $r = 2$ we may even assume (after a suitable bordism) that N is the total space of a canonical circle bundle over the double point manifold D of f_2. Reading off the corresponding refined data involving D yields the link homotopy invariant $\tilde{\beta}(f)$ (cf. [**Ko3**], 3.9) which is much sharper than $\gamma_2(f)$ and , in particular, than

$$\beta(f) \quad := \quad h_1\gamma_2(f) \quad \in \quad \Omega_{p+2q-2m+2}(P^\infty; (m-q-1)\lambda_2). \tag{3.5}$$

In fact, in a large metastable dimension range $\tilde{\beta}[f]$ turns out to be the only obstruction to finding a representative for $[f]$ which embeds S^q; moreover the $\tilde{\beta}$-invariant fits into an exact sequence of abelian groups which reduces the calculation of $LM^m_{p,q}$ to standard homotopy questions (cf. [**Ko4**]).

Example 3.6. In the special setting $(m, p, q) = (4, 2, 2)$ the invariants

$$\beta[f] \in \Omega_0(P^\infty; \lambda) = \mathbf{Z}_2$$

$$\tilde{\beta}[f] = \sigma_-[f] \quad \in \quad \bigoplus_1^\infty \mathbf{Z}$$

were originally introduced by R. Fenn and D. Rolfsen [**FR**], and by P. Kirk [**Ki**], resp., and shown to be surjective on $LM^4_{2,2}$. □

Procedure II: Resolutions. Here we exploit the dual role of the circle S^1 as an abelian Lie group and as a Thom space in order to replace certain low-dimensional immersions by nearby framed embeddings (of modified source manifolds).

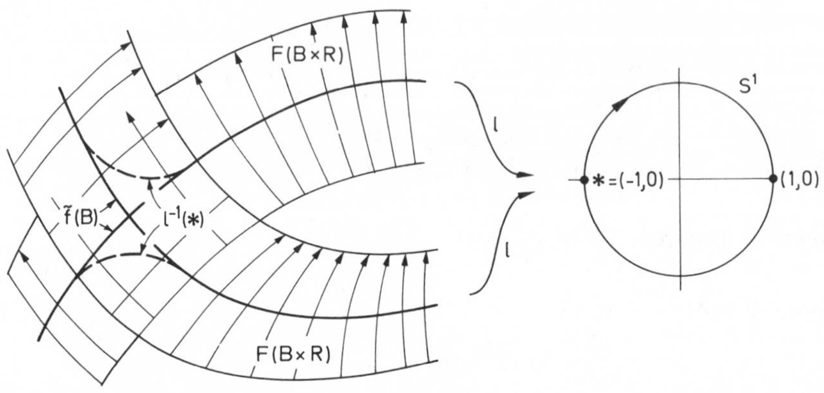

Figure 3.7. Resolving selfintersections of $\tilde{f}$.

Indeed, each selftransverse immersion

$$\tilde{f} : B \looparrowright \mathbf{R}^m$$

of an oriented compact $(m-1)$–manifold B can be extended to an orientation preserving local diffeomorphism

$$F : B \times \mathbf{R} \looparrowright \mathbf{R}^m$$

which maps at most r different elements $(b_1, t_1), ..., (b_r, t_r) \in B \times \mathbf{R}$ to any given point $y \in \mathbf{R}^m$. Thus we can define a map

$$l : F(B \times \mathbf{R}) \longrightarrow S^1$$

by $l(y) := e(t_1) \cdot e(t_2) \cdots e(t_r)$, where the auxiliary smooth map $e : \mathbf{R} \to S^1$ sends 0 to the basepoint $* = (-1, 0)$ of S^1, restricts to an orientation preserving diffeomorphism from a neighborhood of 0 onto $S^1 - (1, 0)$ and maps the whole complement of this neighborhood in $\mathbf{R}$ to the unit element (1,0) of S^1.

Then $l^{-1}(\{*\})$ is a smooth codimension one submanifold of $F(B \times \mathbf{R}) \subset \mathbf{R}^m$ which agrees with $\tilde{f}(B)$ except near the self intersections of $\tilde{f}$.

If B has a boundary ∂B then $l^{-1}(\{*\})$ is a "generalized Seifert surface" of the immersion $\tilde{f}|\partial B$ (in the sense of **[KK]**) and can be "cut off" near the corresponding image $\tilde{f}(\partial B)$ to yield a closed framed codimension 2 submanifold R of $\mathbf{R}^m$, a "resolution of $\tilde{f}|\partial B$".

In particular, this construction produces invariants of $LM^m_{p,q}$ when $p = q = m-2$. Given a link map

$$f = f_1 \amalg f_2 : S^{m-2} \amalg S^{m-2} \longrightarrow \mathbf{R}^m$$

up to link homotopy, we may assume that the component maps extend to selftransverse immersions of framed bordisms of S^{m-2} as above; thus we obtain disjoint (framed) resolutions R_1 and R_2. The α–invariant of $R_1 \amalg R_2$ can be sharpened in a natural way to yield the link homotopy invariant

$$\delta(f) \in \pi_m(S^3) \tag{3.8}$$

(cf. [**KK**]) which does *not* usually suspend to $\alpha(f) \in \pi^S_{m-3}$; e.g. if $m = 9$, then the suspension homomorphism

$$\pi_9(S^3) = \mathbf{Z}_3 \xrightarrow{E^\infty} \pi^S_6 = \mathbf{Z}_2$$

must necessarily vanish (cf. [**T**]), but it is known by an easy precomposition argument that $\alpha(LM^9_{7,7})$ is nontrivial. So what is the precise relationship between α and δ?

The key to answering this question lies in the full invariant

$$\tilde{\delta}(f) \quad = \quad [\varphi|R_2] \quad \in \quad \tilde{\Omega}^{\text{fr}}_{m-2}(\text{Map}(S^{m-2}, S^{m-1})) \tag{3.9}$$

given by procedure II and by (3.1). Clearly there is a canonical homomorphism

$$v_* : \tilde{\Omega}^{\text{fr}}_{m-2}(\text{Map}(S^{m-2}, S^{m-1}) \longrightarrow \tilde{\Omega}^{\text{fr}}_{m-2}(S^1) \cong \pi^S_{m-3}$$

in framed bordism which maps $\tilde{\delta}(f)$ to $E^\infty(\delta(f))$. In addition, in a future paper we will exhibit natural "resolution homomorphisms" $v_{r,s}$, based on the action of the permutation group on the coordinates of suitable tori, such that

$$E^\infty(\delta(f)) = \pm\alpha(f) \quad + \sum_{\substack{2 \leq r,s \\ 4 \leq r+s \leq m-1}} v_{r,s} h_s \gamma_r(f) \tag{3.10}$$

(cf. 3.3 and 3.4). □

The whole recent development of higher dimensional link homotopy was very much inspired by the case $m = 4, p = q = 2$ (compare example 3.6). Here the invariants $\alpha, \beta, \gamma, \delta, E^\infty(\delta)$ and $\tilde{\delta}$ are all equally strong, but weaker than $\tilde{\beta} = \sigma_-$, and the sequence of [**Ko4**] is no longer exact. Hence the following question (cf.[**Ki**], p.42) remains of paramount importance.

Open problem *Is Kirk's full invariant* $\sigma = (\sigma_+, \sigma_-)$ *injective on* $LM^4_{2,2}$*?*

References

[FR] R. Fenn and D. Rolfsen, *Spheres may link homotopically in 4-space*, J. London Math. Soc. (2) **34** (1986), 177–184.

[HL] N. Habegger and X. S. Lin, *The classification of links up to link homotopy*, Jour. of AMS **3** (1990), 389–420.

[HK] N. Habegger and U. Kaiser, *Homotopy classes of two disjoint 4-spheres in R^7*, preprint.

[Ki] P. Kirk, *Link maps in the 4-sphere*, Proc. Siegen Topology Symp., LNM 1350, Springer-Verlag (1988), 31–43.

[KK] P. Kirk and U. Koschorke, *Generalized Seifert surfaces and linking numbers*, preprint.

[Ko1] U. Koschorke, *Higher order homotopy invariants for higher dimensional link maps*, LNM 1172, Springer-Verlag (1985), 116-129.

[Ko2] ——, *Link maps and the geometry of their invariants*, Manuscr. Math. **61** (1988), 383–415.

[Ko3] ——, *Multiple point invariants of link maps*, Proc. Siegen Topology Symp., LNM 1350, Springer-Verlag (1988), 44–86.

[Ko4] ——, *On link maps and their homotopy classification*, Math. Ann. **286** (1990), 753–782.

[Ko5] ——, *Link homotopy with many components*, Topology **30**,2 (1991), 267–281.

[M] J. Milnor, *Link groups*, Ann. of Math. **59** (1954), 177–195.

[MR] W. Massey and D. Rolfsen, *Homotopy classification of higher dimensional links*, Indiana Univ. Math. Jour. **34** (1986), 375–391.

[S] P. Scott, *Homotopy links*, Abh. Math. Sem. Hamburg **32** (1968), 186–190.

[T] H. Toda, *Composition methods in homotopy groups of spheres*, Annals of Math. Stud. 49, Princeton University Press, 1962.

Projective planes in 4-sphere obtained by deform-spinnings

Seiichi Kamada

1. Introduction

For an invertible $(3, 1)$-ball pair with an inverting deformation, a projective plane embedded in S^4, called a deform-spun projective plane, is defined **[P-R]** **[K]**. The construction is analogous to Zeeman's twist-spinning or Litherland's deform-spinning to construct 2-knots. In case the invertible ball pair is non-trivial, there are infinitely many distinct inverting deformations, but many of them often yield the same deform-spun projective plane. In fact, for a torus knot, it is shown in **[K]** that there are at most two deform-spun projective planes obtained from it. In this paper we extend this result to every knot with finite symmetry group. For a 1-knot (S^3, K), we denote by $\mathrm{Homeo}^+(S^3, K)$ the group of all self-homeomorphisms of (S^3, K) which preserve the orientation of S^3. Let (S^3, K) be an invertible 1-knot such that $\mathrm{Sym}^+(S^3, K) = \pi_0\mathrm{Homeo}^+(S^3, K)$ is finite. Then it is a torus knot, a hyperbolic knot or a cable knot with a torus knot companion and $\mathrm{Sym}^+(S^3, K)$ is isomorphic to the dihedral group D_n of order $2n$ for some positive integer n. In particular it is strongly invertible. It is well known that, according as n is odd or even, (S^3, K) admits one or two distinct, up to equivalence, inverting involutions, cf. **[Sa$_1$]**. Hence there are at most two (when n is odd) or at most four (when n is even) distinct quotient knots obtained by inverting involutions, where a quotient knot by an inverting involution h means a constituent knot of the θ-curve in S^3 associated with h **[Sa$_1$]** which is not the image of Fix h.

Theorem. *Let (S^3, K) be an invertible 1-knot with finite $\mathrm{Sym}^+(S^3, K)$. Then any deform-spun projective plane obtained from (S^3, K) is a connected sum of a standard projective plane in S^4 with normal Euler number 2 and the 2-twist spin of a quotient knot of (S^3, K) by an inverting involution. In particular, if n is odd (or even) there are at most two (or at most four), up to equivalence, deform-spun projective planes obtained from (S^3, K).*

Knots 90

For each n $(n = 1, 2, 3, 4)$, there is an invertible 1-knot (S^3, K) with finite $\mathrm{Sym}^+(S^3, K)$ such that just n deform-spun projective planes are obtained from it (§5). Two surfaces in S^4 are said to be equivalent if there exists an orientation-preserving PL self-homeomorphism of S^4 carrying one to the other.

2. Definitions

Let B^3 be the oriented unit 3-ball in the euclidean 3-space R^3 and let β be a properly embedded oriented arc whose boundary points are (0, 1,0) and (0, -1,0) and whose orientation is from (0,-1,0) to (0,1,0). We say that $(B^3,\ \beta)$ is *invertible* if there is a self-homeomorphism h of B^3, called an *inverting homeomorphism of* (B^3, β), such that (1) $h(\beta) = -\beta$ and (2) its restriction to ∂B^3 is the π rotation about the z axis. We denote by $D_+(B^3, \beta)$ (resp. $D_-(B^3, \beta)$) the group of all homeomorphisms of (B^3, β) whose restrictions to $\partial B^3$are the identity (resp. the set of all inverting homeomorphisms of (B^3, β)). Further we denote by $\mathfrak{D}_+(B^3, \beta)$ the *deformation group* $\pi_0 D_+(B^3, \beta)$ (cf. **[L]**) and by $\mathfrak{D}_-(B^3, \beta)$ the *inverting deformation set* $\pi_0 D_-(B^3, \beta)$. Each element of them is called a *deformation* or an *inverting deformation* respectively.

Let g be an inverting deformation and let f be an inverting homeomorphism representing g. Consider the mapping torus for f

$$(W,\ M) = (B^3,\ \beta) \times [0, 1]/(x, 0) \sim (f(x), 1), x \in B^3.$$

Note that W is homeomorphic to $B^3 \times S^1$ and M is a Möbius band. We denote by η the loop (0,0,1)×[0,1]/∼ in W. Let P be a standard projective plane in S^4 of normal Euler number 2 and let ξ be an essential simple loop on P with a regular neighborhood $N(\xi)$ in S^4. Replace $(N(\xi), N(\xi) \cap P)$ in the manifold pair (S^4, P) by (W, M) such that the loop η is null-homotopic in $\overline{S^4 - N(\xi) - P}$. Then we obtain a new projective plane in S^4 and call it the *deform-spun projective plane (in* S^4*) obtained from* $\{(B^3, \beta), g\}$, **[K]**. In case we need not specify the deformation we also call it a *deform-spun projective plane obtained from* (B^3, β) or (S^3, K), where (S^3, K) is a 1-knot associated with (B^3, β).

Let $\{(B^3, \beta_1), g_1\}$ and $\{(B^3, \beta_2), g_2\}$ be invertible ball pairs with inverting deformations. They are said to be *equivalent* if there exist inverting homeomorphisms f_i $(i = 1, 2)$ of (B^3, β_i) representing g_i and a homeomorphism s from (B^3, β_1) to (B^3, β_2) whose restriction to ∂B^3 is the identity such that $s \circ f_1 = f_2 \circ s$. Then deform-spun projective planes obtained from them are equivalent.

3. Reduction of some deformations

Let (S^3, K) be a 1-knot and $N(K)$ be a tubular neighborhood of K in S^3 with the exterior $X = S^3 - \overset{\circ}{N}(K)$. We identify ∂X with the quotient $R^2/ < \vec{\ell}, \vec{m} >$, where vectors $\vec{\ell}$ and $\vec{m} \in R^2$ correspond to a preferred longitude and a meridian on ∂X. $[\vec{x}]$ means the point of $R^2/ < \vec{\ell}, \vec{m} >$ determined by $\vec{x} \in R^2$. A collar of ∂X in X is identified with $\partial X \times [0,1]$ such that $\partial X = \partial X \times \{0\}$, and $N(K)$ is identified with $\partial X \times [-1,0]/\sim$, where $([\vec{x}], \lambda) \sim ([\vec{y}], \mu)$ if $\lambda = \mu = -1$ and $[\vec{x}] - [\vec{y}] = s[\vec{m}]$ $(s \in R)$. For a vector $\vec{\alpha} \in R^2$, we define a self-homeomorphism of (S^3, K) by

$$\begin{cases} H(\vec{\alpha})(x) = x, & \text{for } x \in X - \partial X \times [0,1] \\ H(\vec{\alpha})([\vec{x}], \lambda) = ([\vec{x} + (1-\lambda)\vec{\alpha}], \lambda), & \text{for } ([\vec{x}], \lambda) \in \partial X \times [0,1] \\ H(\vec{\alpha})([\vec{x}], \lambda) = ([\vec{x} + \vec{\alpha}], \lambda), & \text{for } ([\vec{x}], \lambda) \in \partial X \times [-1,0]/\sim. \end{cases}$$

(3.1) Twisting and rolling deformations. Let (B^3, β) be a (3,1)-ball pair associated with (S^3, K), i.e. it is homeomorphic to a complementary ball pair $(\widetilde{D^3}, \widetilde{D^1})$ of a standard ball pair (D^3, D^1) in (S^3, K). We assume that $D^3 \subset N(K)$. Let ϕ be an identification map from $(\widetilde{D^3}, \widetilde{D^1})$ to (B^3, β). Then $r_\tau = \phi \circ H(-\vec{m}) \circ \phi^{-1}$ and $r_\rho = \phi \circ H(-\vec{\ell}) \circ \phi^{-1}$ are elements of $D_+(B^3, \beta)$ and determine deformations of (B^3, β). We call them the *twisting* and the *rolling deformations* and denote them by τ and ρ respectively, **[L]**.

(3.2) Symmetry deformations. Let Homeo$^*(S^3, K)$ be the group of all self-homeomorphisms of (S^3, K) which preserve both of orientations of S^3 and K. In this paper a *symmetry* means a finite subgroup of Homeo$^*(S^3, K)$. If a symmetry G is given, we can take a G-invariant tubular neighborhood $N(K) = \partial X \times [-1,0]/\sim$ and a collar $\partial X \times [0,1]$ parametrized such that each $g \in G$ acts on them by $g([\vec{x}], \lambda) = ([\vec{x} + \vec{g}], \lambda)$, where $\vec{g}$ is a vector on R^2. The deformation of (B^3, β) represented by $r_g = \phi \circ g \circ H(-\vec{g}) \circ \phi^{-1}$ is called a *g-symmetry deformation* and denoted by σ_g. It depends on the choice of $\vec{g}$. Let n be the order of G and q be a unique integer for G such that the slope of G is $< \ell + qm >$ (for the definition, see **[Sa$_2$]**) and $0 \le q < n$. The *canonical generator f of G* is a unique element of G satisfying that $f([\vec{x}], \lambda) = ([\vec{x} + \vec{f}], \lambda)$, where $\vec{f} = (\vec{\ell} + q\vec{m})/n \in R^2$. We can choose $\vec{g}$ such that $\vec{g} = p\vec{f}$ for some unique integer p with $0 \le p < n$. We especially denote by $\sigma_{g \in G}$ the g-symmetry deformation determined by $\vec{g} = p\vec{f}$.

Remark. (1) For the canonical generator f of G, we have that $(\sigma_{f \in G})^n = \rho \circ \tau^q$.

(2) If for every element g of G, the fixed-point set of g is an unknotted circle disjoint from K, we say that G gives a *cyclic periodicity of* (S^3, K). Then its slope is $< \ell >$ **[Sa$_2$]**, so $\sigma_{g \in G}$ is an untwisted deformation in the sense of **[L]**. (In **[L]**, symmetries which give cyclic periodicities are treated.)

Lemma 1. *Let g and h be a deformation and an inverting deformation of (B^3, β) respectively. If they have representatives r_g and r_h such that $r_h \circ r_g \circ r_h^{-1} = r_g^{-1}$, then for any deformation d, we have that $\{(B^3, \beta), h \circ d\}$ and $\{(B^3, \beta), h \circ g \circ d \circ g\}$ are equivalent.*

Proof. Let $r_d \in D_+(B^3, \beta)$ be a representative of d and take $r_h \circ r_d$ and $r_h \circ r_g \circ r_d \circ r_g \in D_-(B^3, \beta)$ as representatives of $h \circ d$ and $h \circ g \circ d \circ g$ respectively. By a map $s = r_g^{-1} : (B^3, \beta) \to (B^3, \beta)$, we have $s \circ (r_h \circ r_d) = (r_h \circ r_g \circ r_d \circ r_g) \circ s$. This completes the proof.

Corollary 2 (cf. [**K**,4]). *Let h be any inverting deformation of (B^3, β). For any deformation d, we have that $\{(B^3, \beta), h \circ d\}$, $\{(B^3, \beta), h \circ \tau^2 \circ d\}$ and $\{(B^3, \beta), h \circ \rho^2 \circ d\}$ are equivalent.*

Proof. We use the same notation as in (3.1). Let r_h be an inverting homeomorphism representing h and let u_h be a homeomorphism of (S^3, K) such that $r_h = \phi \circ u_h \circ \phi^{-1}$. We may assume that u_h acts on $N(K) \cup \partial X \times [0,1]$ by $u_h([\vec{x}], \lambda) = ([-\vec{x}], \lambda)$. Then $r_h \circ r_\tau \circ r_h^{-1} = \phi \circ u_h \circ H(-\vec{m}) \circ u_h^{-1} \circ \phi^{-1} = \phi \circ H(\vec{m}) \circ \phi^{-1} = r_\tau^{-1}$ and $r_h \circ r_\rho \circ r_h^{-1} = r_\rho^{-1}$. Since τ and ρ are elements of the center of the deformation group [**L**], we obtain, by Lemma 1, the result.

4. Dual ball pairs of strongly invertible ball pairs

A (3,1)-ball pair is said to be *strongly invertible* if the π rotation of B^3 about the z axis is an inverting homeomorphism of it. Throughout this section (B^3, β) is strongly invertible and h is the π rotation of B^3 about the z axis (we denote by the same symbol the inverting deformation represented by h.) Let (S^3, K) be a 1-knot associated with (B^3, β) which has a decomposition $(D^3, D^1) \cup (\widetilde{D^3}, \widetilde{D^1})$ with $\phi : (\widetilde{D^3}, \widetilde{D^1}) \to (B^3, \beta)$ as in § 3 such that $D^3 \subset N(K)$. Let h' be an involution of (S^3, K) associated with h, i.e. it satisfies $h = \phi \circ h' \circ \phi^{-1}$. We may assume that it acts on $N(K) \cup \partial X \times [0,1]$ by $h'([\vec{x}], \lambda) = ([-\vec{x}], \lambda)$.

Strongly invertible ball pairs (B^3, β) and (B^3, γ) are said to be *equivalent* if $\{(B^3, \beta), h\}$ and $\{(B^3, \gamma), h\}$ are equivalent.

(4.1) τ-dual, ρ-dual and $\tau\rho$-dual ball pairs. The intersection $\partial\widetilde{D^3} \cap \operatorname{Fix} h'$ is a pair of points. Let A be one of them such that $A = \phi^{-1}((0,0,1))$, where $(0,0,1)$ is the "North Pole" of B^3 and let B be the other one. There is a strongly invertible ball pair (B^3, β_τ), uniquely determined up to equivalence, such that there is an orientation-preserving homeomorphism $\phi_B : (\widetilde{D^3}, \widetilde{D^1}) \to (B^3, \beta_\tau)$ satisfying that $\phi_B(B) = (0,0,1)$ and $\phi_B \circ h' \circ \phi_B^{-1} = h$. We call it the *$\tau$-dual ball pair of*

(B^3,β). Put $(\widetilde{D_1^3},\widetilde{D_1^1}) = H(\vec{\ell}/2)(\widetilde{D^3},\widetilde{D^1})$. It is also h'-invariant and $\partial\widetilde{D_1^3}\cap$ Fix h' is a pair of points $C = H(\vec{\ell}/2)(A)$ and $D = H(\vec{\ell}/2)(B)$. Let (B^3,β_ρ) (resp. $(B^3,\beta_{\tau\rho})$) be a unique, up to equivalence, strongly invertible ball pair such that there is an orientation-preserving homeomorphism ϕ_C (resp. ϕ_D) from $(\widetilde{D_1^3},\widetilde{D_1^1})$ to it satisfying that $\phi_Y(Y) = (0,0,1)$ and $\phi_Y \circ h' \circ \phi_Y^{-1} = h$ $(Y = C, D)$. We call it the *ρ-dual* (resp. *τρ-dual*) *ball pair of* (B^3,β), **[K]**.

(4.2) σ_g-dual ball pair. Suppose that (S^3,K) has a symmetry $G \neq 1$ such that for each $g \in G$, $h' \circ g \circ h' = g^{-1}$. Let σ_g be a g-symmetry deformation and $\vec{g}$ be the vector for σ_g. Put $(\widetilde{D_2^3},\widetilde{D_2^1}) = H(\vec{g}/2)(\widetilde{D^3},\widetilde{D^1})$, which is kept under the involution $g \circ h'$ of (S^3,K). The intersection $\partial\widetilde{D_2^3}\cap\mathrm{Fix}(g\circ h')$ is a pair of points $E = H(\vec{g}/2)(A)$ and $F = H(\vec{g}/2)(B)$. There is a unique strongly invertible ball pair (B^3,β_g) (resp. $(B^3,\beta_{\tau g})$) such that there is an orientation-preserving homeomorphism ϕ_E (resp. ϕ_F) from $(\widetilde{D_2^3},\widetilde{D_2^1})$ to it satisfying that $\phi_Y(Y) = (0,0,1)$ and $\phi_Y \circ (g\circ h')\circ \phi_Y^{-1} = h$ $(Y = E, F)$. We call it the *σ_g-dual* (resp. $\tau\circ\sigma_g$-dual) *ball pair of* (B^3,β).

Lemma 3 ([**K**, 3]).

(1) $\{(B^3,\beta), h\circ\tau\}$ *and* $\{(B^3,\beta_\tau), h\}$ *are equivalent.*
(2) $\{(B^3,\beta), h\circ\rho\}$ *and* $\{(B^3,\beta_\rho), h\}$ *are equivalent.*
(3) $\{(B^3,\beta), h\circ\tau\circ\rho\}$ *and* $\{(B^3,\beta_{\tau\rho}), h\}$ *are equivalent.*

Lemma 4.

(1) $\{(B^3,\beta), h\circ\sigma_g\}$ *and* $\{(B^3,\beta_g), h\}$ *are equivalent.*
(2) $\{(B^3,\beta), h\circ\tau\circ\sigma_g\}$ *and* $\{(B^3,\beta_{\tau,g}), h\}$ *are equivalent.*

Proof. Put $s = \phi_E \circ g \circ H(-\vec{g}/2)\circ\phi^{-1} : (B^3,\beta) \to (B^3,\beta_g)$, and take $h\circ r_g = \phi\circ h'\circ g\circ H(-\vec{g})\circ\phi^{-1} \in D_-(B^3,\beta)$ and $h = \phi_E\circ(g\circ h')\circ\phi_E^{-1} \in D_-(B^3,\beta_g)$ as representatives of $h\circ\sigma_g$ and h respectively. Then $s\circ(h\circ r_g) = h\circ s$ and therefore (1) holds. We see (2) by use of a map $s = \phi_F\circ g\circ H(-(\vec{m}+\vec{g})/2)\circ\phi^{-1}$: $(B^3,\beta) \to (B^3,\beta_{\tau,g})$. This completes the proof of Lemma 4.

5. Proof of Theorem

Let $\mathrm{Homeo}^*(S^3,K)$ be as in (3.2) and $\mathrm{Sym}^*(S^3,K)$ stand for $\pi_0\mathrm{Homeo}^*(S^3,K)$.

Lemma 5. *For a 1-knot* (S^3,K), *there exists an exact sequence,*

$$1 \to \mathfrak{P}(S^3,K) \to \mathfrak{D}_+(S^3,K) \xrightarrow{j} \mathrm{Sym}^*(S^3,K) \to 1,$$

*where $\mathfrak{D}_+(S^3,K)$ is the deformation group of (S^3,K) and $\mathfrak{P}(S^3,K)$ is the subgroup generated by τ and ρ (see [**L**], for definitions).*

Remark. $\mathfrak{D}_+(S^3,K)$ is isomorphic to $\mathfrak{D}_+(B^3,\beta)$, where (B^3,β) is an associated ball pair.

Proof. If K is trivial, then $\mathfrak{D}_+(S^3,K) = \mathrm{Sym}^*(S^3,K) = 1$. Hence we suppose K is nontrivial. Let $\mathrm{Homeo}^*(X)$ and $\mathrm{Homeo}^*(X, rel\partial)$ be the group of all orientation-preserving self-homeomorphisms of the exterior $X = S^3 - \overset{\circ}{N}(K)$ whose restrictions to ∂X are isotopic to the identity and its subgroup consisting of all elements keeping ∂X fixed, respectively. We regard $\pi_0\mathrm{Homeo}^*(X, rel\partial)$ as $\mathfrak{D}_+(S^3,K)$. There exist natural epimorphisms $j' : \mathfrak{D}_+(S^3,K) \to \pi_0\mathrm{Homeo}^*(X)$ and $k : \pi_0\mathrm{Homeo}^*(X) \to \mathrm{Sym}^*(S^3,K)$. We show that the kernel of $j = k \circ j'$ is $\mathfrak{P}(S^3,K)$. Fix a point x on ∂X and consider the group $G = \pi_1(X,x)$ and the peripheral subgroup $i_*(\pi_1(\partial X,x))$ of G. We denote by $\mathrm{Aut}_\partial(G)$ and by $\mathrm{Inn}_\partial(G)$ the group of automorphisms of G whose restrictions to the peripheral subgroup is the identity and its subgroup consisting of inner-automorphisms of G by peripheral elements,respectively. We note that $\mathrm{Inn}_\partial(G) = \mathrm{Aut}_\partial(G) \cap \mathrm{Inn}(G)$, cf.[**Si**]. By arguments of Waldhausen [**W**], we see that $\mathfrak{D}_+(S^3,K) = \pi_0\mathrm{Homeo}^*(X, rel\partial) \cong \mathrm{Aut}_\partial(G)$ [**L**]. There exist natural homomorphisms $\pi_0\mathrm{Homeo}^*(X) \to \mathrm{Aut}_\partial(G)/\mathrm{Inn}_\partial(G)$ and $\mathrm{Sym}^*(S^3,K) \to \mathrm{Aut}(G)/\mathrm{Inn}(G)$, which make the following diagram commutative :

$$\begin{array}{ccccc} \mathfrak{D}_+(S^3,K) & \xrightarrow{j'} & \pi_0\mathrm{Homeo}^*(X) & \xrightarrow{k} & \mathrm{Sym}^*(S^3,K) \\ \downarrow & & \downarrow & & \downarrow \\ \mathrm{Aut}_\partial(G) & \longrightarrow & \mathrm{Aut}_\partial(G)/\mathrm{Inn}_\partial(G) & \longrightarrow & \mathrm{Aut}(G)/\mathrm{Inn}(G). \end{array}$$

Since $\mathfrak{P}(S^3,K) \cong \mathrm{Inn}_\partial(G)$ (cf. [**L**]) and $\mathfrak{P}(S^3,K) \subset \mathrm{Ker}\ j'$, we see that $\mathfrak{P}(S^3,K) = \mathrm{Ker}\ j'$ and that $\pi_0\mathrm{Homeo}^*(X) \to \mathrm{Aut}_\partial(G)/\mathrm{Inn}_\partial(G)$ is an isomorphism. Therefore, noting that $\mathrm{Inn}_\partial(G) = \mathrm{Aut}_\partial(G) \cap \mathrm{Inn}(G)$, we have that k is injective (bijective) and hence the kernel of $j = k \circ j'$ is $\mathfrak{P}(S^3,K)$. This completes the proof of Lemma 5.

Proof of Theorem. By [**Zi**], there exists a finite subgroup $\langle f, h' | f^n = h'^2 = 1, h'fh' = f^{-1}\rangle$ of $\mathrm{Homeo}^+(S^3,K)$ which is isomorphic, via the natural projection, to $\mathrm{Sym}^+(S^3,K) \cong D_n$ and whose element h' is an inverting involution. Let (D^3,D^1) be a regular neighborhood pair in (S^3,K) of a point in $K \cap \mathrm{Fix}h'$. Let $(\widetilde{D^3},\widetilde{D^1})$ be the complementary ball pair and (B^3,β) be a strongly invertible ball pair with a homeomorphism $\phi : (\widetilde{D^3},\widetilde{D^1}) \to (B^3,\beta)$ such that $\phi \circ h' \circ \phi^{-1} = h$, the π rotation about the z axis. By [**K, Lemma 7**], it is sufficient to consider only deform-spun projective planes obtained from (B^3,β).

(Case 1) Suppose that $\mathrm{Sym}^+(S^3, K) \cong D_1$. Then $\mathrm{Sym}^*(S^3, K) = 1$ and hence, by Lemma 5, $\mathfrak{D}_+(B^3, \beta)$ is generated by τ and ρ. Since all inverting deformations of (B^3, β) are peripheral, we see the result by **[K]**.

(Case 2) Suppose that $\mathrm{Sym}^+(S^3, K) \cong D_n \quad (n > 1)$. We may assume that f is the canonical element of $G = \langle f | f^n = 1 \rangle$. Noting that ρ is generated by τ and $\sigma_{f \in G}$ (see Remark (1)), we see, by Lemma 5, that $\mathfrak{D}_+(B^3, \beta)$ is generated by them. Hence any inverting deformation is expressed as $h \circ d$ for some d which is generated by τ and $\sigma_{f \in G}$. Since $h' f h' = f^{-1}$, we can apply Lemma 1 for $\sigma_{f \in G}$ and h. And by Corollary 2, we have that $\{(B^3, \beta), h \circ d\}$ is equivalent to one of $\{(B^3, \beta), h\}, \{(B^3, \beta), h \circ \tau\}, \{(B^3, \beta), h \circ \sigma_{f \in G}\}$ and $\{(B^3, \beta), h \circ \tau \circ \sigma_{f \in G}\}$. By Lemmas 3 and 4 and Price and Roseman's result **[K]**, we see the result. This completes the proof of Theorem.

Examples. (1) 8_5-knot. $\mathrm{Sym}^+(S^3, K)$ is isomorphic to D_2, which has a realization $\langle f, h' | f^2 = h'^2 = 1, h' f h' = f^{-1} \rangle$ in $\mathrm{Homeo}^+(S^3, K)$ as in **[Sa$_1$]**. Four quotient knots of (S^3, K) by inverting involutions h' and fh' are distinct and their π-orbifold groups are D_3, D_7, D_{21} and Z_2, where the π-orbifold group of a 1-knot is the quotient of the knot group by the normal closure of the square of the meridian. Since the knot group of the connected sum of a standard projective plane and the 2-twist spin of a 1-knot is isomorphic to the π-orbifold group of the 1-knot, we see that exact four deform-spun projective planes are obtained from a 8_5-knot.

(2) 9_{35}-knot. $\mathrm{Sym}^+(S^3, K) \cong D_6$. Two of the quotient knots are a trefoil knot whose π-orbifold group is D_3. The π-orbifold groups of others are D_9 and the group represented by $\langle a, b, c | a^2 = b^2 = c^2 = 1, (ab)^2 = (bc)^{-3} = (ca)^3 \rangle$ (order 144). Just three deform-spun projective planes are obtained from a 9_{35}-knot.

(3) Torus knot of type (p, q) $(p \neq 1)$. There is exact one inverting involution. If both p and q are odd integers, the quotient knots are trivial, hence only the standard projective plane is obtained from the torus knot. If p is odd and q is even, then the quotient knots are a trivial knot and a 2-bridge knot of type (p, q), **[K]**. The π-orbifold group of the latter is D_p. Hence just two deform-spun projective planes are obtained; one is standard and the other is the connected sum of the standard one and 2-twist spin of a 2-bridge knot of type (p, q).

References

[K] S. Kamada, *On deform-spun projective planes in 4-sphere obtained from peripheral inverting deformations*, Algebra and Topology, Proc. KAIST Math. Workshop **5,** Korea Adv. Inst. Sci. Tech., Taejon (1990), 197–203.

[L] R. A. Litherland, *Deforming twist-spun knots*, Trans. Amer. Math. Soc. **250** (1979), 311–331.

[P-R] T. M. Price and D. M. Roseman, *Embeddings of the projective plane in four space*, preprint.

[Sa_1] M. Sakuma, *On strongly invertible knots*, Algebraic and Topological Theories, Proc. Miyata Memorial Conf., Kinokuniya Co. Ltd., 1985, pp. 176–196.

[Sa_2] ——, *Uniqueness of symmetries of knots*, Math. Z. **192** (1986), 225–242.

[Si] J. Simon, *Roots and centralizers of peripheral elements in knot group*, Math. Ann. **222** (1976), 205–209.

[W] F. Waldhausen, *On irreducible 3-manifolds which are sufficiently large*, Ann. of Math. **87** (1968), 56–88.

[Ze] E. C. Zeeman, *Twisting spun knots*, Trans. Amer. Math. Soc. **115** (1965), 471–495.

[Zi] B. Zimmermann, *Das Nielsensche Realisierungsproblem für hinreichend große 3-mannigfaltigkeiten*, Math. Z. **180** (1982), 349–359.

Universal abelian coverings of spatial theta-curves

Toshio Harikae

Let L be a spatial θ-curve with exterior X and $\tilde{X}$ the universal abelian covering of X. Litherland [**5**] constructed the Alexander module and the Alexander polynomial of L based on $\tilde{X}$. In the paper we consider some properties of a presentation matrix of $H_1(\tilde{X})$, in particular its relation to the Alexander matrix of L.

1. Introduction

A *θ-curve* is a graph with two vertices and three edges, each edge having distinct endpoints. Let L be a spatial θ-curve with an orientation; each of three edges e_1, e_2 and e_3 is oriented from one vertex A to the other vertex B. A knot $K_{ij} = e_i \cup e_j, 1 \leq i < j \leq 3$, is called a *constituent knot* of L. Let $N = N(L)$ be a regular neighborhood of L in S^3 and $X = X(L) = cl(S^3 - N)$ an exterior of L. Then the fundamental group G of X has a Wirtinger presentation, where $x_1, \ldots, x_\ell; y_1, \ldots, y_m; z_1, \ldots, z_n$ are generators, each of which corresponds to an overpass of e_1, e_2 and e_3, respectively. Note that $H_1(X)$ is the (multiplicative) free abelian group on two generators t_1 and t_2, each of which is the homology class corresponding to a meridian of e_1 and e_2, respectively. Let α be the Hurewicz map, that is, the abelianizer of G to $H_1(X)$ such that

$$\begin{aligned} \alpha(x_i) &= t_1, & 1 \leq i \leq \ell, \\ \alpha(y_j) &= t_2, & 1 \leq j \leq m, \\ \alpha(z_k) &= t_1^{-1}t_2^{-1}, & 1 \leq k \leq n. \end{aligned}$$

Then the covering $\tilde{X}$ of X associated with α is called the *universal abelian covering* of X. Since the covering transformation group of $\tilde{X}$ can be identified with $H_1(X)$, $H_1(\tilde{X})$ is a module over Λ_2, where Λ_2 is the integral group ring of $H_1(X)$, that is, the ring of Laurent polynomials in variables t_1 and t_2. We

Knots 90

call $H_1(\tilde{X})$ the *θ-curve module* of L. Choose any point p of X and let f be the covering projection $\tilde{X} \to X$. Then $H_1(\tilde{X}, f^{-1}(p))$ is also a Λ_2-module. We denote by $A_L = A_L(t_1, t_2)$, the Alexander matrix of L (see [**4**]). Note that A_L is a presentation matrix of $H_1(\tilde{X}, f^{-1}(p))$. Let $M_L = M_L(t_1, t_2)$ be a presentation matrix of $H_1(\tilde{X})$.

If matrices M and M' present an isomorphic Λ_2-module, we call M and M' are *equivalent* and write $M \sim M'$. Note that M and M' are equivalent if and only if M can be transformed to M' by a finite number of elementary operations for matrices. Let M be an infinite rowed n column matrix over Λ_2 such that only the first m rows contain non-zero entries. The *d-th elementary ideal* $E_d(M)$ of M is defined as follows:

(1) For $0 < n - d \leq m$, $E_d(M)$ is the ideal generated by the determinants of all $(n-d) \times (n-d)$ submatrices of M.

(2) If $n - d > m$, $E_d(M) = 0$.

(3) If $n - d \leq 0$, $E_d(M) = \Lambda_2$.

We have the following

Theorem 1.1. $E_1(M_L) = E_2(A_L)$.

2. Proof of Theorem 1.1

We may deform a diagram of any spatial θ-curve so that there is no crossing on e_3 (See Figure 2.1). In Figure 2.1 let T be a 2-string tangle. Then G has generators $x_1, \ldots, x_\ell; y_1, \ldots, y_m; z$ and relations, each of which can be expressed as one of the following six types

$$
\begin{aligned}
r_1 &= x_1 y_1 z, \\
r_2 &= x_\ell y_m z, \\
r_3 &= x_i x_j x_i^{-1} x_k^{-1}, \\
r_4 &= y_i y_j y_i^{-1} y_k^{-1}, \\
r_5 &= x_i y_j x_i^{-1} y_k^{-1}, \\
r_6 &= y_i x_j y_i^{-1} x_k^{-1},
\end{aligned}
$$

where r_1 and r_2 correspond to vertices A and B and r_3, r_4, r_5 and r_6 correspond to four types of crossings as shown in Figure 2.2, respectively. Since one of the relations is derivable from others, the defficiency of G is always two.

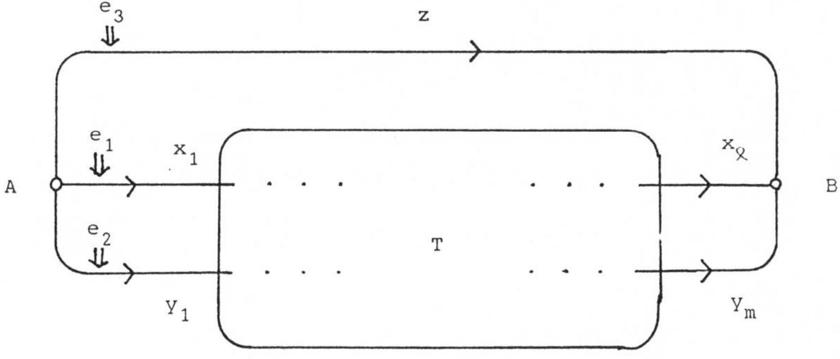

Figure 2.1

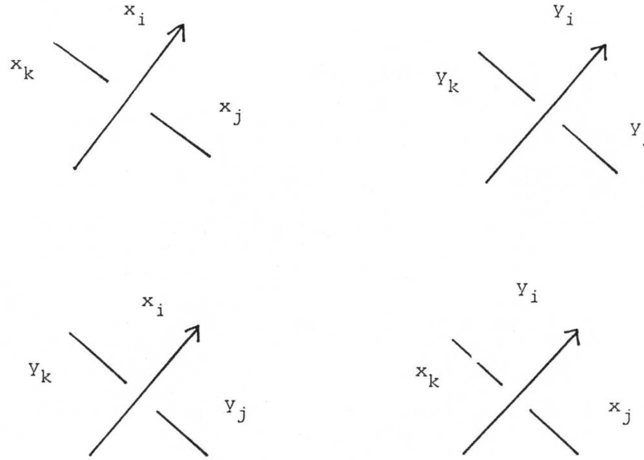

Figure 2.2

We shall express the presentation of $\pi_1(\tilde{X})$ by using the Reidemeister-Schreier method as follows: Generators of $\pi_1(\tilde{X})$:

$$x_{i,pq} = (x_1^p y_1^q) x_i (x_1^{p+1} y_1^q)^{-1},$$
$$y_{j,pq} = (x_1^p y_1^q) y_j (x_1^p y_1^{q+1})^{-1},$$
$$z_{pq} = (x_1^p y_1^q) z (x_1^{p-1} y_1^{q-1})^{-1}.$$

Relations of $\pi_1(\tilde{X})$:

$$
\begin{aligned}
r_1' &= x_{1,pq}y_{1,p+1q}z_{p+1q+1},\\
r_2' &= x_{\ell,pq}y_{m,p+1q}z_{p+1q+1},\\
r_3' &= x_{i,pq}x_{j,p+1q}x_{i,p+1q}^{-1}x_{k,pq}^{-1},\\
r_4' &= y_{i,pq}y_{j,pq+1}y_{i,pq+1}^{-1}x_{k,pq}^{-1},\\
r_5' &= x_{i,pq}y_{j,p+1q}x_{i,pq+1}^{-1}y_{k,pq}^{-1},\\
r_6' &= y_{i,pq}x_{j,pq+1}y_{i,p+1q}^{-1}x_{k,pq}^{-1},
\end{aligned}
$$

each of which corresponds to each of $r_1, \dots, r_6$, respectively.

Let $[g]$ denote the image of g in $H_1(\tilde{X})$ by the canonical projection. Then Λ_2 acts on $H_1(\tilde{X})$ by

$$
\begin{aligned}
t_1[g] &= [x_1 g x_1^{-1}],\\
t_2[g] &= [y_1 g y_1^{-1}].
\end{aligned}
$$

If we put

$$
\begin{aligned}
\bar{x}_i &= [x_{i,00}] = [x_i x_1^{-1}],\\
\bar{y}_j &= [y_{j,00}] = [y_j y_1^{-1}],\\
\bar{z} &= [z_{00}] = [zyx],\\
\bar{\gamma} &= [x_{1,01}] = [y_1 x_1 y_1^{-1} x_1^{-1}],
\end{aligned}
$$

we have

$$
\begin{aligned}
[x_{i,pq}] &= t_1^p t_2^q \bar{x}_i + t_1^p \frac{1-t_2^q}{1-t_2}\bar{\gamma},\\
[y_{j,pq}] &= t_1^p t_2^q \bar{y}_j,\\
[z_{pq}] &= t_1^p t_2^q \bar{z} - t_1^{p-1}\frac{1-t_2^q}{1-t_2}\bar{\gamma},
\end{aligned}
$$

where $\bar{x}_1 = \bar{y}_1 = 0$. Hence $H_1(\tilde{X})$ is generated by $\bar{x}_i$, $\bar{y}_j$, $\bar{z}$ and $\bar{\gamma}$ with operators t_1 and t_2. Further $H_1(\tilde{X})$ has relations:

$$
\begin{aligned}
r_1'' &= t_1^p t_2^q(\bar{x}_1 + t_1\bar{y}_1 + t_1 t_2 \bar{z} - \bar{\gamma}),\\
r_2'' &= t_1^p t_2^q(\bar{x}_\ell + t_1\bar{y}_m + t_1 t_2 \bar{z} - \bar{\gamma}),\\
r_3'' &= t_1^p t_2^q\{(1-t_1)\bar{x}_i + t_1\bar{x}_j - \bar{x}_k\},\\
r_4'' &= t_1^p t_2^q\{(1-t_2)\bar{y}_i + t_2\bar{y}_j - \bar{y}_k\},\\
r_5'' &= t_1^p t_2^q\{(1-t_2)\bar{x}_i + t_1\bar{y}_j - \bar{y}_k - \bar{\gamma}\},\\
r_6'' &= t_1^p t_2^q\{(1-t_1)\bar{y}_i + t_2\bar{x}_j - \bar{x}_k + \bar{\gamma}\},
\end{aligned}
$$

each of which corresponds to each of $r'_1, \ldots, r'_6$, respectively, and two other relations $\bar{x}_1$ and $\bar{y}_1$. Hence the defficiency of M_L is always one. From now on, we shall identify $\bar{x}_1, \ldots, \bar{x}_\ell; \bar{y}_1, \ldots, \bar{y}_m; \bar{z}$ with $x_1, \ldots, x_\ell; y_1, \ldots, y_m; z$.

On the other hand, A_L is defined as follows (See [**4**]): Let ϕ be the canonical map from $(x_1, \ldots, x_\ell, y_1, \ldots, y_m, z)$ to G. If we denote the k-th relation of G by r, then the k-th row of A_L can be expressed as

$$\left(\left(\frac{\partial r}{\partial x_i}\right)^{\alpha\phi} \left(\frac{\partial r}{\partial y_j}\right)^{\alpha\phi} \left(\frac{\partial r}{\partial z}\right)^{\alpha\phi} \right),$$

where ∂ means the free derivative (See [**1**]). For example we have for six types of relations of G

$$\left(\frac{\partial r_1}{\partial x_1}\right)^{\alpha\phi} = 1, \qquad \left(\frac{\partial r_1}{\partial y_1}\right)^{\alpha\phi} = t_1, \qquad \left(\frac{\partial r_1}{\partial z}\right)^{\alpha\phi} = t_1 t_2;$$

$$\left(\frac{\partial r_2}{\partial x_\ell}\right)^{\alpha\phi} = 1, \qquad \left(\frac{\partial r_2}{\partial y_m}\right)^{\alpha\phi} = t_1, \qquad \left(\frac{\partial r_2}{\partial z}\right)^{\alpha\phi} = t_1 t_2;$$

$$\left(\frac{\partial r_3}{\partial x_i}\right)^{\alpha\phi} = 1 - t_1, \qquad \left(\frac{\partial r_3}{\partial x_j}\right)^{\alpha\phi} = t_1, \qquad \left(\frac{\partial r_3}{\partial x_k}\right)^{\alpha\phi} = -1;$$

$$\left(\frac{\partial r_4}{\partial y_i}\right)^{\alpha\phi} = 1 - t_2, \qquad \left(\frac{\partial r_4}{\partial y_j}\right)^{\alpha\phi} = t_2, \qquad \left(\frac{\partial r_4}{\partial y_k}\right)^{\alpha\phi} = -1;$$

$$\left(\frac{\partial r_5}{\partial x_i}\right)^{\alpha\phi} = 1 - t_2, \qquad \left(\frac{\partial r_5}{\partial y_j}\right)^{\alpha\phi} = t_1, \qquad \left(\frac{\partial r_5}{\partial y_k}\right)^{\alpha\phi} = -1;$$

$$\left(\frac{\partial r_6}{\partial y_i}\right)^{\alpha\phi} = 1 - t_1, \qquad \left(\frac{\partial r_6}{\partial x_j}\right)^{\alpha\phi} = t_2, \qquad \left(\frac{\partial r_6}{\partial x_k}\right)^{\alpha\phi} = -1.$$

The defficiency of A_L is always two. The following theorem states an elementary property of A_L.

Theorem 2.1. *We have*

$$A_L(1,1) \sim (1 \quad 0 \quad 0).$$

Proof. Since $A_L(1,1)$ is a presentation matrix of $H_1(X) \cong Z \oplus Z$, we obtain the theorem.

We shall compare the structure of M_L and A_L. Let A'_L be a matrix constructed from A_L by deleting columns of x_1 and y_1. Note that A'_L is a square matrix.

Lemma 2.2. *We have*

(1) $$M_L \sim (A'_L \quad *)$$

and

$$A_L \sim \begin{pmatrix} A'_L & * & 0 & 0 \\ 0 & 1 & t_1 - 1 & 1 - t_2 \end{pmatrix}, \tag{2}$$

where $*$ *is the one column matrix, whose entry is 0, -1 or 1, as the row corresponds to* $r_1, \dots, r_4; r_5$ *or* r_6, *respectively.*

Proof. Since $x_1 = y_1 = 0$ for $H_1(\tilde{X})$, we know M_L is constructed by adding to A'_L the column of $\bar{\gamma}$. Adding the column of z multiplied by $t_1^{-1}t_2^{-1}$ to that of $\bar{\gamma}$, we have the column $*$, which proves (1).

In A_L, we add the columns of $x_2, \dots, x_\ell$ and that of z multiplied by $-t_1^{-1}t_2^{-1}$ to that of x_1, and add the columns of $y_2, \dots, y_m$ and that of z multiplied by $-t_2^{-1}$ to that of y_1. Then we have

$$(A'_L \quad ** \quad ***),$$

where each of $**$ and $***$ is one column matrix $*$ multiplied by $1 - t_1$ and $t_2 - 1$. Therefore we have (2) by a simple elementary operation.

Proposition 2.3. *We have*

$$\det A'_L(1,1) = \pm 1.$$

Proof. By Lemma 2.2 we have

$$A_L(1,1) \sim (A'_L(1,1) \quad 0 \quad 0),$$

which completes the proof.

Theorem 2.4.

$$M_L(1,1) \sim (1 \quad 0).$$

Proof. By Lemma 2.2 (1) and Proposition 2.3 we can prove the theorem.

Proof of Theorem 1.1. Let M'_i be a matrix constructed by deleting the i-th column of A'_L and adding $*$, $i = 1, \dots, \ell + m - 1$. Then we can see that $E_1(M_L)$ is generated by $\det M'_i$, $i = 1, \dots, \ell + m - 1$, and $\det A'_L$ by Lemma 2.2 (1). On the other hand, $E_2(A_L)$ is generated by $(t_1 - 1)\det M'_i$, $(1 - t_2)\det M'_i$, $i = 1, \dots, \ell + m - 1$, and $\det A'_L$ by Lemma 2.2 (2). Hence we have $E_1(M_L) \supseteq E_2(A_L)$. Further, since $\det A'_L(1,1) = \pm 1$ by Proposition 2.3, there exist polynomials P and Q in t_1 and t_2 such that

$$\det A'_L = (1 - t_1)P + (t_2 - 1)Q \pm 1.$$

Hence we have

$$P(t_1 - 1)\det M_i' + Q(1-t_2)\det M_i' + \det M_i' \det A_L' = \pm \det M_i'.$$

Thus $\det M_i'$ is generated by $(t_1-1)\det M_i'$, $(1-t_2)\det M_i'$ and $\det A_L'$. Therefore we have $E_1(M_L) \subseteq E_2(A_L)$, which completes the proof of Theorem 1.1.

3. Some examples

In [2] we introduced two classes of spatial θ-curves called *rational* (Figure 3.1) and *pseudo-rational* (Figure 3.2). Each of rational and pseudo-rational θ-curves is constructed by adding an unknotting tunnel to a two-bridge knot. Hence its exterior is homeomorphic to a genus two handlebody.

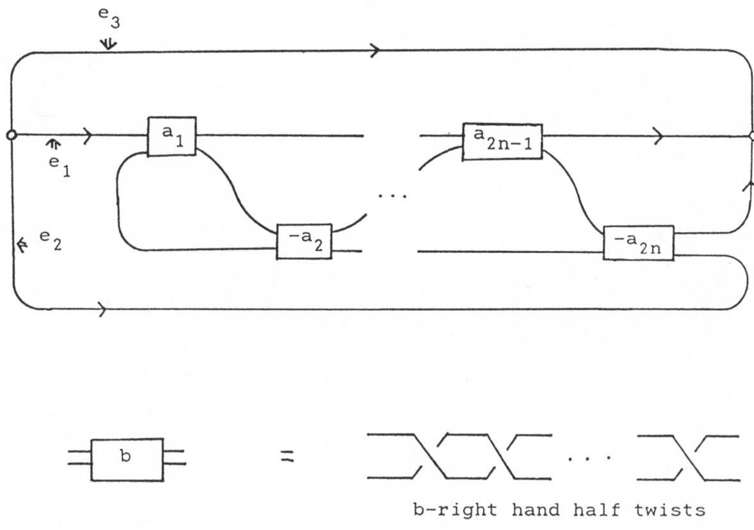

Figure 3.1

Remark. If L is rational or pseudo-rational, then

$$M_L \sim (1 \quad 0)$$

and

$$A_L \sim (1 \quad 0 \quad 0).$$

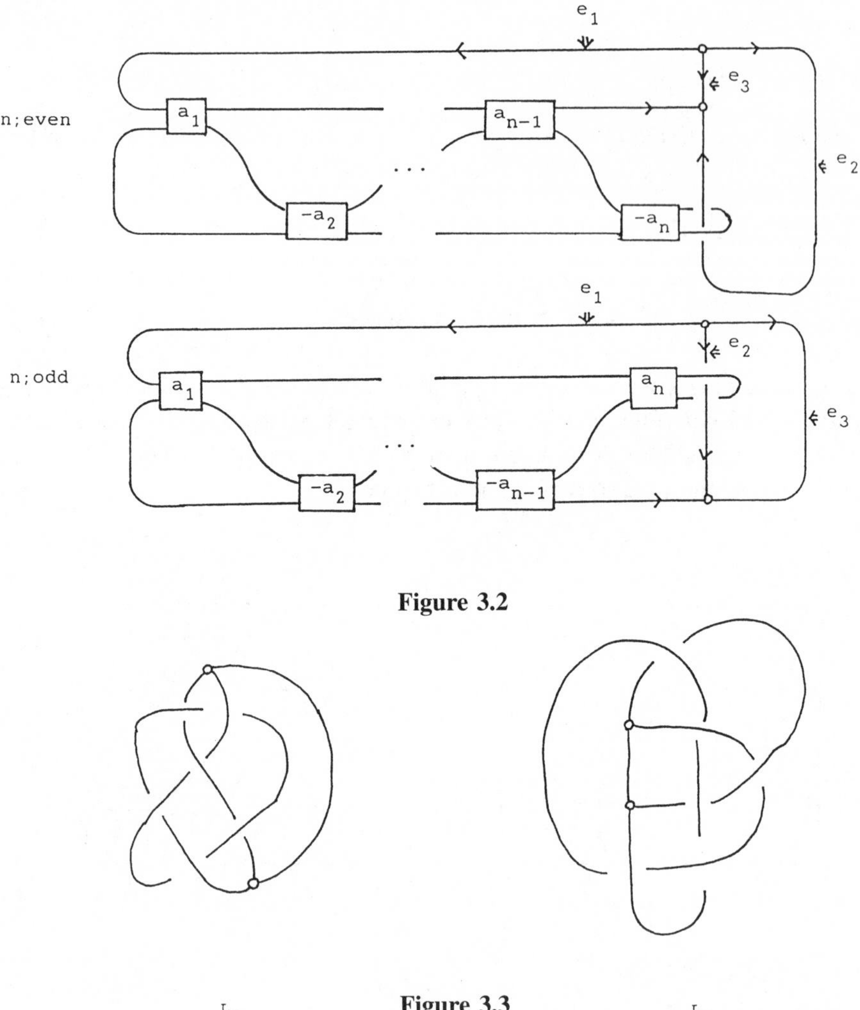

Figure 3.2

Figure 3.3

Remark. Each of two spatial θ-curves L_1 and L_2 as shown in Figure 3.3 is neither rational nor pseudo-rational. For L_1 or L_2 we have

$$M_L \sim (1 \quad 0)$$

and

$$A_L \sim (1 \quad 0 \quad 0).$$

Since L_1 and L_2 are neighborhood equivalent, their exterior have an isomorphic fundamental group which is

$$\langle a, b, c | aba^{-1}b^{-1}cbc^{-2} \rangle.$$

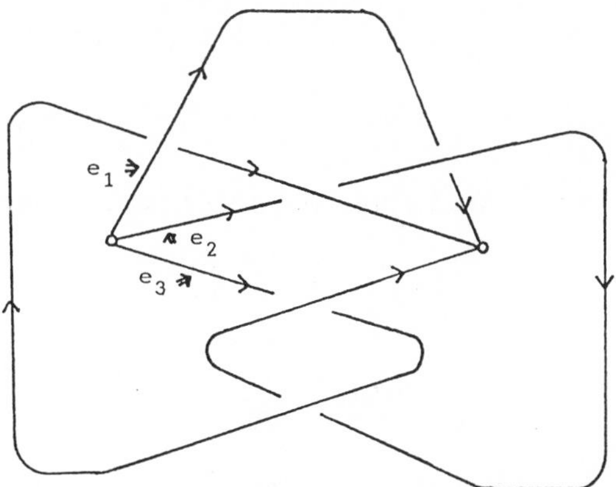

Figure 3.4

Example. The Kinoshita's θ-curve L as shown in Figure 3.4 has three trivial constituent knots. We know that L is not planar by calculating M_L or A_L, where

$$M_L \sim (\, t_1t_2 - t_2 + 1 \quad 2\,)$$

and

$$A_L \sim (\, t_1t_2 - t_2 + 1 \quad 2 \quad 0\,).$$

In [**3**] we consider 3-fold irregular branched coverings of spatial θ-curves. Let μ be a monodromy map of G onto S_3, the symmetric group on three letters $\{1,2,3\}$. We may assume the following conditions without losing generality:

$$\begin{aligned} &\mu(x_i) \in \{(12),(13),(23)\}, \\ (*) \qquad &\mu(y_j) \in \{(12),(13),(23)\}, \\ &\mu(z_k) \in \{e,(123),(132)\}. \end{aligned}$$

The branched covering $\tilde{M}_\mu(L)$ of L associated with μ is called the 3-fold irregular branched covering of L. We know that $\tilde{M}_\mu(L)$ is always an orientable 3-manifold. Further, let ν be the nullity of $M_L(-1,-1)$ in Z_3-coefficients.

Theorem 3.1. *The number of conjugacy classes of S_3-representations, each of which satisfies $(*)$, is equal to*

$$\frac{3^{\nu+1}-3}{3!}.$$

Proof. By Lemma 2.2 in Z_3-coefficients we have

$$A_L(-1,-1) \sim (\, M_L(-1,-1) \quad 0\,).$$

Applying [**3, Theorem 2.1**], we can prove the statement of the theorem.

Since the deficiency of $H_1(\tilde{X})$ is one, we know that ν is at least one.

Collorary 3.2. *For any spatial θ-curve, there exists at least one S_3-representation which satisfies $(*)$.*

References

[1] R. H. Fox, *A quick trip through knot theory,* in "Topology of 3-Manifolds and Related Topics" (ed. M.K. Fort jr.), Prentice-Hall, Englewood Cliffs, New Jersey, 1962, 120–167.

[2] T. Harikae, *On rational and pseudo-rational θ-curves in the 3-sphere*, Kobe J. Math. **7** (1990), 125–138.

[3] T. Harikae, *Three-fold irregular branched coverings of some spatial graphs*, Osaka J. Math. (to appear).

[4] S. Kinoshita, *Elementary ideals in knot theory*, Kwansei Gakuin Univ. Ann. Studies **35** (1986), 183–208.

[5] R. Litherland, *The Alexander module of a knotted theta-curve*, Math. Proc. Camb. Phil. Soc. **106** (1989), 95–106.

Special hermitian forms for Seifert surfaces of boundary links and algebraic invariants

Shinji Fukuhara

Abstract. We introduce algebraic invariants for Seifert surfaces of boundary links. First we see that any Seifert surface is obtained from a standard Seifert surface by special kind of surgery. It is shown that a stable diffeomorphism class of the trace of the surgery depends only on the equivalence class of Seifert surface. The special hermitian form is defined on the 2-dimensional homology group of its universal covering space. Then its determinant, signature etc. are invariants of the Seifert surface.

1. Introduction

An m-component link $L = l_1 \cup \cdots \cup l_m$ is a disjoint union of m simple closed oriented curves in S^3.

Definition 1. *A link $L = l_1 \cup \cdots \cup l_m$ is called a boundary link if there are compact orientable surfaces S_i in S^3 for $i = 1, \ldots, m$ such that $S_i \cap S_j = \emptyset$ for $i \neq j$ and $\partial S_i = l_i$.*

The ordered union $S = S_1 \cup \cdots \cup S_m$ is called a Seifert surface of the boundary link L.

Let F_m denote the free group of rank m. An m-component link L is a boundary link if and only if there is an epimorphism $\theta : \pi_1(S^3 - L) \to F_m$ which carries some set of meridian to a basis of F_m (Smythe[7],Gutierrez[3],Cappell and Shaneson[1], Ko[5]). Such epimorphism θ is called a splitting map for the link L.

Knots 90

Examples.

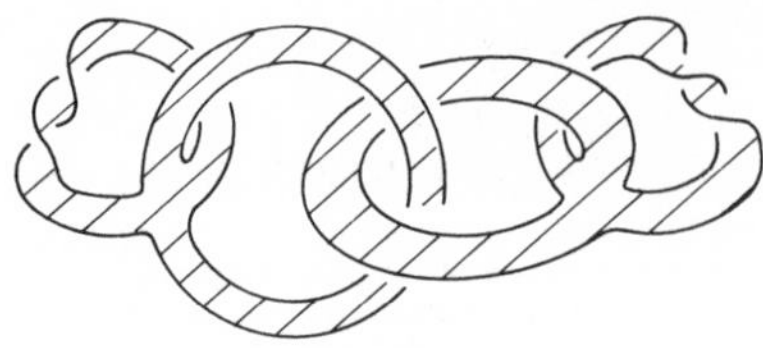

This is a Seifert surface of a boundary link.

Figure 1

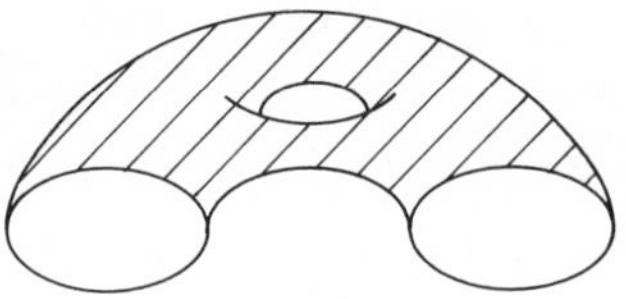

This is not a Seifert surface of a boundary link.

Figure 2

Definition 2. *Seifert surfaces $S = S_1 \cup \cdots \cup S_m$ and $T = T_1 \cup \cdots \cup T_m$ of boundary links are said to be equivalent if there is an orientation preserving diffeomorphism of S^3 which carries S_i to T_i for $i = 1, \ldots, m$.*

2. Surgery descriptions of Seifert surfaces of boundary links

We show that every Seifert surface of a boundary link can be produced by performing special surgery on complement of a standard Seifert surface.

For m non negative integers $g_1, \cdots, g_m$ let $S(g_1, \cdots, g_m)$ denote the surface in S^3 with m components depicted in Figure 3.

Each component is a disk with g_i pairs of bands and has g_i as its genus for $i = 1, \ldots, m$. Let $L(g_1, \cdots, g_m)$ be the boundary of the surface $S(g_1, \cdots, g_m)$. Note that $L(g_1, \cdots, g_m)$ is unlinked.

A (integral) framed link $FL = (C, N)$ is a link $C = c_1 \cup \cdots \cup c_k$ with integers $N = (n_1, \ldots, n_k)$. The integer n_i is assigned to the i-th component of C and called a framing of c_i.

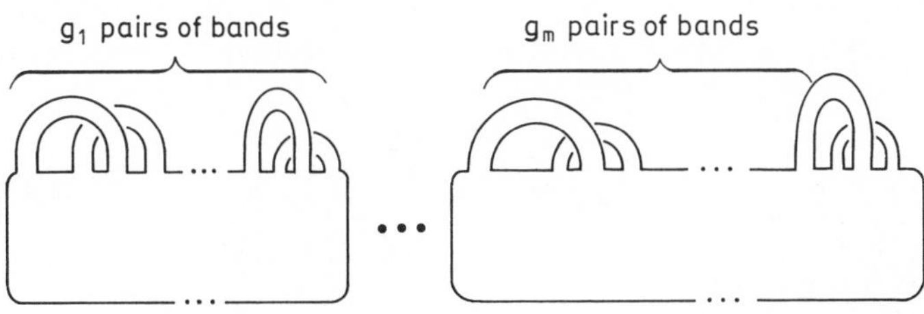

Figure 3

Definition 3. *A framed link $FL = (C, N)$ such that $C = c_1 \cup \cdots \cup c_k$ and $N = (n_1, \ldots, n_k)$ is called special for $S(g_1, \cdots, g_m)$ if it satisfies the following conditions:*

(1) $C \cap S(g_1, \cdots, g_m) = \emptyset$,

(2) C is unlinked and

(3) C has an even number of components, say $k = 2k_0$, and a framing $n_i = 1$ for $i = 1, \ldots, k_0$ and $n_i = -1$ for $i = k_0 + 1, \ldots, k$.

Suppose that FL is a special framed link for $S(g_1, \cdots, g_m)$. Let M_{FL} denote the manifold obtained from S^3 by performing surgery along C with framings N. Since C is unlinked and the framings are ± 1, M_{FL} is diffeomorphic to S^3. Let S_{FL} and L_{FL} be the surface and the link in M_{FL} which correspond to $S(g_1, \cdots, g_m)$ and $L(g_1, \cdots, g_m)$ respectively. Identifying M_{FL} with S^3 we obtain a boundary link L_{FL} and its Seifert surface S_{FL}.

Our first lemma shows that this construction of Seifert surfaces is general.

Lemma 1. *For every Seifert surface S of a boundary link there is a special framed link $FL = (C, N)$ for $S(g_1, \cdots, g_m)$ so that S is equivalent to S_{FL} where g_i is the genus of the i-th component of S.*

Proof of Lemma 1. Any Seifert surface of a boundary link looks like Figure 4 where bands are tangled and twisted.

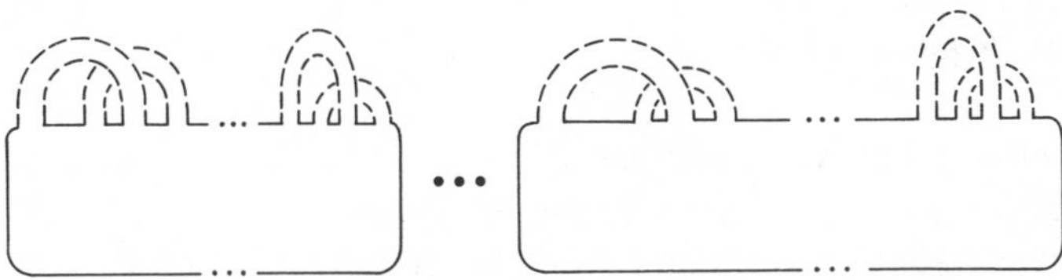

Figure 4

To untangle and untwist these bands we only need the following two moves which are realized by Dehn surgery on the complement of the surface.

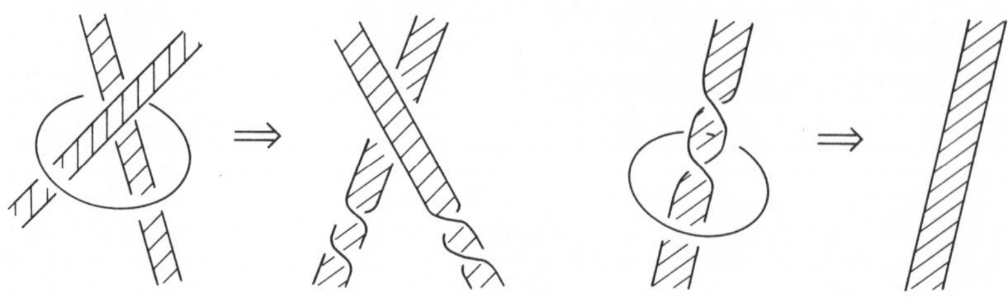

Figure 5

We do surgeries as many times as we need. Finally we obtain a standard surface $S(g_1, \cdots, g_m)$ and surgery curves which miss $S(g_1, \cdots, g_m)$ and form a trivial link in S^3. Conversely we can obtain S from $S(g_1, \cdots, g_m)$ by doing surgeries on the complement of the surface. The condition (3) can be satisfied if we add trivial surgeries. So the result follows.

Consider a special framed link $FL = (C, N)$ for $S(g_1, \cdots, g_m)$. We can choose a tubular neighborhood $N(L(g_1, \cdots, g_m))$ of $L(g_1, \cdots, g_m)$ such that $C \cap N(L(g_1, \cdots, g_m)) = \emptyset$ because $C \cap L(g_1, \cdots, g_m) = \emptyset$. Let $E(g_1, \cdots, g_m)$ denote the closure of $S^3 - N(L(g_1, \cdots, g_m))$. The surgery on S^3 with respect to FL induces surgery on $E(g_1, \cdots, g_m)$. By $X(FL; g_1, \cdots, g_m)$ we denote the trace of this surgery. Then $X(FL; g_1, \cdots, g_m)$ is an orientable 4-manifold which is diffeomorphic to $E(g_1, \cdots, g_m) \times I \cup$2-handles where the 2-handles are attached to $E(g_1, \cdots, g_m) \times \{1\}$. Next we consider how this 4-manifold will change when we replace FL with another special framed link $FL' = (C', N')$ for $S(g_1, \cdots, g_m)$.

Under the notations above, we have:

Lemma 2. *If S_{FL} is equivalent to $S_{FL'}$ then*

$$X(FL; g_1, \cdots, g_m) \sharp pCP(2) \sharp p\overline{CP(2)}$$

is diffeomorphic to

$$X(FL'; g_1, \cdots, g_m) \sharp qCP(2) \sharp q\overline{CP(2)}$$

for some non negative integers p and q.

Proof of Lemma 2. Let $X = X(FL; g_1, \cdots, g_m) = E(g_1, \cdots, g_m) \times I \cup$2-handles and $X' = X(FL'; g_1, \cdots, g_m) = E(g_1, \cdots, g_m) \times I \cup$ 2-handles. Note that the 2-handles are attached in different ways on X and X'. Let $\theta : \pi_1(X) \to F_m$ and $\theta' : \pi_1(X') \to F_m$ be isomorphisms which carry the i-th meridians of $E(g_1, \cdots, g_m) \times \{0\}$ in X and X' to an i-th generator of F_m. We construct a cobordism W fixing the boundary between X and X' with an isomorphism $\Theta : \pi_1(W) \to F_m$ such that Θ satisfies $\theta = \Theta \cdot i$ and $\theta' = \Theta \cdot i'$ for homomorphisms $i : \pi_1(X) \to \pi_1(W)$ and $i' : \pi_1(X') \to \pi_1(W)$ induced from inclusion maps.

Let $\partial_0 X$ and $\partial_1 X$ be submanifolds of X defined by

$$\partial_0 X = E(g_1, \cdots, g_m) \times \{0\}$$

$$\partial_1 X = E(g_1, \cdots, g_m) \times \{1\} - N(L) \cup (\bigcup_i (\partial D_i^2) \times D_i^2).$$

Also let $\partial_0 X'$ and $\partial_1 X'$ be submanifolds of X' defined by

$$\partial_0 X' = E(g_1, \cdots, g_m) \times \{0\}$$

$$\partial_1 X' = E(g_1, \cdots, g_m) \times \{1\} - N(L) \cup (\bigcup_j (\partial D_j^2) \times D_j^2).$$

Let $\Sigma = E(g_1, \cdots, g_m) \cap S(g_1, \cdots, g_m)$. By the assumption that S_{FL} is equivalent to $S_{FL'}$, there is a diffeomorphism $h : \partial_1 X \to \partial_1 X'$ which carries $\Sigma \times \{1\}$ in $\partial_1 X$ to $\Sigma \times \{1\}$ in $\partial_1 X'$.

We consider a 4-manifold $X \cup \partial_0 X \times J \cup \partial_1 X \times J \cup X'$ where $\partial_0 X$ in X is identified with $\partial_0 X \times \{0\}$ in $\partial_0 X \times J$, $\partial_1 X$ in X is identified with $\partial_1 X \times \{0\}$ in $\partial_1 X \times J$ and $\partial_0 X'$ in X' is identified with $\partial_0 X \times \{1\}$ in $\partial_0 X \times J$ by the obvious identifications while $\partial_1 X'$ in X' is identified with $\partial_1 X \times \{1\}$ in $\partial_1 X \times J$ by the diffeomorphism h.

To simplify the argument we assume that $h|\Sigma \times \{1\} = id$. The proof of general case is done by replacing a manifold $N(\Sigma) \times I \times J$ below by an appropriate 5-manifold. But we will omit it.

Let $N(\Sigma)$ be a tubular neighborhood of Σ in $E(g_1, \cdots, g_m)$. We attach $N(\Sigma) \times I \times J$ to the manifold above with the following identification to obatin W_0. Let $N(\Sigma) \times I$ in X be identified with $N(\Sigma) \times I \times \{0\}$, $N(\Sigma) \times I$ in X' with $N(\Sigma) \times I \times \{1\}$, $N(\Sigma) \times J$ in $\partial_0 X \times J$ with $N(\Sigma) \times \{0\} \times J$ and $N(\Sigma) \times J$ in $\partial_1 X \times J$ with $N(\Sigma) \times \{1\} \times J$.

Let $Y = \overline{E^3 - N(\Sigma)}$, $Z = \overline{X - N(\Sigma) \times I}$ and $Z' = \overline{X' - N(\Sigma) \times I}$. Consider a closed 4-manifold

$$M = \partial Y \times I \times J \cup Y \times \partial I \times J \cup Z \cup Z'$$

which is a subspace of W_0. It is easily proved that the signature $\sigma(M)$ of M is just $\sigma(X) - \sigma(X')$ and the condition (3) on FL and FL' assures $\sigma(M) = 0$. Let W_1 be an compact orientable 5-manifold which have M as its boundary. Construct a 5-manifold W_2 from W_0 by attaching W_1 along M. Then W_2 is a cobordism from X to X' fixing the boundary. Since W_2 contains a cobordism of codimension 1 fixing the boundary from $\Sigma \times I$ in X to $\Sigma \times I$ in X', using the Thom-Pontrjagin construction([1],[5]), we can show that there is an epimorphism $\Theta_2 : \pi_1(W_2) \to F_m$ such that Θ_2 satisfies $\theta = \Theta_2 \cdot i$ and $\theta' = \Theta_2 \cdot i'$. Now we apply the 5-dimensinal surgery theory. Performing surgery on W_2 we obtain a cobordism W fixing the boundary which has a required isomorphism $\Theta : \pi_1(W) \to F_m$. Then there is a handle decomposition on W such that X' is obtained from X by surgeries with indices 2 and 3. This means we obtain the same 4-manifold both from X and X' by doing surgeries with index 2. The conditions that Θ satisfies assure that surgery curves of the surgeries are trivial. Hence after being connected with certain number of $CP(2) \sharp \overline{CP(2)}$, X and X' are diffeomorphic. This completes the proof of Lemma 2.

3. Special hermitian forms of Seifert surfaces

Note that $\pi_1(X(FL; g_1, \cdots, g_m))$ is isomorphic to F_m. Let $\widetilde{X}(FL; g_1, \cdots, g_m)$ and $\widetilde{E}(g_1, \cdots, g_m)$ denote the universal covering space of $X(FL; g_1, \cdots, g_m)$ and $E(g_1, \cdots, g_m)$. The 2-dimensinal homology group $H_2(\widetilde{X}(FL; g_1, \cdots, g_m))$ is a free $\mathbf{Z}F_m$-module. The special hermitian form

$$\mathcal{S}_{(FL;g_1,\cdots,g_m)} : H_2(\widetilde{X}(FL; g_1, \cdots, g_m)) \times H_2(\widetilde{X}(FL; g_1, \cdots, g_m)) \to \mathbf{Z}F_m$$

is defined as

$$\mathcal{S}_{(FL;g_1,\cdots,g_m)}(x, y) = \sum_{g \in F_m} (gx, y)g^{-1}$$

where (,) denote ordinary intersection pairing. It is shown that the image from $H_2(\widetilde{E}(g_1, \cdots, g_m))$ is included in the null space of $\mathcal{S}_{(FL;g_1,\cdots,g_m)}$. Let

$$\mathcal{H}(FL; g_1, \cdots, g_m) = H_2(\widetilde{X}(FL; g_1, \cdots, g_m))/H_2(\widetilde{E}(g_1, \cdots, g_m)).$$

Then $\mathcal{S}_{(FL;g_1,\cdots,g_m)}$ induces a special hermitian form

$$\overline{\mathcal{S}}_{(FL;g_1,\cdots,g_m)} : \mathcal{H}(FL; g_1, \cdots, g_m) \times \mathcal{H}(FL; g_1, \cdots, g_m) \to \mathbf{Z}F_m.$$

Definition 4. *The special hermitian forms $\mathcal{S}_1$ and $\mathcal{S}_2$ are called stably equivalent if there is an isomorphism between $\mathcal{S}_1 \oplus (\oplus_p < 1 >) \oplus (\oplus_p < -1 >)$ and $\mathcal{S}_2 \oplus (\oplus_q < 1 >) \oplus (\oplus_q < -1 >)$ for some non negative integers p and q.*

From Lemmas 1 and 2 we obtain our main theorem:

Theorem 1. *Let $S = S_1 \cup \cdots \cup S_m$ be a Seifert surface such that genus of S_i is g_i for $i = 1, \ldots, m$. Then*

(1) there is a special framed link FL for $S(g_1, \cdots, g_m)$ such that S_{FL} is equivalent to S and

(2) the stable equivalence class of the special hermitian form $\mathcal{S}_{(FL;g_1,\cdots,g_m)}$ is an invariant of the equivalence class of S.

Definition 5. *Let $S = S_1 \cup \cdots \cup S_m$ and $T = T_1 \cup \cdots \cup T_m$ be Seifert surfaces of boundary links such that $\partial S = \partial T$. Then S and T are said to be cobordant fixing the boundary if a surface $V = S \times \{0\} \cup (\partial S) \times I \cup T \times \{1\}$ in $S^3 \times I$ bounds m disjoint orientable submanifolds in $S^3 \times I$.*

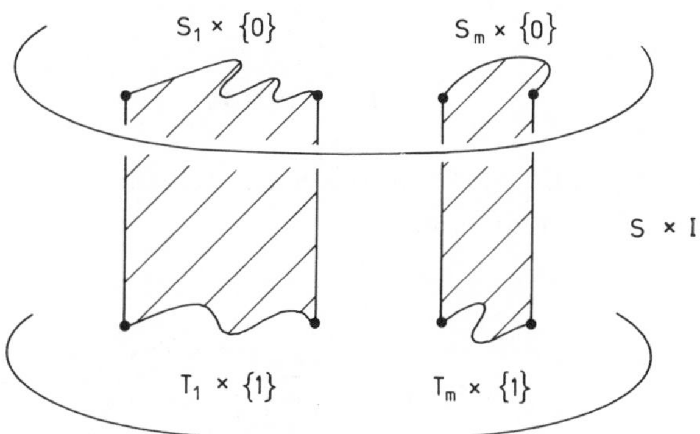

Figure 6

Theorem 1′. *Let $S = S_1 \cup \cdots \cup S_m$ be a Seifert surface such that genus of S_i is g_i for $i = 1, \ldots, m$. Then*

(1) there is a special framed link FL for $S(g_1, \cdots, g_m)$ such that S_{FL} is equivalent to S and

(2′) the stable equivalence class of the special hermitian form $\mathcal{S}_{(FL;g_1,\cdots,g_m)}$ is an invariant of the cobordism class of S fixing the boundary.

Let $\alpha : \mathbf{Z}F_m \to \mathbf{Z}[t_1^{\pm 1}, \ldots, t_m^{\pm 1}]$ be the abelianization. Then the determinants $\det \overline{\mathcal{S}}^{\alpha}_{(FL;g_1,\cdots,g_m)}$ is defined and denoted by $D_S(t_1, \ldots, t_m)$.

Theorem 2. *An integral polynomial $D_S(t_1, \ldots, t_m)$ is an invariant of equivalence class of a Seifert surface S.*

4. Properties of invariants

Our polynomials have the following properties:

(1) $D_S(1, \ldots, 1) = \pm 1$.

(2) $D_S(t_1^{-1}, \ldots, t_m^{-1}) = D_S(t_1, \ldots, t_m)$.

(3) $D_S(t)$ is the normalized Alexander polynomial of ∂S when $m = 1$.

(4) If a Seifert surface $S = S_1 \cup \cdots \cup S_m$ splits into $T_1 = S_1 \cup \cdots \cup S_{m_0}$ and $T_2 = S_{m_0+1} \cup \cdots \cup S_m$ then

$$D_S(t_1, \ldots, t_{m_0}, t_{m_0+1}, \ldots, t_m) = D_{T_1}(t_1, \ldots, t_{m_0}) \cdot D_{T_2}(t_{m_0+1}, \ldots, t_m).$$

5. An example

The following is a link called (once) iterated Whitehead link. Our polynomial $D_S = -2t + 5 - 2t^{-1}$ in this case.

Figure 7

References

[1] S.E. Cappell and J.L. Shaneson, *Link cobordism*, Comment. Math. Helv. **55** (1980), 20-49.

[2] T.D. Cochran and W.B. R. Lickorish, *Unknotting information from 4-manifolds*, Trans. Amer. Math. Soc. **297** (1987), 125-142.

[3] M. Gutierrez, *Boundary links and an unlinking theorem*, Trans. Amer. Math. Soc. **171** (1972), 471-499.

[4] J.A. Hillman, *Alexander ideals of links*, Springer-Verlag, New York, 1981.

[5] K.H. Ko, *Seifert matrices and boundary link cobordisms*, Trans. Amer. Math. Soc. **299** (1987), 675-681.

[6] D. Rolfsen,"Knots and links", Publish or Perish Inc., Berkeley, California, 1976.

[7] N.F. Smythe, *Boundary links*, Topology Seminar (Wisconcin, 1965), (R.H. Bing and R. Bean, eds.), Princeton Univ. Press, Princeton, N.J., 1966, 69-72.

Casson's invariant for branched cyclic covers over iterated torus knots

Amy Davidow

This paper concerns the calculation of the Casson λ invariant in its original formulation as an integer valued invariant of oriented homology 3-spheres [A-M]. Let $K_{p,q;r,s}$ denote the interated torus knot. Start with $K_{p,q}$, the (p,q) torus knot sitting on a standardly embedded solid torus T in S^3. Let h be an embedding of $K_{r,s}$. Applying h to the core of T and extending radially to all of T, we have that $h(K_{p,q})$ is $K_{p,q;r,s}$. Let $B^n_{p,q;r,s}$ be the n-fold branched cyclic cover of S^3 with branch set $K_{p,q;r,s}$ where $n = \pm 1 \bmod pq$. It will be shown below that this restriction is sufficient to guarantee a homology 3-sphere. Then we show that

Theorem.

$$\lambda(B^n_{p,q;r,s}) = \lambda(B^n_{p,q}) + \frac{p}{q}(n \mp 1)\phi_1(K_{r,s}; S^3)$$

where ϕ_1 is the coefficient of t^2 of the Conway polynomial of $K_{r,s} \subset S^3$.

We will begin by constructing a Dehn surgery presentation for $B^n_{p,q;r,s}$. Recall the following

Theorem [D]. *Let $K_{p,q}$ be the (p,q) torus knot and $n = \pm 1 \bmod pq$. Then the n-fold branched cyclic cover with branch set $K_{p,q}$, which will be denoted by $B^n_{p,q}$, can be constructed by doing $\frac{pq}{n \mp 1}$ Dehn surgery along $K_{p,q} \subset S^3$.*

We will also need a "nice" surgery description of $K_{r,s}$: one where all surgery curves have framing $+1$ and all pairwise linking numbers are zero.

Knots 90

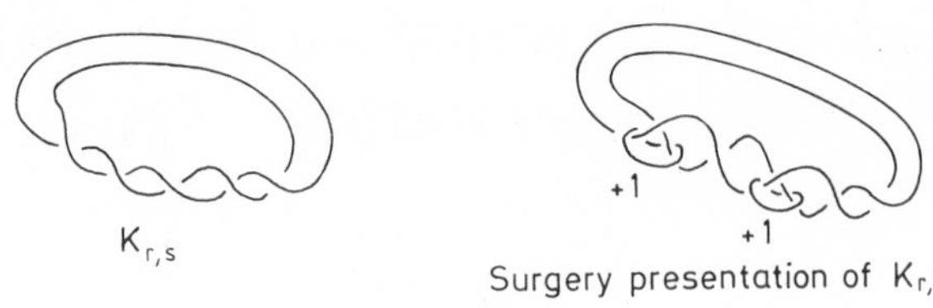

Figure 1

Applying an isotopy to S^3, we can arrange that the resultant framed link is contained in a solid torus which we will henceforth refer to as $S^3 - K_{r,s}$.

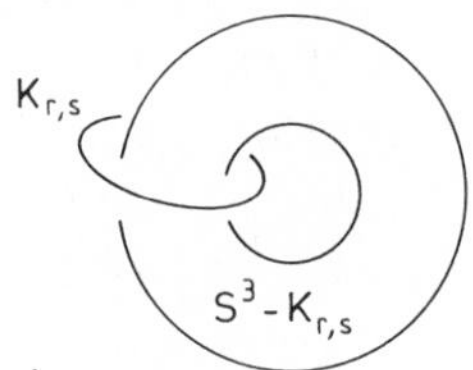

Figure 2

We can now describe $B^n_{p,q;r,s}$ by the following surgery presentation

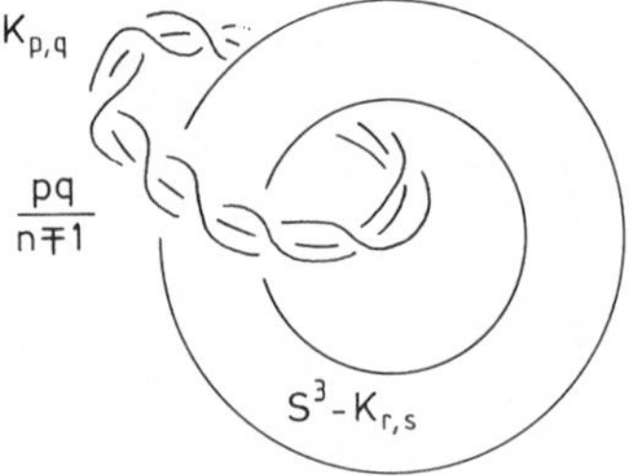

Figure 3

This surgery presentation makes it clear that if $n = \pm 1 \bmod pq$, then $B^n_{p,q;r,s}$ is a homology 3-sphere. Recall the definition of a linking matrix of a surgery presentation as a matrix whose diagonal elements are given by the surgery coefficients and off diagonal elements by the linking numbers. If such a matrix has determinant equal to $\frac{1}{k}$ where k is an integer, then the surgery presentation is that of a homology 3-sphere.

We now recall Hoste's theorem which shows how to compute the Casson invariant from a surgery description given by a framed link. This theorem is a generalization of Casson's original formula which computes the λ invariant of a homology 3-sphere which is presented as the result of Dehn surgery along a single knot.

Theorem [H]. *Let M be a homology sphere, and let L be a framed link in M. Let $\chi(L;M)$ be the manifold resulting from surgery along L in M. Then if all surgery coefficients of L in M are of the form $\frac{1}{q_i}$ where q_i is an integer, and all pairwise linking numbers of L in M are zero, then*

$$\lambda(\chi(L;M)) = \lambda(M) + \sum(\prod q_i)\phi_1(L';M) \tag{1}$$

where the sum is taken over all 1, 2 *and* 3 *component sublinks L' of L, and the product is taken over all components of a sublink L'.*

In order to apply this theorem to the link $S^3 - K_{r,s}$ sitting in the homology sphere $B^n_{p,q}$, we need to check that this link satisfies the conditions of Hoste's theorem. To that end, we call upon the following useful technical lemma, also due to Hoste:

Lemma [H]. *Let $lk(,;M)$ denote linking number in M, where M is described by a framed link $J \subset S^3$. Let K_1 and K_2 be two knots in $S^3 - J$. Let B be the linking matrix of J. Then*

$$lk(K_1, K_2; M) = lk(K_1, K_2; S^3) - lk(K_1, J; S^3) \cdot B^{-1} \cdot lk(K_2, J; S^3)$$

As the link $S^3 - K_{r,s}$ was constructed so that all components linked $K_{r,s}$ in zero, then it is also true that $lk(\cdot, K_{p,q}; S^3) = 0$. It should be noted that this lemma can also be extended from linking numbers to surgery coefficients. Therefore, as all surgery coefficients are $+1$ and all pairwise linking numbers are 0 in S^3, they are also $+1$ and 0 respectively in $B^n_{p,q}$. It follows that

$$\lambda(B^n_{p,q;r,s}) = \lambda(B^n_{p,q}) + \sum \phi_1(L'; B^n_{p,q})$$

Thus in order to prove the formula, it suffices to show that

$$\sum \phi_1(L'; B^n_{p,q}) = \frac{p}{q}(n \mp 1)\phi_1(K_{r,s})$$

We will consider first, the slightly simpler situation below:

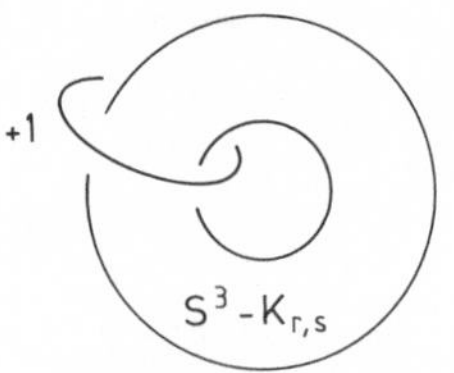

Figure 4

Let $S^3/\{+1\}$ denote the result of doing $+1$ surgery along the unknot above. This is of course S^3 again, but it is important to distinguish between S^3 and $S^3/\{+1\}$ in order to understand how the link $S^3 - K_{r,s}$ sits in $S^3/\{+1\}$.

Lemma 1.

$$\sum \phi_1(L'; S^3/\{+1\}) = \phi_1(K_{r,s})$$

where the sum is taken over all components L' of $S^3 - K_{r,s}$ (rather than over all 1, 2, *and* 3 *component sublinks of $S^3 - K_{r,s}$).*

Proof. Using the following surgery description of B^N_{DK}, the n-fold branched cyclic cover over an untwisted double knot DK, Hoste [H] computes that

$$\lambda(B^n_{DK}) = 2n\phi_1(K)$$

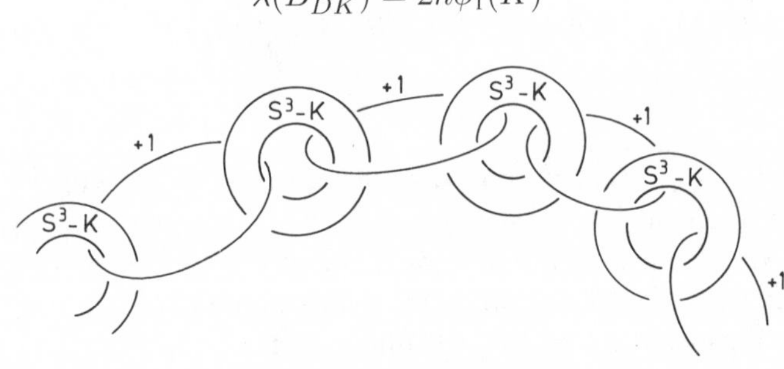

Figure 5

Let $S^3/\{+1\}^n$ denote the result of doing $+1$ surgery along n standardly embedded, geometrically unlinked unknots. Then we can also compute $\lambda(B^n_{DK})$ by applying (1) to the n copies of $S^3 - K$ sitting in $S^3/\{+1\}^n$:

$$\lambda(B^n_{DK}) = \lambda(S^3/\{+1\}^n) + \sum(\prod q_i)\phi_1(L'; S^3/\{+1\}^n)$$

As $S^3/\{+1\}^n$ is homeomorphic to S^3, the first term drops out. Let K be the knot $K_{r,s}$. Picking our surgery description for $K_{r,s}$ as above, we get that

$$\sum \phi_1(L'; S^3/\{+1\}^n) = 2n\phi_1(K)$$

Claim 1.

$$\sum \phi_1(L'; S^3/\{+1\}^n) = n \sum \phi_1(L'; S^3/\{+1\}^2)$$

This means that one can compute $\sum \phi_1(L'; S^3/\{+1\}^n)$ from the following framed link:

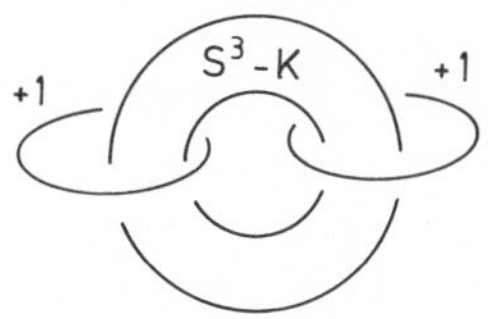

Figure 6

It follows that

$$2\phi_1(K) = \sum \phi_1(L'; S^3/\{+1\}^2)$$

In addition, **Claim 2.**

$$\sum \phi_1(L'; S^3/\{+1\}^2) = 2 \sum \phi_1(L'; S^3/\{+1\})$$

where the sum on the right hand side is taken only over components L' of $S^3 - K$. These two claims prove the lemma.

The following arguments about the calculation of ϕ_1 will be based on the fact that $\phi_1 = 1/2\Delta''(1)$ where $\Delta(t)$ is the Alexander polynomial in "normalized" form: $\Delta(t) = a_0 + a_1(t + t^{-1}) + \cdots + a_k(t^k + t^{-k})$. As the Alexander polynomial can be computed as $det(V - tV^T)$ where V is a Seifert form, we will concern ourselves with the construction of Seifert surfaces.

In order to prove Claims 1 and 2, we will need the following technical result.

Lemma 2. *Let K be a knot. Construct a standard surgery presentation for K by changing the necessary crossings and placing an appropriately framed unknot at each changed crossing in the requisite way - i.e. so that it links K algebraically zero and geometrically twice. Let $L = \cup L_i$ be this collection of unknots. Let L' be an n-component sublink of L where $n =$ 1, 2, or 3. Then L' bounds a Seifert surface in $S^3 - K$ which consists of n twicely punctured disks and n tubes running between the n pairs of punctures.*

Proof. The case $n = 1$ is clear: as L' is an unknot, choose a disk $D^2 \subset S^3$ that it bounds in S^3, puncture it twice where it intersects K and run a single tube from one puncture to the other. We examine next the case where $n = 2$. Consider the following subset of the link $K \cup L'$:

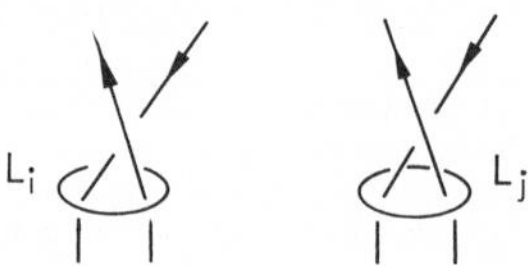

Figure 7

(The four oriented arcs represent a subset of the knot K). There are up to isotopy only two distinct ways that the arcs can be joined together into a knot. We exhibit a choice of punctured disks and tubes for each case.

Figure 8

For the case $n = 3$, we build on our results for $n = 2$, by examining how the two tubes in the above pictures can puncture the disk bounded by the third surgery curve L_k. There are 3 possible cases:
(1) Neither of the tubes punctures the disk bounded by a 3rd surgery curve. (2) Only one of the tubes punctures the disk bounded by a 3rd surgery curve. (3) Both of the tubes puncture the disk bounded by a 3rd surgery curve.
For Case 1, up to isotopy, there are only 4 possibilities, which are shown below. In the second drawing the white tube is nested inside the solid black tube.

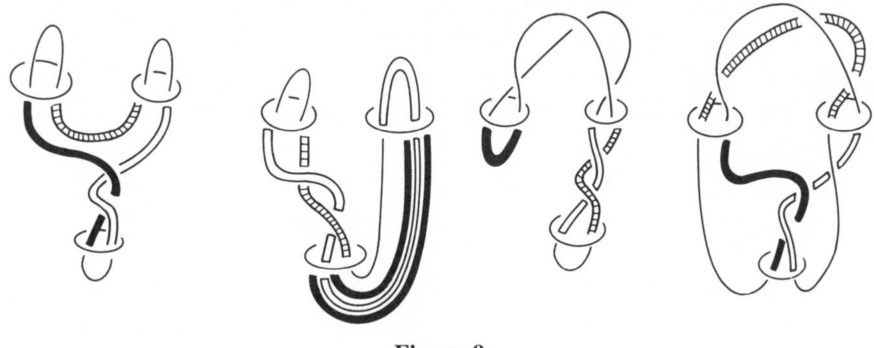

Figure 9

Cases 2 and 3 can be handled similarly. Thus Lemma 2 is proved.

Proof of Claim 1. Recall Hoste's surgery description for B^n_{DK} which is pictured above. Let U_i where $i = 1, 2, \ldots, n$ denote the unknotted components of the link $S^3/\{+1\}^n$. Let L' be a 2 component sublink of $(S^3 - K) \times Z_n$. Then either

(1) each component is in a different copy of $S^3 - K$ i.e. $L' = L_j \cup L_k$ where $L_i \subset (S^3 - K) \times i$ for $i = j, k$ or

(2) both components are in the same copy of $S^3 - K$

If Case 1 holds and either $k > j + 1$ or $k < j - 1$, then construct a Seifert surface for each component according to Lemma 2. By running tubes from one surface to the other, one constructs a Seifert surface with degenerate Seifert form. Hence, this kind of link contributes zero to the sum. If $k = j + 1$ or $j - 1$, then we need to construct a Seifert surface for L' in $S^3 - (U_{j-1} \cup U_j \cup U_{j+1})$. We will first construct a Seifert surface for L_j in $S^3 - (U_{j-1} \cup U_j)$ and one for L_{j+1} in $S^3 - (U_j \cup U_{j+1})$ separately. To this end, we must vary the construction of Lemma 2 slightly: instead of puncturing each disk twice and running one tube between the punctures, we puncture each disk four times and run two tubes between the punctures. The resulting Seifert surfaces - S_j amd S_{j+1} - are both of genus 2.

Figure 10

By cutting along a meridian of the tube of each surface which corresponds to the unknot U_j and glueing the cut ends of S_j to the cut ends of S_{j+1}, one gets a connected Seifert surface S for L' in $S^3 - (U_{j-1} \cup U_j \cup U_{j+1})$ which is twice punctured and of genus 3.

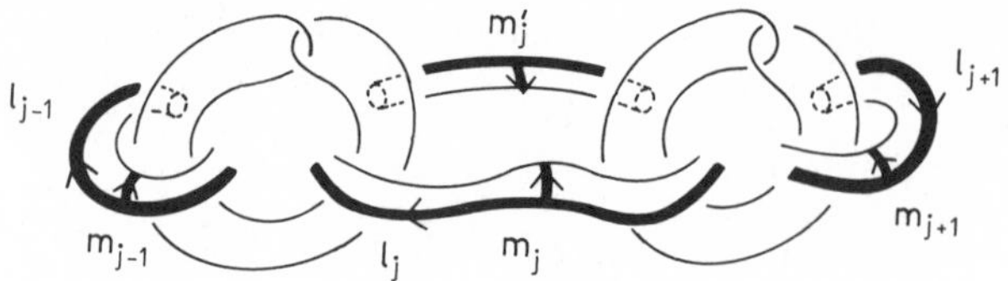

Figure 11

(In order to simplify notation in that which follows, we will suppress the symbols S^3 in the expressions $lk(\, , \, ; S^3)$ and $\phi_1(\, ; S^3)$.) Pick generators for $H_1(S)$ as above. Then as $lk(m_i, U_i) = 1$, we get that

$$lk(m_i, m_i^+; S^3/\{+1\}^3) = -1 \text{ for all } i = j-1, \, j, \, j+1$$
$$lk(m_j', m_j'^+; S^3/\{+1\}^3) = -1$$
$$lk(m_j, m_j'^+; S^3/\{+1\}^3) = -1$$
$$lk(m_j', m_j^+; S^3/\{+1\}^3) = -1$$

Ordering the basis for $H_1(S)$ as m_{j-1}, m_j, m_j', m_{j+1}, l_{j-1}, l_j, and l_{j+1}, we have that a Seifert form for L' is given by

$$\begin{pmatrix} -1 & 0 & 0 & 0 & 0 & 0 & 0 \\ 0 & -1 & -1 & 0 & 0 & 0 & 0 \\ 0 & -1 & -1 & 0 & 0 & 0 & 0 \\ 0 & 0 & 0 & -1 & 0 & 0 & 0 \\ 1 & 0 & 0 & 0 & c_{j-1} & 0 & 0 \\ 0 & 1 & 1 & 0 & 0 & c_j & 0 \\ 0 & 0 & 0 & 1 & 0 & 0 & c_{j+1} \end{pmatrix}$$

where c_i is a constant depending on L', $i = j-1, \, j, \, j+1$. A simple calculation shows that $det(V - tV^T) = 0$ and hence $\phi_1(L'; S^3/\{+1\}^n) = 0$. A similar argument holds if L' is a 3 component link. Thus the only possible contributions will come from those sublinks all of whose components are in the same copy of $S^3 - K$.

Proof of Claim 2. First we will show that all 2 and 3 component sublinks of $S^3 - K$ contribute zero to the sum. Of the Seifert surfaces constructed for a 2 component sublink, there is one case which results in a disconnected surface. Tubing the surfaces together again results in a degenerate Seifert form, and hence $\phi_1 = 0$. For the other case, the Seifert surface is necessarily connected. The resultant twice punctured surface S has genus 1 and hence $H_1(S)$ has three generators: m_1, m_2 , l (see Figure 8). As $lk(m_i, U_{j-1}) = 1$ and $lk(m_i, U_j) = 1$ then

$lk(m_i, m_j^+; S^3/\{+1\}^2) = -2$. Hence the Seifert form looks like

$$V = \begin{pmatrix} -2 & -2 & 0 \\ -2 & -2 & 0 \\ 1 & 1 & c \end{pmatrix}$$

where c is some integer that depends on the particular knot. Again, $det(V - tV^T) = 0$ and so $\phi_1 = 0$.

Of the Seifert surfaces constructed for a 3 component link (see Figure 9), there are 3 cases:

(1) disconnected Seifert surface which has ϕ_1 necessarily 0

(2) connected Seifert surface with three unnested tubes: the generators of $H_1(S)$ are given by the "meridians" of the three tubes and a single "longitude" which intersects each "meridian" once. This has Seifert form

$$V = \begin{pmatrix} -2 & -2 & -2 & 0 \\ -2 & -2 & -2 & 0 \\ -2 & -2 & -2 & 0 \\ 1 & 1 & 1 & c \end{pmatrix}$$

(3) connected Seifert surface with three tubes, two of which are nested. This Seifert surface also has four homology generators: three "meridians" and one "longitude". Its Seifert form is given by

$$V = \begin{pmatrix} -2 & -2 & -2 & 0 \\ -2 & -2 & -2 & 0 \\ -2 & -2 & -2 & 0 \\ 1 & 1 & 0 & c \end{pmatrix}$$

For both cases 2 and 3, $det(V - tV^T) = 0$ and hence $\phi_1 = 0$. Thus we need sum only over each component of $S^3 - K$ rather than over all 1, 2 and 3 component sublinks of $S^3 - K$.

Now we will finish the claim's proof by showing that

$$2\phi_1(L'; S^3/\{+1\}) = \phi_1(L'; S^3/\{+1\}^2)$$

where L' is any single component of the link $S^3 - K$. The punctured torus that L' bounds in $S^3 - (U_{j-1} \cup U_j)$ has Seifert form

$$= \begin{pmatrix} -2 & 0 \\ 1 & c(L') \end{pmatrix}$$

On the other hand, L' bounds a Seifert surface in $S^3 - U_j$ with Seifert form

$$= \begin{pmatrix} -1 & 0 \\ 1 & c(L') \end{pmatrix}$$

Then a simple calculation shows that

$$2\phi_1(L'; S^3/\{+1\}) = \phi_1(L'; S^3/\{+1\}^2).$$

In fact,

$$\phi_1(L'; S^3/\{+1\}) = -c(L')$$

Thus Claim 2 is proved.

Thus we have proved Lemma 1:

$$\phi_1(K) = \sum \phi_1(L'; S^3/\{+1\})$$

or

$$\phi_1(K) = -\sum c(L')$$

Now recall the surgery presentation for $B^n_{p,q;r,s}$ from the beginning of this section. It is easy to establish that no 2 or 3 component sublink of $S^3 - K_{r,s}$ contributes to the sum $\sum \phi_1(L'; B^n_{p,q})$ by a procedure similar to the proof of Claim 2. Thus we need only calculate $\phi_1(L'; B^n_{p,q})$ for each of the individual components of $S^3 - K_{r,s}$. Consider a Seifert surface for L' in $S^3 - K_{p,q}$ which is a punctured torus. Then the Seifert form has two generators m and l. As $lk(m, K_{p,q}) = lk(m^+, K_{p,q}) = p$ and the framing of $K_{p,q}$ is $\frac{pq}{n \mp 1}$, then by Hoste's lemma, we have that

$$\begin{aligned} lk(m, m^+; B^n_{p,q}) &= -p \cdot \left(\frac{n \mp 1}{pq}\right) \cdot p \\ &= -\frac{p}{q}(n \mp 1) \end{aligned}$$

Therefore a Seifert form for L' is given by

$$\begin{pmatrix} -\frac{p}{q}(n \mp 1) & 0 \\ 1 & c(L') \end{pmatrix}$$

which gives that $\phi_1(L'; B^n_{p,q}) = -\frac{p}{q}(n \mp 1)c(L')$. Hence

$$\begin{aligned} \sum \phi_1(L'; B^n_{p,q}) &= -\frac{p}{q} \sum (n \mp 1) \cdot c(L') \\ &= \frac{p}{q}(n \mp 1)\phi_1(K_{r,s}) \end{aligned}$$

which proves the theorem.

References

[A-M] Akbulut, S. and McCarthy, J., *Casson's Invariant for Oriented Homology* 3-*spheres,* Mathematical Notes, Princeton University Press , 1989.

[D] Davidow, Amy, "On Casson's Invariant of Branched Cyclic Covers over S^3", preprint, 1990.

[H] Hoste, Jim, "A Formula for Casson's Invariant", Transactions of the AMS, Vol. 297, No. 2, Oct. 1986, pp. 547 -562.

[K] Kauffman, Louis H., *On Knots,* Annals of Mathematics Studies No. 115, Princeton University Press, NJ, 1987.

[R] Rolfsen, Dale, *Knots and Links,* Publish or Perish, Inc., California, 1976.

Symmetry-spun tori in the four-sphere

Masakazu Teragaito

Introduction

In this note we will study knotted tori in S^4 which are obtained from classical links or knots having cyclic periods by symmetry-spinning. By spinning and twist-spinning a knot in a manner similar to the classical methods of Artin and Zeeman, we get the corresponding knotted tori. It has been studied by several authors [1],[2],[10],[11]. It is easy to construct many examples of tori in S^4 having infinite cyclic knot groups by symmetry-spinning. For example, let K be a 2-bridge knot in S^3. Then K has cyclic period 2. Removing an invariant 3-ball disjoint from K gives a pair (B^3, K). Think of S^4 as $B^3 \times S^1 \cup S^2 \times D^2$. If during the rotation B^3 through the factor S^1 its cyclic period acts on K, then K sweeps out a knotted torus $T(K)$ which is called the *symmetry-spun torus of* K. In this case, $\pi_1(S^4 - T(K))$ is infinite cyclic, so it may be unknotted. There is a conjecture that a surface in S^4 with infinite cyclic knot group is necessarily unknotted, that is, it bounds a handlebody [5]. In fact, in Section 3 we will prove that any symmetry-spun obtained from a periodic link or knot is equivalent to the spun of its factor link or knot. In the case that K is a 2-bridge knot, its factor knot is trivial. Hence $T(K)$ is indeed unknotted.

1. Preliminaries

We will work in the smooth category. All manifolds will be oriented, and all submanifolds are assumed to be locally-flat. The circle is taken to be the quotient space $S^1 = R^1/(\theta \sim \theta + 2\pi$ for all $\theta \in R^1)$. We will write $\theta \in S^1$. $(a, b, \cdots)$ stands for the g.c.d. of the integers $a, b, \cdots$. A *knotted surface* is a pair (S^4, F), where F is a closed oriented (and connected or not) surface in S^4. Two knotted surfaces (S^4, F) and (S^4, G) are *equivalent* if there exists a diffeomorphism $f : (S^4, F) \to (S^4, G)$ preserving the orientations of S^4 and F. Then we use the

Knots 90

notation $(S^4, F) \cong (S^4, G)$. Changing one or both of the orientations, we have three new knotted surfaces $(S^4, -F)$, $-(S^4, -F)$ and $-(S^4, F)$. (Note that we are considering a surface itself, not its embedding map.) A knotted surface (S^4, F) is called (-)*amphicheiral* if $(S^4, F) \cong -(S^4, F)$.

Let U be a standardly embedded torus in S^4 and let $D^2 \times U$ be a tubular neighbourhood of U in S^4. We can assume that its framing is canonical, that is, the homomorphism induced by the inclusion map $H_1(0 \times U; Z) \to H_1(p \times U; Z) \to H_1(S^4 - U; Z)$ where $p \in \partial D^2$, is zero (cf. [9,Lemma 1]). Let $\overline{l} = \partial D^2 \times 0 \times 0, \overline{s} = 0 \times S^1 \times 0, \overline{r} = 0 \times 0 \times S^1$ be curves on $\partial D^2 \times U = \partial D^2 \times S^1 \times S^1$.

Let $E^4 = cl(S^4 - D^2 \times U)$, which is a (trivial) twin (see [6,Lemma 2.1],[12]). Let l, s, r be canonical curves on ∂E^4, which are identified with $\overline{l}, \overline{s}, \overline{r}$, respectively under the natural identification map $i : \partial D^2 \times U \to \partial E^4$. Then l, s, r represent a basis of $H_1(\partial E^4; Z)$.

Let $f : \partial E^4 \to \partial E^4$ be a diffeomorphism with $f_\sharp [l \quad s \quad r] = [l \quad s \quad r] A^f$, where $A^f \in GL(3, Z) \cong \pi_0 Diff(\partial E^4)$. Then f can be extended to a diffeomorphism $\widetilde{f} : E^4 \to E^4$ if and only if $A^f \in H$, where

$$H = \left\{ \begin{bmatrix} \pm 1 & 0 & 0 \\ * & \alpha & \gamma \\ * & \beta & \delta \end{bmatrix} \in GL(3, Z) \middle| \quad \alpha + \beta + \gamma + \delta \equiv 0 \pmod 2 \right\}$$

(see [12,Theorem 5.3]).

Let $\sigma : \partial E^4 \to \partial E^4$ be a diffeomorphism of matrix $\begin{bmatrix} 1 & 0 & 0 \\ 0 & 1 & 1 \\ 0 & 0 & 1 \end{bmatrix}$. Then $E^4 \cup_{\sigma i} D^2 \times S^1 \times S^1$ is diffeomorphic to S^4 (cf. [3],[12]).

2. Symmetry-spun tori

Let K_1 be a knot in $D^2 \times S^1$, which may be geometrically inessential. (We should exclude cases where K_1 bounds a disk in $D^2 \times S^1$ and K_1 is ambient isotopic to the core.) Let $p_a : D^2 \times S^1 \to D^2 \times S^1$ be the a-fold cyclic cover given by $(x, \theta) \to (x, a\theta)$ for $a \in Z \backslash \{0\}$. Let $r_\phi : D^2 \times S^1 \to D^2 \times S^1$ be the rotation map given by $(x, \theta) \to (x, \theta + \phi)$ for $\phi \in S^1$. Let $K_a = p_a^{-1}(K_1) \subset D^2 \times S^1$, which may be a link, and K_a is given the orientation induced by K_1 if $a > 0$, and is given the opposite orientation if $a < 0$. Then K_a is invariant under the rotation $r_{2\pi/a}$. Note that the pairs $(D^2 \times S^1, K_a)$ and $(D^2 \times S^1, K_{-a})$ are diffeomorphic by a diffeomorphism $(x, \theta) \to (x, -\theta)$, which is orientation-reversing both on $D^2 \times S^1$ and K_a.

Lemma 1. *Let $\rho : D^2 \times S^1 \to D^2$ be the projection map. Then $\rho(K_a \cap (D^2 \times \theta)) = \rho(K_b \cap (D^2 \times a\theta/b))$ for $a, b \in Z \backslash \{0\}, \theta \in S^1$.*

Proof. Since the covering $p_a : D^2 \times S^1 \to D^2 \times S^1$ is induced by a covering $S^1 \to S^1$ given by $\theta \to a\theta, \rho(K_a \cap (D^2 \times \theta)) = \rho(K_1 \cap (D^2 \times a\theta)) = \rho(K_b \cap (D^2 \times a\theta/b))$.

Let $i_\theta : D^2 \times S^1 \to D^2 \times S^1 \times \theta \subset D^2 \times S^1 \times S^1$ be the inclusion map $(x, \phi) \to (x, \phi, \theta)$.

Definition. Let $a, b \in Z$ with $b \neq 0$. We define a surface $T^a(K_b)$ in $D^2 \times S^1 \times S^1$, which satisfies

$$T^a(K_b) \cap (D^2 \times S^1 \times \theta) = i_\theta r_{a\theta/b}(K_b).$$

Then we get two knotted surfaces in S^4, called the *symmetry-spun tori of* K_b, identifying ∂E^4 and $\partial D^2 \times S^1 \times S^1$ using the natural identification i and the twisted identification σi (see Section 1). We denote $(S^4, T^a(K_b)), (S^4, \widetilde{T}^a(K_b))$, respectively.

Each connected component of $T^a(K_b)$ is a torus. It is easy to see that K_b has (b, w) components and $T^a(K_b)$ has (a, b, w) components, where w is the winding number of K_1 in $D^2 \times S^1$. Note that $T^0(K_b)$ is just a *spun torus of* K_b (cf. [2],[10]).

Lemma 2. *If* $a \equiv b \pmod{2c}$, *then* $(S^4, T^a(K_c)) \cong (S^4, T^b(K_c)), (S^4, \widetilde{T}^a(K_c)) \cong (S^4, \widetilde{T}^b(K_c))$.

Proof. Define $f : D^2 \times S^1 \times S^1 \to D^2 \times S^1 \times S^1$ by $(x, \phi, \theta) \to (x, \phi + 2\theta, \theta)$. Then

$$f(T^a(K_c)) \cap (D^2 \times S^1 \times \theta) = f i_\theta r_{a\theta/c}(K_c) = i_\theta r_{(a+2c)\theta/c}(K_c),$$

since $f i_\theta = i_\theta r_{2\theta}$ on $D^2 \times S^1$. The diffeomorphism $\tau = i(f|_{\partial D^2 \times S^1 \times S^1}) i^{-1} : \partial E^4 \to \partial E^4$ has a matrix $\begin{bmatrix} 1 & 0 & 0 \\ 0 & 1 & 2 \\ 0 & 0 & 1 \end{bmatrix}$, so τ extends to E^4 (see Section 1). Hence $(S^4, T^a(K_c)) \cong (S^4, T^{a+2c}(K_c))$, which gives the first equivalence. Since $\tau\sigma = \sigma\tau$ on ∂E^4, the second follows.

Lemma 3. $(S^4, T^a(K_b)) \cong (S^4, \widetilde{T}^{a \pm b}(K_b))$.

Proof. . Define $f : D^2 \times S^1 \times S^1 \to D^2 \times S^1 \times S^1$ by $(x, \phi, \theta) \to (x, \phi - \theta, \theta)$. Then

$$f(T^a(K_b)) \cap (D^2 \times S^1 \times \theta) = f i_\theta r_{a\theta/b}(K_b) = i_\theta r_{(a-b)\theta/b}(K_b).$$

The diffeomorphism $\tau = i(f|_{\partial D^2 \times S^1 \times S^1}) i^{-1} : \partial E^4 \to \partial E^4$ has a matrix $\begin{bmatrix} 1 & 0 & 0 \\ 0 & 1 & -1 \\ 0 & 0 & 1 \end{bmatrix}$, so $\sigma^{-1} = \tau$. From this the result follows.

Lemma 4.
(1) $(S^4, T^a(-K_b)) \cong (S^4, -T^a(K_b))$, *where* $-K_b$ *is obtained from* K_b *by reversing its orientation.*
(2) $(S^4, T^{-a}(K_b)) \cong -(S^4, T^a(K_b))$.
(3) $(S^4, T^a(K_{-b})) \cong -(S^4, T^a(K_b))$.
And also, the corresponding equivalences on the twisted cases hold.

Proof. It is easy to see (1). For (2) define $f : D^2 \times S^1 \times S^1 \to D^2 \times S^1 \times S^1$ by $(x, \phi, \theta) \to (x, \phi, -\theta)$. Then

$$f(T^{-a}(K_b)) \cap (D^2 \times S^1 \times \theta) = f i_{-\theta} r_{(-a)(-\theta)/b}(K_b) = i_\theta r_{a\theta/b}(K_b),$$

and f induces an orientation-reversing diffeomorphism on $T^{-a}(K_b)$. The diffeomorphism $\tau = i(f|_{\partial D^2 \times S^1 \times S^1})i^{-1} : \partial E^4 \to \partial E^4$ has a matrix $\begin{bmatrix} 1 & 0 & 0 \\ 0 & 1 & 0 \\ 0 & 0 & -1 \end{bmatrix}$, so τ extends to E^4. Hence $(S^4, T^{-a}(K_b)) \cong -(S^4, T^a(K_b))$.

For (3) define $g : D^2 \times S^1 \to D^2 \times S^1$ by $(x, \phi) \to (x, -\phi)$, and let $h = g \times id : D^2 \times S^1 \times S^1 \to D^2 \times S^1 \times S^1$. Then

$$\begin{aligned} h(T^a(K_{-b})) \cap (D^2 \times S^1 \times \theta) &= h i_\theta r_{a\theta/(-b)}(K_{-b}) = i_\theta g r_{a\theta/(-b)}(K_{-b}) \\ &= i_\theta r_{a\theta/b} g(K_{-b}) = i_\theta r_{a\theta/b}(-K_b). \end{aligned}$$

Since $i(h|_{\partial D^2 \times S^1 \times S^1})i^{-1}$ extends to E^4, we have the equivalence

$$(S^4, T^a(K_{-b})) \cong (-S^4, T^a(-K_b)) \cong -(S^4, T^a(K_b)).$$

For the twisted cases, by Lemma 3 and the above equivalences

$$(S^4, \widetilde{T}^{-a}(K_b)) \cong (S^4, T^{-a+b}(K_b)) \cong -(S^4, T^{a-b}(K_b)) \cong -(S^4, \widetilde{T}^a(K_b)),$$

$$(S^4, \widetilde{T}^a(K_{-b})) \cong (S^4, T^{a-b}(K_{-b})) \cong -(S^4, T^{a-b}(K_b)) \cong -(S^4, \widetilde{T}^a(K_b)).$$

Corollary 5. *Spun tori* $(S^4, T^0(K_b)), (S^4, \widetilde{T}^0(K_b))$ *are (-)amphicheiral.*

It is known that every ribbon 2-knot is (-)amphicheiral and in particular every spun 2-knot is so (cf.[4],[13]). In the case $a = 0$, $T^0(K_b)$ is obtained by spinning K_b, so it has a symmetric normal form. Hence $T^0(K_b)$ is a ribbon surface (cf.[8]).

Corollary 6. *Let* $a \equiv 0 \pmod b$. *Then*

$$(S^4, T^a(K_b)) \cong \begin{cases} (S^4, T^0(K_b)) & \text{if } a/b \equiv 0 \pmod 2, \\ (S^4, \widetilde{T}^0(K_b)) & \text{if } a/b \equiv 1 \pmod 2. \end{cases}$$

Proof. This is an immediate consequence of Lemmas 2 and 3.

Remark. It is clear that $\pi_1(S^4 - T^0(K_b)) \cong \pi_1(S^4 - \widetilde{T}^0(K_b))$. If K is a nontrivial knot or a link with no separated trivial component, then the exteriors of spun tori $(S^4, T^0(K))$ and $(S^4, \widetilde{T}^0(K))$ have different diffeomorphism types. This can be proved using the Z_2−intersection number (see [7,Lemma 2.8] or [10]).

Lemma 7. $(S^4, T^a(K_b)) \cong (S^4, T^b(K_{-a}))$ *for* $a, b \in Z\backslash\{0\}$.

Proof. Define an orientation-preserving diffeomorphism $f : D^2 \times S^1 \times S^1 \to D^2 \times S^1 \times S^1$ by $(x, \phi, \theta) \to (x, \theta, -\phi)$. By Lemma 1,

$$\begin{aligned}
T^a(K_b) \cap (D^2 \times (-\phi) \times S^1) &= \bigcup_{\theta \in S^1} [T^a(K_b) \cap D^2 \times (-\phi) \times \theta] \\
&= \bigcup_{\theta \in S^1} i_\theta|_{D^2 \times (-\phi)} [r_{a\theta/b}(K_b) \cap D^2 \times (-\phi)] \\
&= \bigcup_{\theta \in S^1} i_\theta|_{D^2 \times (-\phi)} \Big(\rho(K_b \cap D^2 \times (-\phi - a\theta/b)), -\phi\Big) \\
&= \bigcup_{\theta \in S^1} i_\theta|_{D^2 \times (-\phi)} \Big(\rho(K_{-a} \cap D^2 \times (b\phi/a + \theta)), -\phi\Big) \\
&= \bigcup_{\theta \in S^1} i_\theta|_{D^2 \times (-\phi)} \Big(\rho(r_{b\phi/(-a)}(K_{-a}) \cap D^2 \times \theta), -\phi\Big).
\end{aligned}$$

Hence

$$\begin{aligned}
f(T^a(K_b)) \cap (D^2 \times S^1 \times \phi) &= i_\phi \Big[\bigcup_{\theta \in S^1} \Big(\rho(r_{b\phi/(-a)}(K_{-a}) \cap D^2 \times \theta), \theta\Big)\Big] \\
&= i_\phi r_{b\phi/(-a)}(K_{-a}),
\end{aligned}$$

and f induces an orientation-preserving diffeomorphism on $T^a(K_b)$. The diffeomorphism $\tau = i(f|_{\partial D^2 \times S^1 \times S^1})i^{-1}$ has a matrix $\begin{bmatrix} 1 & 0 & 0 \\ 0 & 0 & 1 \\ 0 & -1 & 0 \end{bmatrix}$, so τ extends to E^4. Therefore $(S^4, T^a(K_b)) \cong (S^4, T^b(K_{-a}))$.

Thus we have $(S^4, T^a(K_b)) \cong -(S^4, T^b(K_a))$ by Lemmas 3 and 7. Note that a similar equivalence on the twisted case does not hold, because the above equivalence permutes the cores of the twin E^4 (see [12]).

Take a symmetric Wirtinger presentation of a periodic link or knot K_b. By van Kampen Theorem we can see that $\pi_1(S^4 - T^a(K_b)) \cong \pi_1(S^4 - T^0(K_d)) \cong \pi_1(S^3 - K_d)$, where $d = (a, b)$.

Theorem 8. *For $a, b \in Z\backslash\{0\}$, let $a = 2^p a', b = 2^q b'$ with $p, q \geq 0$ and a', b' odd. Then*

$$(S^4, T^a(K_b)) \cong \begin{cases} (S^4, T^0(K_d)) & \text{if} \quad p \neq q, \\ (S^4, \widetilde{T}^0(K_d)) & \text{if} \quad p = q, \end{cases}$$

where $d = (a, b) > 0$.

We will give a proof of Theorem 8 in the next section.

Corollary 9. *Let K_1 be a trivial knot. Let $a, b \in Z$ with $b \neq 0$. If $(a, b) = 1$, then $(S^4, T^a(K_b))$ and $(S^4, \widetilde{T}^a(K_b))$ are unknotted, that is, these bound solid tori in S^4.*

Example. Let K_b be the pretzel link (or knot) $p(n, \cdots b \cdots, n)$ in S^3. It is clear that K_b has cyclic period b. Let $g : (S^3, K_b) \to (S^3, K_b)$ be the diffeomorphism of period b. Let z be a point on $Fix(g)$ and B_0 an invariant regular neighbourhood of z disjoint from K_b. Removing $Int B_0$ from S^3 gives the pair (B^3, K_b). We may assume that $g|_{\partial B^3} = id$. Choose an integer a with $(a, b) = 1$. Then

$$(S^4, T^a(K_b)) \cong \partial B^3 \times D^2 \cup_\partial (B^3, K_b) \times_{g^a} \partial D^2,$$
$$(S^4, \widetilde{T}^a(K_b)) \cong \partial B^3 \times D^2 \cup_\tau (B^3, K_b) \times_{g^a} \partial D^2,$$

where

$$(B^3, K_b) \times_{g^a} \partial D^2 = (B^3, K_b) \times [0, 1] \Big/ \Big((x, 0) \sim (g^a(x), 1) \quad \text{for all } x \in B^3\Big),$$

and $\tau : S^2 \times S^1 \to S^2 \times S^1$ represents the nontrivial element of $\pi_1(SO(3))$ (cf.[3]). By Corollary 9 symmetry-spun tori $(S^4, T^a(K_b))$ and $(S^4, \widetilde{T}^a(K_b))$ are unknotted.

3. Proof of Theorem 8

In this section, we prove Theorem 8. To do this, we use the generalized Euclidean algorithm.

For $a, b \in Z\backslash\{0\}$ with $|b| \geq 2$, let $x_{-1} = a, x_0 = b$. Then

$$x_{-1} = x_0 y_1 + x_1, \qquad \text{where } y_1 \text{ even and } 2 \leq |x_1| \leq |x_0|.$$

If $|x_1| < |x_0|$, then

$$x_0 = x_1 y_2 + x_2, \qquad \text{where } y_2 \text{ even and } 2 \leq |x_2| \leq |x_1|.$$

If $|x_2| < |x_1|$, then proceed to the next step. Repeating this, we have that there exists an integer k such that

$$x_{k-1} = x_k y_{k+1} + x_{k+1}, \qquad \text{where} \quad |x_i| \geq 2 \quad (0 \leq i \leq k),$$

and one of the following cases holds,

(i-1) $x_{k+1} = 0$, y_{k+1} is even,
(i-2) $x_{k+1} = 0$, y_{k+1} is odd,
(ii-1) $|x_{k+1}| = 1, y_{k+1}$ is even, x_k is even,
(ii-2) $|x_{k+1}| = 1, y_{k+1}$ is even, x_k is odd.

As in the (usual) Euclidean algorithm, we can prove that $(a,b) = |x_k| \geq 2$ (Cases i-1,i-2), $(a,b) = |x_{k+1}|$ (Cases ii-1,ii-2). In particular, if $(a,b) = 1$ then neither (i-1) nor (i-2) occurs.

Lemma 10. *Let $a, b \in Z\backslash\{0\}$ with $|b| \geq 2$. Let $a = 2^p a', b = 2^q b'$ with $p, q \geq 0$ and a', b' odd. In the generalized Euclidean algorithm, we have the following.*
(i) Let $(a,b) \neq 1$. If $p \neq q$, then (i-1) occurs, otherwise (i-2).
(ii) Let $(a,b) = 1$. If $p \neq q$, then (ii-1) occurs, otherwise (ii-2).

Proof. Suppose that $(a,b) = 1$. Then as stated before, either (ii-1) or (ii-2) occurs. If x_k is even, then either a or b is even, so $p \neq q$. Assume that x_k is odd. Since x_{k-1} is odd, both a and b are odd, so $p = q = 0$. Thus in the case $(a,b) = 1$, if $p \neq q$ then (ii-1) occurs, otherwise (ii-2).

Next suppose that $d = (a,b) \geq 2$. Let $a = a''d, b = b''d$. If $|b''| = 1$, then $a = \pm ba''$. Hence if a'' is even, then (i-1) occurs, otherwise (i-2). So if $p \neq q$ then (i-1) occurs, otherwise (i-2). If $|b''| \geq 2$, then we apply the generalized Euclidean algorithm to the pair $\{a'', b''\}$. Since $(a'', b'') = 1$, either (ii-1) or (ii-2) occurs. Multiplying the equations by d, we have

$$a = by_1 + dx_1, \cdots, dx_{k-1} = dx_k y_{k+1} \pm d.$$

Since $d \geq 2$, we must proceed to the next step $dx_k = (\pm d)(\mp x_k)$. Hence if x_k is even, then (i-1) occurs, otherwise (i-2). But if x_k is even, then either a'' or b'' is even, so $p \neq q$. If x_k is odd, then both a'' and b'' are odd, so $p = q$. Thus in the case $(a,b) \geq 2$, if $p \neq q$ then (i-1) occurs, otherwise (i-2). The proof is complete.

Proof of Theorem 8. If $|b| = 1$, then the result is a consequence of Corollary 6. Suppose that $|b| \geq 2$. We apply the generalized Euclidean algorithm. Then by Lemma 2,

$$(S^4, T^{x_{-1}}(K_{x_0})) \cong (S^4, T^{x_1}(K_{x_0})),$$

and by Lemmas 3 and 7,

$$(S^4, T^{x_1}(K_{x_0})) \cong -(S^4, T^{x_0}(K_{x_1})).$$

Hence

$$(S^4, T^{x_{-1}}(K_{x_0})) \cong -(S^4, T^{x_0}(K_{x_1})).$$

Repeating this, we have

$$(S^4, T^{x_{-1}}(K_{x_0})) \cong \pm(S^4, T^{x_{k-1}}(K_{x_k})).$$

In each case, we have the following;

(i-1) $(S^4, T^{x_{k-1}}(K_{x_k})) \cong (S^4, T^{x_{k+1}}(K_{x_k})) \cong \pm(S^4, T^0(K_d))$.

(i-2) $(S^4, T^{x_{k-1}}(K_{x_k})) \cong (S^4, T^{x_k}(K_{x_k})) \cong \pm(S^4, \widetilde{T}^0(K_d))$.

(ii-1) $(S^4, T^{x_{k-1}}(K_{x_k})) \cong (S^4, T^{x_{k+1}}(K_{x_k})) \cong \pm(S^4, T^1(K_{x_k}))$
$\cong \pm(S^4, T^{x_k}(K_1)) \cong \pm(S^4, T^0(K_1))$.

(ii-2) $(S^4, T^{x_{k-1}}(K_{x_k})) \cong \pm(S^4, T^{x_k}(K_1)) \cong \pm(S^4, \widetilde{T}^0(K_1))$.

Thus $(S^4, T^a(K_b)) \cong \pm(S^4, T^0(K_d))$ or $\pm(S^4, \widetilde{T}^0(K_d))$. By Corollary 5,

$$(S^4, T^a(K_b)) \cong (S^4, T^0(K_d)) \text{ or } (S^4, \widetilde{T}^0(K_d)).$$

The result follows from Lemma 10.

Acknowledgement. I would like to thank the members of KOOK Seminar for several helpful discussions.

References

[1] K. Asano: *A note on surfaces in 4-spheres*, Math.Sem.Notes Kobe Univ., **4** (1976), 195-198.

[2] J. Boyle: *Classifying 1-handles attached to knotted surfaces*, Trans.Amer.Math.Soc., **306** (1988), 475-487.

[3] H. Gluck: *The embedding of two-spheres in the four-sphere*, Trans.Amer.Math.Soc., **104** (1962), 308-333.

[4] C. McA. Gordon: *A note on spun knots*, Proc.Amer.Math.Soc., **58** (1976), 361-362.

[5] F. Hosokawa and A. Kawauchi: *Proposals for unknotted surfaces in four-spaces*, Osaka J.Math., **16** (1979), 233-248.

[6] Z. Iwase: *Good torus fibrations with twin singular fibers*, Japan J.Math., **10** (2) (1984), 321-352.

[7] Z. Iwase: *Dehn-surgery along a torus T^2-knot*, Pacific J.Math., **133** (2) (1988), 289-299.

[8] A. Kawauchi, T.Shibuya and S. Suzuki: *Descriptions on surfaces in four-space, I. Normal forms*, Math.Sem.Notes Kobe Univ., **10** (1982), 75-125; ibid, *II. Singularities and cross-sectional links*, ibid. **11** (1983), 31-69.

[9] R.A. Litherland: *The second homology of the group of a knotted surface*, Quart.J.Math. Oxford (2), **32** (1981), 425-434.

[10] C. Livingston: *Stably irreducible surfaces in S^4*, Pacific J.Math., **116** (1) (1985), 77-84.

[11] C. Livingston: *Indecomposable surfaces in 4-space*, Pacific J.Math., **132** (2) (1988), 371-378.
[12] J.M. Montesinos: *On twins in the four-sphere I*, Quart.J.Math.Oxford (2), **34** (1983), 171-199.
[13] S. Suzuki: *Knotting problems of 2-spheres in 4-sphere*, Math.Sem.Note Kobe Univ., **4** (1976), 241-371.

From a view of localized link theory

Yasutaka Nakanishi

Dedicated to Professor Fujitsugu Hosokawa on his 60th birthday

One of central problems in the knot theory is to find invariants of knots and links, and characterize them geometrically. Finding an invariant is identical to extracting a structure from complicated structures of knots and links. To study structures of knots and links through local moves is a motivation of localized link theory. In other words, the localized link theory has the following two sides. (1) For a given structure, which kind of local move is corresponding to? (2) For a given local move, which kind of structure is corresponding to? For example, we consider the number of components of knots and links. It is a very poor structure of knots and links, however. We can say that this structure is corresponding to an unknotting operation. Two links have the same number of components if and only if the two links can be deformed into each other by a finite sequence of unknotting operations. There are many local moves corresponding to the structure of numbers of components, but an unknotting operation can be regarded to be the simplest move. Next, we take a Δ-unknotting operation [MN] as a local move. Then, the structure of link-homology is corresponding to it [MN], which is a little rich structure of knots and links.

In this note, we consider the Fox congruence modulo n, q [F2] which is an equivalence relation generated by $\pm 1/n$-surgery along a trivial knot such that the linking number of the trivial knot and the link is a multiple of q. It is clear that the number of components is an invariant for congruence. Congruence modulo $0, q$ is just the link equivalence. And two links are congruent modulo $1, q$ if the numbers of components are equal. It can be seen that there are infinitely many congruence classes modulo n, q of μ-component links for each μ, if $n \geq 2$ and $(n, q) \neq (2,1), (2,2)$. For the remaining cases $(n, q) = (2,1), (2,2)$, there are a finite number of congruence classes modulo n, q of μ-component links for each μ. Especially, there is exactly one congruence class of knots modulo $2, 1$ and $2, 2$.

Knots 90

1. Definitions and theorems

In this note we only consider a μ-component link $L = K_1 \cup \cdots \cup K_\mu$, that is an ordered collection of μ disjoint simple closed oriented curves K_i's in a three dimensional oriented sphere S^3. Two links are said to be *equivalent*, if there is an orientation preserving homeomorphism of S^3 onto itself, which maps one link onto the other preserving the orientation and order of components. And such an equivalence class of links is called a *link type*. A μ-component link $L = K_1 \cup \cdots \cup K_\mu$ is called *trivial*, if there exist μ disjoint disks $D_1 \cup \cdots \cup D_\mu$ in S^3 with $\partial D_i = K_i (i = 1, \ldots, \mu)$. Especially, we call a 1-component link a *knot*, a 1-component link type a *knot type*, and a 1-component trivial link a *trivial knot*.

R. H. Fox introduced the notion of congruence classes of knots in [F2], which can be generalized for links as follows ([N]).

Definition. Let n and q be non-negative integers. Two μ-component link types κ and λ are said to be *congruent modulo* n, q, if there are μ-component links $L_0, L_1, L_2, \ldots, L_l$, integers $c_1, c_2, \ldots, c_l$, and trivial knots $m_1, m_2, \ldots, m_l$ such that

(1) L_{i-1} and m_i are disjoint,

(2) L_i is obtained from L_{i-1} by $1/c_i n$-surgery along m_i (see [R] for a/b-surgery),

(3) the sum of the linking numbers $\sum_{j=1}^{\mu} lk(K_{i-1,j}, m_i) \equiv 0 \pmod q$ where $L_{i-1} = K_{i-1,1} \cup \cdots \cup K_{i-1,\mu}$, and

(4) L_0 represents κ, and L_l represents λ.

We remark that if the numbers of components are distinct, then the two link types are incongruent modulo n, q.

The congruence modulo $0, q$ is just the link equivalence.

Any two μ-component link types are congruent modulo $1, q$. In other words, the congruence modulo $1, q$ is corresponding to the structure of numbers of components.

From an Alexander matrix $A_\kappa(t_1, t_2, \ldots, t_\mu)$ of a μ-component link type κ in the sense of Fox [F1], we obtain a reduced Alexander matrix $\tilde{A}_\kappa(t)$ by rewriting t_i's $(i = 1, 2, \ldots, \mu)$ in entries of the matrix to the same t. Similarly, we obtain the reduced Alexander polynomial $\tilde{\Delta}_\kappa(t)$ and the reduced elementary ideals. In the following, $\sigma_n(t)$ means $(1-t^n)/(1-t) = 1+t+t^2+\cdots+t^{n-1}$, and $f(t) \equiv g(t)$ mod $\{h_1(t), h_2(t), \ldots, h_j(t)\}$ means that $f(t)$ and $g(t)$ are in the same class of the quotient $Z\langle t\rangle/(h_1(t), h_2(t), \ldots, h_j(t))$, where $(h_1(t), h_2(t), \ldots, h_j(t))$ is the ideal generated by $h_1(t), h_2(t), \ldots, h_j(t)$ in $Z\langle t\rangle$.

Theorem 1 ([N], cf. [NS]). *If two link types κ and λ are congruent modulo n, q, then, for properly chosen $\tilde{A}_\kappa(t)$ and $\tilde{A}_\lambda(t)$, we have*

$$\tilde{A}_\kappa(t) \equiv \tilde{A}_\lambda(t) \bmod \left\{ \begin{array}{r} n(1-t) = (1-t)\sigma_n(t^{0\times q}), (1-t)\sigma_n(t^{1\times q}), \\ (1-t)\sigma_n(t^{i_1\times q}), \dots, (1-t)\sigma_n(t^{i_*\times q}) \end{array} \right\},$$

and hence,

$$\tilde{\Delta}_\kappa(t) \equiv \pm t^r \tilde{\Delta}_\lambda(t) \bmod \left\{ \begin{array}{r} n(1-t) = (1-t)\sigma_n(t^{0\times q}), (1-t)\sigma_n(t^{1\times q}), \\ (1-t)\sigma_n(t^{i_1\times q}), \dots, (1-t)\sigma_n(t^{i_*\times q}) \end{array} \right\},$$

where $i_1, \dots, i_*$ are all divisors of n and $1 < i_1 < \dots < i_* < n$. Furthermore, we have similar statements for the reduced elementary ideals of deficiency greater than 1.

Applying Theorem 1, we can find infinitely many link types that are incongruent modulo n, q.

Theorem 2 ([N]). *Let n be an integer greater than 1 and q a non-negative integer such that $(n, q) \neq (2, 1), (2, 2)$. For congruence modulo n, q, there exist infinitely many distinct classes of μ-component link types for each μ.*

To consider the case $(n, q) = (2, 1)$, we use the following notion. Two links $L = K_1 \cup \dots \cup K_\mu$ and $L' = K'_1 \cup \dots \cup K'_\nu$ are said to be Z_2- *link-homologous* if $\mu = \nu$ and $lk(K_i, K_j) \equiv lk(K'_i, K'_j) \pmod 2$ for every $1 \leq i < j \leq \mu$.

Theorem 3 ([N]). *Two given link types are congruent modulo $2, 1$ if and only if two links representing the link types are Z_2-link-homologous.*

Corollary ([N]). *All knots types are congruent modulo $2, 1$.*

At the present time, the author does not know the precise statement for congruence modulo $2, 2$. He knows the finiteness of congruence classes modulo $2, 2$ as follows.

Proposition 4. *All knot types are congruent modulo $2, 2$.*

Proposition 5. *For 2-component link types, two given link types are congruent modulo $2, 2$ if and only if the linking numbers of the link types have the same parity.*

There is no difference between congruence $2, 1$ and congruence $2, 2$ for knot types (and 2-component link types). But we can see the difference for 3-component link types as follows.

Proposition 6 ([N]). *The Borromean rings and a 3-component trivial link are congruent modulo* $2, 1$, *but incongruent modulo* $2, 2$.

Proposition 7. *Let* μ *be an integer greater than or equal to three, and* C_μ *the number of congruence classes modulo* $2, 2$ *for* μ*-component link types. Then, we have* $2^{\mu(\mu-1)/2} < C_\mu < 2^{\mu(\mu-1)/2} \times 4^{2^{\mu-1}-\mu}$.

The proofs of Propositions 4, 5, and 7 will be given in the next section. The estimate in Proposition 7 is very rough, however. For 3-component link types with all linking numbers even, there are at most three congruence classes modulo $2, 2$: the 3-component trivial link type, the link type of Borromean rings, and the partially double of the Whitehead link type as in Fig. 1. If the partially double of the Whitehead link type is incongruent to the 3-component trivial link type modulo $2, 2$, then these examples are a counter-example to the conjectures as mentioned in [N].

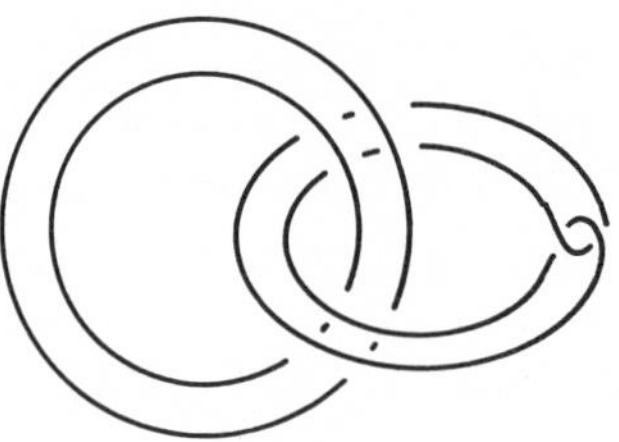

Figure 1

2. Proofs

In order to prove Propositions, it is necessary to recall the proofs in [MN] and [N]. Two links $L = K_1 \cup \cdots \cup K_\mu$ and $L' = K'_1 \cup \cdots \cup K'_\nu$ are said to be *link-homologous* if $\mu = \nu$ and $lk(K_i, K_j) = lk(K'_i, K'_j)$ for every $1 \leq i < j \leq \mu$. In [MN], we proved that a finite sequence of Δ-unknotting operations as in Fig. 2 implies an exchange of a self-crossing of a component. And two links L and L' are link-homologous if and only if L can be deformed into L' by a finite sequence of Δ-unknotting operations.

In [N], we proved that a Δ-unknotting operation can be realized by $\pm 1/2$-surgeries along trivial knots, i.e. Δ-unknotting operations never change congruence classes modulo $2, 1$. By $\pm 1/2$-surgery along a trivial knot which has the linking number 1 with two components of a link, we can change the linking number of the two components by ± 2, i.e. if two links are Z_2-link-homologous, then we can assume that the two links are link-homologous after suitable $\pm 1/2$-surgeries along

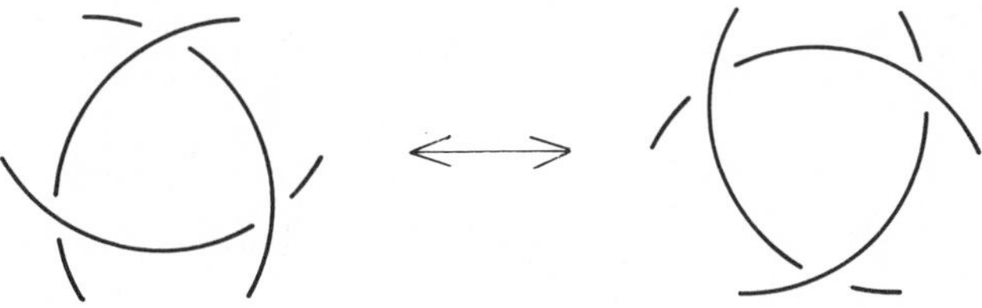

Figure 2

trivial knots. And two given link types are congruent modulo $2, 1$ if and only if two links representing the link types are Z_2-link-homologous.

We make the parallel version of these arguments in the above to Propositions 4, 5, and 7.

Definition. A $D(\Delta)$-operation is a local move on a link diagram as in Fig. 3. If a diagram of a link L' is a result of a $D(\Delta)$-operation on a diagram of L, then we say that L' is obtained from L by a $D(\Delta)$-operation.

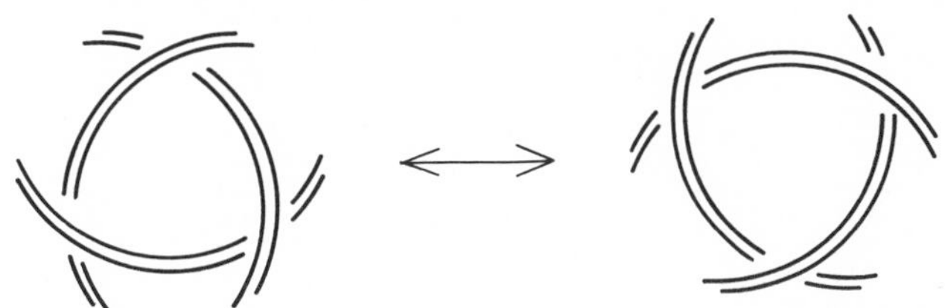

Figure 3

Lemma. *A $D(\Delta)$-operation can be realized by $\pm 1/2$-surgeries along trivial knots such that the linking numbers of the trivial knots and a link are even, i.e. $D(\Delta)$-operations never change congruence classes modulo $2, 2$.*

Proof. Watch Fig. 4. Take three trivial knots m_1, m_2, and m_3 as in Fig. 4 (a). By $-1/2$-surgery along m_1, $+1/2$-surgery along m_2, and $+1/2$-surgery along m_3, we obtain (b) from (a). Take four trivial knots m_4, m_5, m_6, and m_7 as in Fig. 4 (c). By $+1/2$-surgery along m_4, $-1/2$-surgery along m_5, $-1/2$-surgery along m_6, and $-1/2$-surgery along m_7, we obtain (d) from (c). The deformation from (a) to (d) is just a $D(\Delta)$-operation, completing the proof.

We consider a standard form of non-orientable surfaces bounded by a given μ-component link, which consists of a big disk, several non-orientable bands, and $\mu - 1$ orientable bands like as in Fig. 5. And we use the notation $u_1, \ldots, u_n$

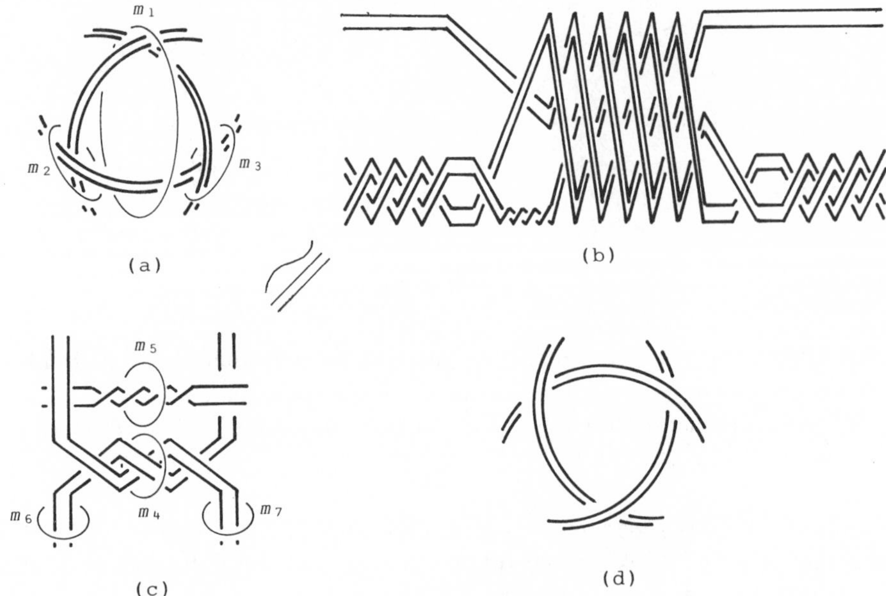

Figure 4

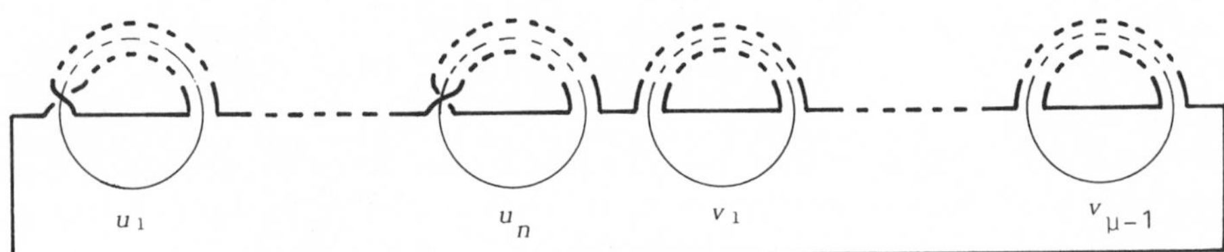

Figure 5

$(v_1, \ldots, v_{\mu-1}$, resp.) to represent a center lines of non-orientable bands (orientable bands, resp.) as in Fig. 5.

We remark that a finite sequence of $D(\Delta)$-operations implies an exchange of self-crossing of a band by the parallel version of the argument in [MN].

If u_1 is knotted, then we apply $D(\Delta)$-operations to change u_1 into a trivial knot. If necessary, we take a trivial knot winding around the band containing u_1, and change the number of half-twists into ± 1 by $\pm 1/2c$-surgery along the trivial knot for a certain integer c. If we have a disregard for the other bands, the band containing u_1 is a trivial non-orientable band like as in Fig. 6. Let δ_1 be a disk bounded by u_1.

If $lk(u_1, u_2)$ is even, we take a trivial knot winding around the bands containing $u_1$and u_2, and change $lk(u_1, u_2)$ into 0 by $\pm 1/2c$-surgery along the trivial knot for a certain integer c. If $lk(u_1, u_2)$ is odd, we slide the root of the band containing u_2

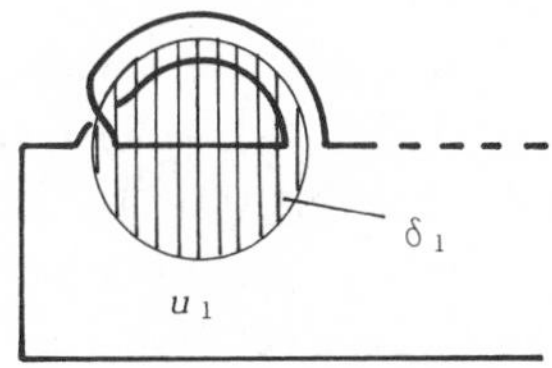

Figure 6

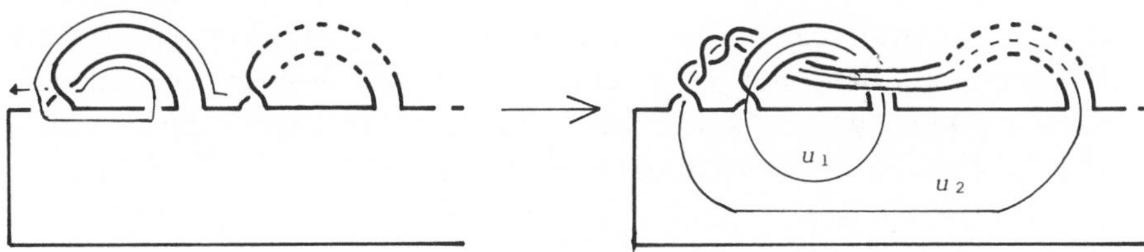

Figure 7

along the boundary of the band containing u_1 to change $lk(u_1, u_2)$ by one. (See Fig. 7.)

Then, we take a trivial knot winding around the bands containing u_1 and u_2, and change $lk(u_1, u_2)$ into 0 by $\pm 1/2c$-surgery along the trivial knot for a certain integer c. If u_2 intersects δ_1, then we apply the parallel version of the argument in the proof of Corollary in [MN] to eliminate all intersection of δ_1 and the band containing u_2 by a finite sequence of $D(\Delta)$-operations. Then, we slide the whole δ_1 and the band containing u_1 along the boundary of the band containing u_2 to be located between the bands containing u_2 and u_3. Inductively, we perform the same argument in order to eliminate all intersection of δ_1 and the bands containing u_i's ($i = 2, 3, \ldots, n$). Here, the band containing u_1 is located between the bands containing u_n and v_1. We apply the above argument for the band containing u_i ($i = 1, 2, \ldots, n$), inductively. Now, we can assume that all the non-orientable bands are trivial if we have a disregard for the orientable bands.

For the case that the given link is a knot, there is no orientable band. Therefore, the result knot is a trivial knot, completing the proof of Proposition 4.

For the case that the given link has two components, there is exactly one orientable band. By the same argument, we can change v_1 into a trivial knot and the number of half-twists of the band containing v_1 can be changed into 0 or 1. And all intersection of δ_1 and the band containing v_1 can be eliminated. Similarly, all the non-orientable bands are trivial, even if we have a regard for the orientable band containing v_1. Therefore, the result link is a 2-component trivial link or a 2-component Hopf link. Since the linking number modulo 2 is an invariant under $\pm 1/2$-surgery along a trivial knot, we complete the proof of Proposition 5.

From now, we consider the case $\mu \geq 3$. By the same argument, we can assume the following: The non-orientable bands are trivial if we have a disregard for the orientable bands. Each v_i $(i = 1, 2, \ldots, \mu - 1)$ is a trivial knot and the number of half-twists of each orientable band is 0 or 1. The linking number $lk(v_i, v_j)$ is 0 or 1 for every $1 \leq i < j \leq \mu - 1$. We remark that we can ignore the linking of three bands or more up to $D(\Delta)$-operations, from the parallel version of the work in [MN].

Since the non-orientable bands are trivial if we have a disregard for the orientable bands, we can change the order of non-orientable bands in any way. We divide the non-orientable bands into groups such that consecutive non-orientable bands are linking with the same orientable bands. For one group of non-orientable bands, we perform the following operation. Since we can ignore the linking of three bands or more, we can assume that the linking of the group of non-orientable bands and the group of orientable bands is like as in Fig. 8.

Figure 8

If there are distinct types of non-orientable bands with $+1$ and -1 half-twist, we can reduce the number of non-orientable bands by $\pm 1/2$-surgeries along trivial knots, that is indicated in Fig. 9. We take three trivial knots m_1, m_2, and m_3. By $-1/2$-surgery along m_1, $+1/2$-surgery along m_2, and $+1/2$-surgery along m_3, we can reduce two non-orientable bands.

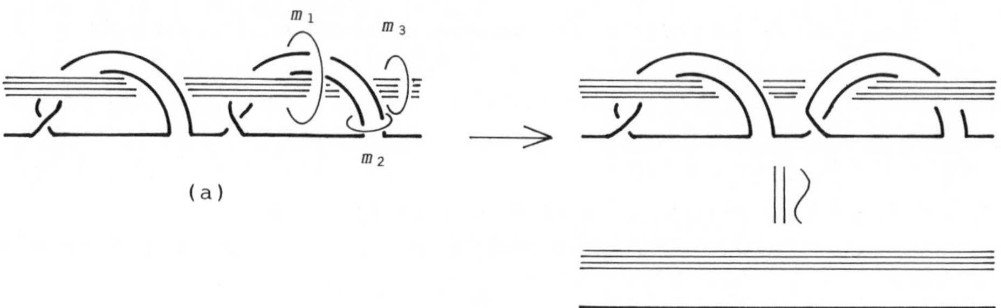

Figure 9

If there are four non-orientable bands or more in the same group, we can reduce the number of non-orientable bands by $\pm 1/2$-surgeries along trivial knots, that is indicated in Fig. 10. Like as in Fig. 10 (i), we take three trivial knots m_1, m_2, and m_3. By $-1/2$-surgery along m_1, $+1/2$-surgery along m_2, and $+1/2$-surgery along m_3, we obtain (ii) from (i). By an ambient isotopy, we obtain (iii) from (ii).

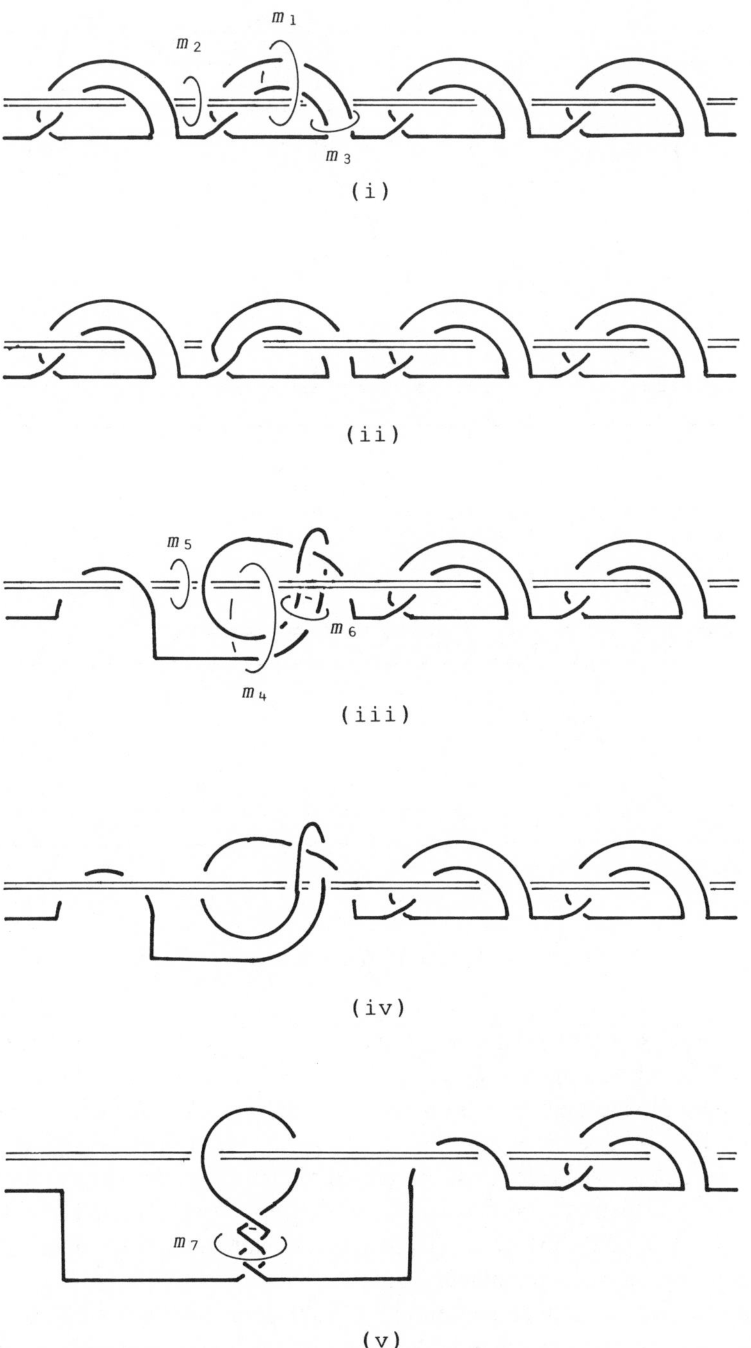

Figure 10

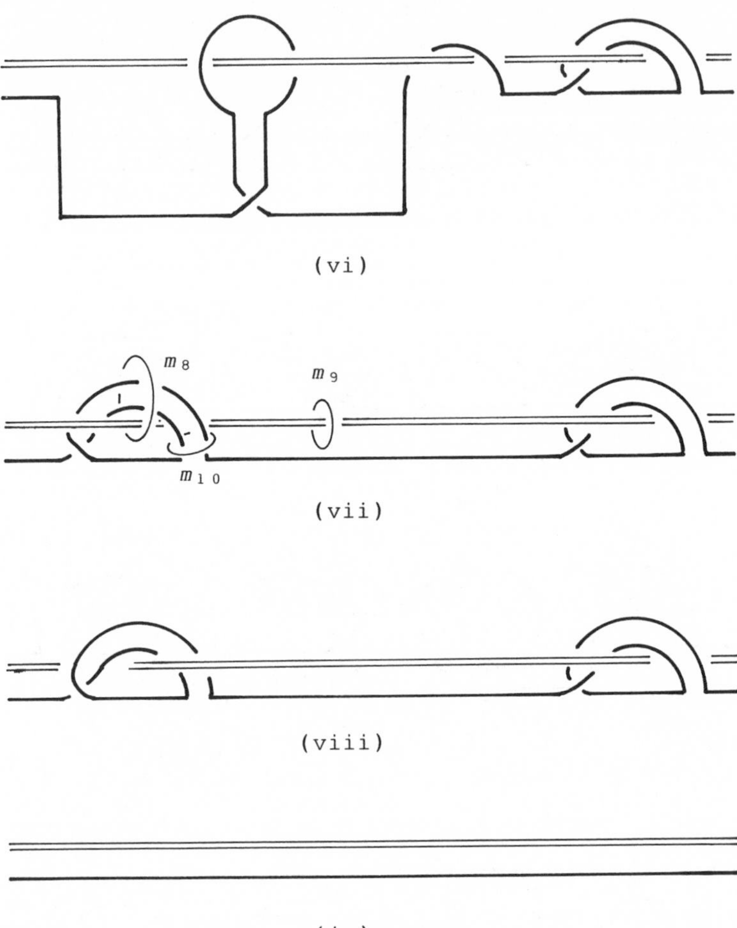

Figure 10 (continued)

We take three trivial knots m_4, m_5, and m_6 as in (iii). By $-1/2$-surgery along m_4, $+1/2$-surgery along m_5, and $+1/2$-surgery along m_6, we obtain (iv) from (iii). By an ambient isotopy, we obtain (v) from (iv). We take a trivial knot m_7 as in (v). By $+1/2$-surgery along m_7, we obtain (vi) from (v). By an ambient isotopy, we obtain (vii) from (vi). We take three trivial knots m_8, m_9, and m_{10} as in (vii). By $-1/2$-surgery along m_8, $+1/2$-surgery along m_9, and $+1/2$-surgery along m_{10}, we obtain (viii) from (vii). By an ambient isotopy, we obtain (ix) from (viii), reducing four non-orientable bands.

Therefore, we can reduce the number of non-orientable bands into $0, 1, 2$, or 3 for each group of orientable bands. We can eliminate non-orientable bands which are linking exactly one orientable band by the same argument for the case of 2-component links. Since the number of groups of two orientable bands or more is

$2^{\mu-1}-\mu$, the number of linking patterns of v_i's is $2^{(\mu-1)(\mu-2)/2}$, and the number of twisting patterns of orientable bands is $2^{\mu-1}$, the number of congruence classes modulo $2,2$ for μ-component link types, C_μ, is at most $2^{\mu(\mu-1)}\times 4^{2^{\mu-1}-\mu}$. On the other hand, if two link types are congruent modulo $2,2$, then the two link types are congruent modulo $2,1$. The number of congruence classes modulo $2,1$ for μ-component link types is $2^{\mu(\mu-1)/2}$ from Theorem 3. And the μ-component trivial link type and the split sum of the $(\mu-3)$-component trivial link type and the link type of the Borromean rings are incongruent modulo $2,2$, that is shown by comparing elementary ideals from Theorem 1. Therefore, C_μ is greater than $2^{\mu(\mu-1)/2}$. The proof of Proposition 7 is completed.

Acknowledgements. This work was partly done while the author was visiting at the University of Toronto, Canada, with a financial support by the Japanese Ministry of Education in the form of an Overseas Research Fellowship. The author would like to express his gratitude to Professor Kunio Murasugi for his hospitaity and valuable discussions.

References

[F1] R.H. Fox, *Free differential calculus, I*, Ann. of Math. **57** (1953), 547-560; *ibid., II*, ibid. **59** (1954), 196-210.

[F2] R.H. Fox, *Congruence classes of knots*, Osaka Math. J. **10** (1958), 37-41.

[MN] H. Murakami and Y. Nakanishi, *On a certain move generating link-homology*, Math. Ann. **284** (1989), 75-89.

[N] Y. Nakanishi, *On Fox's congruence classes of knots, II*, Osaka J. Math. **27** (1990), 207-215.

[NS] Y. Nakanishi and S. Suzuki, *On Fox's congruence classes of knots*, Osaka J. Math. **24** (1987), 217-225.

[R] D. Rolfsen, *Knots and Links*, Publish or Perish Inc., Berkeley, 1976.

Adequate links and the Jones polynomial of unlinks

G.T. Jin, B.K. Kim, and K.H. Ko

1. The Jones polynomial of unlinks

Since the discovery of the Jones polynomial, people have been asking whether there exist nontrivial links having the Jones polynomial of unlinks [APR, LM1, LM2]. Recall that $V_{O^\ell}(t) = (-t^{1/2} - t^{-1/2})^{\ell-1}$ where O^ℓ is the ℓ-component unlink. Since an ℓ-component link L satisfies $V_L(1) = (-2)^{\ell-1}$, we cannot have $V_L(t) = V_{O^m}(t) = (-t^{1/2} - t^{-1/2})^{m-1}$ if $m \neq \ell$.

Definition. Let L be an ℓ-component link. We will say that L is *Jones-trivial* if $V_L(t) = (-t^{1/2} - t^{-1/2})^{\ell-1}$. The *breadth* (or *span*) β_L of L is the difference between the highest degree and the lowest degree of t in $V_L(t)$.

Because of the connected sum formula and the distant union formula for the Jones polynomial, it is enough to investigate only nonsplit prime links.

Lemma 1. *Let D be a link diagram with n crossings, ℓ components and m split components. Then $n \geq 2(\ell - m)$.*

Proof. The lemma is trivially true for $\ell = 1$. Let $\ell > 1$. Suppose the lemma is true for links with less than ℓ components. Choose a component K of D. Let n_K and m_K denote the number of crossings and the number of split components of $D \setminus K$, respectively. Then $n_K \geq 2(\ell - 1 - m_K)$. It is easy to see that K has crossings with $m_K - m + 1$ split components of $D \setminus K$. Therefore

$$n \geq n_K + 2(m_K - m + 1) \geq 2(\ell - m). \qquad \square$$

The first partial answer to our question is here:

Proposition 2. *No nontrivial alternating link is Jones-trivial.*

Knots 90

Proof. Let D be a reduced alternating diagram of a nontrivial nonsplit alternating diagram. If D has n crossings, we have

$$\begin{aligned} \beta_D &= n && \text{by [K, M, T1]} \\ &\geq \max\{2\ell - 2, 1\} && \text{by Lemma 1} \\ &> \ell - 1 \\ &= \beta_{O^\ell} . \quad \square \end{aligned}$$

Given a diagram D, the new diagrams obtained by nullifying all crossings by the rules ⤫ $\Rightarrow$)(and ⤫ $\Rightarrow$ ≍ are denoted by s_+D and s_-D, respectively. The diagrams $s_\pm D$ consist of disjoint simple closed curves. Let $|s_\pm D|$ denote the number of components in $s_\pm D$.

Definition. A link diagram D is said to be *adequate* if, in $s_\pm D$, every pair of corners obtained from any crossing of D belongs to two distinct components of $s_\pm D$. A link is said to be *adequate* if it admits an adequate diagram.

Figure 1 shows that the trefoil knot is adequate. In fact every alternating link is adequate, since every reduced alternating diagram is adequate.

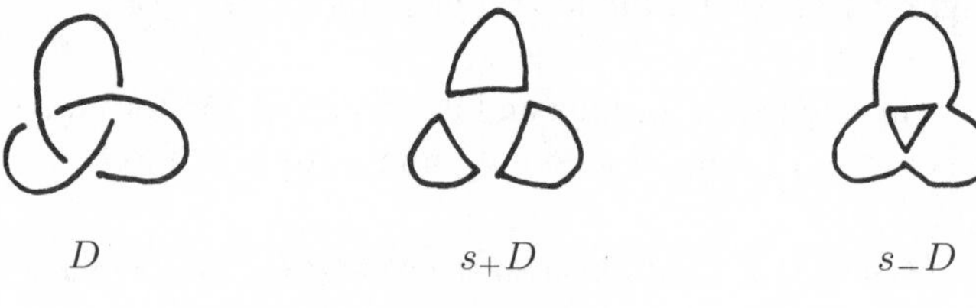

Figure 1

Theorem 3 (Lickorish-Thistlethwaite [LT]). *Let D be an adequate diagram with n crossings. Then*

$$\beta_D = \frac{1}{2}(n + |s_+D| + |s_-D|) - 1.$$

As a corollary to this theorem, Lickorish and Thistlethwaite gave the second partial answer:

Corollary 4. *No nontrivial adequate knot is Jones-trivial.*

Proof. Any nontrivial knot diagram D has $n \geq 1$ and $|s_\pm D| \geq 1$. Therefore $\beta_D > 0 = \beta_O$ if D is adequate. $\square$

Theorem 5. *The breadth of a nontrivial nonsplit adequate link is not smaller than the number of components.*

Proof. Let D be an adequate diagram of a nontrivial nonsplit link. Since D has at least one crossing, $|s_\pm D| \geq 2$. If D has n crossings, then, by Lemma 1 and Theorem 3,

$$\begin{aligned}\beta_D &= \frac{1}{2}(n + |s_+ D| + |s_- D|) - 1 \\ &\geq \frac{1}{2}(n+4) - 1 \\ &\geq \ell. \quad \square\end{aligned}$$

Since a Jones-trivial link must have the breadth equal to the number of components minus one, we have the third partial answer to our question.

Corollary 6. *No nontrivial adequate link is Jones-trivial.*

2. The greatest lower bound for the breadth

It is natural to ask when "$\beta_D = \ell$" can occur in Theorem 5. If $\beta_D = \ell$, then we must have $n = 2\ell - 2$. According to the following lemma, this implies that D is a connected sum of $\ell - 1$ Hopf links. Let H denote the Hopf link. Then

$$\ell = \beta_D = (\ell - 1)\beta_H = 2\ell - 2,$$

which implies $\ell = 2$. Therefore $\beta_D = \ell$ if and only if D is the Hopf link.

Lemma 7. *Let D be a diagram of an ℓ-component nonsplit link. Then there are $2\ell - 2$ crossings in D if and only if D is a connected sum of $\ell - 1$ Hopf links.*

Proof. It is obvious that any connected sum of $\ell - 1$ Hopf links has $2\ell - 2$ crossings. Suppose that D has $2\ell - 2$ crossings. We claim that no component of D has a self-crossing. If any component of D has a self-crossing, we can nullify the crossing without increasing the number of components. Then the new diagram has $2\ell - 3$ crossings, which is a contradiction to Lemma 1. Lemma 7 is trivially true if $\ell = 1$. Assume $\ell > 1$ and that the lemma is true for links with fewer than ℓ components. Let D be a diagram of an ℓ-component nonsplit link with $2\ell - 2$ crossings. Remove any component K of D. Suppose $D \setminus K$ has m split components $D_1, \ldots, D_m$. Let n_i and ℓ_i be the number of crossings and the number of components of D_i.

Since K must cross each D_i at least twice, Lemma 1 implies

$$\begin{aligned} 0 &\leq \sum_{i=1}^{m}(n_i - 2\ell_i + 2) \\ &= (n-c) - 2(\ell - 1) + 2m \\ &= 2m - c \\ &\leq 0 \end{aligned}$$

where c is the number of crossings between K and $D \setminus K$. Therefore, $n_i = 2\ell_i - 2$ and $c = 2m$. By the induction hypothesis, D_i is a connected sum of $\ell_i - 1$ Hopf links. $c = 2m$ implies that each D_i crosses K exactly twice. Therefore D is obtained by a connected sum of the D_i's and a connected sum of m Hopf links. The total number of Hopf links in the connected sum decomposition of D is

$$\sum_{i=1}^{m}(\ell_i - 1) + m = \sum_{i=1}^{m}\ell_i = \ell - 1. \qquad \square$$

We are going to study the greatest lower bound for the breadth of adequate links.

By Lemma 1, Lemma 7 and the fact that the number of crossings in a reduced alternating nonsplit diagram equals the breadth, we obtain

Proposition 8. *If L is an ℓ-component nonsplit alternating link, then $\beta_L \geq 2\ell - 2$. The equality holds if and only if L is a connected sum of $\ell - 1$ Hopf links.*

We want to generalize Proposition 8 for adequate links.

Conjecture. If L is an ℓ-component nonsplit adequate link, then $\beta_L \geq 2\ell - 2$. The equality holds if and only if L is a connected sum of $\ell - 1$ Hopf links.

Since $\beta + 1$ is additive under distant union of links, the conjecture is equivalent to :

> *If L is an ℓ-component adequate link having m split components, then $\beta_L \geq 2\ell - m - 1$. The equality holds if and only if each split component of L is an unknot or a connected sum of Hopf links.*

We will decompose a non-alternating diagram into *maximal alternating tangles* as Thistlethwaite did in [T3].

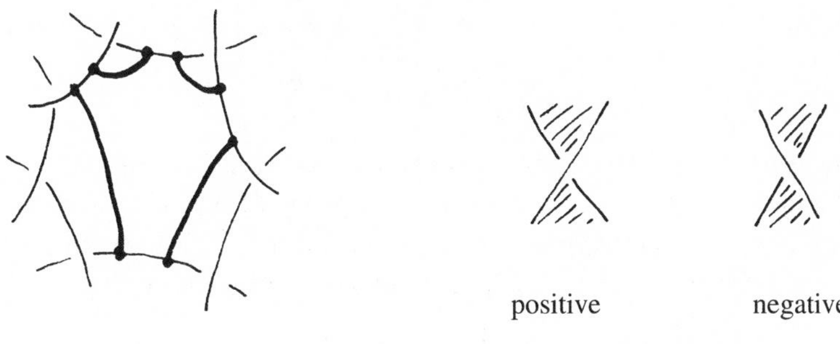

Figure 2 **Figure 3**

Definitions. Let D be a nonsplit link diagram on S^2. An edge of D is said to be *non-alternating* if its ends are both overpasses or both underpasses. Others will be called *alternating edges*. In the interior of every non-alternating edge, we choose two points. In each region of D having non-alternating edges on the boundary, we draw non-intersecting arcs joining these points as in Figure 2. The union of these arcs is a collection of disjoint simple closed curves and is denoted by Γ. Each component of $S^2 \setminus \Gamma$ which does not contain any crossings of D is called a *channel region*. Others will be called *tangle regions*. Then for every tangle region Δ, $\Delta \cap D$ is an alternating tangle which is maximal among connected ones. Suppose the regions of D have black-and-white checkerboard coloring. In each maximal alternating tangle, the crossings are all *positive* or all *negative* in the sense of Figure 3. A maximal alternating tangle is said to be *positive*(*negative*) if its crossings are positive (negative). Notices that $\Gamma = \Gamma_+ \amalg \Gamma_-$ where $\Gamma_+(\Gamma_-)$ is the union of boundary components of positive (negative) tangle regions. Let B and W be the union of the black regions and that of the white regions, respectively, and let $P_\pm$ be the union of $\Gamma_\pm \cap W$, $\Gamma_\mp \cap B$ and the middle pieces of the non-alternating edges. Notice that P_+ and P_- are disjoint unions of simple closed curves. Let $\delta = |\Gamma|$, $p = |P_+| + |P_-|$ and let n, r, and ν denote the number of crossings, the number of maximal alternating tangles and the number of non-alternating edges, respectively.

Theorem 9 (Thistlethwaite [T3]). *Given a nonsplit diagram D, $\beta_D \leq n + r - (\nu + \delta - p)/2 - 1$. The equality holds if D is adequate.*

Another proof of Theorem 5. Let D be a nontrivial nonsplit link diagram with ℓ components. We have three inequalities:
(1) $n + r \geq \nu$
(2) $p \geq 2(\delta - r + 1)$
(3) $n \geq 2\ell - 2$.

To see (1), let n_T, ℓ_T and ν_T be the number of crossings, the number of components (circles and arcs) and the number of non-alternating edges, respectively, in a maximal alternating tangle T. Then, $n_T \geq \ell_T - 1 \geq \frac{1}{2}\nu_T - 1$. Therefore

$$n + r = \sum (n_T + 1) \geq \frac{1}{2} \sum \nu_T = \nu$$

where the sums are taken over all maximal alternating tangles of D. For each channel region, there are at least one component of P_+ (P_-). Since there are $\delta - r + 1$ channel regions, the inequality (2) follows. The inequality (3) is from Lemma 1. The inequalities (1), (2), (3) and Theorem 9 imply that

$$\beta_D \geq n - \frac{1}{2}\nu + \frac{1}{2}\delta \geq \frac{1}{2}n \geq \ell - 1$$

if D is adequate. Notice that $\beta_D = \ell - 1$ implies $n = 2\ell - 2$. Therefore D is a connected sum of $\ell - 1$ Hopf links, and hence $\beta_D = 2\ell - 2$. This is a contradiction if $\ell \neq 1$. If $\ell = 1$, then $\beta_D = 0$. This is also a contradiction to Corollary 4, since D is nontrivial. Therefore $\beta_D \geq \ell$. □

Let us return to the conjecture. Let D be an adequate diagram and let $E_D = \beta_D - (2\ell - 2)$. We want to show that $E_D \geq 0$. Suppose there is a component of D having a self-crossing. Replace that crossing by a clasp as in Figure 4 so that the new diagram D' has one more crossings and one more components than D.

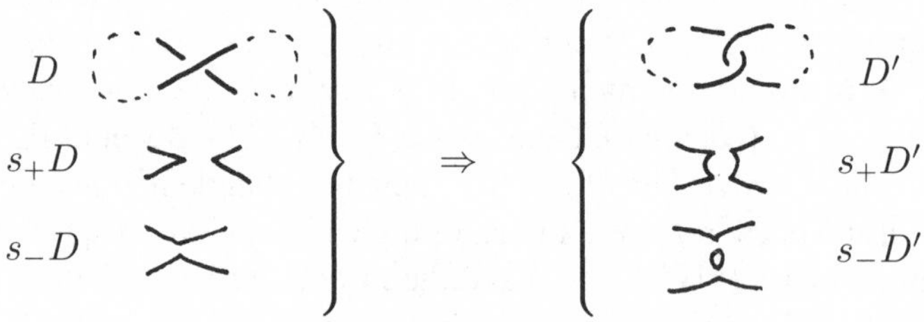

D, s_+D, s_-D $\Rightarrow$ D', s_+D', s_-D'

Figure 4

Notice that D' is also adequate and we have

$$\begin{aligned} \beta_{D'} &= \frac{1}{2}(n + 1 + |s_+D| + |s_-D| + 1) - 1 \\ &= \beta_D + 1, \end{aligned}$$

and hence

$$E_{D'} = \beta_D + 1 - 2\ell < E_D.$$

Therefore, if the conjecture is true for diagrams without self-crossing components, then we are done.

Theorem 10. *Let L be a nonsplit ℓ-component link admitting an adequate diagram D in which*

(1) *no components have self-crossings*

(2) *no two components having non-alternating edges cross.*
Then $\beta_L \geq 2\ell - 2$. *The equality holds if and only if* L *is a connected sum of* $\ell - 1$ *Hopf links.*

At first, we tried to prove the conjecture by induction on the number of self-crossings and the number of crossings between non-alternating components. But we have not yet found any induction argument for the crossings between non-alternating components.

Definition. A crossing in a tangle is called a *nugatory crossing* if there is a simple closed curve in the plane (or S^2) intersecting the tangle only at that crossing. (See Figure 5.)

Lemma 11. *Let* D *be a diagram in which no maximal alternating tangle has nugatory crossings, then* D *is adequate.*

The converse of Lemma 11 is not true as Figure 6 shows.

Proof. Suppose D is not adequate, then there is a crossing such that the corresponding corners in s_+D or s_-D are contained in the same component, say C. Starting from the inside of one of the corners, one draws a path along C on the left hand side of C, until there is a non-alternating edge. In the middle of this edge, the path crosses C and proceeds on the right hand side of C until the next non-alternating edge, and so on. Between any two consecutive non-alternating edges, C has only right-turn corners or only left-turn corners, according to the positiveness or negativeness of the maximal alternating tangle between the two non-alternating edges. It is not hard to see that the path comes back to the inside of the corner opposite to the starting point, as in Figure 7. Joining the two ends of the path, we obtain a simple closed curve intersecting the crossing but no other part of the maximal alternating tangle containing the crossing. □

Figure 5

Figure 6

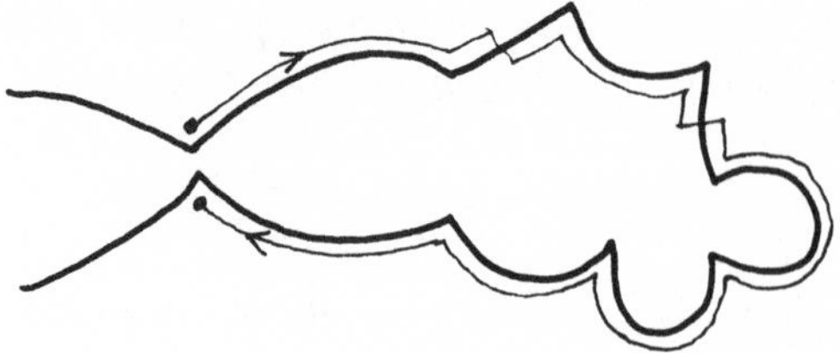

Figure 7

Proof of Theorem 10. By (2), one of the two strands in any crossing of D is contained in an alternating component. This component is contained entirely in one maximal alternating tangle. Therefore under the conditions (2), the only case that a crossing is nugatory is that it is a self-crossing of an alternating component, which is prohibited by (1). Therefore there is no nugatory crossing in D.

We may assume that D is prime. Choose an innermost non-alternating component K. Notice that any region R inside K having non-alternating edges cannot have non-alternating edges from other non-alternating components. We will divide K into two components as follows:

(a) If there are only two non-alternating edges in R, remove the middle parts of them and join the ends of each arc as in Figure 8.

(b) If there are more than two non-alternating edges, then do the same operation on a pair of the same type of non-alternating edges, i.e., an underpassing edge with another underpassing edge or an overpassing edge with another overpassing edge, as in Figure 9.

Notice that under these operations on non-alternating edges, no new crossings are created and no crossings become nugatory. Therefore, the new diagram D' is again adequate by Lemma 11. Furthermore D' satisfies the conditions (1) and (2), and is nonsplit. Under this operation n, r and δ are unchanged. But p, ν and ℓ may change. It is not hard to see that $p' \in \{p, p \pm 2\}$, $\nu' \in \{\nu, \nu - 2\}$ and $\ell' = \ell + 1$. Therefore $\beta_{D'} \in \{\beta_D, \beta_D \pm 1, \beta_D + 2\}$ and hence $E_{D'} \in \{E_D, E_D - 1, E_D - 2, E_D - 3\}$. Therefore $E_D \geq E_{D'}$. Repeating this operation, we can remove all the non-alternating edges. Since adequate diagrams without non-alternating edges are reduced alternating diagrams, the theorem is proved by induction on the number of non-alternating edges. Notice that whenever the operation results in a nonprime diagram, we must divide it into prime diagrams before the next operation. □

If one finds a nice way of handling the crossings between non-alternating components, one can prove the conjecture by using Theorem 10.

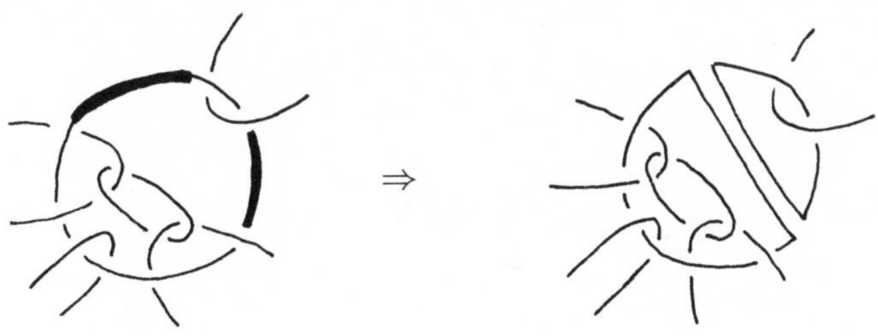

Figure 8

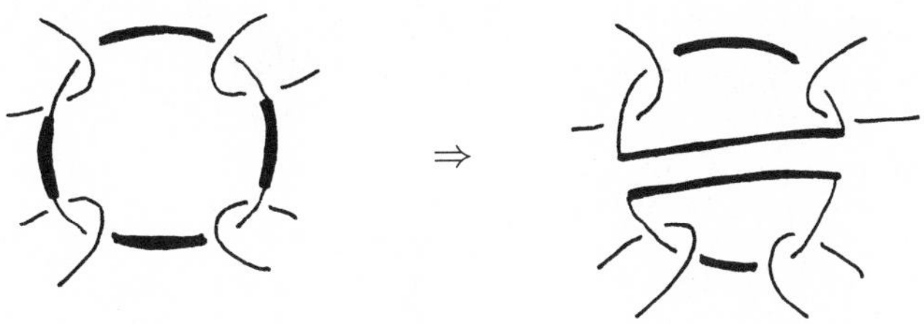

Figure 9

References

[APR] R.P. Anstee, J.H. Przytycki, and D. Rolfsen, *Knot polynomials and generalized mutation*, Topology Appl. 32(1989), 237–249

[J] V.F.R. Jones, *A polynomial invariant for knots via von Neumann algebras*, Bull. Amer. Math. Soc. 12(1985), 103–111

[K] L.H. Kauffman, *State models and the Jones polynomial*, Topology 26(1987), 395–407

[LM1] W.B.R. Lickorish and K.C. Millett, *A polynomial invariant of oriented links*, Topology 26(1987), 107– 141

[LM2] W.B.R. Lickorish and K.C. Millett, *The new polynomial invariant of knots and links*, Math. Magazine 61(1988), 3–23

[LT] W.B.R. Lickorish and M.B. Thistlethwaite, *Some links with non-trivial polynomials and their crossing-numbers*, Comment. Math. Helvetici 63(1988), 527–539

[M] K. Murasugi, *Jones polynomial and classical conjectures in knot theory*, Topology 26(1987), 187–194

[T1] M.B. Thistlethwaite, *A spanning tree expansion of the Jones polynomial*, Topology 26(1987), 297–309

[T2] M.B. Thistlethwaite, *On the Kauffman polynomial of an adequate link*, Invent. Math. 93(1988), 285–296

[T3] M.B. Thistlethwaite, *An upper bound for the breadth of the Jones polynomial*, Math. Proc. Camb. Phil. Soc. 103(1988), 451–456

Chern-Simons theory, link invariants and quasi-Hopf algebras

E. Guadagnini, M. Martellini and M. Mintchev

1. BRS Quantization of the Chern-Simons Theory

The recent interest in the Chern-Simons (C-S) model has its origin mainly in the fact that this model represents a meeting point of nowadays fastly developing areas in mathematics and physics like the theory of invariants, quantum groups, conformal field theory and two-dimensional integrable models. The three-dimensional C-S action reads

$$S_{CS} = \frac{k}{4\pi} \int_{M^3} d^3x \epsilon^{\mu\nu\rho} \mathrm{Tr} \left(A_\mu \partial_\nu A_\rho + i\frac{2}{3} A_\mu A_\nu A_\rho \right) , \tag{1.1}$$

where A_μ is a gauge connection associated with a non-Abelian gauge group G, M^3 is a three-manifold and the coupling constant k is an integer as a consequence of gauge invariance.

The relevance of the C-S model for knot theory has been suggested by A. S. Schwarz [1] and M. Atiyah [2] and has been further explored by E. Witten [3]. Being the integral of a three-form on a three-manifold, S_{CS} is independent of the choice of any metric on M^3. This fact implies that the vacuum expectation value $< \mathcal{O} >$ of any metric-independent (and gauge invariant) observable $\mathcal{O}$ represents a topological invariant, provided that a consistent and diff-invariance preserving quantization of (1.1) exists. Consider in particular the Wilson line operator

$$W_R(C) = \mathrm{Tr\ P\ exp} \left(i \oint_C A^a_\mu(x) R^a \, dx^\mu \right) , \tag{1.2}$$

where the path ordering is performed along the oriented knot C and R is a finite dimensional irreducible representation of G. Then, $< W_R(C) >$ is a knot invariant. More generally, the expectation value

$$< W(L) > = < W_{R_1}(C_1) \cdots W_{R_n}(C_n) > \quad , \tag{1.3}$$

where the link L is a union of non-intersecting knots, is a link invariant.

Knots 90

In order to perform the quantization we adopt the standard Faddeev-Popov procedure. In the Landau gauge the total action is

$$S = S_{CS} + S_{GF} + S_{FP} \, , \tag{1.4}$$

where

$$S_{GF} = \frac{k}{4\pi} \int_{M^3} d^3x \, \sqrt{g} g^{\mu\nu} \, A^a_\mu \partial_\nu B^a \, , \tag{1.5}$$

$$S_{FP} = - \int_{M^3} d^3x \, \sqrt{g} g^{\mu\nu} \, \partial_\mu \bar{c}^a (D_\nu c)^a \, , \tag{1.6}$$

and

$$(D_\mu c)^a = \partial_\mu c^a - f^{abc} A^b_\mu c^c \, . \tag{1.7}$$

In eqs.(1.5,6) $g_{\mu\nu}$ is some metric on M^3, B^a are Lagrange multipliers enforcing the gauge constraint $\partial_\mu(\sqrt{g} g^{\mu\nu} A^a_\nu) = 0$ and c and $\bar{c}$ are the usual Faddeev-Popov ghosts. The total action S is invariant under the BRS transformations [4]

$$\delta A^a_\mu = (D_\mu c)^a \quad , \quad \delta B^a = 0 \, , \tag{1.8}$$

$$\delta c^a = \frac{1}{2} f^{abc} c^b c^c \quad , \quad \delta \bar{c}^a = \frac{k}{4\pi} B^a \, , \tag{1.9}$$

and the asssociated BRS charge operator Q is off-shell nilpotent because there are no gauge anomalies in three dimensions. The cohomology defined by Q determines in the usual way the physical space of the C-S model.

The term $S_{GF} + S_{FP}$ unavoidably spoils the general covariance. Since the metric dependence enters only in the gauge-fixing procedure however, general covariance is maintained on the physical space. In fact, the symmetric energy-momentum tensor $\Theta_{\mu\nu}$ associated with (1.4) is

$$\Theta_{\mu\nu} = \frac{k}{4\pi} (A^a_{\{\mu} \partial_{\nu\}} B^a - g_{\mu\nu} A^a_\rho \partial^\rho B^a) - \partial_{\{\mu} \bar{c}^a (D_{\nu\}} c)^a + g_{\mu\nu} \partial_\rho \bar{c}^a (D^\rho c)^a \tag{1.10}$$

and can be rewritten in the form

$$\Theta_{\mu\nu} = \{ Q, \partial_{\{\mu} \bar{c}^a A^a_{\nu\}} - g_{\mu\nu} \partial^\rho \bar{c}^a A^a_\rho \} \, . \tag{1.11}$$

Since Q annihilates the physical states, the mean value of $\Theta_{\mu\nu}$ between physical states vanishes and general covariance is indeed preserved. Consequently, the expectation value of any gauge invariant and metric independent observable is a topological invariant in M^3. The simplest example of such an observable is the Wilson line operator defined by eq.(1.2).

In what follows we will be interested in quantizing the C-S model and computing $< W(L) >$ for $M^3 = R^3$ equipped with the flat metric $g_{\mu\nu} = \delta_{\mu\nu}$. In this case the theory possesses [5,6] a peculiar supersymmetry; indeed, the total action

(1.4) is invariant under the infinitesimal transformations

$$\delta_\mu A_\nu = \epsilon_{\nu\mu\rho}\partial^\rho c \quad , \quad \delta_\mu B = D_\mu c \, , \tag{1.12}$$

$$\delta_\mu c = 0 \quad , \quad \delta_\mu \overline{c} = -\frac{k}{4\pi} A_\mu \, , \tag{1.13}$$

and

$$\overline{\delta}_\mu A_\nu = \epsilon_{\nu\mu\rho}\partial^\rho \overline{c} \quad , \quad \overline{\delta}_\mu B = \partial_\mu \overline{c} \, , \tag{1.14}$$

$$\overline{\delta}_\mu c = -\frac{k}{4\pi} A_\mu \quad , \quad \overline{\delta}_\mu \overline{c} = 0 \, , \tag{1.15}$$

The generators associated with the above transformations, combined with the BRS and anti-BRS charges give rise [6] to a $N = 2$ supersymmetry algebra. The corresponding Ward identities play a crucial role in proving [7] the finiteness of the C-S model.

In conclusion, we display [8] a discrete symmetry I which is useful in analysing $< W(L) >$. The inversion operator I acts as

$$\begin{gathered} A^a_\mu(x) \mapsto -A^a_\mu(-x) \quad , \quad B(x) \mapsto -B(-x) \quad , \\ c(x) \mapsto c(-x) \quad , \quad \overline{c}(x) \mapsto \overline{c}(-x) \quad , \\ k \mapsto -k \quad . \end{gathered} \tag{1.16}$$

The above transformations have a simple geometrical meaning. The I-invariance states that a change in the sign of the coupling constant k is equivalent to (or can be compensated by) reversing the orientation of the three-manifold M^3. The I-symmetry implies that $< W(L) >$ computed with the coupling constant k is equal to $< W(\widetilde{L}) >$ computed with the coupling constant $-k$, where the link $\widetilde{L}$ is the mirror image of the link L. In fact, consider the transformation of the Wilson line operator (1.2) under (1.16). One has

$$A_\mu(x)dx^\mu \mapsto -A_\mu(-x)dx^\mu = A_\mu(y)dy^\mu \quad , \tag{1.17}$$

where

$$y^\mu = -x^\mu \quad . \tag{1.18}$$

Notice that in three dimensions the transformation (1.18) is equivalent, up to a π-rotation, to the inversion of one of the axes. Therefore, the change of variables (1.18) in (1.2) implies

$$W_R(C) \mapsto W_R(\widetilde{C}) \quad , \tag{1.19}$$

where $\widetilde{C}$ is the mirror (and actually π-rotated) image of the knot C. Combining (1.19) with the rotational and I-invariance of the total action one finds

$$< W(L) > |_k =< W(\widetilde{L}) > |_{-k} \quad . \tag{1.20}$$

Eq.(1.20) provides a non-trivial check on the computation of $< W(L) >$ and gives some restrictions on the k-dependence of the variables entering in the S_L polynomials. We shall return to this point in sect.2. Note that eq.(1.20) still holds on a general background $g_{\mu\nu}$ because of the metric-independence of $< W(L) >$.

It is worth mentioning that the C-S theory makes sense for both signs of k because the Hamiltonian vanishes on the physical space. In other words, with a given orientation of the three-manifold M^3, both the positive and negative values of k are permitted. These two possibilities are simply related by the exchange of the links, associated with the Wilson line operators, with their mirror images. For this reason, it is sufficient to consider positive values of k.

2. Universal link invariants

The expectation value $< W(L) >$ has an expansion in powers of $\lambda = \frac{2\pi}{k}$ which coincides with the standard perturbative expansion. It turns out that each term of this expansion is finite and, in contrast to the usual situation in quantum field theory, the power series in λ is convergent. The computation of $< W(L) >$ presents two aspects. First, one has to construct well-defined correlation functions of the C-S gauge potential. Second, one has to give a definite meaning of the composite operator $W(L)$. The first aspect has been discussed in [9]. It is shown there that, in the framework of the BRS quantization, the correlation functions $\{< A_{\mu_1}(x_1) \cdots A_{\mu_j}(x_j) >\}$ are finite at any order in λ and that the general covariance is preserved on the physical subspace. The explicit two-loop calculation [9] confirms that the theory is finite, as is expected from the constraint on k to be an integer. A general proof of this fact, based on the supersymmetry (1.12-15), is given in [7].

The next step in the construction of $< W_R(C) >$ involves integration of correlation functions of the type $< A_{\mu_1}(x_1) \cdots A_{\mu_j}(x_j) >$ over the same path C. It is worth stressing that, differently from what one could naively expect, no divergences occur in performing these integrations in the C-S theory [8]. In order to preserve the topological character of the Wilson line $W_R(C)$, however, one must give an operative meaning to the framing procedure for knots in the context of quantum field theory. The explicit computations at order λ^2 show [8] that in the C-S model the necessity of framing is related to the well-known self-linking problem. This problem was analyzed [10] some time ago by mathematicians who found that the framing procedure assigns a topological meaning to the self-linking for knots in R^3 by fixing the ambiguities due to a certain direction-dependent but finite limit. In practice, the framing of a knot $C = \{x(t) : 0 \leq t \leq 1\}$ in R^3

consists in introducing a framing contour C_f defined by

$$x^\mu(t) \mapsto y^\mu(t) = x^\mu(t) + \varepsilon n^\mu(t) \quad , \quad (\varepsilon > 0) \quad , \tag{2.1}$$

where n^μ is a vector field orthogonal to C of unit norm. There are two particular framings playing a distinguished role - the so-called "standard" and "vertical" framings. The standard framing is fixed by the requirement that the Gauss integral (linking number)

$$\chi(C, C_f) = \frac{1}{4\pi} \oint_C dx^\mu \oint_{C_f} dy^\nu \epsilon_{\mu\nu\rho} \frac{(x-y)^\rho}{|x-y|^3} \tag{2.2}$$

vanishes, whereas in the vertical framing the field n^μ can be identified simply with the principle normal in the Frenet basis associated with C.

Summarising, $< W_R(C) >$ has a well-defined expansion in $\lambda = \frac{2\pi}{k}$ and the classical framing procedure (2.1) consistently extends [8] to all orders in λ. As a result, $< W_R(C) >$ is entirely expressed in terms of purely topological quantities. The first few terms of the λ-expansion read

$$< W_R(C) >= dim R \Big[1 - i\lambda c_2(R) \chi(C, C_f)$$

$$- \frac{1}{2} \lambda^2 c_2^2(R) \chi^2(C, C_f) + \lambda^2 c_v c_2(R) \varrho(C) \Big] + O(\lambda^3). \tag{2.3}$$

Here C_f is an arbitrary framing contour, $c_2(R)\mathbf{1} = R^a R^a$, $\delta^{ab} c_v = f^{acd} f^{bcd}$ and

$$\varrho(C) = \varrho_1(C) + \varrho_2(C) \quad , \tag{2.4}$$

with

$$\varrho_1(C) = -\frac{1}{32\pi^3} \oint_C dx^\mu \int^x dy^\nu \int^y dz^\rho \epsilon^{\alpha\beta\gamma} \epsilon_{\mu\alpha\sigma} \epsilon_{\nu\beta\lambda} \epsilon_{\rho\gamma\tau} I^{\sigma\lambda\tau}(x, y, z) \quad , \tag{2.5}$$

$$I^{\sigma\lambda\tau}(x, y, z) = \int d^3 w \frac{(w-x)^\sigma}{|w-x|^3} \frac{(w-y)^\lambda}{|w-y|^3} \frac{(w-z)^\tau}{|w-z|^3}, \tag{2.6}$$

and

$$\varrho_2(C) = \frac{1}{8\pi^2} \oint_C dx^\mu \int^x dy^\nu \int^y dz^\rho \int^z dw^\sigma \epsilon_{\sigma\nu\alpha} \epsilon_{\rho\mu\beta} \frac{(w-y)^\alpha}{|w-y|^3} \frac{(z-x)^\beta}{|z-x|^3} \,. \tag{2.7}$$

The explicit form of $\varrho(C)$ and its framing-independence have been analyzed in [8], where it has been shown that $\varrho(C)$ is a knot invariant; it represents the second coefficient of the Alexander-Conway polynomial.

Notice that λ entering in eq.(2.3) is the renormalized coupling constant; i.e. $\frac{2\pi}{\lambda}$ multiplies the renormalized action (1.1). We emphasize that the renormalized coupling constant should not be confused with the bare coupling constant. The expression (2.3) is universal in the sense that it depends neither on the regularization method nor on any finite renormalization, as it should be for any physical quantity . The variable λ appearing in what follows always represents the renormalized coupling constant.

Eq.(2.3) can be easily generalized [8] to the case of a generic link L. In vertical framing (VF) and in the fundamental representation, one has

$$\left.\frac{< W(L) >}{< W(U_0) >}\right|_{VF} = S_L \quad , \tag{2.8}$$

where U_0 is the unknot and S_L is a regular isotopy invariant polynomial introduced in [11]. S_L can be constructed recursively by means of the skein relations

$$S_{\widehat{L}_+} = \alpha S_{\widehat{L}_0} \quad , \quad S_{\widehat{L}_-} = \alpha^{-1} S_{\widehat{L}_0} \quad , \tag{2.9}$$

$$\beta S_{L_+} - \beta^{-1} S_{L_-} = z S_{L_0} \quad , \tag{2.10}$$

where the siagrams associated with $\{L_\pm, L_0, \widehat{L}_\pm, \widehat{L}_0\}$ differ only at the site of one crossing as shown in Fig.1 and the values of the skein parameters are

$$\alpha = \exp\left(-i\lambda \frac{N^2-1}{2N}\right) \quad , \tag{2.11}$$

$$\beta = \exp\left(-i\lambda \frac{1}{2N}\right) \quad , \quad z = -2i \sin\frac{\lambda}{2} \quad . \tag{2.12}$$

Eqs.(2.8) and (2.11,12) have been checked in the manifestly covariant BRS approach at order λ^2. By using the Schrödinger representation in the temporal gauge, we shall verify the skein relation (2.10) and the values (2.12) to all orders in λ.

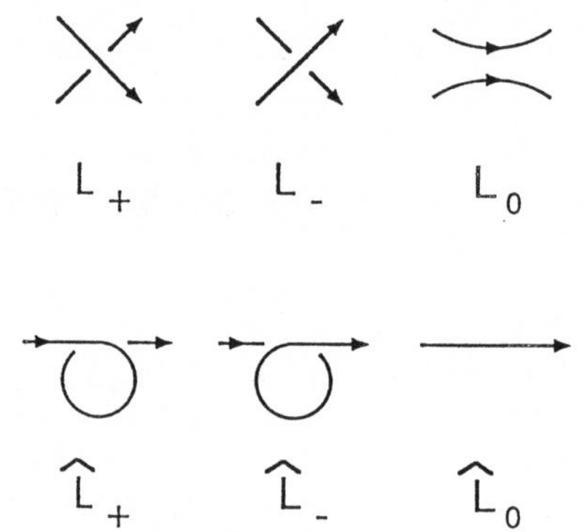

Figure 1

Note that the parameters (2.11,12) are consistent with the relation (1.20) following from the I-symmetry (1.16). Indeed, from the skein relations (2.9,10) it follows that

$$S_L(\alpha,\beta,z) = S_{\widetilde{L}}(\alpha^{-1},\beta^{-1},-z) \quad , \tag{2.13}$$

where $\widetilde{L}$ is the mirror image of the link L. Combining eqs.(1.20), (2.8) and (2.13), one gets

$$\begin{aligned} \alpha(\lambda) = \alpha^{-1}(-\lambda) \quad , \quad \beta(\lambda) = \beta^{-1}(-\lambda) \quad , \\ z(\lambda) = -z(-\lambda) \quad , \end{aligned} \tag{2.14}$$

which are in agreement with eqs.(2.11,12).

Eq.(2.8) is quite remarkable because it provides a field theoretical interpretation of the link polynomial S_L. At this stage one can make also contact with the HOMFLY polynomial P_L, which is by definition [12] an ambient isotopy invariant. Indeed, one can show [11] that

$$P_L(t=\alpha\beta, z) = \alpha^{-w(L)} S_L(\alpha,\beta,z) \quad , \tag{2.15}$$

where the writhe $w(L)$ is given by

$$w(L) = \sum_p \epsilon(p) \quad . \tag{2.16}$$

The sum in (2.16) is performed over all crossings of L and $\epsilon(L_\pm) = \pm 1$. Now, using the relation (2.8) one obtains for a knot C

$$\left. \frac{< W_T(C) >}{< W_T(U_0) >} \right|_{SF} = P_C(t,z) \quad , \tag{2.17}$$

where the Wilson lines in the L.H.S. are computed in the fundamental representation $R^a = T^a$ of $SU(N)$ and "SF" stands for standard framing.

From eq.(2.17) one can extract new information concerning the knot invariants described by P_C. Comparing the coefficients of the λ-expansion of both sides of eq.(2.17), one finds [8]

$$\frac{1}{2}\left[\varrho(C) + \frac{1}{12}\right] \text{ mod } 2 = \text{Arf}(C) \quad , \tag{2.18}$$

where Arf(C) is the Arf-invariant. One also gets

$$\frac{1}{2}\left[\varrho(C) + \frac{1}{12}\right] = \epsilon(C_{n+1}) - \epsilon(C_n) \quad , \tag{2.19}$$

where $\epsilon(C_n)$ is the Casson invariant associated with the $\frac{1}{n}$-th Dehn surgery constructed by the knot C. Eqs.(2.18,19), combined with (2.4-7), are in our knowledge the first analytic formulae for the computation of the Arf- and Casson invariants.

For an arbitrary representation R, the L.H.S. of eq.(2.17) has the following general structure

$$\left.\frac{< W_R(C) >}{< W_R(U_0) >}\right|_{SF} = \sum_{m=0}^{\infty} \lambda^m \left[\sum_{n \in A_m} M_n(R) u_n(C) \right] \quad , \tag{2.20}$$

where M_n is a monomial of the Casimir operators of the group G in the R-representation and A_m is a finite set of indices. The group structure is entirely factorized in $\{M_n(R)\}$ and the set $\mathcal{U} = \{u_n(C)\}$ describes universal knot invariants. Standard perturbation theory provides explicit analytic expressions for $u_n(C)$.

Consider furthermore the coefficients $b_{ij}(C)$ appearing in

$$P_C(t,z) = \sum_{i \geq 0} \sum_{j \in Z} b_{ij}(C) z^{2i} t^{2j} \quad . \tag{2.21}$$

We recall that $b_{ij}(C)$ are integers and represent knot invariants. Inserting in eq.(2.21) the values of the parameters t and z obtained from eq.(2.11,12) and using that [13] $P_C(t, t - t^{-1}) = 1$, one gets the following relations

$$\sum_{p=0}^{n} Q_{n,p} = \delta_{n0} \quad . \tag{2.22}$$

Here for even q, $0 \leq q \leq n$

$$Q_{n,n-q} = \binom{n}{n-q} \sum_{r \geq 0}^{q/2} \sum_{m=0}^{2r} (-1)^m \binom{2r}{m} (m-r)^q \sum_{j \in Z} j^{n-q} b_{rj}(C) \tag{2.23}$$

and $Q_{n,n-q} = 0$ for odd q. From eq.(2.22) it follows that the knot invariants $b_{ij}(C)$ are not independent. It is remarkable that the elements of $\mathcal{U}$ automatically take into account the relations (2.22).

The equation (2.20) can be generalized to the case of links. It is known [14] that $< W_R(L) >$ in the vector representation of $SU(N)$ distinguishes some links which have the same HOMFLY polynomial. Therefore the set of universal invariants $\mathcal{U}$ is more selective than $\{b_{ij}(C)\}$. It is a challenging open problem if $\mathcal{U}$ provides a complete classification of links.

3. Quantization in the temporal gauge

The C-S action in a (1+2)-decomposition of R^3 reads

$$S_{CS} = \frac{k}{4\pi} \int d^3x \left(A_2^a \dot{A}_1^a + A_0^a F_{12}^a \right) \quad , \tag{3.1}$$

where $F^a_{12} = \partial_1 A^a_2 - \partial_2 A^a_1 - f^{abc} A^b_1 A^c_2$ and f^{abc} are the structure constants of G. In our notations $x^\mu = (x^0, \vec{x}) = (x^0, x^i)$ with $i = 1, 2$. We shall use also the two-dimensional antisymmetric tensors ϵ_{ij} and ϵ^{ij} defined by $\epsilon_{12} = \epsilon^{21} = 1$, $\epsilon_{21} = \epsilon^{12} = -1$. The time-components A^a_0 appear in (3.1) without time derivatives and can be viewed as Lagrange multipliers enforcing the Gauss law, which in the absence of sources reads

$$F^a_{12} = 0 \quad . \tag{3.2}$$

The quantization is performed in the usual way: we first impose the canonical commutation relations

$$\left[A^a_1(\vec{x}), A^b_2(\vec{y})\right] = i\delta^{ab} 2\lambda \delta(\vec{x} - \vec{y}) \quad , \tag{3.3}$$

and then select the physical states by means of the Gauss constraint. As is well known,

$$G(\epsilon) = -\frac{k}{4\pi} \int d^2x F^a_{12}(\vec{x}) \epsilon^a(\vec{x}) \quad , \tag{3.4}$$

where $\epsilon^a(\vec{x})$ has compact support, generates the residual local gauge transformations. In what follows we use the standard Schrödinger representation of (3.3). It is convenient [15] to introduce the field variable $U(\vec{x})$ defined by

$$T^a A^a_1(\vec{x}) = -iU^{-1}(\vec{x}) \partial_1 U(\vec{x}) \quad . \tag{3.5}$$

The wave function Ψ_0 of the vacuum takes a simple form in terms of $U(\vec{x})$ [15], namely

$$\Psi_0 = \exp\left[-i\frac{k}{4\pi} \int d^2x \mathrm{Tr}(U^{-1}\partial_1 U U^{-1} \partial_2 U) \right.$$
$$\left. -i\frac{k}{12\pi} \int_D d^3y \epsilon^{\alpha\beta\gamma} \mathrm{Tr}(U^{-1}\partial_\alpha U U^{-1}\partial_\beta U U^{-1}\partial_\gamma U)\right] \quad , \tag{3.6}$$

where the integral of the Wess-Zumino three-form is performed on a three-dimensional manifold D whose boundary is R^2. The requirement of single valuedness of Ψ_0 implies that k takes integer values (our normalization refers to the case $G = SU(N)$). One can easily verify that Ψ_0 satisfies the Gauss law (3.2), i.e. $G(\epsilon)\Psi_0 = 0$.

We would like to describe now the states corresponding to the case when there are some distinguished points (punctures) on the plane R^2 with a definite non-trivial representation of the local gauge group assigned to each puncture. Such states appear as intermediate states at fixed time $t = t_0$ in computing the expectation values $< W(L) >$ of Wilson line operators. The punctures arise as intersection points of the link L with the plane $t = t_0$ [3]. The corresponding wave functions can be constructed as follows [16,17]. Since the matrix valued variable $U(\vec{x})$ takes values in G, we can define $U(\vec{x}, R)$ to be the representative of $U(\vec{x})$

in the R-representation of G. Consider then the wave function

$$\Psi[\vec{y}, R]_{IJ} = U(\vec{y}, R)_{IJ}\Psi_0 \quad , \tag{3.7}$$

where the indices I and J label the matrix elements of $U(\vec{y}, R)$. By making use of eqs.(3.3,4) one finds

$$(iG(\epsilon)\Psi[\vec{y}, R])_{IJ} = i\Psi[\vec{y}, R]_{IK} R^a_{KJ} \epsilon^a(\vec{y}) \quad . \tag{3.8}$$

Eq.(3.8) shows that $\Psi[\vec{y}, R]_{IJ}$ describes a state associated with a single puncture located in $\vec{y}$ and in the R-representation of G. The different vectors of the R-representation on which $G^a(\vec{y})$ act are labelled by the second index J of $\Psi[\vec{y}, R]_{IJ}$. Analogously, the wave function

$$\Psi[\vec{y}, R]^{IJ} = U^{-1}(\vec{y}, R)_{IJ}\Psi_0 \tag{3.9}$$

also describes a puncture in $\vec{y}$ associated with the R-representation of the group G. Differently from the wave function (3.7), however, the vectors of the R-representation associated with $G^a(\vec{y})$ are labelled in this case by the first index I of $\Psi[\vec{y}, R]^{IJ}$. Clearly, multiplying Ψ_0 by several matrix elements of the variable U (and/or of its inverse U^{-1}) in different points and in different representations, one obtains a wave function of the type

$$\Psi[\vec{y}_1, R_1]_{IJ}[\vec{y}_2, R_2]^{KL} \cdots = U(\vec{y}_1, R_1)_{IJ} U^{-1}(\vec{y}_2, R_2)_{KL} \cdots \Psi_0 \,, \tag{3.10}$$

which describes several punctures (located in $\vec{y}_1, \vec{y}_2, \ldots$) associated with the representations $R_1, R_2, \ldots$ of G respectively.

Observe that the first lower index and the second upper index of the wave function (3.10) are free; they label the states of the charges "at infinity" and $G(\epsilon)$ does not act on them. We call these indices q-indices. As we shall see later, the q-indices give rise to the degrees of freedom on which the "quantum deformation" $\mathcal{U}_q(\mathcal{G})$ of G is acting. In the wave functions corresponding to Wilson lines defined in a finite region, all q-indices are contracted. Let us consider indeed the wave function $\Psi(\Gamma)$,

$$\Psi_{IJ}(\Gamma) = W_{IJ}(\Gamma)\Psi_0 \quad , \tag{3.11}$$

obtained by applying the quantum holonomy operator

$$W(\Gamma) = \text{ P exp } \left(i \int_\Gamma A^a_i(\vec{x}) R^a dx^i \right) \tag{3.12}$$

on the vacuum. The path-ordered integral in eq.(3.12) is performed along an oriented non-intersecting smooth path Γ in R^2 connecting the point $\vec{y}_1$ with $\vec{y}_2$. Since $G^a(\vec{x})$ generate gauge transformations, it follows that

$$(iG^a(\vec{x})\Psi(\Gamma))_{IJ} = -i\delta(\vec{x} - \vec{y}_1) R^a_{IK} \Psi_{KJ}(\Gamma) + i\delta(\vec{x} - \vec{y}_2) \Psi_{IK}(\Gamma) R^a_{KJ} \,. \tag{3.13}$$

Therefore, $\Psi(\Gamma)$ describes two punctures in $\vec{y}_1$ and $\vec{y}_2$ associated with the same representation R. Eq.(3.13) is exactly the *operator* Gauss law constraint in the presence of (in this case two) punctures. As usual, we call *physical* the states satisfying the operator Gauss law.

In deriving the explicit form of $\Psi(\Gamma)$, some care is needed. In order to preserve general covariance, a framing procedure has to be introduced [3,8]. For Wilson lines defined on a plane, one can adopt the so-called constant framing [17]. In constant framing one finds [17]

$$\Psi_{IJ}(\Gamma) = \Psi[\vec{y}_1, R]^{IK}[\vec{y}_2, R]_{KJ}\,, \tag{3.14}$$

which shows that in wave functions corresponding to Wilson lines, the q-indices are indeed constracted.

We want to study now the monodromy properties of the physical states. Let us consider for example the wave function Ψ associated with a four-puncture configuration,

$$\Psi = \Psi[\vec{y}_1, R_1]^{IM}[\vec{y}_2, R_1]_{MJ}[\vec{y}_3, R_2]^{KN}[\vec{y}_4, R_2]_{NL} \quad . \tag{3.15}$$

The parallel transport, say, of the puncture $\vec{y}_2$ along a loop C defines a monodromy on the four-puncture physical states. This operation gives the state

$$\begin{aligned} \mathcal{B}\Psi = U^{-1}(\vec{y}_1, R_1)_{IM} U(\vec{y}_2, R_1)_{MQ} \left(\text{P} \exp i \int_C dx^i A_i^a(\vec{x}) R_1^a \right)_{QJ} \\ \times U^{-1}(\vec{y}_3, R_2)_{KN} U(\vec{y}_4, R_2)_{NL} \Psi_0 \quad . \end{aligned} \tag{3.16}$$

For a clockwise-oriented C enclosing the puncture $\vec{y}_4$ only, see Fig.2, one gets [17] (in constant framing)

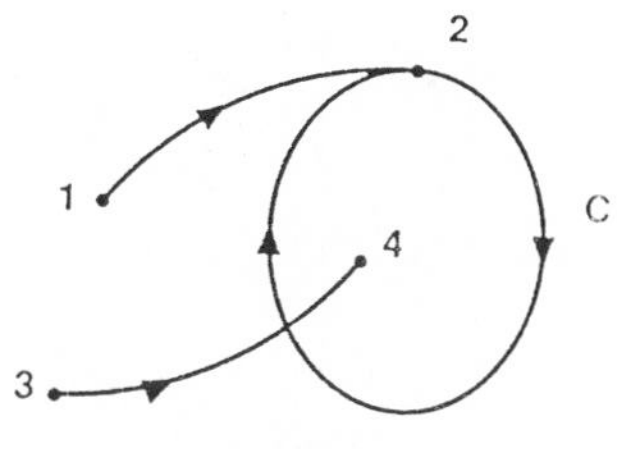

Figure 2

$$\mathcal{B}\Psi = B_{M,N}^{S,Q}\Psi[\vec{y}_1, R_1]^{IM}[\vec{y}_2, R_1]_{SJ}[\vec{y}_3, R_2]^{KN}[\vec{y}_4, R_2]_{QL}\,, \tag{3.17}$$

where $B_{M,N}^{S,Q}$ are the matrix elements of

$$B = \exp\left(i\frac{4\pi}{k} R_1^a \otimes R_2^a \right) \quad . \tag{3.18}$$

The matrix B acts on the tensor product of the two representations R_1 and R_2 of the group G, whose vectors are labelled by the q-indices S and Q in the expression $U(\vec{y}_2, R_1)_{SJ} U(\vec{y}_4, R_2)_{QL}$. Moreover, all the indices of B are contracted with q-indices. Eqs.(3.17,18) can easily be generalized to the case of several punctures. Apart from some phase ambiguities related to the framing, i.e. to the "self-interaction" of the quantum holonomy operator, the results (3.17,18) are invariant under smooth deformations of the contour C on the *punctured* plane. For a loop enclosing several punctures the associated monodromy is the ordered product of B-matrices of the type (3.18).

We illustrate now the important role of the B-matrix by deriving the skein relation (2.10). The starting point is represented by the so-called "half-monodromies" associated with the exchange of two punctures in the same representation, as shown in Fig.3.

Figure 3

Twice this operation must reproduce the matrix B; therefore, the matrix associated to the oriented exchange operation $\vec{y}_1 \leftrightarrow \vec{y}_2$ is taken to be

$$\mathcal{R} = \exp(i\lambda R^a \otimes R^a) \quad . \tag{3.19}$$

In a three-dimensional picture, B and $\mathcal{R}$ can be interpreted as follows [3]. The B-matrix describes the effect of the double exchange, shown in Fig.4, of the two

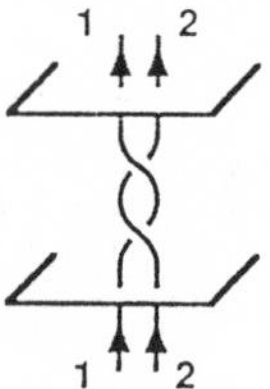

Figure 4

punctures. The exchange operation of the punctures in Fig.3 corresponds to the replacement of the configuration in Fig.5(a) with the configuration in Fig.5(b).

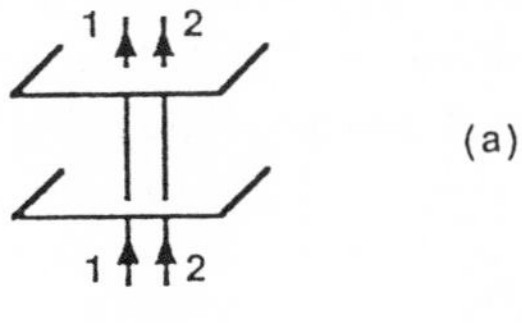

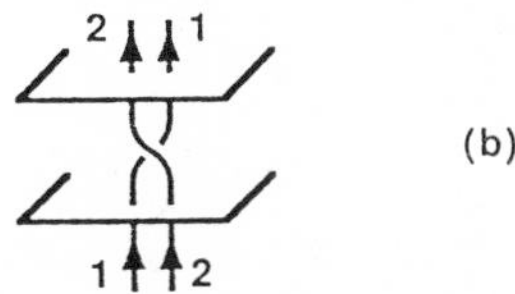

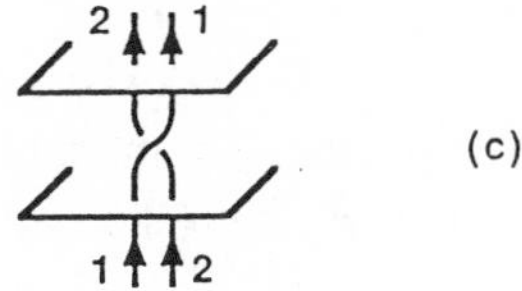

Figure 5

Vice versa, the inverse matrix $\mathcal{R}^{-1}$ is associated with the operation shown in Fig.5(c). At the level of states, one has the following situation. The initial state $|L_0>$, Fig.5(a), is

$$|L_0>=|I>|J> \quad , \tag{3.20}$$

where $|I>|J>\in H_1\otimes H_2$. Under the exchange shown in Fig.5(b), one obtains the state

$$|L_->=\Sigma\mathcal{R}|L_0> \quad , \tag{3.21}$$

whereas for the case shown in Fig.5(c), one has

$$|L_+>=\Sigma\mathcal{R}^{-1}|L_0> \quad , \tag{3.22}$$

where Σ is the permutation operator: $\Sigma|I>|J>=|J>|I>$.

In the fundamental representation of $G=SU(N)$, the above formulae take the simple form

$$|L_0>=|s>+|a> \quad , \tag{3.23}$$

$$|L_\mp>=\exp(\pm\lambda\gamma_-)|s>-\exp(\mp\lambda\gamma_+)|a> \quad , \tag{3.24}$$

$$\gamma_\pm=i\frac{N\pm 1}{2N} \quad , \tag{3.25}$$

where $|s>$ ($|a>$) is the symmetric (antisymmetric) component of the state $|L_0>$.

By means of eqs.(3.23-25) one obtains the skein relation

$$\beta|L_+>-\beta^{-1}|L_->=z|L_0> \quad , \tag{3.26}$$

where β and z are given by eq.(2.12). The values of β and z agree also with Witten's result [3] based on the two-dimensional WZW model, if one makes the identification

$$k = \ell + c_v \quad , \tag{3.27}$$

where ℓ is the level of the WZW model.

4. Braid group representation

On general grounds [3] one expects the existence of a braid group representation acting on the physical space of the C-S model. In what follows we indicate how this representation can be derived from the half-monodromy matrices (3.19). We start with some definitions.

Let $R^2_n = R^2 \backslash \{\vec{y}_1, \vec{y}_2, \dots, \vec{y}_n\}$ be the punctured plane and let us consider the spaces $X_n = \{\vec{y}_1, \vec{y}_2, \dots, \vec{y}_n : \vec{y}_p \neq \vec{y}_q \text{ for } p \neq q\}$ and $Y_n = X_n / S_n$, S_n being the permutation group of n elements. The fundamental groups $P_n = \pi_1(X_n)$ and $B_n = \pi_1(Y_n)$ are called the pure braid group and the Artin braid group respectively. Recall that the generators $\{\sigma_i : i = 1, \dots, n-1\}$ of B_n satisfy

$$\begin{aligned} \sigma_i \sigma_j = \sigma_j \sigma_i \quad &, \quad |i-j| \geq 2 \quad , \\ \sigma_i \sigma_{i+1} \sigma_i = \sigma_{i+1} \sigma_i \sigma_{i+1} \quad &, \quad 1 \leq i \leq n-2 \quad . \end{aligned} \tag{4.1}$$

In sect.3 we have derived the skein parameters by means of the matrix $\Sigma\mathcal{R}$ which, in view of the interpretation illustrated by Figs.4 and 5, gives just a representation

$$\sigma_1 \mapsto \Sigma\mathcal{R} \tag{4.2}$$

of B_2. Our aim now is to find the braiding matrices in the general case of $n \geq 2$ punctures. The problem is well posed; a representation of B_n certainly exists on the physical space of the C-S model because of general covariance. Differently from what in might appear at first sight however, the construction of this representation is not immediate. For instance, the naive extension of the procedure used for $n = 2$ does not work for $n \geq 3$. The difficulties appear in satisfying the relations (4.1) which become non-trivial for $n \geq 3$. In order to construct the B_n-representation, we shall use some results of Drinfel'd [18] and Kohno [19].

We recall first the definition of a quasi-triangular quasi-Hopf (QTQH) algebra given by Drinfel'd [18]. Let A be an associative unital algebra with a homomorphism $\Delta : A \to A \otimes A$ called co-multiplication. In general Δ is neither co-commutative, nor co-associative. In terms of the permutation mapping $\sigma(a_1 \otimes a_2) = a_2 \otimes a_1$ one may construct another co-product, namely the composition $\Delta' = \sigma \circ \Delta$. Suppose that there exists an invertible element $\mathcal{R} \in A \otimes A$, such that $\forall a \in A$,

$$\Delta'(a) = \mathcal{R}\Delta(a)\mathcal{R}^{-1} \quad . \tag{4.3}$$

Assume moreover that there exists an invertible element $\phi \in A \otimes A \otimes A$, obeying:

$$(\mathrm{id} \otimes \Delta)(\Delta(a)) = \phi(\Delta \otimes \mathrm{id})(\Delta(a))\phi^{-1} \quad , \tag{4.4}$$

$$(\mathrm{id} \otimes \mathrm{id} \otimes \Delta)(\phi)(\Delta \otimes \mathrm{id} \otimes \mathrm{id})(\phi) = (1 \otimes \phi)(\mathrm{id} \otimes \Delta \otimes \mathrm{id})(\phi)(\phi \otimes 1) \quad . \tag{4.5}$$

Then the quadruple $\{A, \Delta, \phi, \mathcal{R}\}$ is called a QTQH algebra if the following relations are satisfied:

$$(\Delta \otimes \mathrm{id})(\mathcal{R}) = \phi_{312}\mathcal{R}_{13}\phi_{132}^{-1}\mathcal{R}_{23}\phi \quad , \tag{4.6}$$

$$(\mathrm{id} \otimes \Delta)(\mathcal{R}) = \phi_{231}^{-1}\mathcal{R}_{13}\phi_{213}\mathcal{R}_{12}\phi^{-1} \quad . \tag{4.7}$$

Here the notation is as follows. Denote by $\phi \equiv \phi_{123} = \sum_i a_i \otimes b_i \otimes c_i$; then $\phi_{312} = \sum_i c_i \otimes a_i \otimes b_i$ and so on, and analogously for $\mathcal{R}_{ij}$.

As a consequence of eqs.(4.4-7), the $\mathcal{R}$-matrix satisfies the quasi-Yang-Baxter equation

$$\mathcal{R}_{12}\phi_{312}\mathcal{R}_{13}\phi_{132}^{-1}\mathcal{R}_{23}\phi = \phi_{321}\mathcal{R}_{23}\phi_{231}^{-1}\mathcal{R}_{13}\phi_{213}\mathcal{R}_{12} \quad . \tag{4.8}$$

Given a QTQH algebra and an invertible element $F \in A \otimes A$, one can define a new QTQH algebra $\{A, \Delta_F, \phi_F, \mathcal{R}_F\}$ where

$$\Delta_F(a) = F\Delta(a)F^{-1} \quad , \tag{4.9}$$

$$\phi_F = F_{23}(\mathrm{id} \otimes \Delta)(F)\phi(\Delta \otimes \mathrm{id})(F^{-1})(F_{12})^{-1} \quad , \tag{4.10}$$

and

$$\mathcal{R}_F = F_{21}\mathcal{R}F^{-1} \quad . \tag{4.11}$$

The algebra $\{A, \Delta_F, \phi_F, \mathcal{R}_F\}$ is called a twisting of $\{A, \Delta, \phi, \mathcal{R}\}$ by F.

Starting from a Lie algebra $\mathcal{G}$, one can naturally construct a QTQH algebra $\mathcal{A}_\lambda(\mathcal{G})$ as follows. Take for A the algebra whose elements are power series in some parameter λ with coefficients belonging to the universal enveloping algebra of $\mathcal{G}$. With the standard co-product $\Delta(a) = 1 \otimes a + a \otimes 1$, the $\mathcal{R}$-matrix (3.19) satisfies eq.(4.3). The non-trivial point is to find ϕ satisfying (4.4-7). This problem has been solved by Drinfel'd who has constructed ϕ by using solutions of the Knizhnik-Zamolodchikov [20] equation with apropriate boundary conditions.

Summarising, we have learned that the half-monodromy (3.19) represents actually the $\mathcal{R}$-matrix of a QTQH algebra. In order to make contact with braid groups, we introduce also the so-called quasi-triangular Hopf (QTH) algebras. A QTH algebra is simply a QTQH algebra with $\phi = 1$. In this case the $\mathcal{R}$-matrix satisfies the ordinary Y-B equation

$$\mathcal{R}_{12}\mathcal{R}_{13}\mathcal{R}_{23} = \mathcal{R}_{23}\mathcal{R}_{13}\mathcal{R}_{12} \tag{4.12}$$

and naturally defines a representation θ of B_n given by

$$\theta : \sigma_i \mapsto 1 \otimes \cdots \otimes \Sigma\mathcal{R} \otimes \cdots \otimes 1 \quad . \tag{4.13}$$

The most familiar examples of QTH algebras are the quantum groups $\mathcal{U}_q(\mathcal{G})$. For

$$q = \exp(-i\lambda) \tag{4.14}$$

the algebra A entering in both $\mathcal{A}_\lambda(\mathcal{G})$ and $\mathcal{U}_q(\mathcal{G})$ is the same, but $\mathcal{U}_q(\mathcal{G})$ is equipped with a different co-product, the well-known quantum group co-product [21]. Consequently the relative $\mathcal{R}$-matrices are also different. For example the quantum group $\mathcal{R}$-matrix for $SU(N)$ in the fundamental representation reads [21]

$$\mathcal{R} = \sum_{i=1}^{N} E_{ii} \otimes E_{ii} + \sqrt{q} \sum_{i \neq j}^{N} E_{ii} \otimes E_{jj} + (1-q) \sum_{i>j}^{N} E_{ij} \otimes E_{ji} \quad , \tag{4.15}$$

where E_{ij} are the Weyl matrices. It is also worth mentioning that the representation θ associated with $\mathcal{U}_q(\mathcal{G})$ commutes with the diagonal action of $\mathcal{U}_q(\mathcal{G})$.

At this stage we are ready to find the braid group representation associated with the monodromy matrices (3.19). In fact we can use a theorem by Drinfel'd [18], which states that $\mathcal{U}_q(\mathcal{G})$ is isomorphic to $\mathcal{A}_\lambda(\mathcal{G})$ up to twisting and that the relative B_n-representations are equivalent. Therefore the B_n-representation defined by the monodromy matrices (3.19) is equivalent to θ given by eq.(4.13).

The same conclusion has been reached also in references [16,17,20] by using a diferent method. One has first to observe that the Knizhnik-Zamolodchikov connection w satisfies the C-S Gauss constraint. Then one can apply a theorem of Kohno [19] showing that the holonomies, constructed with w, define a braid group representation which is equivalent to (4.13).

Summarising, we have established the braid group representation carried by the physical space of the C-S model. Using the deep relation of this representation with quantum groups, one can show [16,17] that the C-S model admits a hidden quantum group symmetry. Such symmetries are fundamental for integrable systems and are expected to emerge also in certain conformal field theories.

5. Conclusions

The results reported above confirm that the quantized C-S model can be successfully used for the study of link invariants and the geometry of three-manifolds. The set $\mathcal{U} = \{u_n(C)\}$ of universal knot invariants deserves further investigation. First of all, the possibility to construct explicit analytic expressions for $\{u_n(C)\}$ may provide a differential-geometrical and topological interpretation of these invariants, which is one of the main open problems in knot theory [22]. Second, the elements of $\mathcal{U}$ take into account all the relations between the HOMFLY coefficients and are probably more convenient in the study of the classical conjectures of knot theory. Finally, since the C-S model makes sense on an arbitrary three-manifold M^3, following the above scheme one could in principle produce universal link invariants on M^3.

Acknowledgements. One of the authors (M. Martellini) thanks the organizers of the International Conference on Knot Theory for the kind invitation.

References

[1] A.S. Schwarz, Lett. Math. Phys. 2 (1978) 247; Comm. Math. Phys. 67 (1979) 1; Abstracts of the Baku Intern. Topological Conf. 1987.

[2] M.F. Atiyah, *New Invariants of Three- and Four-Dimensional Manifolds*, in *The Mathemetical Heritage of Hermann Weyl*, Proc. Symp. Pure Math. 48, ed. R. Wells (Amer. Math. Soc., 1988).

[3] E. Witten, Comm. Math. Phys. 121 (1989) 351; Nucl. Phys. B322 (1989) 629.

[4] C. Becchi, A. Rouet and R. Stora, Comm. Math. Phys. 42 (1975) 127, and Ann. Phys. 98 (1976) 287.

[5] D. Birmingham, M. Rakowski and G. Thompson, Nucl. Phys. B229 (1990) 83.

[6] F. Delduc, F. Gieres and S.P. Sorella, Phys. Lett. B225 (1989) 367.

[7] F. Delduc, C. Lucchesi, O. Piguet and S. P. Sorella, *Exact Scale Invariance of the Chern-Simons Theory in the Landau Gauge*, University of Geneva preprint, UGVA-DPT 1990/2-653.

[8] E. Guadagnini, M. Martellini and M. Mintchev, Phys. Lett. B228 (1989) 489; Nucl. Phys. B330 (1990) 575.

[9] E. Guadagnini, M. Martellini and M. Mintchev, Phys. Lett. B227 (1989) 111.

[10] G. Galugareanu, Czechoslovak Math. J. 11 (1961) 588.

[11] P. Cotta-Ramusino, E. Guadagnini, M. Martellini and M. Mintchev, Nucl. Phys. B330 (1990) 557.

[12] P. Freyd, D. Yetter, J. Hoste, W.B.R. Lickorish, K. Millett and A. Ocneanu, Bull. Amer. Math. Soc. 12 (1985) 239.
J.H. Przytycki and P.Traczyk, Kobe J. Math. 4 (1987) 115.

[13] W.B.R. Lickorish and K. Millett, Topology 26 (1987) 107.

[14] M. Wadati, T. Deguchi and Y. Akutsu, Phys. Rep. 180 (1989) 247.

[15] G.V. Dunne, R. Jackiw and C.A. Trugenberger, Ann. Phys. 149 (1989) 197.

[16] E. Guadagnini, M. Martellini and M. Mintchev, Phys. Lett. B235 (1990) 275.

[17] E. Guadagnini, M. Martellini and M. Mintchev, Nucl. Phys. B336 (1990) 581.

[18] V.G. Drinfel'd, *Quasi-Hopf Algebras and Knizhnik-Zamolodchikov Equations*, in *Problems of Modern Quantum Field Theory*, ed. A.A. Belavin, A.U. Klimyk and A.B. Zamolodchikov (Springer-Verlag, 1990).

[19] T. Kohno, Ann. Inst. Fourier, Grenoble 37 (1987) 139.

[20] J. Fröhlich, *"Statistics of Fields, the Yang-Baxter Equation, and the Theory of Knots and Links"*, in *"Nonperturbative Quantum Field Theory"*, ed. by G. 't Hooft et al, (Plenum Press, New York 1988).

[21] V.G. Drinfel'd, Soviet Math. Dokl. 32 (1985) 254.
L.D. Faddeev, N.Yu. Reshetikhin and L.A. Takhtajan, *Quantization of Lie Groups and Lie Algebras*, preprint LOMI E-14-87.
M. Jimbo, Commun. Math. Phys 102 (1986) 537.
N. Yu. Reshetikhin, *Quantized Universal Enveloping Algebras, the Yang-Baxter Equation and Invariants of Links I and II*, preprints LOMI E-4-87 and E-17-87.

[22] W.B.R. Lickorish, Bull. London Math. Soc 20 (1988) 558.

A note on incompressible surfaces in solid tori and in lens spaces

Chichen M. Tsau

Abstract. A classification of simple closed curves on the boundary of a solid torus which bound non-orientable incompressible surfaces is given. Viewing a lens space as being obtained by gluing two solid tori along their boundaries, we study the relationship between incompressible (hence non-orientable) surfaces in a lens space and properly embedded non-orientable incompressible surfaces in one of the solid tori Heegaard splitting of the lens space.

Closed non-orientable surfaces which can be embedded in lens spaces were completely determined in [B-W]; in there a recursive formula for the genus of such surfaces was given, and it was shown that if F is a closed non-orientable surface of minimal genus in a lens space, then F can be isotoped so as to intersect one of the solid tori Heegaard splitting of the lens space in a single meridional disk. Later Rubinstein showed in [R] that this result is true if F is incompressible in the lens space. Indeed, any non-orientable surface of minimal genus in a lens space is incompressible. In [R] it was also proved that the converse of the above statement is true by showing any two incompressible surfaces in a lens space are isotopic. In this note, we first classify the simple closed curves on the boundary of a solid torus which bound non-orientable incompressible surfaces; using this classification we give new proofs of certain well-known results concerning (incompressible) non-orientable surfaces embedded in lens spaces; we show that when an incompressible surface in a lens space is isotoped so as to intersect one of the solid tori Heegaard splitting of the lens space in a single meridional disk, then the portion of the surface in the other solid torus is incompressible in that solid torus, and conversely, if F is a properly embedded incompressible surface in a solid torus, then the surface $\hat{F}$ obtained from F by capping off ∂F with a meridional disk of another solid torus which is glued to the first one along their boundaries is incompressible in the resulting lens space. As a consequence, we

Knots 90

show that, similar to the case of lens spaces, a properly embedded non-orientable surface in a solid torus V bounded by a simple closed curve α is incompresible in V if and only if it is a non-orientable surface of minimal genus in V with boundary α.

By a *surface* we shall mean a compact connected 2-manifold with or without boundary. A surface F properly embedded in a 3-manifold is *compressible* in M if either $F \cong S^2$ and F bounds a 3-ball in M or there exists a 2-disk D in M such that $D \cap F = \partial D$ and ∂D is not contractible in F (equivalently $[\partial D] \neq 1$ in $\pi_1(F)$); otherwise, the surface F is *incompressible* in M. Let V be a solid torus, a pair (m, l) of simple closed curves in ∂V is a *meridian-longitude pair for* V if m bounds a meridional disk in V and l represents a generator of $\pi_1(V)$. Recall that if α is a simple closed curve in ∂V, then $\alpha = pl + qm$ in $\pi_1(\partial V)$, where either $p = q = 0$ or (p, q) is a pair of relative prime integers. We shall adopt the definition of lens space as being obtained by gluing two solid tori along their boundaries; let V_1 and V_2 be solid tori with meridian and longitude pairs (m_1, l_1) and (m_2, l_2), respectively, and let $h : \partial V_1 \mapsto \partial V_2$ be a homeomorphism such that $h(pl_1 + qm_1) = m_2$, then the 3-manifold $V_1 \underset{h}{\cup} V_2$ obtained by gluing V_1 and V_2 along their boundaries via h is the lens space $L(p, q)$. We assume $0 < q < p$ to exclude the degenerate lens spaces S^3 and $S^2 \times S^1$ and to avoid homeomorphic duplicates.

The follwing proposition classifies the simple closed curves on the boundary of a solid torus which bound surfaces in the solid torus.

Proposition 1. *Let V be a solid torus with meridian-longitude pair (m, l), and let α be a simple closed curve in ∂V such that $\alpha = pl + qm$ in $\pi_1(\partial V)$. Then (1) α bounds a (properly embedded) non-orientable surface in V iff α is nontrival in ∂V and p is even. (2) α bounds a (properly embedded) orientable surface iff $p = 0$.*

Proof. (1) Assume α bounds a non-orientable surface in V. Consider the following exact sequence of the $\mathbb{Z}_2$-homology groups of the pair $(V, \partial V)$: $\to H_2(V; \mathbb{Z}_2) \to H_2(V, \partial V; \mathbb{Z}_2) \xrightarrow{\Delta} H_1(\partial V; \mathbb{Z}_2) \xrightarrow{i_*} H_1(V; \mathbb{Z}_2) \to$, where Δ is the connecting homomorphism and i_* is the inclusion-induced homomorphism. We have $H_1(\partial V; \mathbb{Z}_2) = H_1(\partial V) \otimes \mathbb{Z}_2 = (\langle l \rangle \oplus \langle m \rangle) \otimes \mathbb{Z}_2 = \langle l \rangle \otimes \mathbb{Z}_2 \oplus \langle m \rangle \otimes \mathbb{Z}_2 = \langle \bar{l} \rangle \oplus \langle \bar{m} \rangle = \mathbb{Z}_2 \oplus \mathbb{Z}_2$, and $H_1(V; \mathbb{Z}_2) = H_1(V) \otimes \mathbb{Z}_2 = \langle l \rangle \otimes \mathbb{Z}_2 = \langle \bar{l} \rangle = \mathbb{Z}_2$. Now if F is a non-orientable surface in V bounded by α, then $[F] \epsilon H_2(V, \partial V; \mathbb{Z}_2)$ and $\Delta([F]) = [\partial F] = [\alpha]$. This implies $[\alpha] \epsilon Im\Delta = Ker i_* < H_1(\partial V; \mathbb{Z}_2)$. Since $[\alpha] = pl + qm$ in $H_1(\partial V)$, $[\alpha] = \bar{p}\bar{l} + \bar{q}\bar{m}$ in $H_1(\partial V; \mathbb{Z}_2)$, where $\bar{p} = 0$ or 1, $\bar{q} = 0$ or 1, and $p \equiv \bar{p}(mod 2)$, $q \equiv \bar{q}(mod 2)$. Since $i_*([\alpha]) = 0$ in $H_1(V; \mathbb{Z}_2) = \langle \bar{l} \rangle$, it follows $\bar{p} = 0$, so $p \equiv 0(mod 2)$, therefore p is even. Now if α is trivial in ∂V, then α bounds a disk D in the solid torus V, and $F \cup D$ is a closed non-orientable surface contained in V, which is impossible. Conversely, assume α is nontrivial in ∂V and p is even, we show that α bounds a non-orientable surface in V by construction. First we note that since p is even, q must be odd. Let $p = 2k$ with $k \epsilon \mathbb{Z}$. If $k = 0$, then $p = 0$, so $q = \pm 1 (q \neq 0$ since α is nontrivial in $\partial V)$. Thus α is a meridian for V bounding a meridional disk of V. Then α bounds a once-puctured Klein bottle K in V, see Figure 1. For $|k| \geq 1$, we construct a surface F with $\partial F = \alpha$ (viewing V as the standardly embedded solid torus $D^2 \times S^1$ in $\mathbb{R}^3$ and using a projection of V) as shown in the examples in Figure 2; it is not hard to dipict a general construction for arbitrary curve $\alpha = (2k)l + qm$. Note that F can be made to be properly embedded in V, and since q is odd F contains an orientation reversing loop and therefore is non-orientable. This proves (1).

(2) The "only if" part can be proved by considering a similar exact sequence of homology groups of the pair $(V, \partial V)$ but with integer coefficients, and using a similar arguments as in the proof of the "only if" part of (1). The "if" part is trivial. □

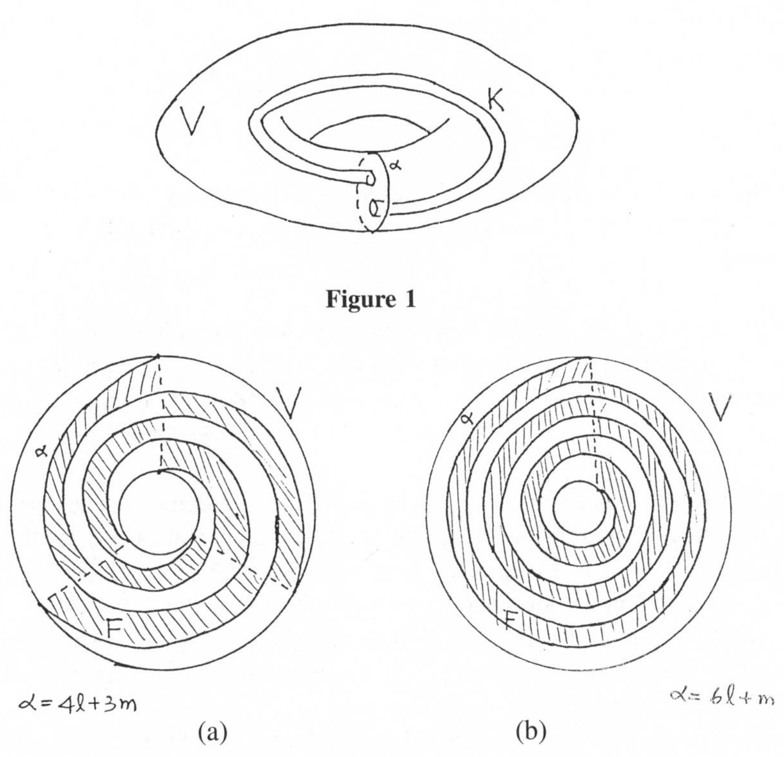

Figure 1

(a)

(b)

Figure 2

[Remark 1]. In the constructive proof of (1) of Proposition 1, if $\alpha = (2k)l + qm$, then α bounds a non-orientable surface F which is a "band connected sum" of $|k|$ Möbius bands (each has $|q|$ half-twists) such that each consecutive pair of

Möbius bands are connected by $|q|$ bands. See Figure 3 for an example. Therefore $\chi(F) = -(|k| - 1)|q|$ and $\text{genus}(F) = 2 - [1 - (|k| - 1)|q|] = 1 + (|k| - 1)|q|$. In the case of $|q| = 1$, $\chi(F) = 1 - |k|$ and $\text{genus}(F) = |k|$; thus in this case F is a band connected sum of $|k|$ Möbius bands(each has one half-twist) such that each consecutive pairs are connected by one band.

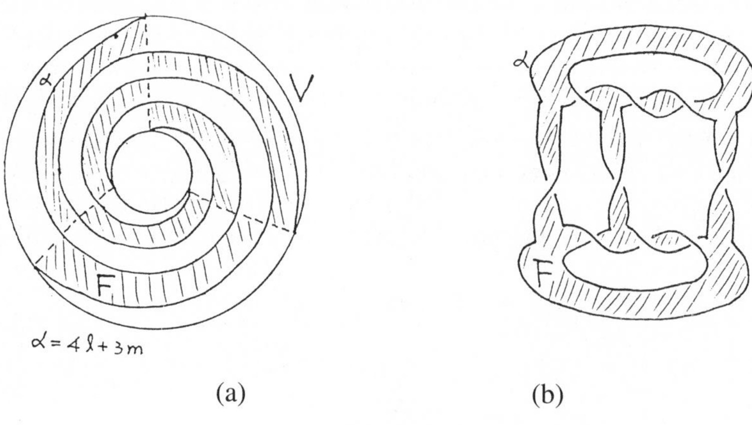

(a) (b)

Figure 3

We now classify the simple closed curves on the boundary of a solid torus which bound non-orientable incompressible surfaces in the solid torus. We need the following

Lemma 2. *Let V be a solid torus with meridian-longitude pair (m, l) and let $\alpha = pl + qm$ be a simple closed curve in ∂V. Assume p is even and $p \neq 0$. If F is a non-orientable surface in V of minimal genus with boundary α, then F is incompressible in V.*

Proof. If F is compressible in V, then there exists a compressing disk D for F in V. Performing surgery on D, we obtain either a new surface F' with $\chi(F') = \chi(F) + 2$ or two new surfaces F' and F'' with $\chi(F') + \chi(F'') = \chi(F) + 2$, according to ∂D does not separate or separate F. In the first case F' must be non-orientable, otherwise $\partial F' = \partial F = \alpha$ bounds an orientable surface in V, by Proposition 1, $p = 0$, contradicting to the assumption. Therefore F' is a non-orientable surface in V with $\partial F' = \alpha$ and $\chi(F') = \chi(F) + 2 > \chi(F)$. In the second case, $F = F' \sharp F''$, so one of the surfaces F' and F'' is non-orientable. Assume F' is non-orientable; then $\partial F' = \alpha$, for otherwise the solid torus V contains a closed non-orientable surface, which is impossible. Now $F'' \ncong S^2$, for otherwise ∂D bounds a disk in F and so D would not be a compressing disk; therefore $\chi(F'') \leq 0$, and it follows $\chi(F') \geq \chi(F) + 2 > \chi(F)$. In either case, we obtain a non-orientable surface F' in V with $\partial F' = \alpha$ and having genus less than that of F, contradicting the assumption of the minimality of the genus of F. □

[Remark 2]. We note that the reason for the assumption $p \neq 0$ in Lemma 2 is to exclude the case when α is a meridian of V bounding a once-punctured Klein bottle K in V, as was shown in the proof of Proposition 1; it is not hard to observe that in such case K is a non-orientable surface of minimal genus contained in V with boundary α but is compressible in V.

Corollary 3. *Let V be a solid torus with meridian-longitude pair (m, l) and let $\alpha = pl + qm$ be a simple closed curve in ∂V. Then α bounds a (properly embedded) non-orientable incompressible surface in V iff p is even and $p \neq 0$.*

Proof. Assume α bounds a non-orientable incompressible surface F in V. Then by Proposition 1, α is nontrivial in ∂V and p is even. If $p = 0$, then $q = 1$, and so α is a meridian for V and hence bounds a meridian disk D of V. By a standard cutting and pasting arguments using the fact that F is incompressible, we may assume $\mathring{F} \cap \mathring{D} = \phi$, thus $F \cup D$ is a closed non-orientable surface in V, which is impossible, concluding that $p \neq 0$.

Conversely, assume p is even and $p \neq 0$. Then by Proposition 1, α bounds a properly embedded non-orientable surface in V, therefore bounds a properly embedded non-orientable surface F in V of minimal genus, by Lemma 2, F is incompressible in V. □

Since solid tori are the building blocks for lens spaces, we may apply Proposition 1 and Corollary 3 to obtain the followig well-known results (compare to [R] and [J, III.6]).

Proposition 4. *The lens space $L(p, q)$ contains a closed non-orientable surface iff p is even.*

Proof. It is well-known that a connected closed $(n-1)$-submanifold K of a closed orientable n-manifold M separates M iff $[K] = 0$ in $H_{n-1}(M; \mathbb{Z}_2)$. If p is odd, then $H_2(L(p, q); \mathbb{Z}_2) = 0$, hence F separates $L(p, q)$ and therefore orientable, a contradiction. Thus p is even. Conversely, if p is even, then by Proposition 1 the curve $\alpha = pl_1 + qm_1$ on the boundary of solid torus V_1 with meridian-longitude pair (m_1, l_1) bounds a non-orientable surface F in V_1, so $L(p, q) = V_1 \underset{h}{\cup} V_2$, where $h : \partial V_1 \mapsto \partial V_2$ is a homeomorphism taking $pl_1 + qm_1$ to m_2, contains the non-orientable surface $\hat{F} = F \cup D_2$, where D_2 is a meridional disk for V_2 bounded by m_2. □

Using the well-known fact that any closed non-orientable surface of minimal genus embedded in a closed orientable 3-manifold with $H_2(M) = 0$ is incompressible, we have

Corollary 5. *The lens space $L(p,q)$ contains an incompressible (hence non-orientable) surface iff p is even.*

The following lemma was proved in [R, Lemma 9]:

Lemma 6. *If F is a closed imcompressible surface in the lens space $L(2k,q) = V_1 \underset{h}{\cup} V_2$, then F can be isotoped to a surface F' such that $F' \cap V_2$ is a single meridional disk D_2 for V_2.*

[Remark 3]. A similar result for closed non-orientable surface F of minimal genus embedded in $L(p,q)$ was obtained in [BW, §7]. Note that, by a previously mentioned well-known result and by [R, Remarks(1), page 194], a closed non-orientable surface in a lens space has minimal genus iff it is incompressible.

Proposition 7. *With the assumption and conclusion of Lemma 6, $F' - \overset{\circ}{D}_2$ is a non-orientable incompressible surface in V_1.*

Proof. Let D be a disk in V_1 with $D \cap (F' - \overset{\circ}{D}_2) = \partial D$. We may assume $D \subset \overset{\circ}{V}_1$ and $\partial D \subset F' - D_2$. Since $F' = (F' - \overset{\circ}{D}_2) \cup D_2$ is incompressible in $L(2k,q) = V_1 \underset{h}{\cup} V_2$ and $\partial D \subset F'$, ∂D bounds a disk σ in F', and so ∂D separates F' into $F - \overset{\circ}{\sigma}$ and σ. Since $\partial D \cap D_2 = \phi$, either $D_2 \subset \overset{\circ}{\sigma}$ or $D_2 \subset F' - \sigma$. If $D_2 \subset \overset{\circ}{\sigma}$, then ∂D and ∂D_2 cobound the annulus $\sigma - D_2$, and since $\sigma \subset F'$ and $F' \cap V_2 = D_2$, the annulus $\sigma - \overset{\circ}{D}_2$ is contained in V_1, it follows that $(\sigma - \overset{\circ}{D}_2) \cup D$ is a disk in V_2 bounded by ∂D_2, so $\partial D_2 \simeq 0$ in V_1. But $\partial D_2 = (2k)l_1 + qm_1$ in ∂V_1, with $k \neq 0$, hence $\partial D_2 \simeq (2k)l_1$ in V_1, a contradiction. Therefore $D_2 \subset F' - \sigma$ or equivalently $\sigma \subset F' - D_2$, so ∂D bounds the disk σ in $F' - D_2$, as required. □

Now we show that when $p = 2k \neq 0$ and $q = \pm 1$, the properly embedded non-orientable surface F constructed in the proof of (1) of Proposition 1 is incompressible in the solid torus V. Figure 4 contains a picture of F when $\alpha = 4l \pm m$. For convenience we shall call such a surface a *standard non-orientable surface bounded by* α. We first observe that in general if S is a properly embedded incompressible surface in V with $\partial S = (2k)l + qm$, where $k \neq 0$, in ∂V, then by a standard cutting and pasting argument, there is a meridional disk D of V such that D intersects S in k disjoint properly embedded arcs in S, and if S' is the 2-manifold obtained from S by cutting along D, then $\chi(S) = \chi(S') - |k|$.

Lemma 8. *Let V be a solid torus with meridian-longitude pair (m, l) and let $\alpha = (2k)l \pm m$, where $k \neq 0$, be a simple closed curve in ∂V. If F is a standard non-orientable surface bounded by α, then F is a (properly embedded) non-orientable surface in V of minimal genus with boundary α and therefore is incompressible in V.*

Proof. We view V as $D^2 \times S^1$. From Remark 1, F is a band connected sum of $|k|$ Möbius bands such that each consecutive pair of Möbius bands are connected by one band, and if F' is the surface obtained from F by cutting along the standard meridional disk $D^2 \times \{(1,0)\}$ of V, then F' is a properly embedded disk in the resulting 3-ball and $\chi(F) = \chi(F') - |k|$. Let S be a non-orientable surface properly embedded in V of minimal genus with boundary α, by Lemma 2, S is incompressible in V. Let S' be the 2-manifold obtained from S by cutting along a meridional disk D of V which intersects S in k disjoint properly embedded arcs in S (such a meridional disk exists by the observation stated in the paragraph that precedes Lemma 8). By an isotopy of V we may assume that D is the standard meridional disk of V. To show that the genus of F is minimal in V with boundary α, we show that genus$(F) \leq$genus(S), and for this it suffices to show $\chi(F) \geq \chi(S)$; since $\chi(F) = \chi(F') - |k|$ and $\chi(S) = \chi(S') - |k|$, it in turn suffices to show $\chi(F') \geq \chi(S')$.

Claim. *$\partial S'$ is a simple closed curve. Therefore S' is a surface (orientable or non-orientable) with a single boundary component and hence $\chi(S') \leq 1 = \chi(F')$.*

Without loss of generality we may assume $k > 0$. Consider $D^2 \times [0,1]$ as the 3-ball resulting from cutting the solid torus $V = D^2 \times S^1$ along the meridional disk $D^2 \times \{(1,0)\}$; since $\partial S = \alpha = (2k)l \pm m$, there are $2k$ consecutive points $a_1, a_2, ..., a_{2k}$ in $\partial D^2 \times \{0\}$ and k pairwise disjoint arcs $\alpha_1, \alpha_2, ..., \alpha_k$ in $D^2 \times \{0\}$, and there are $2k$ consecutive points $b_1, b_2, .., b_{2k}$ in $\partial D^2 \times \{1\}$ and k disjoint arcs $\beta_1, \beta_2, .., \beta_k$ in $D^2 \times \{1\}$, such that for every i, $1 \leq i \leq k$, $\partial\alpha_i \subset \{a_1, .., a_{2k}\}$ and $\partial\beta_i \subset \{b_1, .., b_{2k}\}$, satisfying the condition that for every i, if $\partial\alpha_i = \{a_{i_1}, a_{i_2}\}$, then $\partial\beta_i = \{b_{i_1}, b_{i_2}\}$, where $1 \leq i_1, i_2 \leq 2k$. Also there are $2k$ consecutive disjoint generating arcs $\gamma_1, \gamma_2, .., \gamma_{2k}$ in the annulus $\partial D^2 \times [0,1]$ such that $\partial\gamma_i = \{a_i, b_{i+1}\}$ for $1 \leq i \leq 2k-1$ and $\partial\gamma_{2k} = \{a_{2k}, b_1\}$, and $(\bigcup_{i=1}^{k} \alpha_i) \cup (\bigcup_{i=1}^{k} \beta_i) \cup (\bigcup_{i=1}^{k} \gamma_i) = \alpha = \partial S'$. See Figure 4(a). Now if $\partial S'$ contains more than one simple closed curve, then there exists j, $1 \leq j \leq 2k$, such that γ_j and γ_{j+1} belong to distinct components of $\partial S'$. Let α_l be the arc in $D^2 \times \{0\}$ with $\partial\alpha_l = \{a_{j+1}, a_s\}$ for some $1 \leq s \leq 2k$ and let β_l be the arc in $D^2 \times \{1\}$ with $\partial\beta_l = \{b_{j+1}, b_s\}$.

Without loss of generality we may assume $s > j+1$. Then it is clear $s > j+2$, see Figure 4(b). Now since α_i's and β_i's are pairwise disjoint, the path containing arcs α_l and γ_{j+1} can not be completed to a simple closed curve, a contradiction. □

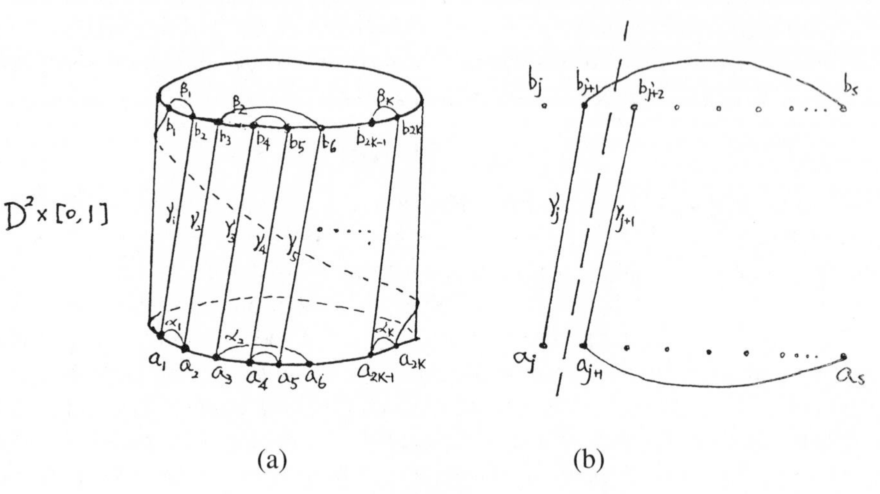

(a) (b)

Figure 4

As a consequence, we obtain the following well-known result (compare to [J,III.2(f)]):

Corollary 9. *The lens space $L(2k,1)$ with $k > 0$ contains an incompressible (hence non-orientable) surface of Euler characteristic $2-k$, or equivalently, of genus k.*

Proof. Let $L(2k,1) = V_1 \underset{h}{\cup} V_2$, where V_1 and V_2 are solid tori with meridian-longitude pairs (m_1, l_1) and (m_2, l_2), respectively, and $h : \partial V_1 \mapsto \partial V_2$ is a homeomorphism taking $\alpha = (2k)l_1 + m_1$ to m_2. Let F be a standard non-orientable surface in V_1 bounded by α, then by Lemma 8, F is a properly embedded non-orientable surface in V of minimal genus with boundary α. Let $\hat{F} = F \cup D_2$, where D_2 is a meridional disk for V_2 bounded by m_2. Then $\hat{F}$ is a closed non-orientable surface in $L(2k,1)$.

Claim. *$\hat{F}$ is a closed non-orientable surface in $L(2k,1)$ of minimal genus.*

If not, then let K be a closed non-orientable surface in $L(2k,1)$ of minimal genus. Then by Remark 3, K can be isotoped to a surface K' so that $K' \cap V_2$ is a single meridional disk Δ of V_2. Now $K' - \overset{\circ}{\Delta}$ is a non-orientable surface in V_1 with $\partial(K' - \overset{\circ}{\Delta}) = (2k)l_1 + m_1$ (when properly oriented), so we may isotope $K' - \overset{\circ}{\Delta}$ in V_1 so that $\partial(K' - \overset{\circ}{\Delta}) = \alpha = \partial F$. Since genus$(K')$ <genus$(\hat{F})$,

genus$(K' - \overset{\circ}{\Delta})$ <genus(F); but this contradicts the fact that F is a non-orientable surface of minimal genus in V_1 with boundary α. The proof of Claim is completed.

Now since $\hat{F}$ is a closed non-orientable surface in $L(2k, 1)$ of minimal genus, $\hat{F}$ is incompressible in $L(2k, 1)$; also by Remark 1, $\chi(\hat{F}) = 2 - k$, hence genus$(\hat{F}) = k$. □

We shall now prove the converse of Proposition 7, the following lemma will be needed in the proof.

Lemma 10. *Let F be a properly embedded non-orientable incompressible surface in a solid torus V. If A is an annulus in V such that $A \cap F = \partial A$, and one boundary component of A is ∂F and the other is contained in $\overset{\circ}{F}$, then ∂A bounds an annulus in F.*

Proof. Assume a_1 and a_2 are the boundary components of A with $a_1 = \partial F$ and $a_2 \subset \overset{\circ}{F}$. Choose a meridional disk D of V that is in general position with A and F. Then $D \cap (F \cup A)$ consists of a collection Γ of properly embedded arcs in D which are the arcs in $D \cap F$, a collection Λ of arcs in D which are the arcs in $D \cap A$, and a collection (may be empty) of circles in $\overset{\circ}{D}$ disjoint from the arcs in $\Gamma \cup \Lambda$, which are the circles in $\overset{\circ}{D} \cap \overset{\circ}{F}$ or in $\overset{\circ}{D} \cap \overset{\circ}{A}$. Note that Γ, therefore Λ, is non-empty by Corollary 3.

We first assert that we may assume that the last collection of circles in $\overset{\circ}{D}$ is empty, that is, $D \cap (F \cup A)$ is the union of arcs in $\Gamma \cup \Lambda$. For if not, then let c be an innermost such circle in $\overset{\circ}{D}$ and thus c is either in $\overset{\circ}{D} \cap \overset{\circ}{F}$ or in $\overset{\circ}{D} \cap \overset{\circ}{A}$. Let δ be the subdisk of D bounded by c. If c is a circle in $\overset{\circ}{D} \cap \overset{\circ}{F}$, then since F is incompressible in V, c bounds a disk Δ in F, thus $\delta \cup \Delta$ is a 2-sphere and hence bounds a 3-ball B in V, therefore by an isotopy of V which has support in a small neighborhood of B, we may eliminate c to obtain a new meridional disk of V without changing ∂A (therefore ∂F) and ∂D. If c is a circle in $\overset{\circ}{D} \cap \overset{\circ}{A}$, then c must be a trivial loop in A, for otherwise $a_1 = \partial F$ would bound a disk in V and hence trivial in V, contradicting Corollary 3. Thus c bounds a disk d in A, and so $\delta \cup d$ is a 2-sphere bounding a 3-ball C in V; again by an isotopy of V which has support in a neighborhood of C, we may eliminate c to obtain a new meridional disk of V without changing ∂A and ∂D. Continue this process we may eliminate all the circles of $D \cap (F \cup A)$ in $\overset{\circ}{D}$.

Next we assert that we may assume that each of the arcs in Λ is a generating arc of A in D. For if an arc τ of Λ is not a generating arc of A, then there is an "outermost" non-generating arc $\hat{\tau}$ of A such that $\hat{\tau}$ together with an arc contained in ∂A bound a subdisk of A containing τ. By using a small regular neighborhood of A in V, we may isotope D, fixing A and ∂D, to change $\hat{\tau}$ into two generating arcs of A. See Figure 5. Repeat this process until τ becomes outermost and may be changed into two generating arcs of A by an isotopy. Then since $\partial F = a_1$ is a boundary component of A, it follows that in D, each arc of Λ has one endpoint at an endpoint of an arc in Γ and has the other endpoint at an interior point of an arc in Γ; and at each endpoint of an arc in Γ, there is an exactly one arc in Λ "issuing" from the endpoint and ending in the interior of an arc in Γ. See Figure 6.

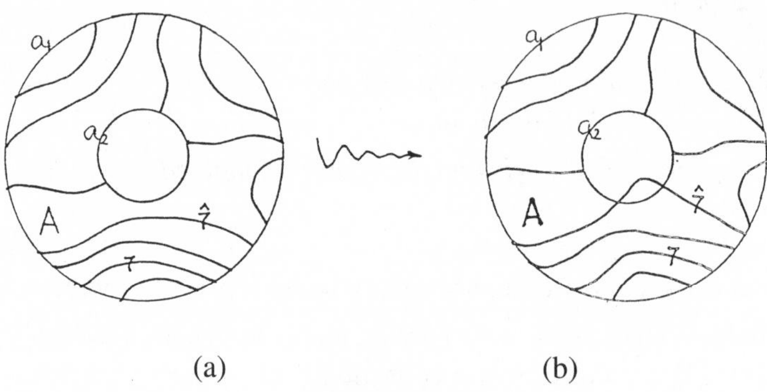

(a) (b)

Figure 5

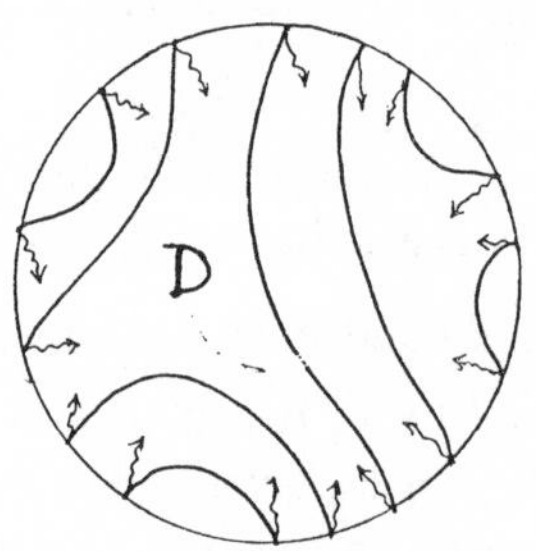

Figure 6

Finally we assert that in the disk D, there exists an arc $\lambda \epsilon \Lambda$ whose endpoints are in a same arc $\gamma \epsilon \Gamma$; that is, one endpoint of λ is at an endpoint of γ and the other endpoint of λ is at an interior point of γ. We shall prove a more general assertion: Given D and Γ as before and let Λ' be a collection of arcs in D satisfying the property that one endpoint of each arc in Λ' is at an endpoint of an arc in Γ and the other endpoint at an interior point of an arc in Γ, such that for every arc in Γ, there is exactly one arc in Λ' issuing from each endpoint of that arc, except possibly for one arc in Γ, for which there is no arc in Λ' issuing from one of its two endpoints of the arc, then there exists an arc $\lambda \epsilon \Lambda'$ whose endpoints are in a same arc $\gamma \epsilon \Gamma$.

We prove by induction on the number n of the arcs in Γ. If $n = 1$, then we have either one of the two cases in Figure 7, and it is clear the results follows. Assume the assertion is true for $n = k$ with $k \geq 1$. Now suppose $n = k + 1$. There are two cases to consider:

Case 1. Every arc in Γ is innermost in D; that is, bounds a subdisk of D that contains no other arcs in Γ. Consider the following two subcases:
Subcase 1.1. At each endpoint of every arc in Γ there is an exactly one arc in Λ' issuing from that endpoint. If so, let β be any arc of Γ in D and let α be an arc in Λ' with one endpoint at an endpoint of β. Let e be the other endpoint of α which lies in the interior of an arc β' in Γ. If $\beta' = \beta$, then we are done; if $\beta' \neq \beta$, then extend α from e, inside the subdisk of D bounded by β', to an endpoint of β', to obtain a new arc α' properly embedded in D, and consider the subdisk D' of D obtained by dividing D along α'. See Figure 8(a). Let Γ' be the collection of arcs in D' consisting of those arcs of Γ in D' and the arc in D' obtained by the splitting of β' by α', and let Λ'' be the arcs of Λ' whose endpoints are endpoints of the arcs in Γ'. Note that the arcs in Λ'' are exactly those arcs in Λ' that are confined by α' in D'. Now the number of arcs in $\Gamma' \leq (k+1) - 1 = k$, thus by indution hypothesis, the result holds for subdisk D' and therefore holds for D.
Subcase 1.2. There is no arc in Λ' issuing from one of the two endpoints of an arc in Γ. If so, we proceed as in Subcase 1.1 except in this subcase we choose β to be the "exceptional" arc of Γ; that is, the arc with only one of its endpoints admits an arc in Λ'; and we choose α to be that arc in Λ'.

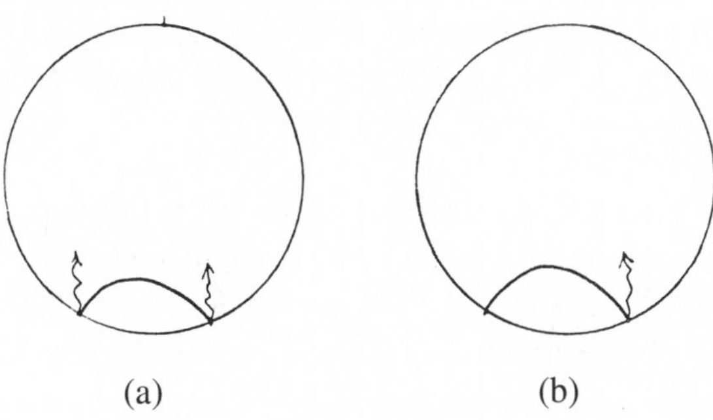

(a) (b)

Figure 7

Case 2. There exists an arc in Γ which is not innermost in D; that is, an arc which separartes D into two subdisks each contains other arcs of Γ. Let b be such an arc in Γ. Then b separates D into 2 subdisks D_1 and D_2, each contains l_i arcs of Γ, $1 \leq l_i \leq k-1$, $i = 1, 2$, excluding b itself. Again we consider two subcases:
Subcase 2.1. At each endpoint of b there is an exactly one arc in Λ'. If both arcs of Λ' at the endpoints of b are in the same subdisk, say D_1, then by considering an enlarged D_1' containing D_1, one checks that the conditions in the assertion are

satisfied, so the result follows by induction hypothesis. If arcs of Λ' at endpoints of b are in different D_i's, then at least one of D_1 and D_2 satisfies the conditions of the assertion and the result follows by the induction hypothesis again. See Figure 8(b).

Subcase 2.2. There is exactly one arc in Γ' at only one endpoint of b. In this case, consider the enlarged disk D_1' containing this arc and use the induction hypothesis.

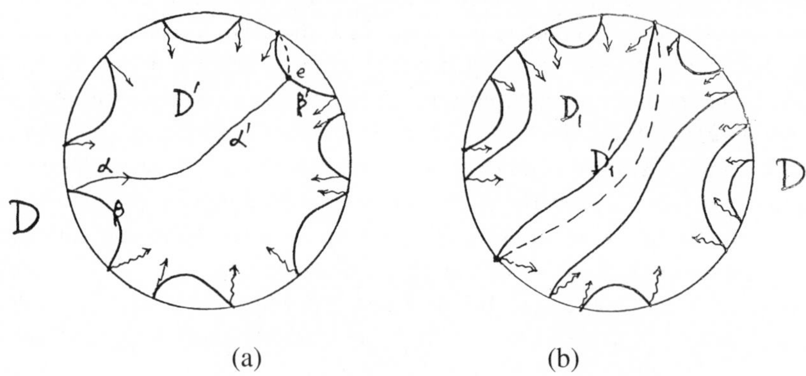

Figure 8

Now by the last assertion, there is an arc $\lambda \epsilon \Lambda$ and there is a subdisk σ of D such that $\partial\sigma = \lambda \cup \gamma'$, where γ' is a subarc of an arc in Γ. Consider a product neighborhood $\sigma \times [-1, 1]$ of σ and consider the disk $\delta = (A - \partial\sigma \times [-1, 1]) \cup \sigma \times \{-1\} \cup \sigma \times \{1\}$. Since $\delta \cap F = \partial\delta$ and F is incompressible, $\partial\delta$ bounds a disk η contained in F, then $\eta \cup (\gamma \times [-1, 1])$ is an annulus contained in F whose boundary is $\partial\delta \cup (\partial\gamma \times [-1, 1]) = \partial A$, therefore ∂A bounds an annulus in F. The proof is completed. □

Theorem 11. *Let V_1 and V_2 be solid tori, and let F be a properly embedded non-orientable incompressible surface in V_1 with a single boundary component and D_2 a meridional disk of V_2. If $h : \partial V_1 \mapsto \partial V_2$ is a homeomorphism such that $h(\partial F) = \partial D_2$, then the closed non-orientable surface $\hat{F} = F \underset{h}{\cup} D_2$ is incompressible in the lens space $V_1 \underset{h}{\cup} V_2$.*

Proof. Let D be a disk in the lens space $L = V_1 \underset{h}{\cup} V_2$ such that $D \cap \hat{F} = \partial D$, we show that ∂D bounds a disk in $\hat{F}$. We may assume that D is in general position with respect to ∂V_2 so that ∂D intersects D_2 in a collection $\mathcal{B}$ of properly embedded arcs in D_2, each arc divides D_2 into two subdisks. We call an arc of $\mathcal{B}$ an "innermost" arc if one of the two subdisks of D_2 divided by it contains no other arcs in $\mathcal{B}$. Let b be an innermost arc of $\mathcal{B}$, and let Δ_b be one of the two subdisks of D_2 divided by b that contains no other arcs of $\mathcal{B}$. Let a be the arc contained in ∂D_2 such that $\partial\Delta_b = a \cup b$. There is a properly embedded arc c in D such that $c \subset D \cap \partial V_2$ and $\partial c = \partial b (= \partial a)$, and there is a subdisk δ of D such that $\partial\delta = b \cup c$. Let $\sigma = \delta \cup \Delta_b$, then σ is a disk in V_2 with $\partial\sigma = a \cup c \subset \partial V_2$.

Now since every properly embedded disk in a solid torus is either a disk whose boundary is homotopically trivial in the solid torus or a meridional disk, we have the following two cases:

Case 1. $\partial\sigma \simeq 0$ in ∂V_2. Then $\partial\sigma$ bounds a disk ϕ in ∂V_2. Let $D_b = (D - \delta) \cup \phi$, then D_b is a disk, $\phi \subset D_b \cap \partial V_2$, and $\partial D_b \subset \hat{F}$. See Figure 9(a). Now by using a regular neighborhood of ∂V_2, we may push ϕ of D_b away from V_2 to obtain a new disk D_b' such that $\partial D_b' \subset \hat{F}$ and $\partial D_b' \cap D_2$ is the collection of circles in $\mathcal{B} - \{b\}$. See Figure 9(b).

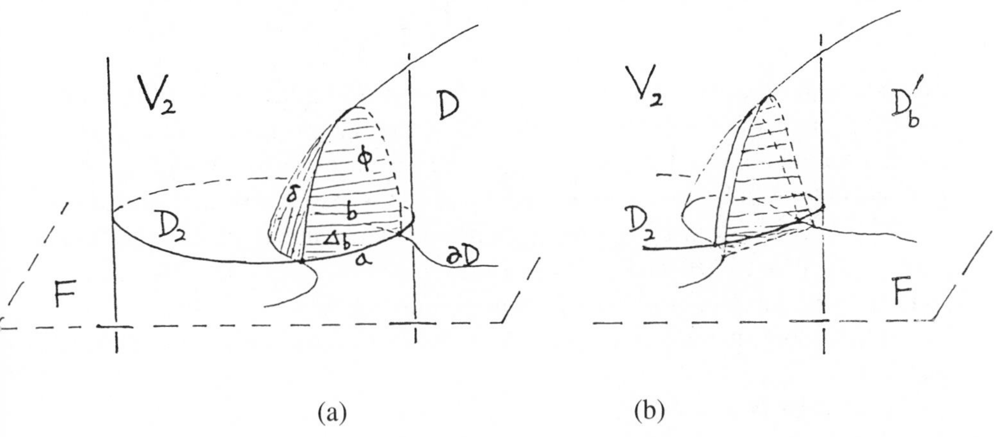

(a) (b)

Figure 9

Case 2. σ is a meridional disk of V_2. Then $\partial\sigma$ bounds a meridional disk ϕ of V_2 with $\phi \cap D_2 = a$. Let $D_b = (D - \delta) \cup \phi$, then D_b is a disk, $\phi \subset D_b \cap V_2$, and $\partial D_b \subset \hat{F}$. See Figure 10(a). Again by using a regular neighborhood of ∂V_2, we may push D_b to obtain a new disk D_b' such that $\partial D_b' \subset \hat{F}$ and $\partial D_b' \cap D_2$ is the collection of circles in $\mathcal{B} - \{b\}$. See Figure 10(b).

Claim. *∂D bounds a disk in $\hat{F}$ iff $\partial D_b'$ does.*

For D_b is obtained from D by replacing the arc b in ∂D by the arc a, and $a \cup b$ bounds a disk in $\hat{F}$, it is not hard to see that ∂D bounds a disk in $\hat{F}$ iff ∂D_b does, therefore it is also true for $\partial D_b'$, so the Claim is proved.

Now apply this procedure to every innermost arc of $\mathcal{B}$, we may obtain a new disk D' such that $\partial D' \subset \hat{F}$ and a subcollection $\mathcal{B}_1$ of $\mathcal{B}$ consisting of properly embedded disjoint arcs of $\partial D' \cap D_2$ in D_2 which are not innermost arcs of $\mathcal{B}$. We

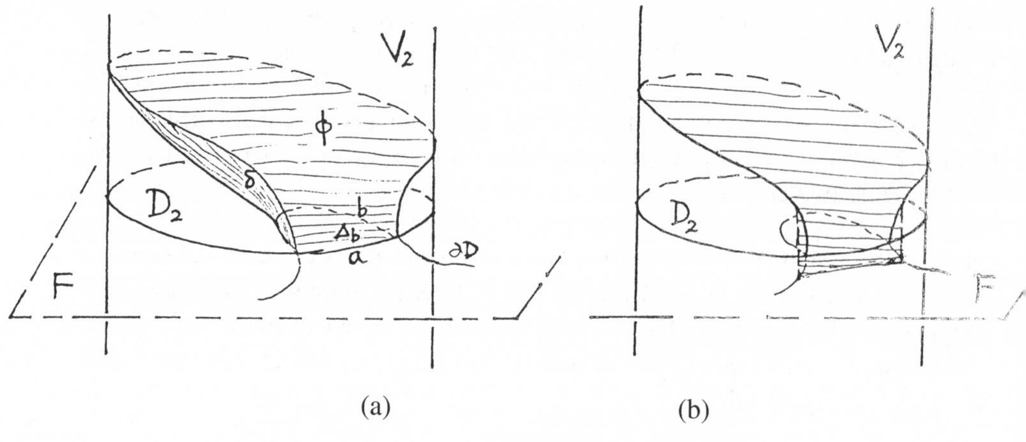

Figure 10

can then apply the previous procedure to $\mathcal{B}_1$ to obtain a new disk D'' such that $\partial D'' \subset \hat{F}$ and $\partial D'' \cap D_2$ consists of arcs which are not innermost arcs in $\mathcal{B}$ or $\mathcal{B}_1$. Using a similar arguments in the above claim, we have that $\partial D'$ bounds a disk in $\hat{F}$ if and only if $\partial D''$ does, thus we have that ∂D bounds a disk in $\hat{F}$ if and only if $\partial D''$ bounds a disk in $\hat{F}$. We may continue this procedure until all the arcs of $\mathcal{B}$ are eliminated and a new disk $\hat{D}$ is obtained such that $\hat{D} \cap \hat{F} = \partial\hat{D}, \partial\hat{D} \cap D_2 = \phi$, and satisfying the property that $\partial\hat{D}$ bounds a disk in $\hat{F}$ if and only if ∂D does. Therefore to show that ∂D bounds a disk in $\hat{F}$, it suffices to show that $\partial\hat{D}$ bounds a disk in $\hat{F}$.

We consider two cases:

Case 2.1. $\hat{D} \cap V_2 = \phi$. If so, then $\hat{D} \subset V_1$ and $\partial\hat{D} \subset F$. Since F is incompressible in V_1, $\partial\hat{D}$ bounds a disk in F and therefore in $\hat{F}$.

Case 2.2. $\hat{D} \cap V_2 \neq \phi$. Since $\partial\hat{D} \cap D_2 = \phi$, $\partial\hat{D} \subset \mathring{F}$, hence $\partial V_2 \cap \hat{D} = \partial V_2 \cap \mathring{\hat{D}}$; therefore ∂V_2 intersects $\hat{D}$ in a collection of circles in $\mathring{\hat{D}}$ (and in ∂V_2). Let u be a circle of $\partial V_2 \cap \hat{D}$ which is an innermost circle in $\mathring{\hat{D}}$. Then $u \subset \partial V_2$ and u bounds a disk τ in $\mathring{\hat{D}}$ which lies either in V_2 or V_1. Now if τ is contained in V_2, then u is either a meridian of ∂V_2 and τ is a meridional disk of V_2 bounded by u or u is a trivial loop in ∂V_2; if u is a trivial loop in ∂V_2, then u bounds a disk ρ in ∂V_2, so $\tau \cup \rho$ is a 2-sphere in V_2 and thus bounds a 3-ball in V_2. Let $u_1, u_2, .., u_n$ be the circles of $\partial V_2 \cap \hat{D}$ such that $u_i \subset \mathring{\rho}$ for $1 \leq i \leq n$. By standard cutting and pasting arguments, we may eliminate the circles u and u_i for $1 \leq i \leq n$, to obtain a new disk $\hat{D}_u$ such that $\partial V_2 \cap \hat{D}_u = (\partial V_2 \cap \hat{D}) - \cup\{u, u_1, u_2, ..., u_n\}$

and $\partial\hat{D}_u \subset \overset{\circ}{F}$. If τ is contained in V_1, then u is either a meridian of ∂V_1 and τ is a meridional disk of V_1 bounded by u or u is a trivial loop in ∂V_1; it can not be the first case, for otherwise u intersects ∂F in at least $|p| > 0$ points, where $\partial F = pl_1 + qm_1$ in ∂V_1, but $\hat{D} \cap \overset{\circ}{F} = \phi$, a contradiction. In the second case, u bounds a disk θ in $\partial V_1 = \partial V_2$, then $\tau \cup \theta$ is a 2-sphere in V_1 and hence bounds a 3-ball in V_1, then by cutting and pasting, we may eliminate u and all the circles of $\partial V_2 \cap \hat{D}$ contained in $\overset{\circ}{\theta}$. Continuing this process, we may eventually obtain a disk $\hat{\hat{D}}$ such that $\partial\hat{\hat{D}} = \partial\hat{D} \subset \overset{\circ}{F}$ and $\hat{\hat{D}}$ intersects V_2 only in meridional disks $\Delta_1, \Delta_2, .., \Delta_r$ of V_2. Let $\partial_i = \partial\Delta_i$ for $1 \leq i \leq r$.

First assume $r = 1$. Let A_1 be one of the two annuli contained in ∂V_2 bounded by ∂_1 and ∂D_2, then $A = (\hat{\hat{D}} - \Delta_1) \cup A_1$ is an annulus in V_1 with $A \cap F = \partial A$, and one boundary component of A is ∂D_2 in L, hence is ∂F in V_1, and the other component is contained in $\overset{\circ}{F}$. By Lemma 10, ∂A bounds an annulus $\hat{A}$ in F, so $\hat{A} \cup D_2$ is a disk in $\hat{F}$ bounded by $\partial\hat{\hat{D}} = \partial\hat{D}$.

Then assume $r > 1$. Let D_1 be a meridional disk of V_1 in general position with $\hat{\hat{D}}$ and F. Since F is incompressible in V_1, we may assume that D_1 intersects F only in properly embedded arcs in D_1. Thus in D_1, $D_1 \cap \hat{\hat{D}}$ consists of a collection Ω of properly embedded arcs, a (possibly empty) collection of circles in $\overset{\circ}{D}_1$, and a (possibly empty) collection of non-properly embedded arcs. See Figure 11. Note that in this case $\Omega \neq \phi$.

Now if α is an innermost circle of $D_1 \cap \hat{\hat{D}}$ in $\overset{\circ}{D}_1$, then α bounds a disk d in D_1 and also bounds a disk d' in $\hat{\hat{D}}$, so $d \cup d'$ is a 2-sphere and therefore bounds a 3-ball B in L. We may then push all the intersections $B \cap (D_1 - d)$ and $B \cap (\hat{\hat{D}} - d')$ away from B, and then by a standard cutting and pasting argument we may eliminate α to obtain a new meridional disk D'_1 of V_1 such that $D'_1 \cap \hat{\hat{D}} \subset (D_1 \cap \hat{\hat{D}}) - \alpha$. Continuing this process we may eliminate all the circles of $D_1 \cap \hat{\hat{D}}$ in $\overset{\circ}{D}_1$, therefore we may assume that $D_1 \cap \hat{\hat{D}}$ does not contain any circles in $\overset{\circ}{D}_1$.

We call an arc of Ω "innermost" if one of the two subdisks of D_1 divided by the arc contains no other arcs of Ω.

Claim. *If ζ is an innermost arc of Ω and if d_1 is one of the two subdisks of D_1 divided by ζ which contains no other arcs of Ω, then d_1 contains no non-properly embedded arcs of $D_1 \cap \hat{\hat{D}}$ in D_1.*

Let $\partial\zeta = \{v_1, v_2\}$; then $v_1 \epsilon \partial D'_2$ and $v_2 \epsilon \partial D''_2$, for some meridional disks D'_2 and D''_2 of V_2. If β is a non-properly embedded arc of $D_1 \cap \hat{\hat{D}}$ in D_1 whose endpoints lie in $\overset{\circ}{D}_1$, then since $\partial\beta \subset \partial\hat{\hat{D}} \subset \overset{\circ}{F}$, there is a properly embedded arc γ in D_1 such that $\gamma \subset D_1 \cap F$ and $\beta \cap \gamma = \partial\beta$. Let $\partial\gamma = \{w_1, w_2\}$; since $\partial\gamma \subset \partial F = \partial D_2$, thus $\{w_1, w_2\} \subset \partial D_2$, and this implies that the arc w in ∂D_1 between w_1 and w_2 runs at least once through the longitude of V_1, it follows that w contains points v_1' and v_2' with $v_1' \epsilon D_2'$ and $v_2' \epsilon D_2''$, and there is a properly embedded arc $\zeta' \epsilon \Omega$ in d_1 (and in D_1) with $\partial\zeta' = \{v_1', v_2'\}$. See Figure 11(a). This contradicts the assumption that d_1 contains no arcs of Ω. Now assume η is a non-properly embedded arc of $D_1 \cap \hat{\hat{D}}$ in D_1 with only one endpoint lying in ∂D_1. Let $\partial\eta = \{z_1, z_2\}$ with $z_2 \epsilon \partial D_1$; then since $z_2 \epsilon \partial\hat{\hat{D}} \subset \overset{\circ}{F}$, there is a properly embedded arc μ in D_1 such that $\mu \subset D_1 \cap F$. Let $\partial\mu \subset \partial F = \partial D_2$, $\{x_1, x_2\} \subset \partial D_2$, thus the arc x in ∂D_1 between x_1 and x_2 runs at least once through the longitude of V_2, and it follows that x contains points x_1' and x_2' with $x_1' \epsilon D_2'$ and $x_2' \epsilon D_2''$, and there is a properly embedded arc $\zeta' \epsilon \Omega$ in d_1 (and in D_1) with $\partial\zeta' = \{x_1', x_2'\}$. See Figure 11(b). This again contradicts the assumption that d_1 contains no arcs of Ω. The proof of Claim is completed.

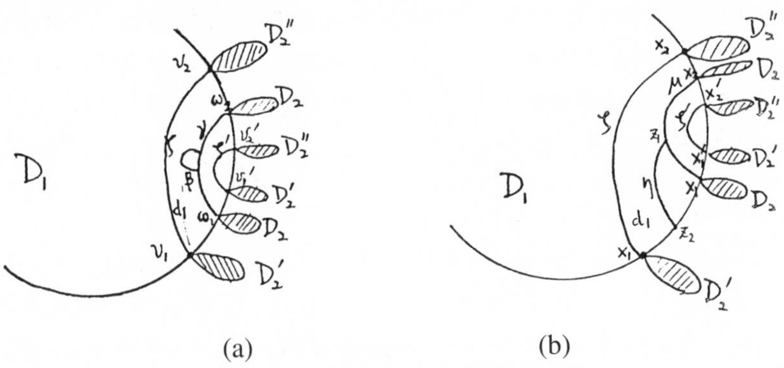

(a) (b)

Figure 11

Finally, ifζ is an innermost arc of Ω and if d_1 is a subdisk of D_1 divided by ζ which contains no other arcs of Ω, then by the preceding Claim, d_1 contains no other arcs of $D_1 \cap \hat{\hat{D}}$. Let $\partial\zeta = \{v_1, v_2\}$ with $v_1 \epsilon D_2'$ and $v_2 \epsilon D_2''$ as before, and let η be the arc contained in ∂D_1 such that $\partial d_1 = \zeta \cup \eta$, then D_2' and D_2'' cobound a 3-ball B_2 containing η. If $N(d_1)$ is a regular neighborhood of d_1 in L, then $N(d_1) \cup B_2$ is a 3-ball, and we can isotope $\hat{\hat{D}}$ by pulling it away from $N(d_1) \cup B_2$ so that at least ζ is eliminated from Ω, and the number r of the meridional disks Δ_i of V_2 of the intersection $\hat{\hat{D}}$ is reduced by at least 2. Continuing this process we may either eliminate all the meridional disks Δ_i of V_2 in the intersection $\hat{\hat{D}} \cap V_2$ and in which case the arguments in Case 2.1 applies, or $\hat{\hat{D}} \cap V_2$ is a single meridional disk of V_2 and the result follows by the arguments in the case when $r = 1$. This completes the proof. □

By Proposition 1 and Lemma 2, and by Remark 3 and Theorem 11, we have

Corollary 12. *Let V be a solid torus, α a nontrivial simple closed curve in ∂V which is not a meridian of V, and F a properly embedded non-orientable surface in V with boundary α. Then F is incompressible in V iff F is a non-orientable surface of minimal genus in V with boundary α.*

Acknowledgement. The author wishes to thank John Kalliongis for helpful conversations in the preparation of this paper.

References

[B-W] G. Bredon and J. Wood, *Non-orientable surfaces in 3-manifolds,* Invent. Math. **7** (1969), pp. 83–110.

[J] W. Jaco, *Lectures on 3-manifold topology,* CBMS Regional Conference Series in Mathematics **no.3**, AMS.

[R] J.H. Rubistein, *One-sided Heegaard splittings of 3-manifolds,* Pacific J. Math. **76, no.1** (1978), pp. 185–200.

Certain abelian subgroups of two-knot groups

Katsuyuki Yoshikawa

Dedicated to Professor Junzo Tao on his 60th birthday

1. Introduction

An *n-knot* K is a smooth submanifold of the $(n+2)$-sphere S^{n+2} which is homeomorphic to the n-sphere S^n. The fundamental group $\pi_1(S^{n+2}-K)$ of the complement is called the *group* of K. Let $H =< S : R >$ and C be groups, and let ϕ_1, ϕ_2 be monomorpihsms of C to H. Then the *HNN extension (Higman-Neumann-Neumann extension)* of H with respect to ϕ_1 and ϕ_2 [LS] is the group defined by

$$G =< S, t : R, t^{-1}\phi_1(c)t = \phi_2(c), c \in C > .$$

The group H is called the *base* of G, and $\phi_1(C), \phi_2(C)$ are called the *associated* subgroups. We denote the group G by $< H, t : t^{-1}\phi_1(C)t = \phi_2(C) >$.

Concerning abelian commutator subgroups of 2-knot groups, it is known that if the commutator subgroup of a 2-knot group is abelian, then it is isomorphic to Z_α, where α is odd, Z^3 (the free abelian group of rank 3) or $Z[1/2]$ (the additive group of the dyadic rationals) [Hi1], [Hi2], [L], [Y4]. Each of these groups can be realized as the commutator subgroup of a 2-knot group. Thus the abelian commutator subgroups of 2-knot groups have been completely determined. We now consider two generalizations of this result. First, since the commutator subgroup is normal, consider abelian normal subgroups of 2-knot groups in §3. Second, since a metabelian 2-knot group can be expressed as an ascending HNN extension of an abelian base H, consider abelian bases of 2-knot groups in §4. In particular, we will discuss the case of 2-knots with minimal Seifert manifold.

Knots 90

2. Preliminaries

Let K be an n-knot. Then there exists a compact orientable $(n+1)$-submanifold V in S^{n+2} such that $\partial V = K$. This is called a *Seifert manifold* for K. Let Y be the $(n+2)$-manifold obtained from S^{n+2} by cutting along V. Then Y is a compact manifold with boundary $\partial Y = V_1 \cup V_2$, where V_1 and V_2 are two copies of V with $V_1 \cap V_2 = \partial V_1 \cap \partial V_2 = K$. Let $\nu_i : V \overset{\approx}{\to} V_i \hookrightarrow Y, i = 1,2$, be the inclusions. If both of the induced homomorphisms $\nu_{i*} : \pi_1(V) \to \pi_1(Y), i = 1,2$, are monomorphisms, we say that V is a *minimal* Seifert manifold for K. If an n-knot K has a minimal Seifert manifold V, then the group G of K is expressed as an HNN extension $< H, t : t^{-1}\nu_{1*}(C)t = \nu_{2*}(C) >$, where $H = \pi_1(Y)$ and $C = \pi_1(V)$.

We will prove some lemmas.

Lemma 2.1. *Let G be an HNN extension $< H, t : t^{-1}\phi_1(C)t = \phi_2(C) >$ of a group H. If A is a finitely generated abelian normal subgroup of G, then we have*

$$A \cap << H >>^G \subset \phi_1(C) \cap \phi_2(C),$$

where $<< H >>^G$ is the normal closure of H in G.

Proof. Let $\tilde{A} = A \cap << H >>^G$. Then, since A and $<< H >>^G$ are normal in G, so is $\tilde{A}$. Hence, if $\tilde{A} \subset \phi_1(C)$, then $\tilde{A} = t^{-1}\tilde{A}t \subset t^{-1}\phi_1(C)t = \phi_2(C)$. Therefore we will prove that $\tilde{A} \subset \phi_1(C)$. Since $G/ << H >>^G \cong Z$, $<< H >>^G$ is the amalgamated free product of infinitely many factors

$$\dots *_{C_{-1,2}=C_{0,1}} H_0 *_{C_{0,2}=C_{1,1}} H_1 *_{C_{1,2}=C_{2,1}} \dots,$$

where $H_i = t^{-i}Ht^i, C_{i,j} = t^{-i}\phi_j(C)t^i, j = 1,2$. Let $H(i,j) = H_i *_{C_{i,2}=C_{i+1,1}} \dots *_{C_{j-1,2}=C_{j,1}} H_j, i \leq j$. Then, since $\tilde{A}$ is finitely generated, there exist integers r, s $(r \leq s)$ such that $\tilde{A} \subset H(r,s)$. Since $\tilde{A}$ is normal, it follows that

$$\tilde{A} = t^{-(s-r+1)}\tilde{A}t^{s-r+1} \subset t^{-(s-r+1)}H(r,s)t^{s-r+1} = H(s+1, 2s-r+1).$$

Thus we see that

$$\tilde{A} \subset H(r,s) \cap H(s+1, 2s-r+1) = C_{s+1,1} = t^{-(s+1)}\phi_1(C)t^{s+1}.$$

The proof is completed.

Lemma 2.2. *Let G be an HNN extension $< H, t : t^{-1}\phi_1(C)t = \phi_2(C) >$, where H is a non-finitely generated group. If G is finitely presented, then there exists a finitely generated subgroup of H which is a base of G. More precisely, G can be described as an HNN extension $< \tilde{H}, t : t^{-1}\tilde{\phi}_1(\tilde{C})t = \tilde{\phi}_2(\tilde{C}) >$, where $\tilde{H}$ and $\tilde{C}$ are finitely generated subgroups of H, C respectively, and $\tilde{\phi}_i = \phi_i|\tilde{C}, i = 1,2$.*

Proof. Let $A = \phi_1(C), B = \phi_2(C)$ and $\phi = \phi_2 \cdot \phi_1^{-1}$. Let $< h_1, h_2, ... : R_1, R_2, ... >$ be a presentation of H, where R_i are words on $h_1, h_2, \cdots$. Then G is presented by

$$(2.1) \qquad < h_1, h_2, ..., t : t^{-1}A_1t = B_1, t^{-1}A_2t = B_2, ..., R_1, R_2, ... >,$$

where $\{A_i\}$ is a generating set of the associated subgroup A, and $B_i = \phi(A_i)$. Since G is finitely generated, G is generated by only finitely many of the generators $h_1, h_2, ..., t$. Thus we may assume that G can be generated by $h_1, ..., h_n, t$. Therefore, for each $j > n$, h_j can be represented by a word W_j on $h_1, ..., h_n, t$. Deleting generators h_j $(j > n)$ in (2.1), we obtain a presentation for G

$$(2.2) \qquad < h_1, h_2, ..., h_n, t : t^{-1}A_1't = B_1', t^{-1}A_2't = B_2', ..., R_1', R_2', ... >,$$

where A_i', B_i' and R_i' are words obtained by substituting W_j for h_j $(j > n)$ in A_i, B_i and R_i, respectively. Moreover, since G is finitely related, G can be related by only finitely many of the relators $\{t^{-1}A_i'tB_i'^{-1}, R_i'\}$ in (2.2) (cf.[B]). Hence we have

$$G =< h_1, h_2, ..., h_n, t : t^{-1}A_1't = B_1', ..., t^{-1}A_m't = B_m', R_1', ..., R_s' > .$$

Let k be the maximum subscript j such that h_j occurs in $A_1, ..., A_m, B_1, ..., B_m, R_1, ..., R_s$. (Note that $k \geq n$.) Then we obtain

$$G =< h_1, h_2, ..., h_n, ..., h_k, t : t^{-1}A_1't = B_1', ..., t^{-1}A_m't = B_m', R_1', ..., R_s',$$
$$h_{n+1} = W_{n+1}, ..., h_k = W_k >,$$
$$(2.3) \qquad =< h_1, h_2, ..., h_k, t : t^{-1}A_1t = B_1, ..., t^{-1}A_mt = B_m, R_1, ..., R_s,$$
$$h_{n+1} = W_{n+1}, ..., h_k = W_k > .$$

In (2.3), we see that A_i, B_i and R_j $(1 \leq i \leq m,\ 1 \leq j \leq s)$ involve no t-generator. We now assume that W_{n+1} involves t or t^{-1}. Let $W_{n+1} =$

$$(2.4) \qquad g_0t^{\varepsilon_1}g_1t^{\varepsilon_2}...t^{\varepsilon_p}g_p,$$

where $\varepsilon_j = \pm 1, p > 0$, and g_i is a word on $h_1, ..., h_n, i = 0, ..., p$. Then, since $p > 0$ and W_{n+1} defines an element of H, it follows from Britton's lemma [LS,p.181] that (2.4) is not reduced relative to the HNN extension $< H, t : t^{-1}\phi_1(C)t = \phi_2(C) >$. That is, there exists a subword $t^{\varepsilon_i}g_it^{\varepsilon_{i+1}}$ such that $\varepsilon_i = -\varepsilon_{i+1} = -1, g_i \in A$ or $\varepsilon_i = -\varepsilon_{i+1} = 1, g_i \in B$.

In (2.3), we introduce a new generator $\tilde{h}_1$ and add a relation $t^{-1}C_1t = D_1$, where C_1 and D_1 are given by $C_1 = \tilde{h}_1, D_1 = g_i$ or $C_1 = g_i, D_1 = \tilde{h}_1$ according as $\varepsilon_i = 1$ or -1. (It should be noted that $\tilde{h}_1$ defines an element in A or B according as $\varepsilon_i = 1$ or -1.) Furthermore, substitute $\tilde{h}_1$ for the subword $t^{\varepsilon_i}g_it^{\varepsilon_{i+1}}$ in W_{n+1}. Thus we can decrease the number of t-generators in W_{n+1}. Using induction on the number of t-generators occuring in $W_{n+1}, ..., W_k$, we can prove that G has a presentation of the following form:

(2.5)

$$< h_1, h_2, ..., h_k, \tilde{h}_1, \tilde{h}_2, ..., \tilde{h}_q, t : t^{-1}A_1t = B_1, ..., t^{-1}A_mt = B_m, R_1, ..., R_s,$$

$$h_{n+1} = W^*_{n+1}, ..., h_k = W^*_k, t^{-1}C_1t = D_1, ..., t^{-1}C_qt = D_q >,$$

where W^*_i, C_j and D_j are words on $h_1, h_2, ..., h_k, \tilde{h}_1, \tilde{h}_2, ..., \tilde{h}_q$, and $C_j \in A, D_j \in B, n+1 \leq i \leq k, 1 \leq j \leq q$.

Let $\tilde{H} = gp\{h_1, h_2, ..., h_k, \tilde{h}_1, \tilde{h}_2, ..., \tilde{h}_q\}, \tilde{A} = gp\{A_1, ..., A_m, C_1, ..., C_q\}$ and $\tilde{B} = gp\{B_1, ..., B_m, D_1, ..., D_q\}$ in H. Then $\tilde{H}, \tilde{A}$ and $\tilde{B}$ are contained in H, A, B, respectively. Let

$$< h_1, h_2, ..., h_k, \tilde{h}_1, \tilde{h}_2, ..., \tilde{h}_q : S_1, S_2, ... >$$

be a presentation for $\tilde{H}$ with generators $h_1, h_2, ..., h_k, \tilde{h}_1, \tilde{h}_2, ..., \tilde{h}_q$. Then the relators R_i and $h_jW^{*-1}_j$ $(1 \leq i \leq s, n+1 \leq j \leq k)$ can be derived from the defining relators $\{S_1, S_2, ...\}$. Thus, from (2.5), we obtain

$$G =< h_1, h_2, ..., h_k, \tilde{h}_1, \tilde{h}_2, ..., \tilde{h}_q, t : t^{-1}A_1t = B_1, ..., t^{-1}A_mt = B_m,$$

$$t^{-1}C_1t = D_1, ..., t^{-1}C_qt = D_q, S_1, S_2, ... >,$$

$$=< \tilde{H}, t : t^{-1}A_1t = B_1, ..., t^{-1}A_mt = B_m, t^{-1}C_1t = D_1, ..., t^{-1}C_qt = D_q > .$$

In the group G, the mapping $\tilde{\phi}$ of $\tilde{A}$ to $\tilde{B}$ given by $A_i \to B_i$ and $C_j \to D_j$ is induced by conjugation of t, i.e., $\tilde{\phi} = \phi|\tilde{A}$. Therefore, $\tilde{A}$ and $\tilde{B}$ are isomorphic under this mapping. Thus G is an HNN extension of the finitely generated subgroup $\tilde{H}$ of H. Hence the proof is completed.

Lemma 2.3. *Let H be a group, and let A, B be subgroups of H. If $A \neq H$ and $B \neq H$, then $A \cup B \neq H$.*

Proof. Suppose that $A \cup B = H$. Then, since $A \neq H$ and $B \neq H$, there exist $g_1 \in A$ and $g_2 \in B$ such that $g_1 \notin B, g_2 \notin A$. However, the product g_1g_2 of g_1 and g_2 is not contained in $A \cup B$.

3. Abelian normal subgroups of two-knot groups

In this section, we will discuss abelian normal subgroups of 2-knot groups. It is known that an abelian normal subgroup of a 1-knot group is trivial or infinite cyclic and that any finitely generated abelian group can be realized as an abelian normal subgroup of an n-knot group $(n \geq 3)$ (cf.[HK]). On the other hand, J. A. Hillman has obtained many results on 2-knot groups with non-trivial abelian normal subgroups [Hi3]. In particular, he proved

Theorem 3.1 (Hillman). *If A is an abelian normal subgroup of a 2-knot group, then $r(A) \leq 4$, where $r(A)$ is the rank of A. Especially, if $r(A) = 3, 4$, then A is free abelian.*

Remark 3.2. There are the following examples of 2-knots with abelian normal subgroup A.

(1) $A \cong Z^4$; the 3-twist spin of the figure-eight 1-knot,
(2) $A \cong Z^3$; 2-knots with commutator subgroup isomorphic to Z^3,
(3) $A \cong Z^2$; r-twist spins of (p,q)-torus 1-knots, where $1/p + 1/q + 1/r \leq 1$,
(4) $A \cong Z$; spins of torus 1-knots,
(5) $A \cong Z \oplus Z_m$; 2-knots with finite commutator subgroup,
(6) $A \cong Z_m$; some satellite 2-knots obtained from 2-knots in (5),
(7) $A \cong Z[1/2]$; the 2-knot with commutator subgroup isomorphic to $Z[1/2]$,
where $m = 2l + 1$ or $2(2l + 1)$, $l \geq 0$.

When $r(A) \leq 2$, it remains open to determine A. In case of 2-knots with minimal Seifert manifold, we will prove the following:

Theorem 3.3. *If A is a finitely generated abelian normal subgroup of the group of a 2-knot K with minimal Seifert maifold, then A is isomorphic to one of the following groups:*

(1) Z_n,
(2) $Z \oplus Z_m$ or
(3) Z^k,
where n is an integer, $m = 2l + 1$ or $2(2l + 1)$, $l \geq 0$, and $k = 2, 3, 4$.

To prove Theorem 3.3, we give two lemmas.

Lemma 3.4. *Let A be an abelian normal subgroup of an n-knot group $(n \geq 1)$. Then we have*

$$A \cong A \cap [G, G] \text{ or } (A \cap [G, G]) \oplus Z.$$

Proof. The proof is elementary.

Lemma 3.5. *If A is a finitely generated abelian normal subgroup of the group G of an n-knot $(n \geq 1)$ with minimal Seifert manifold V, then we have*

$$A \cap [G, G] \subset \nu_{1*}(\pi_1(V)) \cap \nu_{2*}(\pi_1(V)).$$

Proof. The group G can be expressed as an HNN extension $< H, t : t^{-1}\nu_{1*}(C)t = \nu_{2*}(C) >$, where $H = \pi_1(Y)$ and $C = \pi_1(V)$. Therefore, since $<< H >>^G = [G, G]$, the lemma immediately follows from Lemma 2.1.

Proof of Theorem 3.3. Let G be the group of K. By Lemma 3.5, we see that $A \cap [G, G]$ is ismorphic to a subgroup $\hat{A}$ of the group C of the orientable compact 3-manifold V. Therefore, from [He], $\hat{A}$ is isomorphic to Z_n or Z^k, where n is an integer and $1 \leq k \leq 3$. If $\hat{A} \cong Z^k$ or $A \subset [G, G]$, then the assertion follows from Lemma 3.4. Hence we assume that $\hat{A} \cong Z_n$ $(n \neq 1)$ and $A \not\subset [G, G]$, i.e., $A \cong Z \oplus Z_n$. Since $A \not\subset [G, G]$, the commutator subgroup $[G, G]$ is finitely generated (cf.[Hi3]). Therefore we see that

$$[G, G] = H = \nu_{1*}(C) = \nu_{2*}(C) \cong C.$$

Since $\hat{A}$ is a finite subgroup of the 3-manifold group C, it follows from [He] that

(1) $\hat{A} \cong Z_2$ or

(2) $C = W_1 * W_2$, where W_1 is finite and $\hat{A}$ is conjugate to a subgroup $\hat{W}_1$ of W_1.

In case (1), the assertion holds. In case (2), since $\hat{A}$ is normal, we have

$$\hat{A} = \hat{W}_1 \subset W_1 \text{ and } W_2 \cong 1.$$

Therefore, the commutator subgroup $[G, G] (\cong C)$ is finite. Thus, from [Hi1], [Y1], we obtain $n = 2l + 1$ or $2(2l + 1)$, where $l \geq 0$. The proof is completed.

Corollary 3.6. *If A is a finitely generated abelian normal subgroup of a fibered 2-knot group, then A is isomorphic to one of the following groups:*

(1)' Z_m,

(2) $Z \oplus Z_m$ or

(3) Z^k,

where $m = 2l + 1$ or $2(2l + 1)$, $l \geq 0$, and $k = 2, 3, 4$.

Proof. A fibered 2-knot has a minimal Seifert manifold and the commutator subgroup of its group is finitely generated. Therefore, from the proof of Theorem 3.3, we obtain this corollary.

Remarks 3.7. (1) Each of the above groups can be realized as an abelian normal subgroup of a fibered 2-knot group. Thus abelian normal subgroups of fibered 2-knot groups are completely determined.

(2) In [N], L. Neuwirth showed that the center of a 1-knot group is trivial or infinite cyclic. For $n \geq 3$, J. C. Hausmann and M. Kervaire [HK] proved that any finitely generated abelian group is the center of an n-knot group. On the other hand, though some results on the centers of 2-knot groups are obtained (e.g., [Hi3],[Y1],[Y2],[Z]), it still remains open to characterize the centers of 2-

knot groups. It is known that there exist 2-knots whose groups have the centers $1, Z_2, Z, Z \oplus Z_2$ and Z^2 respectively [Y1],[Y2],[Z]. Conversely, by the similar argument of the proof of Therem 3.3, we can obtain the following [Y3]: A finitely generated abelian group A is isomorphic to the center of some 2-knot group with minimal Seifert manifold if and only if $A = 1, Z_2, Z, Z \oplus Z_2$ or Z^2. In particular, if the center of a fibered 2-knot group is finitely generated, then it is isomorphic to $1, Z, Z \oplus Z_2$ or Z^2.

4. Abelian bases of two-knot groups

In this section, we will consider abelian bases of 2-knot groups. For $n \geq 3$, finitely generated abelian bases of n-knot groups are completely determined [Y6]. The author showed that if a 2-knot group is an HNN exention of a finitely generated abelian group H, then H is isomorphic to Z_α, Z^3, or $Z \oplus Z_\beta$, where α and β are odd [Y6]. More generally, we will prove that

Theorem 4.1. *If a 2-knot group is an HNN exention of an abelian group H, then H is isomorphic to*

(1) Z_α,
(2) Z^3,
(3) $Z \oplus Z_\beta$ or
(4) $Z[1/2]$,

where α and β are odd.

Remark 4.2. We have the following examples of 2-knots with abelian base H:

(1) $H \cong Z_\alpha$; 2-twist spins of 2-bridge 1-knots,
(2) $H \cong Z^3$; 2-knots with commutator subgroup isomorphic to Z^3,
(3) $H \cong Z$; ribbon 2-knots with group $< a, t : t^{-1}a^\lambda t = a^{\lambda+1} >$, where $\lambda > 0$,
(4) $H \cong Z[1/2]$; the 2-knot with commutator subgroup isomorphic to $Z[1/2]$,

It is not known whether or not there exists a 2-knot group with base isomorphic to $Z \oplus Z_\beta, \beta > 1$.

To prove Theorem 4.1, we will show the following lemma (cf. [Y5]):

Lemma 4.3. *Let G be an n-knot group $(n \geq 1)$ with non-finitely generated abelian base H. Then the commutator subgroup $[G, G]$ is (non-finitely generated) abelian.*

Proof. By Lemma 2.2, G can be represented as an HNN extension $< \tilde{H}, t : t^{-1}\tilde{\phi}_1(\tilde{C})t = \tilde{\phi}_2(\tilde{C}) >$, where $\tilde{H}$ is a finitely generated abelian subgroup of H.

Let $\tilde{A} = \tilde{\phi}_1(\tilde{C})$ and $\tilde{B} = \tilde{\phi}_2(\tilde{C})$. Since $\tilde{H} \subsetneqq H$, there exists an element $h \in H$ such that $h \notin \tilde{H}$. Let $h =$

$$g_0 t^{\varepsilon_1} g_1 t^{\varepsilon_2} ... g_{p-1} t^{\varepsilon_p} g_p,\ g_i \in \tilde{H}, \tag{4.1}$$

be reduced relative to the HNN extension $< \tilde{H}, t : t^{-1}\tilde{\phi}_1(\tilde{C})t = \tilde{\phi}_2(\tilde{C}) >$. Then, since $h \notin \tilde{H}$, it follows from Britton's lemma [LS] that $p \geq 1$. We assume that $[G, G]$ is not abelian. Then we have $\tilde{H} \neq \tilde{A}$ and $\tilde{H} \neq \tilde{B}$ (cf.[Y4]). Therefore, from Lemma 2.3, there exists an element u of $\tilde{H}$ such that $u \notin \tilde{A} \cup \tilde{B}$. On the other hand, by (4.1), we see that

$$huh^{-1}u^{-1} = (g_0 t^{\varepsilon_1} ... t^{\varepsilon_p} g_p)u(g_p^{-1} t^{-\varepsilon_p} ... t^{-\varepsilon_1} g_0^{-1})u^{-1},$$

$$= g_0 t^{\varepsilon_1} ... t^{\varepsilon_p} u t^{-\varepsilon_p} ... t^{-\varepsilon_1} g_0^{-1} u^{-1}. \tag{4.2}$$

Since $u \notin \tilde{A} \cup \tilde{B}$ and $p \geq 1$, (4.2) is reduced relative to the HNN extension $< \tilde{H}, t : t^{-1}\tilde{\phi}_1(\tilde{C})t = \tilde{\phi}_2(\tilde{C}) >$. Thus, Britton's Lemma shows that $huh^{-1}u^{-1} \neq 1$. However this contradicts the fact that h and u are elements of the abelian group H. Therefore, we have $\tilde{H} = \tilde{A}$ or $\tilde{H} = \tilde{B}$. Thus this completes the proof (cf.[Y4]).

Proof of Theorem 4.1. Let G be a 2-knot group with abelian base H. If H is finitely generated, then the theorem holds [Y6]. Suppose that H is non-finitely generated. Then by Lemma 4.3, the commutator subgroup $[G, G]$ is non-finitely generated abelian. Therefore, from [Hi2] and [Y4], $[G, G]$ is isomorphic to $Z[1/2]$. Since any non-finitely generated subgroup of $[G, G]$ is isomorphic to $Z[1/2]$, we have $H \cong Z[1/2]$. The proof is completed.

Remark 4.4. From the proof of Theorem 4.1, we see that the group presented by $< a, t : t^{-1}at = a^2 >$ is the only 2-knot group with non-finitely generated abelian base. Furthermore, in case of 1-knot groups, there exists no 1-knot group with non-finitely generated abelian base.

In case of 2-knots with minimal Seifert manifolds, we obtain the following:

Theorem 4.5. *An abelian group H is a base of a 2-knot group G with minimal Seifert manifold V if and only if H is isomorphic to*

(1) Z_α, where α is odd,

(2) Z^3,

(3)' Z or

(4) $Z[1/2]$.

Proof. The sufficiency follows from Remark 4.2. Therefore we will prove the necessity. From Theorem 4.1, it suffices to consider only the case (3)'. We assume that G has an abelian base H isomorphic to $Z \oplus Z_\beta$, where $\beta > 1$. Since G is the group of a 2-knot with minimal Seifert manifold V, G can be expressed as an HNN extension $< H, t : t^{-1}\nu_{1*}(\pi_1(V))t = \nu_{2*}(\pi_1(V)) >$, where $H = \pi_1(Y)$. By [Y6], the torsion part $T(H)$ is normal in G. Hence, from Lemma 3.5, $T(H)$ is contained in $\nu_{1*}(\pi_1(V))$. By the same argument as in the proof of Theorem 3.3, we can see that $\nu_{1*}(\pi_1(V))$ is finite. Then, by [LS], it follows that

$$T(H) = T(G) \supset \nu_{1*}(\pi_1(V)).$$

Therefore $T(H) = \nu_{1*}(\pi_1(V)) = \nu_{2*}(\pi_1(V))$. Hence, we have

$$G/ << t >>^G \supset G/ << t, \nu_{1*}(\pi_1(V)) >>^G = H/T(H) \cong Z \neq 1.$$

This contradicts the fact that t is a weighted element of G. Thus we complete the proof.

Remarks 4.6. (1) It is known that every 1-knot has a minimal Seifert manifold [N]. On the other hand, it remains open whether every 2-knot has a minimal Seifert mainifold. Recently D. Silver announces that, for $n \geq 3$, there exists an n-knot with no minimal Seifert manifold.

(2) Lemma 1 in [G] states that if k is a knot of dimension ≥ 3, there exists a minimal Seifert manifold for k. T. Maeda pointed out the existence of gaps in the proof of this lemma. In [Y4] and [Y6], the author refers to this lemma. However without this lemma, we can prove results (theorems etc.) in [Y4] and [Y6] by using Lemma 2.2 of this paper.

References

[B] G. Baumslag, *A remark on generalised free products*, Proc. Amer. Math. Soc. **13** (1962), 53–54.

[G] M. A. Gutierrez, *An exact sequence calculation for the second homotopy of a knot*, Proc. Amer. Math. Soc. **32** (1972), 571–577.

[HK] J. C. Hausmann and M. Kervaire, *Sur le centre des groupes de noeuds multidimensionnels*, C. R. Acad. Sc. Paris, t. **287** (1978), 699–702.

[He] J. Hempel, *3-manifolds*, Princeton Univ. Press, 1976.

[Hi1] J. A. Hillman, *High dimensional knot groups which are not two-knot groups*, Bull. Austral. Math. Soc. **16** (1977), 449–462.

[Hi2] J. A. Hillman, *On metabelian two-knot groups*, Proc. Amer. Math. Soc. **96** (1986), 372–375.

[Hi3] J. A. Hillman, *2-knots and their groups*, Cambrige Univ. Press, 1989.

[L] J. Levine, *Some results on higher dimensional knot groups*, Lecture Notes in Math., Springer **685** (1978), 243–273.

[LS] R. C. Lyndon and P. E. Schupp, *Combinatorial Group Theory*, Springer, 1977.

[N] L. Neuwirth, *The algebraic determination of the genus of knots*, Amer. J. Math. **82** (1960), 791–798.

[Y1] K. Yoshikawa, *On 2-knot groups with the finite commutator subgroups*, Math. Sem. Notes Kobe Univ. **8** (1980), 321–330.

[Y2] K. Yoshikawa, *On a 2-knot group with nontrivial center*, Bull. Austral. Math. Soc. **25** (1982), 321–326.

[Y3] K. Yoshikawa, *A note on the center of fibered two-knot groups*, preprint (1983).

[Y4] K. Yoshikawa, *On 2-knot groups with abelian commutator subgroups*, Proc. Amer. Math. Soc. **92** (1984), 305–310.

[Y5] K. Yoshikawa, *On n-knot groups which have abelian bases*, preprint (1984).

[Y6] K. Yoshikawa, *Knot groups whose bases are abelian*, J. Pure and Appl. Algebra **40** (1986), 321–335.

[Z] E. C. Zeeman, *Twisting spun knots*, Trans. Amer. Math. Soc. **115** (1965), 471–495.

Universal pretzel links

Yoshiaki Uchida

1. Introduction and Main Theorem

A link or knot L in S^3 is *universal* if every closed orientable 3-manifold can be represented as a covering space of S^3 branched over L [7]. Hilden, Lozano and Montesinos gave an necessary and sufficient condition for a 2-bridge knot and link to be universal in [1]. And the auther gave an necessary and sufficient condition for a chain, which is a special type of pretzel link, is universal in [6]. In this paper we consider the following question *"which pretzel links are universal ?"* We remark that a link, being a union of fibers of a graph-manifold structure on S^3 is not universal. That is because the branched covering space over such a link is a graph-manifold.

1.1. Definition. A *pretzel link* is a link consisting of 2-strand braids with q_1-,q_2-, ...,q_m-half twists, which we denote by $p(q_1, q_2, ..., q_m)$. We assume that $q_i \neq 0$ for $i = 1, 2, ..., m$. For example, $p(3, 6, -2)$ is shown in Figure 1.1.

If $q_i = \pm 1$, then $p(q_1, q_2, ..., q_m)$ is equivalent to $p(q_i, q_1, ..., q_{i-1}, q_{i+1}, ..., q_m)$. So we can deform $p(q_1, q_2, ..., q_m)$ into $p(\varepsilon, \varepsilon, ..., \varepsilon, p_1, p_2, ..., p_n)$, where $\varepsilon = \pm 1$ and $| p_i |> 1$ $(i = 1, 2, ..., n)$. We denote this pretzel link by $p(-\varepsilon b; p_1, p_2, ..., p_n)$, where b is the number of ε. If $b \neq 0$ and $p_i = -2\varepsilon$, then $p(-\varepsilon b; p_1, p_2, ..., p_n)$ is equivalent to $p(-\varepsilon(b-1); p_1, p_2, ..., -p_i, ..., p_n)$. So if $b \neq 0$, we can assume that every p_i is not -2ε. We consider only for the case that $p(-\varepsilon b; p_1, p_2, ..., p_n)$ has two components or more, and so only the following two cases occur:

(I) At least two p_i's are even.
(II) All of the p_i's are odd, and $n + b$ is even. (In this case, the number of components is two.)

We say that $p(-\varepsilon b; p_1, p_2, ..., p_n)$ is type (I) (or type (II), resp.), if it is of the case (I) (or (II), resp.).

Knots 90

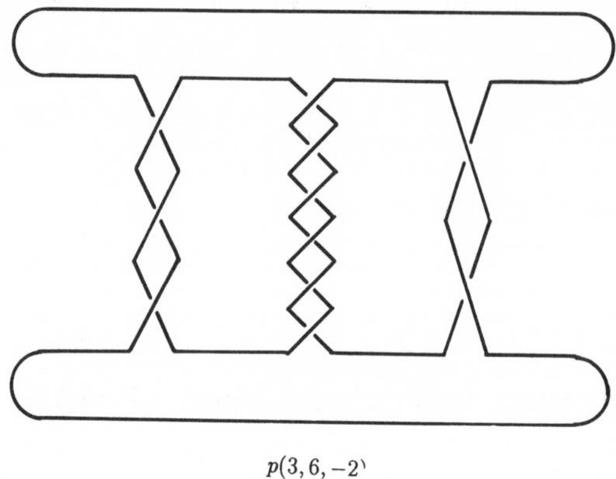

$p(3, 6, -2)$

Figure 1.1

1.2. Theorem. *For the pretzel link* p *of type (I),* p *is univeral if and only if* p *is none of the following:*

$p(2s, 2t), p(2, -2, s), p(0; -2, 3, 4), p(0; 2, -3, -4), p(0; 3, 6, -2), p(0; -3, -6, 2),$ $p(0; 4, 4, -2)$, $p(0; -4, -4, 2)$, $p(0; 2, 2, -2, -2)$, where $s, t \in \mathbf{Z} \setminus \{0\}$.

As a consequence of Theorem 1.2, we have:

1.3. Theorem. *For a pretzel link* p *of type (I),* p *is universal if and only if* p *is not a union of fibers of any graph-manifold structure on* S^3.

Question. Which pretzel links of type (II) are universal ?

2. Preliminaries

We represent the 2-bridge torus knot $T(2, a)$ as in Figure 2.1.

2.1. Lemma. *For an odd integer* a, *the following branched covering space of* S^3 *branched over* $T(2, a)$ *is* S^3.

The monodoromy map $\phi : \pi_1(S^3 \setminus T(2, a)) \longrightarrow S_{|a|}$, *from the knot group of* $T(2, a)$ *to the symmetry group of* $| a |$ *indices, is defined by*

$$\phi(x_1) = (1\ 2\ ...q)$$
$$\phi(x_2) = (1\ q+1\ q+2\ ...2q-1)$$

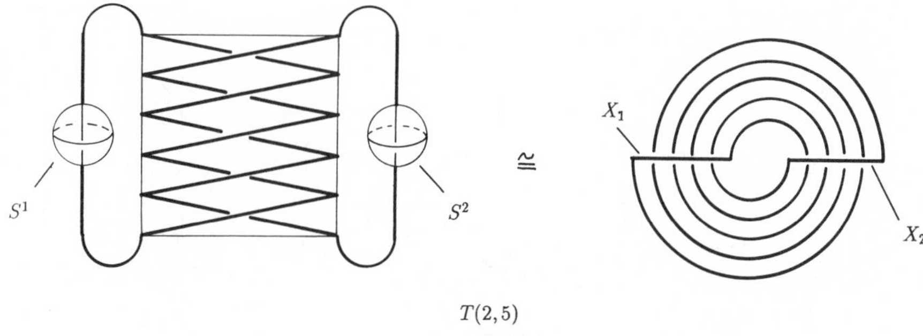

Figure 2.1

where $q = \frac{|a|+1}{2}$ *and* x_1, x_2 *are the meridians as in Figure 2.1.*

We denote this covering by $f : S^3 \longrightarrow S^3$.

Proof. The branched covering space of S^3 branched over a 2-component trivial link, associated with the monodromy as in Figure 2.2, is S^3. And the preimage of this link is a $2q$-component trivial link. Now we consider the 2-disk D as shown in Figure 2.2. The $| \ a \ |$-fold branched covering space of D associated with ϕ is also a 2-disk $\tilde{D}$. By a-half twists at D, this trivial link can be deformed into the torus knot $T(2, a)$. And the a-half twists at D are lifted to one-half twist at $\tilde{D}$ (cf. [4], [5, p.317]).

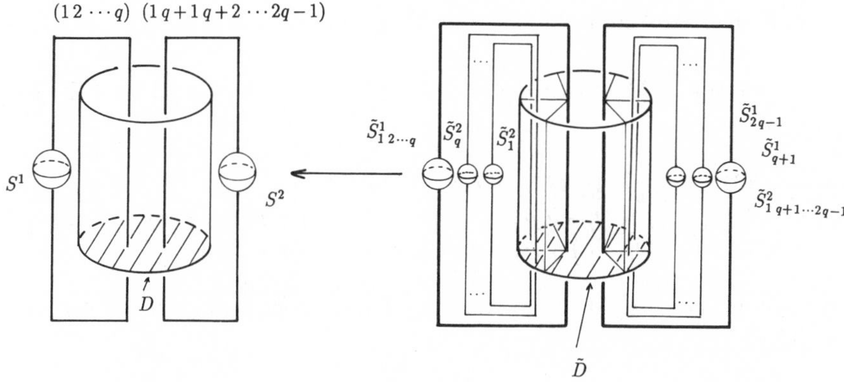

Figure 2.2

Hence the covering space is S^3. And the preimage of $T(2, a)$ is the torus link $T(2q, \varepsilon q) = T(| \ a \ | +1, \varepsilon\frac{|a|+1}{2}) = \frac{|a|+1}{2}T(2, \varepsilon)$ where $\varepsilon = 1$ if $a > 0$ or $\varepsilon = -1$ if

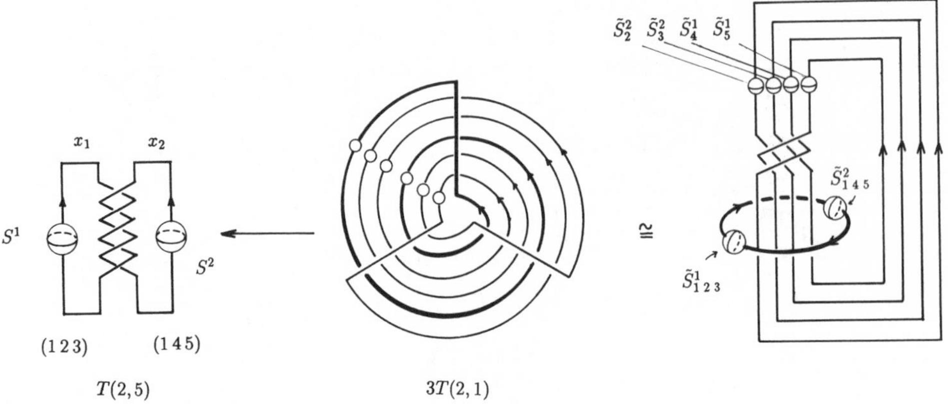

Figure 2.3

$a < 0$. (In Figure 2.3, a = 5.) In Figures 2.2 and 2.3, the bold lines have branched index q. This completes the proof of Lemma 2.1.

2.2. The line with branch index q is a component of $\frac{|a|+1}{2}T(2,\varepsilon)$, and so it is a trivial knot. Then the n-fold cyclic branched covering branched spce over this knot is S^3. The preimage of the other components is $\frac{|a|-1}{2}T(2,\varepsilon n)$.

We denote this covering by $g : S^3 \longrightarrow S^3$.

The monodromy of the covering $f \circ g : S^3 \longrightarrow S^3$ branched over $T(2,a)$, is defined by

$$\begin{aligned} x_1 &\longmapsto (1_1\ 2_1 \cdots q_1\ 1_2\ 2_2\ \cdots q_2\ \cdots\cdots 1_n\ 2_n\ \cdots q_n), \\ x_2 &\longmapsto (1_1\ (q+1)_1\ (q+2)_1\ \cdots (2q-1)_1\ 1_2\ (q+1)_2\ (q+2)_2\ \cdots (2q-1)_2\ \cdot \\ &\quad \cdots\cdots 1_n\ (q+1)_n\ (q+2)_n\ \cdots (2q-1)_n). \end{aligned}$$

We call this covering the (C_1)-covering. Unless confusion, we may use a simple form like as

$$\begin{aligned} x_1 &\longmapsto (1\ 2\ 3\ 4\ 5\ 6\ 7\ 8\ 9)\ (=\ (1_1\ 2_1\ 3_1\ 1_2\ 2_2\ 3_2\ 1_3\ 2_2\ 3_3)), \\ x_2 &\longmapsto (1\ 10\ 11\ 4\ 12\ 13\ 7\ 14\ 15)\ (=\ (1_1\ 4_1\ 5_1\ 1_2\ 4_2\ 5_2\ 1_3\ 4_3\ 5_3)). \end{aligned}$$

2.3. Remark. The length of the cyclic permutation associated with this monodromy is $n \times q$.

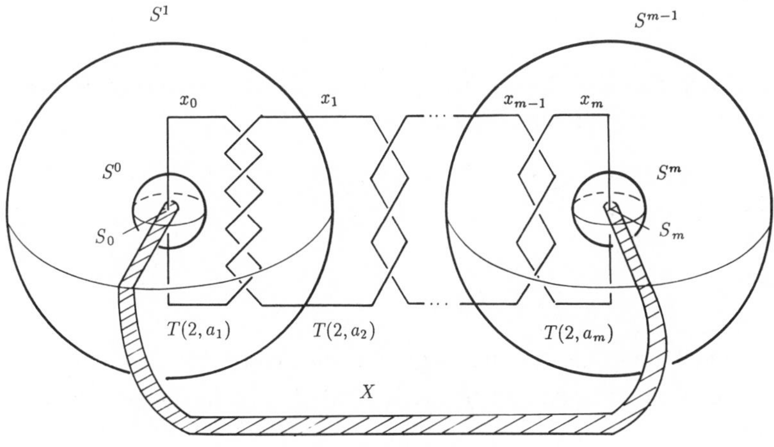

Figure 2.4

2.4. We consider the following branched covering space of S^3 branched over $T(2,a_1)\sharp T(2,a_2)\sharp\cdots\sharp T(2,a_m)$, where a_i is odd for $i=1,2,...,m$.

In Figure 2.4, S^i $(i=0,1,...,m)$ is a 2-sphere and S^i $(i=1,2,...,m-1)$ divides the factors of $T(2,a_i)$'s. And $x_0,x_1,...,x_m$ are the meridians as in Figure 2.4. Let $q_i=\frac{|a_i|+1}{2}$ and q the least common multiple of q_1, q_2,...,q_m. Now we define the permutations corresponding to $x_0,x_1,...,x_m$ as follows:

$x_i\longmapsto(i_1,i_2,...,i_q)$ for $i=0,1,...,m$,
where we identify $i_{1+kq_{i+1}}$ with $(i+1)_{1+kq_{i+1}}$ for $0\le i\le m-1,\ 0\le k<\frac{q}{q_{i+1}}$.

We call this covering the (C_2)-covering for $T(2,a_1)\sharp\cdots\sharp T(2,a_m)$.

2.5. Example. $T(2,5)\sharp T(2,3)\sharp T(2,3)$.

The monodromy of the (C_2)-covering is defined by

$$\begin{array}{l} x_0\longmapsto(0_1\ 0_2\ 0_3\ 0_4\ 0_5\ 0_6),\\ x_1\longmapsto(1_1\ 1_2\ 1_3\ 1_4\ 1_5\ 1_6),\\ x_2\longmapsto(2_1\ 2_2\ 2_3\ 2_4\ 2_5\ 2_6),\\ x_3\longmapsto(3_1\ 3_2\ 3_3\ 3_4\ 3_5\ 3_6),\end{array}\ \text{where}\ \begin{pmatrix} 0_1=1_1=2_1=3_1\\ 0_4=1_4\\ 1_3=2_3=3_3\\ 1_5=2_5=3_5\end{pmatrix}.$$

or

$x_0\longmapsto(1\ 2\ 3\ 4\ 5\ 6)$,
$x_1\longmapsto(1\ 7\ 8\ 4\ 9\ 10)$,
$x_2\longmapsto(1\ 11\ 8\ 12\ 9\ 13)$,
$x_3\longmapsto(1\ 14\ 8\ 15\ 9\ 16)$.

2.6. Lemma. *For $T(2,a_1)\sharp T(2a,a_2)\sharp\cdots\sharp T(2,a_m)$, the (C_2)-covering space is S^3.*

Proof. For each factor $T(2,a_i)$, this covering is the (C_1)-covering. Hence, we can assume that the corresponding monodromy is

$x_{i-1}\longmapsto(1_1\ 2_1\ \cdots(q_i)_1\ 1_2\ 2_2\ \cdots(q_i)_2\ \cdots\cdots 1_r\ 2_r\ \cdots(q_i)_r)$,
$x_i\longmapsto(1_1\ (q_i+1)_1\ \cdots(2q_i-1)_1\ 1_2\ (q_i+1)_2\ \cdots(2q_i-1)_2\ \cdots\cdots 1_r\ (q_i+1)_r\ \cdots(2q_i-1)_r)$,
where $r=\frac{q}{q_i}$.

From 2.2 and Figure 2.3, the preimage is as in Figure 2.5. In Figure 2.5, $\tilde{S}^i_{12\cdots q}$ is the lift of S^i, corresponding to the letter $1,2,...,q$ in the permutation of the monodromy.

Making the connect sum of $T(2,a_i)$ and $T(2,a_{i+1})$ at S^i, is lifted to the following operation in the covering space:

(1) We attach these the (C_1)-covering spaces of $T(2,a_i)$ and $T(2,a_{i+1})$ at $\tilde{S}^i_{1_1\ (q_i+1)_1\ (q_i+2)_1\cdots\cdots(2q_i-1)_r}$.
(2) At the others, $\tilde{S}^i_j$ (unbranched), we attach a copy of the 3-ball bounded by S^i in S^3 which does not contain $T(2,a_i)$.

So, the covering space is S^3. We can perform this operation for $T(2,a_1)\sharp\cdots\sharp T(2,a_{i-1})$ and $T(2,a_i)$ $(i=2,3,...,n)$. Then, the covering space of the (C_2)-covering is S^3. This completes the proof of Lemma 2.6.

2.7. Remark. The knot with branch index q is trivial, because it is a connected sum of trivial knots.

2.8. Example. The preimage of $T(2,5)\sharp T(2,3)\sharp T(2,3)$ under the (C_2)-covering.

The monodromy is given in Example 2.5. Figure 2.6 (a_1) ((a_2), (a_3), resp.) shows the preimage of $T(2,a_1)=T(2,5)$ $(T(2,a_2)=T(2,3), T(2,a_3)=T(2,3)$, resp.) for the (C_2)-covering. The preimage of $T(2,5)\sharp T(2,3)\sharp T(2,3)$ is shown in Figure 2.7.

2.9. Preparation I. We consider the preimage of the 2-disk X as in Figure 2.4 under the (C_2)-covering. Let $X\cap\{T(2,a_1)\sharp T(2,a_2)\sharp\cdots\sharp T(2,a_m)\}=\{S_0,S_m\}$. The preimage of X is the covering space of the disk X branched over $\{S_0,S_1\}$. And {the letter in the permutation corresponding to S_0 } $\cap$ {the letter in the permutation corresponding to S_m } $=\{0_1(=m_1)\}$. So the covering space of X is also a 2-disk $\tilde{X}$. Let X intersect each S^i in the equator. See Figure 2.4.

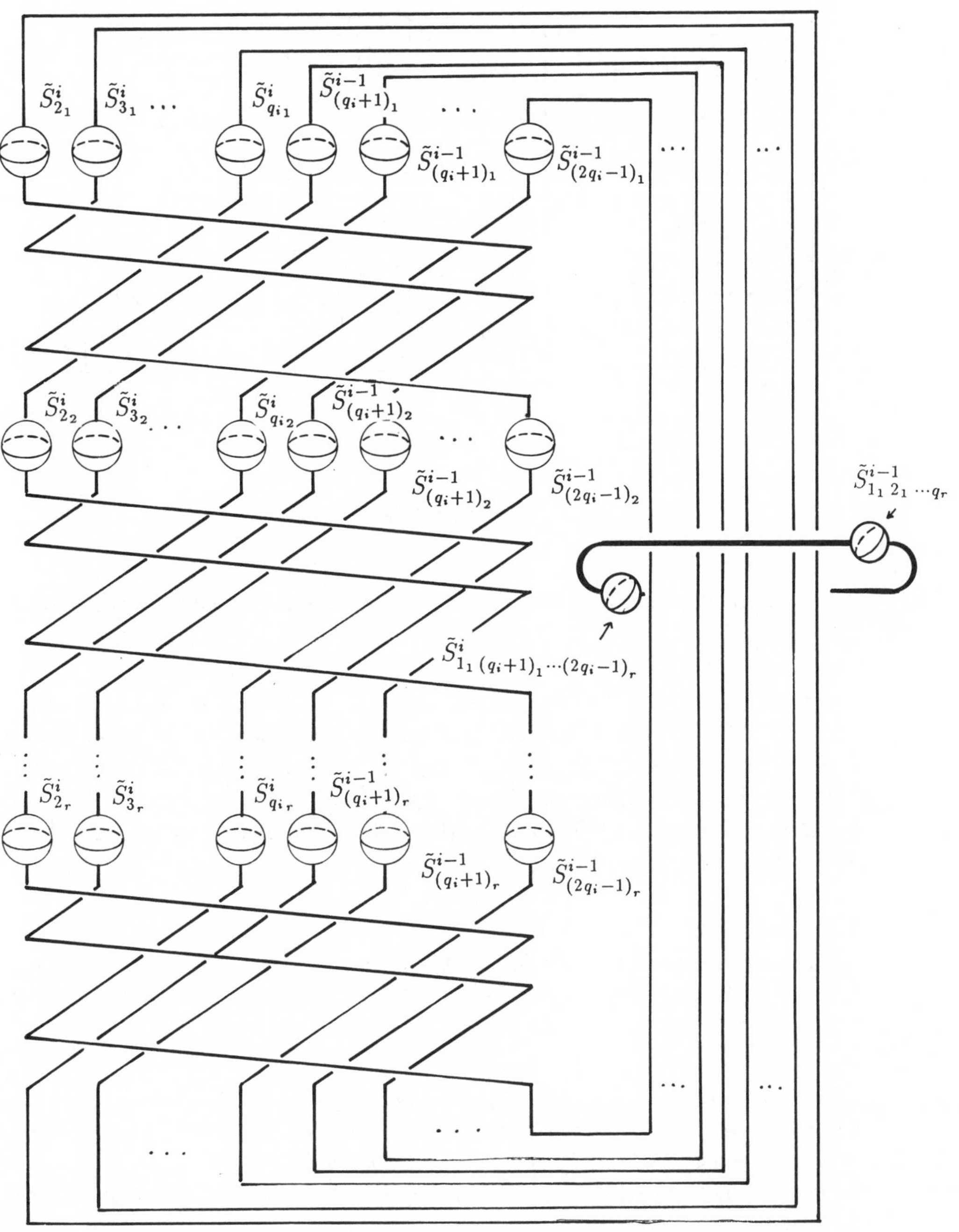

$\tilde{S}^i_{2_1}$
$\tilde{S}^i_{3_1}$ ...
$\tilde{S}^i_{q_{i_1}}$
$\tilde{S}^{i-1}_{(q_i+1)_1}$
...
$\tilde{S}^{i-1}_{(q_i+1)_1}$
$\tilde{S}^{i-1}_{(2q_i-1)_1}$
$\tilde{S}^i_{2_2}$
$\tilde{S}^i_{3_2}$ ...
$\tilde{S}^i_{q_{i_2}}$
$\tilde{S}^{i-1}_{(q_i+1)_2}$
...
$\tilde{S}^{i-1}_{(q_i+1)_2}$
$\tilde{S}^{i-1}_{(2q_i-1)_2}$
$\tilde{S}^{i-1}_{1_1\,2_1\,\cdots q_r}$
$\tilde{S}^i_{1_1\,(q_i+1)_1\cdots(2q_i-1)_r}$
$\tilde{S}^i_{2_r}$
$\tilde{S}^i_{3_r}$
$\tilde{S}^i_{q_{i_r}}$
$\tilde{S}^{i-1}_{(q_i+1)_r}$
$\tilde{S}^{i-1}_{(q_i+1)_r}$
$\tilde{S}^{i-1}_{(2q_i-1)_r}$

Figure 2.5

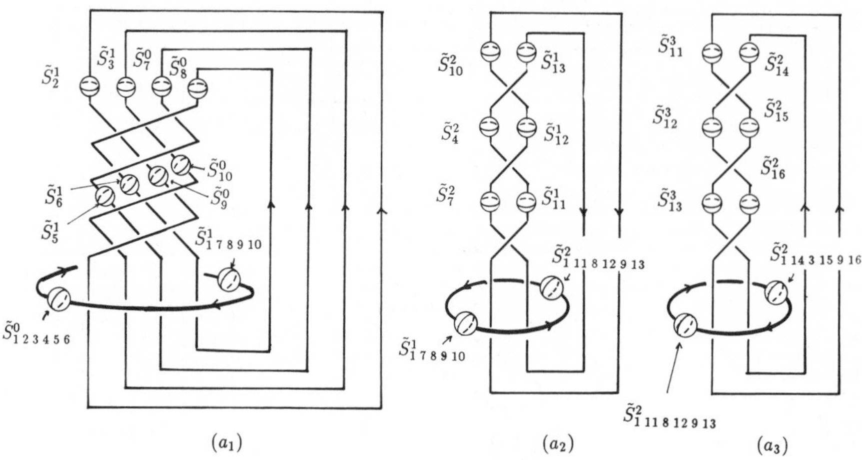

Figure 2.6

2.10. The relation between the preimage of X and the trivial knot with branch index q.

For each $T(2,a_i)$ and X, the preimage of $f : S^3 \longrightarrow S^3$ in Lemma 2.1 is shown in Figure 2.8. We notice the trivial knot with branch index q as in Figure 2.9. In Figure 2.9, the twist of the band "1" depends only on the sign of a_i. (In Figure 2.9, $a_i > 0$.) The preimage of $\tilde{X}$ under the r-fold cyclic covering $g : S^3 \longrightarrow S^3$ is shown in Figure 2.10. Then we perform this operation for each $T(2,a_i)$ and attach them at the lifts of S^i.

2.11. Example. For $T(2,5)\sharp T(2,3)\sharp T(2,3)$, Figure 2.11 ($a_1$) (($a_2$), ($a_3$), resp.) shows the preimage of X for the (C_2)-covering of $T(2,a_1)$ ($T(2,a_2)$, $T(2,a_3)$, resp.), and the knot with branch index q. Figure 2.12 shows the preimage of X and the knot with branch index q.

2.12. Generally, from Example 2.11, the preimage $\tilde{X}$ of X, which is branched, and the trivial knot with branch index q are indicated in Figure 2.13. The twists of the preimage $\tilde{X}$ depend on the sum of signs of a_i's.

For the knot and X as in Figure 2.14, we consider the (C_2)-covering similarly. The preimage is like as in Figure 2.15.

2.13. Preparation II. Let the permutation corresponding to the meridian x_1 for the (C_2)-covering of $T(2,a_1)\sharp T(2,a_2)\sharp\cdots\sharp T(2,a_m)$ be $(1_1\ 2_1\ \cdots (q_1)\ 1_2\ 2_2 \cdots$

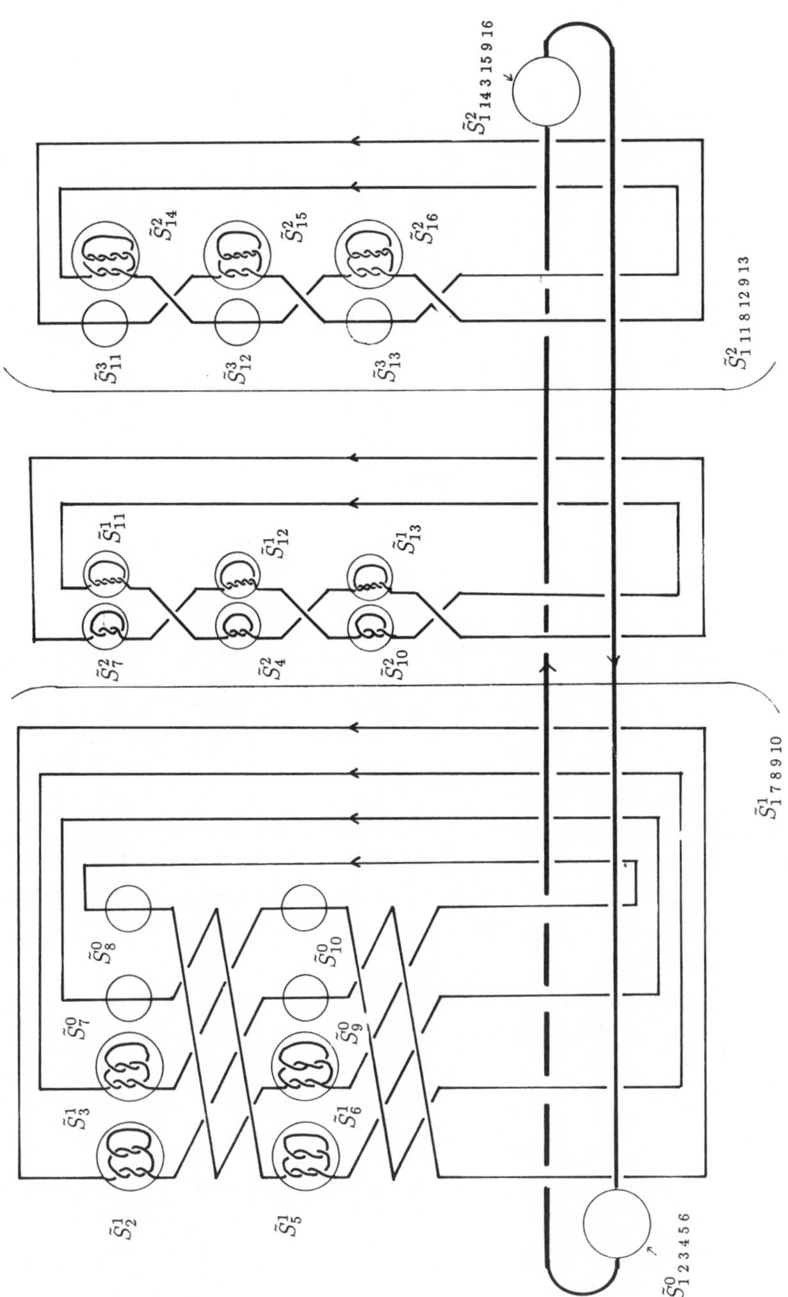
$\tilde{S}^2_{1\,14\,3\,15\,9\,16}$
$\tilde{S}^2_{14}$
$\tilde{S}^2_{15}$
$\tilde{S}^2_{16}$
$\tilde{S}^3_{11}$
$\tilde{S}^3_{12}$
$\tilde{S}^3_{13}$
$\tilde{S}^2_{1\,11\,8\,12\,9\,13}$
$\tilde{S}^1_{11}$
$\tilde{S}^1_{12}$
$\tilde{S}^1_{13}$
$\tilde{S}^2_7$
$\tilde{S}^2_4$
$\tilde{S}^2_{10}$
$\tilde{S}^1_{1\,7\,8\,9\,10}$
$\tilde{S}^0_8$
$\tilde{S}^0_{10}$
$\tilde{S}^0_7$
$\tilde{S}^0_9$
$\tilde{S}^1_3$
$\tilde{S}^1_6$
$\tilde{S}^1_2$
$\tilde{S}^1_5$
$\tilde{S}^0_{1\,2\,3\,4\,5\,6}$

Figure 2.7

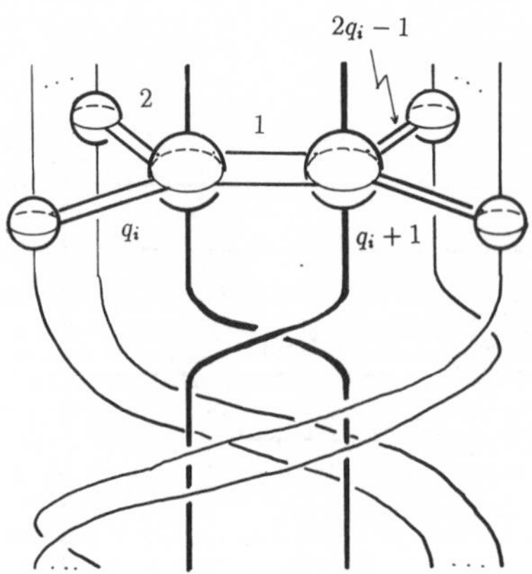

Figure 2.8

$(q_1)_2 \quad \cdots\cdots 1_r \ 2_r \quad \cdots (q_1)_r)$. Now we consider the lifts of X corresponding to 2_1 and $(q_1)_r$. We denote these lifts by $\tilde{X}_{2_1}$ and $\tilde{X}_{(q_1)_r}$.

Remark. 2_1 and $(q_1)_r$ are not the letters contained in the permutation of x_0 and x_2. Hence $g \circ f \mid \tilde{X}_{2_1} : \tilde{X}_{2_1} \longrightarrow X$ and $g \circ f \mid \tilde{X}_{(q_1)_r} : X_{(q_1)_r} \longrightarrow X$ are homeomorphisms. If the permutation corresponding to x_1 is $(1\ 2)$ then $2_1 = (q_1)_r = 2$.

From Figure 2.8, the preimage of $f : S^3 \longrightarrow S^3$ branched over $T(2, a_1)$ is shown in Figure 2.16. And the preimage of $g : S^3 \longrightarrow S^3$ is shown in Figure 2.17. We perform the same operation for $T(2, a_2)$. Then the preimages $\tilde{X}_{2_1}$ and $\tilde{X}_{(q_1)_r}$ are shown in Figure 2.18. Figure 2.19 shows the subset of the preimage under the 2-fold branched covering branched over l in Figure 2.18.

We rewrite the bold lines in Figure 2.19 as in Figure 2.20. In Figure 2.20 K_i $(1 \le i \le 4)$ is the connected sum of 2-bridge torus knots.

Remark. In Figure 2.20, $a_1, a_2 > 0$, but even in the other cases we can consider similar figures.

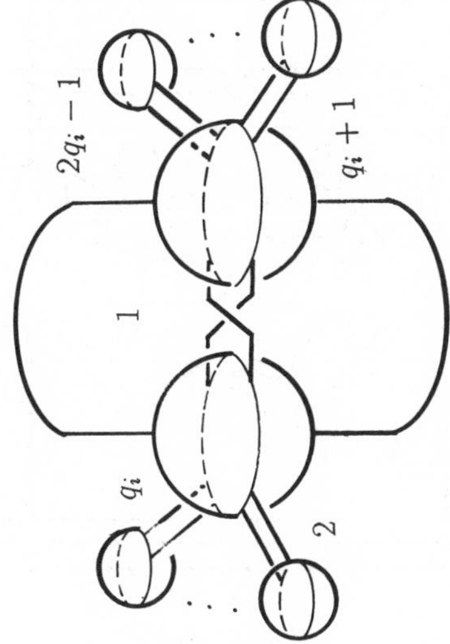

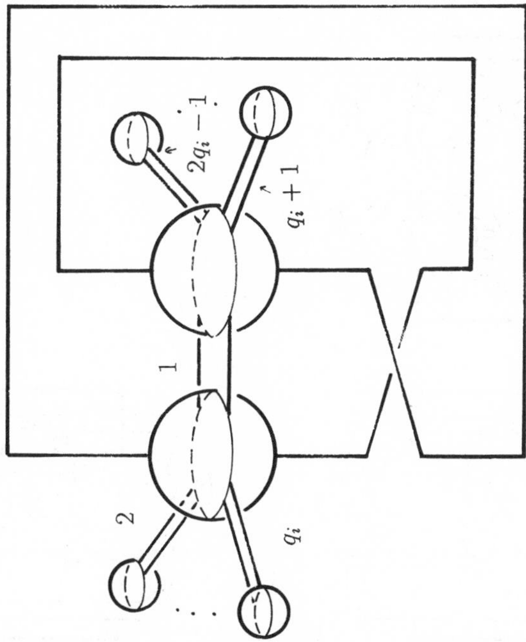

Figure 2.9

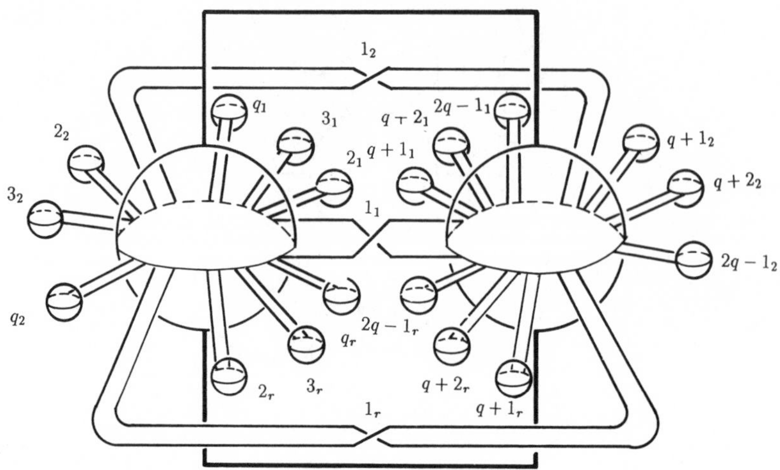

Figure 2.10

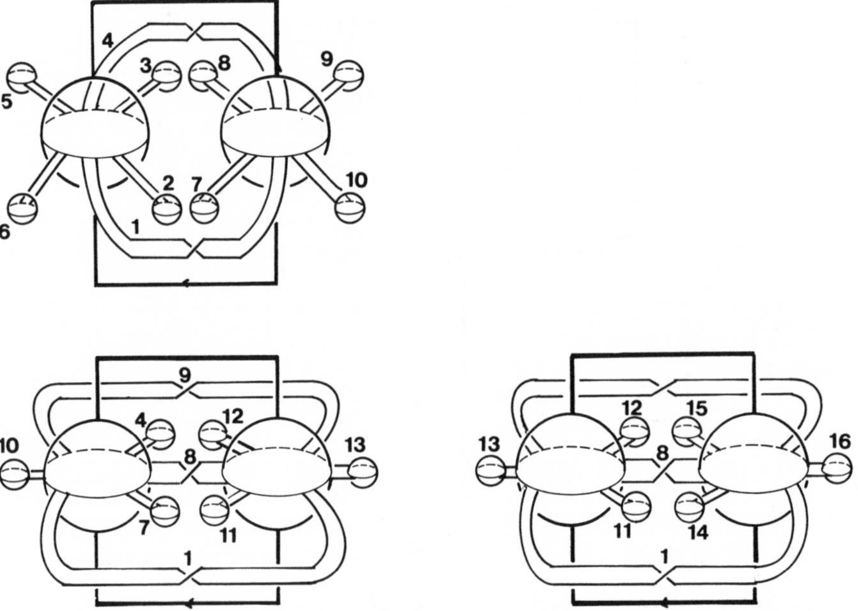

Figure 2.11

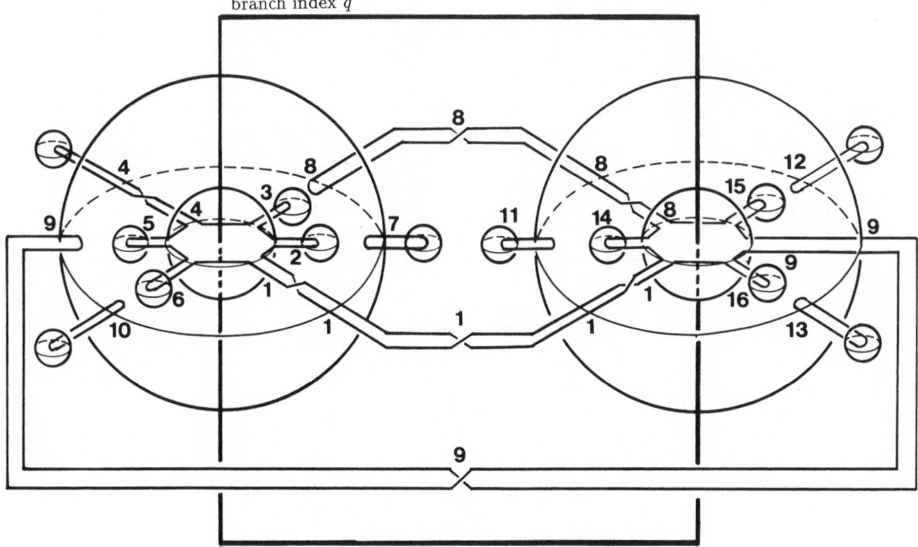

Figure 2.12

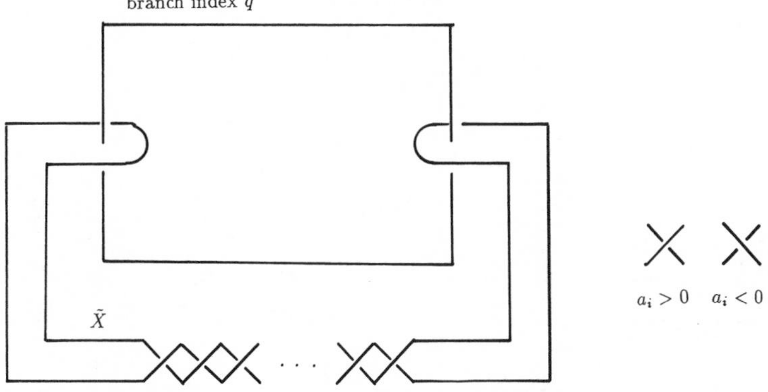

Figure 2.13

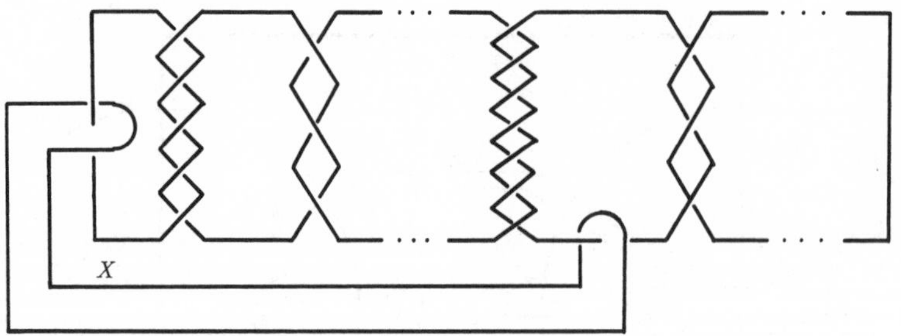

Figure 2.14

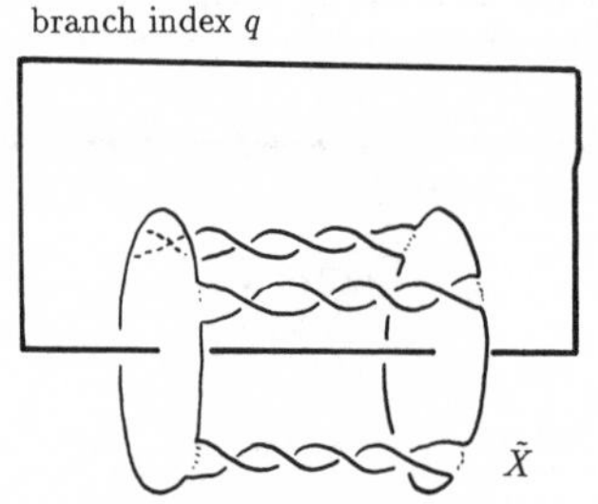

Figure 2.15

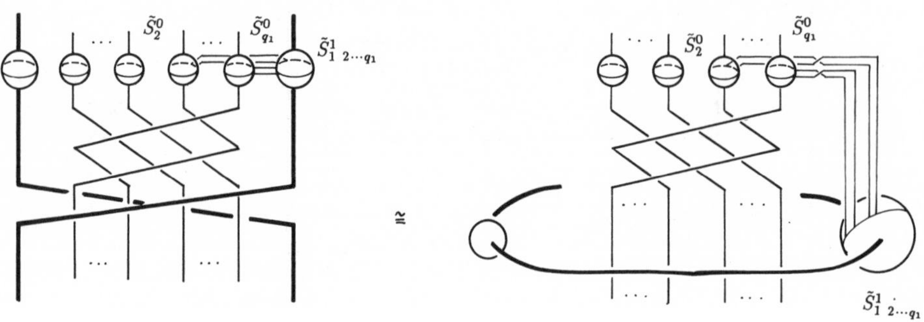

Figure 2.16

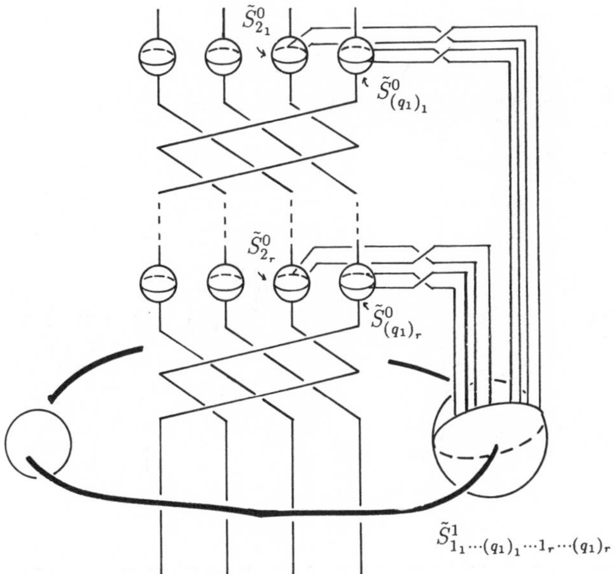

Figure 2.17

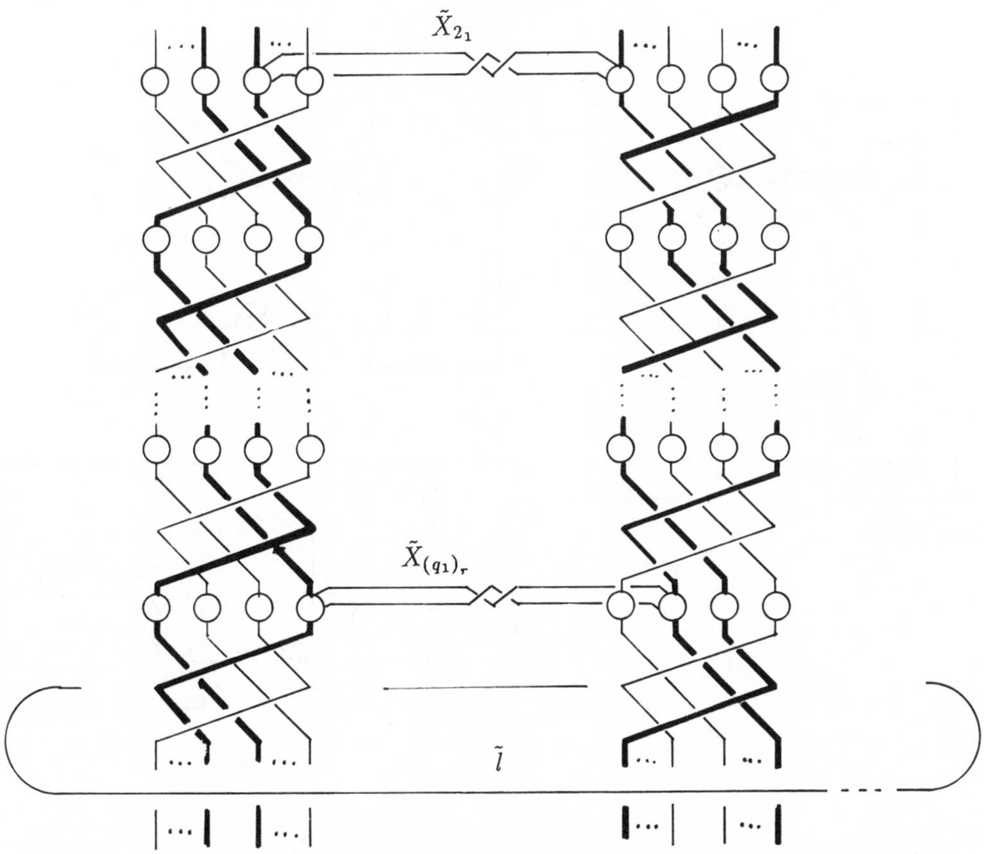

Figure 2.18

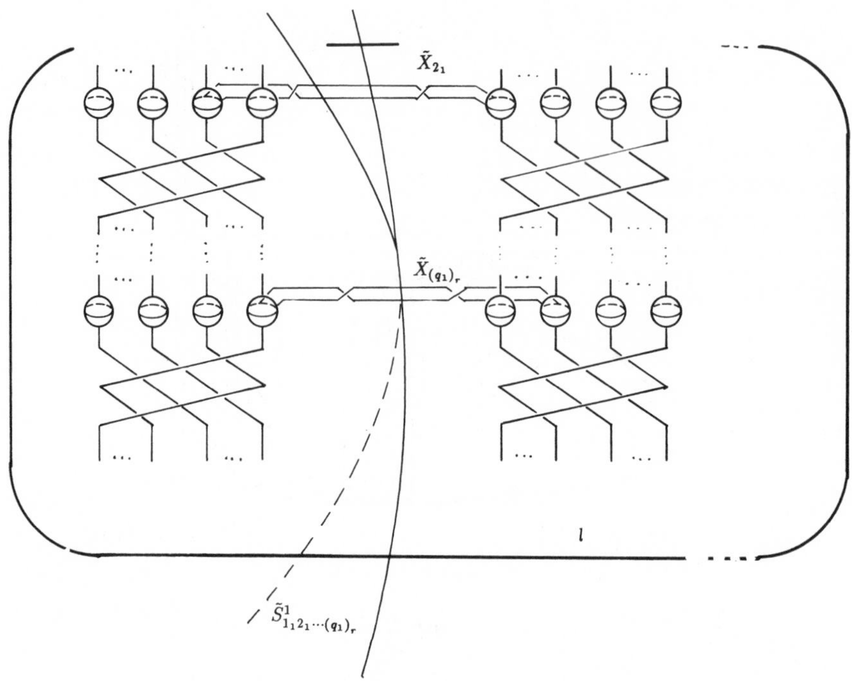

Figure 2.19

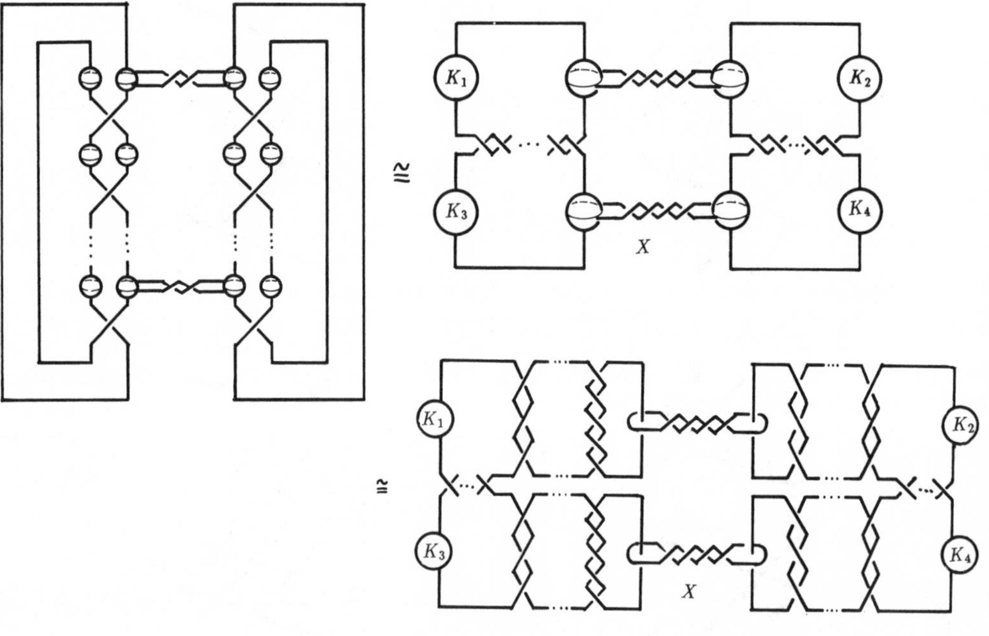

Figure 2.20

3. Proof of Theorem 1.2

In order to show a link L is universal, it is sufficient to find a branched covering space $p : S^3 \longrightarrow S^3$ branched over L such that $p^{-1}(L)$ contains a universal link.

We denote the α-component pretzel link $p(-\varepsilon b; q_1, q_2, ..., q_m)$ by $l_1 \cup l_2 \cup \cdots \cup l_\alpha$. In this case, the pretzel link is type (I) then each l_i is a trivial knot, a 2-bridge torus knot, or a connected sum of 2-bridge torus knots.

We will give the proof on the number of components, $\alpha \geq 5, \alpha = 4, 3$, or 2. The pretzel links in the list of Theorem 1.2 are not universal. Because these links are unions of fibers of graph-manifold structures on S^3. See Figure 3.1.

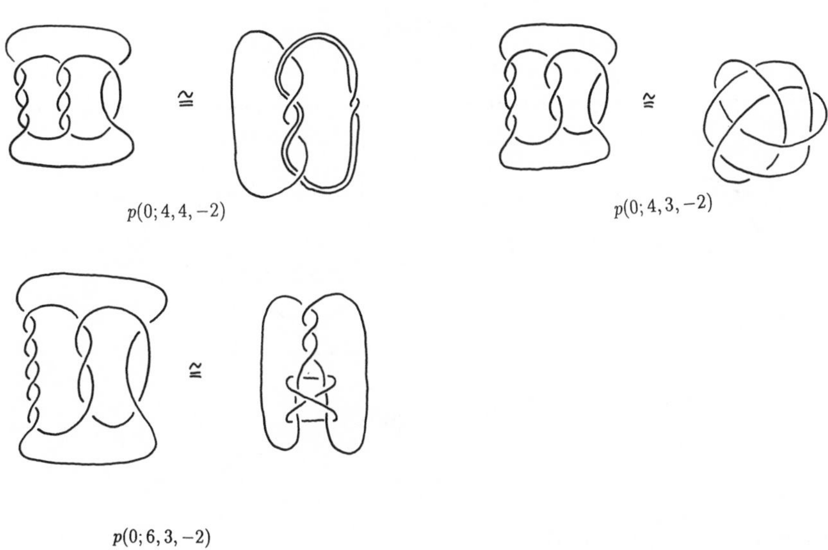

Figure 3.1

3.1. Theorem. *Let α be the number of components of $p(-\varepsilon b; q_1, q_2, ..., q_m)$ and $P = p(-\varepsilon b; q_1, q_2, ..., q_m)\sharp(\sharp_{i=1}^{n} T(2, a_i))$, where $n \in N \cup \{0\}$ and a_i is odd for $i = 1, 2, ..., n$.*

If $\alpha \geq 5$, then P is universal.

3.2. Proposition ([6]). *Let $p = p(-\beta; 2, 2, ..., 2)$ where $\beta \in Z$, $\alpha = \sharp 2 =$ the number of components of p. If $\alpha \geq 5$, then p is universal.*

Proof of Theorem 3.1. Let P be as in Figure 3.2. In Figure 3.2, R is a tangle containg a connected sum of 2-bride link facters as in Figure 3.2.

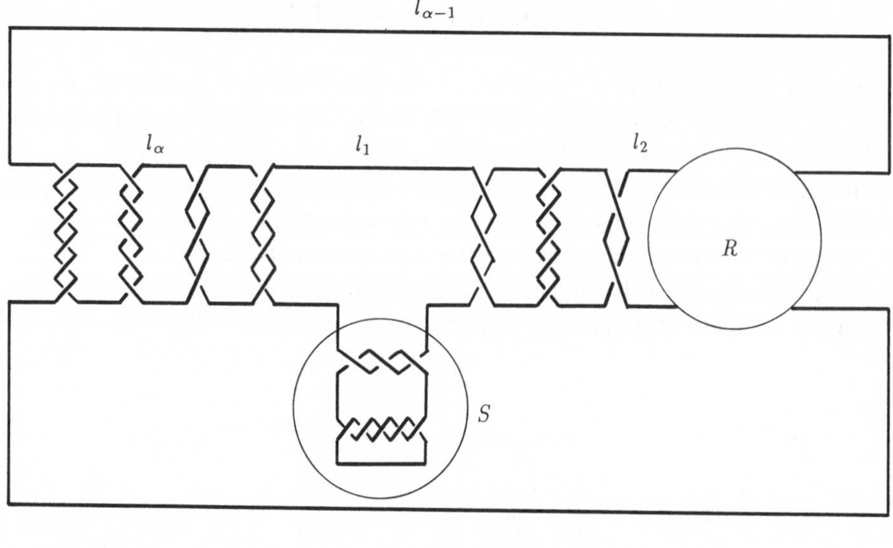

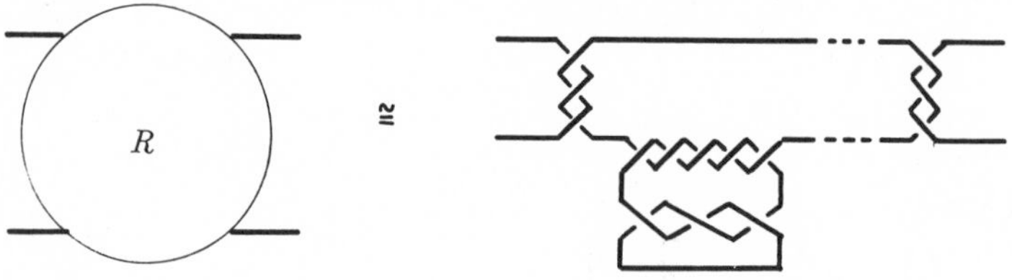

Figure 3.2

(Step I) We divide the Step (I) into two cases.

(Case 1) l_1 is a trivial knot.
(Case 2) Otherwise.

(Case 1) Let $A =| lk(l_1, l_\alpha) |, B =| lk(l_1, l_2) |$ and C the least common multiple of A and B. We consider the C-fold cyclic covering $j : S^3 \longrightarrow S^3$ branched over l_1. See Figure 3.3, where $A = 3, B = 2$, and $C = 6$.

Let $\tilde{l}_i$ $(i = 1, 2, ..., \alpha)$ be the component of $j^{-1}(l_1)$ as in Figure 3.3. Note that R has at least four components. Figure 3.4 shows $\tilde{l}_1 \cup \tilde{l}_2 \cup \cdots \cup \tilde{l}_\alpha$. Then this operation changes each component into one of the following:

$| lk(\tilde{l}_1, \tilde{l}_\alpha) |=| lk(\tilde{l}_1, \tilde{l}_2) |= 1,$
$\tilde{l}_1$ is a trivial knot,

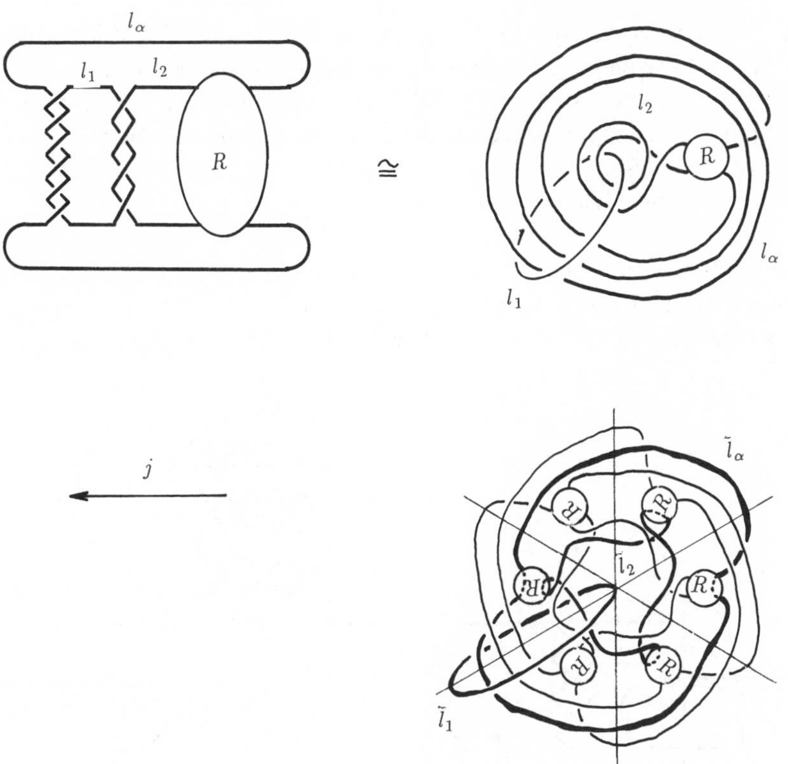

Figure 3.3

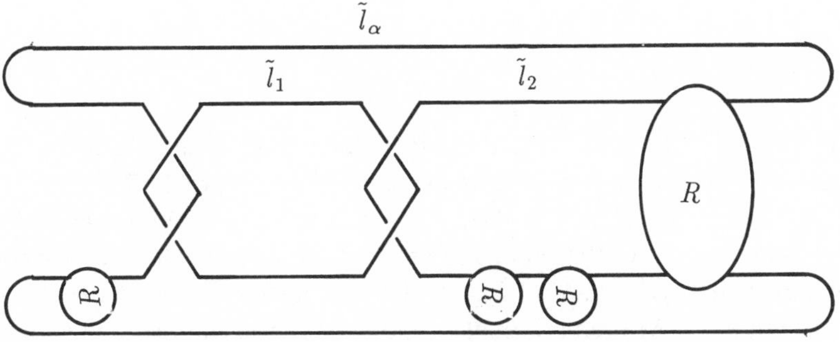

Figure 3.4

$\tilde{l}_\alpha$ is a connected sum of l_α and a connected sum of 2-bridge torus knots (possibilty empty), which is the component contained in R,

$\tilde{l}_2$ is a connected sum of l_2 and a connected sum of 2-bridge torus knots (possibilty empty), which is the component contained in R,

$\tilde{l}_i = l_i$ otherwise.

To simplify we denote $\tilde{l}_1 \cup \tilde{l}_2 \cup \cdots \cup \tilde{l}_\alpha$ by $l_1 \cup l_2 \cup \cdots \cup l_\alpha$.

(Case 2) We consider the (C_2)-covering branched over l_2. Let q be the length of the permutation. To consider the preimage of $l_2 \cup l_3 \cup \cdots \cup l_\alpha$, we can regared this link is contained in the 3-ball $X \times \mathbf{I}$ where $\mathbf{I} = [0, 1]$. See Figure 3.5.

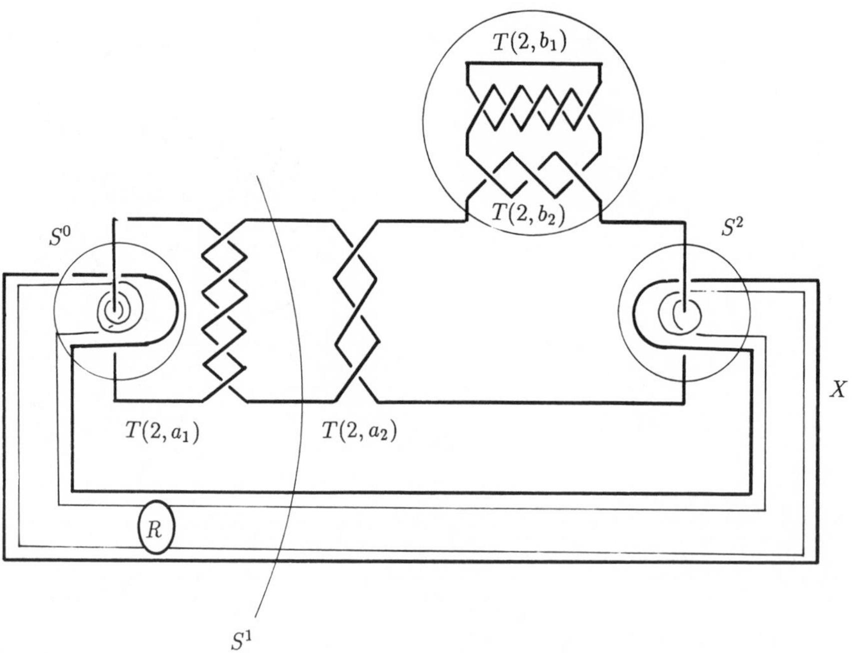

Figure 3.5

From Figures 2.14 and 2.15, the trivial knot with branch index q and the preimage $\tilde{X}$ of X can be regarded as the same as in Figure 2.15.

Let $A = | \, lk(l_1, l_\alpha) \, |$, $B = | \, lk(l_1, l_2) \, |$, C the least common multiple of A and B. Figure 3.6 shows the preimage of $\tilde{X}$ in the C-fold cyclic covering space branched over the trivial knot of Figure 3.6. We rewrite the bold lines in Figure 3.6 as in Figure 3.7. We remark that, $\tilde{l}_1$ and $\tilde{l}_\alpha$ is not inside the tangle R.

Now $\tilde{l}_i$ is the preimage of l_i which is shown in Figure 3.7. Then we have a new link $\tilde{l}_1 \cup \tilde{l}_2 \cup \cdots \cup \tilde{l}_\alpha$ such that

$$| \, lk(\tilde{l}_1, \tilde{l}_\alpha) \, | = | \, lk(\tilde{l}_1, \tilde{l}_2) \, | = 1,$$

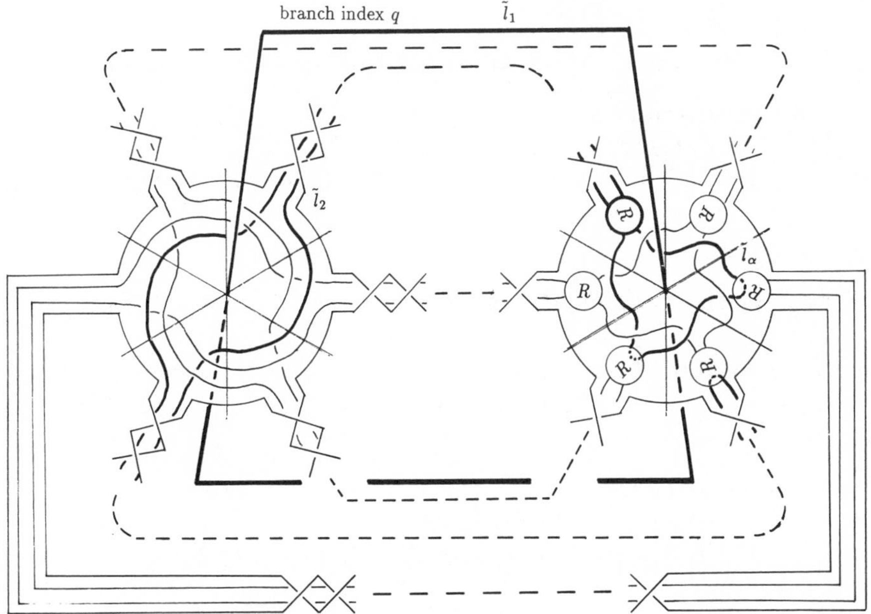

Figure 3.6

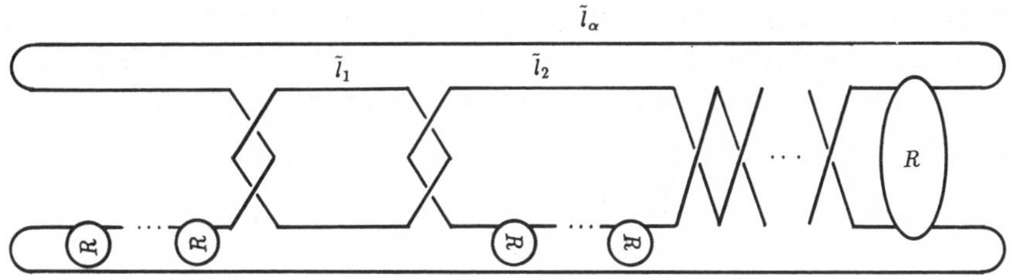

Figure 3.7

$\tilde{l}_1$ is a trivial knot,
$\tilde{l}_2$ ($\tilde{l}_\alpha$ resp.) is the connected sum of l_2 (l_α resp.) and the connected sum of 2-bridge torus knots (possibly empty) which is contained in the tangle R,
$\tilde{l}_i = l_i$ otherwise.

The twists of $\tilde{l}_3$ in Figure 3.7 are the result of the twists on X. For convenience, we use the same simbole $l_1 \cup \cdots \cup l_\alpha$ instead of $\tilde{l}_1 \cup \cdots \cup \tilde{l}_\alpha$.

(Step II) We divide the Step (II) into two cases.

(Case 1) l_2 is trivial.
(Case 2) Otherwise.

(Case 1) We perform the same operation of (Case 1) in (Step I). Since l_1 is trivial and $| lk(l_1, l_\alpha) |=| lk(l_1, l_2) |= 1$, we have a new link $\tilde{l}_1 \cup \cdots \cup \tilde{l}_\alpha$ such that

$\tilde{l}_1$ and $\tilde{l}_2$ are trivial knots,
$| lk(\tilde{l}_1, \tilde{l}_\alpha) |=| lk(\tilde{l}_1, \tilde{l}_2) |=| lk(\tilde{l}_2, \tilde{l}_3) |= 1$,
$\tilde{l}_3 = l_3 \sharp \{T(2, a_1)\sharp \cdots \sharp T(2, a_s)\}$ where $T(2, a_1)\sharp \cdots \sharp T(2, a_s)$ (possibly empty) $\subset R$,
$\tilde{l}_i = l_i$ otherwise.

For convenience, we use the same simbole $l_1 \cup \cdots \cup l_\alpha$ instead of $\tilde{l}_1 \cup \cdots \cup \tilde{l}_\alpha$.

(Case 2) We perform the operation of (Case 2) in (Step I). Since l_1 is trivial and $| lk(l_1, l_\alpha) |=| lk(l_1, l_2) |= 1$, we have a new link $\tilde{l}_1 \cup \cdots \cup \tilde{l}_\alpha$ (see Figure 3.10) such that

$\tilde{l}_1$ and $\tilde{l}_2$ are trivial knots,
$| lk(\tilde{l}_1, \tilde{l}_\alpha) |=| lk(\tilde{l}_1, \tilde{l}_2) |=| lk(\tilde{l}_2, \tilde{l}_3) |= 1$,
$\tilde{l}_3 = l_3 \sharp \{T(2, b_1)\sharp \cdots \sharp T(2, b_s)\}$, where $T(2, b_1)\sharp \cdots \sharp T(2, b_s)$ (possibly empty) $\subset R$,
$\tilde{l}_i = l_i$ otherwise.

For convenience, we use the same simbole $l_1 \cup \cdots \cup l_\alpha$ instead of $\tilde{l}_1 \cup \cdots \cup \tilde{l}_\alpha$.

We perform this operation in order on the number of the index of l_i. Finally, we perform this operation at l_α. Since l_1 and $l_{\alpha-1}$ are trivial and $| lk(l_{\alpha-1}, l_\alpha) |=| lk(l_\alpha, l_1) |= 1$, then the connected sum does not appear in l_1. We have the following link,

l_i $(i = 1, 2, ..., \alpha)$ is trivial knot,
$| lk(l_i, l_{i+1}) |= 1$ $(i = 1, 2, ..., \alpha$ and $l_{\alpha+1} = l_1)$.

This link is the pretzel link $p(-\beta; 2, 2, ..., 2)$ where $\sharp 2 = \alpha \geq 5$, $\beta \in \mathbf{Z}$. From Proposition 3.2, this link is universal. This completes the proof of Theorem 3.1.

3.3. Corollary. *The pretzel link, which has at least five components, is universal.*

3.4. Proposition. *If* $\alpha = 4$, *all pretzel links are universal, but except* $p(0; 2, 2, -2, -2)$.

Proof. Let $p = p(-b\varepsilon; q_1, q_2, ..., q_n) = l_1 \cup l_2 \cup l_3 \cup l_4$. We divide the proof into three cases.

(Case I) $p \supset T(2, a_1)\sharp T(2, a_2)\sharp \cdots \sharp T(2, a_m)$ (a_i is odd, $| a_i |\geq 3$, for $i = 1, 2, ..., m$, and $m \geq 2$). Then p is universal.

(Case II) $p \supset T(2, a)$ (a is odd and $| a |\geq 3$). Then p is universal.

(Case III) All l_i's ($i = 1, 2, 3, 4$) are trivial. Then p is universal, but except $p(0; 2, 2, -2, -2)$.

Proof of Case (I). Let $l_1 = T(2,a_1)\sharp T(2,a_2)\sharp\cdots\sharp T(2,a_m)$. We consider the (C_2)-covering branched over l_1. We regard that $l_2 \cup l_3 \cup l_4 \subset X \times \mathbf{I}$, see Figure 3.5. From Preparation 2.13 and Figure 2.20, we have (a 10-component pretzel link) $\sharp$ (2-bridge torus knots). From Theorem 3.1, this link is universal, completing the proof.

Proof of Case (II). Let $l_1 = T(2,a)$. Figure 3.8 shows the $\mid a \mid$-fold irregular dihedral branched covering space corresponding to $l_2 = T(2,a)$. See [1] and [2].

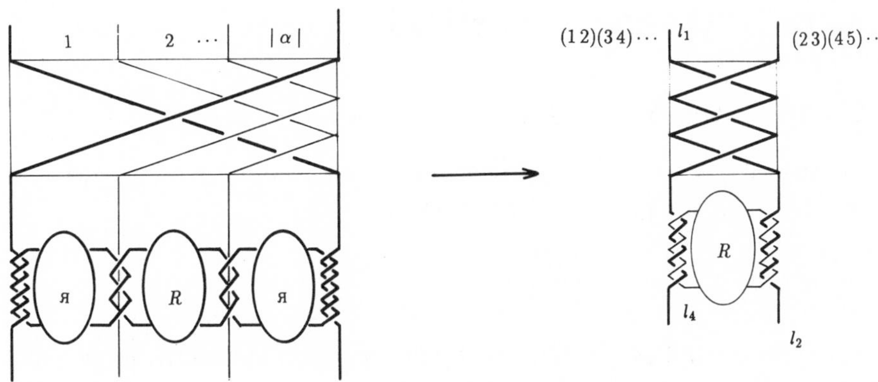

Figure 3.8

The bold line in Figure 3.8 is a pretzel link, which has at least $2 \mid a \mid +2$ components. Since $\mid a \mid\geq 3$, this link is universal, completing the proof.

Proof of Case (III). We divide the proof into two cases.

(Case III-1) $\mid lk(l_i, l_{i+1}) \mid= 1$ $(1 \leq i \leq 4$ and $l_5 = l_1)$.
Then p is universal, but except $p(0;2,2,-2,-2)$.

The proof is in [6].

(Case III-2) $\mid lk(l_i, l_{i+1}) \mid\neq 1$ for some $i(1 \leq i \leq 4)$. Then p is universal.

Proof. Since l_i and l_{i+1} are trivial, $l_i \cup l_{i+1} = T(2,a)$ (a is even and $\mid a \mid\geq 4$). We consider the $\mid a \mid$-fold irregular dihedral covering space branched over $T(2,a)$ (cf. Figure 3.8). Then we have a pretzel link with $\mid a \mid +2$ components or more. Since $\mid a \mid\geq 4$, it is universal by Theorem 3.1. This completes the proof of Proposition 3.6.

3.5. Remark. Suppose that the pretzel link has two (three, resp.) components. If it contains a connected sum of 2-bridge torus knots as a sublink, then we perform the operation of Case (I) in the proof of Proposition 3.4. We have { a six (eight resp.)-component pretzel link } $\sharp$ { 2-bridge torus knots}. Then p is universal.

3.6. Proposition. *If* $\alpha = 3$, *the pretezl link p is universal, but except* $p(0; 4, 4, -2)$, $p(0; -4, -4, 2)$, *and* $p(2, -2s)$ *(s is even).*

Proof. Let $p = l_1 \cup l_2 \cup l_3$. We divide the proof into three cases.

(Case I) $p \supset T(2, a_1)\sharp T(2, a_2)\sharp \cdots \sharp T(2, a_m)$ (a_i is odd, $\mid a_i \mid \geq 3$ and $m \geq 2$). Then p is universl.

(Case II) $p \supset T(2, a)$ (a is odd and $a \geq 3$). Then p is universal.

(Case III) All l_i, $(i = 1, 2, 3)$ are trivial. Then p is universal, but except $p(0; 4, 4, -2)$, $p(0; -4, -4, 2)$, and $p(-2, -2, s)$ (s is even).

Proof of Case (I). See Remark 3.5.

Proof of Case (II). Let $l_1 = T(2, a)$. By the operation in Case(II) in the proof of Propositon 3.4, we have a pretzel link with $\mid a \mid +2$ components or more. Since $\mid a \mid \geq 3$, this link is universal, completing the proof.

Proof of Case (III). We divide the proof into two cases.

(Case III-1) $\mid lk(l_i, l_{i+1}) \mid = 1$ for $1 \leq i \leq 3$. Then p is universal, but except $p(0; 2, -2, -2)$ and $p(0; -2, 2, 2)$.

Proof. See [6].

(Case III-2) $\mid lk(l_i, l_{i+1}) \mid \neq 1$ for some i $(i = 1, 2, 3)$. Then p is universal, but except $p(2, -2, s)$ (s : even), $p(0; 4, 4, -2)$ and $p(0; -4, -4, 2)$.

Proof. $l_i \cup l_{i+1}$ is equivalent to $T(2, a)$ where a is even and $\mid a \mid \geq 4$. Let $l_1 \cup l_2 = l_i \cup l_{i+1}$ and $a_i = \mid lk(l_i, l_3) \mid$.

(Case III-2-1) $a_i \equiv 1 \pmod 2$ for $i = 1, 2$. We will show that p is universal, but except $p(2, -2, s)$.

(i) In the case of $a_1 = a_2 = 1$.

We can represent p by $p(-\beta; 2, -2, a)$ ($\beta \in \mathbf{Z}$). Then we have $p(-(\mid a \mid \beta - 1) \pm 1; 2, 2)$ as a sublink of the preimage by the$\mid a \mid$-fold irregular dihedral covering space branched over $T(2, a)$ (cf. Figure 3.8). Now $p(-\beta'; 2, 2)$ is universal, but except $\beta' = -2, 1$, and 0. If $\mid a \mid \beta - 1 \pm 1 = -2, 1$ or 0, then $\beta = 0$, from $\mid a \mid \geq 4$. Hence, if $\beta \neq 0$, then this link is universal.

(ii) In the case of $a_1 \neq 1$ or $a_2 \neq 1$.

We can assume that $a_1 \neq 1$. Let x_i ($i = 1, 2$) be the meridian of l_i. Then we consider the $\mid a \mid$-fold irregular dihedral covering space branched over $T(2, a)$ with the monodromy defined by

$x_1 \longmapsto$ (1 2)(3 4)...
$x_2 \longmapsto$ (2 3)(4 5)...

There is a 2-component pretzel link in the preimage (see Figure3.8.). Since a_1 is odd, $a_1 \neq 1$ and $\mid a \mid \geq 4$, this pretzel link contains a connected sum of 2-bridge torus knots $T(2, a_1) \sharp \cdots \sharp T(2, a_1)$. From Remark 3.5, p is universal.

(Case III-2-2) $a_1 \equiv 0$ or $a_2 \equiv 0 \pmod 2$. Then p is universal, but except $p(0; 4, 4, -2)$ and $p(0; -4, -4, 2)$.

Proof. We can assume that $a_1 \equiv 0 \pmod 2$. We consider the $\mid a \mid$-fold irregular dihedral covering space branched over $T(2, a)$. (See Figure 3.8.) Then we have a new pretzel link. If $\mid a \mid \geq 6$ ($a_2 \equiv 0 \pmod 2$ resp.), then the new pretzel link has at least $\frac{|a|}{2} + 2$ ($\mid a \mid +1$, resp.) components. Thus we have a pretzel link with five components or more. Then it is universal.

We consider the case of $\mid a \mid = 4$ and $a_2 \equiv 1 \pmod 2$. The new pretzel link has four-components. If $a_1 \geq 4$ or $a_2 \neq 1$, this pretzel knot is not $p(0; 2, 2, -2, -2)$, then it is universal. (See Proposition 3.4.)

We consider the case of $a_1 = 2$ and $a_2 = 1$. This pretzel link is equivalent to $p(-\beta; 4, 4\varepsilon, 2)$ or $p(-\beta; -4, 4\varepsilon, 2)$ ($\varepsilon = \pm 1$). If $p = p(-\beta; 4, 4\varepsilon, 2)$, then we consider the $\mid a \mid$-fold irregular dihedral covering space branched over $T(2, a)$. See Case III-2-1-(ii) in the proof of Proposition 3.6. Then we have a new pretzel link $p(-(4\beta + 1 + \varepsilon); 2, 2, 2, 2)$. This link is universal, if $4\beta + 1 + \varepsilon \neq -2$. So p is universal, but except $\varepsilon = 1$ and $\beta = -1$. Hence p is universal except $p(0; 4, 4, -2)$. If $p = p(-\beta; -4, 4\varepsilon, 2)$, we perform the same operation. Then p is universal, but except $p(0; -4, -4, 2)$. This completes the proof of Proposition 3.6.

3.7. Proposition. *If* $\alpha = 2$, *the pretzel link* p *is universl, but except* $p(0; 2s, 2t)$ ($s, t \in \mathbf{Z} \backslash \{0\}$), $p(0; 3, 6, -2)$, $p(0; -3, -6, 2)$, $p(0; -2, -3, 4)$, $p(0; 2, 3, -4)$, $p(2, -2, s)$ (s *: odd*).

Proof. Let $p = l_1 \cup l_2$. We divide the proof into three cases.

(Case I) $p \supset T(2, a_1)\sharp \cdots \sharp T(2, a_m)$ (a_i is odd, $| a_i | \geq 3$ and $m \geq 2$). Then p is universal.
(Case II) $p \supset T(2, a)$ (a is odd and $| a | \geq 3$). Then p is universal, but except $p(2, -2, s)$ (s : odd), $p(0; 3, 6, -2)$, $p(0; -3, -6, 2)$, $p(0; 3, 4, -2)$, $p(0; -3, -4, 2)$.
(Case III) Both l_1 and l_2 are trivial. Then p is universal, but except $p(2s, 2t)$ ($s, t \in \mathbf{Z}\backslash\{0\}$).

Proof of Case (I). See Remark 3.5.

Proof of Case (II). Let $l_1 = T(2, a)$. From Case (I), we can assume that l_2 is a torus knot or a trivial knot.

Step 1. l_2 is a torus knot. We can denote p by $p(-\beta; 2p_1, a, 2p_3, p_4)$ ($p_1, p_3 \in \mathbf{Z}\backslash\{0\}$, a, p_4 are odd, and $| a |$, $| p_4 | \geq 3$).

We consider the $| a |$-fold dihedral covering space branched over $l = T(2, a)$. See Figure 3.8. If both p_1 and p_3 are even, then we have a ($| a | + 1$)-component pretzel link. If $| a | = 3$, then we have a 4-component pretzel link and it contains $T(2, p_4)$ ($| p_4 | \geq 3$), and so it is not $p(0; 2, 2, -2, -2)$. From Proposition 3.4, p is universal. If $| a | \geq 4$, we have a pretzel link which has at least five components. So it is universal.

If p_1 or p_3 is odd, then we have a pretzel link with two components or more (cf. Figure 3.8) and this pretzel link contains $T(2, p_4)\sharp \cdots \sharp T(2, P_4)$. From Theorem 3.1 and Remark 3.5, it is universal.

Step 2. l_2 is a trivial knot. We can denote p by $p(-\beta; 2p_1, a, 2p_3) = l_1 \cup l_2$ (a : odd, p_1, $p_3 \in \mathbf{Z}\backslash\{0\}$). We divide the proof into two cases.

(Case II-1) p_1, $p_3 \equiv 1 \pmod 2$.

(Case II-1-i) $| p_1 | = | p_3 | = 1$. Then p is universal, but except $p(2, -2, s)$ (s : odd).

(Case II-1-ii) $| p_1 | \neq 1$ or $| p_3 | \neq 1$. Then p is universal, but except $p(0; 3, 6, -2)$ and $p(0; -3, -6, 2)$.

(Case II-2) $p_1 \equiv 0$ or $p_3 \equiv 0 \pmod 2$. Then p is universal, but except $p(0; 3, 4, -2)$ and $p(0; -3, -4, 2)$.

Proof of (Case II-1-i). We can denote p by $p(-\beta; 2, a, -2)$. We consider the $| a |$-fold dihedral covering space branched over $l_1 = T(2, a)$ (cf. Figure 3.8). Then we have a 2-component pretzel link $p(- | a | \beta - 1 + \frac{a}{|a|}; 2, 2)$. If $p(-\beta'; 2, -2)$ is universal, then $\beta' \neq -2, -1$, and 0 (cf. [6]). Since $| a | \geq 3$, if $\beta \neq 0$, then

$p(-\beta; 2, a, -2)$ is universal. We conclude that $p(-\beta; 2, a, -2)$ is universal, but except $p(2, s, -2)$ $(= p(2, -2, s))$ $(s$: odd), completing the proof.

Proof of (Case II-1-ii). We can assume that $| p_1 | \neq 1$. We divide the proof into four cases.

(1) $| a | \geq 5$ — Then p is unversal.
(2) $| a | = 3$, $| p_3 | \neq 1$ — Then p is universal.
(3) $| a | = 3$, $| p_3 | = 1$, $| p_1 | \geq 5$ — Then p is universal.
(4) $| a | = 3$, $| p_3 | = 1$, $| p_1 | = 3$ — Then p is univesal, but except $p(0; 3, 6, -2)$ and $p(0; -3, -2, 6)$.

We consider the $| a |$-fold dihedral covering space branched over $l_1 = T(2, a)$. Then we have a 2-component pretzel link.

(1) Since $| p_1 | \neq 1$, the pretzel link contains $T(2, p_1)\sharp \cdots \sharp T(2, p_1)$. From Case (I) in the proof of Proposition 3.7, this pretzel link is universal.

(2) Since $| a | = 3$ and $| p_3 | \neq 1$, this pretzel link contains $T(2, p_1)\sharp T(2, p_3)\sharp \cdots$. From Case (I), this link is universal.

(3) We consider the 3 $(= | a |)$-fold dihedral covering branched over $l = T(2, a)$. Then we have $p(-(3\beta + \frac{a}{|a|} + p_3); 2p_1, p_1, 2p_3)$. (cf. Figure 3.8.) Since $| p_1 | \geq 5$, this pretzel link is universal from (1).

(4) It is sufficient to consider $p(-\beta; 6, 3, 2)$ and $p(-\beta; 6, -3, 2)$.
For $p(-\beta; 6, 3, 2)$, the 3-fold covering associated with Figure 3.9 is the lens space $L(2, 1)$ by [3]. And the 2-fold unbranched covering of $L(2, 1)$ is S^3.

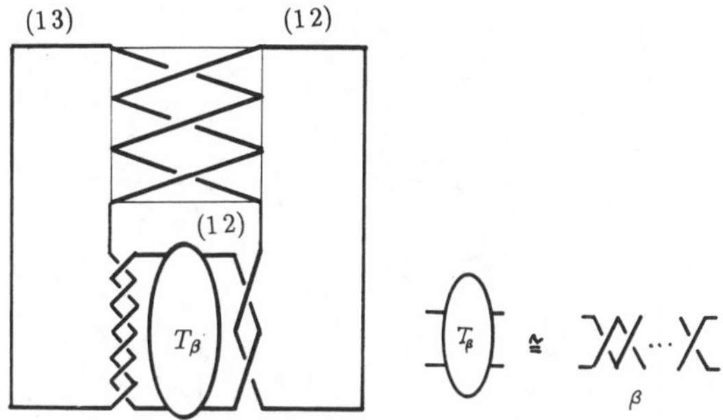

Figure 3.9

First, we consider $p(-\beta; 6, 3, 2)$. We will show that, in this covering $S^3 \longrightarrow S^3$ the preimage of this pretzel link contains $p(-(6\beta + 4); 2, 2, 2, 2)$. First we consider the link L with the permutation shown in Figure 3.10. If we perform 3 (6 resp.)-half twists at the 3-ball **3** (**6** resp.) in Figure 3.10, then we have $p(-\beta; 6, 3, 2)$. There is a sphere S^2 dividing S^3 into two balls **A** and **B** such that **A** and **B** are shown in Figure 3.10 (b). The ball **B** does not contain the permutation

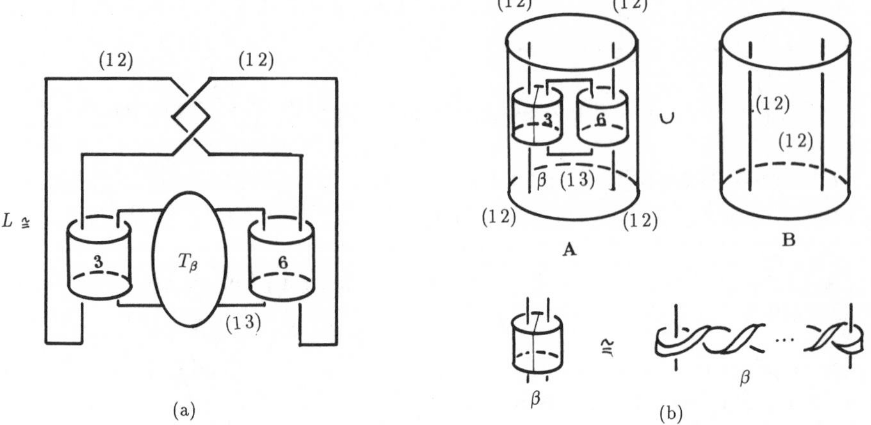

Figure 3.10

(13). Thus the covering space of S^3 branched over L is divided by the preimage of S^2 into two part $\tilde{\mathbf{A}}$ and $\tilde{\mathbf{B}}$, the lifts of **A** and **B** respectively. We note that $\tilde{\mathbf{A}}$ and $\tilde{\mathbf{B}}$ are tori. Figure 3.11 indecates the lift $\tilde{\mathbf{A}}$. In Figure 3.11, the bold line has branched index two, and the others have one.

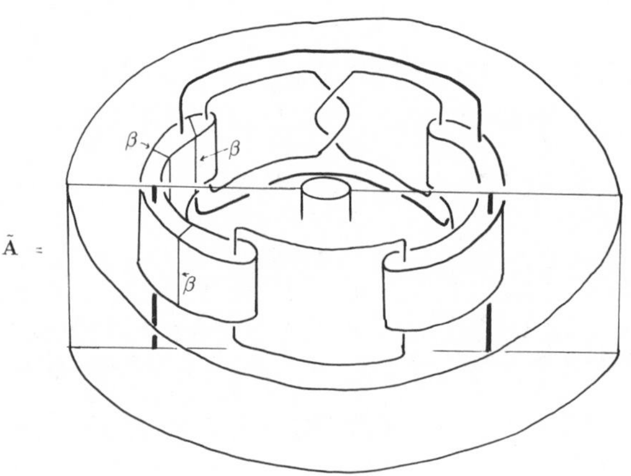

Figure 3.11

The link with branch index 1 is contained in $\tilde{\mathbf{A}}$. We denote this link by l. So we can show this link in $L(2,1)$ as Figure 3.12(a). In Figure 3.12, the bold circle denotes the surgery link. By the twist at the ball **3** and **6**, we have the

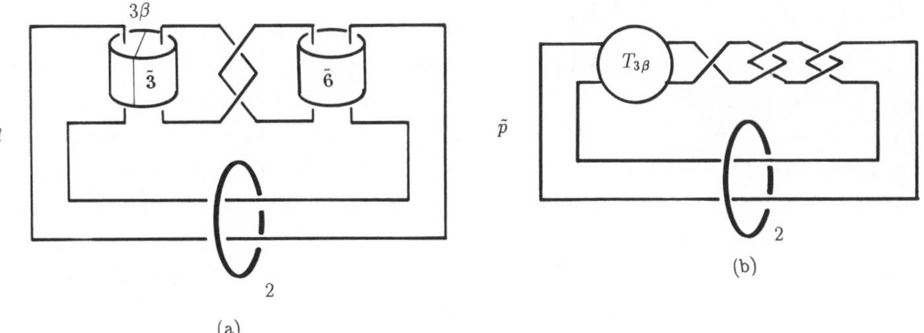

Figure 3.12

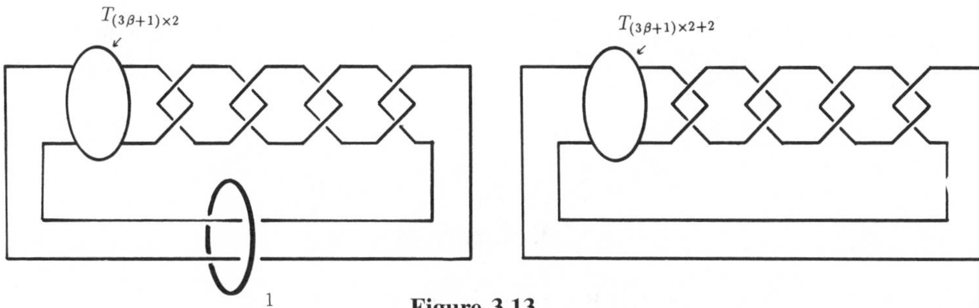

Figure 3.13

link as in Figure 3.12 (b).(See [5; p. 317.].) We denote this link by $\tilde{p}$, which is a sublink of the preimage of $p(-(6\beta+4);2,2,2,2)$. Figure 3.13 shows the preimage of $\tilde{p}$ under the 2-fold unbranched covering $S^3 \longrightarrow L(2,1)$. Now we have $p(-(6\beta+4);2,2,2,2)$. Unless $6\beta+4=-2$, i.e. $\beta=-1$, this pretzel link is universal. (See Proposition 3.4.) So $p(-\beta;6,3,2)$ is universal, but except $p(0;6,3,-2)$ $(=p(-1;6,3,2))$.

For $p(-\beta;6,-3,2)$, we perform the same operation. And we have the pretzel link $p(-(-6\beta+5);2,2,2,2)$. From Proposition 3.4, since $-6\beta+5\neq-2$, this pretzel link is universal.

Then p is universal, but expect $p(0;6,3,-2)$ and $p(0;-6,-3,2)$, completing the proof.

Proof of (Case II-2). We divide the proof into two cases.

(i) $p_i \equiv 0 \pmod 2$ for $i=1,3$.

Since $|a| \geq 3$ and $p_1 \equiv 0 \pmod 2$, we can have a 4-component pretzel link, which is not $p(0;2,2,-2,-2)$, or a pretzel link with five or more components link, by the $|a|$-fold dihedral covering space branched over $l_1=T(2,a)$. So this pretzel link is universal.

(ii) $p_1 \equiv 0,\ p_2 \equiv 1 \pmod 2$

If $\mid a \mid \geq 5$, then we perform the same operation of (i). Thus it is universal. In the case of $\mid a \mid = 3$. If $a = 3$, we denote this pretzel link by $p = p(-\beta; 4m, 3, 2n)$ $(m, n \in \mathbf{Z}\backslash\{0\})$. And we consider the 3-fold dihedral covering space branched over $l_1 = T(2,3)$. (See Figure 3.12.) Then we have a new pretzel link $p(-(3\beta+1); 2n, 4m, n, 2m)$ — (1). This link has three components. The 3-component pretzel link, which is not universal, is $p(2, -2, s)$ (s:even) and $p(4, 4, -2)$. From (1), if $(m, n, \beta) \neq (1, 1, -1)$, then p is universal. Hence p is universal, but except $p(0; 4, 3, -2)$ ($= p(0; 3, 4, -2)$).

If $a = -3$, then we perform same operation. Thus p is universal, but except $p(0; -4, -3, 2)$ $(= p(0; -3, -4, 2))$, completing the proof.

Proof of Case (III). This pretzel link is a 2-bridge link. From [1], it is universal, but except $p(0; 2s, 2t)$ $(s, t \in \mathbf{Z}\backslash\{0\})$. This completes the proof of Proposition 3.7.

References

[1] H. Hilden, M. Lozano and J. Montesinos: *On knots that are univeral*, Topology, **24** (1985), 499-504.

[2] H. Hilden, M. Lozano and J. Montesinos: *Non-simple universal knots*, Math. Proc. Camb. Phlos. Soc., **102** (1987), 87-95.

[3] F. Hosokawa and Y. Nakanishi: *On 3-fold irregular branched covering spaces of pretzel knots*, Osaka J. Math., **23**(1986), 249-254.

[4] S. Kinoshita: *On the branched points in the branched coverings of links*, Canada. Math. Bull., **28**(2) (1985), 165-173.

[5] D. Rolfsen: Knots and links, Math. Lecture Series ♯7, Publish or Perish Inc., Berkely, 1976.

[6] Y. Uchida: *Universal chains*, to appear in Kobe J. Math.

[7] W. Thurston: *Universal links*. (preprint)

Knots, spin networks and 3-manifold invariants

Louis H. Kauffman

I. Introduction

The purpose of this paper is to connect the basic ideas and techniques of the Penrose theory of spin networks [**31**] to a knot theoretic context. By doing this via the bracket polynomial [**7**] model for the Jones polynomial, we obtain a theory of q-spin nets (where $q = \sqrt{A}$, A the variable in the bracket) that is appropriate for the representation theory of the quantum group $SL(2)_q$. In particular, we obtain generalizations of the spin network formulas for anti-symmetrizers, $3j$ and $6j$ symbols. These generalizations make possible a knot theoretic basis for three manifold invariants of Reshetikhin and Turaev, and also for the invariants of Turaev and Viro (via $q - 6j$ symbols).

Details will appear elsewhere. This paper indicates the generalizations and shows how the "magic weaving elements" (I am indebted to S. Yamada for this fine terminology.) in the Temperley Lieb algebra are nothing but the q-deformed anti-symmetrizers in the q-spin network theory. This means concretely that there is a specific knot theoretic formula for the magic weave. It really is a weave! The formula is

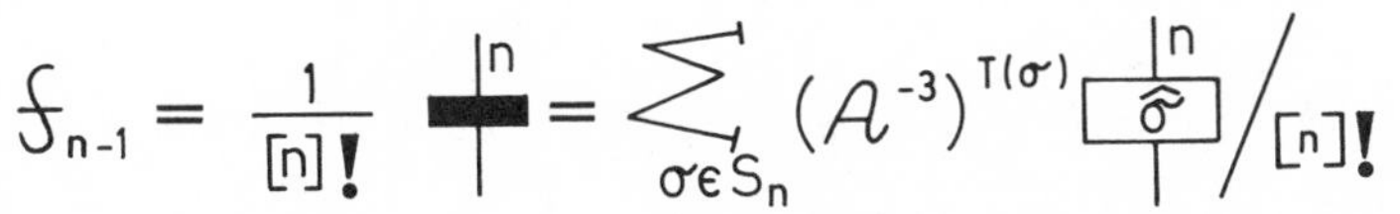

$$f_{n-1} = \frac{1}{[n]!} \;\; \text{(diagram: strand labeled } n \text{ through a black box)} \;\; = \sum_{\sigma \in S_n} (A^{-3})^{T(\sigma)} \;\; \text{(diagram: strand labeled } n \text{ through box } \hat{\sigma}) \Big/ [n]!$$

where $T(\sigma)$ is the minimal number of transpositions needed to write the element σ of the symmetric group S_n, and $\hat{\sigma}$ is a braid projecting to σ with all negative crossings. Here

$$[n]! = \sum_{\sigma \in S_n} (A^{-4})^{T(\sigma)}.$$

Thus

Knots 90

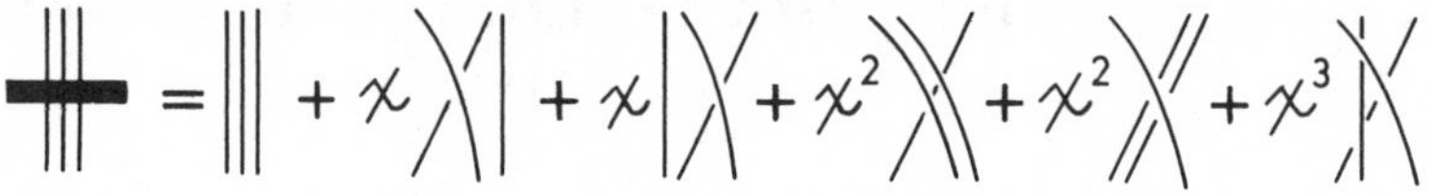

where $x = A^{-3}$. To obtain the corresponding element of the Temperley Lieb algebra, expand these braids via the bracket identity

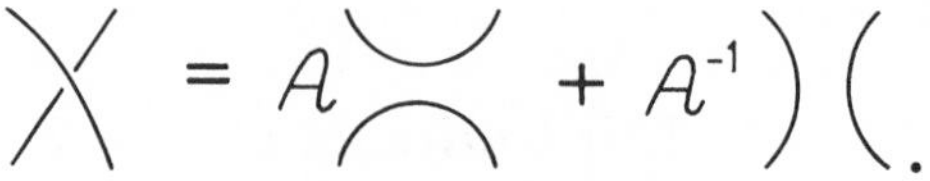

The paper is organized as follows. Section 2 discusses the bracket polynomial and the quantum group $SL(2)_q$. Section 3 discusses classical spin networks. Section 4 discusses the q-deformed spin networks and the magic weave.

It gives the author great pleasure to thank the organizers of Knots 90 in Osaka for a very stimulating conference. This rearch was supported in part by NSF Grant Number DMS-8822602.

II. The bracket polynomial and $SL(2)_q$

First recall the construction of the bracket polynomial **[7]**, giving an elementary picture of the Jones polynomial as a state summation. The bracket is defined by the equations:

1.  $< \times > = A < \asymp > + A^{-1} <)(>$
2. $< \bigcirc \quad K > = \delta < K > \quad \text{with} \quad \delta = -A^2 - A^{-2}.$
3. $< \bigcirc > = \delta.$

The diagrams stand for larger diagrams that, in a given equation, are identical except as indicated by the small diagrams. For any unoriented knot or link K, $< K >$ belongs to the ring of Laurent polynomials $Z[A, A^{-1}]$ and is a regular isotopy invariant of K. Regular isotopy **[6]** is the equivalence relation generated by Reidemeister moves II and III. Note that it follows from these axioms that the bracket behaves multiplicatively on curls (Reidemeister I):

$$\langle \text{curl} \rangle = -A^{3} \langle \sim \rangle , \langle \text{curl} \rangle = -A^{-3} \langle \sim \rangle .$$

The bracket provides a model for the original Jones polynomial, $V_K(t)$ via the formula

$$V_K(t) = (-A^3)^{-w(K)} < K > (A) / < \bigcirc > (A)$$

where $A = t^{-1/4}$ [7]. Here $w(K)$ is the writhe of the oriented link K; that is, $w(K)$ is the sum of the signs of the crossing of K where the sign conventions are as indicated below:

A Yang-Baxter model for the bracket is obtained as follows [**16**]: Let M denote the matrix

$$M = \sqrt{-1}\tilde{\epsilon}$$

where

$$\tilde{\epsilon} = \begin{bmatrix} 0 & A \\ -A^{-1} & 0 \end{bmatrix}.$$

Associate to each diagram a height function so that the diagram is decomposed into crossings, minima and maxima. Delineate each of these critical points by placing **nodes** on the diagram as shown below:

For example,

A **state** of the diagram is a choice of assignments of the indices $\{1,2\}$ to the nodes of the diagram. With a given state, associate matrix elements to each maxima, minima or crossing as indicated below

$$\text{(crossing)} \Longleftrightarrow R^{ab}_{cd}\,, \quad \text{(crossing)} \Longleftrightarrow \bar{R}^{ab}_{cd}$$

$$\text{(cap)} \Longleftrightarrow M_{ab}\,, \quad \text{(cup)} \Longleftrightarrow M^{ab}.$$

Here the lower and upper index entries of M denote the same elements of M - its ab entry, and the matrices R and $\bar{R}$ are defined by the formulas

$$R^{ab}_{cd} = A\,\text{(cup over cap)} + A^{-1}\,\text{(two vertical arcs)} = AM^{ab}M_{cd} + A^{-1}\delta^a_c\delta^b_d$$

$$\bar{R}^{ab}_{cd} = A^{-1}M^{ab}M_{cd} + A\delta^a_c\delta^b_d.$$

The Kronecker delta corresponds to an arc that is free of critical points with respect to this height function. To each state $S : Nodes(K) \longrightarrow \{1,2\}$, let $< K|S >$ denote the product of the matrix entries that are produced by S at the critical points of the diagram. Then $< K >= \sum < K|S >$ where the summation is taken over all possible states. For example,

$$\langle \bigcirc \rangle = {}_1\bigcirc_1 + {}_1\bigcirc_2 + {}_2\bigcirc_1 + {}_2\bigcirc_2$$
$$= M^2_{11} + M^2_{12} + M^2_{21} + M^2_{22} = -A^2 - A^{-2}.$$

It is easy to see from the definition of the matrices that $< K >$ satisfies the identities 1., 2., 3. above. Abstract properties of the bracket make it easy to deduce that R satisfies the Yang-Baxter equation (see **[13]** and **[16]** for a discussion of this point.).

Special value $A = -1$ and the Penrose Binor Calculus

With this model of the bracket in front of us, lets note how it behaves at the special value $A = -1$. Here we have

0. $< \times >=< \times >=$ (by def) $< \times >$.
1. $< \times > + < \asymp > + <)(>= 0.$
2. $< \bigcirc >= -2.$
3. $< \gamma >=< \sim >$.

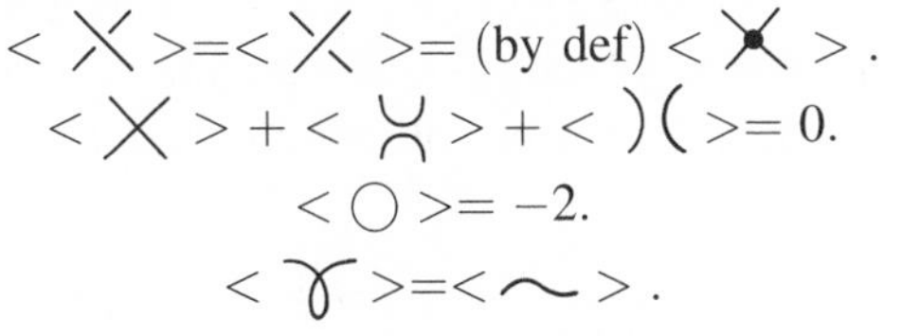

In this special case, the matrix $M = \sqrt{-1}\epsilon$ where ϵ is the matrix

$$\epsilon = \begin{bmatrix} 0 & 1 \\ -1 & 0 \end{bmatrix}.$$

This matrix, epsilon, is the defining invariant for the group $SL(2)$ (read $SL(2, C)$ if you like). That is, $SL(2)$ is the set of matrices of determinant 1, and epsilon has the property that for any matrix with commuting entries $P\epsilon P^T = DET(P)\epsilon$. In this context, we can interpret the equation 1. above as a matrix identity where the crossed arcs are Kronecker deltas:

$$\delta^a_d \, \delta^b_c + (\sqrt{-1})^2 \, \varepsilon^{ab} \varepsilon_{cd} + \delta^a_c \, \delta^b_d = 0$$

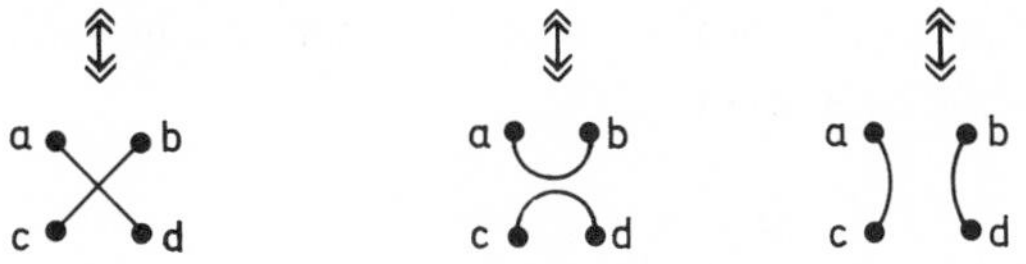

a b a b a b
c d c d c d

Thus the network calculus associated with the value $A = -1$ corresponds to a diagrammatic calculus of tensor identities for $SL(2)$ invariant tensors. This calculus was originally invented by Roger Penrose [**31**] in order to investigate the foundations of spin, angular momentum and the structure of space-time.

I wish to show how the calculus of link evaluations via the bracket polynomial provides a significant generalization of the Penrose spin networks, and how this is obtained by shifting from $SL(2)$ at $A = -1$ to $SL(2)_q$ at $q = \sqrt{A}$ for arbitrary A, where $SL(2)q$ denotes the quantum group in the sense of Drinfeld.

In the next section we discuss how the generalization works for spin network evaluations. Here I indicate how $SL(2)_q$ arises in this context. The idea is simply this: Since $SL(2)$ is characterized by matrices P (with commutative entries) such that $P\epsilon P^T = \epsilon$, what sorts of matrices will satisfy the equations

$$(*) \qquad \begin{aligned} P\tilde{\epsilon}P^T &= \tilde{\epsilon} \\ P^T\tilde{\epsilon}P &= \tilde{\epsilon} \end{aligned} \qquad ?$$

Attempting to generalize invariance in this way leads to well-known difficulties. One finds that it is necessary to assume that the entries of P do not necessarily commute with one another. Thus, we shall assume that P has the form

$$P = \begin{pmatrix} a & b \\ c & d \end{pmatrix}$$

where a, b, c, d belong to an associative (not necessarily commutative) ring. It is then an exercise in elementary algebra to see that the equations (*) are equivalent to the system of relations shown below.

$$\begin{aligned} ba &= qab \qquad ca = qac \\ dc &= qcd \qquad db = qbd \\ bc &= cb \\ ad - da &= (q^{-1} - q)bc \\ ad - q^{-1}bc &= 1 \end{aligned}$$

where $q = \sqrt{A}$.

The entries of the matrix P form a non-commutative algebra, $\mathcal{A}$. The algebra $\mathcal{A}$ is a Hopf algebra with coproduct

$$\Delta : \mathcal{A} \longrightarrow \mathcal{A} \otimes \mathcal{A}$$

given by the formula

$$\Delta(P^i_j) = \sum_k P^i_k \otimes P^k_j$$

The antipode is determined by the fact that P is invertible with respect to the algebra of its own elements. We have $s : \mathcal{A} \longrightarrow \mathcal{A}$, the antipode with

$$s(P) = P^{-1} = \begin{pmatrix} d & -qb \\ -q^{-1}c & a \end{pmatrix}.$$

This Hopf algebra is called the quantum group $SL(2)_q$.

Thus we see how $SL(2)_q$ arises quite naturally from the bracket model of the Jones polynomial. See [**13**],[**16**],[**23**] for a more complete discussion of this point of view.

III. Spin networks

Classical Penrose spin networks ([**31**],[**3**]) are based on the binor calculus (see section 2), and they are designed to facilitate calculations about angular momentum and $SL(2)$. I will not go into this background here, but it is worth noting part of the motivation. A spinor is a vector in two complex variables, denoted by Ψ^A, $A = 1, 2$. The spinor space is acted on by elements U in $SL(2)$ so that

$$(U\Psi)^A = \sum_B U^A_B \Psi^B$$

(Einstein summation convention). A natural $SL(2)$ invariant inner product on spinors is given by the formula $\Psi\Psi^*$ where

$$\Psi^*_A = \epsilon_{AB}\Psi^B$$

so that $\Psi\Psi^* = \Psi^A\epsilon_{AB}\Psi^B$ (sum on A and B). If we wish to diagram this inner product, then we let

$$\Psi^A \Longleftrightarrow$$

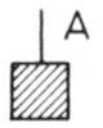

A

It is natural to lower the index via

$$\Psi^*_B \Longleftrightarrow$$

B

Then

$$\Psi\Psi^* \Longleftrightarrow$$

and the fragment  A B should be interpreted as the epsilon, ϵAB.

In order to design a planar calculus for the spinors it would be very pleasant if the calculations did not depend upon the topological types of the diagrams. (Surely, this is one of the thoughts of the "Old One".)

In any case, Penrose was led to make special conventions for maxima or minima (a minus sign for the minima) in order to insure

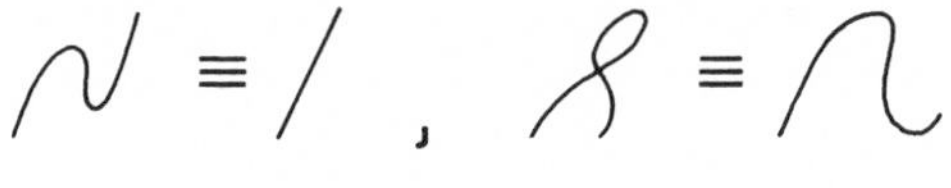

$$\equiv \; , \quad \equiv$$

These conventions are equivalent to choosing to replace ϵ by $\sqrt{-1}\epsilon$, as we have done in section 2. Thus in the classical spin nets

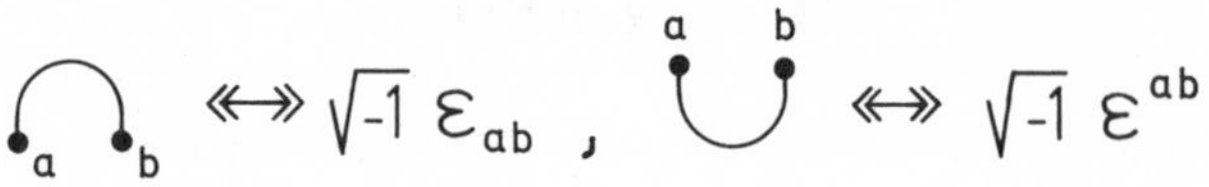

and

$$\epsilon_{ab} = \epsilon^{ab} \Longleftrightarrow \begin{pmatrix} 0 & -1 \\ 1 & 0 \end{pmatrix}.$$

Note that the loop value is

$\sum_{a,b}$  $= (\sqrt{-1})^2 + (-\sqrt{-1})^2 = -2.$

We know from the bracket, that this calculus entails the binor identity

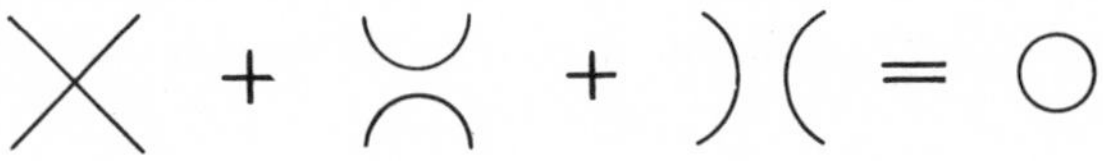

and that it is invariant under the projections of the Reidemeister moves. Therefore, any loop, even with self-crossings, has value -2 – since it is regularly homotopic to a connected sum of curls (Whitney-Graustein Theorem), and

The binor calculus is a unique planar calculus associated with both the bracket polynomial and the representations of $SL(2)$.

The next important spin network ingredient is the **antisymmetrizer**. This is a diagram sum associated to a bundle of lines, and is denoted by

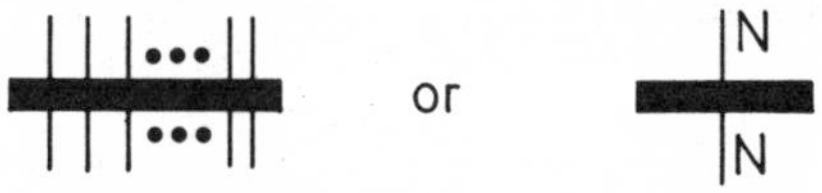

where the N denotes a bundle of parallel strands of multiplicity N.

The antisymmetrizer is defined by the formula

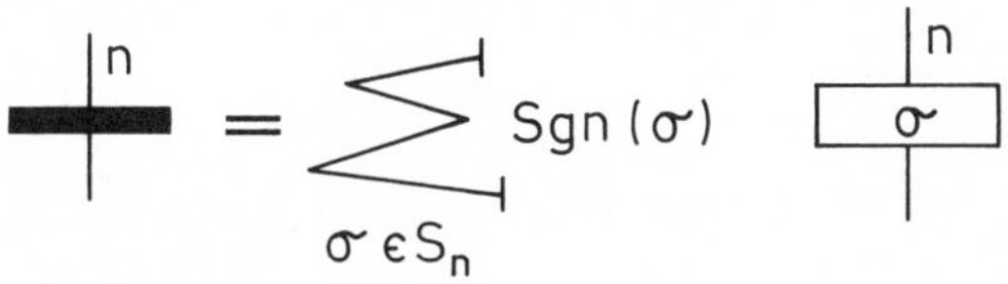

$$= \sum_{\sigma \in S_n} \mathrm{Sgn}(\sigma)$$

where σ runs over all permutations in S_n, $sgn(\sigma)$ is the sign of the permutation, and the σ in the box denotes the diagrammatic representation of this permutation as a braid projection. Thus

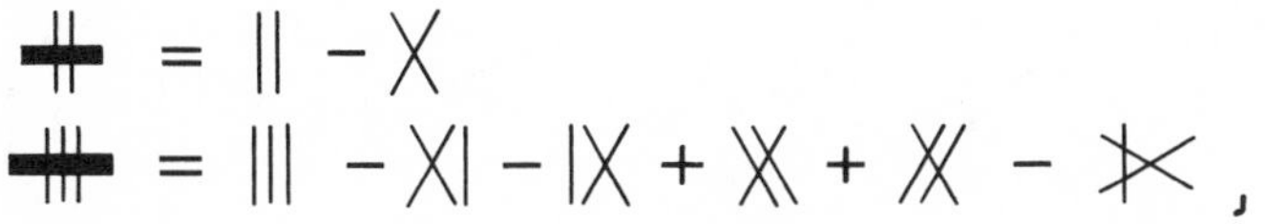

These antisymmetrizers are the basic ingredients for making the spin network calculations of Clebsch-Gordon coefficients, $3j$ and $6j$ symbols, and other apparatus for angular momentum. In this framework, the 3-vertex is defined as follows:

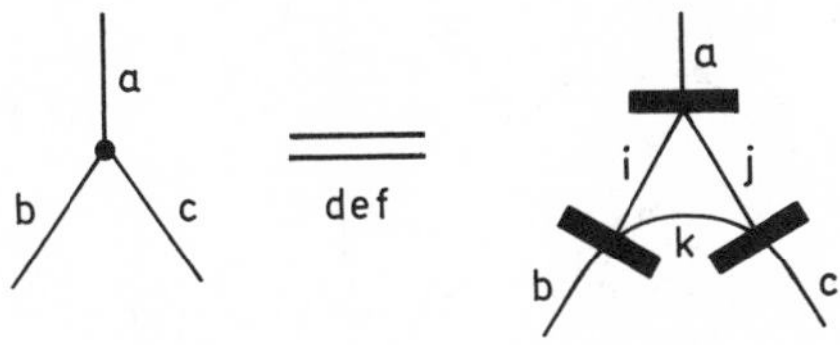

Here a, b, c are positive integers satisfying the condition that the equations $i+j = a, i+k = b, j+k = c$ can be solved in non-negative integers. Think of spins being apportioned in this way in interactions along the lines. This vertex gives the spin-network analog of the quantum mechanics of particles of spins b and c interacting to produce spin a.

Formally, the $6j$ symbols are defined in terms of these three-vertices by the formula

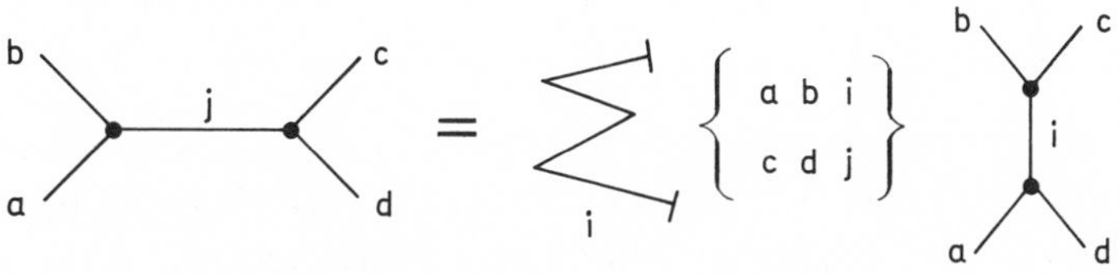

$$= \sum_i \begin{Bmatrix} a & b & i \\ c & d & j \end{Bmatrix}$$

From this it is clear why the $6j$ symbol is called a recoupling coefficient. To obtain a formula for these recoupling coefficients, we proceed diagrammatically as follows:

1°

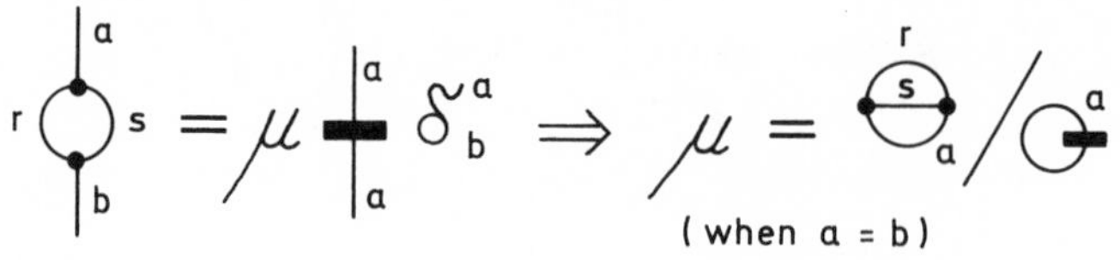

(when a = b)

2°

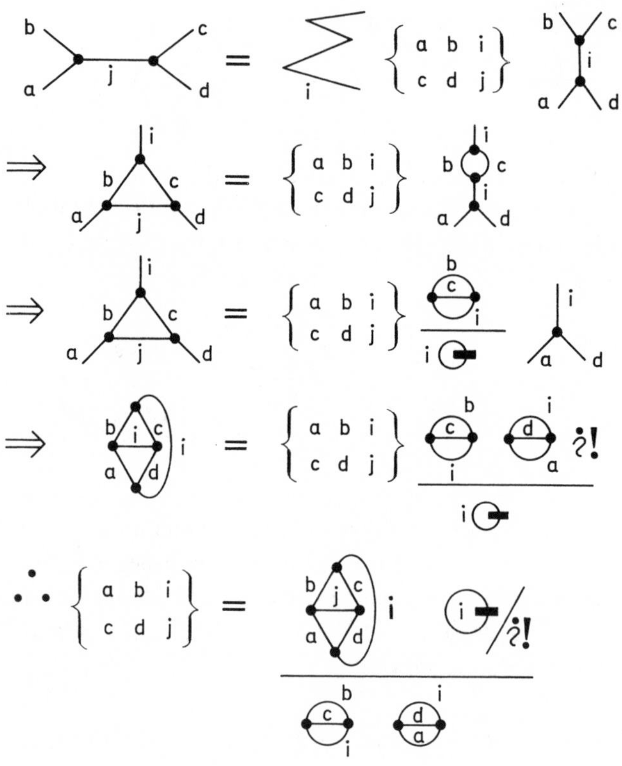

The final formula determines the $6j$ symbol in terms of spin network evaluations of some small nets. These can be handled by combinatorial means, and the various factorials in the resulting formulas are interpreted in terms of counting loop evaluations in the state expansions for the nets (via binor calculus). This same approach can be used to prove various properties of these objects such as orthogonality relations, and pentagon (or Elliot- Biedenharn) identities that arise from the recoupling shown below:

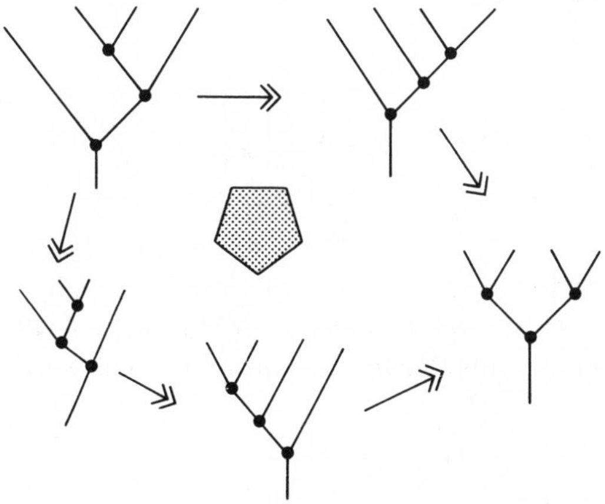

Now, with this sketch of recoupling theory in mind, let us return to the antisymmetrizers. They are **projection operators** in the sense that

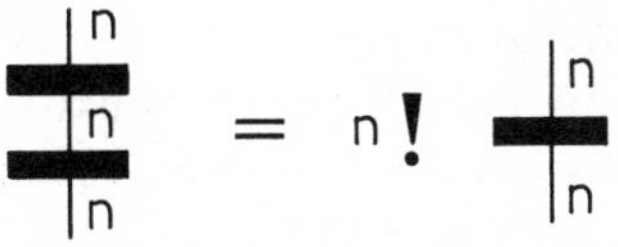

and that they kill off the generators of the Temperley Lieb algebra [**7**] $e_1, e_2, \ldots, e_{n-1}$:

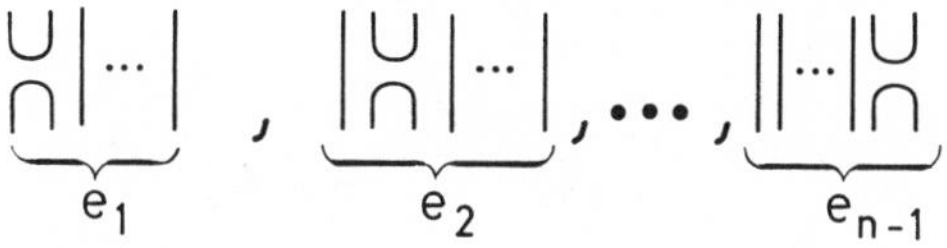

since

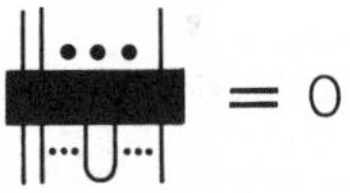

by dint of antisymmetry. These remarks mean that the expanded forms of the antisymmetrizers are special projection operators in the Temperley Lieb algebra. For example

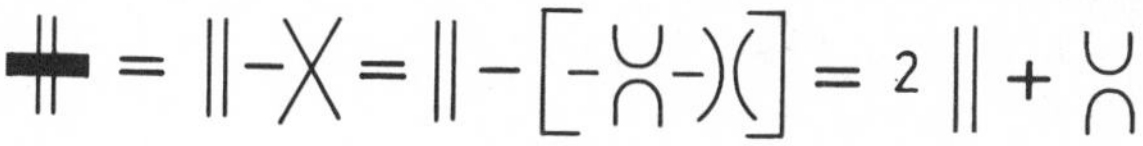

and

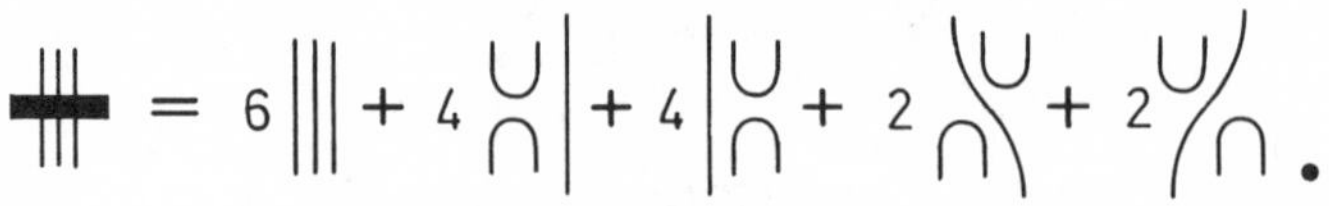

$$= 6 \,|||\, + 4 \ldots + 4 \ldots + 2 \ldots + 2 \ldots .$$

Such projection operators are of great value, and we shall see in the next section, that a suitable generalization of spin-nets produces the analogs of these operators in the full Temperley Lieb algebra with arbitrary loop value.

IV. q-Spin networks and the magic weave

I shall just show how to construct the generalized antisymmetrizers (see section 3). A q-**Spin Network** is nothing more than a link diagram with special nodes that are interpreted as these antisymmetrizers. Thus a q-spin net is no longer confined to the plane, but we will not here discuss formulas for twisting the nodes of these networks. I will discuss elsewhere how the q-spin nets are acted upon by the quantum group corresponding to $SL(2)$ (compare however with section 2), and how the Penrose Spin Geometry Theorem and other results fit into this context.

We take $q = \sqrt{A}$, and use the bracket identity

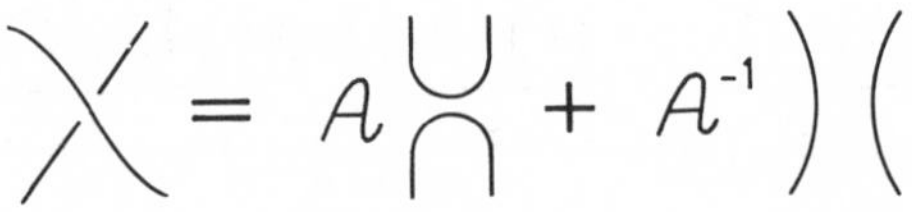

$$= A \ldots + A^{-1} \,)\,(\,$$

in place of the binor identity, with loop value

$$\bigcirc = (-A^2 - A^{-2})$$

Thus any q-spin network computes its own bracket polynomial. Now define the q-antisymmetizer n by the formula

$$= \sum_{\sigma \in S_n} (A^{-3})^{T(\sigma)} \; \hat{\sigma}$$

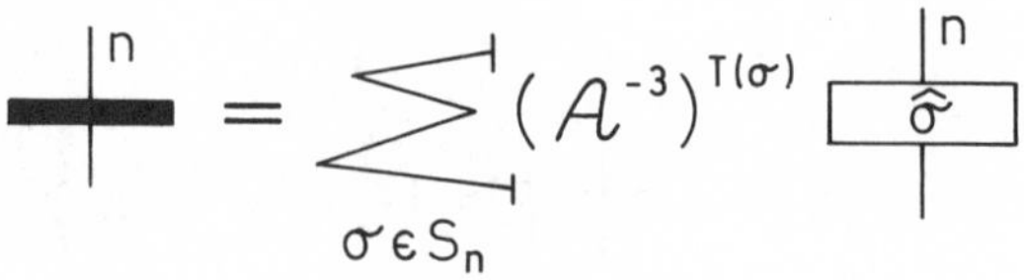

where $T(\sigma)$ is the minimal number of transpositions needed to return s to the identity, and $\hat{\sigma}$ is a minimal braid representing s with all negative crossings, i. e. with all crossings in the form shown below with respect to the braid direction

Example 1.

$$\begin{aligned}
\# &= \| + A^{-3} \times \\
&= \| + A^{-3}\left[A \asymp + A^{-1})(\right] \\
&= \| + A^{-2} \asymp + A^{-4})(\\
&= (1+A^{-4})\left[\| + \frac{A^{-2}}{1+A^{-4}} \asymp\right] \\
&= (1+A^{-4})\left[\| + \frac{1}{A^{2}+A^{-2}} \asymp\right] \\
\# &= (1+A^{-4})\left[\| - \delta^{-1} \asymp\right]
\end{aligned}$$

($\delta = -A^2 - A^{-2}$. Note that $[2]! = 1 + A^{-4}$.)

Note that $f_1 = 1 - \delta^{-1} e_1$ is the first of a sequence of Temperley Lieb projectors f_n defined inductively via

$$\begin{cases} f_0 = 1, & \mu_1 = \delta^{-1} \\ f_{n+1} = f_n - \mu_{n+1} f_n e_{n+1} f_n, & \mu_{n+1} = (\delta - \mu_n)^{-1}. \end{cases}$$

$$\begin{aligned}
\overset{n}{+} \Rightarrow f_1 &= 1 - \delta^{-1} e_1 \\
f_2 &= 1 - \mu_2(e_1 + e_2) + \mu_1\mu_2(e_1e_2 + e_2e_1) \\
\mu_1 &= \delta^{-1} = 1/\delta \\
\mu_2 &= \delta/(\delta^2 - 1).
\end{aligned}$$

(See **[28]** for an exposition of this algebra.)

Theorem. *The Temperley Lieb elements f_n (for loop value $\delta = -A^2 - A^{-2}$) are equivalent to the q-antisymmetrizers. In particular, we have the formula*

$$f_n = \frac{1}{[n]!}$$

where

$$[n]! = \sum_{\sigma \in S_n} (A^{-4})^{T(\sigma)} = \prod_{k=1}^{n} \left(\frac{1 - A^{-4k}}{1 - A^{-4}} \right)$$

is the q-deformed factorial in this framework.

Proof. To appear elsewhere.

Example 2.

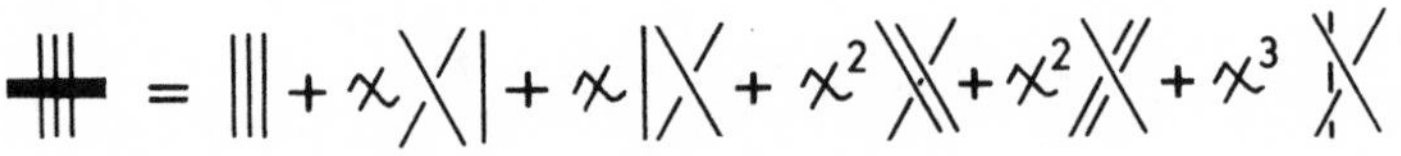

$(x = A^{-3})$

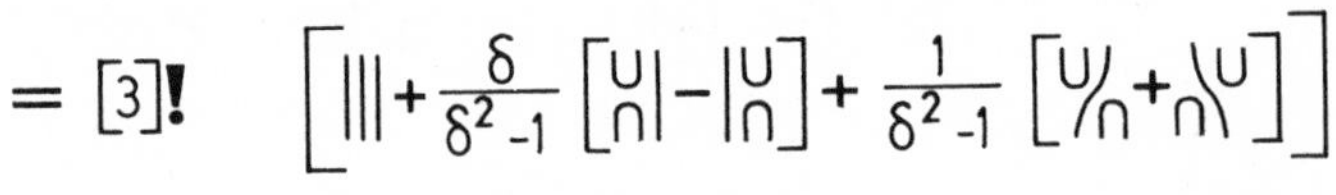

$= [3]!$

$$[3!]_A = 1 + 2A^{-2} + 2A^{-8} + A^{-12}.$$

By a sustained calculation, check that this holds, and that the Temperley Lieb element is given by

$$f_2 = 1 - \frac{\delta}{\delta^2 - 1}(e_1 + e_2) + \frac{1}{\delta^2 - 1}(e_1 e_2 + e_2 e_1).$$

Various features of these operators are clear from the context of the antisymmetrizers. Note for example, that the proper generalization of the antisymmetric property is that

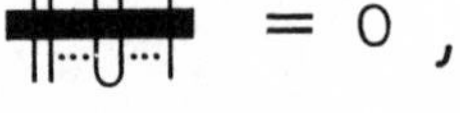

$= 0$,

and we see that this is a curl compensation!

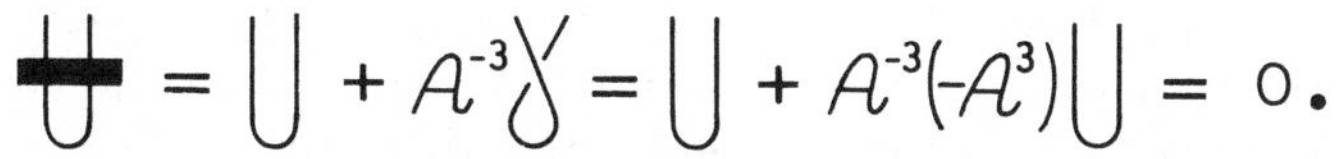

$$= \bigcup + A^{-3} \ldots = \bigcup + A^{-3}(-A^{3}) \bigcup = 0.$$

I have emphasized the q-spin network construction for these magic weaving operators because they have extraordinary use in knot theory and invariants of 3-manifolds. The Lickorish proof of existence of 3-manifold invariants [**28**] depends crucially on subtle properties of these operators. Yamada [**38**] has used them for constructing invariants of graphs. We now see that the q-antisymmetrizers form a special structure for direct generalization of the methods outlined in section 3 to q-angular momentum. Thus spin network techniques can be used as a foundation for the theory of $q-6j$ recoupling and other intricacies of the $SL(2)$ quantum group. In particular, this applies directly to the 3-manifold invariants of Viro, and Turaev. Thus these invariants can also be given a basis in terms of the Jones polynomial and the Temperley Lieb Algebra. It remains to be seen how the theory thus reformulated will reflect on the nature of these invariants.

References

1. M. F. Atiyah, *The Geometry and Physics of Knots*, Cambridge University Press, 1990.
2. L. Crane, *2-D Physics and 3-D Topology*, Comm. Math. Physics **135** (1991), 615–640.
3. B. Hasslacher and M. J. Perry, *Spin networks are simplicial quantum gravity*, Physics Letters **103B, No. 1**, July 1981..
4. V. F. R. Jones, *On knot invariants related to some statistical mechanics models*, Pacific J. Math. **137, No. 2** (1989).
5. L. H. Kauffman, *Formal Knot Theory*, Lecture Notes No. 130, Princeton University Press, 1983.
6. L. H. Kauffman, *On Knots*, Annals Study Number 115, Princeton University Press, 1987.
7. L. H. Kauffman, *State models and the Jones polynomial*, Topology **26** (1987), 395–407.

7.1. L. H. Kauffman, *State Models for Link Polynomials*, L'Enseignment Mathematique **t.36** (1990), 1–37.

8. L. H. Kauffman, *New invariants in the theory of knots*, Amer. Math. Monthly **95** (1988), 195–242.
9. L. H. Kauffman, *Statistical mechanics and the Jones polynomial*, Contemp. Math. vol. 78, AMS, 1988, pp. 263–297.
10. L. H. Kauffman, *Invariants of graphs in three-space*, Trans. Amer. Math. Soc. **311** (1989), 697–710.
11. L. H. Kauffman, *Polynomial invariants in knot theory*, Braid Group, Knot Theory and Statistical Mechanics, ed. by C. N. Yang and M. L. Ge, World Sci., 1989, pp. 27–58.

12. L. H. Kauffman, *An integral heuristic*, Int. J. Mod. Phys. A **5, No.7**, 1363–1367.
13. L. H. Kauffman, *Knots abstract tensors and the Yang-Baxter equation*, Knots, Topology and Quantum Field Theories, Proceedings of the Johns Hopkins Workshop on Current Problems in Particle Theory. Vol. 13 (ed. by L.Lussana), World Sci., 1990, pp. 179–334.
14. L. H. Kauffman, *Statistical mechanics and the Alexander polynomial*, Proceedings of 1988 Conf. on Algebraic Topology, Northwestern University AMS Contemp. Math. Series. Vol. 96, pp. 221–231.
15. L. H. Kauffman, *A Tutte polynomial for signed graphs*, Discrete Applied Math. **25** (1989), 105–127.
16. L. H. Kauffman, *Spin networks and knot polynomials*, Int. J. Theo. Phys. A **5, No.1** (1990), 93-115.
17. L. H. Kauffman, *Knot polynomials and Yang-Baxter models*, IXth International Congress on Math. Phys. 17-27 July 1988, Swansea, Wales. (ed. by Simon, Truman, Davies), Adam, Hilger Pub..
18. L. H. Kauffman, *Super Twist Spinning*, Festschrift for Antonio Plans, Universita de Zaragoza Press, 1990, pp. 139–154.
19. L. H. Kauffman, *An invariant of regular isotopy*, Trans. Amer. Math. Soc. **318** (1990), 417-471.
20. L. H. Kauffman, *Problems in Knot Theory*, Open Problems in Topology, ed. by J. van Mill, North Holland, 1990.
21. L. H. Kauffman, *From knots to quantum groups and back*, Proceedings of the fall 1989 Montreal conference on Hamiltonian systems, ed. by Harnad and Marsden, CRM Pub., 1990, pp. 161–176.
22. L. H. Kauffman, *Knots, Spin Networks and Invariants of Three Manifolds*, (to appear in the proceedings of Knots 90 - Conference, August 1990, Osaka, Japan).
23. L. H. Kauffman, *Knots and Physics*, World Scientific, 1991.
24. L. H. Kauffman and H. Saleur, *Free Fermions and the Alexander-Conway polynomial*, (to appear Comm. Math. Phys.).
25. R. Kirby and P. Melvin, *On the 3-manifold invariants of Witten and Reshetikhin-Turaev for sl(2,C)*, (preprint 1990).
26. T. Kohno, *Topological invariants for 3-manifolds using representations of mapping class groups I*, (preprint 1990).
27. M. Kontsevich, *Rational conformal field theory and invariants of three dimensional manifolds*, (preprint 1989).
28. W. B. R. Lickorish, *Invariants for 3-manifolds derived from the combinatorics of the Jones polynomial*, Pacific J. Math. (to appear).
29. G. Moore and N. Seiberg, *Classical and quantum conformal field theory*, Comm. Math. Phys. **123** (1989), 177–254.
30. J. P. Moussouris, *Quantum models of space-time based on recoupling theory*, (Mathematics Thesis, Oxford University - 1983).
31. R. Penrose, *Applications of negative dimensional tensors*, Combinatorial Mathematics and its Applications, ed. by D. J. A. Welsh, Academic Press, 1971.
32. G. Ponzano and T. Regge, *Semiclassical limit of Racah coefficients*, Spectroscopic and Group Theoretical Methods in Physics, North- Holland Publ. Co., Amsterdam, 1968.
33. N. Yu. Reshetikhin, *Quantized universal enveloping algebras, the Yang Baxter equation and invariants of links I and II*, LOMI preprints 1988.
34. N. Yu. Reshetikhin and V. Turaev, *Invariants of Three Manifolds via link polynomials and quantum groups*, (preprint 1989).

35. V. Turaev, *The Yang-Baxter equation and invariants of links*, Invent. Math. **92** (1988), 527–533.
36. V. Turaev and O. Viro, *State sum invariants of 3-manifolds and quantum 6j-symbols*, (preprint 1990).
37. E. Witten, *Quantum field theory and the Jones polynomial*, Comm. Math. Phys **121** (1989), 351–399.
38. S. Yamada, *Talk given at Knots 90*, Osaka, August 1990.

The framed braid group and representations

Ki Hyoung Ko and Lawrence Smolinsky***

Many approaches to the Jones polynomial have emerged in the last several years. Jones' original approach is through group representations. He constructs representations of B_n which are unitary when they arise from $\mathbf{C}^*$ algebras. E. Witten viewed the Jones polynomial through quantum mechanics and generalized it to an invariant of links in 3-manifolds. There is also the combinatorial viewpoint that considers skein relations.

The situation for Witten's 3-manifold invariant is similar. Witten's original approach is through quantum mechanics. This approach has been broadened and reinterpreted by several authors (Reshetikhin-Turaev [**R-T**], Kirby-Melvin [**K-M**], Cappell-Lee-Miller [**C-L-M**]). The Kirby-Melvin interpretation motivated W.B.R. Lickorish to develop a combinatorial approach [**L1**], [**L2**] and [**K-S1**]. The combinatorial approach uses the Kauffman bracket and the evaluation of the Jones polynomial at roots of unity (from Jones' viewpoint, unitary representations of B_n). However there is no development via group representations. In fact there is no group. This paper introduces a candidate for a group, the framed braid group. The second portion discusses its representation theory. In particular it is shown that framing information always separates from braiding information when examining irreducible unitary representations.

Framed Braids. We begin with the viewpoint that any orientable 3-manifold can be produced by surgery on a link [**L**]. The nature of the equivalence relation on links that yields 3-manifolds is described by the Kirby calculus in [**K**]. Unframed oriented links can be described in terms of braids: given a geometric braid β the link $\hat{\beta}$ is gotten by identifying the initial points and the terminal points of β. Every link can be obtained in this manner. If we assign framing numbers to the top of each strand of a braid we obtain a framed braid.

* Partially supported by a Korea Science and Engineering Foundation Postdoctoral Fellowship.

** Partially supported by a Louisiana Education Quality Support Fund grant LEQSF(87-89)-RD-A-8. Speaker at the Knots 1990 conference

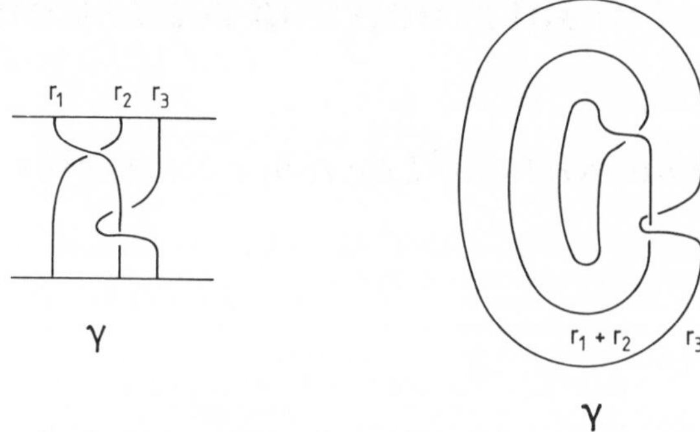

Figure 1

Let π be the natural map from the braid group to the symmetric group, $\pi : B_n \to \Sigma_n$. Then each braid β has a decomposition into cycles given by $\pi(\beta)$. If we begin with a framed braid we can observe which framings correspond to strands in the same cycle of $\pi(\beta)$. Hence we can close a framed braid to obtain a framed oriented link; close the braid and add the framings that correspond to a cycle, i.e., yield a link component (Figure 1). By forgetting the orientation we obtain a framed link description of a 3-manifold.

$$(*) \qquad \begin{matrix}\text{Framed}\\ \text{braids}\end{matrix} \xrightarrow{\text{close}} \begin{matrix}\text{Framed}\\ \text{oriented}\\ \text{links}\end{matrix} \xrightarrow{\text{Forget orientation}} \begin{matrix}\text{Framed}\\ \text{links}\end{matrix} \xrightarrow{\text{surgery}} \text{3-manifolds}$$

Let B_n denote the braid group with the explicit generators $\sigma_1, \sigma_2, \ldots, \sigma_{n-1}$ and relations

(1) $\sigma_i\sigma_j = \sigma_j\sigma_i$ if $|i-j| > 1$;

(2) $\sigma_i\sigma_{i+1}\sigma_i = \sigma_{i+1}\sigma_i\sigma_{i+1}$.

B_n acts on $\{1, 2, \ldots, n\}$ through π, i.e., $\sigma(i) = \pi(\sigma)(i)$ for $\sigma \in B_n$. This paper follows the convention that the symmetric group acts from the right so that $(\sigma\tau)(i) = \tau(\sigma(i))$ for $\sigma, \tau \in B_n$. Note that $\pi(\sigma_i) = (i, i+1)$.

Definition. The framed braid group $\mathfrak{F}_n$ is the group generated by $\sigma_1, \sigma_2, \ldots, \sigma_{n-1}$, $t_1, t_2, \ldots, t_n$ with the relations (1), (2) and additional relations

(3) $t_it_j = t_jt_i$ for all i, j;

(4) $\sigma_it_j = t_{\sigma_i(j)}\sigma_i$.

The group $\mathfrak{F}_n$ is a semidirect product $\mathbf{Z}^n \rtimes B_n$ where the action of B_n on $\mathbf{Z}^n$ is given by $\sigma(r_1, r_2, \ldots, r_n) = (r_{\sigma(1)}, r_{\sigma(2)}, \ldots, r_{\sigma(n)})$. If $t_1^{r_1}t_2^{r_2}\cdots t_n^{r_n}\alpha \in \mathfrak{F}_n$ with $\alpha \in B_n$ then the r_i's are called *framings*. Note that $\sigma t_i = t_{\sigma^{-1}(i)}\sigma$ for $\sigma \in B_n$. The product and the inverse in this notation are given as follows. See Figure 2 for an example.

$$(t_1^{r_1} t_2^{r_2} \cdots t_n^{r_n} \alpha)(t_1^{s_1} t_2^{s_2} \cdots t_n^{s_n} \beta) = t_1^{r_1 + s_{\alpha(1)}} t_2^{r_2 + s_{\alpha(2)}} \cdots t_n^{r_n + s_{\alpha(n)}} \alpha\beta$$

and

$$(t_1^{r_1} t_2^{r_2} \cdots t_n^{r_n} \alpha)^{-1} = t_1^{-r_{\alpha^{-1}(1)}} t_2^{-r_{\alpha^{-1}(2)}} \cdots t_n^{-r_{\alpha^{-1}(n)}} \alpha^{-1}.$$

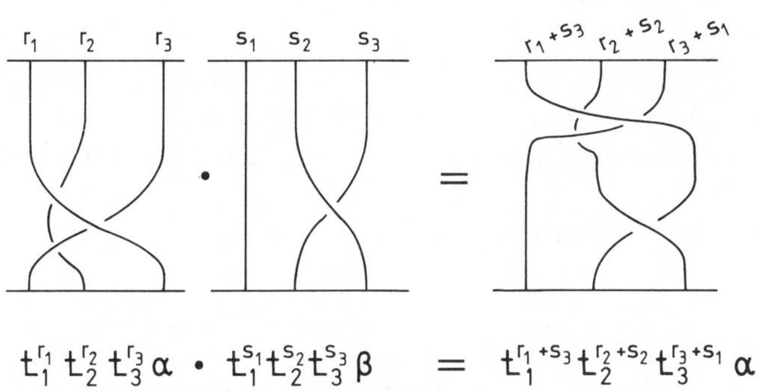

Figure 2

The subgroup $\mathbf{Z}^n$ is the subgroup of framings.

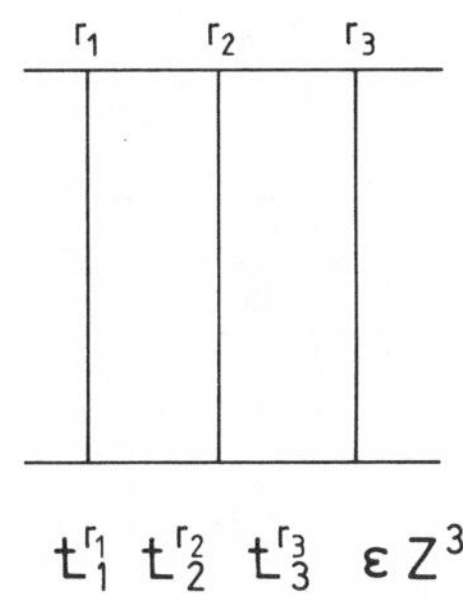

Figure 3

In [**K-S2**] the relationship between framed braids and 3-manifolds via the correspondence $(*)$ is determined. This relationship is given in the theorem below. Framing changes are given in terms of functions on the symmetric group. For $i < n$ let λ_i assign to each permutation in Σ_{n-1} a function from $\{1, 2, \ldots, n-1\}$ to $\{0, 1, \ldots, n-i\}$ as follows. Suppose $\tau \in \Sigma_{n-1}$ decomposes into a product of disjoint cycles $\tau_1 \tau_2 \cdots \tau_k$. If p occurs in some $\tau_j = (j_1, j_2, \ldots, j_m)$ then $\lambda_i(\tau)(p)$ is defined to be the number of elements in the intersection $\{j_1, j_2, \ldots, j_m\} \cap \{i, i+$

$1, \ldots, n-1\}$. Denote by $W_{n,j}$ the braid $\sigma_{n-1}\sigma_{n-2}\cdots\sigma_{j+1}\sigma_j^2\sigma_{j+1}\cdots\sigma_{n-2}\sigma_{n-1}$ in B_n.

Theorem 1. *Two framed braids represent homeomorphic 3-manifolds if and only if they are related by the equivalence relation generated by the following moves:*
(1) conjugation by framed braids;
(2) Markov move : for $\beta \in \mathfrak{F}_n$, $\beta\sigma_n \sim \beta \sim \beta\sigma_n^{-1}$;
(3) blow up : for $\beta \in \mathfrak{F}_n$, $t_{n+1}\beta \sim \beta \sim t_{n+1}^{-1}\beta$;
handle slide : for $\alpha, \beta \in \mathfrak{F}_{n-1}$,
$t_nW_{n,j}\alpha W_{n,i}^{-1}\beta \sim t_{n+1}t_n^{2\lambda+1}(W_{n,j}\sigma_nW_{n,j}\sigma_n^{-1})\alpha(W_{n,i}^{-1}\sigma_n^{-1}W_{n,i}\sigma_n)\beta\sigma_n^{-2}\sigma_{n-1}^{-1}$
where $\lambda = \lambda_i(\pi(\beta\alpha))(\alpha(n-1)) - \lambda_j(\pi(\alpha\beta))(n-1)$;
(3) orientation reversing : for $\alpha, \beta \in \mathfrak{F}_{n-1}$, $W_{n,j}\alpha W_{n,i}^{-1}\beta \sim W_{n,j}^{-1}\alpha W_{n,i}\beta$.

Figures 4, 5, 6 and 7 describe the moves 2, 3, 4 and 5.

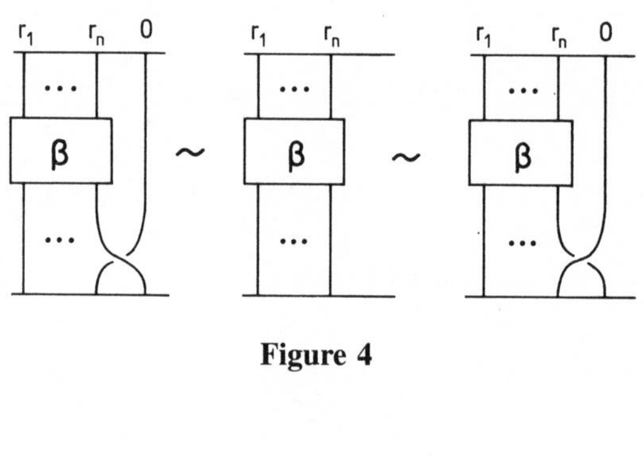

Figure 4

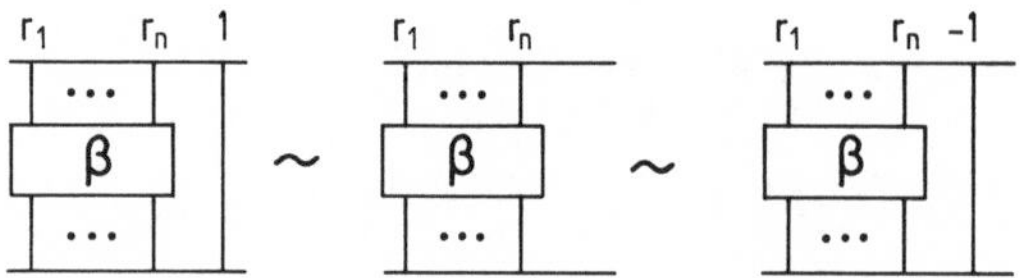

Figure 5

Unitary representations. In the semidirect product $\mathfrak{F}_n$, B_n acts on $\mathbf{Z}^n$ and so gives an action on the characters of $\mathfrak{F}_n$. Let $char(\mathbf{Z}^n)$ be the characters of $\mathbf{Z}^n$ (i.e., homomorphisms to S^1) and take $\chi \in char(\mathbf{Z}^n)$. If $\beta \in B_n$ then $\chi \cdot \beta \in char(\mathbf{Z}^n)$ by $\chi \cdot \beta(z) = \chi(\beta z \beta^{-1})$ As a space, $char(\mathbf{Z}^n)$ is $S^1 \times S^1 \times \cdots \times S^1$ the n-torus. The action of B_n on $\mathbf{Z}^n$ is by permuting the coordinates so B_n acts on $char(\mathbf{Z}^n)$ by permuting the S^1 factors, i.e., $(\omega_1, \omega_2, \cdots, \omega_n) \cdot \beta = (\omega_{\pi(\beta)(1)}, \omega_{\pi(\beta)(2)}, \cdots, \omega_{\pi(\beta)(n)})$. For $\chi \in char(\mathbf{Z}^n)$ let B_χ be the stabilizer of χ in B_n, $B_\chi \subset B_n$. Suppose $\chi = (\omega_1, \omega_2, \cdots, \omega_n)$. The n-tuple χ has k distinct entries $\{\eta_1, \eta_2, \cdots, \eta_k\}$ and η_j occurs with multiplicity n_j. From the action of Σ_n

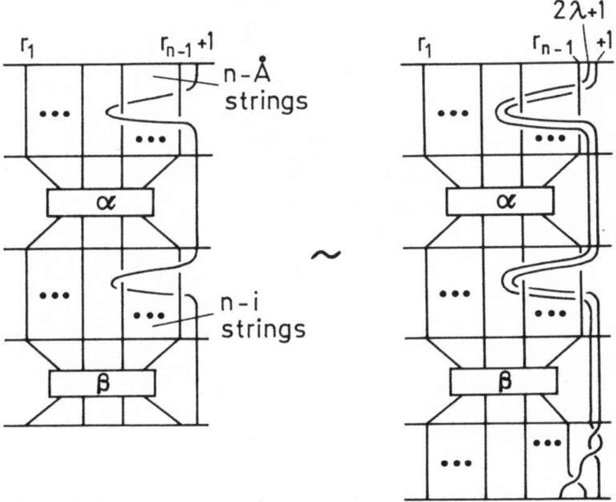

Figure 6

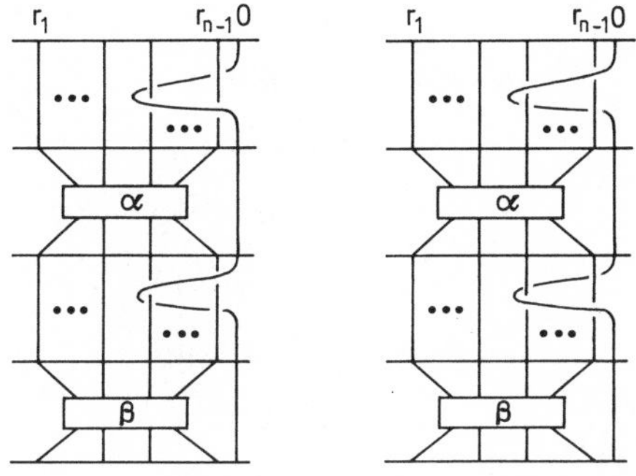

Figure 7

on $char(\mathbf{Z}^n)$ we see that the stablizer of χ in Σ_n is isomorphic to $\Sigma_\chi = \Sigma_{n_1} \oplus \Sigma_{n_2} \oplus \cdots \oplus \Sigma_{n_k}$. Since B_n acts via $\pi : B_n \to \Sigma_n$ we have that $B_\chi = \pi^{-1}(\Sigma_\chi)$. This subgroup consists of elements that preserve some collection of subsets of the initial points. For example if $\chi \in char(\mathbf{Z}^3)$ and $\omega_1 = \omega_2 \neq \omega_3$ then $\beta \in B_\chi$ for the β in Figure 8 preserves the sets $\{1, 2\}$ and $\{3\}$.

The induced map $\pi : B_n/B_\chi \to \Sigma_n/\Sigma_\chi$ is a bijection so that $|B_n/B_\chi| = \binom{n}{n_1, n_2, \cdots, n_k}$ and B_χ is of finite index in B_n.

We now consider the general theory of representations of semidirect products $G = N \rtimes S$ with N abelian. A complete discussion is given in [**B-R**] chapter 17. In the semidirect product G, S acts on N and so on $char(N)$. In order to describe

Figure 8

the unitary representations of G we need G to satisfy a technical condition: G is a regular semidirect product if $char(N)$ contains a countable family $Z_1, Z_2, Z_3, \cdots$ of Borel subsets, each a union of G orbits such that every orbit in $char(N)$ is the intersection of members of the subfamily containing that orbit.

Suppose G is a regular semidirect product of N and S. Let $\chi \in char(N)$ and denote by S_χ the stabilizer of χ under the S action. Let λ be a representation of S_χ. If $a \in N$ and $b \in S$ then let $\chi \cdot \lambda(ab) = \chi(a)\lambda(b)$. If $a' \in N$, $b' \in S$, $g = ab$ and $g' = a'b'$ then

$$\begin{aligned}\chi \cdot \lambda(gg') &= \chi \cdot \lambda(aba'b') \\ &= \chi \cdot \lambda(aba'b^{-1}bb') \\ &= \chi(a)\chi(ba'b^{-1})\lambda(b)\lambda(b') \\ &= \chi(a)\chi(a')\lambda(b)\lambda(b') \\ &= \chi \cdot \lambda(ab)\chi \cdot \lambda(a'b') \\ &= \chi \cdot \lambda(g)\chi \cdot \lambda(g')\end{aligned}$$

so $\chi \cdot \lambda$ is a representation of the subgroup $N \rtimes S_\chi = G_\chi$. We can now form the representation

$$\rho = \mathrm{ind}_{G_\chi}^G(\chi \cdot \lambda)$$

a representation of the whole group G. The following theorems are results from Mackey's theory of induced representations (see [**B-R**] pages 508-509).

Theorem 2. *Let G be a regular semidirect product $N \rtimes S$ of separable, locally compact groups N and S with N abelian.*

(1) Let ρ be an irreducible unitary representation of G. Then there is a character χ of N and an irreducible unitary representation λ of S_χ such that ρ is unitary equivalent to $\mathrm{ind}_{G_\chi}^G(\chi \cdot \lambda)$.

(2) If $\chi \in char(N)$ and λ is an irreducible unitary representation of S_χ then $ind_{G_\chi}^G(\chi \cdot \lambda)$ is an irreducible unitary representation of G.

The framed braid group is a regular semidirect product by,

Proposition 3. *If $G = N \rtimes S$, N a finitely generated abelian group and if S acts on N through a finite group then G is a regular semidirect product.*

Proof. If N is finite then so is $char(N)$ and the Z_i's are the orbits. If not then $char(N)$ is a finite union of tori, $S^1 \times \cdots \times S^1$. Since this space is second countable, let $\{\mathfrak{O}_i\}_{i=1}^{\infty}$ be a countable base and take $Z_i = G(\mathfrak{O}_i)$. □

As a corollary to Theorem 2 we have,

Corollary 4. *If ρ is an irreducible unitary representation of $\mathfrak{F}_n$ then there is a character χ of $\mathbf{Z}^n$ and an irreducible representation λ of B_χ such that*

$$\rho = \operatorname{ind}_{\mathbf{Z}^n \rtimes B_\chi}^{\mathbf{Z}^n \rtimes B_n} (\chi \cdot \lambda).$$

The $\mathbf{Z}^n$ subgroup is purely framing phenomena and the B_χ subgroup is purely braiding phenomena so the representation theory of $\mathfrak{F}_n$ separates these aspects.

Some examples. These examples are motivated by the Kirby-Melvin interpretation of Witten's invariants [**K-M**]. Let $c : \{1, 2, \cdots, n\} \to \{1, 2, \cdots, r\}$ so that $c(i)$ is a natural number and let T_j be a braid of m_j strings. For example see Figure 9.

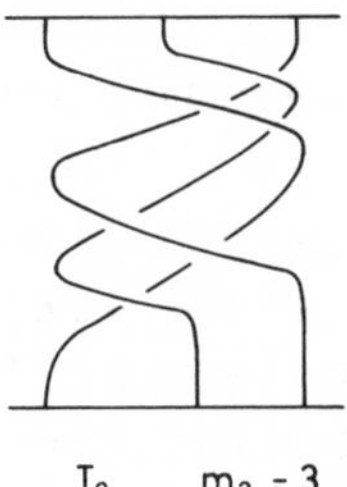

Figure 9

Let A_n be the Temperley-Lieb algebra considered by Jones in [**J**]. This is a finite dimensional $*$-algebra with generators denoted by $e_1, e_2, \cdots, e_n$. Let $\gamma_q : \mathbf{C}[B_q] \to A_q$ be the map $\gamma_q(\sigma_i) = A + A^{-1} e_i$ for A a root of unity. This defines a unitary representation of B_q. In particular, $\gamma_q(T_{c(i)})$ is unitary. Let $\mathfrak{z}_i : \mathbf{Z} \to A_{m_{c(i)}}$ be defined by $\mathfrak{z}_i(k) = \gamma_{m_{c(i)}}(T_{c(i)}^k)$ and define a unitary representation of $\mathbf{Z}^n$ by

$\mathfrak{Z} : \mathbf{Z}^n \to A_{\Sigma_{i=1}^n m_{c(i)}}$ by

$$\mathbf{Z}^n \xrightarrow{\oplus \mathfrak{Z}_i} \oplus A_{m_{c(i)}} \subset A_{\Sigma_{i=1}^n m_{c(i)}}.$$

So if $n = 3$ and $c(1) = c(2) = 1$ and $c(3) = 2$ then $\mathfrak{Z}(1, 3, 2)$ is shown in Figure 10.

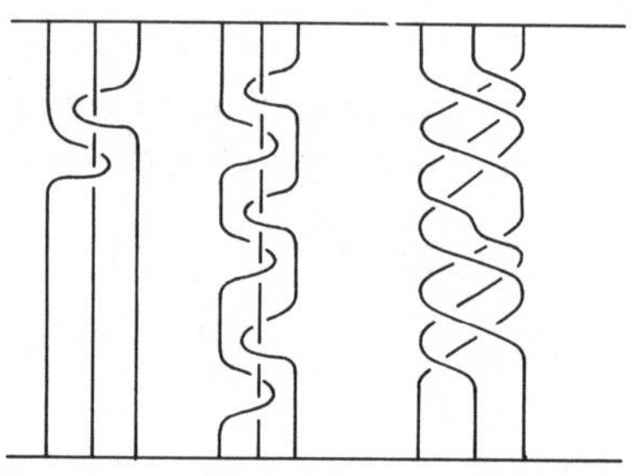

Figure 10

As a representation of $\mathbf{Z}^n$, $\mathfrak{Z} = \Sigma a_j \chi_j$ for $\chi_j \in char(\mathbf{Z}^n)$. Let P be the subset of the powerset of $\{1, \cdots, n\}$ given by $\{c^{-1}(1), c^{-1}(2), \cdots, c^{-1}(r)\}$ and $B_P \subset B_n$ the subgroup that preserves these sets of points, i.e., $\{\alpha \in B_n |$ if $x \in c^{-1}(i)$ then $\pi(\alpha)(x) \in c^{-1}(i)\}$. Now $B_P \cong \pi^{-1}(\Sigma_{|c^{-1}(1)|} \oplus \Sigma_{|c^{-1}(2)|} \oplus \cdots \oplus \Sigma_{|c^{-1}(r)|})$. The functions c and c' yield conjugate subgroups if and only if there is a $\delta \in \Sigma_n$ with $|c^{-1}(i)| = |c'^{-1}(\delta(i))|$, i.e., c and c' determine the same partition of the number n. If $G = \beta B_P \beta^{-1}$ for $\beta \in B_n$ then let $c' = c \circ \pi(\beta)$ and P' be the resulting set of subsets of $\{1, \cdots, n\}$. We have that $G = B_{P'}$. So the collection of subgroups conjugate to B_P is $\{B_{P'} | P'$ and P determine the same partition of $n\}$. This observation is important if one wishes to consider characters of representations. If $a_j \neq 0$ then $B_{\chi_j} \subset B_P$ and we can construct representations of $\mathbf{Z}^n \rtimes B_P$ by constructing representations λ_P of B_P. Let λ_P be the composition

$$B_n \to B_{\Sigma_{i=1}^n m_{c(i)}} \xrightarrow{\gamma_{\Sigma_{i=1}^n m_{c(i)}}} A_{\Sigma_{i=1}^n m_{c(i)}}$$

where the map between braid groups makes $c(i)$ parallel copies of the ith string as shown below. Then the $a_j \chi_j \cdot \lambda_P$ are representations of $\mathbf{Z}^n \rtimes B_P$ and determine representations of $\mathfrak{F}_n$.

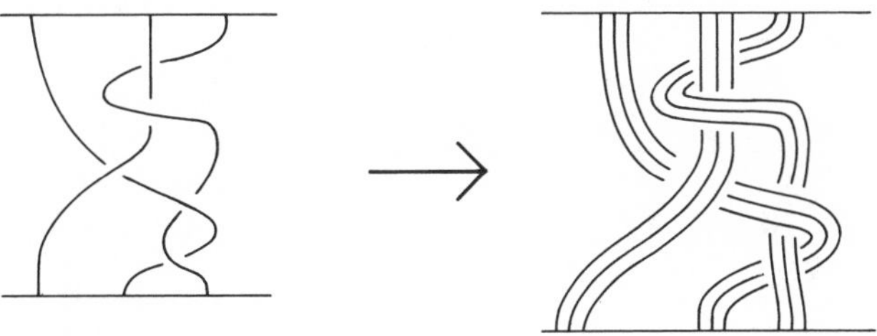

Figure 11

References

[B-R] A.O. Barut and R. Raczka, *Theory of group representations and applications*, World Scientific Publishing Co. Pte. Ltd., Singapore, 1986.

[B] J. Birman, *Braids, Links and Mapping Class Groups*, Ann. Math. Stud., vol. 82, Princeton University Press, Princeton, New Jersey, 1975.

[C-L-M] S. Cappell, R. Lee and E. Miller, *Invariants of 3-manifolds from conformal field theory, preprint.*

[J] V.F.R. Jones, *Index for subfactors*, Inventiones Math. **72** (198), 1–25.

[K] R. C. Kirby, *A calculus for framed links in S^3*, Inventiones Math. **45** (1978), 35–56.

[K-M] R. C. Kirby and P. Melvin, *On the 3-manifold invariants of Witten and Reshetikhin-Turaev for $sl(2, \mathbf{C})$, preprint.*

[K-S1] K.H. Ko and L. Smolinsky, *A Combinatorial matrix in 3-manifold theory*, Pacific J. Math., to appear.

[K-S2] ——, *The framed braid group and 3-manifolds, preprint.*

[L] W. B. R. Lickorish, *A representation of orientable, combinatorial 3-manifolds*, Ann. Math. **76** (1962), 531–540.

[L1] ——, *3-Manifold invariants from the combinatorics of the Jones polynomial*, Pacific J. Math., to appear.

[L2] ——, *3-Manifolds and the Temperley-Lieb algebra, preprint.*

[R-T] N. Yu. Reshetikhin and V.G. Turaev, *Invariants of 3-manifolds via link polynomials and quantum groups*, Inventiones Math., to appear.

[W] E. Witten, *Quantum field theory and the Jones polynomial*, Comm. Math. Phys. **121** (1989), 351–399.

On the decompositions of elliptic surfaces

Masaaki Ue

Freedman and Taylor [FT] shows that every smooth closed 1-connected 4-manifold whose intersection form is a direct sum of some quadratic forms has a decomposition along some homology 3-spheres such that the intersection form of each piece corresponds to a direct summand of the intersection form. But the decompositions are far from unique and seem difficult to obtain explicitely in general. The purpose of this paper is to give two kinds of decompositions of 1-connected elliptic surfaces along some Brieskorn homology spheres both of which correspond to the same splitting of the intersection forms and do not depend on the choices of the gluing maps. We will also construct involutions of the elliptic surfaces which leaves the decompositions of the first type invariant (Theorem 2). Furthermore these two kinds of decompositions contain the same 1-connected exotic pieces with second betti number 2 (which we will denote by $Z_{p,q}^k$ below). The same exotic manifolds have been obtained independently by Gompf [G] (and also by Viro [V] for the Dolgachev surfaces). In this paper we will describe their structures together with involutions acting on them (which will be extended to those on the whole spaces) and will discuss the extendability of the self-diffeomorphisms of their boundaries (§ 2, Proposition 3 and Corollary). Now we fix some notations. The intersection form of a 4-manifold M is denoted by $q(M)$. The quadratic form over $\mathbf{Z}$ corresponding to the dual graph in Fig. 1 in which each vertex has self-intersection number -2 is denoted by E_i. We will denote the compact 4-manifold obtained by the plumbing construction associated to E_i also by the same symbol E_i. The quadratic form over $\mathbf{Z}$ of rank 2 corresponding to the matrix $\begin{pmatrix} 0 & 1 \\ 1 & 0 \end{pmatrix}$ is denoted by H. The Brieskorn homology 3-sphere of type p, q, r is denoted by $\Sigma(p, q, r)$. The orientation of $\Sigma(p, q, r)$ will be that induced from the Brieskorn variety in the complex space.

Now our theorems are stated as follows.

Knots 90

E_i

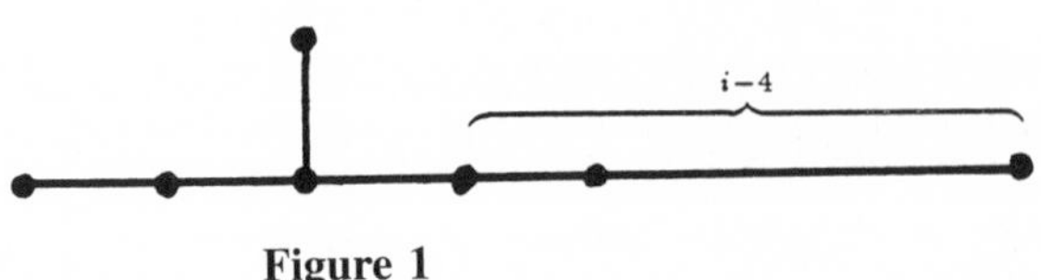

Figure 1

Theorem 1. *Let $S^k_{p,q}$ be a relatively minimal 1-connected elliptic surface with euler number $12k$ ($k \geq 1$) and with multiple fibers of type p and q ($p, q \geq 1$, $\gcd(p,q) = 1$). Then $S^k_{p,q}$ is decomposed in two ways as follows.*

(1) $S^k_{p,q} \cong \bigcup_{i=1}^{k} (E_8)_i \bigcup Y \bigcup Z^k_{p,q}$,
(2) $S^k_{p,q} \cong \bigcup_{i=1}^{k-1} (E_8)_i \bigcup Y' \bigcup E' \bigcup Z^k_{p,q}$ (in case $k \geq 2$).
Here $(E_8)_i$ is an E_8-plumbing with $\partial(E_8)_i = \Sigma(2,3,5)$, and the other pieces are 1-connected compact 4-manifolds satisfying the following conditions.

$$\partial Y = (k \text{ copies of } -\Sigma(2,3,5)) \cup \Sigma(2,3,6k-1), \quad q(Y) = \oplus^{2k-2} H$$
$$\partial Y' = (k-1 \text{ copies of } -\Sigma(2,3,5)) \cup \Sigma(2,3,6k-5), q(Y') = \oplus^{2k-2} H$$
$$\partial E' = -\Sigma(2,3,6k-5) \cup \Sigma(2,3,6k-1), \quad q(E') = E_8.$$

Finally $Z^k_{p,q}$ is a 1-connected 4-manifold with $\partial Z^k_{p,q} = -\Sigma(2,3,6k-1)$ and $q(Z^k_{p,q}) = H'$ where

$$H' = \begin{cases} H & \text{if } p, q \text{ are odd and } k \text{ is even} \\ (1) \oplus (-1) & \text{otherwise.} \end{cases}$$

(3) Both decompositions do not depend on the choices of the gluing maps on the boundaries of any pieces and correspond to the same splitting of $q(S^k_{p,q})$ of the form

$$q(S^k_{p,q}) = \oplus^k E_8 \oplus (\oplus^{2k-2} H) \oplus H'.$$

Remark 1. If $k = 1$ the decomposition of the first type is chosen so that $Y = \emptyset$ (cf. [Ma]). Also if $k = 1$ the decomposition of the second type degenerates to $E' \cup Z^1_{p,q} \cup$ (a 4-ball) where $\partial E' = \Sigma(2,3,5) \cup S^3$, $q(E') = E_8$.

Theorem 2. *There is an smooth involution τ on $S^k_{p,q}$ which acts both on the base and on the general fiber as a reflection and leaves each piece of the decomposition of $S^k_{p,q}$ of the first type in Theorem 1 invariant. The involution τ reverses the orientation of the maximal positive subspace H_2^+ of $H_2(S^k_{p,q}; \mathbf{R})$ (or the homology orientation of $S^k_{p,q}$ in Donaldson's sense [D]) if and only if k is odd. For the*

Donaldson invariant γ_d *of* $S^k_{p,q}$ *of degree* $d = 4c - 3k$ *for* $k \geq 2$ *we have* $\tau^* \gamma_d = (-1)^k \gamma_d$.

Remark 2. (1) The involution in Theorem 2 is equivalent to that constructed in [FKV] if $k = 1$. (2) It is easy to modify the argument in §1 to construct a similar involution of any (not necessarily 1-connected) elliptic surface with positive euler number. (3) It should be noted that according to Friedman and Morgan's theorem [FM2] there are no self-diffeomorphisms of $S^k_{p,q}$ which reverse the homology orientation if k is even, $k > 2$ (see [D] for the $K3$-surfaces).

§1. Constructions of involutions

An elliptic surface is a closed complex surface with projection to a nonsingular complex curve whose general fiber is an elliptic curve. Hereafter we will only consider the 1-connected and relatively minimal ones (that is, those which do not contain (-1)-curves in any fiber). Hence the base curve is $\mathbf{CP}^1$ and the number of the multiple fibers are at most 2. We can assume that any multiple fiber is a multiple torus ([M]). Put $k = 1 + p_g$ where p_g is the gemetric genus of the elliptic surface. Then its euler number is $12k$ ($k \geq 1$). It is known that once k and the pair of multiplicities (p, q) of the multiple tori are fixed then the diffeomorphism type of the corresponding elliptic surface is uniquely determined. Therefore we denote such a surface by $S^k_{p,q}$. Note that $\gcd(p, q) = 1$ if and only if $S^k_{p,q}$ is 1-connected. Here we assume that $p,q \geq 1$. A multiple fiber of multiplicity 1 means a regular fiber. Now we start with $S^k_{1,1}$ (with no multiple fibers). Then as is well known we can assume that the elliptic fibration $\pi\colon S^k_{1,1} \to \mathbf{CP}^1$ has a smooth cross section $B \subset S^k_{1,1}$ with self-intersection number $-k$ (by adjunction formula). Since the diffeomorphism type of $S^k_{1,1}$ does not depend on the choice of the singular fibers we can assume that $S^k_{1,1}$ contains exactly k copies of fibers of type II^* and just $2k$ copies of singular fibers of type I_1 (a rational curve with one node). Here the fiber of type II^* (or $\tilde{E}_8$) is the core of E_9 in our notation ([K]). To describe the fibers of type I_1 in $S^k_{p,q}$ fix a pair of curves ℓ, h in the general fiber $T^2 = \mathbf{R}^2/\mathbf{Z}^2$ defined by $\ell = \mathbf{R} \times */\mathbf{Z}^2$ and $h = * \times \mathbf{R}/\mathbf{Z}^2$. Let I_1^+ (resp. I_1^-) be the regular neighborhood of the fiber of type I_1 which has a vanishing cycle attached along h (resp. ℓ). Decompose $D^2 = \{z \in \mathbf{C} \mid |z| \leq 1\}$ as $D^2 = D^+ \cup D^-$ where $D^+ = \{z \in D^2 \mid Re\, z \geq 0\}$ and $D^- = \{z \in D^2 \mid Re\, z \leq 0\}$. Then I_1^+ and I_1^- are diffeomorphic to $D^+ \times T^2 \cup h_1$ and $D^- \times T^2 \cup h_2$ respectively where h_1 (resp. h_2) is a 2-handle attached along a curve parallel to h (resp. ℓ) in $\{1\} \times T^2 \subset \partial D^+ \times T^2$ (resp. $\{-1\} \times T^2 \subset \partial D^- \times T^2$) with framing -1. If we put $m^\pm = \partial D^\pm$, then $m^{+}(\ell,h)(m^{+})^{-1} = (\ell,h)\begin{pmatrix}1 & 0\\ 1 & 1\end{pmatrix}$, $m^{-}(\ell,h)(m^{-})^{-1} = (\ell,h)\begin{pmatrix}1 & -1\\ 0 & 1\end{pmatrix}$ in $\pi_1(\partial I_1^{\pm})$ where the above matrices correspond to the monodromies of $\partial I_1^{\pm}$ as T^2-bundles over S^1. Furthermore I_1^{+} and I_1^{-} are patched together to form

$$Z^0 = I_1^{+} \cup I_1^{-} = D^2 \times T^2 \cup h_1 \cup h_2.$$

Note that Z_0 is diffeomorphic to the regular neighborhood of the fiber of type II (a rational curve with one cusp) (see Fig. 2 and [M]). We can assume that $S_{1,1}^{k}$ contains just k-copies of Z^0.

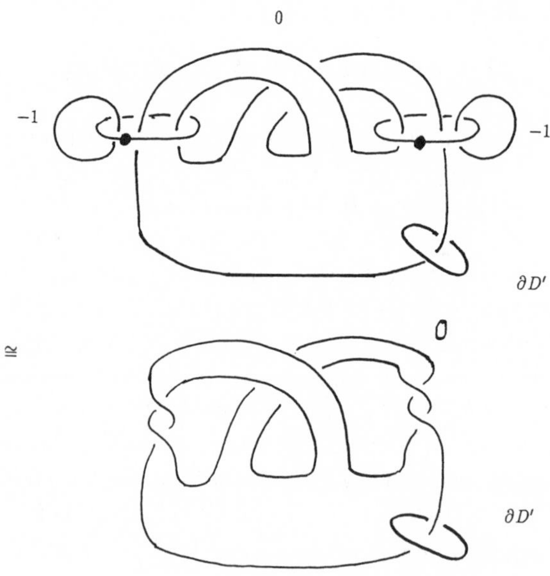

Figure 2

Now we will describe the involution ι acting on Z^0. Let ι_0 be the involution of $D^2 \times T^2$ defined by $\iota_0(z,w,u) = (\overline{z},w,\overline{u})$ for $z \in D^2 \subset \mathbf{C}$, $w, u \in S^1 \subset \mathbf{C}$. Put $\partial_0 h_i = h_i \cap \partial D^2 \times T^2$. If $\partial D^2 \times T^2 \cong T^3$ is represented by a cube with corresponding faces identified then $\iota_0|_{\partial D^2 \times T^2}$ is induced by the rotation through angle π in the axis A in Fig. 3. We can assume that $\partial_0 h_1$ (resp. $\partial_0 h_2$) is attached to an ι_0-invariant tubular neighborhood $\partial_0 h_1'$ (resp. $\partial_0 h_2'$) of $\{1\} \times \{-1\} \times S^1$ (resp. $\{-1\} \times S^1 \times \{1\}$) in $\partial D^2 \times T^2$. Furthermore we can take $D' = D^2 \times * \times \{-1\} \subset D^2 \times T^2$ (disjoint from $\partial_0 h_i$) as a cross section for the Lefshetz fibration for Z^0 over D^2 (cf. Fig. 2). We can also take an ι_0-invariant tubular neighborhood $N(D') \cong D^2 \times D^2$ of D' in $D^2 \times T^2$. Moreover $\partial_0 h_i'$'s are identified with $D^2 \times S^1$

so that $\iota_i = \iota_0|_{\partial_0 h'_i}$ satisfies

$$\iota_1(z,w) = (\overline{z}, \overline{w}), \quad \iota_2(z,w) = (-z, w)$$

for $z \in D^2$, $w \in S^1$. On the other hand $h_i \cong D^2 \times D^2$ is attached via the identification map ϕ between $\partial_0 h_i \cong D^2 \times S^1$ and $\partial_0 h'_i \cong D^2 \times S^1$ defined by $\phi(z,w) = (zw^{-1}, w)$. Since ϕ commutes with ι_1 and ι_2 the involutions ι'_i of h_i defined by $\iota'_1(z,w) = (\overline{z}, \overline{w})$, $\iota'_2(z,w) = (-z,w)$ for $z, w \in D^2$ and ι_0 induce the involution ι of Z^0.

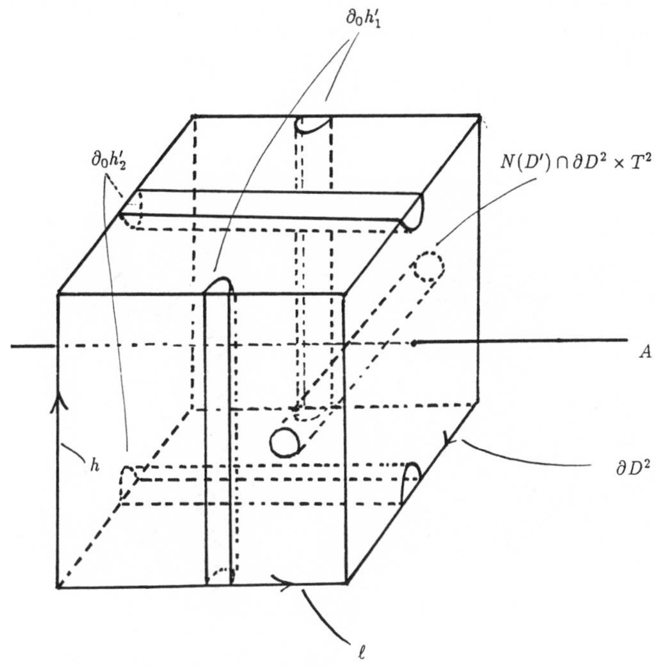

Figure 3

Let $Z^k_{1,1} \subset S^k_{1,1}$ be the union of Z^0 and the regular neighborhood $N(B)$ of the cross section B. Then $Z^k_{1,1} \cong Z^0 \cup h_3$ where h_3 is a 2-handle attached along $m_0 = \partial D'$ in $\partial(D^2 \times T^2)$ with framing $-k$. Then we have

Proposition 1. $\partial Z^k_{1,1} \cong -\Sigma(2,3,6k-1)$.

Proof. $\partial Z^k_{1,1}$ is represented by a framed link in Fig. 2 and is the manifold obtained by the $1/k$- surgery along the $(2,3)$-torus knot which is $-\Sigma(2,3,6k-1)$.

Put $\partial_0 I_1^{\pm} = \partial Z^0 \cap I_1^{\pm}$. Then ι leaves $\partial_0 I_1^{\pm}$ invariant. Furthermore we can see that $(\partial_0 I_1^+/\iota, Fix\,\iota/\iota)$ is obtained from $(D^2 \times S^1, L_0)$ where L_0 consists of two circles parallel to $(* \times S^1)$ by $-1/2$-disk twist among two components of L_0. Likewise $(\partial_0 I_1^-/\iota, Fix\,\iota/\iota)$ is obtained from $(D^2 \times S^1, L_0)$ by performing -2-surgery along one component of L_0 (the other one is left as $Fix\,\iota \cap \partial_0 I_1^-/\iota$). Furthermore $(Z^0/\iota, Fix\,\iota/\iota)$ is a union of $(D^2 \times T^2/\iota_0, Fix\,\iota_0/\iota_0) \cong (B^3 \times S^1, A_1 \cup A_2)$ (where A_1 and A_2 are annuli), $(h_1/\iota_1, Fix\,\iota_1/\iota_1) \cong (D^2 \times D^2, I \times I)$ (where $I = \{z \in D^2 \mid Im\, z = 0\}$) and $(h_2/\iota_2, Fix\,\iota_2/\iota_2) \cong (D^2 \times D^2, D^2 \times \{0\})$ such that $Z^0/\iota \cong D^4$ and $Fix\,\iota/\iota$ consists of $A_1 \cup A_2$, the twisted band connecting the one components of ∂A_1 and ∂A_2, and the core of 2-handle $(Fix\,\iota_2/\iota_2)$ attached along the other component of ∂A_2 with framing -2. The band b in Fig. 4 connecting two components of $(Fix\,\iota \cap \partial Z^0)/\iota$ is the quotient of $\partial D' \times I \subset \partial D' \times D^2 \cong N(D') \cap \partial Z^0$ by ι where $I = \{z \in D^2 \mid Im\, z = 0\}$.

$(Z^0/\iota, Fix\,\iota \cap \partial Z^0/\iota)$

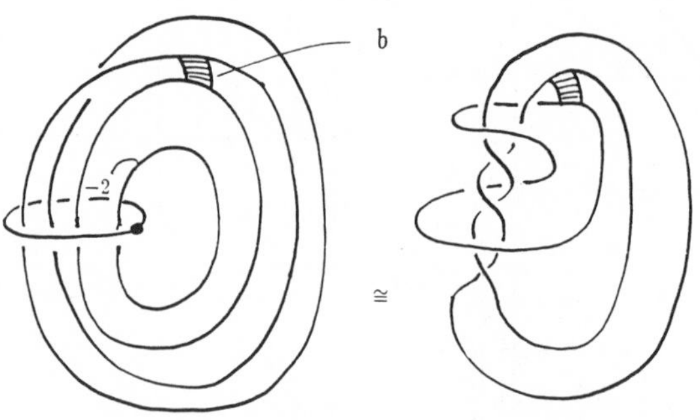

Figure 4

Next consider E_9 which is represented by the framed link in Fig. 5. The rational elliptic surface $S^1_{1,1}$ is the union of E_9 and Z_0. Decompose the cross section B of $S^1_{1,1}$ as $B = D' \cup D''$ where $D' = B \cap Z_0$ and $D'' = B \cap E_9$ (we can assume that D' coincides with that constructed above). Let $N(D')$ (resp. $N(D'')$) $\cong D^2 \times D^2$ be the tubular neighborhood of D' (resp. D'') in Z_0 (resp. E_9). Thus $S^1_{1,1} - (Z^0 \cup N(D'')) = E_9 - N(D'')$ whose boundary is $\Sigma(2,3,5)$ by Proposition 1. Since we can assume that D'' intersects the core of E_9 transversely at one point other than the singularities of the divisors in E_9 we can see that $D'' \cap$ the E_9-fiber is in the divisor corresponding to the component on the right end of the framed link in Fig. 5 in which $\partial D''$ is represented by a circle C_0. (This can be proved by the fact that 0-surgery of ∂E_9 along $\partial D''$ must yield the homology 3-sphere $\Sigma(2,3,5)$.) On the other hand the framed link together with C_0 in Fig. 5 is arranged so that each component is symmetric with respect to the axis A'. Thus the rotation through angle π in the axis A' defines the involution ι' on E_9 such that $E_9/\iota' \cong D^4$ and $Fix\iota'$ is the surface with nine bands with -1 twists whose boundaries are represented by the second picture in Fig. 5 (see [HKK]). As in ι, we can arrange $N(D'') \cong D^2 \times D^2$ so that it is invariant by ι' which acts on $N(D'')$ by $\iota'(z,w) = (\overline{z}, \overline{w})$ for $z, w \in D^2$. The band b' in Fig. 5 is the quotient of $\partial D'' \times I \subset \partial D'' \times D^2 \cong N(D'') \cap \partial E_9$ by ι' where $I = \{z \in D^2 \mid Im\, z = 0\}$.

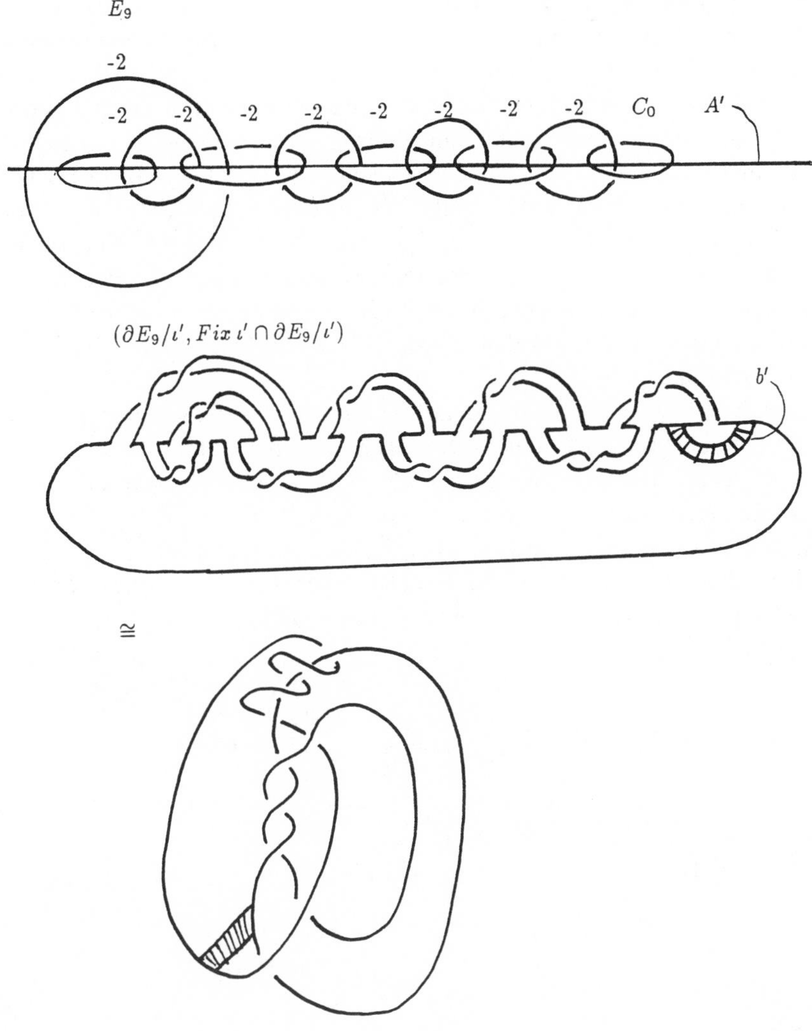

Figure 5

Then we can see that there is a diffeomorphism ϕ from $(\partial(E_9/\iota'), L_2)$ to $(-\partial Z_0/\iota, L_1)$ with $\phi(\partial D''/\iota') = \partial D'/\iota$ where $L_1 = (Fix\,\iota \cap \partial Z^0)/\iota$ and $L_2 = (Fix\,\iota' \cap \partial E_9)/\iota'$ (Fig. 6). Then as usual ϕ is lifted to the diffeomorphism $\tilde{\phi}\colon \partial E_9 \to -\partial Z_0$ such that $\iota' = \tilde{\phi}^{-1}\iota\tilde{\phi}$. This gives the equivalence of ι and ι' via some identifications of ∂E_9 and $-\partial Z_0$ with the T^2-bundle M over S^1 with monodromy $\begin{pmatrix} 0 & 1 \\ -1 & 1 \end{pmatrix}$. It should be noted that the orientation preserving involution on M which reverses the orientation of the base is unique up to equivalence ([S]). In fact the above identification is chosen so that $\tilde{\phi}_* = id$ on $\pi_1 M$ and $\tilde{\phi}$ is isotopic to the identity (since M is Haken). Thus $\tilde{\phi}$ extends to the self diffeomorphism Φ of E_9 and $\Phi \cup id|_{Z^0}$ maps $E_9 \cup_{id} Z^0$ diffeomorphically to $E_9 \cup_{\phi} Z^0$ and the involution $\Phi^{-1}\iota'\Phi \cup \iota$ on $E_9 \cup_{id} Z^0$ is equivalent to $\iota' \cup \iota$ on $E_9 \cup_{\phi} Z^0$. Here we note that if the orientations of L_1 and L_2 are induced from those of $Fix\iota/\iota$ and $Fix\iota'/\iota'$ respectively then ϕ reverses the orientation of just one component of L_1 and hence the band b' is mapped by ϕ to the band obtained from b by adding extra one half twist. It follows that if $\partial_0 N(D'') = N(D'') \cap \partial E_9$ and $-\partial_0 N(D') = -N(D') \cap \partial Z^0$ are identified with $S^1 \times D^2$ via $N(D'')$ and $-N(D')$ then $\tilde{\phi}\colon \partial_0 N(D'') \to -\partial_0 N(D')$ is identified with the map $(z, w) \to (zw^1, w)$ for $z \in D^2$, $w \in S^1$. Thus the global involution $\tau = \iota \cup \iota'$ on $S^1_{1,1}$ is constructed so that τ leaves the decomposition $E_9 \cup Z_0$ and the tubular neighborhood $N(B) = N(D') \cup N(D'')$ of the cross section B invariant. It follows that $Z^1_{1,1} = Z^0 \cup N(D'')$ and $E_9 - N(D') \cong E_8$ are also τ-invariant.

In case $k = 1$ we can perform logarithmic transformations within $D^2 \times T^2 \subset D^2 \times T^2 \cup h_1 \cup h_2 = Z_0$ to obtain the surfaces with multiple fibers as follows. Take small two 2-disks D_1 and D_2 in $D^2 \subset \mathbf{C}$ which are invariant under the complex conjugation $z \to \bar{z}$. (We can take them in $D^+ \subset D^+ \cup D^- = D^2$.) Put $D_0 = D^2 - (D_1 \cup D_2)$ and $m_0 = \partial D^2$, $m_1 = \partial D_1$, $m_2 = \partial D_2$ which are oriented so that $m_0 = m_1 + m_2$ in $H_1(D_0)$. Then $S^1_{p_1,p_2}$ is a union of E_9 and $Z^0_{p_1,p_2}$ which is defined as follows. Take a copy $(D^2 \times T^2)_i$ of $D^2 \times T^2$ and put $M_i = (\partial D^2 \times * \times *)_i$, $L_i = (* \times S^1 \times *)_i$, $H_i = (* \times * \times S^1)_i$ in $(\partial D^2 \times T^2)_i$. Then

$$Z^0_{p_1,p_2} = D_0 \times T^2 \cup_{\phi_1} (D^2 \times T^2)_1 \cup_{\phi_2} (D^2 \times T^2)_2 \cup h_1 \cup h_2$$

where h_i's are attached in the same way as in Z^0 and the identification map ϕ_i between $m_i \times T^2$ and $(\partial D^2 \times T^2)_i$ is defined by the following matrix with respect to the framings (m_i, ℓ, h) on $\partial D_i \times T^2$ and (M_i, L_i, H_i) on $\partial(D^2 \times T^2)_i$.

$$(\ell, h, m_i) \begin{pmatrix} 1 & 0 & 0 \\ 0 & 0 & 1 \\ 0 & -1 & p_i \end{pmatrix} = (L_i, H_i, M_i) \tag{1}$$

Remark 3. The choice of ϕ_i which gives the multiple fiber of type p_i is not unique. But the diffeomorphism type of $Z^0_{p_1,p_2}$ depends only on p_1 and p_2. To

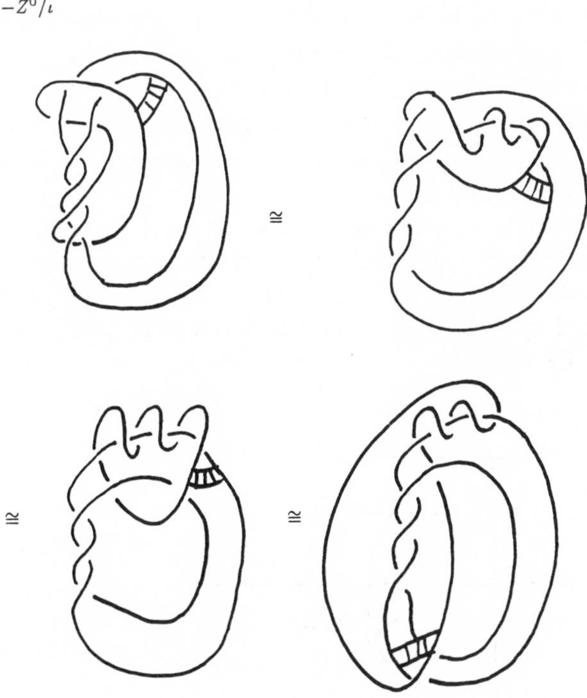

Figure 6

prove this note that Z^0 contains the fibers of type I_1 with monodromy $\begin{pmatrix} 1 & -1 \\ 0 & 1 \end{pmatrix}$ and $\begin{pmatrix} 1 & 0 \\ 1 & 1 \end{pmatrix}$. These matrices (and their inverses) generate $SL_2\mathbf{Z}$ and then the proofs of Lemmas in [U] and of Lemma 7 in [M], Part II can be modified so that they are applicable to $Z^0_{p_1,p_2}$. This leads to the above conclusion. In particular $Z^0_{1,1}$ is identified with Z^0.

Since ϕ_i commutes with the involution defined by $(z, w, u) \to (\overline{z}, w, \overline{u})$ on $\partial D^2 \times S^1 \times S^1$ we can modify τ to obtain the new involution $\tilde{\tau}$ of $Z^0_{p_1,p_2}$ such that $\tilde{\tau} = \tau$ on the complement of $\cup(D^2 \times T^2)_i$ and $\tilde{\tau}(z, w, u) = (\overline{z}, w, \overline{u})$ on $(D^2 \times T^2)_i$. Thus $\tilde{\tau}$ naturally extends to the involution of $S^1_{p_1,p_2}$ (which is also denoted by τ for brevity). It follows that $S^1_{p_1,p_2}$ is decomposed into the union of $Z^1_{p_1,p_2} = Z^0_{p_1,p_2} \cup N(D'')$ and $E_8 = E_9 - N(D'')$ both of which are τ-invariant. These observations and Proposition 2 in §2 prove Theorems 1 and 2 in case $k = 1$ since every self-diffeomorphism of $\Sigma(2, 3, 5)$ is isotopic to the identity ([BO]).

Remark 4. Decompose $S^1_{1,1}$ as $S^1_{1,1} = (E_9 \cup N(D')) \cup (Z^0 - N(D'))$ which is also preserved by τ. Then $E_9 \cup N(D')$ is obtained by -1 surgery along the curve C_0 in Fig. 5 and is diffeomorphic to $\mathbf{CP}^2 \sharp 9\overline{\mathbf{CP}}^2 - D^4$. Moreover $Z^0 - N(D') \cong D^4$ and $(D^4/\tau, Fix\,\tau/\tau) \cong (D^4, D^2)$. To get $Fix\,\tau/\tau$ in D^4/τ remove the band from the annulus A_1 in $Fix\,\iota/\iota \subset Z^0/\iota$ described above. It is not difficult to see that $Fix\,\tau/\tau \cong D^2$ can be isotoped to a 2-disk on ∂D^4 and hence is unknotted in D^4. On the other hand we can see that $Fix\,\tau|_{E_{\hat{q}} \cup N(D')}/\tau$ is obtained from $Fix\,\iota'|_{E_9}/\iota'$ by adding a band with $-1/2$ twist along C_0/τ in Fig. 5. It follows that τ on $S^1_{1,1}$ (before the logarithmic transformations) is equivalent to the complex conjugation (slide the bands on $Fix\,\tau \cap (E_9 \cup N(D'))/\tau$ to see this) and hence our τ is obtained by "real logarithmic transformations" and is equivalent to that constructed in [FKV].

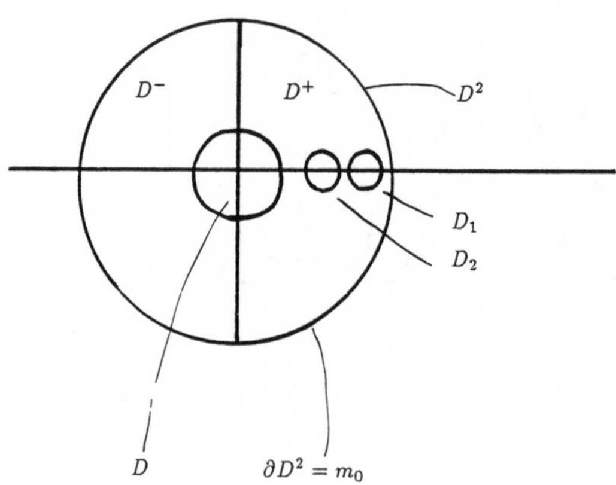

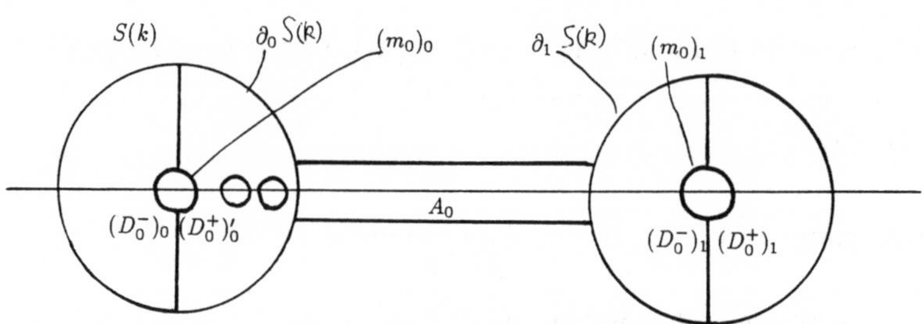

Figure 7

Now we go on to the cases with $k \geq 2$. First take a small 2-disk D in $D_0 \subset D^2 = D^+ \cup D^-$ which is divided into two pieces $(D \cap D^+) \cup (D \cap D^-)$

and is invariant by the reflection $z \to \overline{z}$ (Fig. 7). Remove $D \times T^2$ from $Z^0_{p_1,p_2}$ to get $T^0_{p_1,p_2} = Z^0_{p_1,p_2} - D \times T^2$ and $S^0_{p_1,p_2} = S^1_{p_1,p_2} - D^2 \times T^2 = T^0_{p_1,p_2} \cup E_9$ together with the involution $\tau^0 = \tau|_{S^0_{p_1,p_2}}$ which acts on $\partial S^0_{p_1,p_2} \cong \partial D^2 \times T^2$ by $\tau^0(z,w,u) = (\overline{z}, w, \overline{u})$ for $z,w,u \in S^1$. Likewise we get $T^0 = Z^0 - D \times T^2$ and $S^0 = T^0 \cup E_9$ together with $\tau^0 = \tau|_{S^0}$. Here $D \times T^2$ is (up to isotopy) the regular neighborhood of the general fiber of the Lefshetz fibration for $Z^0_{p_1,p_2}$. Also we have the cross section $D'_0 \cup D''$ of S^0 where $D'_0 = D' - D \times T^2$ and the τ^0-invariant tubular neighborhood $N(D'_0) \cup N(D'')$ of $D'_0 \cup D''$. Let $S(k)$ be the k-th punctured 2-sphere whose each boundary component $\partial_i S(k)$ $(i = 0, \cdots, k-1)$ and $S(k)$ itself are preserved by the reflection $\overline{\iota}$ of $S(k)$. Define the involution ι'' of $S(k) \times T^2$ by $\iota''(z,w,u) = (\overline{\iota}(z), w, \overline{u})$ for $z \in S(k)$, $w, u \in S^1$. Also take $k-1$ copies (S^0_i, τ^0_i) $(i = 1, \ldots, k-1)$ of (S^0, τ^0). Then since $\tau^0_i = \iota''$ on $\partial_i S(k) \times T^2 = -\partial S^0_i$ and $\tau^0 = \iota''$ on $\partial_0 S(k) \times T^2 = -\partial S^0_{p_1,p_2}$ these involutions induce the involution of

$$S^k_{p_1,p_2} = S(k) \times T^2 \cup_{\partial_0} S^0_{p_1,p_2} \cup_{\partial_1} (S^0)_1 \ldots \cup_{\partial_{k-1}} (S^0)_{k-1} \tag{2}$$

where S^0_i (or $S^0_{p_1,p_2}$) and $S(k) \times T^2$ are glued along $\partial_i = \partial_i S(k) \times T^2$ by the identity. (If $(p_1, p_2) = (1,1)$ replace $S^0_{p_1,p_2}$ by S^0.) Hereafter for the sake of brevity the involution of $S^k_{p_1,p_2}$ constructed from the above pieces is also denoted by τ since there is no danger of confusion. The restriction of τ to any subspace of $S^k_{p_1,p_2}$ will also be denoted by the same symbol τ. Note that we can take the τ-invariant tubular neighborhood $N(S(k))$ of the cross section $S(k) \times (*) \times (*) \subset S(k) \times T^2$ so that in case of $S^k_{1,1}$, $N(S(k))$ together with the copies $N(D'_0)_i \cup N(D'')_i$ of $N(D'_0) \cup N(D'')$ $(i = 0, \ldots, k-1)$ forms the τ-invariant tubular neighborhood $N(B)$ of the cross section B. Here $N(D'')_i$ and $N(D'_0)_i$ are glued so that the framing is twisted by -1 via the identification map constructed above. This reflects the fact that the self-intersection number of B is $-k$. We will see in the next section that this involution has the properties required in Theorem 2.

§2. Constructions of decompositions

Consider the construction of $(S^k_{p_1,p_2}, \tau)$ in §1. In Fig. 7 $D^2 - D$ is decomposed as $D^2 - D = D^+_0 \cup D^-_0$ where $D^\pm_0 = D^\pm - D$. Put $(D^+_0)' = D^+ \cap D_0 - D$. Let $(D^\pm_0)_i$, $(D^+_0)'_i$ be the i-th copies of $D^\pm_0$ and $(D^+_0)'$ $(i = 0, \ldots, k-1)$. Note that we can define the cross sections $(D')^\pm_0 = D' \cap D^\pm_0 \times T^2$ in D' and their τ-invariant tubular neighborhoods $N((D')^\pm_0)$ as well as their k copies (denoted by subscripts i). On the other hand take an annulus A_0 in $S(k)$ connecting $\partial_0 S(k)$ and $\partial_1 S(k)$ so that it is invariant by the reflection $\overline{\iota}$ (Fig. 7). Consider $(I^+_1)'_0 = (D^+_0)'_0 \times T^2 \cup h_1 \cup (D^2 \times T^2)_1 \cup (D^2 \times T^2)_2$ in $S^0_{p_1,p_2}$ where h_1 and $(D^2 \times T^2)_i$ are attached as in $Z^0_{p_1,p_2}$ and $(I_1^-)_1^0 = (D_0^-)_1 \times T^2 \cup h_2$ in $(S^0)_1$ where h_2 is attached as in Z^0. Put

$$(Z^0_{p_1,p_2})' = (I_1^+)_0' \cup A_0 \times T^2 \cup (I_1^-)_1^0 \subset S^k_{p_1,p_2}.$$

Then $(Z_0^{p_1,p_2})'$ (which is τ-invariant) is naturally diffeomorphic to $Z^0_{p_1,p_2}$ and the action of τ on $(Z^0_{p_1,p_2})'$ is equivalent to that on $Z^0_{p_1,p_2}$ constructed in §1. In case $(p_1,p_2) = (1,1)$ consider $Z^0_{1,1} = (I_1^+)_0^0 \cup A_0 \times T^2 \cup (I_1^-)_1^0$ where $(I_1^+)_0^0 = (D_0^+)_0 \times T^2 \cup h_1$. Likewise $(Z^0_{1,1}, \tau)$ is equivalent to (Z^0, ι) constructed in §1. Next take a τ-invariant tubular neighborhood $N(S(k) - A_0)$ of $(S(k)-A_0) \times * * \times * \subset S(k) \times T^2$ so that the union of $N(S(k) - A_0)$, $N((D')_0^-)_i$ $(i \neq 1)$, $N((D')_0^+)_i$ $(i \neq 0)$, and $N(D'')_i$ (in $(S^0)_i$ or $S^0_{p_1,p_2}$) form the τ-invariant 2-handle h_3 attached along the cross sectional curve on $\partial(Z^0_{p_1,p_2})'$ with framing $-k$. Then we can see that $(Z^k_{p_1,p_2})' = (Z^0_{p_1,p_2})' \cup h_3$ is naturally identified with

$$\begin{aligned} Z^k_{p_1,p_2} &= Z^0_{p_1,p_2} \cup h_3 \\ &= D_0 \times T^2 \cup (D^2 \times T^2)_1 \cup (D^2 \times T^2)_2 \cup h_1 \cup h_2 \cup h_3 \end{aligned} \tag{3}$$

where h_3 on the right hand side is attached along m_0 in $\partial Z^0_{p_1,p_2}$ (in §1) with framing $-k$. On the other hand τ on $Z^0_{p_1,p_2}$ constructed in §1 can be extended to the involution $\tilde{\tau}$ on $Z^k_{p_1,p_2}$ by putting $\tilde{\tau}(z,w) = (\overline{z}, \overline{w})$ for $(z,w) \in D^2 \times D^2 \cong h_3$ since this map commutes with the identification map between $\partial_0 h_3 = h_3 \cap \partial Z^0_{p_1,p_2}$ and the tubular neighborhood of m_0. It is easy to see that τ on $(Z^k_{p_1,p_2})'$ is equivalent to $\tilde{\tau}$ on $Z^k_{p_1,p_2}$. Hence we will identify them. Next take an τ-invariant regular neighborhood $(E_8)_i$ of the divisors in $(E_9)_i - N(D'')_i$ which is diffeomorphic to E_8 and lies within the interior of $(E_9)_i - N(D'')_i$. This is possible since $(E_8)_i$ is represented by the framed link obtained from that in Fig. 5 by removing the component which links C_0 and the action of τ on $(E_8)_i$ is also induced by the rotation through angle π in the same axis A'. Finally put $Y = S^k_{p_1,p_2} - \cup(E_8)_i - (Z^k_{p_1,p_2})'$. Thus we get the τ-invariant decomposition of $S^k_{p_1,p_2}$. Now we will show that this decomposition is the first one required in Theorem 1. Consider $(Z^k_{p_1,p_2})'$ which is identified with $Z^k_{p_1,p_2}$. We use the decomposition (3) and the notations introduced in §1. Fix the integers q_1, q_2 satisfying $q_1 p_1 + q_2 p_2 = 1$ (note that $\gcd(p_1,p_2) = 1$). Since $M_i = p_i m_i + h$ by (1) in §1 we see that $p_2 M_1 + p_1 M_2$ in $(\partial D^2 \times T^2)_1 \cup (\partial D^2 \times T^2)_2$ is homologous in $D_0 \times T^2$ to $p_1 p_2 (m_1 + m_2) + (p_1 + p_2) h$ in $m_0 \times T^2$. Take a 2-chain y_0 in $D_0 \times T^2$ realizing this homologous relation. Since $m_0 = m_1 + m_2$ and the 2-handle h_3 is attached along m_0 we can take the 2-cycle y of the form

$$y = p_2 D_1 + p_1 D_2 + y_0 + (p_1 + p_2) h_1^0 + p_1 p_2 h_3^0$$

where D_i is the meridian disk of $(D^2 \times T^2)_i$, and h_j^0 is the core of the 2-handle h_j. Also define the 2-cycle x by

$$x = (q_2 m_1 + q_1 m_2) \times \ell.$$

Proposition 2. *$H_2(Z^k_{p_1,p_2} : \mathbf{Z}) \cong \mathbf{Z}^2$ which is generated by x and y. The intersection form of $Z^k_{p_1,p_2}$ is determined by*

$$x \cdot x = 0, \quad x \cdot y = 1, \quad y \cdot y = -k(p_1 p_2)^2 - (p_1 + p_2)^2.$$

Proof. A computation using Mayer-Vietoris exact sequences shows the first assertion. Clearly we have $x \cdot x = 0$. We can push $y \cap (Z^k_{p_1,p_2} - \cup h_i)$ along the ℓ-direction. Since h_2 and h_3 are attached with framing -1 and $-k$ respectively the contributions of the self-intersections of h_j^0's to $y \cdot y$ show the claim for $y \cdot y$. Similar arguments show that $x \cdot y = q_2 p_2 + q_1 p_1 = 1$.

Since $\gcd(p_1, p_2) = 1$ we can see that $y \cdot y$ is even if and only if k is even and p, q are odd. This proves the claim for the intersection form of $Z^k_{p_1,p_2}$ in Theorem 1. Note that $x = q_2 m_1 \times \ell + q_1 m_2 \times \ell = q_2 L_1 \times H_1 + q_1 L_2 \times H_2$ by (1). This implies that $p_1 p_2 x$ is homologous to $p_2 q_2 \ell \times h + p_1 q_1 \ell \times h = \ell \times h$. It follows that x is a rational multiple of the canonical divisor of $S^k_{p_1,p_2}$.

Proposition 3. *Any self-diffeomorphism of $\partial Z^k_{p,q} = -\Sigma(2,3,6k-1)$ extends to that of $Z^k_{p,q}$.*

Proof. First we note that $\partial Z^k_{p,q} = \partial Z^k_{1,1} \cong -\Sigma(2,3,6k-1)$ by Proposition 1. Next we will show that $\tilde{\tau}$ on $\partial Z^k_{p_1,p_2}$ constructed above is not isotopic to the identity. In Fig. 8 T^3 is represented by the cube so that the coresponding faces are identified to form $T^3 - \bigcup_{i=1}^3 \partial_0 h_i$ (which is diffeomorphic to the Borromean rings complement). Choose the vertices O_i (all O_i's are identified) as the base point and deform $m = m_0$, ℓ, h as in Fig. 8. Then the oriented meridian μ_i and the longitude λ_i for ∂h_i defined via the path γ_i in Fig. 8 are represented as

$$\begin{aligned}
\mu_1 &= [m, \ell], \quad \lambda_1 = h \\
\mu_2 &= [h, m], \quad \lambda_2 = m \ell m^{-1} \\
\mu_3 &= [\ell, h], \quad \lambda_3 = [m, \ell]^{-1} h m h^{-1}.
\end{aligned}$$

Here $[x, y] = x y x^{-1} y^{-1}$. Thus $\pi_1(\partial Z^k_{p_1,p_2})$ has the following representation.

the generators: ℓ, h, m

the relations: $h = [m, \ell]$ (R_1)

$$m\ell m^{-1} = [h, m] \quad (R_2)$$

$$[m, \ell]^{-1} hmh^{-1} = [l, h]^k \quad (R_3)$$

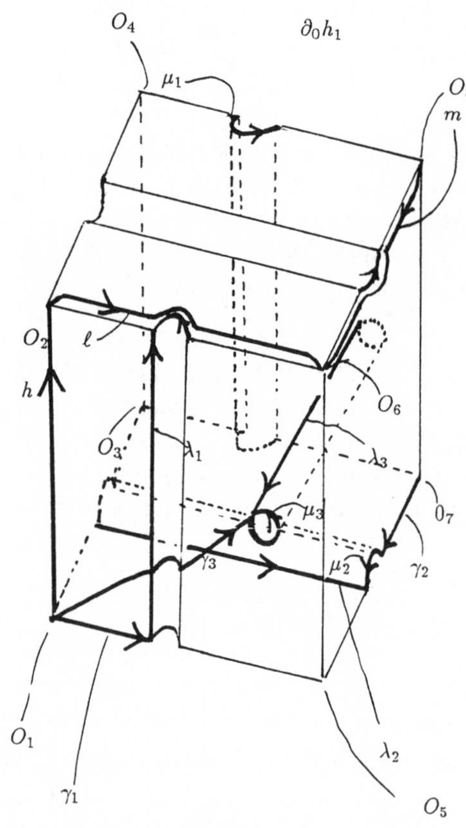

Figure 8

These relations come from the Dehn surgery along the three curves represented by ${\mu_2}^{-1}\lambda_2 = 1$, ${\mu_1}^{-1}\lambda_1 = 1$ and ${\mu_3}^{-k}\lambda_3 = 1$ respectively. (In $\pi_1(T^3 - \cup\partial_0 h_i)$ there are the relations $[\lambda_i, \mu_i] = 1$ coming from the Wirtinger representation of the Borromean rings complement. But they are absorbed in $R_1 \sim R_3$.) (R_2) and (R_3) can be replaced by

$$\ell = [m^{-1}, h] \quad (R_2')$$

$$\ell m = [\ell, h]^k \quad (R_3')$$

since $[m, \ell] = [h, m]\ell^{-1}$.

Here we note that by (R_1) and (R_2')

$$\ell m = h^{-1}m\ell = h^{-1}m[m^{-1},h] = mh^{-1}. \qquad (R_4)$$

Now we deduce some relations from R_1, R_2', R_3' and R_4. First using

$$m = \ell^{-1}[\ell,h]^k = [\ell,h]^k h \qquad (R_5)$$

or equivalently

$$m^{-1} = h^{-1}[\ell,h]^{-k} = [\ell,h]^{-k}\ell$$

we get

$$[\ell,h]^k h[\ell,h]^{-k} = \ell^{-1}. \qquad (R_6)$$

Next from R_2' we deduce

$$\ell = m^{-1}hmh^{-1} = h^{-1}[\ell,h]^{-k}h[\ell,h]^k$$

and hence using R_6 we have

$$[\ell,h]^k\ell[\ell,h]^{-k} = \ell h. \qquad (R_7)$$

Successive applications of R_6 and R_7 show that

$$[\ell,h]^{6k}\ell[\ell,h]^{-6k} = [\ell,h]\ell[\ell,h]^{-1}$$

and

$$[\ell,h]^{6k}h[\ell,h]^{-6k} = [\ell,h]h[\ell,h]^{-1}.$$

Thus together with R_5 we can see that $[\ell,h]^{6k-1}$ lies in the center of $\pi = \pi_1(\partial Z^k_{p_1,p_2})$. We note that if $k \geq 2$, then π has no torsion since $\partial Z^k_{p_1,p_2} \cong -\Sigma(2,3,6k-1)$ is aspherical (its universal cover is $\mathbf{H}\times\mathbf{R}^2$). Therefore if $[\ell,h]^{6k-1} = 1$ then $[\ell,h] = 1$. But this would imply by R_1, R_2' and R_3' that $\pi = 1$ which is a contradiction. Hence $[\ell,h]^{6k-1}$ is not a torsion if $k \geq 2$. On the other hand $\tilde{\tau}_\sharp$ on π induced by $\tilde{\tau}$ maps (m,ℓ,h) to $(hm^{-1}h^{-1}, m\ell m^{-1}, h^{-1})$ where

$$m\ell m^{-1} = [\ell,h]^k h\ell h^{-1}[\ell,h]^{-k} = [\ell,h]^{k-1}\ell[\ell,h]^{-k}$$

by R_4.

Hence using R_7 we deduce

$$\tilde{\tau}_\sharp([\ell,h]) = [\ell,h]^{-1}$$

and

$$\tilde{\tau}_\sharp([\ell,h]^{6k-1}) = [\ell,h]^{-6k+1}. \qquad (R_8)$$

On the other hand by the theorem of Boileau and Otal [BO] we deduce that

$$\pi_0 \mathrm{Diff}\,\Sigma(2,3,6k-1) = \begin{cases} 1 & \text{if } k = 1 \\ \mathbf{Z}_2 & \text{if } k \geq 2. \end{cases}$$

Here $\pi_0\mathrm{Diff}\,\Sigma(2,3,6k-1)$ for $k \geq 2$ is generated by the involution of $\Sigma(2,3,6k-1)$ which preserves the Seifert fibration of $\Sigma(2,3,6k-1)$ and acts both on the base orbifold and on the fiber as a reflection. Then by R_8 it follows that $\tilde{\tau}|_{\partial Z^k_{p_1,p_2}}$ is the generator of the diffeotopy group of $\Sigma(2,3,6k-1)$. This proves Proposition 3.

Corollary (see also [G]). (1) *$Z^k_{p,q}$ is homeomorphic to $Z^k_{p',q'}$ if their intersection forms are the same.* (2) *$Z^k_{p,q}$ is diffeomorphic to $Z^k_{p',q'}$ if and only if $S^k_{p,q}$ is diffeomorphic to $S^k_{p',q'}$. In particular $Z^k_{p,q}$ is not diffeomorphic to $Z^k_{p',q'}$ if $pq \neq p'q'$ in case $k \geq 2$. In case $k = 1$, $Z^1_{2,q}$ is diffeomorphic to $Z^1_{2,q'}$ if and only if $q = q'$.*

Proof. The claim for the homeomorphism types is due to a slight extension of Wall's theorem ([W]) (which shows the existence of the h-cobordism between $Z^k_{p,q}$ and $Z^k_{p',q'}$ which is the product over the boundary) and Freedman's theorem ([F]). The assertion for the diffeomorphism types is deduced from Proposition 3 and Friedman-Morgan's theorem [FM2] if $k \geq 2$. The assertion for the case with $k = 1$ also comes from [FM1], [OV].

Next consider Y. Note that Y does not depend on the choices of (p_1, p_2).

Proposition 4. *Y is 1-connected and ∂Y is the union of $\Sigma(2,3,6k-1)$ and k-copies of $-\Sigma(2,3,5)$. The intersection form $q(Y)$ of Y is isomorphic to $\oplus^{2k-2}H$. The maximal positive subspace H^+ of $H_2(Y{:}\mathbf{Q})$ is isomorphic to a direct sum $H^+_+ \oplus H^+_-$ where H^+_+ (resp. H^+_-) is a subspace of H^+ of rank $k-1$ on which $\tau_* = id$ (resp. $\tau_* = -id$).*

Proof. The claim for ∂Y is obvious. Clearly $\pi_1 Y$ is normally generated by the fundamental groups of the boundary components of Y. As is shown in the proof of Proposition 3 $\pi_1\Sigma(2,3,6k-1)$ is normally generated by the loops ℓ, h in the general fiber. The same statement holds for $-\Sigma(2,3,5)$ (Proposition 3 for the case with $k = 1$). But there still exist the vanishing cycles for ℓ and h within Y if $k \geq 2$ (for $k = 1$ we put $Y = \emptyset$). Therefore $\ell = h = 1$ in $\pi_1 Y$ and hence Y is 1-connected. On the other hand since x in $Z^k_{p_1,p_2}$ is the rational multiple of the canonical divisor and y is its dual the support of the second Stiefel-Whitney class of $S^k_{p_1,p_2}$ lies in $Z^k_{p_1,p_2}$ and hence Y must be spin. Therefore the computation of the euler number and the signature of Y shows that $q(Y) = \oplus^{2k-2}H$. To see the generators of $H_2(Y)$ it suffices to consider $S^k_{1,1}$ since Y is left unchanged even if $Z^k_{1,1}$ is replaced by $Z^k_{p_1,p_2}$. Applying the Mayer-Vietoris sequence to the decomposition (2) of $S^k_{1,1}$ in §1 (where $S^0_{p_1,p_2}$ is replaced by the copy $(S^0)_0$ of S^0) we can see that $H_2(S^k_{1,1})$ is generated by the general fiber $F = \ell \times h$, the cross section B, the divisors in $(E_8)_i$ $(i = 0, \dots, k-1)$, $\partial_i \times \ell$, $\partial_i \times h$ $(i = 1, \dots, k-1)$ (where $\partial_i = \partial_i S(k)$) and the 2-cycles Σ_i, Σ'_i $(i = 1, \dots, k-1)$ constructed as follows. (Here the suffices are counted modulo k.)

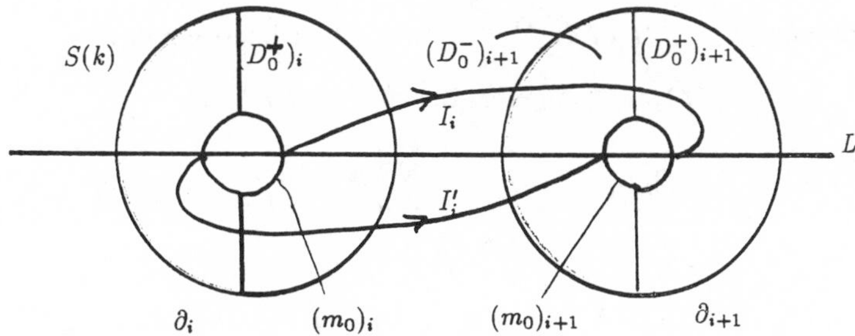

Figure 9

In Fig. 9 the copies $(D_0^{\pm})_i$ $(i = 0, \dots, k-1)$ are attached along ∂_i to $S(k)$ so that the union is invariant by the reflection $\bar{\iota}$ in the line L. Note that $S_{1,1}^k$ contains $(S(k) \cup (\cup (D_0^+)_i) \cup (\cup (D_0^-)_i) \times T^2$ on which τ descends to $\bar{\iota}$. Choose the arcs I_i, I_i' as in Fig. 9. Let $(h_j)_i$ be the i-th copy in $(S^0)_i$ of the 2-handle h_j in S^0. Then Σ_i is the union of $I_i \times h$ and the cores $(h_1)_i^0$, $(h_1)_{i+1}^0$ of $(h_1)_i$, $(h_1)_{i+1}$. Likewise let Σ_i' be the union of $I_i' \times \ell$ and the cores $(h_2)_i^0$, $(h_2)_{i+1}^0$ of $(h_2)_i$, $(h_2)_{i+1}$. Then we have

$$\Sigma_i \cdot \Sigma_j' = \Sigma_i \cdot \partial_j \times h = \Sigma_i' \cdot \partial_j \times \ell = 0$$
$$\partial_i \times \ell \cdot \partial_j \times \ell = \partial_i \times h \cdot \partial_j \times h = \partial_i \times \ell \cdot \partial_j \times h = 0$$

and moreover

$$\Sigma_i \cdot \Sigma_j = \Sigma_i' \cdot \Sigma_j' = \begin{cases} -2 & \text{if } i = j \\ 1 & \text{if } |i-j| = 1 \\ 0 & \text{if } |i-j| \geq 2 \end{cases}$$

$$\Sigma_i \cdot \partial_j \times \ell = \begin{cases} 1 & \text{if } i = j \\ -1 & \text{if } j = i+1,\ i < k-1 \\ 0 & \text{otherwise.} \end{cases}$$

$$\Sigma_i' \cdot \partial_j \times h = \begin{cases} -1 & \text{if } i = j \\ 1 & \text{if } j = i+1,\ i < k-1 \\ 0 & \text{otherwise.} \end{cases}$$

Here we recall that $((D_0^+)_0 \cup (D_0^-)_1 \cup A_0) \times T^2$ together with $(h_1)_0 \cup (h_2)_1 \cup N(B)$ forms the part $Z_{1,1}^1$ of the first decomposition of $S_{1,1}^k$. The 2-cycles $\partial_1 \times \ell$, $\partial_1 \times h$, Σ_i, Σ_i' for $i = 1$ and $i = k-1$ intersect $Z_{1,1}^1$ (the others are contained in Y). But each one is homologous to the cycle in Y due to the fact that $H_2(Z_{1,1}^1)$ is generated by F and B. Then we can see that $\sigma_i = \sum_{t=i}^{k-1} (\Sigma_t + \partial_t \times \ell)$ and $\partial_i \times \ell$ $(i = 1, \dots, k-1)$ generate the first $k-1$-summands $\oplus^{k-1} H$ of $H_2(Y)$ and $\sigma_i' = \sum_{t=i}^{k-1} (\Sigma_t' - \partial_t \times h)$ and $-\partial_i \times h$ $(i = 1, \dots, k-1)$ generate the other

$k-1$-summands $\oplus^{k-1}H$ of $H_2(Y)$. On the other hand we have

$$\tau\Sigma_i = -\Sigma_i - \partial_{i+1}\times h \quad (i<k-1)$$
$$\tau\Sigma_{k-1} = -\Sigma_{k-1} + \sum_{t=1}^{k-1}\partial_t \times h$$
$$\tau\Sigma_i' = \Sigma_i' - \partial_i \times \ell$$
$$\tau\partial_i\times h = \partial_i \times h, \quad \tau\partial_i\times\ell = -\partial_i\times\ell.$$

Then an easy computation shows that $\sigma_i' - (\sum_{t=i}^{k-1}\partial_t\times\ell)/2 - \partial_i\times h$ and $\sigma_i - (\sum_{t=1}^{i}\partial_t\times h)/2 + \partial_i\times\ell$ $(i=1,\ldots,k-1)$ give the desired basis over $\mathbf{Q}$ of $H_\pm^+$ on which $\tau_* = \pm 1$.

Note that $\tau_* = -id$ on any other summand of $H_2(S^k_{p_1,p_2})$. Therefore the orientation of the maximal subspace H_2^+ of $H_2(S^k_{p_1,p_2}\colon\mathbf{R})$ is reversed by τ if and only if k is odd. Moreover if $k\geq 2$ we have $\tau^*\gamma_d = (-1)^k\gamma_d$ for any Donaldson polynomial invariant γ_d with $d=4c-3k$ since $\tau^*\kappa = -\kappa$ and $\tau^*q = q$ in $S^*H^2(S^k_{p_1,p_2})$ where κ is the positive rational multiple of the canonical divisor and q is the intersection form of $S^k_{p_1,p_2}$ ([FM2]). This proves the required properties of the decomposition of $S^k_{p_1,p_2}$ of the first type and Theorem 2.

For the second decomposition we use Proposition 5 below. Take a 10-th punctured 2-disk D_{10} and let $\partial_i D_{10}$ be the i-th boundary component of D_{10} $(i=0,\ldots 10)$ oriented so that $\sum_{i=1}^{10}\partial_i D_{10} = \partial_0 D_{10}$. Put $\tilde{s}_1 = \begin{pmatrix}1 & -1\\ 0 & 1\end{pmatrix}$, $\tilde{s}_2 = \begin{pmatrix}1 & 0\\ 1 & 1\end{pmatrix}$. Take $\tilde{t}_i \in SL_2\mathbf{Z}$ $(i=1,\ldots 10)$ which is conjugate in $SL_2\mathbf{Z}$ to $\tilde{s}_1$ (or equivalently $\tilde{s}_2$). Construct the T^2-bundle X over D_{10} with the projection $\pi\colon X\to D_{10}$ such that $\partial_i X = \pi^{-1}(\partial_i D_{10})$ is a T^2-bundle over S^1 with monodromy $\tilde{t}_i$ if $i\neq 0$ and with monodromy $\tilde{t}_1\tilde{t}_2\ldots\tilde{t}_{10}$ if $i=0$ with respect to the fixed framing (ℓ,h). Then 10-copies of the tubular neighborhood of the fiber of type I_1 can be glued to X along $\partial_i X$ $(i\neq 0)$ to get the Lefshetz fibration $X(\tilde{t}_1,\ldots,\tilde{t}_{10})$ over D^2 with projection $\tilde{\pi}$.

Proposition 5. *There is a diffeomorphism from E_9 to $X(\tilde{t}_1,\ldots,\tilde{t}_{10})$ which maps the cross section D'' of E_9 to that of $\tilde{\pi}$ where $\tilde{t}_i = \tilde{s}_2$ if i is odd and $\tilde{t}_i = \tilde{s}_1$ if i is even.*

Proof. According to [M], Part II, Lemma 6 $\pi\colon E_9\to D^2$ is deformed to a sum of 10 copies of fibers of type I_1 and the cross section of E_9 is deformed to that of $\tilde{\pi}$. Hence E_9 is diffeomorphic to $X(\tilde{t}_1,\ldots\tilde{t}_{10})$ with $\prod\tilde{t}_i = (\tilde{s}_2\tilde{s}_1)^5 = \begin{pmatrix}0 & 1\\ -1 & 1\end{pmatrix}$. Note that each $\tilde{t}_i$ is conjugate in $SL_2\mathbf{Z}$ to $\tilde{s}_1$ and we can assume that the cross section of E_9 coincides with D'' in §1. There are isotopic deformations called elementary transformations which maps $X(\tilde{t}_1 \dots \tilde{t}_{10})$ to $X(\tilde{t}'_1 \dots \tilde{t}'_{10})$ and fixes $\partial X(\tilde{t}_1, \dots, \tilde{t}_{10})$ where

$$\tilde{t}'_i = \tilde{t}_{i+1}, \quad \tilde{t}'_{i+1} = \tilde{t}_{i+1}^{-1}\tilde{t}_i\tilde{t}_{i+1}, \quad \tilde{t}'_j = \tilde{t}_j \quad (j \neq i, i+1)$$

or

$$\tilde{t}'_i = \tilde{t}_i\tilde{t}_{i+1}\tilde{t}_i^{-1}, \quad \tilde{t}'_{i+1} = \tilde{t}_i, \quad \tilde{t}'_j = \tilde{t}_j \quad (j \neq i, i+1)$$

for $i = 1, \dots, 9$. Thus if $(\tilde{t}_1 \dots \tilde{t}_{10})$ is transformed to $(\tilde{t}'_1 \dots \tilde{t}'_{10})$ by a finite sequence of elementary transformations then there is a diffeomorphism between $X(\tilde{t}_1 \dots \tilde{t}_{10})$ and $X(\tilde{t}'_1 \dots \tilde{t}'_{10})$ which is the identity on the boundary. Now the proof follows from

Proposition 6. *If $(\tilde{t}_1, \dots, \tilde{t}_{10}) \in \prod SL_2\mathbf{Z}$ satisfies*

$$\tilde{t}_1 \dots \tilde{t}_{10} = \begin{pmatrix} 0 & 1 \\ -1 & 1 \end{pmatrix}$$

and each $\tilde{t}_i$ is conjugate in $SL_2\mathbf{Z}$ to $\tilde{s}_1$. Then $(\tilde{t}_1, \dots, \tilde{t}_{10})$ is transformed by finitely many elementary transformations to $(\tilde{t}'_1, \dots, \tilde{t}'_{10})$ where $\tilde{t}'_i = \tilde{s}_2$ if i is odd and $\tilde{t}'_i = \tilde{s}_1$ if i is even.

Let $p\colon SL_2\mathbf{Z} \to PSL_2\mathbf{Z}$ be the canonical projection and put $s = p(\tilde{s})$ for $\tilde{s} \in SL_2\mathbf{Z}$. Also put $a = s_2s_1$ and $b = s_2(s_2s_1)^{-2}$. Then $s_1 = aba$, $s_2 = ba^2$. Furthermore $PSL_2\mathbf{Z}$ is generated by a, b and is isomorphic to $G = \{a, b \mid a^3 = b^2 = 1\}$. In $PSL_2\mathbf{Z} = G$ we have

Proposition 7. *Suppose that $\mu = t_1 \dots t_{10} = a^{-1} = a^2$ in G where $t_i \in G$ and is conjugate in G to s_1. Then by finitely many elementary transformations (just defined as in the case with $SL_2\mathbf{Z}$), $(t_1, \dots t_{10})$ is changed to $(s_2, s_1, \dots, s_2, s_1)$.*

Assuming Proposition 7 (which will be proved in §3) the rest of the proof of Proposition 6 goes just as that for [M], Lemma 8. The elemetary transformations in G can be lifted to those in $SL_2\mathbf{Z}$ and the lift $\tilde{s}_i \in SL_2\mathbf{Z}$ of s_i is uniquely determined since $\tilde{s}_i$ must be conjugate in $SL_2\mathbf{Z}$ to $\begin{pmatrix} 1 & -1 \\ 0 & 1 \end{pmatrix}$. This proves Proposition 6 (and 5).

Now we turn to the construction of the second decomposition of $S^k_{p_1,p_2}$. Let E be the union of one copy of E_9 in $S^k_{1,1}$ and the τ-invariant tubular neighborhood $N(B)$ of the cross section B of $S^k_{1,1}$ constructed in §1. Then E is represented by the framed link obtained from that in Fig. 3 by performing $-k$ surgery on C_0 and τ on E is also induced by the rotation through angle π in the same axis A' in Fig. 3 . Therefore $\partial E = -\Sigma(2, 3, 6k-5)$ and τ maps the regular fiber of the Seifert fibration of $-\Sigma(2, 3, 6k-5)$ to its inverse. It also follows from [BO] as in Proposition 3 that τ induces the generator of $\pi_0\mathrm{Diff}(\Sigma(2,3,6k-5))$ and hence any self-diffeomorphism of $\Sigma(2,3,6k-5) = \partial(S^k_{1,1} - E)$ extends to that of $S^k_{1,1} - E$. As in the decomposition of the first type $S^k_{1,1} - E$ is decomposed as the union of $k-1$ copies of E_8 (which come from the other E_9-fibers) and $Y' = S^k_{1,1} - E - \cup E_8$. On the other hand according to Proposition 5 we can take $Z^k_{1,1}$ (which is the union of I^+_1, I^-_1 and the regular neighborhood $N(B)$ of B) in the interior of E. Then after performing the logarithmic transformations within $Z^k_{1,1} \subset E$ we get the decomposition of $S^k_{p_1,p_2}$ of the form

$$S^k_{p_1,p_2} = Z^k_{p_1,p_2} \cup Y' \cup E' \cup (\cup_{i=1}^{k-1}(E_8)_i)$$

where $E' = E - Z^k_{1,1}$ with $\partial E' = \Sigma(2,3,6k-1) \cup -\Sigma(2,3,6k-5)$. Just as in the first decomposition we can observe that both E' and Y' are 1-connected and the intersection forms of E' and Y' are E_8 and $\oplus^{2k-2}H$ respectively. Proposition 3 and the above observation (together with [BO]) show that this decomposition does not depend on the choice of the gluing maps. This proves the required properties of the second decomposition.

§3. Proof of Proposition 7

The proof of Proposition 7 is a modification of that of Livine's theorem and its extension in [M] and consists of two steps. Put $G = PSL_2\mathbf{Z} = \{a, b \mid a^3 = b^2 = 1\}$ and $s_0 = a^2b$, $s_1 = aba$, $s_2 = ba^2$ (which are called "short" words in [M]).

Claim. *By finitely many elementary transformations $(y_1, y_2, \cdots, y_{10})$ is changed to $(y'_1, y'_2, \cdots, y'_{10})$ where each y'_i is one of s_0, s_1 and s_2.*

Assuming this claim we start with $(y_1, \cdots, y_{10})$ where each y_i is short. For any product X of short words let $\ell'(X)$ be the length of X with respect to s_i's. Then $\ell'(\mu) = 10$. Since $s_0^m = (a^2b)^m \neq a^2$ in G for $m \neq 0$, it follows that μ contains s_1 or s_2. Write

$$\mu = s_0^{m_1} X_1 s_0^{m_2} X_2 \cdots s_0^{m_\nu} X_\nu$$

with $m_i \geq 0$ where

$$X_j = \prod_{i=0}^{N_j} s_1^{k_{ij}} s_2^{\ell_{ij}}$$

with k_{ij}, $\ell_{ij} \geq 0$ satisfying the following conditions.

$$m_i \geq 1 \qquad \text{for } i \geq 2 \text{ if } \nu \geq 2$$
$$k_{ij} \geq 1, \ell_{i-1,j} \geq 1 \qquad \text{for } 1 \leq i \leq N_j.$$

We assume that the exponent sum $\sum_{i=1}^{\nu} m_i$ of s_0 in μ is minimal under the elementary transformations of μ. If a product x of s_0, s_1, s_2 can be changed to another product y by finitely many elementary transformations we will write $x \sim y$. Then we have ([M])

$$s_1 s_0 \sim s_2 s_1 \sim s_0 s_2 = a, \quad s_1 s_2 s_1 \sim s_2 s_1 s_2.$$

It follows that

$$s_0 s_1^k s_2 s_1 \sim s_0 s_2 s_1 s_2^k \sim s_2 s_1^2 s_2^k.$$

Then by the assumption we deduce

(1) There are no subproducts in μ of the form

$$s_1 s_0, \quad s_0 s_2, \quad s_0 s_1^k s_2 s_1 \quad \text{for } k \geq 1.$$

Thus we have

(2)
$$k_{0j} \geq 1 \qquad \text{unless } j = 1 \text{ and } m_1 = 0,$$
$$\ell_{N_j,j} \geq 1 \qquad \text{unless } j = \nu.$$

Furthermore we assume that

(3) There is no subproduct of the form $s_2 s_1 s_2$ in any X_j.

(Replace it by $s_1 s_2 s_1$ if it exists). Thus in any X_j we have

(4) $k_{ij} \geq 2 \quad$ for $1 \leq i \leq N_j$ if $N_j \geq 1$

except for the cases when $i = N_j$, $\ell_{N_j,j} = 0$ (and hence $j = \nu$ by (2)). Now we assume for the moment that any X_j does not contain the subproduct $\sim (s_1 s_2)^3$. Then as is shown in [M]

$$(s_1 s_2)^3 \sim (s_2 s_1)^3 \sim s_2 s_1^2 s_2 s_1^2 \sim s_1^2 s_2 s_1^2 s_2$$

and $s_1 s_2 s_1^k s_2 s_1 \sim (s_1 s_2)^3 s_2^{k-2}$ for $k \geq 2$. Hence we have

(5) there are no subproducts in X_j of the form

$$s_2 s_1^2 s_2 s_1^2, \quad s_1^2 s_2 s_1^2 s_2, \quad s_1 s_2 s_1^k s_2 s_1 \quad (k \geq 2).$$

Now we will see how X_j can be reduced. For brevity put $k_{ij} = k_i$, $\ell_{ij} = \ell_i$. We must have the subproduct $s_2 s_1$ of μ to reduce it to a^2 in G. Hence if $N_j = 0$ the reduction cannot occur at X_j. Suppose that $N_j \geq 1$. Then from the above conditions and the fact that $\ell'(\mu) = 10$ we deduce that X_j has one of the following forms.

Case I. $\ell_0 = 1$. In this case $j = 1$ and $m_1 = 0$ by (1).

(1) $\mu = X_j = s_1^{k_0} s_2 s_1^{k_1}$ with $k_0 + k_1 = 9$.

(2) $\mu = X_j = s_2 s_1^{k_1} s_2 s_1^{k_2}$ where either $k_2 \geq 2$ and $k_1 \geq 3$, or $k_1 \geq 2$ and $k_2 \leq 1$.

(3) $X_j = s_1^{k_0} s_2 s_1^{k_1} s_2$ where either $k_0 \geq 2$ and $k_1 \geq 3$, or $k_0 \leq 1$ and $k_1 \geq 2$.

(4) $\mu = X_j = s_2 s_1^3 s_2 s_1^3 s_2^2$.

(5) $X_j = s_2 s_1^{k_1} s_2 s_1^{k_2} s_2$ where $k_1, k_2 \geq 3$.

(6) $\mu = X_j = s_1^{k_0} s_2 s_1^{k_1} s_2^{\ell_1} s_1^{k_2}$ where $\ell_1 \geq 2$ and either $k_0 \geq 2, k_1 \geq 3$ or $k_0 \leq 1, k_1 \geq 2$.

(7) $X_j = s_1^{k_0} s_2 s_1^{k_1} s_2^{\ell_1}$ where k_0, k_1, ℓ_1 satisfy the same conditions as in I-(6).

(8) $\mu = X_j = s_1^{k_0} s_2 s_1^{k_1} s_2^{\ell_1} s_1^{k_2} s_2^{\ell_2} s_1$ where $k_1, k_2, \ell_1 \geq 2$, $\ell_2 \geq 1$ and $k_0 \leq 1$.

(9) $X_j = s_1^{k_0} s_2 s_1^{k_1} s_2^{\ell_1} s_1^{k_2} s_2^{\ell_2}$ where k_i's and ℓ_i's satisfy the same conditions as in I-(8).

Case II. $\ell_0 \geq 2$.

(1) $X_j = s_1^{k_0} s_2^{\ell_0} s_1^{k_1}$ where $\nu = j$, $\ell_0 \geq 2$, $k_1 \geq 1$.

(2) $X_j = s_1^{k_0} s_2^{\ell_0} s_1^{k_1} s_2 s_1^{k_2}$ where $j = \nu$, $\ell_0 \geq 2$ and either $k_2 \geq 2$ and $k_1 \geq 3$, or $k_2 = 1$ and $k_1 \geq 2$.

(3) $X_j = s_1^{k_0} s_2^{\ell_0} s_1^{k_1} s_2$ where ℓ_0, $k_1 \geq 2$.

(4) $\mu = X_j = s_2^2 s_1^3 s_2 s_1^3 s_2$.

(5) $X_j = s_1^{k_0} s_2^{\ell_0} s_1^{k_1} s_2^{\ell_1} s_1^{k_2}$ where ℓ_0, ℓ_1, $k_1 \geq 2$, $j = \nu$ and $k_2 \geq 1$.

(6) $X_j = s_1^{k_0} s_2^{\ell_0} s_1^{k_1} s_2^{\ell_1}$ where k_i's and ℓ's satisfy the same conditions as in II-(5).

(7) $\mu = X_j = s_2^2 s_1^2 s_2^2 s_1^2 s_2^2$ or $s_2^2 s_1^2 s_2^2 s_1^2 s_2 s_1$.

(8) $X_j = s_2^{\ell_0} s_1^{k_1} s_2^{\ell_1} s_1^{k_2} s_2$ where ℓ_0, ℓ_1, k_1, $k_2 \geq 2$.

We can see directly that either $\nu \neq a^2$ or X_j has one of the following forms.

(i) $X_j = R s_2^{\ell'}$ where R is a reduced word of the form $\cdots b \cdots a$ and there are no words in μ on the left hand side of R.

(ii) $X_j = R s_2^{\ell'}$ where $R = a^2$, $\ell' \geq 1$ and there are no words in μ on the left hand side of R.

(iii) $X_j = s_1^{k'} R$ where R is a reduced word of the form $b \cdots$ and there are no words in μ on the right hand side of R.

(iv) $X_j = s_1^{k'} R s_2^{\ell'}$ where R is a reduced word of the form $b \cdots a$.

Taking the condition (1) into account we have $\mu \neq a^2$ in G in each case. Therefore μ must contain a subproduct $\mu_0 \sim (s_1 s_2)^3$. Let z_1, z_2, z_3, z_4 be the short words which are the complements to μ_0 and which appear in this order in μ. Again we can assume that the exponent sum of s_0 in μ is minimal in the equivalence classes of μ under the elementary transformations. Then μ satisfies (1). On the other hand

$$\begin{aligned}\mu_0 s_2^2 s_0 &\sim (s_2 s_1)^2 s_1 s_2 s_1 s_2 s_0 \\ &\sim (s_1 s_2)^2 s_1^2 s_1 s_2 s_1 s_0 \sim (s_2 s_1)^2 s_1^3 s_2^2 s_1\end{aligned}$$

Likewise we have $\mu_0 s_1 s_2 s_0 \sim (s_1 s_2)^2 s_1^2 s_2^2 s_1$. These facts and analogous facts for $s_0 s_1^2 \mu_0$ and $s_0 s_1 s_2 \mu_0$ show that each z_i is s_1 or s_2. Furthermore since $\mu_0 = 1$ in G we have $z_1 z_2 z_3 z_4 = a^2$ in G. Since we need the subproduct $s_2 s_1$ to reduce the left hand side we have $(z_1, z_2, z_3, z_4) = (s_2, s_1, s_2, s_1)$, (s_1, s_2, s_1, s_1) or (s_2, s_2, s_1, s_2). Then an easy computation shows that $\mu \sim (s_2 s_1)^5$. This proves Proposition 7.

Proof of Claim. The proof goes as in that given in [M], Appendix. For the reduced product $g = t_1 \cdots t_k$ where $t_i = a$, a^2, or b let $\ell(g)$ be the length of g with respect to a, a^2, b. Since y_i is conjugate to s_1 any y_i is either one of s_i's (which are short) or $y_i = Q_i^{-1} s_1 Q_i = Q_i^{-1} aba Q_i$ where Q_i is expressed in reduced form $t_1 \cdots t_k$ with $t_1 = b$. Fix such representations. For $\mu = y_1 \cdots y_{10}$ define $L(\mu) = \sum \ell(Q_i)$ where the sum is taken over all "long" $y_i = Q_i^{-1} s_1 Q_i$. First suppose that for all i for $1 \leq i \leq 9$ we have $\ell(y_i y_{i+1}) \geq \max(\ell(y_i), \ell(y_{i+1}))$. Consider $y_i = t_k \cdots t_2 t_1$, $y_{i+1} = \tilde{t}_1 \cdots \tilde{t}_\ell$ (in reduced forms). Then there are some maximal subproducts $t_m \cdots t_1$ and $\tilde{t}_1 \cdots \tilde{t}_m$ which can be reduced in $y_i y_{i+1}$ to get $t_m \cdots t_1 \tilde{t}_1 \cdots \tilde{t}_m = r$ with $\ell(r) \leq 1$. The proof of Assertion in [M], Appendix shows the followings. If $\ell(r) = 0$, that is $r = 1$ then $m = 0$ and no reductions occur in $y_i y_{i+1}$. (Here t_1 or $\tilde{t}_1$ is b.) Suppose that $\ell(r) = 1$. If k is even then $y_i = s_2 = ba^2$, $y_{i+1} = a^2 b \cdots$ and no reductions occur in $y_{i-1} y_i$. If ℓ is even then $y_{i+1} = s_0 = a^2 b$, $y_i = \cdots ba^2$ and no reductions occur in $y_{i+1} y_{i+2}$. If k is odd then in $y_i = t_k \cdots t_{k-m+1} t_{k-m} \cdots t_{n+1} t_n \cdots t_1$ the subproduct $t_k \cdots t_{k-m+1}$ (resp. $t_n \cdots t_1$) is the maximal part which can be reduced in $y_{i-1} y_i$ (resp. $y_i y_{i+1}$) where $m \leq (k-1)/2$, $n \leq (k-1)/2$. In this case $t_{n+1} = t_{k-m} = b$ $(n+1 \leq k-m)$. In either case the reduced form of $y_1 \cdots y_{10}$ contains b and not equal to a^2 in G. It follows that for some i $(1 \leq i \leq 9)$ we have $\ell(y_i y_{i+1}) < \max(\ell(y_i), \ell(y_{i+1}))$. Then as in the proof of [M], Appendix some elementary transformations reduce $L(\mu)$. (Note that the proof of this part is valid in our case without any change.) Thus after a finite number of steps we get $y_1' \cdots y_{10}' \sim y_1 \cdots y_{10}$ with $L(y_1' \cdots y_{10}') = 0$ which means that y_i''s are short.

References

[BO] M. Boileau, J. P. Otal, Groupe des difféotopies de certaines variétés de Seifert, C. R. Acad. Sci. Paris **303** (1986), 19–22.

[D] S. K. Donaldson, Polynomial invariants for smooth four-manifolds, Topology **39** (1990), 257–315.

[F] M. H. Freedman, The topology of four-dimensional manifolds, J. Diff. Geometry **17** (1982), 357–453.

[FKV] S. Finashin, M. Kreck, O. Viro, Non-diffeomorphic but homeomorphic knottings of surfaces in the 4-sphere, in "Topology and Geometry-Rohlin Seminar Lect. Notes in Math. 1346," Springer, 1988, pp.157–198.

[FM1] R. Friedman, J. W. Morgan, On the diffeomorphism type of certain elliptic surfaces I, II, J. Diff. Geo. **77** (1988), 297–369, 371–398.

[FM2] R. Friedman, J.W. Morgan, Complex versus differentiable classification of algebraic surfaces, Topology and its appl. **32** (1989), 135–139.

[FT] M. Freedman, L. Taylor, Λ-splitting 4-manifolds, Topology **16** (1977), 181–184.

[G] R. E. Gompf, Nucli of elliptic surfaces, Topology **30** (1991), 479–512.

[HKK] J. Harer, A. Kas, R. Kirby, Handlebody decompositions of complex surfaces, Memoirs of AMS. **62** (1986).

[K] K. Kodaira, On Compact analytic surfaces II, Ann. of math. **77** (1963), 563–626.

[M] B. Moishezon, "Complex Surfaces and Connected Sums of Complex Projective Planes", Lect.Notes in Math. **603**, Springer, 1977.

[Ma] T. Matumoto, Extension problem of diffeomorphisms of a 3-torus over some 4-manifolds, Hiroshima Math. J. **14** (1984), 189–201.

[OV] C. Okonek, A. Van de Ven, Stable bundles and differentiable structures on certain elliptic surfaces, Invent. math. **86** (1986), 357–370.

[S] M. Sakuma, Involutions on torus bundles, Osaka J. Math. **22** (1985), 163–185.

[U] M. Ue, On the diffeomorphism types of elliptic surfaces with multiple fibers, Invent. Math. **84** (1986), 633-643.

[V] O. Ya. Viro, Compact four dimensional exotica with small homology, (in Russian) Algebra and analysis **1.4** (1989), 67–77, (to be translated as Leningrad Mathematical Journal)

[W] C. T. C. Wall, On simply connected 4-manifolds, J. London Math. Soc. **39** (1964), 141–149.

Symmetry groups of prime knots up to 10 crossings

Kouzi Kodama and Makoto Sakuma

Dedicated to Professor Fujitsugu Hosokawa on his 60th birthday

Let K be a smooth oriented knot in S^3. The symmetry group $Sym(S^3, K)$ of K is defined as the mapping class group of the pair (S^3, K); that is, $Sym(S^3, K) = \pi_0 Diff(S^3, K)$. This group contains information on chirality and invertibility of K, and essentially controls the rigid symmetries of K (cf. [BiZ2, Theorem 2.1]).

The purpose of this paper is to study the symmetry groups of prime knots up to 10 crossings. Since symmetries of knots have been subject of extensive literature, most part of them are already determined. In particular, chirality, invertibility, and free periods of these knots are already known (see [Pe, HK; H2; H1, BiZ1, Sa3]). However, cyclic periods of them remain undetermined, though a list of possible cyclic periods are presented by Lüdicke [Lü1] (cf. [BiZ1, Bu, DL, Lü2, Ms1, 2, 3, Pr, Tr1, 2, 3, Yo1, 2]). Here a knot K in S^3 is said to have *cyclic* [resp. *free*] *period* n, if there is an orientation preserving periodic diffeomorphism f on S^3 of period n which preserves K and its orientation, and $Fix(f) \cong S^1$ [resp. $Fix(f^i) = \emptyset \quad (1 \leq i \leq n-1)$]. In this paper, we give some additional methods for the study of symmetries of hyperbolic knots. It should be noted that the prime knots up to 10 crossings are simple, and hence all of them except torus knots are hyperbolic by Thurston's uniformization theorem (cf. [Ri1, 2]). Using this fact and our methods we determine the cyclic periods and the symmetry groups of them except 10_n with $n = 102, 117$, and 119 in the table of [Ro].

For alternating knots, our results can be stated nicely in terms of Conway's notation, which describes a knot diagram by decomposing it into a *basic polyhedron* and *algebraic tangles* (see [Cw]). A *basic polyhedron* is a connected 4-valent graph P^* embedded in $S^2(\subset S^3)$, such that no region of S^2 divided by P^* has just two vertices. An *algebraic tangle* is a (diagram of a) tangle which is obtained from a trivial tangle through certain operations, and it has the property that its double branched covering is a graph manifold. A *rational tangle* is an algebraic

Knots 90

tangle which is homeomorphic to a trivial tangle. Conway observed that every knot diagram is obtained from a basic polyhedron by replacing the regular neighbourhoods of its vertices by algebraic tangles. For example, the knot 10_{101} is obtained from the basic polyhedron (which is named) 6^* by replacing the regular neighbourhoods of their vertices by rational tangles as illustrated in Figure 0.1.

Figure 0.1 and 0.2

If a knot K is obtained from a nontrivial basic polyhedron P^* by replacing its vertices by rational tangles, then a graph automorphism of (S^2, P^*) which "respects" the sabstitution data for K induces a symmetry of K. For example, there is only one nontrivial automorphism of the basic polyhedron 6^* which respects the substitution data for 10_{101}; and it induces a strong inversion of 10_{101} as illustrated in Figure 0.2. A basic polyhedron P^* is called a *Conway graph*, if every simple loop in S^2 meeting P^* in 2 or 4 points bounds a disk D in S^2 such that $(D, P^* \cap D)$ is $\ominus$, $\otimes$ or ⦶. Using the hyperbolic Dehn surgery theorem [Th1], Bonahon and Siebenmann [BnS] have proved the following.

Proposition 0.1. *If K is obtained from a Conway graph P^* by replacing its vertices by "sufficiently complicated" rational tangles, then $Sym(S^3, K)$ is isomorphic to the group of graph automorphisms of (S^2, P^*) which "respect" the substitution data.*

However, no information is known on the actual range of rational tangles for which the conclusion of Proposition 0.1 holds. Our results for alternating knots can be stated as follows:

Result 0.2. *The conclusion of Proposition 0.1 holds for prime alternating knots up to 10 crossings except possibly* 10_n *with* $n = 102$, 117 *and* 119.

Thus, for example, the symmetry group of 10_{101} is the cyclic group of order 2. The authors do not know any alternating knot diagram for which the conclusion of Proposition 0.1 does not hold. For non-alternating knots, there are several diagrams up to 10 crossings for which Claim 0.1 does not hold, since they admit "inequivalent" diagrams of minimal crossings.

This paper is organized as follows. In Section 1, we give some methods for the study of symmetries of hyperbolic knots. In particular, the problem of whether an invertible hyperbolic knot has cyclic or free period 2 is reduced to a problem of whether a spatial θ-curve in S^3 admits a certain symmetry. In Section 2, we study symmetries of spatial θ-curves, and show that certain spatial θ-curves are not "*reversible*". To do this we have used a computer program written by the first author. In Section 3, we give a table of the symmetry groups and cyclic periods of prime knots up to 10 crossings.

The authors would like to thank Professor Kunio Murasugi for teaching us useful information including the work of [DL].

1. Symmetry groups of hyperbolic knots

Let K be a hyperbolic knot in S^3, and $E(K)$ be the exterior of K (that is, the complement of an open tubular neighbourhood of K). Then the interior $intE(K)$ of $E(K)$ admits a complete hyperbolic structure of finite volume, and there is a natural isomorphism $Sym(S^3, K) \cong Isom(intE(K))$. In fact, by Waldhausen's theorem [Wa] and Mostow's rigidity theorem, we see $Sym(S^3, K)$ is isomorphic to a subgroup of $Isom(intE(K))$. On the other hand, by the positive solution of the knot complement conjecture [GLu], we see the above inclusion is an epimorphism (cf. [Si, Proposition 5.1]). Let $\tilde{I}$ be the finite subgroup of $Diff(S^3, K)$ which is obtained by extending the action of $Isom(intE(K))$. Then $\tilde{I}$ gives a realization of $Sym(S^3, K)$, that is, we obtain the following commutative diagram:

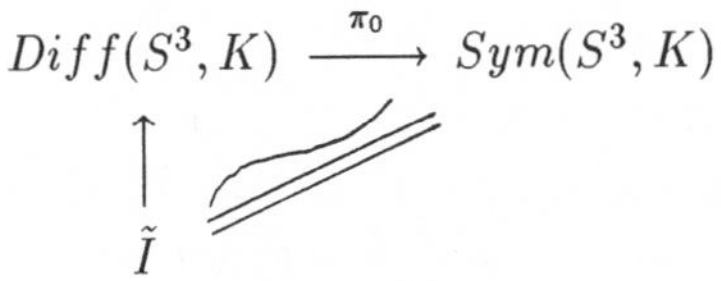

Diagram 1.1.

We denote the subgroups of $Diff(S^3, K)$, $Sym(S^3, K)$, and $\tilde{I}$ generated by the elements which preserve the orientation of S^3 [resp. the orientations of S^3 and K] by putting the symbol $^+$ [resp. *]. By considering the restriction of $\tilde{I}$ to the subset K, we obtain the following lemma by virtue of the Smith conjecture [MB] (see [Ri1, p.124, Sa2, p.186]).

Lemma 1.1. *The restriction of $\tilde{I}$ to K is an effective group action. In particular, we have the followings:*

(1) *$Sym^*(S^3, K)$ is a finite cyclic group Z_n for some positive integer n.*

(2) *$Sym^+(S^3, K)$ is isomorphic to the dihedral group D_n of order $2n$ or Z_n, according to whether K is invertible or not.*

(3) *If K is non-amphicheiral, then $Sym(S^3, K) \cong Sym^+(S^3, K)$.*

(4) *Suppose K is amphicheiral. If K is invertible, then $Sym(S^3, K) \cong D_{2n}$. If K is noninvertible, then $Sym(S^3, K) \cong Z_{2n}$ or D_n according as K is positive or negative amphicheiral.*

Now assume K is hyperbolic and invertible. Then K is strongly invertible; that is, there is an involution h on (S^3, K), such that $Fix(h)$ is a circle intersecting K in two points. We call h a *strong inversion* of K. Let q be the projection $S^3 \to S^3/h \quad (\cong S^3)$, and let $\theta(K, h)$ denote the spatial θ-curve $q(Fix(h) \cup K)$.

We introduce some notions concerning spatial θ-curves. Let θ be a spatial θ-curve, that is, a graph embedded in S^3 with two vertices v_- and v_+ and three edges e_1, e_2 and e_3, each edge having distinct endpoints. We say that θ is *reversible*, if there is an orientation preserving homeomorphism f of S^3 such that $f(v_-) = v_+$ and $f(e_i) = e_i \quad (1 \leq i \leq 3)$. θ is said to be *strongly reversible*, if we can choose the above f as an involution. [In this case $Fix(f)$ is an unknotted circle which intersects each e_i in a point.] We say that θ has *period 2 with center in e_i*, if there is an orientation preserving involution f of S^3 such that (1) $Fix(f)$ is a circle intersecting e_i in a point, and (2) $f(e_j) = e_k$, where $\{1, 2, 3\} = \{i, j, k\}$.

Proposition 1.2. *Let K be an invertible hyperbolic knot, and let h be a strong inversion of K. Then the following holds:*

(1) *K has cyclic period 2, if and only if $\theta(K, h)$ is strongly reversible.*

(2) *K has free period 2, if and only if $\theta(K, h)$ has period 2 with center in $q(K)$.*

Proof. Let h' be the element of $\tilde{I}$ corresponding to the class $[h] \in Sym(S^3, K)$ via the natural isomorphism $Sym(S^3, K) \cong \tilde{I}$. Then by [To], the involutions $h|_{E(K)}$ and $h'|_{E(K)}$ are strongly equivalent. Hence we see that the involutions h and h' on (S^3, K) are strongly equivalent, that is, there is a diffeomorphism φ of (S^3, K) such that $h' = \varphi h \varphi^{-1}$ and φ is isotopic to the identity. Thus we may assume that h is an element of $\tilde{I}$. Now suppose that K has cyclic or free period 2, and let f be an involution of (S^3, K) realizing the period. Then by [Cn], $[f]$ has order 2 in $Sym(S^3, K)$. We denote the order 2 element of $\tilde{I}$ corresponding to $[f]$ by the same symbol f. [By the preceding argument, this involution is strongly equivalent to the original f]. By Lemma 1.1, we see the elements f and h of $\tilde{I}$ are commutative. Hence f induces an involution $\hat{f}$ on S^3/h which preserves $\theta(K, h) = q(Fix(h) \cup K)$. We can observe that the restriction of $\hat{f}$ to the arc $q(K)$ is an inversion, since $f|_K$ is the π-rotation. Thus $\hat{f}$ is an orientation preserving involution on S^3/h such that $Fix(\hat{f})$ is a circle intersecting $q(K)$ in a point. Let e_1 and e_2 be the arcs of $\theta(K,h)$ such that $e_1 \cup e_2 = q(Fix(h))$. Suppose f realizes free period 2. Then $\hat{f}(e_1) = e_2$, since f induces π-rotation on $Fix(h)$. Hence $\theta(K,h)$ has period 2 with center in $q(K)$. Suppose f realizes cyclic period 2. Then we have $\hat{f}(e_i) = e_i \quad (i = 1, 2)$, and hence $\theta(K,h)$ is strongly reversible. In fact, if this is not the case, then $\hat{f}(e_1) = e_2$ and therefore $Fix(f) \cap Fix(h) = \emptyset$. Thus the projection $S^3 \to S^3/ < f, h >$ is a $Z_2 \oplus Z_2$-covering branched over the 2-component link which is the image of $Fix(f) \cup Fix(h)$. Hence fh is a free involution of S^3 and preserves the orientation of K, a contradiction. Conversely, we can see that if $\theta(K,h)$ is strongly reversible [resp. has period 2 with center in $q(K)$] and $\hat{g}$ is an involution of S^3/h realizing the symmetry, then a suitable lift of $\hat{g}$ to S^3 realizes cyclic [resp. free] period 2 of K.

If a spatial θ-curve $e_1 \cup e_2 \cup e_3$ has period 2 with center in e_3, then the constituent knots $e_1 \cup e_3$ and $e_2 \cup e_3$ must be of the same type. By using this fact, we can confirm that 10_{99} and 10_{123} do not have free period 2 (cf. [Sa3]). In fact the above constituent knots of a θ-curve obtained from 10_{99} or 10_{123} are a nonamphicheiral 2-bridge knot and its mirror image.

However, it is not so easy to prove non-reversibility of a spatial θ-curve. Recently, Litherland [Li] has defined a polynomial invariant of a spatial θ-curve, and solved this problem for a certain family of spatial θ-curves. In Section 2, we give a method for solving this problem using an idea similar to that of Wolcott [Wo] (cf. [Sa2]). It enables us to determine whether an invertible knot up to 10 crossings has cyclic period 2 or not. Combined with other results, this achieves the determination of cyclic periods of prime knots up to 10 crossings except three knots. Since other kinds of symmetries of these knots are already known, their symmetry groups are almost decided. However, there remains the following problem.

Problem 1.3. *Let p be a prime number. Suppose a hyperbolic knot K has cyclic or free period $p^r \quad (r \geq 1)$. Then $Sym^*(S^3, K) \cong Z_{p^s m}$, where $s \geq r$ and m is prime to p. How can we determine the integer s?*

In the remainder of this section, we consider the above problem. First, we treat the case where $p = 2$ and K is invertible. Let K be an invertible hyperbolic knot, and suppose there is an element f of $Diff^*(S^3, K)$ of order $2^r \quad (r \geq 1)$. Then $Sym^+(S^3, K) \cong D_{2^s m}$, where $s \geq r$ and m is odd. Let h be a strong inversion of K, such that fh is also a strong inversion. Then we have the following.

Proposition 1.4. *The equality $s = r$ holds, if and only if h and fh are not conjugate in $Diff^+(S^3, K)$.*

Proof. We denote the elements of $\tilde{I}^+ \cong D_{2^s m}$ corresponding to f and h by the same symbols. Then h and fh are conjugate in $\tilde{I}^+$, if and only if $s > r$ (cf. [Sa3, Proposition 3.1]). Since the elements h and fh of $\tilde{I}^+$ are strongly equivalent to the original involutions (see Proof of Proposition 1.1), we obtain the desired result.

There are effective methods for showing inequivalences of strong inversions. For example, there is a polynomial invariant for a strong inversion defined by the second author [Sa2], which is in fact an invariant of the spatial θ-curve $\theta(K, h)$. (For further invariants of spatial θ-curves, see [Li, Ya, Wo].) Either this invariant or the set of the knot types of the three constituent knots of $\theta(K, h)$ is enough for the proof of the following examples.

Example 1.5. (1) 10_{120} has cyclic period 2 and has no other cyclic nor free periods (see [H1, Lü1]). It is invertible, and the strong inversions h and fh are as illustrated in Fig. 1.1. We see they are not conjugate and hence its symmetry group is D_2. The same result holds for 9_{16}, 9_{28}, 10_{78} and 10_{122}.

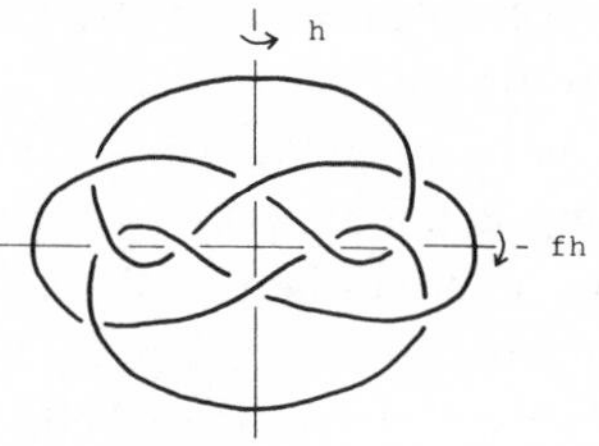

Figure 1.1

(2) 7_4 has cyclic period 2 and has no other cyclic nor free periods. However, the corresponding two strong inversions h and fh are equivalent. In fact, there is an element f_0 of $Diff^+(S^3, 7_4)$ of order 4, such that $f_0^2 = f$ (see [Sa2, pp.187-188] e.g.). Moreover we see the strong inversions h and f_0h are not conjugate. Thus its symmetry group is D_4.

(3) 10_{155} and 10_{157} have free periods 2 and 4 respectively. We can apply our method to them, and see that their symmetry groups are D_2 and D_4 respectively (cf. [Sa2, Example 3.5]).

(4) 9_{40} has cyclic periods 2 and 3, and has no other cyclic nor free periods. By our method, we see its symmetry group is $D_{2\cdot 3^r}$ for some $r \geq 1$.

Next, we consider Problem 1.3, where K has cyclic period n. Let f be a periodic map on (S^3, K) realizing the cyclic period n. Then the quotient space S^3/f is again a 3-sphere, and we obtain a 2-component link $q(Fix(f) \cup K)$, where q is the projection. We denote it by the symbol $L(K, f)$. By an argument similar to that of [Sa1], we obtain the following.

Lemma 1.6. *K is simple, if and only if one of the following holds.*
(1) *$n \geq 3$ and $L(K, f)$ is simple.*
(2) *$n = 2$, $L(K, f)$ is simple, and any 2-sphere in S^3/f intersecting $q(Fix(f))$ in 4 points bounds a trivial tangle in $(S^3/f, L(K, f))$*

Remark 1.7. In [Sa1, Theorem 3], we constructed distinct prime knots J_n and K_n whose n-fold branched cyclic coverings are homeomorphic, for each integer $n \geq 2$. By the above lemma, we see $J_n \quad (n \geq 2)$ and $K_n \quad (n \geq 3)$ are hyperbolic, while K_2 is non-simple.

In particular, it follows that, if K is hyperbolic, then $L(K, f)$ is hyperbolic. Let $Sym^*(S^3, L(K, f))$ be the subgroup of the symmetry group of $L(K, f)$ consisting of the elements which preserve each of the components and orientations of S^3 and the components of $L(K, f)$.

Proposition 1.8. *Suppose K is hyperbolic. Then we have the following exact sequence:*

$$1 \to Z_n \to Sym^*(S^3, K) \to Sym^*(S^3, L(K, f)) \to 1$$

To prove this proposition, we need the following lemma which is proved by using the result of Takeuchi [Ta].

Lemma 1.9. *Let f' be the element of $\tilde{I}$ corresponding to f. Then f and f' are strongly equivalent.*

Proof. Let f_* be the element of $Out(\pi_1(E(K)))$ induced by f. Then by [Cn], f_* has order n and hence f' has order n. We see that the orbifold fundamental groups of the quotient orbifolds $E(K)/f$ and $E(K)/f'$ are isomorphic. In fact, both groups are the extensions of the centerless group $\pi_1(E(K))$ by Z_n with the action of Z_n on $\pi_1(E(K))$ given by f_*; so they are isomorphic to the subgroup of $Aut(\pi_1(E(K)))$ which is the inverse image of the subgroup of $Out(\pi_1(E(K)))$ generated by f_*. Further the above orbifolds are sufficiently large by [D, Theorem 11], and they belong to the class $\mathcal{W}$ considered by Takeuchi [Ta]. Since E(K) is simple, the isomorphism $\pi_1^{orb}(E(K)/f) \cong \pi_1^{orb}(E(K)/f')$ preserves the peripheral structures. Hence, by [Ta], the above orbifolds are isomorphic and therefore f and f' are conjugate in $Diff(S^3, K)$. It follows that they are strongly equivalent since $[f] = [f']$ in $Sym(S^3, K)$.

Remark 1.10. This lemma is also obtained in [BiZ2, Theorem 2.1] using the orbifold uniformization theorem announced by Thurston[Th2].

Proof of Proposition 1.8. By Lemma 1.9, we may assume f belongs to $\tilde{I}$. We show $\tilde{I}^*/<f>$ is isomorphic to $Sym^*(S^3, L(K,f))$. Since $\tilde{I}^*/<f>$ acts effectively on the exterior $E(L(K,f))$, there is a natural injective homomorphism $\tilde{I}^*/<f>\rightarrow Out(\pi_1(E(L(K,f))))$ by [Cn]. The above action preserves the orientation of S^3/f and $q(K)$. Since $L(K,f)$ has a non-zero linking number, the above action also preserves the orientation of $q(Fix(f))$. Hence we obtain a natural injective homomorphism $\tilde{I}^*/<f>\rightarrow Sym^*(S^3, L(K,f))$. Since any homeomorphism of $(S^3, L(K,f))$ preserving the components of $L(K,f)$ lifts to a homeomorphism of (S^3, K), the above injective homomorphism is an epimorphism.

Example 1.11. Let $S(\alpha,\beta)=K_1\cup K_2$ be the 2-bridge link of type (α,β), and let n be a positive integer relatively prime to $lk(K_1,K_2)$. Since K_1 is trivial, the n-fold cyclic covering of S^3 branched along K_1 is S^3, and the inverse image of K_2 forms a cyclically periodic knot of period n. We denote it by the symbol $S_n(\alpha,\beta)$. If $\beta\equiv\pm1 \pmod{\alpha}$, then $S_n(\alpha,\beta)$ is a torus knot, and its symmetry group is isomorphic to D_1. Otherwise, $S_n(\alpha,\beta)$ is hyperbolic by Lemma 1.6, and we obtain the following results by using the calculation of the symmetry groups of 2-bridge links by [BnS] (cf. [Sa6]):

(1) $Sym^*(S^3, S_n(\alpha,\beta))$ is isomorphic to Z_{2n} or Z_n according to whether $\beta^2\equiv1 \pmod{\alpha}$ or not.

(2) $S_n(\alpha,\beta)$ is invertible.

(3) $S_n(\alpha,\beta)$ is amphicheiral, if and only if $\beta^2\equiv-1 \pmod{\alpha}$.

Any prime knot with ≤ 10 crossings which has cyclic period ≥ 3 is of this form. The following is the list of such knots which are not torus knots: $8_{18}\cong S_4(10,3)$, $9_{40}\cong S_3(24,-7)$, $9_{41}\cong S_3(18,5)$, $9_{47}\cong S_3(16,3)$, $9_{49}\cong S_3(14,3)$, $10_{123}\cong S_5(10,3)$.

2. Symmetries of θ-curves

In this section, B denotes a fixed 3-ball in S^3, and B^c denotes S^3-intB. Consider the arcs $\alpha_1^{\pm}$, $\alpha_2^{\pm}$, α_3, β_1 and β_2, and the points $p_i (1\leq i\leq 4)$ and $w_{\pm}$ as illustrated in Fig. 2.1. Here β_1 and β_2 are contained in ∂B, and the union $\alpha_1^-\cup\alpha_1^+\cup\alpha_2^-\cup\alpha_2^+\cup\alpha_3$ forms an H-shaped graph in B; we denote it by the symbol γ. Let ξ_1 and ξ_2 be the π -rotations of B as illustrated in Fig. 2.2, and let δ denote a Dehn twist along the disk D as illustrated in Fig. 2.2. We choose the above diffeomorphisms so that they are commutative and preserve γ.

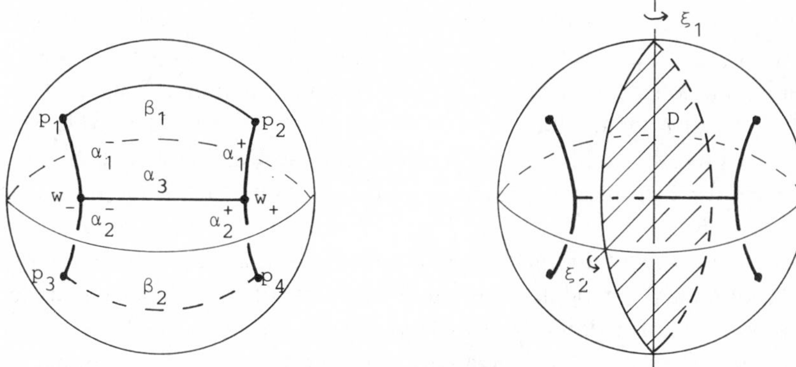

Figure 2.1 and 2.2

We use the term *tangle* to denote the pair (B, t), where t is a one dimensional submanifold of B with $\partial t = \{p_i | 1 \leq i \leq 4\}$. Following [Cw], we denote the tangles $(B, \xi_1(t))$ and $(B, \xi_2(t))$ by the symbols (B, t_v) and (B, t_r) respectively.

Let $\theta = e_1 \cup e_2 \cup e_3$ be a spatial θ -curve in S^3 with vertices v_- and v_+. By an ambient isotopy, we can deform θ so that $B \cap \theta = \gamma$, $B \cap e_i = \alpha_i^- \cup \alpha_i^+ \quad (i = 1, 2)$, $B \cap e_3 = \alpha_3$, and $v_\pm = w_\pm$. Note that the isotopy type of $B^c \cap \theta$ in B^c rel ∂B^c is not uniquely determined by θ (see Fig. 2.3).

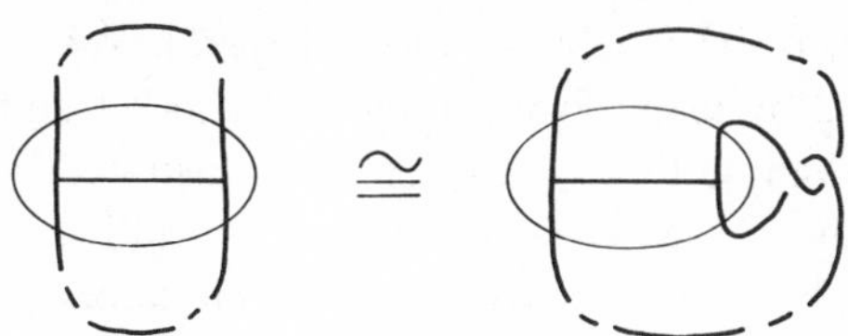

Figure 2.3

In the following, we fix a particular embedding θ in S^3 which satisfy the above conditions. For a tangle (B, t), let $\theta(t)$ denote the link in S^3 given by $(S^3, \theta(t)) = (B, t) \cup (B^c, B^c \cap \theta)$.

Proposition 2.1. (1) *Suppose θ is reversible, then for any tangle (B, t), the links $\theta(t)$ and $\theta(t_v)$ are ambient isotopic.*
(2) *Suppose θ has period 2 with center in e_3, then for any tangle (B, t), the links $\theta(t)$ and $\theta(t_r)$ are ambient isotopic.*

Proof. Let f be a homeomorphism of (S^3, θ) which realizes the reversibility of θ. By the uniquenesses of regular neighbourhoods, we may assume $f(B) = B$. By [Wo, Lemma 3.1], the restriction $f|_{(B,t)}$ is isotopic to $\xi_1 \circ \delta^n$ for some integer n. Put $e_i' = e_i \cap B^c$, and consider the knots $K_i = e_i' \cup \beta_i$ for $i = 1, 2$. Then $f(K_i) = e_i' \cup \delta^n(\beta_i)$, and hence $lk(K_1, K_2) = lk(f(K_1), f(K_2)) = lk(K_1, K_2) \pm n$. Hence $n = 0$, and we may assume $f|_{(B,\gamma)} = \xi_1$. Then we have $f(\theta(t)) = \theta(t_r)$ and obtain (1), since f preserves the orientation of S^3. (2) is proved similarly.

Thus to prove that a spatial θ-curve θ in S^3 is not reversible, we have only to show that the links $\theta(t)$ and $\theta(t_r)$ are not ambient isotopic for some (non-reversible) tangle (B, t). However, skein polynomials and their 2-parallel versions cannot distinguish them, since $\theta(t_r)$ is a mutant of $\theta(t)$ (c.f. [LL, MT, Mk]). To accomplish this, we employ the following "classical" method: Let K be a knot in S^3. Then for each natural number n and a partition $\mathcal{P}$ of n, the followings are invariants of K.

(2.2) The number of all transitive representations of $\pi_1(E(K))$ to the symmetric group S_n of degree n modulo inner-automorphisms of S_n, which send a meridian of K to products of disjoint cycles whose lengths are given by $\mathcal{P}$.

(2.3) The set of homology invariants (i.e., the first homology groups and the covering linkage invariants) of the n-fold coverings of S^3 branched over K corresponding to the above representations.

The first author wrote a program in which we can get the above invariants of a link by drawing its diagram with a mouse. It also calculates Alexander, Conway, Jones, Kauffmann, P- and Q-polynomials, and checks Murasugi's condition [Ms1] for the Alexander polynomials of periodic knots. The program works on NEC PC-9801 series, and is available on request.

The knots 10_n (n = 77, 99, 100, 101, 103, 108, 111, 112, 113, 114, 121, 123, 163, 165 and 166) are strongly invertible and have possibility to admit cyclic period 2 according to the algebraic study of [Lü1]. We show these knots do not have cyclic period 2 by using Propositions 1.2 and 2.1 and the above method. To do this, we use the tangle (B, s) or (B, s') as illustrated in Fig. 2.4, and compare the above invariants of the knots $\theta(t)$ and $\theta(t_v)$ with $t = s$ or s' for a θ-curve arising from each of the above strongly invertible knots.

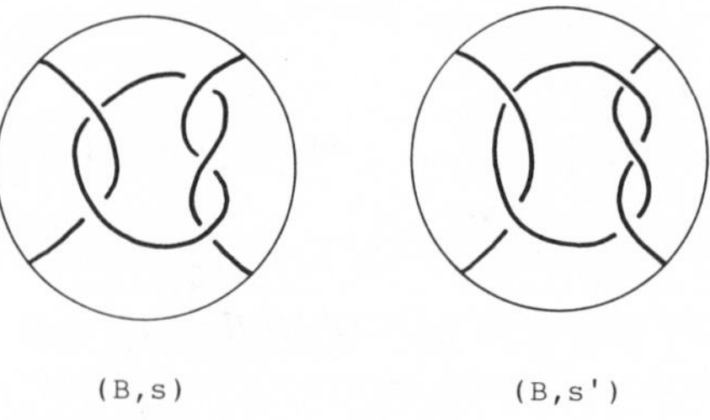

Figure 2.4

In all cases, the first homology groups of 5 or 6-fold branched coverings suffice to distinguish them, and we see the above knots do not have cyclic period 2 by Propositions 1.2 and 2.1. In the following Table 2.1, we present a θ-curve $\theta(K, h)$, a partition $\mathcal{P}$ of a number n (= 5 or 6), and the invariants of $\theta(t)$ and $\theta(t_v)$ described in (2.3). In the construction of these knots, the ball B is taken to be the regular neighbourhood of an arc with an arrow, and we identify $(B, B \cap \theta)$

Table 2.1

10_{77} $\mathcal{P} = \{3,3\}$
$\theta(s)$: $D(2,2,2,34,204)$
$D(3,23616), (4,68,2720)$
$(9,945), (45,45)$
$D(68,68,136), (53436)$
$\theta(s_v)$: $D(2,2,2,34,204)$
$(4,68,2720), D(45,45)$
$D(68,68,136), (53436)$
$D(70200)$

10_{99} $\mathcal{P} = \{4,1\}$
$\theta(s)$: $(5;1), (5;1), (25,25), (105)$
$\theta(s_v)$: $(2;2), (5;1), (25,25), (105)$.

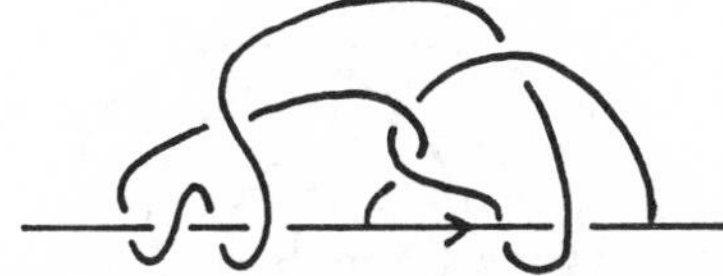

10_{100} $\mathcal{P} = \{4,1\}$
$\theta(s)$: (50)
$\theta(s_v)$: —

10_{101} $\mathcal{P} = \{4,1\}$
$\theta(s)$: (24)
$\theta(s_v)$: (39)

10_{103} $\mathcal{P} = \{5\}$
$\theta(s)$: $(6,6,246,246), (3165)$
$\theta(s_v)$: $(6,6,246,246), (9165)$

10_{108} $\mathcal{P} = \{4,1\}$
$\theta(s)$: $(5), (76)$
$\theta(s_v)$: $(24), (130)$

10_{111} $\mathcal{P} = \{4,1\}$
$\theta(s)$: $(2,10,60)$
$\theta(s_v)$: $(5,5), (5,210), (1424)$

Table 2.1 (cont.)

10_{112} $\mathcal{P} = \{4,1\}$
$\theta(s)$: (30), (95), (810), (1278)
$\theta(s_v)$: (2, 40), (30), (260), (1460)

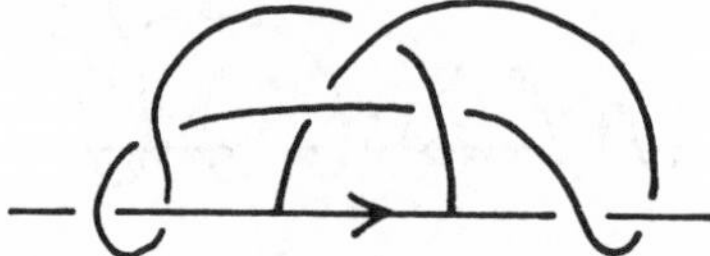

10_{113} $\mathcal{P} = \{4,1\}$
$\theta(s)$: (2; 1), (25), (195), (2474)
$\theta(s_v)$: (2; 1), (5), (5, 520), (150)

10_{114} $\mathcal{P} = \{4,1\}$
$\theta(s)$: (37), (51), (65), (90)
$\theta(s_v)$: (43), (65), (110), (265)

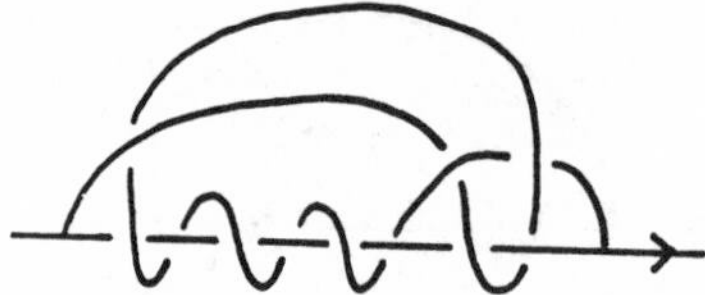

10_{121} $\mathcal{P} = \{3,2\}$
$\theta(s)$: (209), (2764)
$\theta(s_v)$: (43), (65), (110), (265)

10_{123} $\mathcal{P} = \{4,1\}$
$\theta(s)$: (35), (40), (78)
$\theta(s_v)$: (2, 30), (335), (418)

10_{163} $\mathcal{P} = \{4,1\}$
$\theta(s')$: (5), (80), (1858)
$\theta(s'_v)$: (45), (120), (833)

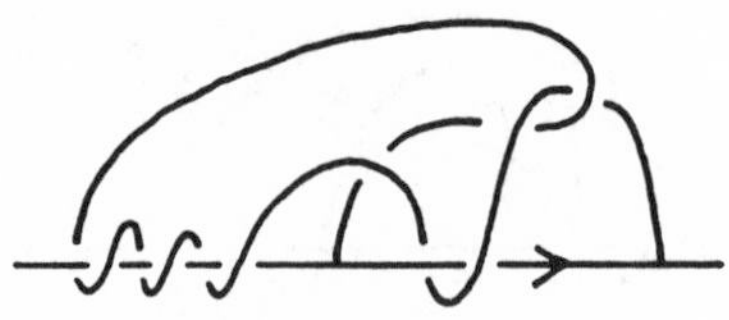

10_{165} $\mathcal{P} = \{4,1\}$
$\theta(s')$: (2, 14)
$\theta(s'_v)$: (357)

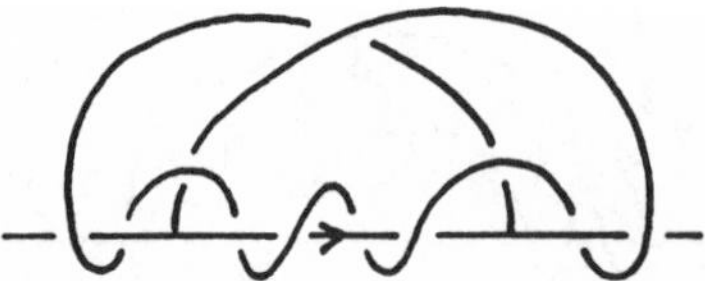

10_{166} $\mathcal{P} = \{4,1\}$
$\theta(s)$: (7), (305)
$\theta(s_v)$: (7), (225)

with (B, γ) "naturally" so that the origin and the target of the arrow correspond to w_- and w_+ respectively. The symbols following $\theta(t)$ [resp. $\theta(t_v)$] give the first homology groups of the coverings of S^3 branched over $\theta(t)$ [resp. $\theta(t_v)$]. Here $(d_1, \ldots, d_r)$ [resp. $(d_1, \ldots, d_r; \beta)$] represents the abelian group $Z_{d_1} \oplus \cdots \oplus Z_{d_r}$ [resp. $Z_{d_1} \oplus \cdots \oplus Z_{d_r} \oplus Z^\beta$], and the symbol '$D$' means that there are two branched coverings whose first homology groups are given by the symbol following it.

3. Table of the symmetry groups

In this section, we give a table of the symmetry groups and the cyclic periods of prime knots up to 10 crossings except three knots. To do this, we need the following facts (1) $\sim$ (6) in addition to our calculation and the results stated in the introduction.

(1) The symmetry group of a nontrivial torus knot is isomorphic to D_1 (see [Sc]).

(2) The symmetry group of a 2-bridge knot $S(\alpha, \beta)$ of type (α, β) is calculated by Conway (see [GLM]) and Bonahon and Siebenmann [BnS] (cf. [Sa6]). Suppose $\beta \not\equiv \pm 1 \pmod{\alpha}$, that is $S(\alpha, \beta)$ is not a torus knot. Then the symmetry group is isomorphic to D_4 or D_2 according to whether $\beta^2 \equiv \pm 1 \pmod{\alpha}$ or not; and the knot is amphicheiral, if and only if $\beta^2 \equiv -1 \pmod{\alpha}$ (cf. Example 1.11).

(3) The symmetry groups of nonelliptic Montesinos knots are determined by Boileau and Zimmermann [BiZ1].

(4) For elliptic Montesinos knots, the isometry groups of the elliptic π-orbifolds coming from them are calculated by the second author (see [Sa6]). If we use [Th2], this gives their symmetry groups. For elliptic Montesinos knots up to 10 crossings, their symmetry groups are determined without using [Th2] except 8_{10} and 8_{20}.

(5) If Conway's notation of a prime knot with ≤ 10 crossings has basic polyhedron 1^* or 6^{**}, then it is algebraic (see [Cw, p.335]). Further if it is not a Montesinos knot, then it is equivalent to the knot K as illustrated in Figure 3.1.

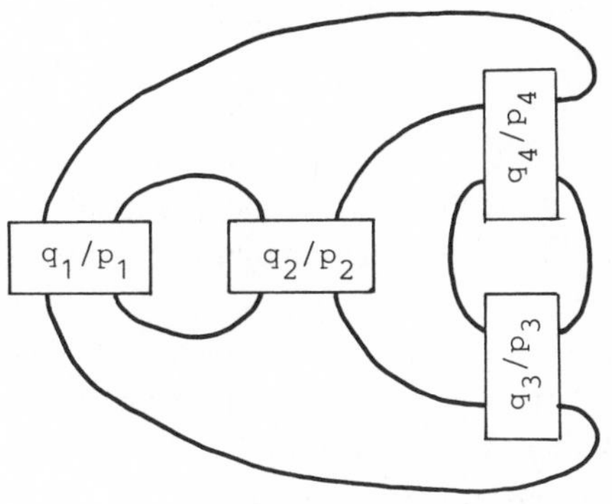

Figure 3.1

Here for each i $(1 \leq i \leq 4)$, p_i and q_i are relatively prime integers, and the box $\boxed{q_i/p_i}$ stands for the rational tangle represented by the rational number q_i/p_i (see [Cw]). To describe its symmetry group, we introduce some notation. Let $\mathcal{S}$ be the subgroup of the symmetric group S_4 of degree 4 consisting of permutations which sends the subset $\{1,2\}$ to either itself or $\{3,4\}$. Let $\mathcal{S}(K)$ be the subgroup of $\mathcal{S}\times\{\pm\}$ consisting of the elements (σ,ϵ) which satisfy the following conditions:

$$\frac{q_i}{p_i} \equiv \epsilon \frac{q_{\sigma(i)}}{p_{\sigma(i)}} \quad \text{in} \quad Q/Z \quad (1 \leq i \leq 4)$$

$$\frac{q_i}{p_i} + \frac{q_{i+1}}{p_{i+1}} = \epsilon \left(\frac{q_{\sigma(i)}}{p_{\sigma(i)}} + \frac{q_{\sigma(i+1)}}{p_{\sigma(i+1)}} \right) \quad \text{in} \quad Q \quad (i = 1,3)$$

Then as a very special case of the results of Bonahon and Siebenmann [BoS], the symmetry group of K is isomorphic to $\mathcal{S}(K)$, if the following conditions are satisfied:

(i) $p_i \geq 2 \quad (1 \leq i \leq 4)$.

(ii) For $i = 1$ or 3, if $\frac{q_i}{p_i} \equiv \frac{q_{i+1}}{p_{i+1}} \equiv \frac{1}{2}$ in Q/Z, then $\frac{q_i}{p_i} + \frac{q_{i+1}}{p_{i+1}} \neq 0$ in Q.

(6) According to [Lü1], 10_{159} has possibility to admit cyclic period 3. However, by using the modified Murasugi's condition presented by Davis and Livingston [DL], we see it is impossible. To see this, we use the fact that the prime factorization of its Alexander polynomial in $Z[\zeta] < t >$, where ζ is the primitive third root of 1, is

$$(t+\zeta)(t+\bar{\zeta})\{t^2+(-1+\zeta)t+1\}\{t^2+(-1+\bar{\zeta})t+1\}.$$

In the following Table 3.1, the columns "Sym" and "CP" give symmetry groups and cyclic periods respectively. [There are some errors in the table of cyclic periods contained in [BuZ, Table I].] Each symmetry group $Sym(S^3, K)$ is presented paying attention to its action on the subset $K \cong S^1$ (see Lemma 1.1). Thus, though Z_2 and D_1 are isomorphic as abstract groups, they have different meanings; that is, Z_2 means that the symmetry group is generated by a diffeomorphism which preserves the orientation of K, while D_1 means that it is generated by one which reverses the orientation of K.

A symbol on the upper right of a symmetry group describes the amphicheirality of the corresponding knot. The symbol i [resp. f] means that the knot is *involutory* [resp. with *full* symmetry]; that is, amphicheiral and non-invertible [resp. amphicheiral and invertible]. If no such symbols are added, then the knot is non-amphicheiral.

A question mark n? in the column "CP" means that there remains possibility for the knot to have cyclic period n. A question mark G? in the column "Sym" means that, if we show that the knot does not have the cyclic period with question mark, then the symmetry group is isomorphic to G.

Table 3.1

Knot	Sym	CP	Knot	Sym	CP	Knot	Sym	CP
3_1	D_1	2, 3	9_8	D_2	2	10_1	D_2	2
4_1	$D_4^{\dagger}$	2	9_9	D_2	2	10_2	D_2	2
5_1	D_1	2, 5	9_{10}	D_4	2	10_3	D_2	2
5_2	D_2	2	9_{11}	D_2	2	10_4	D_2	2
6_1	D_2	2	9_{12}	D_2	2	10_5	D_2	2
6_2	D_2	2	9_{13}	D_2	2	10_6	D_2	2
6_3	$D_4^{\dagger}$	2	9_{14}	D_2	2	10_7	D_2	2
7_1	D_1	2, 7	9_{15}	D_2	2	10_8	D_2	2
7_2	D_2	2	9_{16}	D_2	2	10_9	D_2	2
7_3	D_2	2	9_{17}	D_4	2	10_{10}	D_2	2
7_4	D_4	2	9_{18}	D_2	2	10_{11}	D_2	2
7_5	D_2	2	9_{19}	D_2	2	10_{12}	D_2	2
7_6	D_2	2	9_{20}	D_2	2	10_{13}	D_2	2
7_7	D_4	2	9_{21}	D_2	2	10_{14}	D_2	2
8_1	D_2	2	9_{22}	D_1	-	10_{15}	D_2	2
8_2	D_2	2	9_{23}	D_4	2	10_{16}	D_2	2
8_3	$D_4^{\dagger}$	2	9_{24}	D_1	-	10_{17}	$D_4^{\dagger}$	2
8_4	D_2	2	9_{25}	D_1	-	10_{18}	D_2	2
8_5	D_2	2	9_{26}	D_2	2	10_{19}	D_2	2
8_6	D_2	2	9_{27}	D_2	2	10_{20}	D_2	2
8_7	D_2	2	9_{28}	D_2	2	10_{21}	D_2	2
8_8	D_2	2	9_{29}	D_1	-	10_{22}	D_2	2
8_9	$D_4^{\dagger}$	2	9_{30}	D_1	-	10_{23}	D_2	2
8_{10}	D_1	-	9_{31}	D_4	2	10_{24}	D_2	2
8_{11}	D_2	2	9_{32}	1	-	10_{25}	D_2	2
8_{12}	$D_4^{\dagger}$	2	9_{33}	1	-	10_{26}	D_2	2
8_{13}	D_2	2	9_{34}	D_1	-	10_{27}	D_2	2
8_{14}	D_2	2	9_{35}	D_6	2, 3	10_{28}	D_2	2
8_{15}	D_2	2	9_{36}	D_1	-	10_{29}	D_2	2
8_{16}	D_1	-	9_{37}	D_2	2	10_{30}	D_2	2
8_{17}	D_1^{i}	-	9_{38}	D_1	-	10_{31}	D_2	2
8_{18}	$D_8^{\dagger}$	2, 4	9_{39}	D_1	-	10_{32}	D_2	2
8_{19}	D_1	2,3,4	9_{40}	D_6	2, 3	10_{33}	$D_4^{\dagger}$	2
8_{20}	D_1	-	9_{41}	D_3	3	10_{34}	D_2	2
8_{21}	D_2	2	9_{42}	D_1	-	10_{35}	D_2	2
9_1	D_1	2,3,9	9_{43}	D_1	-	10_{36}	D_2	2
9_2	D_2	2	9_{44}	D_1	-	10_{37}	$D_4^{\dagger}$	2
9_3	D_2	2	9_{45}	D_1	-	10_{38}	D_2	2
9_4	D_2	2	9_{46}	D_2	2	10_{39}	D_2	2
9_5	D_2	2	9_{47}	D_3	3	10_{40}	D_2	2
9_6	D_2	2	9_{48}	D_6	2	10_{41}	D_2	
9_7	D_2	2	9_{49}	D_3	3	10_{42}	D_2	

Table 3.1 (cont.)

Knot	Sym	CP	Knot	Sym	CP	Knot	Sym	CP
10_{43}	$D_4^{\dagger}$	2	10_{85}	1	-	10_{127}	D_1	-
10_{44}	D_2	2	10_{86}	1	-	10_{128}	D_1	-
10_{45}	$D_4^{\dagger}$	2	10_{87}	1	-	10_{129}	D_1	-
10_{46}	D_1	-	10_{88}	D_1^i	-	10_{130}	D_1	-
10_{47}	D_1	-	10_{89}	D_1	-	10_{131}	D_1	-
10_{48}	D_1	-	10_{90}	1	-	10_{132}	D_1	-
10_{49}	D_1	-	10_{91}	1	-	10_{133}	D_1	-
10_{50}	D_1	-	10_{92}	1	-	10_{134}	D_1	-
10_{51}	D_1	-	10_{93}	1	-	10_{135}	D_1	-
10_{52}	D_1	-	10_{94}	1	-	10_{136}	D_2	2
10_{53}	D_1	-	10_{95}	1	-	10_{137}	D_1	-
10_{54}	D_1	-	10_{96}	D_1	-	10_{138}	D_2	2
10_{55}	D_1	-	10_{97}	D_1	-	10_{139}	D_2	2
10_{56}	D_1	-	10_{98}	Z_2	2	10_{140}	D_1	-
10_{57}	D_1	-	10_{99}	$D_2^{\dagger}$	-	10_{141}	D_2	2
10_{58}	D_2	2	10_{100}	D_1	-	10_{142}	D_2	2
10_{59}	D_1	-	10_{101}	D_1	-	10_{143}	D_1	-
10_{60}	D_2	2	10_{102}	1?	2?	10_{144}	D_2	2
10_{61}	D_2	2	10_{103}	D_1	-	10_{145}	D_2	2
10_{62}	D_1	-	10_{104}	D_1	-	10_{146}	D_2	2
10_{63}	D_2	2	10_{105}	D_1	-	10_{147}	Z_2	2
10_{64}	D_2	2	10_{106}	1	-	10_{148}	1	-
10_{65}	D_1	-	10_{107}	1	-	10_{149}	1	-
10_{66}	D_2	2	10_{108}	D_1	-	10_{150}	1	-
10_{67}	Z_2	2	10_{109}	D_1^i	-	10_{151}	1	-
10_{68}	D_2	2	10_{110}	1	-	10_{152}	D_1	-
10_{69}	D_2	2	10_{111}	D_1	-	10_{153}	1	-
10_{70}	D_1	-	10_{112}	D_1	-	10_{154}	D_1	-
10_{71}	D_1	-	10_{113}	D_1	-	10_{155}	D_2	-
10_{72}	D_1	-	10_{114}	D_1	-	10_{156}	D	-
10_{73}	D_1	-	10_{115}	D_1^i	-	10_{157}	D_4	-
10_{74}	D_2	2	10_{116}	D_1	-	10_{158}	D_1	-
10_{75}	D_6	2	10_{117}	1?	2?	10_{159}	1	-
10_{76}	D_2	2	10_{118}	D_1^i	-	10_{160}	D_1	-
10_{77}	D_1	-	10_{119}	1?	2?	10_{161}	D_1	-
10_{78}	D_2	2	10_{120}	D_2	2	10_{162}	= 10_{161}	
10_{79}	D_1^i	-	10_{121}	D_1	-	10_{163}	D_1	-
10_{80}	1	-	10_{122}	D_2	2	10_{164}	D_1	-
10_{81}	D_1^i	-	10_{123}	$D_{10}^{\dagger}$	5	10_{165}	D_1	-
10_{82}	1	-	10_{124}	D_1	3, 5	10_{166}	D_1	-
10_{83}	1	-	10_{125}	D_1	-			
10_{84}	1	-	10_{126}	D_1	-			

Added in proof. The question marks in Table 3.1 can be omitted by the work of C. Adams et al., "Hyperbolic invariants of knots and links" [Trans. A.M.S. 326 (1991), 1–56]. Professor Jeffery Weeks kindly pointed out that the symmetry group of 10_{159} is not 1 (as listed in our paper), but D_1.

References

[BiZ1] M. Boileau and B. Zimmermann, *Symmetries of nonelliptic Montesinos links*, Math. Ann. **277** (1987), 563–584.

[BiZ2] M. Boileau and B. Zimmermann, *The π-orbifold group of a link*, Math. Z. **200** (1989), 187–208.

[BnS] F. Bonahon and L. Siebenmann, *Geometric splittings of links, and Conway's algebraic knots*, Proc. Bangor Conf. 1979, (to appear).

[Bu] G. Burde, *Über periodische Knoten*, Archiv Math. **30** (1978), 487–492.

[BuZ] G. Burde and H. Zieschang, *Knots*, De Gruyter Studies in Mathematics 5, de Gruyter, Berlin, Ney York, 1985.

[Cn] P. E. Conner, *Transformation groups on $K(\pi, 1)$ II*, Mich. Math. J. **6** (1959), 413-417.

[Cw] J. H. Conway, *An enumeration of knots and links and some of their related properties*, Computational problems in abstract algebra, Proc. Conf. Oxford 1967 (ed. J. Leech), Markham, Chicago, pp. 329–359.

[DL] J. F. Davis and C. Livingston, *Alexander polynomials of periodic knots*, preprint.

[D] W. D. Dunbar, *Hierarchies for 3-orbifolds*, Topology and its applications **29** (1988), 267–283.

[GLM] C. McA. Gordon, R. A. Litherland, and K. Murasugi, *Signatures of covering links*, Canad. J. Math. **33** (1981), 381–394.

[GLu] C. McA. Gordon and J. Luecke, *Knots are determined by their complements*, J. Amer. Math. Soc. **2** (1989), 371–415.

[H1] R. Hartley, *Knots with free periods*, Canad. J. Math. **33** (1981), 91–102.

[H2] R. Hartley, *Identifying non-invertible knots*, Topology **22** (1983), 137-145.

[HK] R. Hartley and A. Kawauchi, *Polynomials of amphicheiral knots*, Math. Ann. **243** (1989), 63–70.

[LL] W. B. R. Lickorish and A. S. Lipson, *Polynomials of 2-cable-like links*, Proc. A.M.S. **100** (1987), 355–361.

[Li] R. Litherland, *The Alexander module of a knotted theta-curve*, Math. Proc. Cambridge Phil. Soc. **105** (1989), 95–106.

[Lü1] U. Lüdicke, *Zyklische Knoten*, Archiv Math. **32** (1979), 588–599.

[Lü2] U. Lüdicke, 9_{25} *has no period* 3, C. R. Math. Rep. Acad. Sci. Canada **VI** (1984), 157–158.

[MB] J. Morgan and H. Bass, *The Smith conjecture*, Academic Press, 1984.

[MT] H.R. Morton and P. Traczyk, *The Jones polynomials of satellite links*, Contemp. Math. **78** (1988), 587–592.

[Mk] J. Murakami, *The parallel version of polynomial invariants of links*, Osaka J. Math. **26** (1989), 1–55.

[Ms1] K. Murasugi, *On periodic knots*, Comment. Math. Helv. **46** (1971), 162–174.

[Ms2] K. Murasugi, *On symmetries of knots*, Tsukuba J. Math. **4** (1980), 331–347.

[Ms3] K. Murasugi, *Jones polynomials of periodic links*, Pacific J. Math. **131** (1988), 319–329.

[Pe] A. P. Perko, *On 10-crossing knots*, Portugalliae Math. **38** (1979), 5–9.

[Pr] J.H. Przytycki, *On Murasugi's and Traczyk's criteria for periodic knots*, Math. Ann. **283** (1989), 465–478.

[Ri1] R. Riley, *An elliptic path from parabolic representations to hyperbolic structures*, Topology of low dimensional manifolds, Second Sussex Conf. 1977 (ed. R. Fenn), Lect. Notes in Math. 722 (1979) pp.99-133, Springer-Verlag.

[Ri2] R. Riley, *Seven excellent knots*, London Math. Soc. Lect. Note Series 48 (1982) pp.81-151.

[Ro] D. Rolfsen, *Knots and links*, Math. Lect. Note Series 7, Publish or Perish Inc., 1976.

[Sa1] M. Sakuma, *Periods of composite links*, Math. Sem. Notes, Kobe Univ. **9** (1981), 445–452.

[Sa2] M. Sakuma, *On strongly invertible knots*, Algebraic and Topological Theories, Proc. Miyata memorial Conf., Kinokuniya Co. Ltd., 1985.

[Sa3] M. Sakuma, *Non-free-periodicity of amphicheiral hyperbolic knots*, Homotopy Theory and related topics, Adv. Stud. in Math. 9 (1986), Kinokuniya and North-Holland.

[Sa4] M. Sakuma, *Uniqueness of symmetries of knots*, Math. Z. **192** (1986), 225–242.

[Sa5] M. Sakuma, *Realization of the symmetry groups of links*, Transformation groups (ed. K. Kawakubo), Lect. Notes in Math. 1375 (1989) pp.291-306, Springer-Verlag.

[Sa6] M. Sakuma, *The geometries of spherical Montesinos links*, Kobe J. Math. **7** (1990), 167–190.

[Sc] O. Schreier, *Über die Gruppen* $A^a B^b = 1$, Abh. Math. Sem. Univ. Hamburg **3** (1924), 167–169.

[Si] L. Siebenmann, *On vanishing of the Rochlin invariant and nonfinitely amphicheiral homology* 3-*spheres*, Topology symposium, Siegen, Lect. Notes in Math. 788 (1980), Springer-Verlag, Berlin-Heidelberg-New York.

[Ta] Y. Takeuchi, *Waldhausen's classification theorem for finitely uniformizable* 3-*orbifolds*, preprint.

[Th1] W. Thurston, *The geometry and topology of* 3-*manifolds*, mimeographed note.

[Th2] W. Thurston, *Three manifolds with symmetry*, preliminary report.

[To] J. L. Tollefson, *Involutions of sufficiently large* 3-*manifolds*, Topology **20** (1981), 323–352.

[Tr1] P. Traczyk, 10_{101} *has no period* 7*: a criterion for periodicity of links*, Proc. A.M.S. **108,** (1990), 845–846.

[Tr2] P. Traczyk, *A criterion for knots of period* 3, Topology and its appl. **36** (1990), 275–281.

[Tr3] P. Traczyk, *Periodic knots and the skein polynomial*, preprint.

[Ya] S. Yamada, *An invariant of spatial graphs*, J. of Graph Theory **13** (1989).

[Yo1] Y. Yokota, *The Jones polynomial of periodic knots*, preprint.

[Yo2] Y. Yokota, *The skein polynomial of periodic knots*, preprint.

[Wa] F. Waldhausen, *On irreducible* 3-*manifolds which are sufficiently large*, Ann. of Math. **87**, 56–88.

[Wo] K. Wolcott, *The knotting theta curves and other graphs in* S^3, Geometry and Topology, (1987), Marcel Dekker.

Tangle invariants and centre of the quantum group

H.C. Lee

Abstract. An ambient isotopy $\mathcal{V}$ for (n,n)-tangles $T^{(n)}$, $\mathcal{V} : T^{(n)} \rightarrow \mathcal{A}^{\otimes n}$ is contructed on the quasitriangular Hopf algebra $\mathcal{A}$ of a quantum group. $\mathcal{V}[T^{(0)}]$ is an invariant for links. A theorem stating that the set $\{\mathcal{V}[T^{(1)}] |$ all $T^{(1)}\} \subset$ centre of $\mathcal{A}$ is proved. This implies that $\mathcal{V}[T^{(1)}]$ is also an invariant of links. In particular, if $\pi \in End(V_{\mathcal{A}})$ is an irreducible representation of $\mathcal{A}$, then for all $(1,1)$-tangles $T^{(1)}$ whose closure is L, a link polynomial $Q_\pi[L] \in C[[q]]$ is given by the relation $\pi(\mathcal{V}[T^{(1)}]) = Q_\pi[L]\pi(1)$, and redundantly by $\pi(\mathcal{V}[L]) = Q_\pi[L]Trace_\pi(h)$, where h is a universal element of $\mathcal{A}$ and q denotes a generic deformation parameter of the quantum group. The isotopy is extended to cabled tangles using the coproduct in $\mathcal{A}$. Two examples, when π is the fundamental 2×2 representation of $sl(2;C)_q$ and when it is that of $(gl(2;C)/U(1))_{q,s}$, are discussed. The $Q_\pi[L]$ in these cases are respectively the Jones and Alexander-Conway polynomials.

1. Quantum group and Hopf algebra

We understand a quantum group in the sense of Drinfeld [**1**]. Specifically it is a quantized universal enveloping algebra of a Lie algebra g, denoted by $\mathcal{U}_q(g)$, or simply $(g)_q$ in this paper. It has a Hopf algebra structure composed of the set $\{\mathcal{A}, m, \triangle, u, \epsilon, S\} = \{$*bialgebra, product, coproduct, unit, counit, antipode*$\}$ over the field $V_{\mathcal{A}}$. Often we refer to the Hopf algebra as $\mathcal{A}$. The maps $m, \triangle, u$ and ϵ are homomorphisms; S is an antihomomorphism. The following relations are especially important for our purpose:

$$(id \otimes \triangle)\triangle(a) = (\triangle \otimes id)\triangle(a), \tag{1.1}$$

$$m(id \otimes S)\triangle(a) = m(S \otimes id)\triangle(a) = u\epsilon(a), \quad \forall a \in \mathcal{A}. \tag{1.2}$$

Knots 90

2. Quasitriangular $\mathcal{R}$-matrix

It is assumed that $\mathcal{A}$ is quasitriangular, namely that there exists an invertable universal $\mathcal{R}$-matrix $\mathcal{R} \in \mathcal{A} \otimes \mathcal{A}$, whose inverse is $\mathcal{R}^{-1} = (S \otimes id)\mathcal{R} = (id \otimes S^{-1})\mathcal{R}$, that satisfies the relations

$$\mathcal{T} \circ \triangle(a)\mathcal{R} = \mathcal{R}\triangle(a), \qquad \forall a \in \mathcal{A}, \tag{2.1}$$

$$(\triangle \otimes id)\mathcal{R} = \mathcal{R}_{13}\mathcal{R}_{23}, \quad (id \otimes \triangle)\mathcal{R} = \mathcal{R}_{13}\mathcal{R}_{12}, \tag{2.2}$$

where if $\mathcal{R} = \sum_i a_i \otimes b_i$, then, e.g., $\mathcal{R}_{13} = \sum_i a_i \otimes 1 \otimes b_i$, and so on. The braid relation

$$\mathcal{R}_{12}\mathcal{R}_{13}\mathcal{R}_{23} = \mathcal{R}_{23}\mathcal{R}_{13}\mathcal{R}_{12}, \tag{2.3}$$

is a consequence of the above two relations.

From its property of quantum doubling [**1**], $\mathcal{A}$ has Hopf subalgebras $\mathcal{A}^+$ and $\mathcal{A}^-$ which are dual to each other, so that the (opposite) coproduct on $\mathcal{A}^+$ ($\mathcal{A}^-$) induces the product on $\mathcal{A}^-$ ($\mathcal{A}^+$). This means that if $\{e_\sigma\}$ is a basis for $\mathcal{A}^-$, then [**1,2**] there exists a basis $\{e^\sigma\}$ for $\mathcal{A}^+$ dual to $\{e_\sigma\}$, and

$$\mathcal{R} = \sum_\sigma e_\sigma \otimes e^\sigma. \tag{2.4}$$

For quantum groups that are quantized universal envoloping algebras of simple Lie algebras $\mathcal{A}$ are finitely generated infinite algebras so that the summation over σ in (2.4) is infinite. This makes the derivation of certain properties of a quasitriangular Hopf algebra problematic. (For instance (2.1) and (2.2) express the same relation if the sum were finite.) In the following we take another approach to the description of $\mathcal{R}$ so that the summation in (2.4) may be replaced by the limit of a finite summation.

Consider $\mathcal{A}^\pm$ as linear spaces respectively spanned by $\{e_\sigma\}$ and $\{e^\sigma\}$. We want to construct a series of finite subspaces

$$\mathcal{S}^\pm_{N_1} \subset \mathcal{S}^\pm_{N_2} \ldots \subset \mathcal{S}^\pm_\infty = \mathcal{A}^\pm, \quad N_1 < N_2 \ldots \infty \tag{2.5}$$

each having the property

$$\triangle(\mathcal{S}^\pm_N) \subset \mathcal{S}^\pm_N \otimes \mathcal{S}^\pm_N, \tag{2.6}$$

This is can be achieved because coproduct is homogeneous in the lowering (and raising) generators. In the case of $sl(2;C)_q$ with generators [**2**] $\{k, k^{-1}, X^\pm\}$ and coproduct $\triangle(X^\pm) = X^\pm \otimes k + k^{-1} \otimes X^\pm$ we let $\mathcal{S}^\pm_N$ be spanned by the smallest set containing $(X^\pm)^N$. Then the ideals $\mathcal{I}^\pm_N$ generated by $(X^\pm)^{N+1}$ are $\notin \mathcal{S}^\pm_N$. For example $\mathcal{S}^\pm_1$ is spanned by $\{k^{-1}, k, X^\pm\}$, $\mathcal{S}^\pm_3 \supset \mathcal{S}^\pm_1$ by $\{k^{-3}, k^{-1}, k, k^3, k^{-2}X^\pm, X^\pm, k^2X^\pm, k^{-1}(X^\pm)^2, k(X^\pm)^2, (X^\pm)^3\}$, etc. In general,

$$\forall \mathcal{S}^\pm_N, \quad \exists N' > N \textit{ such that } \mathcal{S}^\pm_N \subset \mathcal{S}^\pm_{N'}. \tag{2.7}$$

The generalization to quantum groups with higher ranks is straightforward.

We suppress the subscript N in the ensuing discussion. Denote $\mathcal{S} \equiv \mathcal{S}^- \cup \mathcal{S}^+$. We now construct a homomorphism π on $\mathcal{A}$ such that

$$\pi(\mathcal{S})\pi(\mathcal{S}) \subset \pi(\mathcal{S}). \tag{2.8}$$

This implies either $\pi((X^{\pm})^{N+1}) \sim \pi(1)$ or $\pi(\mathcal{I}^-) = \pi(\mathcal{I}^+) = 0$. The remainder of π is then completely described by a finite set of expansion coefficients in the product m_π that specify (2.8). These coefficients, which are not all independent because $\mathcal{S}^- \cap \mathcal{S}^+ \neq \emptyset$, may be computed as follows.

Let the basis $\{\tilde{e}_v\} \equiv \mathcal{S}^-$ and the basis $\{\tilde{e}^v\}$ be a linear transformation ϕ of $\mathcal{S}^+$ (we think of $\mathcal{S}^\pm$ as bases) so that it is dual to $\{\tilde{e}_v\}$ in the sense that: (a) The product m_π in $\pi(\{\tilde{e}_v\})$ is given by the coprduct $\triangle \circ \phi$ in $\{\tilde{e}^v\}$; (b) The product $m_\pi \circ \phi$ in $\pi(\{\tilde{e}^v\})$ is given by the opposite coproduct $\mathcal{T} \circ \triangle$ in $\{\tilde{e}_v\}$.
It can be verified that these conditions are sufficient for (1.1) and (1.2) to hold under the image of π provided that $\mathcal{A}$ and m are replaced by $\mathcal{S}$ and m_π. The conditions however may not completely constrain the products $\pi(\mathcal{S}^-)\pi(\mathcal{S}^+)$ and $\pi(\mathcal{S}^+)\pi(\mathcal{S}^-)$. Define

$$\tilde{\mathcal{R}} \equiv \sum_v \tilde{e}_v \otimes \tilde{e}^v, \tag{2.9}$$

Then under the image of $\pi \otimes \pi$, $\tilde{\mathcal{R}}^{-1} \sim (S \otimes id)\tilde{\mathcal{R}} \sim (id \otimes S^{-1})\tilde{\mathcal{R}}$. There may be sufficient degrees of freedom left in m_π after (a) and (b) above for the condition

$$(\pi \otimes \pi)(\mathcal{T} \circ \triangle(a)\tilde{\mathcal{R}} - \tilde{\mathcal{R}}\triangle(a)), \qquad \forall a \in \mathcal{S} \tag{2.10}$$

to be satisfied. We call the set $\{\{\tilde{e}_v\}, \{\tilde{e}^v\}, \pi\}$ a *finite representation* of $\mathcal{A}$ when it satisfies the conditions (a) and (b), and a *finite quasitriangular representation* when it also satisfies (2.10). In this paper we make the following:

Conjecture 2.11. *If $\mathcal{A}$ is quasitriagular, then a quasitriangular representation of it exists for every N.*

An explicit construction is given in [**4**] and examples of some results are given in section 10.

In the limit $N \to \infty$, $\mathcal{I} \to \emptyset$ and $\mathcal{S}_N^\pm \to \mathcal{A}^\pm$, so that by definition $\{\tilde{e}_v\} \to \{e_\sigma\}$, and $\{\tilde{e}^v\} \to \{e^\sigma\}$ to within a linear transformation ϕ_∞. Then the dualities between $\{e_\sigma\}$ and $\{e^\sigma\}$ and between $\{\tilde{e}_v\}$ and $\{\tilde{e}^v\}$ imply

$$m_{\pi_\infty} = m \circ \phi_\infty, \tag{2.12}$$

on $\{e_\sigma\} \otimes \{e_\sigma\}$, and

$$m_{\pi_\infty} = m, \tag{2.13}$$

on $\{e^\sigma\} \otimes \{e^\sigma\}$. Furthermore (2.10) is exactly the same as (2.1) when m_{π_∞} is replaced by m. Thus ϕ_∞ is a linear identity map and m_{π_∞} is isomorphic to m. Therefore π_∞ is an isomorphism. This leads finally to:

Theorem 2.14. *If $\mathcal{A}$ is quasitriangular and has a unique $\mathcal{R}$ and the series $\tilde{\mathcal{R}}_N, N = N_1, N_2 \ldots \infty$, exists, then $\tilde{\mathcal{R}}_\infty$ is isomorphic to $\mathcal{R}$.*

In the following, we assume Conjecture 2.11 to be true and $\mathcal{R}$ to be unique and treat the infinite sum in (2.4) as if it were a finite one. In particular, the results in sections 3 and 5 are derived by first (formally) replacing $\mathcal{R}$ by $\tilde{\mathcal{R}}_N$, and then taking the limit $N \rightarrow \infty$.

3. The h-matrix

The quantity $\Lambda = m\,((id \otimes S)\mathcal{R})m(\mathcal{T}((id \otimes S)\mathcal{R})))$ is central in $\mathcal{A}$ **[2,3,4]**. Then from Schur's lemma, for any irreducible $\pi \in End(V_{\mathcal{A}})$, $\pi(\Lambda) = \lambda\pi(1)$, where $\lambda \in C$. In this sense $\Lambda^{1/2} \sim \lambda^{1/2}$ is meaningful. Define

$$h \equiv \Lambda^{-1/2} m(id \otimes S)\mathcal{R}, \tag{3.1}$$

then $h^{-1} = \Lambda^{-1/2} m(\mathcal{T}((id \otimes S)\mathcal{R})) = S(h)$. Furthmore **[2,3,4]**

$$hah^{-1} = S^{-2}(a), \qquad \forall a \in \mathcal{A} \tag{3.2}$$

and, from the relations (1.2) and (2.1) (repeated Greek indicies are summed),

$$e_\sigma h e^\sigma = e^\sigma h^{-1} e_\sigma = \Lambda^{-1/2}, \tag{3.3}$$

$$e^\sigma h S(e_\sigma) = S(e_\sigma) h^{-1} e^\sigma = \Lambda^{1/2}, \tag{3.4}$$

$$S(e_\sigma) e_\tau \otimes e^\tau h^{-1} e^\sigma = e^\sigma e^\tau \otimes S(e_\tau) h^{-1} e_\sigma = 1 \otimes h^{-1}, \tag{3.5}$$

$$e_\sigma h S(e_\tau) \otimes e^\tau e^\sigma = e^\sigma h e^\tau \otimes e_\tau S(e_\sigma) = h \otimes 1. \tag{3.6}$$

4. Tangles and splices

We consider two-dimensional projection of tangles in R^3, with a certain restriction given below. By a (n, n)-tangle in R^3 we mean the union of n entangled, disjoint, oriented strands all of whose entanglements are enclosed within an imaginary B^3 and all of whose 2n open ends are outside the B^3. The boundary of the B^3 and the space exterior to it is fixed. The boundary is separated by an equator into top and bottom hemispheres and the open ends of strands of the tangle do not enter into or exit from B^3 on the equator. On the two-dimensional projection, the shadows of the two hemispheres are represented by two vertically separated horizontal lines, between which is a finite strip, called the *interior strip*, on which the interior of the B^3 is projected. We consider only tangles all of whose strands enter the strip from above and exit from below. Closing a strand means joining the two ends of a strand in an area external to the interior strip. Closing a tangle means closing all strands of a tangle. The result is the *closure* of the tangle. In this work we restrict our consideration to those tangles that can be closed without generating any new crossings. In what follows, all Reidemeister moves [**5**] are restricted to within the interior strip.

Consider an operation, called splicing, under which a positive (negative) crossing is replaced by an *uncrossing*, represented by a plus (minus) sign and two noncrossing strands, as shown below:

$$\times \sim)+(\quad , \quad \times \sim)-(\tag{4.1}$$

The result of splicing all the crossings in a $T^{(n)}$ is the union of a set of n noncrossing strands, m unknots, called Seifert circles, and uncrossings located at the verticies of $T^{(n)}$. In a spliced tangle, called a *splice*, the relative positions of uncrossings along any strand or Seifert circle may not be altered. An *untangle* is a tangle that has no crossings. The following lemma is self evident:

Lemma 4.2. *Splicing is a bijection between tangles and splices.*

Consider a Seifert circle with s uncrossings around its perimeter which divide its circumference into s sections. If $s > 1$ the Seifert circle may be deformed such that no two uncrossings are allowed to be on the same latitude so that the sections point generally either upward or downward; at least one section will point up and at least one down. If $s = 1$ we consider the sole section to be pointing up and call the Seifert circle trivial. A tangle is said to be trivial if its splice contains at most only trivial Seifert circles; otherwise it is nontrivial. A trivial tangle is equivalent to a braid (by the Reidemeister move I, see below). The following has been proven elsewhere [22]:

Lemma 4.3. *It is possible to deform, without generating any additional crossing, a tangle to one such that each of the Seifert circles in the splice contains one and only one section pointing up.*

5. Reidemeister moves

We construct a homomorphism $\mathcal{V}$ between the complete set of (n, n)-tangles and $\mathcal{A}^{\otimes n}$ and define the Reidemeister moves under $\mathcal{V}$. It is then shown that under $\mathcal{V}$ the tangles are invariant under the Reidemeister moves.

The homomorphism $\mathcal{V}$ is a natural one: each strand in the tangle is mapped to an element in an $\mathcal{A}$ factor in $\mathcal{A}^{\otimes n}$. The actions of $\mathcal{V}$ on a local part of a tangle are defined as

$$\rightarrow \quad 1 \quad \in \mathcal{A}, \tag{5.1a}$$

$$\rightarrow \quad 1 \otimes 1 \quad \in \mathcal{A} \otimes \mathcal{A}, \tag{5.1b}$$

$$\rightarrow \quad e_\sigma \otimes e^\sigma = \mathcal{R} \quad \in \mathcal{A} \otimes \mathcal{A}, \tag{5.1c}$$

$$\rightarrow \quad h \quad \in \mathcal{A}, \tag{5.1e}$$

(in a counterclockwise circle)

$$\rightarrow \quad h^{-1} \quad \in \mathcal{A}. \tag{5.1f}$$

(in a clockwise circle)

Summation over repeated Greek indicies in the target is understood. The factor of $\mathcal{A}$ in the target space onto which an element is mapped is the image of the strand to which that element belongs. It is clear that $\mathcal{V}$ restricted to (5.1a-d) is bijective. Two examples will further illustrate the relation between the domain (strands in a tangle) and the target (factors of $\mathcal{A}$) of $\mathcal{V}$:

$$\rightarrow \quad e_\sigma e^\tau \otimes e^\sigma e_\tau = \mathcal{R}^2, \tag{5.2}$$

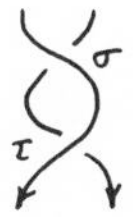

$$\rightarrow \quad e_\sigma S(e_\tau) \otimes e^\sigma e^\tau = \mathcal{R} \otimes \mathcal{R}^{-1} = 1 \otimes 1. \tag{5.3}$$

Define the equivalence relation between tangles as one in which the tangles have the same image under $\mathcal{V}$. Then (5.3) and (5.1b) establishes the Reidemeister move II as an invariant action only when it acts on two strands having the same orientations, that is

$$\sim \qquad \text{Reidemeister move } II_a \tag{5.4a}$$

We call such a move Reidemeister move II_a. A Reidemeister move II_b acts on two strands with opposite orientations. It is not trivially an invariant move because

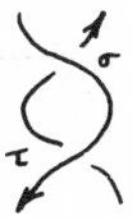

$$\rightarrow \quad S(e_\sigma)e_\tau \otimes e^\tau e^\sigma \neq 1 \otimes 1. \tag{5.5}$$

Examination of the inverse images of relations (3.5 and 6) suggests a type of II_b invariance can be defined by inserting of a factor of h or h^{-1} into the section of strand between the two crossings that is pointed in the *worng* or upward, direction.

This explains the need for (5.1e and f). The general rules for the insertion of h and/or h^{-1} factors on a tangle T is:

(a) Splice T and call the result P. P is the union of a set of noncrossing strands, uncrossings and Seifert circles.

(b) On each counter-clockwise (clockwise) Seifert circle in P, insert a factor of h (h^{-1}) on the upward-pointing section (each Seifert circle has one and only one such). Call the result P'.

(c) Generate from P' a unique tangle T' by inverse splicing (by Lemma 4.1). T' is the same as T except that it has factors of h and/or h^{-1} inserted on all the sections having the wrong directions.

An example suffices to explain these rules.

Henceforth it will be understood that by a tangle T we mean T'.

It follows that (3.5) and (3.6), respectively, induce the equivalence relations

$$\sim \quad \sim \qquad (5.4b)$$

$$\sim \quad \sim \qquad (5.4c)$$

These, together with (5.4a), establish the complete invariance under the Reidemeister move II.

The inverse images of (3.3) and (3.4) under $\mathcal{V}$, respectively, induce the equivalence relations

$$\sim \quad \sim \Lambda^{-1/2} \qquad (5.6a)$$

$$\sim \quad \sim \Lambda^{1/2} \tag{5.6b}$$

which collectively establish invariance under the Reidemeister move I to within regular isotopy.

Invariance of the Reidemeister III move on three downward oriented strands is induced by the braid relation (2.3)

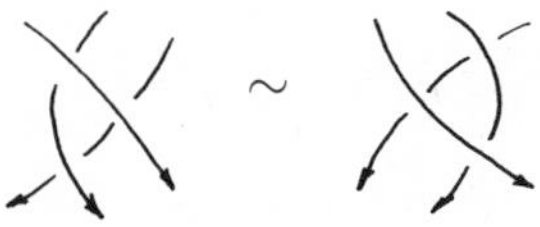

$$\sim \tag{5.7a}$$

Variations of the braid relation involving upward oriented strand(s) can be derived from (5.7a) and moves I and II. An example is

$$\sim \tag{5.7b}$$

This completes our demonstration that $\mathcal{V}$ is invariant (to within regular isotopy) under the actions of all three of Reidemeister's moves.

6. Tangle invariants

We now discuss the global property of $\mathcal{V}$. Define a word in $\mathcal{A}$ as an ordered product of elements in $\mathcal{A}$. Then,

$$\mathcal{V}[\textit{an open strand}] \rightarrow \textit{a word} \in \mathcal{A}. \tag{6.1a}$$

For a closed strand an additional procedure is needed. Define a cyclically symmetric map Tr

$$Tr : \mathcal{A} \rightarrow C\,[[q]], \qquad Tr(a_1 a_2 \ldots a_n) = Tr(a_2 \ldots a_n a_1). \tag{6.2}$$

Then

$$Tr(\mathcal{V}[\textit{a closed strand}]) \rightarrow C\,[[q]]. \tag{6.1b}$$

Recall that (5.6) only gives regular isotopy. In order to establish ambient isotopy the factors of $\Lambda^{\pm 1/2}$ on the right-hand side of (5.6) must be compensated. This is achieved by defining the action of $\mathcal{V}$ on the writhe number w of the tangle as

$$\mathcal{V}[w] \to (\textit{multiplicative factor of})\ \Lambda^{w/2} \in C\,[[q]]. \tag{6.1c}$$

To summarize, a $T^{(n)}$ that is a disjoint union of n open strands $\{o_i\}$ and any number of closed strands $\{c_j\}$ and has a writhe number w is mapped by $\mathcal{V}$ to $\mathcal{A}^{\otimes n}$ by

$$\mathcal{V}[T^{(n)}] = \Lambda^{w/2}(\prod_j Tr(\mathcal{V}(c_j)))(\bigotimes_i(\mathcal{V}(o_i))) \quad \in \mathcal{A}^{\otimes n}. \tag{6.3}$$

By definition $\mathcal{V}$ is a homomorphism:

$$\mathcal{V}[T^{(n)}T^{(n)'}] = \mathcal{V}[T^{(n)}]\mathcal{V}[T^{(n)'}], \tag{6.4}$$

where the product on $\mathcal{V}$ is just the product on $\mathcal{A}$. We now have our first result:

Theorem 6.5. *The map $\mathcal{V}$ gives an ambient isotopy for tangles.*

The proof follows from the local properties, (5.1a-f), and global properties, (6.1a-c), of $\mathcal{V}$, and from Reidemeister's theorem [**5**]. □

An l-component link, denoted generically by L, is a $(0,0)$-tangle with l closed strands, therefore

Corollary 6.6. *The map $\mathcal{V}[L]$ is a link invariant.*

$\mathcal{V}$ has special properties when $n = 1$. For simplicity denote $T^{(1)}$ by T. From the invariance of Reidemeister moves II and III it follows that for any two T and T', $\mathcal{V}[TT'] = \mathcal{V}[T'T]$ so that, from (6.4)

$$[\,\mathcal{V}[T], \mathcal{V}[T']\,] = 0. \tag{6.7}$$

Therefore,

Lemma 6.8. *The set $\{\mathcal{V}[T]|$ all $(1,1)$-tangles $T\}$ is a commutative subset of $\mathcal{A}$.*

Then we have, from Schur's lemma,

Theorem 6.9. *Every irreducible representation of $\mathcal{V}[T]$ is one dimensional.*

In section 7 we prove a more powerful theorem:

Theorem 6.10. *$\mathcal{V}[T]$ is a central element in $\mathcal{A}, \forall T \in \{T^{(1)}\}$.*

From which Theorem 6.9 follows. Then,

Corollary 6.11. *An irreducible representation of $\mathcal{A}$ is also an irreducible representation of $\mathcal{V}[T]$.*

Let $\pi \in End(V_{\mathcal{A}})$ be an irreducible representation of $\mathcal{A}$, then,

$$\pi(\mathcal{V}[T]) = Q_{\pi}[T]1_{\pi}, \tag{6.12}$$

where $Q_{\pi} \in C[[q]]$ and 1_{π} is the unit matrix of π. Let T be a tangle whose closure is L. Then

$$\mathcal{V}[L] = Tr(\mathcal{V}[T]h) = Tr(\mathcal{V}[T]h^{-1}). \tag{6.13}$$

Note that the tangle may be closed anticlockwise or clockwise. From (6.12),

$$\pi(\mathcal{V}[T]) \equiv \mathcal{V}_{\pi}[L] = Q_{\pi}[T]Tr_{\pi}(h) = Q_{\pi}[T]Tr_{\pi}(h^{-1}), \tag{6.14}$$

where Tr_{π} is taken to be the matrix trace in π. The second equality is guaranteed by the cyclic symmetry of Tr and the properties of h and h^{-1} (see (3.1)). Therefore,

Corollary 6.15. *For any T that closes to L, $Q_{\pi}[T]$ is a link invariant for L.*

This implies that the set $\{\mathcal{V}[T]\}$ contains all the information of link invariants and that, if the closures of T and T' are equivalent, then $Q_{\pi}[T] = Q_{\pi}[T']$; it is not necessary that $T \sim T'$. In the following we shall sometimes use, instead of T, its closure L as the argument of Q_{π}.

For other constructions of link invariants equivalent to $\mathcal{V}[L]$ on the quantum group, see for example Reshetikhin [**2**], Turaev [**6**], Lawrence [**7**], Jones [**8**], Rosso [**9**], and Kirillov and Reshetikhin [**10**]. The set $\{\mathcal{R}, h\}$ is equivalent to the *extended Yang-Baxter set* needed for those constructions that are based on Markov's theorem [**11**]. An earlier derivation of $Q[T^{(1)}]$ as a link invariant based on a conjecture of theorem 6.10 was given in [**12**].

7. Invariants of quantum group

We prove Theorem 6.10, restated below:

Theorem 6.10. *$\mathcal{V}[T]$ is a central element in $\mathcal{A}, \forall T \in \{T^{(1)}\}$.*

In the rest of this section, unless otherwise specified, $T \in \{T^{(1)}\}$. The theorem is based on the quasitriangularity property (2.1) of $\mathcal{R}$ and proved topologically using the properties of splices.

Consider a splice. Call the result of replacing all its uncrossings by unsigned dots a *shadow splice*. Consider an equivalence relation whereby two tangles are equivalent if they only differ by at most the signs of the crossings, and call the equivalence classes of tangles under this relation *shadows*. It is clear that splicing a shadow obtains a shadow splice. The first of the following is obvious.

Lemma 7.1. *Splicing is a surjection from tangles to shadow splices and a bijection between shadows and shadow splices.*

Lemma 7.2. *Under splicing, the reverse image of every shadow splice contains a tangle equivalent to an untangle.*

Proof of Lemma 7.2. Consider a shadow splice S whose reverse image the shadow P is composed of a set of closed and open strands. Consider a three dimensional manifold $M^3 \sim R^2 \times Z$ in whose R^2 factor reside S and P. Let $\tilde{T}$ be a tangle in M^3 that is the disjoint union of the components of P, each of which resides in R^2 but with a distinct Z coordinate. Since there are no crossings in $\tilde{T}$, by Reidemeister's theorem its projection T on R^2 is equivalent to an untangle. By definition the shadow of T is equivalent to P. □

$\mathcal{V}[T]$ is central if it commutes with all the generators $\{k_i^{\pm 1}, X_i^{\pm}\}$ of $\mathcal{A}$. We omit the proof of $[\mathcal{V}[T], k_i^{\pm 1}] = 0$, which follows trivially from the relations $[\triangle(k_i^{\pm 1}), \mathcal{R}] = 0$ and $[k_i^{\pm 1}, h^{\pm 1}] = 0$ and is a simpler version of the proof of $[\mathcal{V}[T], X_i^{\pm}] = 0$, given below. Under coproduct

$$\triangle(X_i) = X_i \otimes k_i + k_i^{-1} \otimes X_i, \qquad i = 1 \ldots r, \tag{7.3}$$

where r is the rank of the quantized Lie algebra. Since the index i plays no role in the proof it will be suppressed in the rest of the discussion. On a diagram, represent X by a solid circle, k (k^{-1}) by a hollow arrow pointing along (against) the orientation of the strand. Thus (2.1), with $a = X$, is diagrammatically

$$+ \quad \sim \quad + \tag{7.4a}$$

and its companion relation $\mathcal{R}^{-1}T \circ \triangle(X) = \triangle(X)\mathcal{R}^{-1}$ is

(7.4b)

The splices of the above two relations differ only by the signs of the uncrossings, therefore they combine to give a relation for shadow splices

(7.5)

Similarly, from the relations $\triangle(k)\mathcal{R} = \mathcal{R}\triangle(k)$ one obtains

(7.6)

or with all the directions of the hollow arrows reversed when k is replaced by k^{-1}. These relations mean that, on a shadow splice, a pair (k,k) (or (k^{-1},k^{-1})) may freely cross a dot, while the two pairs (X,k) and (k^{-1},X) must move in tandem to cross a dot. Recall that any dot in these shadow splices may represent a positive or a negative crossing. Let P be the splice of the the tangle T, and denote the spliced $\mathcal{V}[T]X$ and $X\mathcal{V}[T]$, respectively, by PX and XP. Similarly define Pk and kP.

Lemma 7.7. $[\,\mathcal{V}[T], X\,] = 0$ *implies* $PX = XP$*, and vice versa.*

This follows from Lemma 4.1. The lemma does not apply to higher tangles.

To see how X may be commuted through P, consider

(7.8)

obtained by combining (7.5 and 7.6). T being a $(1,1)$-tangle, there is only one open strand in P. By identifying the left-hand strands in the diagram on the left-hand side and the first diagram on the right-hand side of (7.8), the latter can be applied on PX to move X through P, at the cost, due to the last two terms in (7.8), of introducing new diagrams in the process. For example, the result after one application is

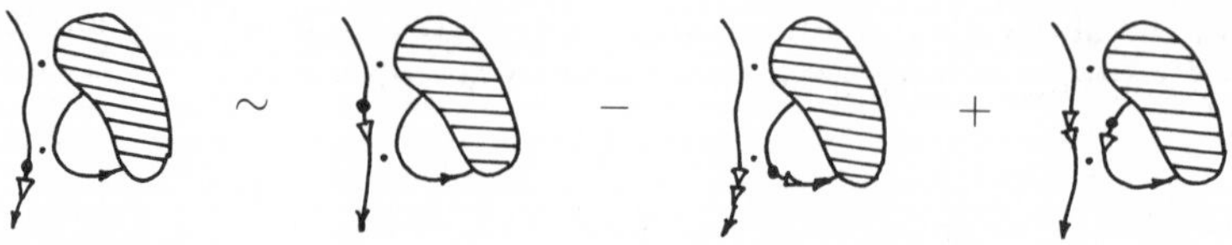

$$\sim \quad - \quad + \tag{7.9}$$

where the shaded area represents P complement to the open strand. Repeated applications of (7.8) will eventually lead to

$$PX = XkPk^{-1} + B = XP + B, \tag{7.10}$$

because P and k commute; B represents a set of splices induced by the application of (7.8).

Lemma 7.11. *The set* $B = 0$ *in (7.10) for any shadow splice* P.

Proof of Lemma 7.11. By Lemma 7.2 there is an untangle T_0 whose shadow splice is P. Then the shadow of $\mathcal{V}[T_0]X$ splices to PX; similarly $X\mathcal{V}[T_0]$ to XP. Since $\mathcal{V}[T_0]X = X\mathcal{V}[T_0] = X$, it follows from Lemma 7.7 that $PX = XP$. Therefore $B = 0$. □

We now prove the main theorem.
Proof of Theorem 6.10. Let P be the splice of T. (7.10) being true for shadow splices implies it is also true for any splice having that shadow. Then, from Lemma 7.11, $PX = XP$, and from Lemma 7.7, $\mathcal{V}T, X = 0$. Since $\mathcal{V}[T]$ already commutes with k, it commutes with all the generators of $\mathcal{A}$. □

8. Representation of tangle invariants

Let $T^{(n)}$ be a tangle with n open strands o_i and m closed strands c_j. Denote their union by the set $\{s_i | i = 1 \dots l = n + m\}$, the first n being the o_i's and the last m being the c_i's. Denote by $\{\pi_i \in End(V_\mathcal{A}) | i = 1 \dots l\}$ a set of l irreducible representations of $\mathcal{A}$; all π_i's need not be different. Let w_i be the number of positive crossings minus the number of negative crossings crossed by the strand

s_i. Then $\sum_i w_i/2 = w$. Define

$$\mathcal{V}_{\{\pi_i\}}[T^{(n)}] \equiv (\prod_i \pi_i)(\mathcal{V}[T^{(n)}]) = (\prod_{i=1}^{l} \lambda_{\pi_i}^{w_i/4})(\prod_{j=n+1}^{l} Tr_{\pi_j}(\mathcal{V}[c_j]))\,(\bigotimes_{i=1}^{n} \mathcal{V}_{\pi_i}[o_i]). \tag{8.1}$$

Theorem 8.2. *$\mathcal{V}_{\{\pi_i\}}[T^{(n)}]$ is an (n,n)-tangle invariant.*

This follows from (6.3) and Corollary 6.11. If $T^{(n-1)}$ is obtained from $T^{(n)}$ by closing, say, the open strand o_v, then the invariant $\mathcal{V}_{\{\pi_i\}}[T^{(n-1)}]$ is given by the right-hand side of (8.1), with the exception that the factor $(\mathcal{V}_{\pi_v}[o_v])$ is replaced by $Tr_{\pi_v}(\mathcal{V}[o_v])Tr_{\pi_v}(h) = Tr_{\pi_v}(\mathcal{V}[o_v])Tr_{\pi_v}(h^{-1})$.

9. Cabled tangle invariants

We use the coproduct in the quantum group on $\mathcal{V}$ to construct invariants for *cabled tangles*. In a cabled tangle, at least one of the strands (closed or open) is replaced by a set of parallel strands. The point is to use $\triangle : \mathcal{A} \rightarrow \mathcal{A} \otimes \mathcal{A}$ for $\mathcal{V}[strand] \rightarrow \mathcal{V}[cable]$. Define a homomorphism $\triangle^{(u)} : \mathcal{A} \rightarrow \mathcal{A}^{\otimes u}$, where u is a positive integer, by

$$\triangle^{(u)} \equiv (1 \otimes \cdots \otimes 1 \otimes \triangle)\triangle^{(u-1)}, \qquad u > 1, \tag{9.1}$$

with $\triangle^{(1)} = 1$. If $\pi \in End(V_{\mathcal{A}})$ is a fundamental representation of $\mathcal{A}$, then from the fusion rule

$$\pi^{\otimes u} = \bigoplus_k \phi_k^{(u)} \quad \in End({V_{\mathcal{A}}}^{\otimes u}) \tag{9.2}$$

irreducible representations $\phi^{(u)}$ of $\mathcal{A}^{\otimes u}$ are obtained. The idea of using the coproduct and the fusion rule for cabling is expressed in the following diagram

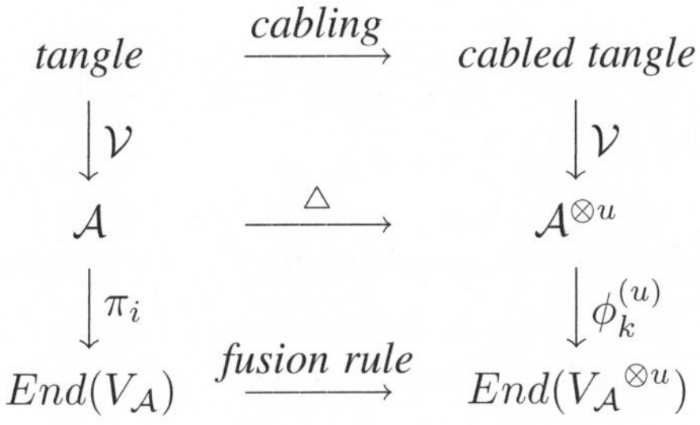

$$\begin{array}{ccc}
\textit{tangle} & \xrightarrow{\textit{cabling}} & \textit{cabled tangle} \\
\downarrow \mathcal{V} & & \downarrow \mathcal{V} \\
\mathcal{A} & \xrightarrow{\triangle} & \mathcal{A}^{\otimes u} \\
\downarrow \pi_i & & \downarrow \phi_k^{(u)} \\
End(V_{\mathcal{A}}) & \xrightarrow{\textit{fusion rule}} & End({V_{\mathcal{A}}}^{\otimes u})
\end{array}$$

Define

$$\mathcal{R}_{ij}^{(u,v)} \equiv (\phi_i^{(u)} \otimes \phi_j^{(v)})(\triangle^{(u)} \otimes \triangle^{(v)})\mathcal{R}, \tag{9.3}$$

$$h_i^{(u)} \equiv \phi_i^{(u)}(\triangle^{(u)} h), \tag{9.4}$$

where in (9.3), $\mathcal{R} = e_\sigma \otimes e^\sigma$, and $\triangle^{(u)} \quad (\triangle^{(v)})$ acts on $e_\sigma \quad (e^\sigma)$. For simplicity we write

$$a' = \triangle^{(u)}(a) \quad \in \mathcal{A}^{\otimes u}, \tag{9.5}$$

for generic $a \in \mathcal{A}$ and u. Similarly the various quantities in (9.3) and (9.4) will respectively be written as $\mathcal{R}'_{ij}$, ϕ'_i and h'_i, etc., with the understanding that, in (9.3), u and v need not be the same. The discussions in section 5 can be repeated for cables, provided the appropriate $\mathcal{R}'_{ij}$, h'_i and λ'_i are substituted for the $\mathcal{R}$, h and λ used there. We omit the discussion, except to point out that

$$\lambda'_i = \phi_i^{(u)}(\Lambda) = (\lambda_i)^u, \tag{9.6}$$

which is easily understood from from the diagram below.

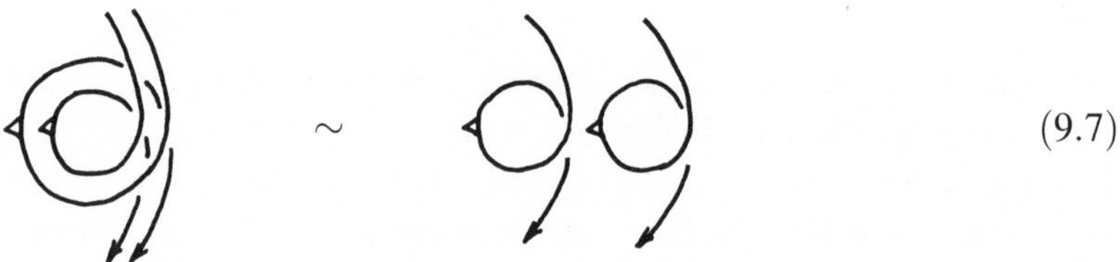

(9.7)

This establishes the ambient invariance of the Reidemeister moves for cables. There is a significant difference between the description here and in section 5: here cables are mapped to *irreducible representations* of $\mathcal{A}^{\otimes u}$, whereas in section 5 single strands were directly mapped to $\mathcal{A}$. Let T' be a cabled tangle. By changing every cable in T' into a single strand one obtains the single-strand tangle T, for which (8.1) gives an invariant. By replacing the quantities T, Tr, λ o_i and c_i in (8.1) by their respective cabled counterparts (denoted by primed symbols) and the set $\{\pi_i\}$ by $\{\phi_k\}$ one obtains

$$\mathcal{V}'_{\{\phi_k\}}[T^{(n)'}] \equiv \bigotimes_i \phi_i(\mathcal{V}[T^{(n)'}]) = (\prod_i (\lambda'_{\phi_i})^{\frac{w_i}{4}})(\prod_j Tr'_{\phi_j}(\mathcal{V}[c'_j]))(\bigotimes_i \phi_i(\mathcal{V}[o'_i])). \tag{9.8}$$

where the indices i and j have the same range as in (8.1). The discussion in section 6 can be repeated to give:

Theorem 9.9. $\mathcal{V}'_{\{\phi_k\}}[T^{(n)'}]$ *is an ambient isotopy for cabled tangles.*

Similarly, Theorem 6.10 holds with $\mathcal{A}$ replaced by $\triangle^{(u)}(\mathcal{A}) \subset \mathcal{A}^{\otimes u}$. Therefore,

Corollary 9.10. *For any cabled* $(1,1)$*-tangle* T', $\mathcal{V}'_{\{\phi_k\}}[T']$ *is proportional to the unit matrix.*

Suppose $\phi_1 \in \{\phi_k\}$ acts on the open cable in T' whose closure is L', then the analog of (6.6) is

$$\mathcal{V}'_{\{\phi_k\}}[T'] = \mathcal{Q}'_{\{\phi_k\}}[T']\phi_1(1), \tag{9.11}$$

so that

Corollary 9.12. $\mathcal{Q}'_{\{\phi_k\}}[T^{(1)'}]$ *is a link polynomial for the closure* L' *of* $T^{(1)'}$.

10. Examples

We use the simplest nontrivial representations of two quantum groups, $sl(2;C)_s$ and $(gl(2;C)/U(1))_{q,s}$, to demonstrate some of the ideas discussed in the preceding sections. q and s are deformation parameters; s instead of the usual q is used for $sl(2;C)_s$ to make the comparison between the two quantum groups more transparent. Correspondingly, the generic algebraic field $C\,[[q]]$ referred to in previous section should be replaced respectively by $C\,[[s]]$ and $C\,[[q,s]]$ for the two quantum groups.

$sl(2;C)_s$. The generators H, X^+ and X^- are constrained by the usual defining relations [**1,13,2**]. Define $k = s^{H/2}$. The minimum set (section 2) $\mathcal{S}^\pm$ closed under coproduct are

$$\mathcal{S}^\pm = \{k^{-1}, k, X^\pm\}, \tag{10.1}$$

whose finite bases are [**12**]

$$\{e_\sigma\} = \{e_1 = k^{-1}, e_2 = k, e_3 = X^-\}, \tag{10.2}$$

$$\begin{aligned}\{e^\sigma\} = \{e^1 &= (s^{-1/2}k^{-1} - s^{1/2}k)/(s^{-1} - s),\\ e^2 &= -(s^{1/2}k^{-1} - s^{-1/2}k)/(s^{-1} - s), \quad e^3 = X^+\},\end{aligned} \tag{10.3}$$

that are dual under the 2×2 representation π':

$$\pi'(H) = \begin{pmatrix}1 & 0\\ 0 & -1\end{pmatrix}, \quad \pi'(X^-) = \eta\begin{pmatrix}0 & 0\\ 1 & 0\end{pmatrix}, \quad \pi'(X^+) = \eta\begin{pmatrix}0 & 1\\ 0 & 0\end{pmatrix}. \tag{10.4}$$

where $\eta^2 \equiv s^{1/2}(s^{-1} - s)$. Then from (2.8) and (3.1)

$$\pi'(\mathcal{R}) = s^{1/2}\left(\begin{pmatrix}s^{-1} & 0\\ 0 & 1\end{pmatrix}\otimes\begin{pmatrix}1 & 0\\ 0 & 0\end{pmatrix} + \begin{pmatrix}1 & 0\\ 0 & s^{-1}\end{pmatrix}\otimes\begin{pmatrix}0 & 0\\ 0 & 1\end{pmatrix}\right.$$

$$+(s^{-1}-s)\begin{pmatrix}0 & 0\\ 1 & 0\end{pmatrix}\otimes\begin{pmatrix}0 & 1\\ 0 & 0\end{pmatrix}\Bigg), \tag{10.5}$$

$$\pi'(\lambda)=s^3, \qquad \pi'(h)=\begin{pmatrix}s^{-1} & 0\\ 0 & s\end{pmatrix}. \tag{10.6}$$

By definition (6.1) $Q_{\pi'}[unknot]=1$. Furthermore it can be shown [**12,18**] that $Q_{\pi'}[L]$ of (6.15) satisfies the skein relation of the Jones polynomial [**14**]. Therefore $Q_{\pi'}[L]$ is precisely the Jones polynomial. (Note that $Tr(h)=0$ when $s^2=-1$. In this case, $\mathcal{V}_{\pi'}[L]$ of (6.14) vanishes identically for every L. On the other hand, the Jones polynomial as derived from the skein relation is well defined and nonvanishing (except accidentally) at $s^2=-1$. This shows that the Jones polynomial is $Q_{\pi'}[L]$, not $\mathcal{V}_{\pi'}[L]$.) It is readily verified that under π' the tangles of (7.12) and (7.13) are proportional to the unit matrix:

$$\mathcal{V}_{\pi'}[\textit{Hopf tangle}]=Q_{Jones}[\textit{Hopf link}]\begin{pmatrix}1 & 0\\ 0 & 1\end{pmatrix}, \tag{10.7}$$

$$\mathcal{V}_{\pi'}[\textit{trefoil tangle}]=Q_{Jones}[\textit{trefoil}]\begin{pmatrix}1 & 0\\ 0 & 1\end{pmatrix}. \tag{10.8}$$

The fusion rule for the u-fold direct product of π' is

$$\pi'^{\otimes u}=\{(u+1)-\textit{dim. rep.}\}\oplus \textit{lower dim. reps.}. \tag{10.9}$$

The link polynomials of Akutsu and Wadati [**15**] derived from the solutions of the Yang-Baxter equation [**16,17**] that are associated with the spin-$u/2$ state models are just the u-cable link polynomials of (9.8) with the ϕ representation belonging to the $\{(u+1)-\textit{dim. rep.}\}$ of (10.9).

$(gl(2;C)/U(1))_{q,s}$. The structure of the this quantum group [**4,18**] is similar to that of $sl(2;C)_q$, except that

$$k=q^{(H-I)/2}s^{I/2}, \tag{10.10}$$

where I is a commuting generator associated with the $U(1)$ factor in $GL(2;C)$. When $s^2=q^2$ this factor is modded out and the quantum group reduces to $sl(2;C)_q$. For generic s and $s^2\neq q^2$, the quantum group has a fundamental m-dimensional representations at $q^2=m^{th}$ root of unity, $m>1$. We consider the case $m=2$, when $q^2=-1$, and call the representation π, under which $(X^{\pm})^2=0$. To avoid dealing with explicitly complex representation, we consider the set

$$k_1=s^{H/2}, \quad k_2=q^{H-I}s^{I-H/2}, \quad Z^+=lX^+, \quad Z^-=X^-l, \tag{10.11}$$

where $l^2=k_1{k_2}^{-1}$. The set $\{{k_1}^{\pm},{k_2}^{\pm},Z^{\pm}\}$ generates the same Hopf algebra as $\{k^{\pm},X^{\pm}\}$ because $k_1k_2=k^2$. Then from $\triangle(X)=X\otimes k+k^{-1}\otimes X$ means

$\triangle(Z) = Z \otimes k_2 + {k_1}^{-1} \otimes Z$. The minimum sets that close under coproduct are

$$\mathcal{S}^{\pm} = \{{k_1}^{-1}, k_2, Z^{\pm}\}, \tag{10.12}$$

with bases [**4**]

$$\{e_\sigma\} = \{e_1 = {k_1}^{-1}, e_2 = k_2, e_3 = Z^-\}, \tag{10.13}$$

$$\{e^\sigma\} = \{e^1 = (s^{1/2}{k_1}^{-1} + s^{-1/2}k_2)/2,$$
$$e^2 = s^{-1}(s^{1/2}{k_1}^{-1} - s^{-1/2}k_2)/2, \quad e^3 = Z^+\}, \tag{10.14}$$

that are dual under the 2×2 representation π

$$\pi(H) = \begin{pmatrix} 1 & 0 \\ 0 & -1 \end{pmatrix}, \quad \pi(I) = \begin{pmatrix} 1 & 0 \\ 0 & 1 \end{pmatrix},$$

$$\pi(Z^-) = \eta \begin{pmatrix} 0 & 0 \\ 1 & 0 \end{pmatrix}, \quad \pi(Z^+) = \eta \begin{pmatrix} 0 & 1 \\ 0 & 0 \end{pmatrix}, \tag{10.15}$$

Then

$$\pi(\mathcal{R}) = s^{1/2}\left(\begin{pmatrix} s^{-1} & 0 \\ 0 & 1 \end{pmatrix} \otimes \begin{pmatrix} 1 & 0 \\ 0 & 0 \end{pmatrix} + \begin{pmatrix} 1 & 0 \\ 0 & -s \end{pmatrix} \otimes \begin{pmatrix} 0 & 0 \\ 0 & 1 \end{pmatrix}\right.$$
$$\left. + (s^{-1} - s)\begin{pmatrix} 0 & 0 \\ 1 & 0 \end{pmatrix} \otimes \begin{pmatrix} 0 & 1 \\ 0 & 0 \end{pmatrix}\right), \tag{10.16}$$

$$\pi(\lambda) = s^{-1}, \quad \pi(h) = \begin{pmatrix} s & 0 \\ 0 & -s \end{pmatrix}. \tag{10.17}$$

Note (10.16) differs from (10.5) only by one matrix element in the second term. By definition $Q_\pi[unknot] = 1$. Furthermore, it can be shown [**4,18**] that $Q_\pi[L]$ is the Alexander-Conway polynomial [**19,20**]. A peculiar property of this representation is that $Tr_\pi(h) = 0$ identically, so that $\mathcal{V}_\pi[L] = 0$ identically. This is the reason why the polynomial vanishes for any split link and the source of a well known difficulty one encounters when attempting to construct the polynomial using the Markov trace method, and explains why the Markov trace method breaks down even for the Jones polynomial at $s^2 = -1$, since at this value of s the Jones and Alexander-Conway polynomials are identical.

It is easy to verify by direct computation that the counterparts of (10.7 and 8) hold [**18**], with Jones replaced by Alexander-Conway. The fusion rule of π is very different from that of its counterpart π', however [**21**]. Here, for generic s,

$$\pi^{\otimes u} = \bigoplus_{k=1}^{2^{u-1}} \{2 - dim.\ reps.\ \phi_k\}. \tag{10.18}$$

ϕ_k are parafermionic representations [**21**]. $(gl(2;C)/U(1))_{q,s}$ do have $m \times m$ representations, but only at $q^2 = e^{-2i\pi/m}$. The associated polynomials form a hierarchy of Alexander-Conway polynomials [**4**]. Since these representations do not appear on the right-hand side of (10.18) for any u, unlike their Jones counterparts, the polynomials are not cabled versions of the fundamental Alexander-Conway polynomial.

This work is partly supported by a Canadian NSERC grant; I thank the organizers of the Osaka Knot Conference, especially Professor A. Kawauchi, for support at the conference and for giving me the opportunity to present it.

Note added in proof. Invariants of tangles were recently discussed in terms of category theory by Turaev [**23**] and Reshetikhin [**24**].

References

[1] V. Drinfeld, *Quantum groups*, Proc. Int. Cong. Math. vol. 1, 798-820 (Berkeley, 1986).

[2] N.Y. Reshetikhin, *Quantum envoloping algebra, the Yang-Baxter equation and the invariants of links*, LOMI preprint E-4-87, I and II (Leningrad, 1988).

[3] S. Majid, Int. J. Mod. Phys. A5 (1990) 1.

[4] H.C. Lee, *Twisted quantum groups of A_n and the Alexander- Conway link polynomial*, CRL preprint TP-90-0220 (submitted to Pac. J. Math).

[5] K. Reidemeister, *Knotentheorie* (Chelsea Publ. Co. 1948).

[6] V.G. Turaev, The Yang-Baxter equations and invariants of links, Invent. Math. 92 (1988) 527-553.

[7] R. Lawrence, *A universal link invariants using quantum groups*, Proc. XVII Int. Cong. Diff. Goem. Meth. in Theo. Phys. 1988 (to be publ. by World Scientific).

[8] V.F.R. Jones, *On knot invariant related to some statistical mechanical models*, Pac. J. Math. 137 (1989) 311-334.

[9] M. Rosso, *Groupes quantiques et modeles a vertex de V. Jones en theorie des noeuds*, C.R.A.S. 307 (1988) 207.

[10] A.N. Kirillov and N.Y. Reshetikhin, *Representations of the algebra $U_q sl(2)$, q-orthogonal polynomials and invariants of links*, LOMI preprint.

[11] A.A. Markov, *Uber die freie aquivalenz geschlossener Zopfe*, Recueil Math. Moscou, 1 (1935) 73-78.

[12] H.C. Lee, *Tangles, links and twisted quantum groups*, in Physics, Geometry and Topology, (Plenum, 1990), 623-655.

[13] M. Jimbo, *A q-difference analogue of $U(gl(n+1))$, Hecke algebras and the Yang-Baxter equation*, Lett. Math. Phys. 11 (1986) 247-252.

[**14**] V.F.R. Jones, *A polynomial invariant for knots via von Neumann algebras*, Bull. Amer. Math. Soc. 12 (1985) 103-110.

[15] Y. Akutsu and M. Wadati, *Exactly solvable models and new link polynomials. I. N-state vertex models*, J. Phys. Soc. Jap. 56 (1987) 3039-3051.

[16] C.N. Yang, Phys. Rev. Lett. 19 (1967) 1312; Phys. Rev. 168 (1967) 1920.

[17] R. Baxter, *Exactly solved models in statistical mechanics*, (Academic Press, London, 1982).

[18] H.C. Lee and M. Couture, *Twisted quantum groups of A_n II. Ribbon links, Chern-Simons theory and graded vertex models*, CRL preprint, TP-90-0505 (submitted to Nucl. Phys. B).

[19] J.W. Alexander, Trans. Amer. Math. Soc. 20 (1928) 275.

[20] J.H. Conway, in *Computational problems in abstract algebra* (Pergamon Press, 1970) 329.

[21] H.C. Lee, *Quantization of* $gl(N,C)/U(1)$ *at roots of unity and parafermions*, Proc. 25th Int. Conf. High Energy Phys. Singapore, 1990 (to be publ. by World Scientific).

[22] H.C. Lee, *On tangles and partially closed braids,* Kyoto U. RIMS preprint 777 (1991) 31pp.

[23] V.G. Turaev, *Operator invariants of tangles, and R-matrices,* Math. USSR Izvestiya 35 (1990) 411-444.

[23] N. Yu. Reshetikhin, *Quasitriangular Hopf algebras and invariants of tangles,* Leningrad Math. J. 1 (1990) 491-513.

A survey of skein modules of 3-manifolds

Jim Hoste and Józef H. Przytycki

Abstract. This paper is a short survey and guide to the literature of the rapidly developing theory of skein modules. These are modules which may be associated to any 3-manifold M^3 and which strive to capture information about M^3 based on the knot theory that M^3 supports. There are many variations on the basic definition of a skein module and hence many different skein modules that can be defined for a given manifold. What all the various definitions have in common however, is their attempt to capitalize on, and generalize, the so-called skein theory of the Alexander/Conway polynomial that was first promoted by Conway and later made popular by the emergence of the Jones polynomial. In some sense the theory of skein modules is the natural extension of the HOMFLY, Kauffman and Jones polynomials to links in arbitrary 3-manifolds.

Introduction

Given an oriented 3-manifold M^3 and a ring R, it is possible to associate to M^3 a host of different modules each loosely described as a "skein" module of M^3. The various possible definitions each strive to capture information about M^3 based on the knot theory that M^3 supports. The fact that one can consider links in M^3 either as oriented or unoriented, framed or unframed, and up to isotopy, homotopy or homology for example, is responsible, in part, for the multitude of different possible skein modules that one can associate to M^3. What all the various definitions have in common however, is their attempt to capitalize on, and generalize, the so-called skein theory of the Alexander polynomial that was first noted by Alexander himself [A], later developed by Conway [C] and then Giller [G], and most recently thrust into prominence via the emergence of the Jones and related polynomials [J], [FYHLMO], [PT_1]. (See also [L_1] and [LM_1] for treatments of skein theory.) In short, the theory of skein modules, as introduced in [P_1] and [T_2], can be seen as the natural extension of the skein (HOMFLY), Kauffman and Jones polynomial link invariants to links in arbitrary 3-manifolds.

Knots 90

We intend this paper to be a short survey and guide to the literature of the rapidly developing theory of skein modules. We have tried to give a reasonable overview of the major results at a moderate level of detail. We do not include proofs of any theorems, but rather hope to inform the reader of the various directions that research in this area has taken. We begin by describing the basic definition of a skein module involving oriented links considered up to isotopy. This definition can then be varied in several important ways to produce other skein modules that are similar in spirit. Rather than presenting material in historical order, we have grouped together the various examples according to their type, e.g. those that employ oriented links as opposed to unoriented links, etc.

Definitions

Before giving a general definition of a skein module of a 3-manifold, it is worth recalling the original skein relation satisfied by the Alexander polynomial $\nabla(z)$ (as normalized by Conway [C]). If L_+, L_- and L_0 are three oriented links in S^3 which have identical diagrams, except near a single right handed crossing of L_+, where they appear instead as shown in Figure 1, then their Alexander polynomials are related by

$$\nabla_{L_+}(z) - \nabla_{L_-}(z) = z\nabla_{L_0}(z). \tag{1}$$

This relation, together with the fact that the unknot has value equal to 1, uniquely determines the Alexander polynomial.

Figure 1

Following the discovery of the Jones polynomial, it was found by several people ([FYHLMO], [PT$_1$]) that the skein relation (1) could be replaced with the more general relation

$$v^{-1}P_{L_+} - vP_{L_-} = zP_{L_0} \tag{2}$$

giving rise to a two variable polynomial invariant $P_L(v, z)$ of oriented links. The Alexander polynomial and the Jones polynomial $V_L(t)$ occur as the special cases $v = 1$, and $v = t, z = t^{1/2} - t^{-1/2}$ respectively.

A skein module of a 3-manifold is a module that is defined essentially by trying to generalize the skein relation (2). Various ways in which we can attempt to generalize this formula are:

(1) Replace S^3 with an arbitrary 3-manifold M^3.
(2) Replace the ring $\mathbb{Z}[v^{\pm 1}, z^{\pm 1}]$ with an arbitrary ring.
(3) Maintain a linear relation but increase (or decrease) the number of terms, using links which differ only locally in prescribed ways.
(4) Consider links as oriented or unoriented, framed or unframed, and up to various types of equivalence. If M^3 has boundary, allow links to have components which are properly embedded arcs.
(5) Use different skein relations in different cases distinguished by global topological data. For example the number of components in L_+ relative to L_0, or some information based on linking numbers.

In order to replace S^3 with an arbitrary 3-manifold M^3, one must move away from projections and diagrams and into a more general setting. This is easily done by considering links which are identical outside some ball B^3 contained in M^3 and which appear inside the ball as various prescribed tangles. However, one often wants to be able to distinguish the links from one another based on their local appearance in B^3, and in order to do this it will usually be necessary to assume that M^3 is an oriented manifold.

If an arbitrary ring R is used then distinguished elements of R will be needed to take the place of v and z. In other words, to serve as the coefficients in the skein relation. Without assuming anything special about these elements, for example that they are invertible in R, the worth of the associated module, as well as its tractability, come into question.

Following closely on the heels of the Jones polynomial was the discovery of the Brandt–Lickorish–Millett/Ho polynomial as well as the Kauffman polynomial [BLM], [Ho], [K$_3$]. These examples point to the need to consider unoriented links.

We can now give a fairly general definition of a skein module of a 3-manifold M^3. The definition will incorporate most of the ideas mentioned above. But the reader will soon see that there seems to be no end to the amount of generality that can be introduced. For the present it seems clear that there can be no ultimate definition. At this point in the development of the theory there is simply too big a gap between what modules can be defined and what special cases of these modules can actually be computed and understood for reasonably interesting manifolds.

The Basic Definition (Isotopy Skein Module of Oriented Links). Let M^3 be an oriented 3-manifold, R a commutative ring with unit, $\mathcal{L}$ the set of all oriented links in M^3 considered up to isotopy, and $R\mathcal{L}$ the free R-module generated by $\mathcal{L}$. Let $T_0, T_1, \ldots, T_{k-1}$ be a sequence of k oriented tangles in a 3-ball B^3. (An oriented tangle is a properly embedded oriented 1-manifold in B^3 having a prechosen, and henceforth fixed, set of inputs and outputs on ∂B^3. Thus $\partial T_i = T_i \cap \partial B^3$ is the same oriented 0-manifold for each i.) Let $r_0, \ldots, r_{k-1}$ be a chosen sequence of elements of R. Now let S be the submodule of $R\mathcal{L}$ generated by all expressions of the form $r_0 L_{T_0} + \cdots + r_{k-1} L_{T_{k-1}}$ where $L_{T_0}, \ldots, L_{T_{k-1}}$ are k oriented links in M^3 which are identical outside some ball B^3 contained in M^3 but inside of B^3 they appear as the tangles $T_0, \ldots, T_{k-1}$. We will often abbreviate this "skein relation" as $r_0T_0 + \cdots + r_{k-1}T_{k-1}$. Finally the quotient module $\vec{\mathcal{S}} = R\mathcal{L}/S$ is called a skein module of M^3. It should be denoted by something like $\vec{\mathcal{S}}(M^3; R; r_0T_0 + \cdots + r_{k-1}T_{k-1}; \text{isotopy})$ to fully incorporate all the ingredients of its definition, but this is obviously unwieldy, and we shall always drop at least the last qualifier, and sometimes the last two or even three qualifiers, from the notation. The arrow over the $\mathcal{S}$ is used to indicate that oriented links are used to define the module.

Notice that this definition is topological and 3-dimensional, not dependent on the existence of any particular projection of the 3-manifold to a 2-dimensional submanifold.

The most straightforward way to alter this definition, while still retaining a certain degree of generality, is to consider unoriented links, or to consider links only up to link homotopy or some other relation. Another important variant is to consider framed links. In this case the tangles must also be framed. It is now possible, and usually desirable, to introduce additional relations between any link and another which differ only by their framing. Many manifolds, for example homology spheres or products, possess natural framings on links. For these manifolds it is not clear how much more can be gained by considering framings (compare [He]). But for an arbitrary 3-manifold it seems most natural to consider framed links [K$_3$], [HP$_3$], [T$_2$].

Note that this definition makes no attempt to incorporate generalities of the kind described in (5) above. Yet we will see that the Kauffman 2-variable polynomial can be associated to such a module. Most of the research involving skein modules so far has focussed on some subset of the tangles shown in Figure 2. Note that $\vec{\tau}_j$ is oriented and has j right-handed half twists. The tangle τ_∞ is shown as unoriented since there is no cannonical way to orient it in a way that is compatible with the other tangles. Nevertheless, there will still be occasions in which we employ τ_∞ with some orientation. The tangles $\vec{\tau}_j$ are also used without their orientations in defining skein modules associated to unoriented links. In these cases the tangles will be denoted by τ_j. (Our general convention will be to place an arrow over oriented links or tangles and to place no such accent over unoriented objects. Occasionally however, we will place a bar over an unoriented object to emphasize that it is not oriented.)

Notice that the tangles of Figure 1 are simply $\vec{\tau}_1, \vec{\tau}_{-1}$ and $\vec{\tau}_0$. In cases like this where k "consecutive" tangles from $\{\vec{\tau}_j\}$ are used, we will often denote the skein module as $\vec{\mathcal{S}}_k(M^3; R)$. If $\vec{\tau}_\infty$ is also used (with some orientation) we will write $\vec{\mathcal{S}}_{k,\infty}(M^3; R)$. (In the case of unoriented links we'll write $\overline{\mathcal{S}}_k(M^3; R)$ or $\overline{\mathcal{S}}_{k,\infty}(M^3; R)$.)

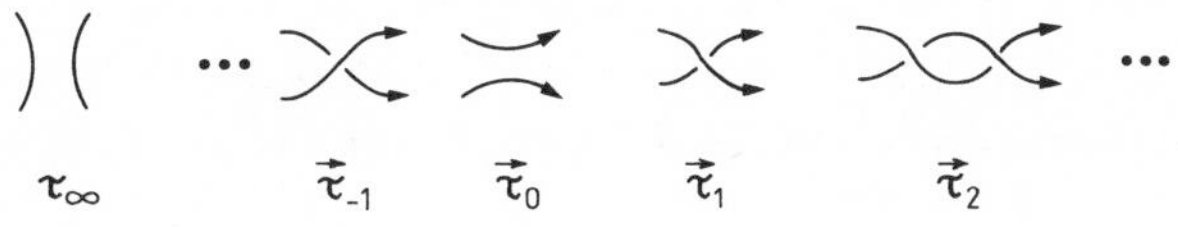

τ_∞ $\vec{\tau}_{-1}$ $\vec{\tau}_0$ $\vec{\tau}_1$ $\vec{\tau}_2$

Figure 2

Examples, results and conjectures

Oriented links up to isotopy

Example 1. Consider oriented links in an arbitrary 3-manifold M^3 and the tangles $\vec{\tau}_0$ and $\vec{\tau}_1$ of Figure 2 together with the skein relation $\vec{\tau}_0 - \vec{\tau}_1$. Then the skein module $\vec{\mathcal{S}}_2(M^3; R; \vec{\tau}_0 - \vec{\tau}_1) \cong RH_1(M^3, \mathbb{Z})$ (McCarthy-Przytycki). This is relatively easy to prove and yet can be used to prove a not so trivial formula of Lickorish and Millett concerning the Kauffman polynomial F of links in S^3 [LM$_2$]. This formula, which was also proven by Turaev [T$_1$], states that

$$F_L(a, -(a + a^{-1})) = [(-1)^{c(L)} \sum_{S \subset L} a^{-4lk(S, L-S)}]/2$$

where $c(L)$ is the number of components of L and the summation is taken over all sublinks S of L (including $S = \emptyset$).

Example 2. Consider oriented links in an arbitrary oriented 3-manifold M^3 and the tangles $\vec{\tau}_{-1}$ and $\vec{\tau}_1$ of Figure 2 with the skein relation $\vec{\tau}_{-1} - \vec{\tau}_1$. Then the skein module $\vec{\mathcal{S}}(M^3; R; \vec{\tau}_{-1} - \vec{\tau}_1)$ is equal to $\bigoplus_1^\infty R\hat{\pi}$, where $R\hat{\pi}$ is the free R-module generated by $\hat{\pi}$, the set of conjugacy classes of the fundamental group $\pi = \pi_1(M^3)$. If we allow the empty link then $\vec{\mathcal{S}}(M^3; R; \vec{\tau}_{-1} - \vec{\tau}_1)$ becomes an algebra where multiplication is given by $L_1 \cdot L_2 = L_1 \cup L_2$. As an algebra it is R-isomorphic to the symmetric tensor algebra over $R\hat{\pi}$, denoted by $S(R\hat{\pi})$, or equivalently to the algebra of R-polynomials with variables from $\hat{\pi}$.

Example 3. Consider oriented links in an arbitrary oriented 3-manifold M^3, $R = \mathbb{Z}[v^{\pm 1}, z^{\pm 1}]$ and the tangles $\vec{\tau}_{-1}, \vec{\tau}_0$ and $\vec{\tau}_1$ of Figure 2 with the skein relation $v^{-1}\vec{\tau}_1 - v\vec{\tau}_{-1} - z\vec{\tau}_0$. The skein module

$$\vec{\mathcal{S}}_3(M^3) = \vec{\mathcal{S}}_3(M^3; \mathbb{Z}[v^{\pm 1}, z^{\pm 1}]; v^{-1}\vec{\tau}_1 - v\vec{\tau}_{-1} - z\vec{\tau}_0)$$

has been computed in several cases described below.

The original result that the skein (HOMFLY) polynomial $P_L(v, z)$ is well defined can be interpreted in this language as the following theorem.

Theorem 4 (Freyd–Yetter; Hoste; Lickorish–Millett; Ocneanu; Przytycki–Traczyk [FYHLMO], [PT$_1$]). *$\vec{\mathcal{S}}_3(M^3) \cong \mathbb{Z}[v^{\pm 1}, z^{\pm 1}]$ with basis the unknot $\mathcal{U}_1$. For each link L, the element it represents in the skein module is $P_L(v, z)\mathcal{U}_1$.*

Theorem 5 (Hoste–Kidwell [HK], Turaev [T$_2$]). *$\vec{\mathcal{S}}_3(S^1 \times D^2)$ is a free infinitely generated $\mathbb{Z}[v^{\pm 1}, z^{\pm 1}]$ – module. The basis will be described in the more general setting of the following theorem.*

Theorem 6 (Przytycki [P$_3$]). *Let $M^3 = F \times I$ where F is a compact oriented surface. Then the skein module $\vec{\mathcal{S}}_3(F \times I)$ is a free infinitely generated $\mathbb{Z}[v^{\pm 1}, z^{\pm 1}]$-module. In this case we allow the empty link $\emptyset$ and the skein relation $v^{-1}\emptyset - v\emptyset - z\mathcal{U}_1$ where $\mathcal{U}_1$ is the unknot. The module has the structure of an algebra where $L_1 \cdot L_2$ is defined to be the link obtained by placing L_1 over L_2. The empty link is the multiplicative unit. Let $\hat{\pi}^0$ denote the set of conjugacy classes of nontrivial elements of $\pi = \pi_1(F)$ and let $R\hat{\pi}^0$ be the free $\mathbb{Z}[v^{\pm 1}, z^{\pm 1}]$-module over $\hat{\pi}^0$. Finally, let $S(R\hat{\pi}^0)$ be the tensor symmetric algebra over $R\hat{\pi}^0$. Then there is an R-module isomorphism $i : S(R\hat{\pi}^0) \to \vec{\mathcal{S}}_3(F \times I)$ such that for $w \in \hat{\pi}^0$, $i(w)$ is represented by a knot K in $F \times I$ such that the class of K in $\hat{\pi}^0$ equals w.*

The proof of Theorem 6 consists of a delicate multistep induction which incorporates the ideas of [BM], [FYHLMO], [PT$_1$], [HK], [HP$_2$] and [T$_3$], and the method used in the classical proof of the Poincaré–Birkhoff–Witt Theorem on universal enveloping algebras of Lie algebras [Ab]. The case of a closed surface is different and relies on a recent (and as yet uncirculated) result of Hass–Scott–Grayson on curvature flow on hyperbolic surfaces [HS]. The theorem itself can be thought of as a variant of the Poincaré–Birkhoff–Witt Theorem.

The assumption that F be compact is not really necessary because for any surface F, Reidemeister moves suffice to realize (ambient) isotopies of links in $F \times I$ (see [Hu]). Furthermore, the methods that are used in the case where F has boundary also work for noncompact F.

In the case where F is a planar surface, so that $F \times I$ is a handlebody, the module $\vec{\mathcal{S}}_3(F \times I)$ also has the structure of a Hopf algebra (i.e. quantum group). (See [Ab] for definitions.) This result, stated below, solves a conjecture of Turaev [T$_4$].

Theorem 7 (Przytycki [P$_4$]; Cotta-Ramusino–Rinaldi [CR]). *If F is a planar surface then the module $\vec{\mathcal{S}}_3(F \times I)$ also has the structure of a Hopf algebra (i.e. quantum group). Co-multiplication is that given by Turaev [T$_4$] (based on an idea of Jaeger [Ja]) and the antipode map is related to reflecting in the I factor.*

Example 8. Consider oriented links in an arbitrary oriented 3-manifold M^3, any ring R and the tangles T_0 and T_1 of Figure 3 together with the skein relation $T_0 - T_1$. The module $\vec{\mathcal{S}}(M^3; R; T_0 - T_1)$ has been determined in the case of $M^3 = S^3$ as described in the following theorem.

Figure 3

Theorem 9 (Nakanishi–Murakami [NM]). *The module $\vec{\mathcal{S}}(M^3; R; T_0 - T_1)$ is free on infinitely many generators. A generating set consists of a set of links representing all possible linking matrices (up to reordering of the link components) without repetition.*

Example 10. Consider oriented links in an arbitrary oriented 3-manifold M^3, where additionally each component of the link has been "colored" with a natural number $n \in \mathbb{N}$. Let $\{t_i\}_{i \in \mathbb{N}}$ be indeterminates and let R be the ring of rational functions in $\{t_i\}$. We may now consider skein modules based on skein relations that employ the colorings. At least two important theorems have been proven in this setting.

Theorem 11 (Nakanishi [N_1]). *Consider oriented colored links in S^3 employing exactly three colors. Let R be the ring of rational functions in t_1, t_2, t_3. Then the colored skein module $\vec{\mathcal{S}}(S^3; R; r_1, r_2, r_3)$, where r_1, r_2, r_3 are the skein relations shown in Figure 4, is finitely generated by the links shown in Figure 5.*

Theorem 12 (J. Murakami [Mu]). *Consider oriented colored links in S^3 employing any number of colors. Let R be the ring of rational functions in $\{t_i\}$. Then the colored skein module $\vec{\mathcal{S}}(S^3; R; r_1, r_2, r_3, r_4, r_5)$, where r_1, r_2, r_3, r_4, r_5 are the skein relations shown in Figure 4 is free on one generator $\mathcal{U}_1$, which can be taken to be an unknot of color c. If L is any colored link then, in the skein module, $L = (t_c - t_c^{-1})\Delta_L \mathcal{U}_1$, where Δ_L is the multi-variable Alexander polynomial of L as normalized by Conway.*

It is worth pointing out that Murakami derived Theorem 12 by first describing the multi-variable Alexander polynomial Δ in terms of a solution to a "color enhanced" Yang–Baxter equation thereby producing a states model for Δ. Turaev has considered "enhanced" Yang–Baxter operators in great generality and has

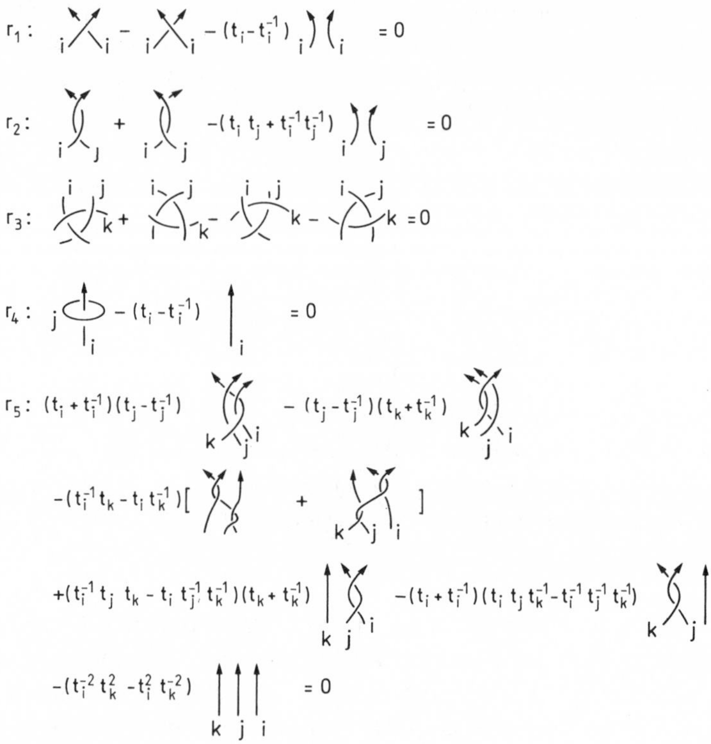

Figure 4

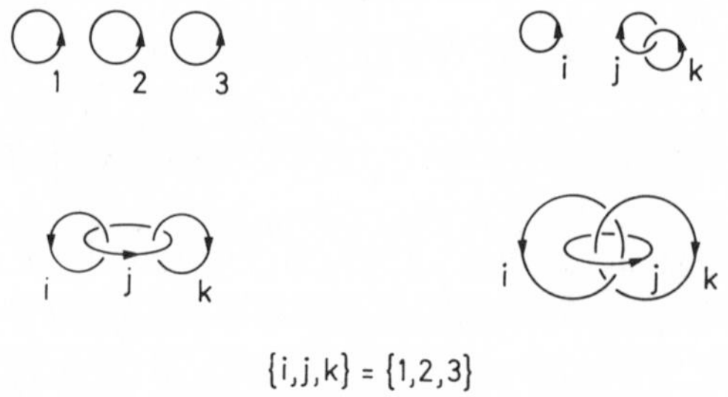

$\{i,j,k\} = \{1,2,3\}$

Figure 5

shown that they always give rise to link invariants [T_1]. Moreover these link invariants always satisfy skein relations employing the tangles of Figure 2.

Another variation on the basic definition is to make use of skein relations that depend on global criteria. A good example of this is the module associated to the Kauffman polynomial $F(a,x)$. The next theorem follows from the original work of Kauffman [K_3] together with the work of Lickorish [L_2].

In order to state the theorem we must describe a way to orient the tangle τ_∞ of Figure 2 relative to the tangles $\vec{\tau}_j$ of Figure 2. Let $c(L)$ denote the number of components in the link L. Now define $\vec{\tau}_\infty$ as in Figure 6a if $c(L_{\vec{\tau}_1}) < c(L_{\vec{\tau}_0})$ and as in Figure 6b if $c(L_{\vec{\tau}_1}) > c(L_{\vec{\tau}_0})$. In case $c(L_{\vec{\tau}_1}) < c(L_{\vec{\tau}_0})$ let λ be the linking number of the bottom component of $L_{\vec{\tau}_0}$ in Figure 2 with the remainder of $L_{\vec{\tau}_0}$. If $c(L_{\vec{\tau}_1}) > c(L_{\vec{\tau}_0})$ let μ be the linking number of the bottom-left to top-right component of $L_{\vec{\tau}_1}$ in Figure 2 with the remainder of $L_{\vec{\tau}_1}$.

Theorem 13. *Let $M^3 = S^3$ and $R = \mathbb{Z}[a^{\pm 1}, x^{\pm 1}]$. Consider oriented links in S^3 up to isotopy and modulo the following skein relations of two different types.*

(1) $a\vec{\tau}_1 + a^{-1}\vec{\tau}_{-1} - x\vec{\tau}_0 - xa^{-4\lambda}\vec{\tau}_\infty$ *if* $c(L_{\vec{\tau}_1}) < c(L_{\vec{\tau}_0})$
(2) $a\vec{\tau}_1 + a^{-1}\vec{\tau}_{-1} - x\vec{\tau}_0 - xa^{-4\mu}\vec{\tau}_\infty$ *if* $c(L_{\vec{\tau}_1}) > c(L_{\vec{\tau}_0})$

Then the skein module $\vec{\mathcal{S}}$ is freely generated by the unknot $\mathcal{U}_1$. If L is a link then, in the module $\vec{\mathcal{S}}$, $L = F_L(a, x)\mathcal{U}_1$.

Figure 6a and 6b

It is not necessary to define the Kauffman polynomial as above. In particular if one works with unoriented framed links it is possible to use only local skein relations. This approach has merit in that it allows for an obvious generalization to arbitrary oriented 3-manifolds. We describe this in greater detail later in this paper.

Oriented links up to link anti-homotopy

A crucial step in the proof of Theorem 6 is the computation of the "anti-homotopy" skein module of $F \times I$. This module (which is called anti-homotopy in contrast to the homotopy skein modules described later in this paper) is defined as follows.

Definition (Anti-homotopy skein module of oriented links). Let $R = \mathbb{Z}[v^{\pm 1}]$. Consider oriented links in M^3, the tangles $\vec{\tau}_{-1}$, $\vec{\tau}_0$ and $\vec{\tau}_1$ of Figure 2 and the skein relations

(1) $v^{-1}\vec{\tau}_1 - v\vec{\tau}_{-1} - \vec{\tau}_0$ if the crossing of $\vec{\tau}_1$ is a self-crossing
(2) $v^{-1}\vec{\tau}_1 - v\vec{\tau}_{-1}$ otherwise.

Furthermore, include the relation $v^{-1}\emptyset - v\emptyset - \mathcal{U}_1$. Then the associated skein module is called the anti-homotopy skein module and denoted $\mathcal{A}\vec{\mathcal{S}}(M^3)$.

Theorem 14 (Hoste–Przytycki [HP$_2$]). *If F is a compact planar surface then $\mathcal{A}\vec{\mathcal{S}}(F \times I)$ is isomorphic to $S(R\hat{\pi}^0)$. Moreover, $\mathcal{A}\vec{\mathcal{S}}(F \times I)$ corresponds to the part of $\vec{\mathcal{S}}_3(F \times I)$ associated to the smallest power of z.*

The role of the anti-homotopy skein module in the proof of Theorem 6 is to start an induction argument that essentially inducts on that part of $\vec{\mathcal{S}}(F \times I)$ associated to a given power of z. This is the similarity to [BM]. To complete the inductive step requires the construction of a "Lie bracket" and it is this part of the proof that is similar to the classic Poincaré–Birkhoff–Witt Theorem.

Oriented links including arcs up to isotopy

One simple way to modify the basic definition of skein module is to work inside a manifold with boundary and to allow links to include components which are properly embedded arcs. One result in this category is the following theorem.

Theorem 15 (Morton, Traczyk [MT]). *Let $\vec{\mathcal{S}}_3(B^3)$ be the oriented skein module of links in B^3 consisting of n properly embedded arcs and any number of embedded circles. Consider these links up to isotopy rel endpoints. Let $R = \mathbb{Z}[v^{\pm 1}, z^{\pm 1}]$ and use the skein relation $v^{-1}\vec{\mathcal{T}}_1 - v\vec{\mathcal{T}}_{-1} - z\vec{\mathcal{T}}_0$, where $\vec{\mathcal{T}}_{-1}, \vec{\mathcal{T}}_0$ and $\vec{\mathcal{T}}_1$ are as in Figure 2. Then $\vec{\mathcal{S}}_3(B^3)$ is free on $n!$ generators. Each generator consists of n arcs which connect the $2n$ boundary points in one of the $n!$ ways.*

Oriented links up to link homotopy

Considering links up to link homotopy rather than isotopy gives rise to the following definition.

Definition (Homotopy skein module). Let $R = \mathbb{Z}[z]$. Consider oriented links in M^3, the tangles $\vec{\mathcal{T}}_{-1}, \vec{\mathcal{T}}_0$ and $\vec{\mathcal{T}}_1$ of Figure 2, and the skein relations

(1) $\vec{\mathcal{T}}_1 - \vec{\mathcal{T}}_{-1}$ if the crossing of $\vec{\mathcal{T}}_{-1}$ is a self-crossing
(2) $\vec{\mathcal{T}}_1 - \vec{\mathcal{T}}_{-1} - z\vec{\mathcal{T}}_0$ otherwise.

The resulting skein module is called the homotopy skein module of M^3 and is denoted $\mathcal{H}\vec{\mathcal{S}}(M^3)$.

Theorem 16 (Hoste–Przytycki [HP$_1$]; Turaev [T$_4$]).

(1) *For a compact surface F, $\mathcal{H}\vec{\mathcal{S}}(F \times I)$ is an infinitely generated free R-module isomorphic to the symmetric tensor product $S(R\hat{\pi})$.*

(2) *A description of $\mathcal{H}\vec{\mathcal{S}}(F \times I)$ analogous to the Poincaré–Birkhoff–Witt Theorem can be made. $\mathcal{H}\vec{\mathcal{S}}(F \times I)$ has a natural algebra structure with product $L_1 \cdot L_2$ obtained by placing L_1 above L_2. The empty link $\emptyset$ is the multiplicative unit. On the other hand, $R\hat{\pi}$ has a Lie algebra structure (the Goldman–Wolpert Lie algebra [Go]). The Lie bracket $[\,,\,] : R\hat{\pi} \times R\hat{\pi} \to R\hat{\pi}$ is defined as follows. Suppose that α and β are immersed loops in F in general position. At a double point p let $sgn(\alpha, \beta, p) = \pm 1$ as illustrated in Figure 7. Let $\alpha_p\beta_p$ denote the product of α_p and β_p in $\pi_1(F, p)$ as shown in Figure 7. Then $[\alpha, \beta] \in R\hat{\pi}$ is defined as $[\alpha, \beta] = z \sum_{p \in \alpha \cap \beta} sgn(\alpha, \beta, p)|\alpha_p\beta_p|$ where $|\gamma|$ denotes the class of γ in $\hat{\pi}$. Then $\mathcal{H}\vec{\mathcal{S}}(F \times I)$ is algebra-isomorphic to the universal enveloping algebra of the Goldman–Wolpert Lie algebra $U(R\hat{\pi}) = T(R\hat{\pi})/(\alpha \otimes \beta - \beta \otimes \alpha - [\alpha, \beta])$.*

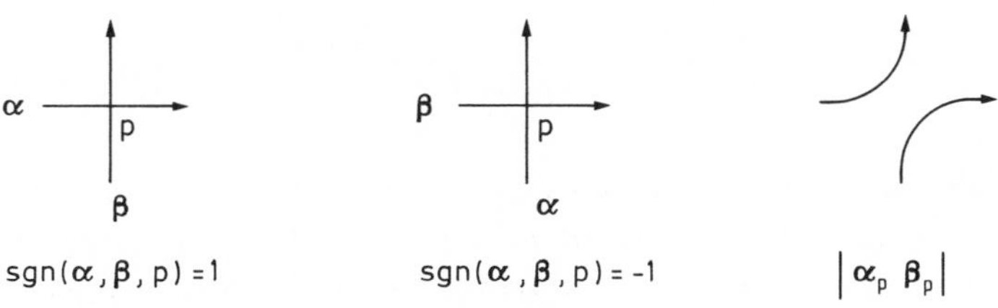

Figure 7

Unoriented links up to isotopy

Example 17. Consider unoriented links in an arbitrary 3-manifold M^3, any ring R and the unoriented tangles τ_0 and τ_3 of Figure 2, with the skein relation $\tau_0 - \tau_3$. In the case $M^3 = S^3$, Montesinos and Nakanishi have conjectured that the skein module $\overline{\mathcal{S}}(S^3; R; \tau_0 - \tau_3)$ is freely generated by the set of unlinks [Mo].

The well-definedness of the Brandt–Lickorish–Millett/Ho polynomial $Q(x)$ translates into the following theorem.

Theorem 18 (Brandt–Lickorish–Millett [BLM]; Ho [Ho]). *Let $\tau_\infty, \tau_{-1}, \tau_0$ and τ_1 be the unoriented tangles of Figure 2. The module $\overline{\mathcal{S}}_{3,\infty}(S^3; \mathbb{Z}[x]; \tau_1 + \tau_{-1} - x\tau_0 - x\tau_\infty)$ is isomorphic to $\mathbb{Z}[x]$. Here the bar over the $\mathcal{S}$ is to indicate that unoriented links are used, considered up to isotopy.*

The Jones polynomial is a special case of the skein (HOMFLY) polynomial and so can be described in terms of an $\vec{\mathcal{S}}_3$ skein module. But it can also be described via Kauffman's bracket polynomial and hence in terms of an $\overline{\mathcal{S}}_{2,\infty}$ skein module. We state this in Theorem 22 within the context of unoriented framed links.

Example 19. Consider unoriented links in an arbitrary 3-manifold M^3, any ring R and the tangles T_0^k and T_1^k for any k shown in Figure 8, together with skein relations $T_0^k - T_1^k$. For the case $M^3 = S^3$, Nakanishi [N2] proved the following result, the proof of which is based on Theorem 9.

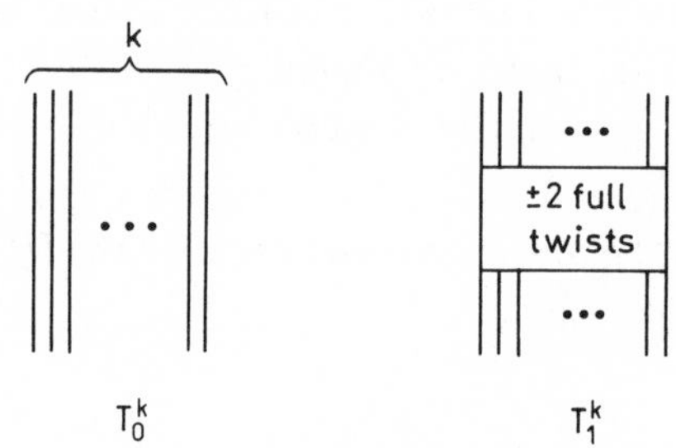

Figure 8

Theorem 20 (Nakanishi). *The module $\overline{\mathcal{S}}(S^3; R; T_0^k - T_1^k)$ is free on infinitely many generators. A generating set consists of a set of links representing all possible linking matrices reduced mod 2 (up to reordering of the link components) without repetition.*

Theorem 20 is crucial in Akbulut and King's proof of the 3-dimensional Nash conjecture which states that any two closed connected smooth 3-manifolds can be made diffeomorphic after topologically blowing up along submanifolds [Na], [AK].

Unoriented framed links up to isotopy

Example 21. Let $R = \mathbb{Z}[A^{\pm 1}]$. Consider unoriented framed links in an oriented 3-manifold M^3 and the unoriented tangles τ_∞, τ_0 and τ_1 of Figure 2 together with the skein relation $\tau_1 - A\tau_0 - A^{-1}\tau_\infty$. (The framing of these tangles is taken to be perpendicular to the plane of projection of Figure 2.) Consider also the framing relation $L^{(1)} + A^3 L$ where $L^{(1)}$ denotes the framed link obtained from L by adding one positive twist to the framing of L. Then the associated skein module, denoted $\overline{\mathcal{S}}_{2,\infty}^{fr}(M^3; \mathbb{Z}[A^{\pm 1}]; \tau_1 - A\tau_0 - A^{-1}\tau_\infty, L^{(1)} + A^3 L)$, or more simply $\overline{\mathcal{S}}_{2,\infty}^{fr}(M^3)$, is called the $(2,\infty)$-skein module of M^3.

Theorem 22 (Kauffman [K$_2$]). *$\overline{S}^{fr}_{2,\infty}(S^3) \cong \mathbb{Z}[A^{\pm 1}]$ with basis the unknot. This is equivalent to saying that the Jones polynomial of links is well defined.*

Theorem 23 (Hoste–Przytycki [HP$_1$]; Ju. Drobotuchina [Dr]). *Suppose that M^3 is an oriented 3-manifold which is diffeomorphic to an I-bundle over a surface F. Then $\overline{S}^{fr}_{2,\infty}(M^3)$ is a free $\mathbb{Z}[A^{\pm 1}]$-module with basis $B(F)$ consisting of the unknot together with links in F without trivial components. In particular $\overline{S}^{fr}_{2,\infty}(\mathbb{RP}^3) \cong \mathbb{Z}[A^{\pm 1}] \oplus \mathbb{Z}[A^{\pm 1}]$. (Notice that $\mathbb{RP}^3 - B^3$ is an I-bundle over $\mathbb{RP}^2$.)*

Theorem 24 (Hoste–Przytycki [HP$_3$]). *For the lens space $L(p,q), p \geq 1$, $\overline{S}^{fr}_{2,\infty}(L(p,q))$ is a free $\mathbb{Z}[A^{\pm 1}]$-module generated by $[p/2]+1$ generators where $[y]$ is the greatest integer less than or equal to y. The module $\overline{S}^{fr}_{2,\infty}(S^1 \times S^2)$ is infinitely generated, but $\overline{S}^{fr}_{2,\infty}(S^1 \times S^2) \otimes \mathcal{F}(A^{\pm 1}) \cong \mathcal{F}(A^{\pm 1})$, where $\mathcal{F}(A^{\pm 1})$ is the field of rational functions in the variable A. (The case $q=1$ was independently proven by Turaev.)*

Theorem 25 (Hoste–Przytycki [HP$_4$]). *Let W^3 be the classic Whitehead manifold. Then $\overline{S}^{fr}_{2,\infty}(W^3)$ is infinitely generated, torsion free, but not free. In fact, $\overline{S}^{fr}_{2,\infty}(W^3)$ is infinitely generated, torsion free but not free for uncountably many, perhaps all, genus 1 Whitehead type open contractible manifolds. (These are the manifolds that are constructed by the obvious variation of Whitehead's classic construction.)*

Conjecture 26. *If M^3 is compact then $\overline{S}^{fr}_{2,\infty}(M^3)$ is a free $\mathbb{Z}[A^{\pm 1}]$-module if and only if M^3 contains no nonseparating 2-sphere.*

If S^2 is a separating 2-sphere in M^3 then one can easily show that any link cutting S^2 transversely in a single point represents a torsion element in $\overline{S}^{fr}_{2,\infty}(M^3)$.

It may seem curious to the reader that we have passed from unframed to framed links. After all, Theorems 22-25 generalize the Jones polynomial in S^3 which is an invariant of unframed links. The answer lies in the fact that we have approached the Jones polynomial via the Kauffman bracket. The bracket is an invariant of regular isotopy in S^3 which really can be interpreted as saying that it is an invariant of framed diagrams in the plane with framings perpendicular to the plane. Links which differ by a Type I Reidemeister move now differ by framing as well, and so we should not expect their bracket polynomials to agree. The manifold S^3 is special because $\mathbb{R}^3$ is the product $\mathbb{R}^2 \times \mathbb{R}$, and there is a natural framing perpendicular to the $\mathbb{R}^2$ factor. It is probably correct to think of all the known skein modules of links in S^3 as actually involving framed links, with the special (product) nature of $\mathbb{R}^3$ making it unnecessary to explicitly mention framings.

Example 27. Consider unoriented framed links in an arbitrary oriented 3-manifold M^3, $R = \mathbb{Z}[a^{\pm 1}, x^{\pm 1}]$, the unoriented tangles $\tau_\infty, \tau_{-1}, \tau_0$ and τ_1 of Figure 2 and the skein relation $\tau_1 + \tau_{-1} - x\tau_0 - x\tau_\infty$. Consider additionally the framing relations $L^{(1)} - aL$.The module $\overline{\mathcal{S}}^{fr}_{3,\infty}(M^3; R; \tau_1 + \tau_{-1} - x\tau_0 - x\tau_\infty, L^{(1)} - aL)$, or more simply $\overline{\mathcal{S}}^{fr}_{3,\infty}(M^3)$, is called the Kauffman skein module of M^3. We call it this because the original statement that the Kauffman polynomial is well defined can be reformulated as the following theorem.

Theorem 28 (Kauffman [K$_3$]). *The module $\overline{\mathcal{S}}^{fr}_{3,\infty}(S^3)$ is free on one generator which can be taken to be the unknot $\mathcal{U}_1$ with trivial framing. If $\overline{L}$ is an unoriented framed link and $\vec{L}$ is $\overline{L}$ with some chosen orientation, and moreover the framing of $\overline{L}$ agrees with the preferred framing of $\vec{L}$ (i.e. the one induced by a Seifert surface of $\vec{L}$), then $\overline{L} = F_{\vec{L}}(a, x)\mathcal{U}_1$.*

Theorem 29 (Hoste–Kidwell; Turaev [T$_2$]). *The module $\overline{\mathcal{S}}^{fr}_{3,\infty}(S^1 \times D^2)$ is free on infinitely many generators.*

Future directions

P. Traczyk proposes to use the theory of skein modules to find a counterexample to the Poincaré conjecture. In particular, he hopes to show that the skein module of a potential counterexample is not free on one unknotted generator by making actual computations involving specific links in the manifold. His investigations have suggested the following conjecture.

Conjecture 30 (Traczyk [Tr]). *A homotopy 3-sphere M^3 is homeormorphic to S^3 if and only if $\vec{\mathcal{S}}_3(M^3)$ is generated by the unknot.*

Associated to every skein module $\mathcal{S}$ is its co-skein module $\mathcal{S}^* = Hom(\mathcal{S}, R)$. Various invariants of links in 3-manifolds can be interpreted as elements of certain co-skein modules. In particular the Witten invariant is related to the co-skein module $\vec{\mathcal{S}}_3(M^3)$ of Example 3.

The existence of the Witten invariant together with the fact that it satisfies a kind of Van Kampen-Seifert theorem suggests the possibility of such a theorem for skein modules.

Finally, in order to develop the theory of skein modules along the lines of traditional algebraic topology, for example like homotopy or homology theory, it would appear necessary to extend the definition in a meaningful way to other dimensions. The obvious generalization is to consider codimension 2 embedded (or immersed) submanifolds of an n-dimensional manifold M^n up to isotopy (or regular homotopy). One should then form the free R-module generated by equivalence classes of these submanifolds and mod out by some set of local skein relations.

References

[Ab] E. Abe, *Hopf Algebras*, Cambridge University Press, 1977.

[AK] S. Akbulut and H. King, *Rational structures on 3-manifolds*, preprint.

[ADW] Akutsu, Deguchi and Wadati, *Exactly solvable models and new link polynomials I, II, III and IV*, J. Phys. Soc. Japan **56 and 57** (1987).

[A] J. W. Alexander, *Topological invariants of knots and links*, Trans. Amer. Math. Soc. **30** (1928), 275–306.

[BM] R. Ball and R. L. Metha, *Sequence of invariants of knots and links*, J. Physique **42** (1981), 1193–1199.

[BLM] R.D. Brandt, W.B.R. Likorish and K. Millett, *A polynomial invariant for unoriented knots and links*, Invent. Math. **84** (1986), 563–573.

[C] J. H. Conway, *An enumeration of knots and links*, Computational problems in abstract algebra (J. Leech, ed.), Pergamon press, 1969, pp. 329–358.

[CR] P. Cotta-Ramusino and M. Rinaldi, *On the algebraic structure of link-diagrams on a 2-dimensional surface*, preprint.

[D] V. G. Drinfeld, *Quantum groups*, Proc. Internat. Congress Math. Berkeley, Am. Math. Soc., 1986, pp. 798–820.

[Dr] Ju. Drobotuchina, *An analogue of the Jones polynomial for links in $\mathbb{RP}^3$ and a generalization of the Kauffman-Murasugi Theorem*, Algebra i Analiz **2** no. 3 (1990), 171–191.

[FYHLMO] P. Freyd, D. Yetter, J. Hoste, W.B.R. Lickorish, K. Millett and A. Ocneanu, *A new polynomial invartiant of links*, Bull. Amer. Math. Soc. **12 (2)** (1985), 239–246.

[G] C. A. Giller, *A family of links and the Conway calculus*, Trans. Amer. Math. Soc. **270** (1982), 75–109.

[Gr] M. Grayson, Annals of Math. (to appear).

[Go] W. Goldman, *Invariant functions of Lie groups and Hamiltonian flows of surface group representations*, Invent. Math. **85** (1986), 263–302.

[HS] J. Hass and P. Scott, *Flows and intersections of curves on surfaces*, Lecture of J. Hass at Ohio State Univ. April 10, 1990.

[He] M. A. Hennings, *A polynomial invariant for banded links*, preprint.

[Ho] C.F. Ho, *A new polynomial invariant for knots and links, preliminary report*, Abstracts of the Amer. Math. Soc. 6,4 (1985), 300.

[H] J. Hoste, *A polynomial invariant of knots and links*, Pacific J. Math. **124 (2)** (1986), 295–320.

[HK] J. Hoste and M. Kidwell, *Dichromatic link invariants*, Trans. Amer. Math. Soc. **321** no. 1 (1990), 197–229.

[HP_1] J. Hoste and J.H. Przytycki, *Homotopy skein modules of oriented 3-manifolds*, Math. Proc. Camb. Phil. Soc. **108** (1990), 475–488.

[HP_2] ——, *Anti-homotopy skein modules of 3-manifolds*, in preparation; Abstracts of the Am. Math. Soc **10 (2)** (1989), 165.

[HP_3] ——, *The $(2,\infty)$-skein module of lens spaces: a generalization of the Jones polynomial*, preprint; Abstracts of the Am. Math. Soc. **11 (3)** (1990), 29.

[HP_4] ——, *The skein module of genus 1 Whitehead type manifolds*, preprint.

[Hu] J.F.P. Hudson, *Piecewise linear topology*, Benjamin Inc. N. Y., 1969.

[Ja] F. Jaeger, *Composition products and models for the HOMFLY polynomial*, preprint.

[J] V.F.R. Jones, *A polynomial invariant for knots via von Neumann algebras*, Bull. Amer. Math. Soc. **12** (1985), 103–111.

[K_1] L. Kauffman, *Formal Knot Theory*, Math. Notes 30, Princeton Univ. Press, 1983.

[K_2] ——, *State models and the Jones polynomial*, Topology **26** (1987), 395–401.

[K_3] ——, *New invariants in the theory of knots*, Amer. Math. Monthly **95** (1988), 195-242.

[L_1] W.B.R. Lickorish, *Linear skein theory and link polynomials*, Topology and its Applications **27** (1987), 265–274.

[L_2] ——, *A relationship between link polynomials*, Math. Proc. Camb. Phil.Soc. **100** (1986), 109–112.

[L_3] ——, *The panorama of polynomials for knots, links and skeins*, Contemp. Math. (J.S. Birman and A. Libgober, eds.) **78** (1988).

[L_4] ——, *Polynomials for links*, Bull. London Math. Soc. **20** (1988), 558–588.

[LM_1] W.B.R. Lickorish and K. Millett, *A polynomial invariant of oriented links*, Topology **26** (1987), 107–141.

[LM_2] ——, *An evaluation of the F-polynomial for links*, Proc. of Siegen topology meeting.

[Mo] H. R. Morton, *Problems*, Contemp. Math. (J.S. Birman and A. Libgober, eds.) **78** (1988), 557–574.

[MT] H. Morton and P. Traczyk, *Knots, skeins and algebras*, preprint.

[MN] H. Murakami and Y. Nakanishi, *On a certain move generating link-homology*, Math. Ann. **284** (1989), 75–89.

[Mu] J. Murakami, *A state model for the multi-variable Alexander polynomial*, preprint.

[N_1] Y. Nakanishi, *Three-variable Conway potential functions of links*, Tokyo J. Math. **13** (1990), 163–177.

[N_2] ——, *On Fox's congruence classes of knots,2*, Osaka J. Math. **27** (1990), 207–215.

[Na] J. Nash, *Real algebraic manifolds*, Ann. of Math. **56** (1952), 405–421.

[P_1] J.H. Przytycki, *Skein modules of 3-manifolds*, Bull. Polo. Acad. Sci.: Math. (to appear).

[P_2] ——, *Survey on recent invariants in classical knot theory*, Warsaw Univ. preprint.

[P_3] ——, *Skein module of links in a handlebody*, Proc. of the OSU research seminar (to appear).

[P_4] ——, *Quantum group of links in a handlebody*, preprint.

[PT_1] J.H. Przytycki and P. Traczyk, *Invariants of links of Conway type*, Kobe J. Math. **4** (1987), 115–139.

[PT_2] ——, *Conway algebras and skein equivalence of links*, Proc. Amer. Math. Soc. **100(4)** (1987), 744–748.

[Tr] P. Traczyk, private communication.

[T_1] V.G. Turaev, *The Yang-Baxter equations and invariants of links*, Invent. Math. **92** (1988), 527–553.

[T_2] ———, *The Conway and Kauffman modules of the solid torus*, Zap. Nauchn. Sem. LOMI ;English transl. in J. Soviet Math. **167** (1988), 79–89.

[T_3] ———, *Algebras of loops on surfaces, algebra of knots, and quantization, in Braid groups, knot theory and statistical mechanics*, Advanced Series in Math. Physics **9** (1989), 59–95.

[T_4] ———, *Skein quantization of Poisson algebras of loops on surfaces*, preprint.

Representations of link groups

Pierre Vogel

Many different invariants of manifolds like Reidemeister torsion, high signature invariants, Atiyah-Patodi-Singer η invariants, Casson-Gordon invariants of classical knots, Casson invariants of three dimensional homology spheres etc ... are constructed by considering representations of the fundamental group of some manifold to a compact Lie group (U_n in general). If the fundamental group G of the manifold is finite or commutative or free, the set $R(G)$ of unitary representations of G is very well understood and we can define different invariants.

A particular case is the following: If L is a high dimensional boundary link, the manifold M_L obtained by surgery along L has a free fundamental group F. Then the η-invariant induces a map on the set $R(F)$ of unitary representations of F [**L4**]. If we want to extend this construction for arbitrary links, we have to understand a little bit the set of representations of the fundamental group G of the complement of the link L. Unfortunately this group is unknown. It depends strongly on the link. If L is not a boundary link, G is not necessarily free. Another point is the following: the group G depends on L and not only on its concordance class. So it is difficult to define an invariant on $R(G)$ which is invariant under concordance.

The group G contains the free group F and this map induces an isomorphism on H_1 and H_2. Furthermore the canonical inclusion from F to its algebraic closure F' [**LD**], [**L1**] extends in a unique way to G. Therefore every representation from F' to a Lie group induces a representation of the fundamental group of the complement of any link. But now there is another problem: if F and $R(F)$ are very well understood, more or less nothing is known about F' except universal properties [**L1,2,3**] and the existence of a map from $\mathbb{Z}[F']$ to a finite dimensional algebra over an algebraic extension of a polynomial ring [**V**].

In this paper we'll show that there exist a lot of representations from F' to U_n. More precisely, for every connected compact Lie group Γ, every representation from the free group F to Γ extends to F'. As a corollary every invariant of boundary links defined by a unitary representation of the free group F may be extended to all links.

Knots 90

It is not clear if the compactness condition on Γ is necessary, but the connectivity condition is very important. For instance, a surjective representation from F to the symmetric group $\mathfrak{S}_n$ never extends to F'.

An important point in the proof concerns the Magnus expansion of elements of F' in the ring of series in non commuting variables. We prove that the coefficients of the Magnus expansion of every element of F' are not too big in some precise sense. Hence every representation ρ of F in a Lie group Γ which is sufficiently near the trivial representation sends the Magnus expansion of a given element of F' to a convergent series in any finite dimensional $\mathbb{C}$-algebra containing Γ. The limit of this series lies in Γ and the representation ρ is extended in this way to a big subgroup of F'. The rest of the argument uses some techniques from algebraic geometry.

§1 The algebraic closure of the free group and the Magnus expansion

Let F be the free group generated by $x_1, \cdots, x_n$. Denote by F' the algebraic closure of F [**L1**]. This group contains F and satisfies the following properties:

i) F' is the direct limit of a sequence of finitely presented groups G_p such that $G_0 = F$ and for every $p \geq 0$:

$$H_1(G_p, \mathbb{Z}) \xrightarrow{\sim} H_1(G_{p+1}, \mathbb{Z}), \quad H_2(G_p, \mathbb{Z}) = 0$$

and the map $G_p \to G_{p+1}$ is normally surjective.

ii) F' is universal with respect to property i).

Let $A = A(t_1, \cdots, t_n)$ be the algebra of formal series in non commuting variables t_i with integral coefficients. Every element u of A can be written as an infinite sum $\sum_\alpha f(\alpha)\alpha$ where f is a map from the set of monomials in the t_i's to $\mathbb{Z}$.

The group F may be embedded in A by the Magnus embedding $x_i \mapsto 1 + t_i$. This map extends uniquely to the nilpotent completion of F and therefore to F'. The series corresponding to an element ω of F' is called the Magnus expansion of ω.

Definition 1-1. Let $u = \sum_\alpha f(\alpha)\alpha$ be an element of A and c be a scalar in $[0, \infty]$. We'll say that the size of u is less than or equal to c if there exists a scalar a such that:

for every monomial α of length m: $|f(\alpha)| \leq ac^m$.

The series u is called convergent if its size is finite and such a couple (a, c) is called a bound for u.

Remark. The reason for this definition is the following: if B is a Banach algebra and if $u = u(t_1, \cdots, t_n)$ is a convergent series in A there exists $\varepsilon > 0$ such that: $\forall b_1, \cdots, b_n \in B \parallel b_1 \parallel < \varepsilon, \cdots, \parallel b_n \parallel < \varepsilon \Rightarrow$ the series $u(b_1, \cdots, b_n)$ converges in B.

The main result of this section is the following:

Theorem 1-2. *The Magnus expansion of every element of F' is convergent.*

Lemma 1-3 Let $c > 0$. *Then the set A_c of series of size less than c is a sub-algebra of A.*

Proof. The only non trivial point to check is that A_c is stable under multiplication. Let u and v be two elements of A_c of the following form:

$$u = \sum_{\alpha} f(\alpha)\alpha \ , \qquad v = \sum_{\alpha} g(\alpha)\alpha.$$

There exist a real $c' < c$ and a real a such that (a, c') is a bound for u and v.

The coefficient $h(\alpha)$ of α in the product uv has the following form:

$$H(\alpha) = \sum_{\alpha = \beta\gamma} f(\beta)g(\gamma).$$

So, for every monomial α of length m, we get the following inequality:

$$|h(\alpha)| < (m+1)a^2 c'^m$$

and the size of uv is less then c'' for every $c'' > c'$.

Lemma 1-4. *Let $v, u_1, \cdots, u_n$ be convergent elements in A. Suppose that $u_1, \cdots, u_n$ have no constant term. Then the composite series $v(u_1, \cdots, u_n)$ is convergent.*

Proof. For every monomial α, denote by $f_1(\alpha), \cdots, f_n(\alpha), g(\alpha)$ the coefficient of α in $u_1, \cdots, u_n, v$. Let $w = v(u_1, \cdots, u_n) = \Sigma_\alpha h(\alpha)\alpha$. The coefficient $h(\alpha)$ is given by:

$$h(\alpha) = \sum_{p>0} \sum_{\alpha = \beta_1 \cdots \beta_p} \sum_{i_1, \cdots, i_p \leq n} g(t_{i_1} \cdots t_{i_p}) f_{i1}(\beta_1) \cdots f_{ip}(\beta_p).$$

Suppose that $v, u_1, \cdots, u_n$ have size less than c. There exists $a > 0$ such that (a, c) is a bound for $v, u_1, \cdots, u_n$.

Therefore, for every monomial α of length m, we get the following inequality:

$$|h(\alpha)| < \sum_{p>0} \sum_{\substack{m_1, \cdots, m_p > 0 \\ m_1 + \cdots + m_p = m}} n^p a^{p+1} c^p c^{m_1} \cdots c^{m_p}.$$

Let ω_m be the right part of this inequality. We have the following in $\mathbb{R}[[t]]$:

$$\begin{aligned}\sum_{m>0} \omega_m t^m &= a \sum_{p>0} (\sum_{m>0} nac^{m+1} t^m)^p \\ &= a \sum_{p>0} (\frac{nac^2 t}{1-ct})^p = a \frac{nac^2 t}{1-ct} \frac{1}{1 - \frac{nac^2 t}{1-ct}} \\ &= \frac{na^2c^2 t}{1 - ct - nac^2 t}.\end{aligned}$$

So we deduce the value of ω_m:

$$\omega_m = na^2 c^{m+1} (1 + nac)^{m-1}$$

and w is convergent.

Lemma 1-5. *Let w be a convergent series in the ring $A(t_1, \cdots, t_n, \theta)$. Suppose that w has no constant term and that the coefficient of θ in w is 1. Then there exists a unique series u in $A(t_1, \cdots, t_n)$ without constant term, suth that $w(t_1, \cdots, t_n, u) = 0$. Moreover u is convergent.*

Proof. Let $f(\alpha)$ be the coefficient of α in $w - \theta$. By assumption, $f(1) = f(\theta) = 0$. Let $u = \sum_\alpha g(\alpha)\alpha$ be any series in A without constant term ($g(1) = 0$). We have the following:

$$\begin{aligned} w(t_1, \cdots, t_n, u) &= \sum_\alpha g(\alpha)\alpha \\ &+ \sum_{p \geq 0} \sum_{\beta_0, \cdots, \beta_p} \sum_{\alpha_1, \cdots, \alpha_p} f(\beta_0 \theta \beta_1 \cdots \theta \beta_p) g(\alpha_1) \cdots g(\alpha_p) \beta_0 \alpha_1 \cdots \alpha_p \beta_p. \end{aligned}$$

Therefore $w(t_1, \cdots, t_n, u)$ vanishes if and only if:

$$\sum_\alpha g(\alpha)\alpha = -\sum_{p \geq 0} \sum_{\beta_0, \cdots, \beta_p} \sum_{\alpha_1, \cdots, \alpha_p} f(\beta_0 \theta \beta_1 \cdots \theta \beta_p) g(\alpha_1) \cdots g(\alpha_p) \beta_0 \alpha_1 \cdots \alpha_p \beta_p$$

and $g(\alpha)$ can be uniquely determined by induction.

Let (a, c) be a bound for $w - \theta$. Define numbers a_m, $m > 0$, such that the following formula holds in $\mathbb{R}[[t]]$:

$$\sum_{m>0} a_m t^m = \sum_{p \geq 0} \sum ac^{p+q_0+\cdots+q_p} a_{m_1} \cdots a_{m_p} t^{q_0+m_1+\cdots+m_p+q_p}$$

where the indices run over all tuples $(q_0, \cdots, q_p, m_1, \cdots, m_p)$ of integers satisfying:

$$\begin{array}{ll} \forall i = 0, \cdots, p \ q_i \geq 0 & \forall i = 1, \cdots, p \ m_i > 0 \\ \text{if } p = 0, \ q_0 > 0 & \text{if } p = 1, \ q_0 + q_1 > 0 \end{array}$$

For every monomial α of length m, $|g(\alpha)|$ is clearly less than a_m.

Denote by φ the series $\sum_{m>0} a_m t^m$. We have:

$$\begin{aligned}\varphi &= \sum_{q>0} ac^q t^q + ac((\sum_{q\geq 0} c^q t^q)^2 - 1)\varphi + \sum_{p\geq 2} ac^p (\sum_{q\geq 0} c^q t^q)^{p+1} \varphi^p \\ &= \sum_{p\geq 0} ac^p (\sum_{q\geq 0} c^q t^q)^{p+1} \varphi - a - ac\varphi \\ &= \sum_{p\geq 0} \frac{ac^p \varphi^p}{(1-ct)^{p+1}} - a - ac\varphi = \frac{a}{1-ct-c\varphi} - a - ac\varphi\end{aligned}$$

and we get the following formula:

$$c(1+ac)\varphi^2 + (ct(1+ac)-1)\varphi + act = 0.$$

Thus φ is analytic near 0 and there exist a' and c' such that a_m is less than $a'c'^m$ for every m. Therefore u is convergent.

1-6 Proof of theorem 1-2. Let $\omega \in F'$. This element is contained in some group G_p. This group is finitely presented and normally generated by the x_i's. Since G_p and F have the same H_1, G_p is generated by $x_1, \cdots, x_n, y_1, \cdots, y_q$ where $y_1, \cdots, y_q$ are contained in $[G_p, G_p]$.

Let $F(x,y)$ be the free group generated by $x_1, \cdots, x_n, y_1, \cdots, y_q$. The elements $y_1, \cdots, y_q$ may be written as products of commutators. Hence there exist elements $\omega_1, \cdots, \omega_q \in [F(x,y), F(x,y)]$ such that y_j and ω_j have the same image in G_p. Let v_j be the Magnus expansion of $y_j\omega_j^{-1} - 1$ in the algebra $A(t_1, \cdots, t_n, \theta_1, \cdots, \theta_q)$ (via $y_j \mapsto 1 + \theta_j$). The constant term in v_j is trivial and θ_j is the only non trivial term in v_j of degree 1. By virtue of lemma 1-5, we can solve the equation $v_q = 0$. We get:

$$\theta_q = u_q(t_1, \cdots, t_n, \theta_1, \cdots, \theta_{q-1})$$

where u_q is a convergent series in $A(t_1, \cdots, t_n, \theta_1, \cdots, \theta_{q-1})$.

If we substitute θ_q by u_q in $v_1, \cdots, v_{q-1}$, we get $q-1$ new convergent series $v_1', \cdots, v_{q-1}'$, in $A(t_1, \cdots, t_n, \theta_1, \cdots, \theta_{q-1})$. Now we are able to solve the equation $v_{q-1} = 0$ etc. ... We proceed this way q times and we construct convergent series u_j in the algebre $A(t_1, \cdots, t_n, \theta_1, \cdots, \theta_{j-1})$ such that $\theta_j = u_j$ is the solution of

$$v_j(t_1, \cdots, t_n, \theta_1, \cdots, \theta_j, u_{j+1}, \cdots, u_q) = 0.$$

Therefore the solution in A of:

$$v_1 = 0, \cdots, v_q = 0$$

is: $$\theta_1 = u_1, \cdots, \theta_q = u_q(t_1, \cdots, t_n, u_1, \cdots, u_{q-1})$$

and the Magnus expansions of $y_1, \cdots y_q$ are convergent series in A.

Since $y_1, \cdots, y_q$ generate G_p, the Magnus expansions of all elements of G_p are contained in some sub-algebra A_c of A. In particular the Magnus expansion of ω is convergent.

§2 Representations of F'

Lemma 2-1. *Let G be a finitely generated subgroup of F' containing F and let m be a integer. Then there exists a neighbourhood V of 1 in $GL_m(\mathbb{C})$ such that every representation ρ from F to $GL_m(\mathbb{C})$ satisfying:*

$$\forall i = 1, \cdots, n \quad \rho(x_i) \in V$$

extends to all the group G.

Proof. There exists $c > 0$ such that the Magnus expansion of every element of G is contained in the algebra A_c. There exists a neighbourhood V' of 0 in $M_m(\mathbb{C})$ such that for every $u_1, \cdots, u_n$ in V' and every series v in A_c the series $v(u_1, \cdots, u_n)$ converges in $M_m(\mathbb{C})$. Hence every representation ρ from F to $GL_m(\mathbb{C})$ sending $x_1, \cdots, x_n$ to $1 + V'$ extends to G.

Theorem 2-2. *Every representation from the free group F to a connected compact Lie group extends to the algebraic closure F' of F.*

Proof. Let Γ be a connected compact Lie group. For every group G, $R(G)$ will denote the set of representations from G to Γ. If G is finitely generated, $R(G)$ is a compact algebraic variety and the component of the trivial representation ρ_0 in $R(G)$ will be denoted by $R_0(G)$.

The variety $R(F) = R_0(F)$ is a product of copies of Γ; hence it is smooth. The set $R(F)$ is the inverse limit of varieties $R(G_p)$ (cf. §1). It contains the set $R_0(F') = \varprojlim R_0(G_p)$. To prove the theorem, it is enough to show that the restriction map from $R_0(F')$ to $R_0(F)$ is surjective. Since $R_0(G_p)$ is compact, it is enough to prove that $R_0(G_p) \to R_0(F)$ is surjective for every p.

Consider the elements $y_1, \cdots, y_q$ and $\omega_1, \cdots, \omega_q$ defined in 1-6. Let G' be the group defined by the presentation:

$$G' = < x_1, \cdots, x_n, y_1, \cdots, y_q; y_1 = \omega_1, \cdots, y_q = \omega_q > .$$

The group G' contains F and maps onto G_p. Let $F(x, y)$ be the free group generated by elements $x_1, \cdots, x_n, y_1, \cdots, y_q$. We have a map φ from $R(F(x, y))$ to $R(F) \times \Gamma^q$ defined by:

$$\rho \mapsto (\rho|_{R(F)}, \rho(y_1\omega_1^{-1}), \cdots, \rho(y_q\omega_q^{-1}))$$

This map is algebraic. Moreover $\varphi^{-1}(\rho_0, 1)$ is the trivial representation ρ_0' of $R(F(x,y))$ and the differential of φ at ρ_0' is bijective. But $\varphi^{-1}(R(F) \times 1)$ is exactly the variety $R(G')$. Therefore there exists a neighbourhood V of ρ_0 in $R(F)$ such that the restriction ψ from $R(G')$ to $R(F)$ induces a homeomorphism from $W = \psi^{-1}(V)$ to V.

Consider an embedding from Γ to $GL_m(\mathbb{C})$. The system of equations:

$$y_1 = \omega_1 \qquad y_2 = \omega_2 \qquad \cdots \qquad y_q = \omega_q$$

has a unique solution $(y_1, \cdots, y_q)$ in A_c for some c. But it also has a unique solution in $M_m(\mathbb{C})$ if $x_1, \cdots, x_p$ are chosen in a small neighbourhood of 1 in $M_m(\mathbb{C})$. Therefore if V is a small neighborhood of ρ_0 in $R(F)$, for every ρ in V, the representation $\rho' = \psi^{-1}(\rho)$ is given by the continuous extension f of ρ from A_c to $M_m(\mathbb{C})$. Hence f is an extension of ρ from G_q to Γ.

On the other hand, the map from $R_0(G')$ to $R_0(G_p)$ is injective. Hence the restriction map ψ' from $R_0(G_p)$ to $R_0(F)$ has the following property: for some neighbourhood V of ρ_0 in $R_0(F)$, ψ' is a homeomorphism from $\psi'^{-1}(V)$ to V. Consider the complexification $\psi'_{\mathbb{C}}$ of ψ'. Over ρ_0 we have only one real point and hence an odd number of points. Therefore over every point in $R_0(F)$ we have an odd number of points and hence some real point. Thus ψ' is surjective.

Another way to prove the surjectivity of ψ' is the following:

$R_0(G_p)$ has a well defined fundamental class in mod 2 homology [**BCR**], and because of the property of ψ' near ρ_0, the degree of ψ', well defined module 2, is odd. Hence ψ' is surjective.

So the restriction map from $\varprojlim R_0(G_p)$ to $R_0(F)$ is surjective and every representation from F to Γ extends to F'.

References

[BCR] J. Bochnak, M. Coste and M-F. Roy, *Géométrie algébrique réelle*, Ergebnisse der Mathematik und ihrer Grenzgebiete (3) 12, Springer Verlag 1987.

[LD] J. Y. Le Dimet, *Cobordisme d'Enlacement de Disques*, Memoire de la S. M. F. **32** (1988).

[L1] J. Levine, *Algebraic Closure of Groups, Combinatorial Group Theory*, Contemporary Mathematics 109, A. M. S., Providence, R. I., 1990.

[L2] J. Levine, *Link Concordance and algebraic Closure of Groups*, Comm. Math. Helv. **64** (1989), 236–255.

[L3] J. Levine, *Link Concordance and algebraic Closure II*, Invent. Math. **96** (1989), 571–592.

[L4] J. Levine, *Signature Invariants of Homology Bordism with Applications to Links*, preprint.

[V] P. Vogel, 2 × 2-*Matrices and Application to Link Theorey*, Proceedings Göttingen 1987, Lecture Notes in Math. 1361, 1988, pp. 269–298.

Applications of Morse theory to link theory

Nathan Habegger

In the last 10 years a lot of progress has been made towards the classification of links. The seminal paper of le Dimet [D] established an exact sequence, using surgery theory, for the classification of higher dimensional disk links. That paper's consequences are still being explored. On another front, Habegger and Lin [HL1] obtained the classification of classical links up to homotopy. The ideas there have been pushed further to give information on link concordance in the classical dimension [HL2]. Koschorke [Ko] has obtained an exact sequence for the classification of higher dimensional links up to homotopy. In conclusion, a lot of progress has been made, but the study of links remains a rich field of investigation.

The purpose of this note is to point out how the study of links up to concordance has been enriched by the use of a little Morse theory. Specifically, we wish to survey 3 results whose proof all involve the central observation that a concordance comes equipped with a natural Morse function. We will refer to the original papers for more complete details. The point being made here is that Morse theory is the common thread among these results.

The 3 results are: 1) the theorem of Goldsmith [Go] and Giffen [Gi] that concordance implies link homotopy in the classical dimension. 2) Lin's proof of the Cochran-Orr conjecture that k-nulcobordism implies the Milnor $\bar{\mu}$ invariants of length less than or equal to 2k vanish. 3) The solution of the basing problem for classical links. (Regarding this latter, the author would like to thank Smolinsky for pointing out at the conference that the proof extends to higher dimensions as well.)

Elementary Morse theory [M1] shows that every smooth manifold admits a Morse function. On closer observation, it is observed that the critical points can be rearranged so that they are in order of increasing index. At the heart of the h- and s-cobordism theorems [M2] [Ke] are the still deeper observations of how the critical points can be made to cancel.

Even though the strongest applications of Morse theory (or handlebody theory) are to the higher dimensions, it should not go unnoticed that there are applications to low dimensions as well. (For example, it is an exercise for every good student

Knots 90

of topology to show that handlebody theory can be used to obtain the classification of closed smooth surfaces.)

Throughout the rest of this paper we will constantly make use of the following elementary result (see [HL2] for a proof):

Proposition 1. *Suppose* $f : W^{n+1} \longrightarrow Q^{n+q} \times I$ *is a proper embedding of smooth manifolds. Let* $\Pi : Q \times I \longrightarrow I$ *denote the projection.*

If $q \geq 1$, *there is an isotopy of f (rel boundary) to* f' , *so that the composite map* $t = \Pi \circ f'$ *is a Morse function. Furthermore, one may take t to have critical points of increasing index.*

If $q \geq 2$, *the order of the critical points of the same index may be rearranged as well.*

Investigation of the steps of the proof of the s-cobordism theorem reveals that they can be carried through in the embedded setting, provided the codimension is strictly greater than 2.

Proposition 2. *In the setting of Proposition 1, suppose in addition that* $n \geq 5$, $q \geq 3$ *and that W is an s-cobordism between* $f^{-1}(0) = M$ *and* $f^{-1}(1) = M'$. *Then there is an isotopy of f to a map* f' *so that the map t has no critical points. In particular M and* M' *are isotopic as submanifolds.*

Remarks. The above result, applied to a concordance $M \times I \longrightarrow Q \times I$ yeilds a map $t : M \times I \longrightarrow I$ without critical points. This falls short of a level preserving map (which would yeild an isotopy of M in W) by a self diffeomorphism of $M \times I$ (pseudoisotopy of M). Thus the above is weaker than the well-known theorem (see [H]) that concordance implies isotopy in codimension greater than 2.

However, in the classical setting, i.e. dim M = 1, dim Q = 3, there is no difference between isotopy of embeddings and isotopy of submanifolds, since any diffeomorphism of a 1-manifold homotopic to the identity is isotopic to the identity.

Embedded handle addition

Let $M^n \subset Q^{n+q}$ be an embedding. Suppose $S^{p-1} \times D^{n-p+1}$ is embedded in M and this embedding extends to an embedding of $D^p \times D^{n-p+1}$ in Q. This yields an embedding of $M' = M - S^{p-1} \times int(D)^{n-p+1} \cup D^p \times S^{n-p}$ in Q. We say that M' is obtained from M by an embedded handle addition.

The manifold $W = M \times I \cup D^p \times D^{n-p+1}$ admits a Morse function with a single critical point of index p. Combining this with the map to $M \cup D^p \times D^{n-p+1} \subset Q$ yields an embedding of the cobordism W in $Q \times I$.

Conversely, it is not difficult to show that every embedded cobordism, whose projection to I yields a Morse function, arizes in this way out of a sequence of embedded handle additions, up to a level preserving isotopy of $Q \times I$. Proposition 1 then follows from general position of the embedded handles. Proposition 2 follows from the observation that in codimension greater than 2, a pair of handles which cancel can be pieced together in Q to form an embedded ball attached to M along a ball in its boundary. This ball can then be used to isotope M to M'', where M'' is obtained from M after both handle additions. We call the pair of handles embedded cancelling handles.

Remark. In codimension 2, cancelling embedded handle additions of index i and i+1 cannot generally be made into embedded cancelling handles, because the interior of the core of the index i+1 handle will meet the interior of the cocore of the index i handle. Nevertheless:

Theorem 3 (Goldsmith[Go], Giffen[Gi]). *Two 1-manifolds M_1 and M_2 in Q^3 which are concordant are homotopic, i.e. there is a homotopy of M_1 to M_2 during which disjoint components of M_1 remain disjoint.*

Proof. . Arrange for the 0-handle additions from M_1 to M_2 to be cancelled by 1-handles and let M be the result obtained from M_1 by addition of these cancelling pairs. After a homotopy of M, induced by an isotopy of each 1 handle off the cocore (a 2-disk) of the 0-handle which it cancels, the cancelling pairs can be made into embedded cancelling pairs and thus eliminated, one pair at a time. Thus M is homotopic to M_1 by a homotopy in which distinct components do not intersect.

Now M_2 is obtained from M by embedded handle additions of index 1 and 2 which must algebraically cancel, since we began with a product cobordism. Hence M may be obtained from M_2 by embedded handle additions of index 0 and 1 which may be arranged in cancelling pairs. As above, M is then homotopic to M_2, and hence M_1 is homotopic to M_2.

K-cobordism and Lins's theorem

We will describe one of the main geometric ingrediants in Lin's proof [L] of the Cochran-Orr conjecture. We refer to [C] and [O] for the definition of k-cobordism. For our purposes, it suffices to consider just an embedded cobordism of 1-manifolds. Call an embedded cobordism special, if all critical points of the corresponding Morse function lie on a single component. Call a cobordism 2-sided if every component has exactly one boundary component in $M = f^{-1}(0)$ and one in $M' = f^{-1}(1)$. Call an embedded cobordism elementary, if it is special, 2-sided and the corresponding Morse function has only two critical points of index 1. (The nontrivial component is then just a twice punctured torus.)

Proposition 4. *Every 2-sided cobordism of 1-manifolds in $Q^3 \times I$ is a product of concordances and elementary cobordisms.*

Proof. As in the proof of concordance implies homotopy, arrange for the 0-handles to be cancelled by 1-handles and dually, the 2 handles to be cancelled by 1-handles. These two cobordisms, consisting of canceling pairs, are then diffeomorphic to $M \times I$ and $M' \times I$ and hence yeild concordances. What remains is an embedded cobordism consisting only of 1-handles which can be rearranged freely. Grouping these handles by components factors this cobordism into a product of special 2-sided cobordisms. Furthermore, since the cobordisms are 2-sided, attaching the first 1-handle will increase the number of components by one. Therefore there is a second 1-handle connecting the two components back together. Continuing in this way to pair up the 1-handles factors the cobordisms into elementary pieces.

Lin's analysis of k-cobordisms then reduces to a study of elementary k-cobordisms. We refer the reader to [L] for further details.

Basing unbased links

The third result we'd like to discuss is the solution of the basing problem for unbased links.

Given a link $M_1^{p_1} \cup M_2^{p_2} \cup \cdots \cup M_k^{p_k} \subset Q^m$, one would like to know how to base the link in a natural way. There are many proposals for this, but all of them run into the following problem: Suppose two links are concordant. How can one propagate the basing along the concordance? We call this the basing problem.

Consider the most naive way to base a link, i.e. choose a base point on each component and paths joining these to the base point of Q. Given a concordance of the link, the problem is how to push this basing along the concordance. The first obstruction is that the base point on each component many not respect the level as one moves along the concordance. Considering the restriction of the concordance to the base points then gives a concordance of 0-manifolds in $Q \times I$. The projection to I then yeilds a Morse function whose critical points come in cancelling pairs. If $m = dimQ \geq 3$, the cancelling pairs can be made into embedded cancelling pairs. Thus up to isotopy of $Q \times I$, we may assume the base points decend in time. Given this, it is natural to try to push the curves down in time as well. However, if there are codimension 2 components, the trace of the path will in general intersect the concordance in its interior in points. We are unhappily obstructed from achieving our goal.

The solution of the basing problem is given as follows: Suppose that the concordance from $M_1 \cup M_2 \cup \cdots \cup M_k$ to $M_1' \cup M_2' \cup \cdots \cup M_k'$ is obtained by embedded handle additions of increasing index. Arrange for the 0-handles to be cancelled by 1-handles. This then gives two concordances, one from M to M'' and another from M' to M'', neither of which has handles of index of codimension 2 in Q. The handles can therefore be added away from the arcs basing M (and M'). We have proven the

Proposition 5. *Suppose $M \subset Q$ is concordant to M', and the components have codimension greater than or equal to 2. Given basings of M, M', there is a link M'' and "based" concordances of M and M' to M''.*

The above result reduces questions of different basings for concordant links to questions of different basings for a fixed link. It is at the heart of the solution of the problem of the indeterminacies of the Milnor $\bar{\mu}$ invariants. It is also one of the main geometric ingrediants in the discovery that the set of links up to concordance possesses the structure of an orbit space for a certain group action. See [HL2] for details.

References

[C] T. Cochran, *Derivatives of links: Milnor's concordance invariants and Massey's products,* Mem. Amer. Math. Soc., vol. 84, Amer. Math. Soc., Providence, RI, 1990

[D] J.Y. le Dimet, *Cobordisme d'enlacements de disques.* Mém. Soc. Math. France. no. 32, Supplément Bull. Soc. Math. France **116** (1988).

[Gi] C. Giffen, *Link concordance implies link homotopy,* Math. Scand. **45** (1979), 243-254.

[Go] D. Goldsmith, *Concordance implies homotopy for classical links in M^3,* Comment. Math. Helv. **54** (1979), 347-355.

[H] J.F.P. Hudson, *Piecewise Linear Topology,* Benjamin (1969).

[HL1] N. Habegger, X.S. Lin, *The Classification of Links up to Homotopy,* Journal of the American Mathematical Society, Volume 3, No. 2, April 1990, 389-419.

[HL2] N. Habegger, X.S. Lin, *On Milnor's $\bar{\mu}$ invariants and the concordance classification of links,* (in preparation).

[Ke] M. Kervaire, *Le théorème de Barden-Mazer-Stallings,* Commentarii Math. Helv. **40** (1965), 31-42.

[Ko] U. Koschorke, *On link maps and their homotopy classification* (to appear in Math. Annalen).

[L] X.S. Lin, *Artin-type representation theorems and Milnor's $\bar{\mu}$ invariants,* Ph.D. thesis, Univ. of California, San Diego, 1988.

[M1] J. Milnor, *Morse Theory,* Princeton University Press,(1963).

[M2] J. Milnor, *Lectures on the h-cobordism Theorem,* (notes by L. Siebenmann and J Sondow), Princeton University Press (1965).

[0] K. Orr, *Homotopy invariants of links,* Invent. Math. **95** (1989), 379-394.

Signature invariants of homology bordism with applications to links

*J.P. Levine**

In [**APS**], an invariant $\rho_\alpha(M)$ is defined for an odd-dimensional closed oriented manifold M equipped with a unitary representation $\alpha : \pi_1(M) \to U(n)$ of its fundamental group, by considering the spectrum of a certain differential operator. In general, $\rho_\alpha(M)$ is a real number but if $M = \partial V$ and α extends to a unitary representation $\alpha' : \pi_1(V) \to U(n)$, then it is shown that $\rho_\alpha(M) = n\sigma(V) - \sigma_{\alpha'}(V)$, where $\sigma(V)$ is the ordinary signature of V and $\sigma_{\alpha'}(V)$ is the signature using local coefficients defined by α'. Thus, $\rho_\alpha(M)$ is rational if some multiple of (M, α) bounds — in particular, if α factors through some unitary representation of a finite group.

We will consider pairs (M, α), where $\alpha : \pi_1(M) \to G$ for some group G and M is an oriented manifold. Two closed *G-manifolds* (M, α), (M', α') *are homology G-bordant* if there is an oriented G-manifold (V, β) with $\partial V = M' - M$ and β restricting to α, α' on M, M' (up to conjugacy in G) and $H_*(V, M) = H_*(V, M') = 0$. In general M and V may be disconnected and we require α to be defined on each component. Let $R_n(G)$ denote the set of unitary representations $G \to U(n)$. If (M, α) is a closed odd-dimensional G-manifold, we can define a function $\sigma(M, \alpha) : R_n(G) \to \mathbb{R}$ by $\sigma(M, \alpha) \cdot \theta = \rho_{\theta\alpha}(M)$; $\sigma(M, \alpha)$ is obviously a diffeomorphism invariant of (M, α). We will investigate the extent to which $\sigma(M, \alpha)$ measures the homology G-bordism class of (M, α), when G is infinite. We will also investigate the use of signature functions $R_n(G) \to \mathbb{Z}$ as invariants of the Cappell-Shaneson homology surgery groups [**CS**] of G. The main application will be to obtain invariants of link concordance which, in some cases, extend earlier signature invariants of links. In particular we obtain:

(i) a general link invariant $\sigma : R_1(\mathbb{Z}^m) = T^m \to \mathbb{Z}$, defined on the m-torus $T^m = S^1 \times \cdots \times S^1$ (for m component odd-dimensional links)

* Partially supported by NSF Grant No. 59500 and by the U.S.-Israel Binational Science Foundation (Jerusalem, Israel), Grant No. 88-00114.

Knots 90

(ii) an invariant $\sigma_F : R_n(F) = U(n) \times \cdots \times U(n) \to \mathbb{Z}$, where F is the free group of rank m, for odd-dimensional m-component boundary links.
(iii) an invariant $\bar{\sigma} : R_n(\bar{F}) \to \mathbb{Z}$, where $\bar{F}$ is the "algebraic closure" of F in its nilpotent completion, for odd-dimension links with vanishing $\bar{\mu}$-invariants in dimension one.

These will all measure, to some extent, the concordance class of the link.

These signature functions will be seen to have the following type of continuity: $R_n(G)$ admits an "analytic" stratification such that $\sigma(M, \rho)$ is continuous on the open strata. The invariance under homology G-bordism (or under concordance, in the application to links) is of the following sort. If (M, α) or (M', α') are homology G-bordant then there is an "analytic" hypersurface Σ of a special type in $R_n(G)$ such that $\sigma(M, \alpha) = \sigma(M', \alpha')$ except on Σ.

The invariants σ, $\bar{\sigma}$ include earlier signature invariants: The Levine-Tristram invariants [**L**], [**T**] and the p-signatures and nilpotent signature of Smolinsky [**Sm**], [**Sm1**].We will give some examples, though, of non-concordant links detected by σ but not these earlier invariants. We will also show that $\bar{\sigma}$ can be used to detect links not concordant to boundary links — a phenomenon first detected by Cochran-Orr [**CO**]. σ_F has the potentiality to distinguish boundary link concordance from ordinary link concordance, but I have not yet been able to realise this potentiality.

This contains no proofs. These will appear in a subsequent exposition.

1. Let G be a group and $R_n(G)$ the set of all representations $G \to U(n)$. We give $R_n(G)$ the weak topology with respect to the collection of evaluation maps $e_g : R_n(G) \to U(n)$ defined by $e_g(\theta) = \theta(g)$, for all $g \in G$. We say a function $f : R_n(G) \to \mathbb{R}$ is (real) analytic if $^{\exists}g_1, \ldots, g_k \in G$ and an analytic function $h : U(n) \times \cdots \times U(n) \to \mathbb{R}$ such that $f(\theta) = h(\theta(g_1), \ldots, \theta(g_k))$ for all $\theta \in R_n(G)$. If V is a (real) analytic manifold then $\phi : V \to R_n(G)$ is analytic if $e_g \circ \phi$ is analytic for all $g \in G$ — alternatively, if $f \circ \phi$ is analytic for all analytic $f : R_n(G) \to R$.

Example A.

(i) $G = \mathbb{Z}^m$: then $R_1(G) = T^m$

(ii) $G = F$, free group on m-generators: $R_n(G) = U(n) \times \cdots \times U(n) = U(n)^m$

(iii) $G = D$, the infinite dihedral group $\{x, t : txt^{-1} = x^{-1}, t^2 = 1\}$
An analytic function $\phi : S^1 \to R_2(D)$ is defined by $\phi(\theta) \cdot t = \begin{pmatrix} 0 & 1 \\ 1 & 0 \end{pmatrix}$, $\phi(\theta) \cdot x = \begin{pmatrix} e^{2\pi i\theta} & 0 \\ 0 & e^{-2\pi i\theta} \end{pmatrix}$.

(iv) $G = \bar{D}$, the algebraic closure of D. $\bar{D}$ can be represented (see [**L**], [**CO**]) as a semi-direct product $\mathbb{Z}_{(2)} \times \mathbb{Z}/2$, where the generator t of $\mathbb{Z}/2$ acts, by $tqt^{-1} = q^{-1}$, $q \in \mathbb{Z}_{(2)}$. ($\mathbb{Z}_{(2)}$ is the rational 2-adic integers).

An analytic function $\bar{\phi} : \mathbb{R} \to R_2(\bar{D})$ is defined by $\bar{\phi}(s) \cdot t = \begin{pmatrix} 0 & 1 \\ 1 & 0 \end{pmatrix}$, $\hat{\phi}(s) \cdot q = \begin{pmatrix} e^{2\pi isq} & 0 \\ 0 & e^{-2\pi isq} \end{pmatrix}$ if $q \in Z_{(2)} \subseteq \mathbb{Q} \subseteq \mathbb{R}$.

(v) A similar example to (iii), (iv) which we shall have use for, is the group $B = \{x, y, t : t^2 = 1, txt^{-1} = y, [x, y] = 1\}$, which is a semi-direct product $A \times \mathbb{Z}/2$, A the free abelian group generated by x, y. An analytic map $\phi : S^1 \to R_2(B)$ is defined by $\phi(\theta) \cdot t = \begin{pmatrix} 0 & 1 \\ 1 & 0 \end{pmatrix}$, $\phi(\theta)(x^m y^n) = \begin{pmatrix} e^{2\pi im\theta} & 0 \\ 0 & e^{2\pi in\theta} \end{pmatrix}$.

The algebraic closure $\bar{B}$ is a semi-direct product $\bar{A} \times \mathbb{Z}/2$ where $\bar{A} \subseteq \mathbb{Z}_{(2)} \times \mathbb{Z}_{(2)}$ consists of all pairs (p, q), where $p + q \in \mathbb{Z}$ ($A \to \bar{A}$ is defined by $x^m y^n \to (m, n)$). An analytic map $\bar{\phi} : \mathbb{R} \to R_2(\bar{B})$ is defined by $\bar{\phi}(s) \cdot t = \begin{pmatrix} 0 & 1 \\ 1 & 0 \end{pmatrix}$, $\bar{\phi}(\theta)(p, q) = \begin{pmatrix} e^{2\pi ips} & 0 \\ 0 & e^{2\pi iqs} \end{pmatrix}$.

If (M, α) is a closed odd-dimensional G-manifold, then $\sigma(M, \alpha) : R_n(G) \to \mathbb{R}$ is defined in the introduction.

We first discuss the continuity of $\sigma(M, \alpha)$. For this we need to introduce the following "stratification" of $R_n(G)$, which depends on (M, α). Let $\Sigma_r^1 \subseteq R_n(G)$, for a non-negative integer r, be the set of all θ such that $\sum_i \dim H_i(M; \theta\alpha) \geq r$. $H_*(M; \theta\alpha)$ is the homology of M using local coefficients in $\mathbb{C}^n$ with the action of $\pi_1 M$ defined by the representation $\theta\alpha$. Clearly $\Sigma_{r+1}^1 \subseteq \Sigma_r^1$, $\Sigma_0^1 = R_n(G)$, and Σ_r^1 is empty for large enough r. We can reindex the $\{\Sigma_i^1\}$, setting $\Sigma_r = \Sigma_{r+k}^1$ for all r, where k is the maximum such that $\Sigma_k^1 = R_n(G)$.

Theorem 1.

(i) *Each Σ_r is the zero set of an analytic function on $R_n(G)$.*
(ii) *$\sigma(M, \alpha)$ is continuous on each open stratum $\Sigma_r - \Sigma_{r+1}$.*

If some non-zero multiple of (M, α) bounds, in the oriented bordism group of G — a condition which is well-known to be equivalent to the vanishing of "characteristic numbers" — then $\sigma(M, \alpha)$ is defined, according to the Atiyah-Patodi-Singer theorm [**APS**], by signatures and takes rational values. In this case we conclude that $\sigma(M, \alpha)$ is locally constant on each open stratum.

We will refer to $\Sigma_0 - \Sigma_1$, or, in general, any non-empty subset X of $R_n(G)$ on which $\sigma(M,\alpha)$ is continuous and which is the complement of the zero set of an analytic function, as a *domain of continuity*.

Suppose (λ_{ij}) is a square matrix with entries in $\mathbb{Z}G$ such that the integral matrix $(\epsilon(\lambda_{ij}))$, where $\epsilon : \mathbb{Z}G \to \mathbb{Z}$ is the usual augmentation, is non-singular. Then $g : R_n(G) \to \mathbb{Z}$, defined by $g(\theta) = \mid \det(\theta(\lambda_{ij})) \mid^2$ is analytic. Any such function will be called *special*. The complement of the zero set of a special function, will be called *large*.

Proposition 2. *If $g : R_n(G) \to \mathbb{R}$ is special, then:*

(i) $g(\theta_1\theta_2\theta_1^{-1}) = g(\theta_2)$

(ii) *If P is a group of prime-power order and $\phi : G \to P$ a homomorphism, then $g(\theta\phi) \neq 0$ for any $\theta \in R_n(P)$.*

Denote by $P_n(G) \subseteq R_n(G)$, the set of all θ which factor through groups of prime-power order, as in (ii). So every large subset of $R_n(G)$ contains $P_n(G)$.

Proposition 3. *If (M,α) and (M',α') are homology G-bordant, then $\sigma(M,\alpha)$ and $\sigma(M',\alpha')$ agree on some large subset of $R_n(G)$.*

Corollary. *If V is an analytic manifold and $\phi : V \to R_n(G)$ is an analytic map such that $\phi^{-1}(P_n(G))$ is non-empty, then $\sigma(M,\alpha)\circ\phi$ and $\sigma(M',\alpha')\circ\phi$ agree on an open dense subset of V.*

In many cases of interest $\sigma(M,\alpha)$ has a large domain of continuity. It then follows that no information about the homology bordism class of (M,α) can be obtained outside of this domain. However we will see that, in some cases, homology bordism information is carried by the values of $\sigma(M,\alpha)$ on the lower strata of continuity.

Proposition 4. *Suppose M is parallelizable, $H_1(M)$ free abelian and $H_i(M) = 0$ if $1 < i < n-1$. Suppose there is a finite G-complex (P,β) and a G-map $\phi : (M,\alpha) \to (P,\beta)$ such that $H_i(P) = 0$ for $i \geq q$ (where $n = 2q-1$), $H_{q-1}(P)$ torsion-free and $H_1(M) \to H_1(P)$ is an isomorphism. Then $\sigma(M,\alpha)$ is continuous on a large set.*

2. Let $\mathcal{F} : \mathbb{Z}G \to \Lambda$ be an epimorphism of rings, where G is any group. In **[CS]** an abelian group $\Gamma_n(\mathcal{F})$ is defined, for any positive integer n, as the group of obstructions to homology surgery. We will consider a special case corresponding to $\Lambda = \mathbb{Z}$, $\mathcal{F} = \epsilon$, n even, and ignore the quadratic function μ which is part of the structure of the objects which define $\Gamma_n(\mathcal{F})$. Thus we consider a finite-generated $\mathbb{Z}G$-module A, equipped with an ϵ-Hermitian sesquilinear form $< , >$ $(\epsilon = \pm 1)$

with values in $\mathbb{Z}G$, such that $\mathbb{Z} \otimes_{\mathbb{Z}G} A$ is free abelian and the induced integral form is non-singular. $(A, < , >)$ is an ϵ-form. A subkernel of $(A, < , >)$ is a submodule $B \subseteq A$ such that $< , >| B = 0$ and Image$\{\mathbb{Z} \otimes B \to \mathbb{Z} \otimes A\}$ is a direct summand of half the rank. If $(A, < , >)$ contains a subkernel, it is a kernel. Two ϵ-form are equivalent if, after adding kernels to both, they became isomorphic. The set of equivalence classes $\Gamma_\epsilon(G)$ is an abelian group and there is an obvious homomorphism $\Gamma_{2q}(\epsilon) \to \Gamma_\epsilon(G)$, whose kernel and cokernel are 2-torsion. Every element of $\Gamma_\epsilon(G)$ can be represented by an ϵ-form $(A, < , >)$ in which A is free.

If $(A, < , >)$ is an ϵ-form, $\theta \in R_n(G)$, then one can obtain an associated complex Hermitian form $< , >_\theta$ on $\mathbb{C}^n \otimes_\theta A$. Define $\sigma(a, < , >) \cdot \theta =$ signature $< , >_\theta$ and $\Delta(A, < , >) \cdot \theta = | \det < , >_\theta|^2$.

Proposition 5. *$\Delta(A, < , >)$ is a special function and $\sigma(A, < , >)$ is continuous on the large subset $V(A, < , >)$ which is the complement of the zero set of $\Delta(A, < , >)$. If $(A', < , >')$ represents the same element of $\Gamma_\epsilon(G)$, then $\sigma(A', < , >') = \sigma(A, < , >)$ on a large subset of $R_n(G)$.*

Corollary. *If V is an analytic manifold and $\phi : V \to R_n(G)$ is analytic and $\phi^{-1}(P_n(G))$ is non-empty, then $\sigma(A, < , >) = \sigma(A', < , >')$ on an open dense subset of V.*

We can define $L_\epsilon(G)$ as the analogous simplification of the Wall group. If $(A, < , >)$ represents an element of $L_\epsilon(G)$, then $\sigma(A, < , >)$ is continuous on *all* of $R_n(G)$.

Let $\mathcal{F}_n(G)$ be the abelian group whose elements are equivalence classes of function $R_n(G) \to \mathbb{Z}$ which are continuous on some large subset, where two such are equivalent if they agree on some large subset addition; is defined pointwise. Then $\sigma(A, < , >)$ defines, by Proposition 5, a homomorphism $\sigma : \Gamma_\epsilon(G) \to \mathcal{F}_n(G)$.

Example B.

(i) $G = \mathbb{Z}^m$: If $f(z_1, \ldots, z_m)$ is any Laurent polynomial satisfying $f(1, \ldots, 1) = \pm 1$ and $f(z_1^{-1}, \ldots, z_m^{-1}) = f(z_1, \ldots, z_m)$, then there is an element $\alpha \in \Gamma_\epsilon(\mathbb{Z}^m)$, $\epsilon = \pm 1$, such that $\sigma(\alpha) : R_1(\mathbb{Z}^m) = T^m \to Z$ satisfies

$$\sigma(z_1, \ldots, z_m) = \begin{Bmatrix} +1 & \epsilon = +1 \\ +2 & \epsilon = -1 \end{Bmatrix} \quad \text{if} \quad f(z_1, \ldots, z_m) > 0 \quad \text{and}$$

$$\sigma(z_1, \ldots, z_m) = \begin{Bmatrix} -1 & \epsilon = +1 \\ -2 & \epsilon = -1 \end{Bmatrix} \quad \text{if} \quad f(z_1, \ldots, z_m) < 0.$$

(ii) $G = F/F_q$, where F is the free group of rank m and F_q is the subgroup generated by commutators of weight q (i.e. $F_1 = F$, $F_q = [F, F_{q-1}]$). Then there exists an infinite set $\{\alpha_k\} \subseteq \mathrm{Ker}\{\Gamma_\epsilon(F) \to \Gamma_\epsilon(F/F_q)\}$ such that $\{\sigma(\alpha_k)\} \subseteq \mathcal{F}_2(F)$ is linearly independent. α_k is represented by the matrix $(g_q^k + g_q^{-k} - 1)$ if $\epsilon = +1$ or $\begin{pmatrix} g_q^k - g_q^{-k} & 1 \\ -1 & g_q^k - g_q^{-k} \end{pmatrix}$ if $\epsilon = -1$, where $g_q \in F$ is defined successively by $g_1 = x$ and $g_q = [y, g_{q-1}]$ (x, y are two generators of F — in fact, we can assume $m = 2$). Now use the analytic map $S^1 \xrightarrow{\phi} R_1(D) \longrightarrow R_2(F)$ of Example (iii), for $\epsilon = +1$ or $S^1 \to R_2(B) \to R_2(F)$ of $A(v)$ if $\epsilon = -1$ where $F \to D$ and $F \to B$ are defined by $y \mapsto t$, $x \mapsto x$ and Corollary of Proposition 5.

(iii) $G = D$. We see that $\Gamma_+(D)$ contains an infinitely-generated set. Let $\beta_k' \in \Gamma_+(D)$ be represented by the matrix $(x^k + x^{-k} - 1)$ and again use the analytic map ϕ of Example A (iii).

(iv) Let $f : D \to D$ be the homomorphism $f(t) = t$, $f(x) = x^a$ for an odd integer $a > 1$. Then $f_* : \Gamma_+(D) \to \Gamma_+(D)$ is not onto. In fact, if $\beta \in \mathrm{Im} f_+$ then $\sigma = \sigma(\beta) \circ \phi$ (ϕ as above) is easily seen to have the following periodicity property: $\sigma(z) = \sigma(zw)$ for all $z \in S^1$ and w any a-th root of unity. The elements β_k of (iii) have this property if and only if k (or $2k + 1$, if $t = -1$) is a multiple of a.

Similar considerations show that the homomorphism $F \to F$ defined by $x \mapsto x[x, y]^k$, $y \mapsto y$ does not induce an epimorphism $\Gamma_+(F) \to \Gamma_+(F)$, if $k > 0$.

(v) $G = \bar{D}$: Let $i : D \to \bar{D}$ be the standard inclusion. Then $i_* : \Gamma_+(D) \to \Gamma_+(\bar{D})$ is not onto. Let $\bar{\phi} : \mathbb{R} \to R_2(\bar{D})$ be the analytic map of Example A (iv). Then $i^*\bar{\phi}(t+1) = i^*\bar{\phi}(t)$ where $i^* : R_2(\bar{D}) \to R_2(D)$ and so if $\alpha \in \Gamma_\epsilon(D)$, then $\sigma = \sigma(\alpha) \circ \bar{\phi}$ exhibits the periodicity $\sigma(t+1) = \sigma(t)$. If α is represented by the element $(r + r^{-1} - 1)$ where $r = \frac{1}{a} \in \mathbb{Z}_{(2)} \subset \bar{D}$, this periodicity does not hold ($a > 1$).

Using the natural commutative diagram: $\begin{array}{ccc} F & \to & \bar{F} \\ \downarrow & & \downarrow \\ D & \to & \bar{D} \end{array}$ where the vertical maps are epimorphisms, one sees that $\Gamma_+(F) \to \Gamma_+(\bar{F})$ is not onto.

(vi) $G = \bar{B}$. Then $i_* : \Gamma_-(B) \to \Gamma_-(\bar{B})$ is not onto. If $\bar{\phi} : \mathbb{R} \to R_2(\bar{B})$ is as in Example A (v), then $\bar{\phi}(t+1) = \bar{\phi}(t)$ on $R_2(\bar{B})$ and so $\sigma = \sigma(\alpha) \circ \bar{\phi}$ exhibits this same periodicity for $\alpha \in \Gamma_\epsilon(B)$. If we consider the element α represented by the matrix $\begin{pmatrix} r-r^{-1} & 1 \\ -1 & r-r^{-1} \end{pmatrix}$ where $r = \left(\frac{k+1}{2k+1}, \quad \frac{k}{2k+1}\right)$ for any $k \geq 1$, then $\sigma(\alpha) \circ \bar{\phi}$ does not have this periodicity.

If we use the epimorphism $F \to B$ ($u \mapsto x, v \mapsto t$) we conclude that $\Gamma_-(F) \to \Gamma_-(\bar{F})$ is not onto.

Example (iii), (iv) and (v) are, of course, just an alternative method for obtaining the result of Cochran-Orr [**C-O**].

3. An *n-link* will be an imbedding $\ell : S^n + \cdots + S^n \to S^{n+2}$. $M(\ell)$ is the closed oriented n-manifold obtained by doing surgery on S^{n+2} along ℓ with the 0-framing; i.e. the unique framing such that, if $n = 1$, the linking number of each component with its translate along the normal field is zero. *We assume*, when $n = 1$, that the linking number of any two components of ℓ is zero.

We will consider three different ways of making $M(\ell)$ into a G-manifold. Let $\pi(\ell) = \pi_1(M(\ell))$.

(i) $G = \mathbb{Z}^m$; $\alpha : \pi(\ell) \to \mathbb{Z}^m$ ($m =$# components of ℓ), the abelianization which sends the i-th meridian $\mu_i \in \pi(\ell)$ to $(0, \ldots, 0, 1, 0, \ldots, 0)$, where 1 is the i-th coordinate.

(ii) $G = F$ (the free group on $x_1, \ldots, x_m$); $\alpha : \pi(\ell) \to F$ defined by $\alpha(\mu_i) = x_i$. Of course, this can be defined if ℓ is a *boundary link*. Two such *F-structures* α, α' on ℓ differ by an element $\phi \in A(F) =$ the group of automorphisms of F which conjugate each x_i.

(iii) $G = \bar{F}$; $\alpha : \pi(\ell) \to \bar{F}$ defined by $\alpha(\mu_i) = x_i$. This requires that $n > 1$, or the $\bar{\mu}$-invariants (see [**L2**]) vanish if $n = 1$. Two such *$\bar{F}$-structures* α, α' on L differ by an element of $A(\bar{F}) =$ the group of automorphisms $\bar{F}$ which conjugate each x_i.

For each of these categories of links, the notion of G-concordance is clear. $\mathbb{Z}^m$-concordance is the same as concordance. If two $\bar{F}$-links are concordant, then, after changing the $\bar{F}$-structure on one of them, they are $\bar{F}$-concordant. If two F-links are *boundary concordant*, then, after changing the F-structure on one of them, they are F-concordant.

Proposition 6. *If G-links ℓ, ℓ' are G-concordant then $M(\ell)$ and $M(\ell')$ are homology G-bordant.*

We define signature invariants for links in each of these categories. If ℓ is a G-link, then we define $\sigma_G(\ell) = \sigma(M(\ell)) : R_k(G) \to \mathbb{R}$ where $k = 1$ when $G = \mathbb{Z}^m$. For convenience we write $\sigma_G = \begin{Bmatrix} \sigma & G = \mathbb{Z}^m \\ \bar{\sigma} & G = \bar{F} \end{Bmatrix}$.

Integrality of these functions is described by the following theorem.[1]

Theorem 2.

(i) *If L is any odd-dimensional link, then*

$$\sigma(L)\cdot(z_1,\dots,z_n)\equiv -2\sum_i s_i(L)\arg z_i \mod \mathbb{Z}$$

(where $z=\exp\big(2\pi(\arg z)\sqrt{-1}\big)$*)*

(ii) *If L is an SHB link, and* $\theta\in R_n(\bar F)$*, then*

$$\bar\sigma(L)\cdot\theta\equiv -2\sum_i s_i(L)\arg\det\theta(x_i).$$

This formula also holds for $\sigma_F(L)$ *when L is an F-link.*

It would be interesting to know whether (ii) is true for any $\bar F$-link, but, in fact, it is still an open question whether every $\bar F$-link is (or, at least, is concordant to) an SHB-link. There are purely homotopy theoretic obstructions which measure this difference (see [**Le**], [**L3**]) but it is not known whether the groups in which these obstructions live are zero.

As a consequence of Proposition 6 and 3, we see that $\sigma_G(\ell)=\sigma_G(\ell')$ on some large set if ℓ and ℓ' are G-concordant links. We also have

Proposition 7. *If* ℓ *is an SHB-link, then* $\sigma(\ell)$ *and* $\bar\sigma(\ell)$ *have a large domain of continuity. For every boundary link* $\sigma_F(\ell)$ *has a large domain of continuity.*

Proposition 7 is probably true for all $\bar F$-links. We will see that Proposition 7 is false, in general, for σ, and, in fact, σ can contain concordance information outside of any domain of continuity.

σ and $\bar\sigma$ are related to some previous signature invariants as follows:

(i) The Levine-Tristram signature [**L**] $\tau(\ell)\colon S^1\to\mathbb{Z}$, defined using a connected Seifert surface, is related to $\sigma(\ell)$ by $\tau(\ell)\cdot z=\sigma(\ell)\cdot(z,\dots,z)$. The signatures of the knot components are given by the values of $\tau(\ell)$ on the m circles $(1,\dots,1,z,1,\dots,1)$.

(ii) The Smolinsky signature [**Sm**] of ℓ associated to a sequence $(w_1,\dots,w_m)$ of p-th power roots of unity is the same as $\sigma(w_1,\dots,w_m)$.

---

1 We use the following terminology and notation. If L is a k-dimensional link of multiplicity m, let $S_i(L)$, $i=1,\dots,m$, denote the signature of the i-th component of L as a knot, when $k\equiv 3 \bmod 4$, and be zero if $k\equiv 1 \bmod 4$. Also recall the notion of homology boundary link and sublink of homology boundary link (SHB link) (see [**C**]).

(iii) If N is any group of prime-power order (necessarily nilpotent), $\phi : F/F_q \to N$ an epimorphism and $\rho : N \to U(n)$ a unitary representation, then Smolinsky [**Sm1**] defines an associated signature of a link ℓ. The composite: $\bar{F} \to F/F_q \xrightarrow{\phi} N \xrightarrow{\rho} U(n)$ defines $\theta \in R_n(\bar{F})$ and the Smolinsky signature coincides with $\bar{\sigma}(\ell) \cdot \theta$.

That these are all concordance invariants follows from our results.

4. To construct links exhibiting certain signatures we will use two realization theorems.

Theorem 3. *Let G be a finitely-generated group with $H_1(G) \approx \mathbb{Z}^m$ and $\{g_1, \dots, g_m\} \subseteq G$ a set of normal generators. Let (λ_{ij}) be a $(-1)^{q+1}$-Hermitian matrix over $\mathbb{Z}G$ such that $\epsilon(\lambda_{ij}) = \pm\delta_{ij}$ if $q = 1,3,7$, or is an even, non-singular matrix over $\mathbb{Z}$ otherwise, and assume the coefficients, in each λ_{ij}, of the elements of G of order 2 are even. Then there exists a $(2q-1)$-link ℓ with G-structure $\alpha : M(\ell) \to G$ such that $\alpha(\mu_i) = g_i$, for suitable meridians $\{\mu_i\}$, and so that $\sigma(M(\ell), \alpha) \cdot \theta = n$ signature$\big(\epsilon(\lambda_{ij})\big)$ – signature$\big(\theta(\lambda_{ij})\big)$ for all θ in some large subset of $R_n(G)$.*

(If the Whitehead conjecture on the asphericity of 2-complexes is true, or if $q > 1$, or if G has no finitely-generated perfect subgroups, then the formula holds for *all* θ.)

The second realization theorem deals with a much more restricted situation but obtains a wider realization result within this class. In particular, we will be able to contstruct examples whose domain of continuity is not large.

Theorem 4. *Let (λ_{ij}) be a Hermitian matrix over $\mathbb{Z}(\mathbb{Z}^2) = \mathbb{Z}[x, x^{-1}, y, y^{-1}]$ satisfying:*

(i) $\epsilon(\lambda_{ij}) = \pm\delta_{ij}$ *for* $i, j \geq 2$

(ii) $\lambda_{11} = 0$

Then there exists a 2-component *one-dimensional* link ℓ such that $\sigma(\ell) \cdot \theta =$ signature$(\theta(\lambda_{ij}))$ – signature$(\epsilon(\lambda_{ij}))$ for all $\theta \in R_1(\mathbb{Z}^2)$.

This result and proof is similar to [**L4 appendix**]. The Alexander polynomial of ℓ is $(x-1)(y-1)\Delta$ and the Alexander polynomials of the component knots are $\Phi(x,1)$ and $\Phi(1,y)$, where $\Delta = \det(\lambda_{ij})$, $\Phi = \det(\lambda'_{ij})$. The general picture of the continuity of $\sigma(\ell)$ is very simple. If $\Sigma_1 \subseteq T^2 = R_1(\mathbb{Z}^2)$ is the zero set of Δ and Σ_2 the zero set of Φ — so $T^2 - \Sigma_2$ is large — then $\sigma(\ell)$ is continuous on $T^2 - \Sigma_1$ and $\Sigma_1 - (\Sigma_1 \cap \Sigma_2)$. If Δ is primitive but $\Delta(1,1) \neq \pm 1$, then $T^2 - \Sigma_1$ is not large. Thus $\sigma(\ell) \mid \Sigma_1 - (\Sigma_1 \cap \Sigma_2)$ can contain concordance information.

5. *Example C:*

(1) Consider the (2×2)-matrix $\tau = \begin{pmatrix} 0 & \rho \\ \bar{\rho} & \lambda \end{pmatrix}$ over $\mathbb{Z}[x, x^{-1}, y, y^{-1}]$, where

$$\rho = p(x + x^{-1}) - q$$
$$\lambda = [1 - N_0(2 - x - x^{-1})(2 - y - y^{-1})][1 - N_1(2 - x - x^{-1})(2 - y - y^{-1})]$$

where p, q, N_0, N_1 are integers to be specified. Let ℓ be a link such that $\sigma(\ell)(z_1, z_2)$ $=\operatorname{sign}\tau(z_1, z_2) - \operatorname{sign}\tau(1, 1)$, from Theorem 4. Let us project T^2 onto the unit square S in the real (u, v)-plane ($|u|, |v| \leq 1$) by $u = \operatorname{Re}(z_1)$, $v = \operatorname{Re}(z_2)$. Then ρ, λ define real polynomials $\rho' = 2pu - q$, $\lambda' = \big(1 - 2N_0(1 - u)(1 - v)\big)\big(1 - 2N_1(1 - u)(1 - v)\big)$. The strata Σ_1, Σ_2 projected onto S, are, respectively, a line and two hyperbolas, as pictured in Figure 1 (assume $4N_i > 1 + q/2p$, $N_0 \neq N_1$).

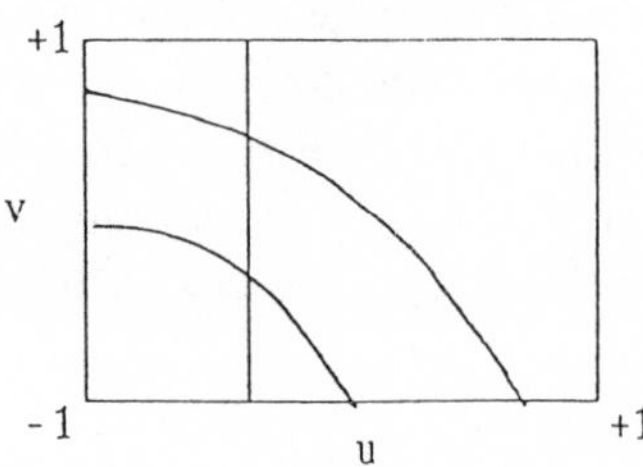

Figure 1

One computes easily that $\sigma(\ell) = 0$ on $T^2 - \Sigma_1$, while

$$\sigma(\ell) = \begin{cases} +1 & \text{on upper and lower segments of } \Sigma_1 - \Sigma_2 \\ -1 & \text{on middle segment } \Sigma \text{ of } \Sigma_1 - \Sigma_2 \end{cases}$$

Now consider a link ℓ' constructed from the (2×2)-matrix $\begin{pmatrix} 0 & \rho \\ \bar{\rho} & \lambda^2 \end{pmatrix}$. Obviously $\sigma(\ell')$ will have the same values as $\sigma(\ell)$ *except* that $\sigma(\ell) = +1$ on Σ. It can be shown that any large subset of T^2 contains all but a finite subset of Σ — and then ℓ and ℓ' are not concordant — if $2p$ and q are relatively prime and $q \neq 2p \pm 1$. By choose N_0, N_1 appropriately we can insure that Σ is disjoint from the diagonal. Furthermore Σ is also disjoint from the axes $z_1 = 1$ and $z_2 = 1$ and all (w_1, w_2) where w_i are prime-power roots of unity (unless $p = q = 1$). Then the Levine-Tristram and Smolinsky signatures as well as the signatures of the corresponding knot components all coincide for ℓ and ℓ'.

2) Let ℓ, ℓ' be constructed from matrices $\begin{pmatrix} 0 & \rho \\ \bar{\rho} & \lambda \end{pmatrix}$ and $\begin{pmatrix} 0 & \rho \\ \bar{\rho} & \lambda^2 \end{pmatrix}$, by Theorem 4, where λ is as in (1) but

$$\rho = [p(x + x^{-1}) - q]^2 + [r(y + y^{-1}) - s]^2$$

for non-zero integers $p,\ q,\ r,\ s$. Now Σ_1 consists of the point $(q/2p,\ s/2r)$ and Σ_2 as in (1). $\sigma(\ell) = \sigma(\ell') = 0$ except at Σ_1,

$$\sigma(\ell) \cdot \Sigma_1 = \begin{cases} -1 & \text{if } \Sigma_1 \text{ lies between the hyperbolas} \\ +1 & \text{otherwise} \end{cases}$$

and $\sigma(\ell') \cdot \Sigma_1 = +1$. We can choose $N_0,\ N_1$ so that $\sigma(\ell) \cdot \Sigma_1 = -1$, for a given $p,\ q,\ r,\ s$. It can be shown that Σ_1 belongs to every large subset of T^2 if $(q, 2p) = 1 = (s, 2r)$ and $(2p - 1, 2r - s) > 1$. Such $p,\ q,\ r,\ s$ are plentiful. Assuming $\sigma/2p \neq s/2r$, Σ_1 avoids all points which define any previous signatures.

3) Let G be the free group on a, b and set $g_1 = [a,b]^k a$, $g_2 = b$, for some integer k. Let λ be the (1×1)-matrix with entry $a + a^{-1} - 1$, if $q = 1,\ 3,\ 7$, or the (2×2)-matrix $\begin{pmatrix} 2(a + a^{-1} - 1) & 3 \\ 3 & 4(a + a^{-1} - 1) \end{pmatrix}$, if q odd $\neq 1,\ 3,\ 7$ or $\begin{pmatrix} a - a^{-1} & 1 \\ -1 & a - a^{-1} \end{pmatrix}$ if g is even. Let ℓ be the link constructed for this data by Theorem 3. The $\bar{F}$-structure on ℓ is defined from the G-structure on ℓ by the map $G \to \bar{F}$, where $g_1 \mapsto u$, $g_2 \mapsto v$. Composing with the homomorphism $\bar{F} \to \bar{D}$ (or $\bar{B}$) of example A, one checks that $b \to t$ and $a \to 1/2k + 1$ (or $[\frac{k+1}{2k+1}, \frac{k}{2k+1}]$) in $\mathbb{Z}_{(2)} \subseteq \bar{D}$ (or $\bar{A} \subseteq \bar{B}$). If we use the analytic map $\bar{\phi} : \mathbb{R} \to \bar{D}$ (or $\bar{B}$) of Example A, we see, as in Example B, that $\bar{\sigma}(\ell) \circ \bar{\phi}$ has period 1 if the F-structure on ℓ comes from an $\bar{F}$-structure. But, as in Example B, for this choice of ℓ such periodicity fails.

To conclude that ℓ is not concordant to a boundary link, we must consider the effect of a change in the $\bar{F}$-structure on ℓ. But it is not hard to check that any special automorphism of $\bar{F}$ commutes under the homomorphism $\bar{F} \to \bar{D}$ (or $\bar{B}$), with an inner automorphism of $\bar{D}$ (or $\bar{B}$). Since inner automorphisms of G do not change the signature function $R_n(G) \to \mathbb{R}$, $\bar{\sigma}(\ell)$ is unchanged by its $\bar{F}$-structure.

References

[APS] M. Atiyah, V. Patodi, I. Singer, *Spectral Asymmetry and Riemannian Geometry II*, Math. Proc. Camb. Phil. Soc. **77** (1975), 405–432.

[C] T. Cochran, *Link Concordance Invariants and Homotopy Theory*, Invent. Math. **90** (1987), 635–646.

[CO] T. Cochran, K. Orr, *Not All Links are Concordant to Boundary Links*, Bull. AMS **23** (1990), 99–106.

[CS] S. Cappell, J. Shaneson, *The Codimension Two Placement Problem and Homology Equivalent Manifolds*, Annals of Math. **90** (1974), 277–348.

[Le] J. Y. LeDimet, *Cobordisme D'Enlacements de Disques*, Mem. de la Soc. Math. de France No. 32, 1988.

[L] J. Levine, *Knot Cobordism Groups in Codimension Two*, Comm. Math. Helv. **44** (1969), 229–244.

[L1] J. Levine, *Algebraic Closure of Groups,*, Combinatorial Group Theory, Contemporary Mathematics vol. 109, vol. 109, AMS Providence, R.I., 1990.

[L2] J.Levine, *Link Concordance and Algebraic Closure of Groups*, Comment. Math. Helv. **64** (1989), 236–255.

[L3] J. Levine, *Link Concordance and Algebraic Closure II*, Invent. Math. **96** (1989), 571–592.

[L4] J. Levine, *The Module of a 2-Component Link*, Comm. Math. Helv. **57** (1982), 377–399.

[Sm] L. Smolinsky, *A Generalization of the Levine-Tristram Link Invariant*, Trans. AMS.

[Sm1] L. Smolinsky, *Invariants of Link Cobordism*, Topology and its Applications **32** (1989), 161–168.

[T] A.G. Tristram, *Some Cobordism Invariants for Links*, Proc. Camb. Phil. Soc. **66** (1969), 251-264.

Lusternik-Schnirelmann category of a fibered knot complement with fiber a punctured torus

Takao Matumoto

Abstract. We showed in [**20**] that a locally flat knot is topologically unknotted if and only if the Lusternik-Schnirelmann category of its complement is one. We showed also in [**21**] that the complement of a ribbon knot is of category two. In this paper we shall prove the following: If there is a fibered knot with fiber a punctured torus, then its complement attains the maximal category and the complement of its iterated spun knot has the same category. Hence, we know now the existence of knot $S^{n-2} \subset S^n$ whose complement is of category k with $2 \leq k \leq 7 \leq n-1$ for any n.

§1. Introduction

The category $\operatorname{cat} X$ of a space X is the least integer n such that X can be covered by $n+1$ open subsets each of which is contractible to a point in X; if there is no such integer, $\operatorname{cat} X = \infty$. In particular, $\operatorname{cat} X = 0$ if and only if X is itself contractible. So, $\operatorname{cat} S^n = 1$. We reduced the number by one from Fox's definition.

One of the most important properties is that $\operatorname{cat} X$ is a homotopy type invariant of X. Using this we can easily prove that $\operatorname{cat} X \leq \dim X$ for a connected CW complex X. To estimate the category of a CW complex X from below we use the property that if $\operatorname{cat} X \leq n$ then the cup product $u_0 \smile \ldots \smile u_n$ vanishes for $u_k \in H^*(X)$ $(k = 0, \ldots, n)$ with $* \geq 1$. We have immediately for example that $\operatorname{cat}(S^{n_1} \times \cdots \times S^{n_k}) = k$.

A closed manifold M is a homotopy sphere if and only if $\operatorname{cat} M = 1$. In fact, if $\operatorname{cat} M = 1$ then M is a homology sphere by the property above and the Poincaré duality theorem. So, it suffices to know that $\pi_1(X)$ is a free group if X is a connected manifold and $\operatorname{cat} X \leq 1$ (cf. Proposition 3.1).

A corresponding theorem in the case of knot complement is

Knots 90

Theorem 1 [20]. *A locally flat knot $S^{n-2} \subset S^n$ is topologically unknotted if and only if* $\text{cat}(S^n - S^{n-2}) = 1$.

By the facts known in geometric topology we have only to show that $\text{cat}(S^n - S^{n-2}) = 1$ implies $S^n - S^{n-2} \simeq S^1$ and a detailed proof is given in [**20**].

Any non-trivial classical knot is an example of knot whose complement is of category two. More examples are given by simple fibered odd-dimensional knots. The ribbon knots are the other examples.

Theorem 2 [21]. *For any non-trivial ribbon knot $S^{n-2} \subset S^n$ we have that* $\text{cat}(S^n - S^{n-2}) = 2$.

We have examples of closed connected n-dimensional manifolds with any integer from 1 to n as their category; $\text{cat}(S^{n-k+1} \times T^{k-1}) = k$. We may ask if there is also an example of knot $S^{n-2} \subset S^n$ whose complement has a given category from 3 to $n-1$.

A knot $S^{n-2} \subset S^n$ is called a fibered knot with fiber a punctured torus if its complement has a structure of fiber bundle over a circle with fiber a punctured torus $T^{n-1} - pt$. The following Theorems 3 and 4 would give an affirmative answer if there is a locally flat fibered knot with fiber a punctured torus for any dimension. But such a knot is now known to exist only for $3 \leq n \leq 8$ as in Proposition 5.

Theorem 3. *The Lusternik-Schnirelmann category of the total space of a fiber bundle over a circle S^1 with fiber a punctured torus $T^m - pt$ is equal to m.*

Theorem 4. *Let $S^{n-2} \subset S^n$ be a locally flat fibered knot with fiber a punctured torus $T^{n-1} - pt$. Then, the complement of its k-iterated spun knot $S^{n+k-2} \subset S^{n+k}$ is of category $n-1$ for any $k \geq 0$.*

Proposition 5. *For $3 \leq n \leq 8$ there is a locally flat fibered knot $S^{n-2} \subset S^n$ with fiber a punctured torus.*

Theorem 3 is an outcome of the knot conference. Professors González-Acuña and Gómez-Larrañaga were interested in the question and suggested a proof of Theorem 3. The author is deeply grateful to them for the kind suggestion.

§2. Covering manifolds with open cells

Any connected n-dimensional manifold is covered by $n+1$ open n-cells (cf. [**18**]). Here an open n-cell stands for an open subset homeomorphic to $\mathbf{R}^n$. If the manifold is non-closed we have the following proposition. We will outline a proof, since the proposition should be known for smooth or piecewise linear manifolds.

Proposition 2.1. *Let M be a non-compact connected n-dimensional manifold without boundary. Then, M is covered by n open n-cells.*

Proof. In case $n \leq 4$ we know that M is smoothable due to [**22**]+[**9**] and hence M has a handle decomposition. If $n \geq 6$, we get a topological handle decomposition of M owing to [**16**, III]. In the remaining case $n = 5$ we note that $\beta_0 k(M) \in H^5(M;\mathbf{Z})$ vanishes where $k(M) \in H^4(M;\mathbf{Z}_2)$ is the Kirby-Siebenmann obstruction and β_0 is the integral Bockstein operation. By [**23**]+[**3**] (cf.[**19**],[**10**];[**6**]) we have a simplicial complex K which is homeomorphic to M, that is, M has a structure of polyhedral homology manifold. So, M has a handle decomposition with homology handles; a homology handle is a product of a cone of homology sphere with a disk. In this case the interior of these homology handles are open 5-cells and homology 4- and 5-handles can be taken by usual handles.

We may assume that (homology) i-handles are attached to the union of (homology) j-handles with $j < i$. We get its associated "Morse" function and gradient-like vector field, where the critical points of index $\leq n-2$ may not be of Morse type. Moreover, since $H^n(M;\mathbf{Z}) = 0$ we can cancel all the critical points of index n. Note that the union of an open regular neighborhood of each remaining handle and its $+$-trajectories along the gradient-like vector field forms an open n-cell which has a locally flat boundary at at least one frontier point. So, connecting these disjoint neighborhoods of all the handles of index i by tubes we get an open n-cell for each i with $0 \leq i \leq n-1$. Clearly they cover M.

§3. Proof of Theorem 3

A subset U of a space X is called π_1-contractible in X if every loop in U is contractible to a point in X. We write $\operatorname{cat}_{\pi_1} X = n$ if n is the least integer such that X can be covered by $n+1$ open subsets $U_0, \dots, U_n$ which are π_1-contractible in X. Again, if no such integer exists then $\operatorname{cat}_{\pi_1} X = \infty$. Clearly $\operatorname{cat}_{\pi_1} X \leq \operatorname{cat} X$.

Proposition 3.1 (Eilenberg & Ganea [**7**], Gómez-Larrañaga & González-Acuña [**11**]). *Let X be a paracompact, connected and locally arcwise connected space. Then, $\operatorname{cat}_{\pi_1} X \leq n$ if and only if there exist a connected CW complex L of dimension n and a map $f : X \to L$ such that $f_* : \pi_1 X \to \pi_1 L$ is an isomorphism.*

By using Proposition 3.1 we will show the following Theorem 3.2 which is a stronger restatement of Theorem 3. We acknowledge here that the idea to use $\hat{X}$ and to prove $H_m(X) = 0$ from the condition $\operatorname{cat}_{\pi_1} X \leq m-1$ is due to Professors González-Acuña and Gómez-Larrañaga.

Theorem 3.2. *Let $q : X \to S^1$ be a fiber bundle with fiber a punctured torus $T^m - pt$. Then,* $\operatorname{cat}_{\pi_1} X = \operatorname{cat} X = m$.

Proof. If $m = 1$, then $X \simeq S^1$ and $\operatorname{cat}_{\pi_1} X = \operatorname{cat} X = 1$. So, we may assume $m \geq 2$ hereafter. Since X is a non-compact connected $(m+1)$-dimensional manifold without boundary, $\operatorname{cat}_{\pi_1} X \leq \operatorname{cat} X \leq m$ by Proposition 2.1. Note that any self-homeomorphism of $T^m - pt$ can be extended to that of T^m. So, $X = \hat{X} - S^1$ for a fiber bundle $\hat{X}$ over S^1 with fiber T^m. Since the closed $(m+1)$-manifold $\hat{X}$ is orientable over $\mathbf{Z}_2$ and S^1 is a compact ANR, $H_{m+1}(\hat{X}, X; \mathbf{Z}_2)$ is isomorphic to $H^0(S^1; \mathbf{Z}_2) \cong \mathbf{Z}_2$ by the duality theorem [**25**, 6.1.10 + 6.2.17]. Of course $H_{m+1}(\hat{X}; \mathbf{Z}_2) \cong \mathbf{Z}_2$. Take a point $y \in \hat{X} - X$ and let $j : X \to \hat{X}$ and $j_y \circ j_{S^1} : (\hat{X}, \emptyset) \to (\hat{X}, X) \to (\hat{X}, \hat{X} - y)$ denote the natural inclusions. Then, by the argument to determine the fundamental class of $\hat{X}$ we see that the composition $(j_y)_* \circ (j_{S^1})_* : H_{m+1}(\hat{X}; \mathbf{Z}_2) \to H_{m+1}(\hat{X}, X; \mathbf{Z}_2) \to H_{m+1}(\hat{X}, \hat{X} - y; \mathbf{Z}_2)$ is an isomorphism. So, $(j_{S^1})_* : H_{m+1}(\hat{X}; \mathbf{Z}_2) \to H_{m+1}(\hat{X}, X; \mathbf{Z}_2)$ is injective and hence an isomorphism. We see that $j_* : H_m(X; \mathbf{Z}_2) \to H_m(\hat{X}; \mathbf{Z}_2)$ is injective in the long exact sequence for the pair $(\hat{X}, X)$.

Now assume that $\operatorname{cat}_{\pi_1} X \leq m - 1$. Then, by Proposition 3.1 we have a map $f : X \to L^{m-1}$ into a connected CW complex L^{m-1} of dimension $m-1$ inducing an isomorphism $f_* : \pi_1 X \cong \pi_1 L$. Since $\hat{X}$ is an aspherical space (i.e., $K(\pi_1, 1)$-space), we can get a map $g : L \to \hat{X}$ such that $j_* = g_* \circ f_* : \pi_1 X \cong \pi_1 L \to \pi_1 \hat{X}$ by extending a map on the 2-skeleton. In fact, there is no obstruction because the coefficients vanish in the obstruction theory. Then, j and $g \circ f$ are seen to be homotopic as well by the obstruction theory. Hence $j_* = g_* \circ f_* : H_m(X; \mathbf{Z}_2) \to H_m(\hat{X}; \mathbf{Z}_2)$ is a zero-map, because it passes through $H_m(L^{m-1}; \mathbf{Z}_2) = 0$. As we have proved that j_* is injective, we see that $H_m(X; \mathbf{Z}_2) = 0$.

Let $\tilde{X}$ be an infinite cyclic covering over X associated to q and $t : \tilde{X} \to \tilde{X}$ a generator of the covering transformations. Take a finite cyclic intermediate k-sheet covering X_k over X for $k \geq 1$ with a projection $\tilde{X} \to X_k$. Let $C_*(\tilde{X}; \mathbf{Z}_2)$ and $C_*(X_k; \mathbf{Z}_2)$ denote the singular chain complexes of $\tilde{X}$ and X_k respectively with the coefficient in $\mathbf{Z}_2$. Then, we have an exact sequence:

$$0 \to C_*(\tilde{X}; \mathbf{Z}_2) \xrightarrow{t^k - 1} C_*(\tilde{X}; \mathbf{Z}_2) \to C_*(X_k; \mathbf{Z}_2) \to 0.$$

Its associated long exact sequence gives the following short exact sequences:

$$0 \to \operatorname{Coker}(t^k - 1)_* | H_i(\tilde{X}; \mathbf{Z}_2) \to H_i(X_k; \mathbf{Z}_2) \xrightarrow{\partial_*} \operatorname{Ker}(t^k - 1)_* | H_{i-1}(\tilde{X}; \mathbf{Z}_2) \to 0.$$

Now take the corresponding k-sheet coverings $\hat{X}_k$ and L_k^{m-1}. Note that $\hat{X}_k$ is also a fiber bundle over S^1 with fiber T^m and $X_k = \hat{X}_k - S^1$. Then, the inclusion $j_k : X_k \to \hat{X}_k$ induces a map $(j_k)_* : H_m(X_k; \mathbf{Z}_2) \to H_m(\hat{X}_k; \mathbf{Z}_2)$ which is injective as before and at the same time a zero-map because $j_k \simeq g_k \circ f_k$ for the induced maps $f_k : X_k \to L_k^{m-1}$ and $g_k : L_k^{m-1} \to \hat{X}_k$. Hence, under the assumption that $\mathrm{cat}_{\pi_1} X \leq m-1$ we have $H_m(X_k; \mathbf{Z}_2) = 0$ for any $k \geq 1$. Moreover, by using the short exact sequence above with $i = m$ we have for any integer $k \geq 1$

$$\mathrm{Ker}(t^k - 1)_* | H_{m-1}(\tilde{X}; \mathbf{Z}_2) = 0.$$

Note that $\tilde{X} \simeq T^m - pt$ and $H_{m-1}(\tilde{X}; \mathbf{Z}_2)$ is an m-dimensional vector space over $\mathbf{Z}_2$. So, we have $(t^\ell - 1)_* | H_{m-1}(\tilde{X}; \mathbf{Z}_2) = 0$ for some ℓ, say $\ell = \Gamma(2^m + 1)$. Taking $k = \ell$ in the above formula, we see $H_{m-1}(\tilde{X}; \mathbf{Z}_2) = 0$. This contradicts that $H_{m-1}(\tilde{X}; \mathbf{Z}_2) \cong H_{m-1}(T^m - pt; \mathbf{Z}_2) \neq 0$.

§4. Proof of Theorem 4

Let $S^{n-2} \subset S^n$ be a locally flat knot. Then, removing an open regular neighborhood of a point of S^{n-2} in S^n, we get a disk pair $B^{n-2} \subset B^n$. The interior of the complement $C = B^n - B^{n-2}$ is homeomorphic to the complement $S^n - S^{n-2}$. The spun knot is defined by $(B^n, B^{n-2}) \times S^1 \cup (\partial B^n, \partial B^{n-2}) \times D^2$, that is, the boundary of $(B^n, B^{n-2}) \times D^2$. So, its complement is homeomorphic to $\partial(C \times D^2) = C \times S^1 \cup_{\partial C \times S^1} S^1 \times \mathbf{R}^{n-2} \times D^2$ where we use a natural identification $\partial C = S^1 \times \mathbf{R}^{n-2}$.

Now we have a natural inclusion $C = C \times \{1\} \to \partial(C \times D^2)$ and a retraction $\partial(C \times D^2) \xrightarrow{incl} C \times D^2 \xrightarrow{proj} C$. Remark that they both induce isomorphisms on the fundamental group π_1. According to Proposition 3.1 we see that $\mathrm{cat}_{\pi_1} C = \mathrm{cat}_{\pi_1} \partial(C \times D^2)$. Since the complement $S^n - S^{n-2}$ is homeomorphic to the interior of C and has the same homotopy type as C, we get the following proposition.

Proposition 4.1. *The π_1-category of the complement of the spun knot is equal to that of the original locally flat knot.*

This together with Theorem 3 shows the following proposition.

Proposition 4.2. *For any $k \geq 0$ let X^{n+k} be the complement of the k-iterated spun knot $S^{n+k-2} \subset S^{n+k}$ of a locally flat fibered knot $S^{n-2} \subset S^n$ with fiber a punctured torus $T^{n-1} - pt$. Then,* $\mathrm{cat}_{\pi_1} X^{n+k} = n - 1$.

Since $\operatorname{cat}_{\pi_1} X \leq \operatorname{cat} X$, it is more than enough to prove the following lemma in order to prove Theorem 4.

Lemma 4.3. *Let X^{n+k} be the same as in Proposition 4.2. Then, X^{n+k} is covered by n open $(n+k)$-cells in X^{n+k}.*

Proof. This will be proved by induction on k. Let C^{n+k} be the complement of the disk pair associated to the k-iterated spun knot for $k \geq 0$. We will find open subsets $U_0, \ldots, U_{n-1}$ in C^{n+k} such that $U_0 \cup \cdots \cup U_{n-1} = C^{n+k}$, U_0 is a half of open $(n+k)$-cell with boundary $U_0 \cap \partial C^{n+k}$, U_1 is a half of open $(n+k)$-cell with boundary $U_1 \cap \partial C^{n+k}$ and U_i for $2 \leq i \leq n-1$ are open $(n+k)$-cells. Here a half of open $(n+k)$-cell means a subset homeomorphic to a half space $\mathbf{R}^{n+k}_+$.

Note that we can extend any self-homeomorphism of $T^{n-1} - pt$ to that of T^{n-1}. Moreover, we see that any orientation preserving self-homeomorphism of (T^{n-1}, pt) is isotopic keeping pt fixed to the one which is identity on a disk neighborhood of pt by using the affirmative answer to the annulus conjecture due to [**15**]+[**9**]. So, for the case of $k = 0$ we have a compactification M of $C = C^n$; M is a fiber bundle over S^1 with fiber $T^{n-1} - \operatorname{Int} D^{n-1}$ and $\operatorname{Int} M$ is homeomorphic to $\operatorname{Int} C$. Take the closure $\overline{\partial C} = S^1 \times D^{n-2}$ of ∂C in ∂M. The collar $\overline{\partial C} \times I$ in M has a handle decomposition with one 0-handle and one 1-handle. Starting from these two handles we can get a handle decomposition of M or of $M - S^1$. Here S^1 is a core of $\partial M - \partial C$. In fact, in case $n \leq 4$ any homeomorphism of $T^{n-1} - \operatorname{Int} D^{n-1}$ to itself is isotopic to a diffeomorphism [**22**] and M is a smooth manifold; if $n \geq 6$ or $n = 5$ we see that M or $M - S^1$ respectively has a structure of piecewise linear manifold relative to $\overline{\partial C} \times I$ by [**16**] because M and $\overline{\partial C} \times I$ are homology circles and $H^4(M, \overline{\partial C} \times I; \mathbf{Z}_2) = 0$. Unless $n = 5$ we may assume that the handle decomposition of M has no n-handles and that i-handles are attached to the union of j-handles with $j < i$. Connecting the open regular neighborhoods of all the handles of index i by tubes and taking intersection with C we get an open subset U_i of C for $0 \leq i \leq n-1$. When $n = 5$ and M is not a piecewise linear manifold, we use a Morse function on $M - S^1$ and trajectories along the gradient-like vector field as in the proof of Proposition 2.1; we can get open subsets U_i of C for $0 \leq i \leq n-1$ in the same way. In case $n = 5$ we can show this also by using a polyhedral homology manifold structure on M relative to ∂M which exists by [**19**]+[**3**] (cf.[**10**],[**23**];[**6**]) because $H^5(M, \partial M; \mathbf{Z})$ has no 2-torsions. These $U_0, \ldots, U_{n-1}$ satisfy the condition stated in the first paragraph. Of course, $U_0 \cap \operatorname{Int} C^n$, $U_1 \cap \operatorname{Int} C^n$, $U_2, \ldots, U_{n-1}$ are open n-cells and they cover $\operatorname{Int} C^n$ which is homeomorphic to X^n.

Now for each i with $2 \leq i \leq n-1$ we take an arc γ_i from a point p_i in the frontier of U_i to a point q_i of ∂C^{n+k} outside the closure of $U_i \cup \partial C^{n+k}$. A carefully chosen neighborhood of $U_i \cup \gamma_i$ gives a half U_i' of open $(n+k)$-cell.

Put $U'_0 = U_0$ and $U'_1 = U_1$. It follows that $\partial(C^{n+k} \times D^2)$ is covered by n open $(n+k+1)$-cells $\partial(U'_0 \times D^2), \ldots, \partial(U'_{n-1} \times D^2)$.

We see that the following subset C_0^{n+k+1} of X^{n+k+1} is a choice for C^{n+k+1}:

$$C_0^{n+k+1} = \partial(C' \times D^2) \cup \partial\overline{\partial C'} \times \operatorname{Int} D^2 \cup (\partial C - \partial C') \times \partial D^2,$$

where C' and $\overline{\partial C'}$ are defined by $\operatorname{Int} C' = \operatorname{Int} C^{n+k}, \partial C' = S^1 \times \operatorname{Int} \mathbf{R}_+^{n+k-2} \subset \partial C^{n+k} = S^1 \times \mathbf{R}^{n+k-2}$ and $\overline{\partial C'} =$ closure of $\partial C'$ in ∂C^{n+k}. Then, for each of $i = 2, \ldots, n-1$ we could have chosen q_i in $\partial C'$ so that $\partial(U'_i \times D^2)$ is an open $(n+k+1)$-cell contained in C_0^{n+k+1}. We see also that each of

$$\partial(U_i \cap C') \times D^2 \cup (U_i \cap \partial\overline{\partial C'}) \times \operatorname{Int} D^2 \cup (U_i \cap (\partial C - \partial C')) \times \partial D^2$$

for $i = 0$ and 1 is a half of open $(n+k+1)$-cell. These two subsets together with $\partial(U'_2 \times D^2), \ldots, \partial(U'_{n-1} \times D^2)$ cover C_0^{n+k+1}. This completes an induction step.

Remark. It is conjectured in general that the category of the complement of the spun knot is equal to or less than that of the original knot.

§5. Proof of Proposition 5 and some comments

When $n = 3$, we have two locally flat fibered knots of genus one; trefoil knot and figure eight knot and there are no others due to [**2**]+[**12**]. The case $n = 4$ is discussed in [**1**] to some extent. The examples for the case $n = 5$ or 6 are given in [**4**] and for the case $n = 7$ or 8 in [**28**]. Although these are enough for Proposition 5, we will give some concrete examples for $4 \leq n \leq 8$.

Given $A \in GL(n-1, \mathbf{Z})$, let $A^{(r)}$ be its r-fold exterior product with itself. Then, A induces a diffeomorphism of T^{n-1} which leaves the origin fixed. So, we get a fiber bundle M_A over a circle with fiber a punctured torus $T^{n-1} - pt$ and monodromy A. If $\det A = +1$, we may assume that the corresponding diffeomorphism is identity on a neighborhood of the origin. The next result is an easy consequence of van Kampen, Mayer-Vietoris and duality theorems (cf. [**4**]).

Lemma 5.1. *Let* $n \geq 4$. *Suppose that for* $1 \leq r \leq [(n-1)/2]$, $\det(E - A^{(r)}) = \pm 1$ *and that* $\det A = +1$. *Then, there is a smooth fibered knot* $S^{n-2} \subset \tilde{S}^n$ *in a smooth homotopy sphere whose complement is diffeomorphic to* M_A.

Remark 5.2. Since any smooth homotopy sphere $\tilde{S}^n$ is homeomorphic to n-sphere for $n \geq 4$ due to [**24**]+[**9**] and any smooth submanifold is locally flat, Lemma 5.1 suffices to give examples for Proposition 5.

By using the relation between the elementary symmetric functions and the symmetric sums of constant powers $\Sigma \xi_i^m$ of the eigenvalues (cf. [**26**, p.39]), we see that the characteristic polynomial of $A^{(r)}$ is determined by the characteristic polynomial of A. Note also that the first condition in Lemma 5.1 is equivalent to say the characteristic polynomial of $A^{(r)}$ takes value ± 1 at $t = 1$. Now given $f(t) = t^m - a_m t^{m-1} - \cdots - a_2 t - a_1$ with $a_i \in \mathbf{Z}$ and $m = n - 1$, we get $\det(tE - A) = f(t)$ for A defined by $Ae_i = e_{i+1}$ for $i = 1, \ldots, m-1$ and $Ae_m = a_1 e_1 + a_2 e_2 + \cdots + a_m e_m$. So, it is enough to write the characteristic polynomial $\det(tE - A)$ to give examples of A satisfying the conditions of Lemma 5.1.

Examples 5.3. $f = t^3 - t - 1$, $g = t^4 - 2t^2 - t + 1$, $h = t^5 + t^2 - 1$, $f_6 = t^6 + t^4 - t^3 - t + 1$ and $f_7 = t^7 - t - 1$.

By Lemma 3.1 of [**13**] and the argument after it we see that $\det(E - A^{(2)}) = \pm 1$ for the matrix A which has any multiple of f, g and h as its characteristic polynomial. So, by Lemma 5.1 and Remark 5.2 f, g and h give the examples for $3 \leq n - 1 \leq 5$. We have to calculate explicitly $\det(E - A^{(2)})$ and $\det(E - A^{(3)})$ for f_6 and f_7. The examples f_6 and f_7 are due to Yamada [**28**].

Remark. One may think that some aspherical knot has a higher category. But the complement of any topologically non-trivial locally flat knot is not aspherical if $n > 3$ [**5**].

References

1. I. R. Aitchison and J. H. Rubinstein, *Fibered knots and involutions on homotopy spheres*, Four-manifold Theory, Contemp. Math. **35** (1984), 1–74.
2. G. Burde and H. Zieschang, *Neuwirthsche Knoten und Flächenabbildungen*, Abh. Math. Sem. Univ. Hamburg **31** (1967), 239–246.
3. J. W. Cannon, *The recognition problem: What is a topological manifold?*, Bull. Amer. Math. Soc. **84** (1978), 832–866.
4. S. E. Cappell and J. L. Shaneson, *There exist inequivalent knots with the same complement*, Ann. of Math. **103** (1976), 349–353.
5. E. Dyer and A. T. Vasquez, *The sphericity of higher dimensional knots*, Canad. J. Math. **25** (1973), 1132–1136.
6. R. D. Edwards, *The topology of manifolds and cell-like maps*, Proc. Int. Congress Math. Helsinki, 1978 (1980), 111-127.
7. S. Eilenberg and T. Ganea, *On the Lusternik-Schnirelmann category of abstract groups*, Ann. of Math. **65** (1957), 517–518.
8. R. H. Fox, *On the Lusternik-Schnirelmann category*, Ann. of Math. **42** (1941), 333–370.

9. M. H. Freedman and F. Quinn, *The Topology of 4-Manifolds*, Princeton Math. Ser. 39, Princeton Univ. Press, Princeton, 1990.
10. D. E. Galewski and R. J. Stern, *Classification of simplicial triangulations of topological manifolds*, Ann. of Math. **111** (1980), 1–34.
11. J. C. Gómez-Larrañaga and F. González-Acuña, *Lusternik-Schnirelmann category of 3-manifolds*, Preprint 1990.
12. F. González-Acuña, *Dehn's construction on knots*, Bol. Soc. Mat. Mexicana **15** (1970), 58–79.
13. J. C. Hausmann and M. Kervaire, *Sous-groupes dérivés des groupes de nœuds*, Enseign. Math. **24** (1978), 111–123.
14. I. M. James, *On category in the sense of Lusternik-Schnirelmann*, Topology **17** (1978), 331–348.
15. R. C. Kirby, *Stable homeomorphisms and the annulus conjecture*, Ann. of Math. **89** (1969), 575–582.
16. R. C. Kirby and L. C. Siebenmann, *Foundational Essays on Topological Manifolds, Smoothings and Triangulations*, Ann. Math. Studies 88, Princeton Univ. Press, Princeton, 1977.
17. L. Lusternik and L. Schnirelmann, *Méthodes Topologiques dans les Problèmes Variationnelles*, Hermann, Paris, 1934.
18. H. Luft, *Covering of manifolds with open cells*, Illinois J. Math. **13** (1969), 321–326.
19. T. Matumoto, *Variétés simpliciales d'homologie et variétés topologiques métrisables; Triangulation of manifolds*, Thèse Univ. de Paris-Sud, 91405 Orsay, 1976; Proc. Symp. Pure Math. **32** (1978), 3–6.
20. ———, *Lusternik-Schnirelmann category and knot complement*, J. Fac. Sci. Univ. Tokyo **37** (1990), 103–107.
21. ——— *Lusternik-Schnirelmann category of ribbon knot complement*, Proc. Amer. Math. Soc., (to appear).
22. E. Moise, *Affine structures on 3-manifolds*, Ann. of Math. **56** (1952), 96–114.
23. L. C. Siebenmann, *Are non-triangulable manifolds triangulable?*, "Topology of Manifolds," edited by Cantrell-Edwards, Markham, 1970, 77–84.
24. S. Smale, *Generalized Poincaré's conjecture in dimension greater than four*, Ann. of Math. **64** (1956), 399–405.
25. E. H. Spanier, *Algebraic Topology*, McGraw-Hill Inc., New York, 1966.
26. H. Weyl, *The Classical Groups, Their Invariants and Representations*, Princeton Math. Ser. 1, Princeton Univ. Press, Princeton, 1939.
27. G. W. Whitehead, *Elements of Homotopy Theory*, Graduate Texts in Math. 61, Springer-Verlag, New York, 1978.
28. T. Yamada, *On fibered knot with fiber* $T^n - intD^n$ (in Japanese), Master Thesis, Hiroshima Univ. 1990.

On 3-fold irregular branched coverings of θ_4 curves in a 3-sphere

Shin'ichi Kinoshita

In this paper we consider 3–fold irregular branched coverings of θ_4-curves in a 3–sphere. We will show how these branched coverings may be identified with those of links. We apply this technique to pretzel θ_4-curves, as examples. Note that the same technique can also be applied to any 4 valent graphs in a 3–sphere.

1.

Let K be a θ_4-curve in a 3–sphere S^3. Consider the fundamental group $G = \pi_1(S^3 - K)$ and present it by using a Wirtinger presentation (see, for instance, Harikae [**2**] or Nakao [**3**]). Assume that there exists a homomorphism ϕ of G onto S_3, the symmetric group of degree 3. The homomorphism ϕ is called a monodromy map of G onto S_3, or simply that of K. Further, we assume that the image of each of the generators in the Wirtinger presentation is one of (12), (13) and (23). For convenience, we put $\{a, b, c\} = \{(12), (13), (23)\}$. The branched covering associated with ϕ is the 3–fold irregular branched covering of K, denoted by $\bar{M}_\phi(K)$. Clearly, the collection of $\bar{M}_\phi(K)$ for all possible ϕ is a topological invariant of the position of K in S^3.

Now we look at the projection of K on a plane. At a vertex of K, the four generators x, y, z and t of G, as shown in Fig. 1, must satisfy one of the following conditions:

I. $\phi(x) = \phi(y) = \phi(z) = \phi(t) = a$,
II. $\phi(x) = \phi(y) = a$ and $\phi(z) = \phi(t) = b$, or
III. $\phi(x) = \phi(z) = a$, $\phi(y) = b$ and $\phi(t) = c$.

Knots 90

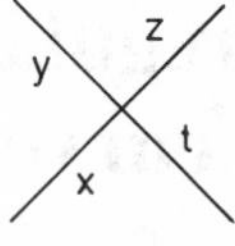

Figure 1

If Case I occurs, $\bar{M}_\phi(K)$ is not locally Euclidean over this vertex. Otherwise, $\bar{M}_\phi(K)$ is locally Euclidean over the vertex. From now on, we restrict our investigation to Cases II and III at each of two vertices of K, i.e. we only consider the case, when $\bar{M}_\phi(K)$ is an orientable 3–manifold.

Then, without changing the branched covering, we can modify the projection of K around two vertices to the one shown in Fig. 2. Note that we use $\phi(x)$, $\phi(y)$, $\phi(z)$ and $\phi(t)$, instead of x, y, z and t, respectively, in Fig. 2. Using this procedure, we can modify K to a link L and the monodromy map ϕ of K to ψ of L. Note that $\bar{M}_\phi(K)$ is homeomorphic to $\bar{M}_\psi(L)$, where the latter may be more easily identified by well-known technique.

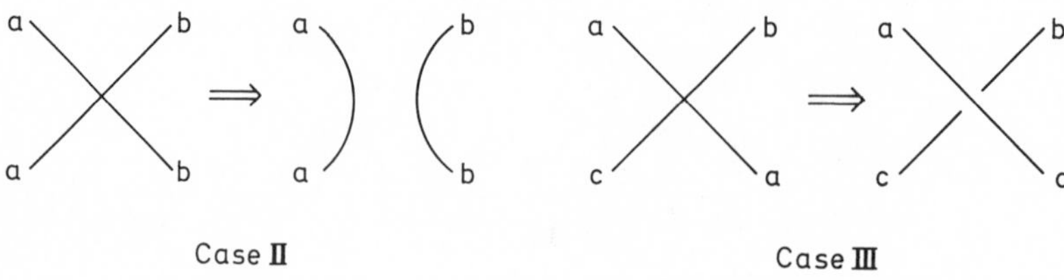

Figure 2

The link L in the above is obtained through a projection of K on a plane. Hence L itself is not uniquely associated with K. This can be seen from Fig. 3,

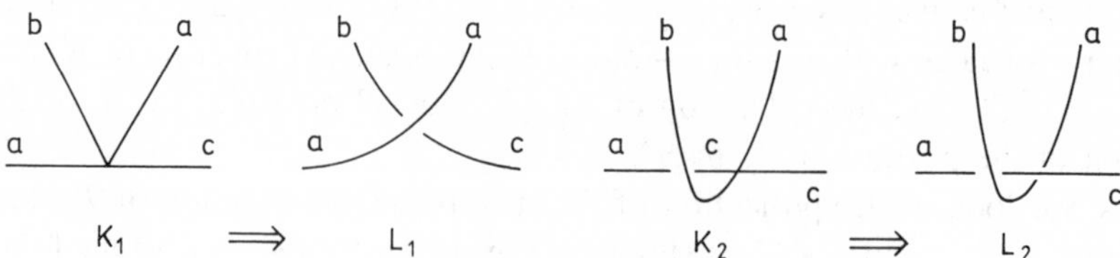

Figure 3

where two links L_1 and L_2 are obtained from two different projections K_1 and K_2 of K, respectively. However, we have $\bar{M}_{\psi_1}(L_1) \approx \bar{M}_{\psi_2}(L_2)$, because the corresponding 3–fold irregular branched coverings are homeomorphic under the operation shown in Fig. 4.

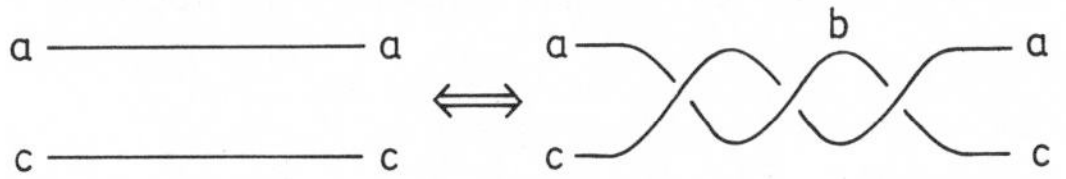

Figure 4

2.

In this section we investigate 3–fold irregular branched coverings of pretzel θ_4-curves, which are defined as follows: Let p_1, p_2, p_3 and p_4 be integers. Then $\theta(p_1, p_2, p_3, p_4)$ is a spatial graph as shown in Fig. 5. We note that if all p_1, p_2, p_3 and p_4 are even, then $\theta(p_1, p_2, p_3, p_4)$ minus one of its branches is planar. On the other hand, if all p_1, p_2, p_3 and p_4 are odd, $\theta(p_1, p_2, p_3, p_4)$ has two components, a θ_4-curve and a knot. Non-triviality and symmetricity of $\theta(p_1, p_2, p_3, p_4)$ are discussed by Hara [**1**] and Scharlemann [**4**].

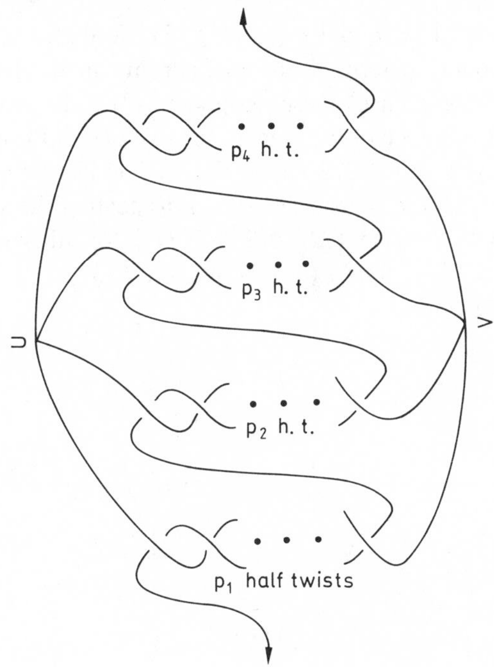

Figure 5

Now we show how to identify $\bar{M}_\phi(\theta(p_1, p_2, p_3, p_4))$ by some examples. We will supply a table of $\bar{M}_\phi(\theta(p_1, p_2, p_3, p_4))$ in the next section.

Example 1. $\theta(2, 2, 2, 2)$ (see Fig. 6).

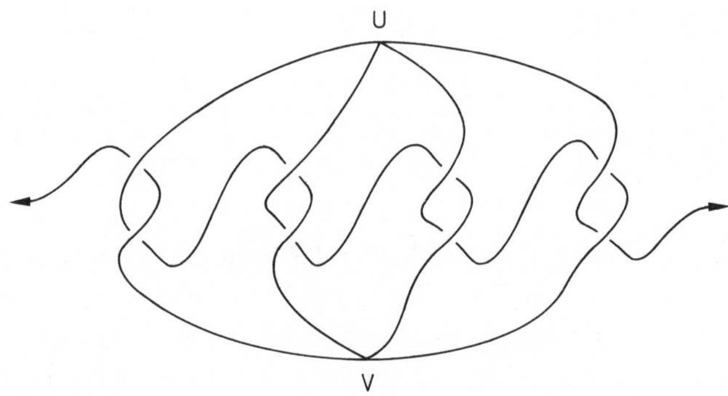

Figure 6

This is the Suzuki's θ_4-curve in [**5**]. At the vertex V, there are four ways of choosing assignments of a, b and c (see Fig. 7). Further, each of these assignments does not uniquely determine the monodromy map. Hence, we will have 3 monodromy maps for each of these assignments, i.e. 12 monodromy maps all together. For the convenience of the reader, we give one of these maps, say ψ, and show how to identify $\bar{M}_\psi(\theta(2, 2, 2, 2))$ in Fig. 8. The last figure in Fig. 8, shows that $\bar{M}_\psi(\theta(2, 2, 2, 2))$ is a lens space $L(3, -1)$. Repeating the same procedure for all monodromy maps, we have $\{\bar{M}_\phi(\theta(2, 2, 2, 2))\}$ for all possible ϕ, which are as follows: two $S^2 \times S^1$'s, four $L(3, 1)$'s and six $L(3, -1)$'s. From this we can see that $\theta(2, 2, 2, 2)$ is chiral.

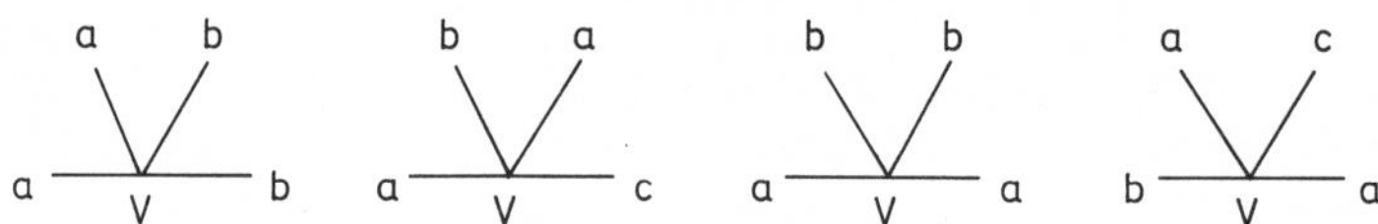

Figure 7

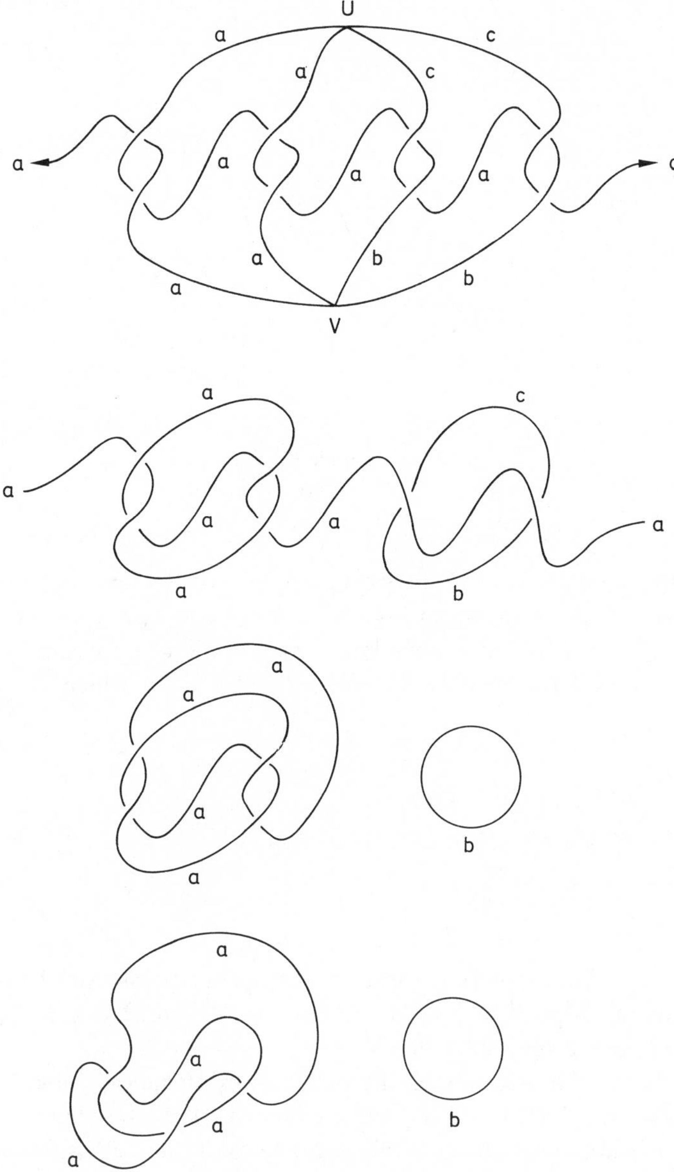

Figure 8

Example 2. $\theta(2, 2, -2, -2)$ (see Fig. 9).

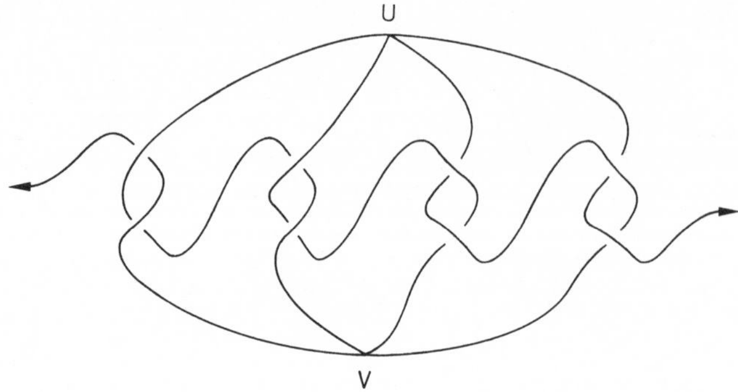

Figure 9

As before, we try to give 12 different assignments. However, we can have only 3 monodromy maps as shown in Fig. 10. Further, the case 3 must be discarded, because, at the vertex U, the monodromy map is the one appeared as Case I in Fig. 1. Hence we have only two $\bar{M}_\phi(\theta(2, 2, -2, -2))$'s, which are $L(3, 1)$ and $L(3, -1)$.

3.

A table of $\bar{M}_\phi(\theta(p_1, p_2, p_3, p_4))$ for all possible ϕ is given. All of them are lens spaces, including $S^2 \times S^1 \approx L(0, *)$. We note that $L(-p, q) \approx L(p, -q)$. Here are some remarks for reading the table.

The entries of the first column are quadruplets of integers mod. 3. Hence, if we want to know $\bar{M}_\phi(\theta(1, 2, 3, 4))$'s, for example, we look at the row containing $(1, 2, 0, 1)$. It's not there, but its cyclic permutation $(0, 1, 1, 2)$ is. Hence, we consider the row containing $(0, 1, 1, 2)$, where $p_1 = 3,\ p_2 = 4,\ p_3 = 1$ and $p_4 = 2$. Therefore, $\{\bar{M}_\phi(\theta(p_1, p_2, p_3, p_4))\}$ consists of one $L(5, 3)$ and three S^3's.

As another example, let us consider $\theta(2, 2, 2, -2)$. We find that neither $(2, 2, 2, 1)$, nor any of its cyclic permutations, is on the list. However, $(-2, -2, -2, 2) = (1, 1, 1, 2)$ mod. 3, and the latter is on the list. So, we first consider $\theta(-2, -2, -2, 2)$, i.e. $p_1 = p_2 = p_3 = -2$ and $p_4 = 2$. Then $\bar{M}_\phi(\theta(-2, -2, -2, 2))$'s are three $L(-5, -2)$'s and one $L(1, 1)$. Reversing the orientations, we have three $L(5, -2)$'s and one S^3 as $\bar{M}_\phi(\theta(2, 2, 2, -2))$'s.

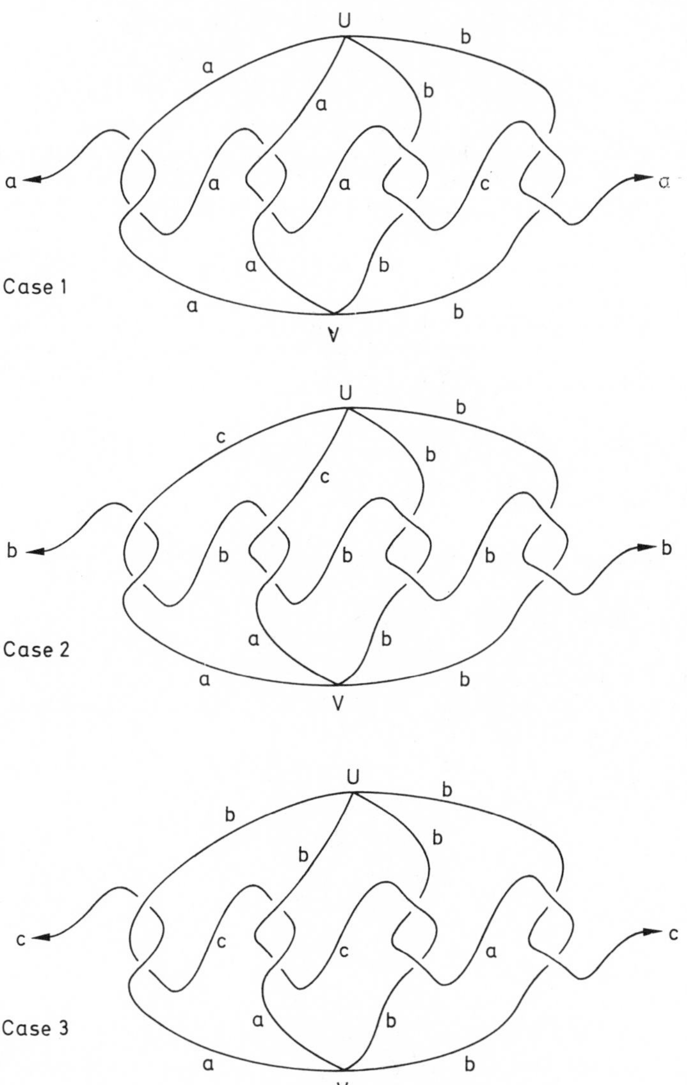
U
b
a
b
a
a
a
a
c
a
a
a
b
Case 1
a
b
V
U
b
c
b
c
b
b
b
b
b
b
a
b
Case 2
a
b
V
U
b
b
b
b
c
c
c
a
c
c
a
b
Case 3
a
b
V

Figure 10

A table of $\bar{M}_\phi(\theta)$ for pretzel $\theta(p_1, p_2, p_3, p_4)$

$(1,1,1,1)$	$S^2 \times S^1,\ S^2 \times S^1,\ L(p_1-1,\ 1),\ L(p_2-1,\ 1),\ L(p_3-1,\ 1),$ $L(p_4-1,\ 1),\ L(p_1p_2-1,\ p_1),\ L(p_1p_3-1,\ p_1),\ L(p_1p_4-1,\ p_1),$ $L(p_2p_3-1,\ p_2),\ L(p_2p_4-1,\ p_2),\ L(p_3p_4-1,\ p_3)$
$(1,1,1,2)$	$L(p_1p_4-1,\ p_1),\ L(p_2p_4-1,\ p_2),\ L(p_3p_4-1,\ p_3),\ L(p_4-1,\ 1)$
$(1,1,2,2)$	$L(p_1p_2-1,\ p_1),\ L(p_3p_4-1,\ p_3)$
$(1,2,1,2)$	$L(p_1p_3-1,\ p_1),\ L(p_2p_4-1,\ p_2)$
$(0,1,1,1)$	$L(p_1p_2-1,\ p_1),\ L(p_1p_3-1,\ p_1),\ L(p_1p_4-1,\ p_1),\ L(p_1-1,\ 1)$
$(0,1,1,2)$	$L(p_1p_4-1,\ p_1),\ S^3,\ S^3,\ S^3$
$(0,1,2,1)$	$L(p_1p_3-1,\ p_1),\ S^3,\ S^3,\ S^3$
$(0,2,1,1)$	$L(p_1p_2-1,\ p_1),\ S^3,\ S^3,\ S^3$
$(0,1,0,1)$	$L(p_1p_3-1,\ p_1),\ S^3,\ S^3,\ S^3$
$(0,0,1,1)$	$L(p_1p_2-1,\ p_1),\ S^3,\ S^3,\ S^3$
$(0,0,1,2)$	$S^3,\ S^3,\ S^3,\ S^3$
$(0,1,0,2)$	$S^3,\ S^3,\ S^3,\ S^3$
$(0,0,0,1)$	$S^3,\ S^3,\ S^3,\ S^3$
$(0,0,0,0)$	$S^3,\ S^3,\ S^3,\ S^3$

References

1. M. Hara, *Symmetry of θ_4-curves*, Tokyo J. Math **14** (1991), 7–16.
2. T. Harikae, *Universal abelian coverings of spatial theta–curves*, in this book.
3. M. Nakao, *On the $Z_2 \oplus Z_2$ branched coverings of spatial K_4–graphs*, in this book.
4. M. Scharlemann, *Some pictorial remarks on Suzuki's Brunnian graph*, preprint.
5. S. Suzuki, *Almost unknotted θ_n-curves in the 3–sphere*, Kobe J. Math. **1** (1984), 19–22.

Spherical space forms and Dehn surgery

*Steven A. Bleiler and Craig D. Hodgson**

This paper concerns those Dehn surgeries on a knot in the 3-sphere which yield manifolds with finite fundamental group. The focus will be on knots whose exterior contains an incompressible torus which is not boundary parallel, a so-called essential torus. Much of what follows holds for torally bounded 3-manifolds M and not just for knot complements in the 3-sphere. Following the notation of [CGLS] we will refer to the isotopy class of an unoriented simple closed curve on ∂M as a *slope*. The *distance* or *difference in slope* $\Delta(r, s)$ for two slopes r and s is taken to be the minimum geometric intersection number of the corresponding simple closed curves. The manifold obtained by Dehn filling along r, *i.e.* gluing a solid torus to ∂M by identifying the boundary of a meridian disc to a curve representing the slope r, will be denoted by $M(r)$, or in the case where M is the exterior of a knot K in the 3-sphere by $K(r)$. In this notation we show:

Theorem 1. *Let K be a knot in the 3-sphere whose exterior contains an essential torus. If r and s are slopes such that the manifolds $K(r)$ and $K(s)$ have finite fundamental group, then $\Delta(r, s) \leq 5$.*

The result is sharp. In §2, we give examples of non-trivial satellite knots and slopes r, s such that $\pi_1(K(r))$ and $\pi_1(K(s))$ are finite and $\Delta(r, s) = 5$. From the bound on the difference in slopes we can immediately draw the following corollary.

Corollary 2. *There exist at most 8 surgeries on a knot in the 3-sphere whose exterior contains an essential torus which yield a manifold with finite fundamental group.*

Part of the proof of the theorem 1 is a complete classification of the finite surgeries on iterated torus knots; this is outlined in §2. Here we will find our examples of finite surgeries a distance 5 apart. The remainder of the proof of the main theorem

* Partially supported by NSF grant DMS 88-05672.

Knots 90

is an analysis in light of the proof of the cyclic surgery theorem [CGLS], recent work of Colin Adams [A], and an unpublished theorem of Thurston and Gromov, the so-called "2π" theorem. In §3 we state and outline a proof of Thurston and Gromov's remarkable theorem. In §4 we give the proof of theorem 1 and show that the number of finite surgeries on a non-torus knot is at most 24, although probably much smaller. Further details of the proofs and new examples of finite surgeries will be given in the paper [BlH].

§1. Introduction

A Dehn surgery on a knot which yields a manifold with finite fundamental group will be called a *finite surgery*, following the convention used for Dehn surgeries which yield a manifold with cyclic fundamental group. Although little seems to be known about general finite surgeries, a great deal is known about cyclic surgeries, and it is useful here to recall some of it.

Torus knots, whose surgeries were classified by Moser [Mo], provided the first examples of cyclic surgeries. In [Mo] it was conjectured that surgery on a non-torus knot could not yield a Seifert fibre space and in particular, could not yield a lens space. Many counter-examples have since been found. For example, Rolfsen and Bailey [BaR] found the first cyclic surgery on a non-torus knot: an example of a lens space arising via an integer surgery on an iterated torus knot. Subsequently, Gordon [Go] classified the manifolds obtained by surgery on iterated torus knots and noted that among these knots there is a single family of cyclic surgeries, the Rolfsen-Bailey example being the "smallest". Fintushel and Stern [FS] found the first examples of a cyclic surgery on a simple non-torus (i.e. hyperbolic) knot and later showed [CGLS] that the $+18$ and $+19$ surgeries on the -2, 3, 7 pretzel knot are both cyclic. Berge [Be] and Gabai [Ga] have independently constructed collections of knots in solid tori such that a certain non-trivial surgery on them yields a solid torus, and therefore yield lens spaces when considered as knots in the 3-sphere.

In 1985 Culler, Gordon, Luecke, and Shalen [CGLS] announced the cyclic surgery theorem. In our notation, the cyclic surgery theorem states that the distance between slopes on a non-torus knot which correspond to cyclic surgeries is no more than one. This has as corollaries that the cyclic surgeries on a non-torus knot are integral and that a non-torus knot can have at most three cyclic surgeries. The -2, 3, 7 pretzel knot is thus a sort of "maximal" example. Of course, torus knots have infinitely many non-integral cyclic surgeries and so must be excluded.

For certain families of knots, the cyclic surgeries are classified. The cyclic surgeries on satellite knots were classified by Bleiler and Litherland [BlL] and are in fact just the iterated torus knots enumerated by Gordon in [Go]. The classification was given independently by Wu [Wu] and by Gonzales-Acuña and Whitten [GW]. Wang [Wa] and Hempel [H1] proved similar theorems.

Passing to finite non-cyclic surgeries, much less is known; although Thurston's hyperbolic Dehn surgery theorem [T1] shows that for a hyperbolic knot the number of such surgeries is bounded. Indeed, the "2π" theorem of Thurston and Gromov, together with a result of Adams gives the bound 24 for the number of finite surgeries on a hyperbolic knot (see §3). Known 3-manifolds with finite non-cyclic fundamental group are all Seifert fibre spaces over the two-sphere with three exceptional fibres. The manifolds have finite fundamental group when the branching indices of the three exceptional fibres form a so-called platonic triple. That is to say, that the branching indices are either 2, 2 and r (the so-called prism manifolds); 2, 3 and 3; 2, 3 and 4; or 2, 3 and 5; see for example [H2]. The spherical space form conjecture asserts that every 3-manifold with finite fundamental group is a Seifert fibre space. There is a unique Seifert fibring with orientable base for each closed Seifert fibred 3-manifold with finite non-cyclic fundamental group; again, see [H2].

As with the lens spaces, these manifolds arise naturally as surgeries on certain torus knots [Mo]. The most familiar example is Dehn's picture of the Poincaré homology sphere arising via $+1$ surgery on the right hand trefoil; the fundamental group is just the binary icosahedral group I_{120}. A less familiar example is $17/2$ surgery on the right hand trefoil, whose fundamental group is the product of the binary icosahedral group with the cyclic group of order 17. It turns out that $+17$ surgery on the -2, 3, 7 pretzel knot is also a finite surgery. In fact, this surgery gives the same oriented manifold as does $17/2$ surgery on the right hand trefoil. The proof uses the Montesinos trick ([M]). The details of this argument and some new examples of finite, non-cyclic surgeries on hyperbolic knots are given in [BlH]. Further examples of finite, non-cyclic surgeries were recently found by Hodgson and Weeks [HW] during their survey of "small" hyperbolic manifolds via Weeks' computer program "SNAPPEA".

§2. The finite surgeries on iterated torus knots

The results of surgery on an iterated torus knot were classified by Gordon in [Go]. While Gordon does not in every case explicitly indicate which manifold arises from a given surgery, the surgeries yielding lens spaces (i.e. the cyclic surgeries) are classified, as are those surgeries yielding Seifert fibre spaces over the 2-sphere with 3 exceptional fibres. He further notes that the Seifert invariants of these manifolds can be worked out from the proof of the classification theorem [Go, 7.5], and that the manifolds obtained by the remaining surgeries on iterated torus knots contain at least one incompressible torus. The strategy we'll use to classify the finite non-cyclic surgeries on iterated torus knots is to begin with a predetermined set of Seifert invariants, work backwards through the proof of [Go, 7.5], and then solve the resulting Diophantine equations. A complete list of solutions is obtained because the numbers involved are small.

To begin we note from [Go, 7.5] that if an iterated torus knot admits a non-trivial surgery yielding a Seifert fibre space over the 2-sphere with 3 exceptional fibres, then the knot is cabled at most twice, *i.e.* the knot is at worst the p_1,q_1 cable of the p_2,q_2 cable of the p_3,q_3 torus knot. Moreover, if m/n surgery on such a knot yields a Seifert fibre space over the 2-sphere with 3 exceptional fibres, then in fact $m = 4p_2q_2 \pm 1$, $n = 1$, $p_1 = 2p_2q_2 \pm 1$, $q_1 = 2$, and the exceptional fibres of the resulting manifold have branching indices p_3, q_3 and d, where $d = |4q_2(p_2 - q_2p_3q_3) \pm 1|$. Since q_2 is at least 2, it follows that d is at least 7, so the triple p_3, q_3, d is never platonic and hence the resulting Seifert fibre space has infinite fundamental group. We conclude that an iterated torus knot admitting a finite surgery is cabled precisely once.

For these knots we have from [Go, 7.5] that if m/n surgery on the p_1, q_1 cable of the p_2, q_2 torus knot yields a Seifert fibre space over the 2-sphere with 3 exceptional fibres, then $m = np_1q_1 + \epsilon$ with $\epsilon = \pm 1$ and the exceptional fibres of the resulting manifold have branching indices p_2, q_2 and d, where $d = |m - np_2q_2q_1^2|$. Again, these indices must form one of the elliptic triples $2, 2, r$; $2, 3, 3$; $2, 3, 4$; or $2, 3, 5$. As p_2 and q_2 are relatively prime, it follows that the pair p_2, q_2 is not $2, 2$; hence $2 \leq d \leq 5$. Further, we can assume that $n \geq 1$, $q_1 \geq 2$.

The equation for d can be rewritten as $\pm d = nq_1(p_1 - p_2q_2q_1) + \epsilon$. Since $d \geq 2$, the sign of the last expression is the same as the sign of $(p_1 - p_2q_2q_1)$, and we have $nq_1|p_1 - p_2q_2q_1| = d - \tilde{\epsilon}$, where $\tilde{\epsilon} = \epsilon\ \mathrm{sign}(p_1 - p_2q_2q_1)$. Since $q_1 \geq 2$ and $d \leq 5$ this gives:

$$1 \leq n \leq \frac{d - \tilde{\epsilon}}{q_1} \leq \frac{5 - \tilde{\epsilon}}{2} \leq 3. \qquad (*)$$

For surgeries giving lens spaces, [Go,7.5] shows that $n = 1$ so $(*)$ still holds. From this we see there are at most 6 finite surgeries on any iterated torus knot. (In fact, there are at most 5 combinations of n and ϵ satisfying $(*)$, as well as the the trivial surgery $m = 1$, $n = 0$ giving the 3-sphere.)

Assume m/n and m'/n' are both non-trivial slopes giving finite surgeries, with $m = np_1q_1 + \epsilon$, $m' = n'p_1q_1 + \epsilon'$. Then the difference in slopes is $\Delta = |mn' - m'n| = |\epsilon n' - \epsilon' n|$. If $\epsilon = \epsilon'$ then we obtain $\Delta = |n' - n| \leq 3 - 1 = 2$, and if $\epsilon = -\epsilon'$ then $\Delta = n + n' \leq \dfrac{5 - \tilde{\epsilon}}{2} + \dfrac{5 - \tilde{\epsilon}'}{2} = 5$.

This leads to the following:

Theorem 3. *On any iterated torus knot, there are at most* 6 *finite surgeries. The maximal distance between finite surgeries is* 5*, and the maximal distance from a* *finite surgery to the meridian or a cyclic surgery is* 3. *All of these situations are realized for the* 11, 2 *and* 13, 2 *cables of the trefoil.*

Remark. In fact, the previous argument shows that the extreme cases occur if and only if $q_1 = 2$, $d = 5$, $\{p_2, q_2\} = \{2, 3\}$ and $|p_1 - p_2q_2q_1| = |p_1 - 12| = 1$, *i.e.* $p_1 = 11$ or 13.

By a similar analysis, we can obtain a complete list of finite surgeries on iterated torus knots. In summary, there are two infinite families of finite surgeries: one yielding lens spaces and one yielding prism manifolds, and 34 "sporadic" instances of finite surgery, most of these on cables of the trefoil. The list includes 10 examples of non-integral finite surgeries; these are the first examples of non-integral finite surgeries on non-torus knots. (See [BlH] for the details.)

§3. Negatively curved Dehn fillings

In this section we sketch a proof of the so-called "2π" theorem of Thurston and Gromov. As usually stated the theorem concerns the number of surgeries that must be excluded from each component of a hyperbolic link complement in order to assure that the manifolds resulting from the remaining surgeries have negatively curved metrics. A well known conjecture in differential geometry would imply that these manifolds are in fact hyperbolic. As a 3-manifold with a negatively curved metric must necessarily have $\mathbf{R}^3$ as its universal cover by the Cartan-Hadamard theorem, it follows that such a manifold has infinite fundamental group. The reader is reminded that the converse is false as the 3-manifolds without negatively curved metrics include the Seifert fibre spaces and manifolds with essential tori and spheres, as well as the manifolds with finite fundamental group (see [S]).

For the purposes of this section we extend our notation for Dehn fillings to 3-manifolds M whose boundary is a union $V_1 \cup \ldots \cup V_v$ of tori and denote by $M(r_1, \ldots, r_v)$ the manifold obtained by Dehn filling along the slopes $(r_1, \ldots, r_v)$.

Theorem 4 (The "2π" Theorem). *Let M be a complete hyperbolic 3-manifold of finite volume and $P_1, \ldots, P_v$ disjoint horoball neighborhoods of the cusps of M. Suppose r_i is a slope on ∂P_i represented by a geodesic α_i with length in the Euclidean metric satisfying* $\text{length}(\alpha_i) > 2\pi$, *for each $i = 1, \ldots, v$. Then $M(r_1, \ldots, r_v)$ has a metric of negative curvature.*

Proof. The idea here is to begin with the complete hyperbolic metric on M, remove large disjoint horoball neighborhoods of the cusps, and try to fill in a negatively curved metric on each filling torus. For this one appeals to the following lemma due to Gromov (compare [GT], [BlH]). It is from this lemma that the theorem gets its name.

Lemma 5. *Let V be a solid torus supplied with a hyperbolic metric near its boundary so that ∂V is the quotient of a horosphere. Then the metric near the boundary can be extended to a negatively curved metric on V provided that the length of the Euclidean geodesic representing the meridian curve on ∂V is at least 2π.*

Remarks. The necessity of this condition easily follows from the Gauss-Bonnet theorem as applied to a meridian disc, noting that the geodesic curvature of a Euclidean geodesic on a horospherical torus is 1. The lemma is proved by constructing a negatively curved metric on V invariant under the action of $S^1 \times S^1$, using the technique of [GT].

Next we outline an argument of Thurston giving an explicit bound on the number of non negatively-curved Dehn fillings on a cusped hyperbolic 3-manifold.

Theorem 6. *Let M be a 3-manifold with boundary components consisting of tori, whose interior admits a complete hyperbolic structure. Then if at most 48 Dehn fillings are excluded for each boundary component, all other Dehn fillings on M admit negatively curved metrics. If the boundary of M consists of a single torus, then there are at most 24 Dehn fillings on M admitting no negatively curved metric.*

Proof. Suppose that M has been decomposed into a compact component and a finite set of cusps, each homeomorphic to $T^2 \times [0, 1)$. Lifting a cusp to hyperbolic space $\mathbf{H}^3$ we obtain an infinite set of disjoint horoballs. Then expand these horoballs equivariantly until two first become tangent. The projection of these expanded horoballs to M is a *maximal cusp* of M. We can arrange that one of these horoballs is centered about the point ∞ in the upper half-space model of $\mathbf{H}^3$ and is the set of points at Euclidean height 1 above the x-y plane. We note that lengths on this horosphere are the same in both the hyperbolic and Euclidean metrics. The other horospheres are, of course, tangent to the boundary plane of upper half space, and some have diameter 1 by maximality. These horospheres are disjoint from their translates under the action of the $\mathbf{Z} \oplus \mathbf{Z}$ subgroup of parabolic isometries fixing the cusp at ∞; so all translations have Euclidean length at least 1. It follows that the shortest geodesic on the corresponding boundary component of M minus the maximal cusps also has length at least 1.

Taking the universal cover of the boundary then, one obtains a lattice in the Euclidean plane $\mathbf{E}^2$ such that the distance between any two lattice points is at least 1, and such that all surgeries corresponding to lattice points outside the circle of radius 2π have negatively curved metrics. The worst possible case (in terms of the number of excluded lattice points) occurs for the equilateral triangular lattice. This gives the bound 48.

In the case of a manifold with one cusp, a recent result of Colin Adams [A,Thm 2] can be used to improve this estimate by a factor of 2. In particular, Adams shows that a fundamental domain for the cusp must contain at least two inequivalent maximal horoballs. Using this information, it is not difficult to show that the worst possible case occurs when the horoball centers form an equilateral triangular lattice with sides of length 1 and the lattice of parabolic translations is an index two subgroup in this. This shows that the number of excluded surgeries per cusp is at most

$$\frac{1}{2}\ \text{card}\ \{(a,b) : a, b \text{ are relatively prime integers, with } |a+\sqrt{-3}b| < 2\pi\} = 24.$$

Remark. Recent work of Cameron Gordon leads to stronger results for 3-manifolds M as in theorem 6, with more than one boundary component. In this case, theorem 6 can be combined with Gordon's work to show that if at most 8 surgeries are excluded for all but one of the boundary tori and at most 24 surgeries are excluded for the other boundary torus then all remaining surgeries admit negatively curved metrics.

§4. A proof of the main theorem

We begin by giving an estimate on the maximum distance between Dehn fillings on a hyperbolic manifold which yield manifolds with no negatively curved metric.

Theorem 7. *Let M be 3-manifold bounded by a torus, whose interior admits a complete hyperbolic structure. Suppose α and β are surgeries on M yielding manifolds with no negatively curved metric. Then $\Delta(\alpha,\beta) < 23$.*

Proof. We consider the maximal cusp picture used in the proof of theorem 6. In particular, note that the geometric intersection number of two (geodesic) curves α, β on a Euclidean torus T is given by dividing the area $A(\alpha,\beta)$ of the parallelogram spanned by the lifts of α and β to the universal cover by the area $A(T)$ of a fundamental domain for the torus. So if α and β correspond to surgeries giving manifolds which do not admit negatively curved metrics, then $A(\alpha,\beta) \leq (2\pi)(2\pi) = 4\pi^2$, and so $\Delta(\alpha,\beta)$ is at most $4\pi^2/A(T)$. The fact that the shortest translation on T has length at least one gives $A(T) \geq \sqrt{3}/2$ with the worst case occuring for the equilateral triangular lattice; hence $\Delta(\alpha,\beta) \leq 45$. Using Adams' result [A] improves this estimate by a factor of 2, giving $A(T) \geq \sqrt{3}$ and $\Delta(\alpha,\beta) < 4\pi^2/\sqrt{3}$. In particular $\Delta(\alpha,\beta) < 23$.

Remark. In fact, this estimate can be improved slightly by examining the configuration of lattice points inside the circle of radius 2π to give $\Delta(\alpha, \beta) < 21$. However, we won't use this refinement and omit the details.

Proof of theorem 1. Assume there is an essential torus T in the knot exterior and finite surgeries α', β' at a distance $d = \Delta(\alpha', \beta') > 5$ apart. From [CGLS, Theorem 2.1] we conclude that T and ∂M cobound a cable space, so our knot is in fact a cable knot with wrapping number q, say, about its companion. From [Go] it follows that α' and β' pass to a pair of finite surgeries α'', β'' on this companion with $\Delta(\alpha'', \beta'') = q^2 d$. Proceeding inductively we come to a simple knot which, by the classification of finite surgeries on iterated torus knots above, must be hyperbolic. As $d \geq 6$ and $q \geq 2$, we then have a hyperbolic knot admitting finite surgeries α, β with $\Delta(\alpha, \beta) > 23$. This contradiction to our previous result establishes the theorem.

Theorem 6 shows that there are at most 24 finite surgeries on a hyperbolic knot. This extends via Corollary 2 to all non-torus knots as follows. Simple non-torus knots are hyperbolic by [T2]. If not simple, the knot exterior contains an incompressible torus, so by Corollary 2 admits at most 8 finite surgeries. So we have proved:

Theorem 8. *There are at most* 24 *finite surgeries on a non-torus knot.*

Remarks. The bounds given by theorems 7 and 8 are undoubtedly not the best possible. Here are some examples giving the "worst" known behavior in the following situations: (a) non-hyperbolic Dehn fillings on hyperbolic manifolds, (b) finite Dehn fillings on hyperbolic manifolds, (c) finite surgeries on hyperbolic knot complements, (d) finite surgeries on non-torus knots.

(a) For the figure eight knot, there are 10 non-hyperbolic surgeries and these satisfy $\Delta(\alpha, \beta) \leq 8$ (see [T1].)

(b) For 5/1 surgery on one component of the (right handed) Whitehead link, there are 5 finite Dehn fillings, satisfying $\Delta(\alpha, \beta) \leq 3$. (This cusped hyperbolic manifold with the same volume as the figure eight complement is discussed extensively in [We].)

(c) For the $-2, 3, 7$ pretzel knot, there are 4 finite surgeries, satisfying $\Delta(\alpha, \beta) \leq 2$.

(d) The examples of iterated torus knots mentioned in theorem 3 admit 6 finite surgeries satisfying $\Delta(\alpha, \beta) \leq 5$.

From this we conjecture that there are at most 6 finite surgeries on a non-torus knot in the 3-sphere. Compare [CGLS, Cor. 2].

Finally, for torally bounded 3-manifolds which contain an essential torus but are not knot complements in S^3 the behavior is more complicated. For example, any manifold obtained by gluing a cable space to the twisted I-bundle over the Klein bottle admits an infinite number of finite Dehn fillings as, with one possible exception, a filling at a distance one from the Seifert fibre of the cable space yields a prism manifold.

References

[A] C. Adams, The noncompact hyperbolic 3-manifold of minimal volume, Proceedings AMS 100, no.4, (1987), 601-606.

[BaR] J. Bailey and D. Rolfsen, An unexpected surgery construction of a lens space, Pacific J. of Mathematics 71 (1977), 295-298.

[Be] J. Berge, The knots in $D^2 \times S^1$ which have nontrivial Dehn surgeries that yield $D^2 \times S^1$, Topology and its Applications 38 (1991), 1-20.

[BlH] S. Bleiler and C. Hodgson, Spherical space forms and Dehn filling, preprint.

[BlL] S. Bleiler and R. Litherland, Dehn surgery and satellite knots, Proceedings AMS 107 No.4 (1990), 1127-1131.

[BuZ] G. Burde and H. Zieschang, Knots, de Gruyter 1985.

[CGLS] M. Culler, C. Gordon, J. Luecke, and P. Shalen, Dehn surgery on knots, Annals of Mathematics 125 (1987), 237-300.

[FS] R. Fintushel and R. Stern, Constructing lens spaces by surgery on knots, Mathematische Zeitschrift 175, (1980), 33-51.

[Ga] D. Gabai, Surgery on knots in solid tori, Topology 28 (1989), 1-6.

[GW] F. Gonzales-Acuña and W. Witten, Imbeddings of three manifold groups, preprint.

[Go] C. Gordon, Dehn surgery on satellite knots, Transactions AMS 275, No.2, (1983), 687-708.

[GT] M. Gromov and W. Thurston, Pinching constants for hyperbolic manifolds, Invent. Math. 89 (1987), 1-12.

[H1] J. Hempel, Dehn fillings of coverings of surface bundles, Topology and its Applications 24 (1986), 63-70.

[H2] J. Hempel, 3-manifolds, Annals of Mathematics Studies 86 (1976), Princeton University Press.

[HW] C. Hodgson and J. Weeks, A census of closed hyperbolic 3-manifolds, in preparation.

[M] J. Montesinos, Surgery on links and double branched coverings of S^3, Annals of Math. Studies 84 (1975), 227-260.

[Mo] L. Moser, Elementary surgery along a torus knot, Pacific J. of Mathematics 38 (1971), 737-745

[S] P. Scott, The geometries of 3-manifolds, Bull. London Math. Soc. 15 (1983), 401-487.

[T1] W. Thurston, The geometry and topology of 3-manifolds, Princeton Univ. Math. Dept., 1979.

[T2] W. Thurston, Hyperbolic structures on 3-manifolds, I, Annals of Math. 124 (1986), 203-246.

[Wa] S. Wang, Cyclic surgery on knots, Proceedings AMS 107 No.4 (1990), 1091-1094.

[We] J. Weeks, Ph.D. Thesis, Princeton University, 1985.

[Wu] P. Wu, Dehn surgery on knots, preprint.

Some remarks on knots and links in spatial graphs

Takashi Kohara and Shin'ichi Suzuki

Abstract. A finite graph G is said to be self-knotted (resp. self-linked) iff for any embedding f of G into the 3-space $\mathbf{R}^3$, there exists a cycle c (resp. a pair of disjoint cycles c_1, c_2) in G such that $f(c)$ is a non-trivial knot (resp. $f(c_1 \cup c_2)$ is a non-trivial link).

Let $\mathcal{F}_{SK}$ (resp. $\mathcal{F}_{SL}$) be the family of all self-knotted (resp. self-linked) graphs, and $\mathcal{F}^c_{SK}$ (resp. $\mathcal{F}^c_{SL}$) the complement of $\mathcal{F}_{SK}$ (resp. $\mathcal{F}_{SL}$) in the family of finite graphs. We discuss $C(\mathcal{F}_{SK})$ (resp. $C(\mathcal{F}_{SL})$), the set of critical graphs with respect to self-knottedness (resp. self-linkedness), and $\Omega(\mathcal{F}^c_{SK})$ (resp. $\Omega(\mathcal{F}^c_{SL})$), the obstruction set for $\mathcal{F}^c_{SK}$ (resp. $\mathcal{F}^c_{SL}$).

1. Introduction

By a *graph* we mean a finite undirected graph. Loops and multiple edges are allowed. We call a *tame* embedding f of a graph G into the 3-dimensional space $\mathbf{R}^3$ a *spatial embedding* of G, and the image $f(G) = G_f$ a *spatial graph* or a *spatial representative* of G. We say that two spatial representatives G_f and G_g of a graph G are *equivalent*, denoted by $G_f \approx G_g$, iff there exists an orientation preserving homeomorphism of $\mathbf{R}^3$ that maps G_f to G_g; that is, G_f and G_g are ambient isotopic.

It is an interesting problem what types of spatial representatives of a graph exist, and there is a lot of characterization problems of spatial graphs.

A *cycle* of a graph G is a subgraph of G which is connected and regular of degree 2; that is, homeomorphic to a circle. A *Hamilton cycle* of G is a cycle which contains every vertex of G. By $\Gamma(G)$ and $\Gamma_H(G)$, we denote the set of all cycles in G and the set of all Hamilton cycles in G, respectively, and by $\Lambda(G)$ we denote the set of all pairs of disjoint cycles in G.

1.1 Definition. A graph G is said to be *self-knotted* (resp. *self-linked*) iff for any spatial embedding f of G, there exists $c \in \Gamma(G)$ (resp. $\{c_1, c_2\} \in \Lambda(G)$) such that

Knots 90

$f(c)$ is a non-trivial knot (resp. $f(c_1 \cup c_2)$ is a non-trivial link). For the definitions and elementary terminology in knot theory, we refer to Rolfsen [7] .

This definition follows from the following memorable results of Sachs [8] and Conway-Gordon [2] .

1.2 Theorem (Sachs [8], cf. Conway-Gordon [2]). *The following eight graphs illustrated in* Figure 1 *are self-linked.* □

Here, we denote by K_n the complete graphs on n vertices, by $K_{n_1,n_2,\cdots,n_k}$ a complete k-partite graph with part sizes $n_1, n_2, \cdots, n_k$, and by $G - e$ a graph obtained from a graph G by removing an edge e.

1.3 Theorem (Conway-Gordon [2]). *The complete graph K_7 is self-knotted.* □

Sachs [8] proved Theorem 1.2 by using the unlinking number of links. Conway-Gordon [2] proved that every spatial representative of K_6 contains an odd number of pairs in $\Lambda(K_6)$ with odd linking number by using the mod 2 linking number of links. By the same method as in Conway-Gordon's proof, we can check that every spatial representative of $K_{4,4}-e, K_{3,3,1}, G_7, G_8, G_9$ and PG contains an odd number of pairs in Λ with odd linking number. To prove Theorem 1.3, Conway-Gordon calculated the sum of Arf-invariants taken over $\Gamma_H(K_7)$, and showed that every spatial representative of K_7 contains an odd number of Hamilton cycles with nonzero Arf-invariant. Shimabara [9] showed that $K_{5,5}$ is self-knotted by using a generalization of Conway-Gordon's method.

We will use the following notation. Let $\mathcal{F}_{SK}$ (resp. $\mathcal{F}_{SL}$) denote the family of self-knotted (resp. self-linked) graphs, and let $\mathcal{F}_{SK}^c$ (resp. $\mathcal{F}_{SL}^c$) the complement of $\mathcal{F}_{SK}$ (resp. $\mathcal{F}_{SL}$) in the family of finite graphs.

1.4 Definition. (1) A graph G is said to be *critical with respect to self-knottedness* (resp. *self-linkedness*), iff $G \in \mathcal{F}_{SK}$ (resp. $G \in \mathcal{F}_{SL}$), the minimum degree $\delta(G) \geq 3$ and any proper subgraph H of G belongs to $\mathcal{F}_{SK}^c$ (resp. $\mathcal{F}_{SL}^c$). Any critical graph is simple and connected.

(2) By $C(\mathcal{F}_{SK})$ (resp. $C(\mathcal{F}_{SL})$) we denote the set of critical graphs with respect to self-knottedness (resp. self-linkedness).

As Sachs asserted in [8] , the proof of Theorems 1.2 and 1.3 actually yield the following:

1.5 Proposition. (1) $C(\mathcal{F}_{SL}) \supset \{K_6, K_{3,3,1}, G_7, G_8, G_9, K_{4,4} - e, PG\}$.

(2) $C(\mathcal{F}_{SK}) \supset \{K_7\}$. □

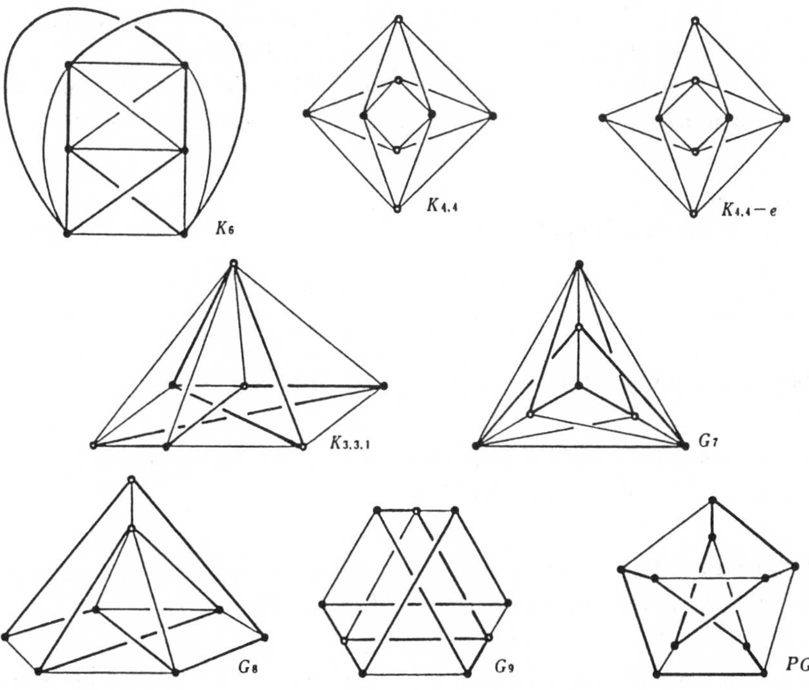

Figure 1

2. Graph minors and obstruction sets

To examine the self-knottedness and self-linkedness of graphs, we need the basic definitions and fact from the theory of graph minors.

2.1 Definition. (1) Let G be a graph and $e = uv$ be an edge of G. The *edge contraction* of G along e is the graph obtained from $G - e$ by indentifying the ends u and v, and is denoted by G/e. We image that the edge e is topologically shrunk to a point.

(2) A graph H is a *minor* of a graph G, denoted by $H \leq G$, iff there exists a subgraph G' of G and its edges $e_1, e_2 \dots e_n$ such that H is isomorphic to the graph which is obtained from G' by a sequence of edge contractions along $e_1, e_2 \dots e_n$. Any subgraph of G is also a minor of G.

2.2 Definition. Let $\mathcal{F}$ denote the family of graphs satisfying the property $P(\mathcal{F})$.

(1) The family $\mathcal{F}$ (or the property $P(\mathcal{F})$) is said to be *closed* under minor-reductions, iff for all $G \in \mathcal{F}$, if $H \leq G$ then $H \in \mathcal{F}$.

(2) An *obstruction set* for $\mathcal{F}$, denoted by $\Omega(\mathcal{F})$, is the set of all minor-minimal graphs which are not in $\mathcal{F}$. Each element of $\Omega(\mathcal{F})$ is called a *forbidden* graph for $\mathcal{F}$.

From the definitions, it holds that

$$\Omega(\mathcal{F}_{SK}^c) \subset C(\mathcal{F}_{SK}),$$
$$\Omega(\mathcal{F}_{SL}^c) \subset C(\mathcal{F}_{SL}).$$

We have the following fundamental properties:

2.3 Proposition (Motwani-Raghunathan-Saran [5]). *The families $\mathcal{F}_{SK}^c$ and $\mathcal{F}_{SL}^c$ are closed under minor-reductions.* □

2.4 Proposition (Robertson-Seymour [6]). *If a family $\mathcal{F}$ of graphs is closed under minor-reduction, then the obstruction set $\Omega(\mathcal{F})$ is finite.* □

By Propositions 2.3 and 2.4, both $\Omega(\mathcal{F}_{SK}^c)$ and $\Omega(\mathcal{F}_{SL}^c)$ are finite. It is easy to see that the seven critical graphs in $C(\mathcal{F}_{SL})$ are members of $\Omega(\mathcal{F}_{SL}^c)$ and $K_7 \in C(\mathcal{F}_{SK})$ is a member of $\Omega(\mathcal{F}_{SK}^c)$; that is,

2.5 Proposition (Motwani-Raghunathan-Saran [5]).
(1) $\Omega(\mathcal{F}_{SL}^c) \supset \{K_6, K_{3,3,1}, G_7, G_8, G_9, K_{4,4} - e, PG\}$,
(2) $\Omega(\mathcal{F}_{SK}^c) \supset \{K_7\}$. □

2.6 Remark. Motwani-Raghunathan-Saran reported in [5, Theorem 2] that $\Omega(\mathcal{F}_{SL}^c)$ consists of exactly these seven graphs, but unfortunately the unproved fundamental lemma [5, Lemma 3] is false and the induction step in the proof of main theorem is incomplete; and so $\Omega(\mathcal{F}_{SL}^c)$ is not yet settled. The following Figure 2 is a counterexample to Lemma 3 of [5] .

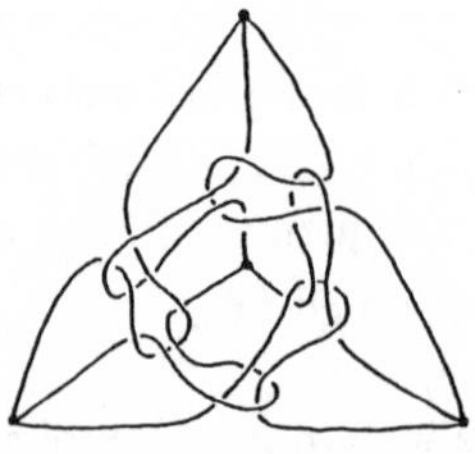

Figure 2

A spatial representative of a planar praph is said to be *unknotted* iff it is equivalent to its planar representative, and to be *almost unknotted* iff it is knotted and every proper subgraph is unknotted. Almost unknotted spatial graphs are given in Kinoshita [4] , Suzuki [10] , Kawauchi [3] etc., and these are also counterexamples.

2.7 Remark. Motwani-Raghunathan-Saran [5] remarked that using similar technique of that of Conway-Gordon [2] , they established that $K_{3,3,1,1}$ is also in $\Omega(\mathcal{F}^c_{SK})$. We cannot however believe this remark, because $\Gamma = \Gamma_H(K_{3,3,1,1})$ or any other significant subset Γ of $\Gamma(K_{3,3,1,1})$ does not satisfy the condition for the sum of Arf-invariants taken over $\Gamma \pmod 2$ to be independent to its spatial representatives. The spatial representative of $K_{3,3,1,1}$ in Figure 3 (a) contains only one trefoil as a Hamilton cycle, and the other in Figure 3 (b) contains exactly two trefoils as Hamilton cycles.

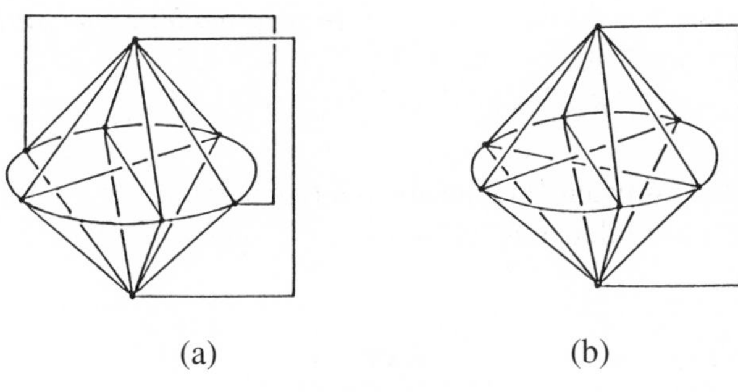

(a) (b)

Figure 3

3. ΔY transformations

Motwani-Raghunathan-Saran [5] used socalled triangle-star transformations of graphs to examine spatial graphs.

3.1 Definition. Let $G = (V, E)$ be a graph having a triangle (i.e. a 3 cycle) $\triangle uvw$. Let G' denote the graph obtained from G by replacing the edges uv, vw and wu of the triangle by a degree three vertex x with the incident edges xu, xv and xw (see Figure 4). We say that G' was obtained from G by a *triangle-star transformation* (or *delta-wye transformation*; in abbreviated notation: ΔY transformations). The inverse transformation will be called the *star-triangle transformation* (or *wye-delta transformation*: YΔ transformations).

To each cycle $c' \in \Gamma(G')$ that contains x, there corresponds a unique cycle $c \in \Gamma(G)$ obtained by the path uxv, vxw or wxu in c' replaced by the edge uv, vw

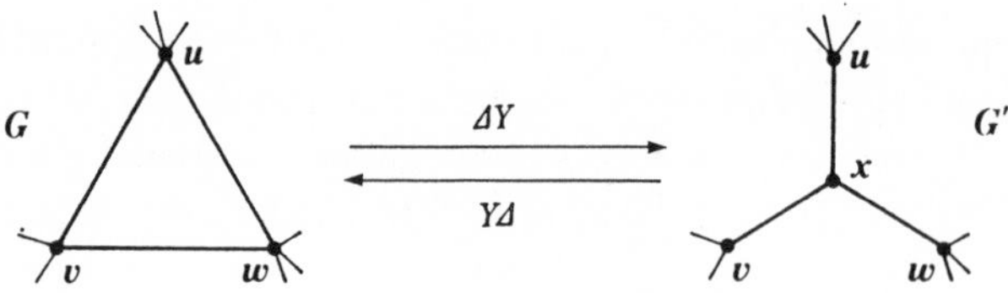

Figure 4

or wu, respectively. Since every cycle $c' \in \Gamma(G')$ that does not contain x is also a cycle in $\Gamma(G)$, this natural correspondence $\gamma : \Gamma(G') \longrightarrow \Gamma(G)$ is clearly an injection. The following fact illustrates the importance of the ΔY transformations.

3.2 Proposition (Motwani-Raghunathan-Saran [5, Lemma 2]). *Suppose that G' be a graph obtained from a graph G by a ΔY transformation. Then, it holds that*

(1) $G \in \mathcal{F}_{SL} \Longrightarrow G' \in \mathcal{F}_{SL}$, $G' \in \mathcal{F}_{SL}^c \Longrightarrow G \in \mathcal{F}_{SL}^c$

(2) $G \in \mathcal{F}_{SK} \Longrightarrow G' \in \mathcal{F}_{SK}$, $G' \in \mathcal{F}_{SK}^c \Longrightarrow G \in \mathcal{F}_{SK}^c$ □

3.3 Remark. As Motwani-Raghunathan-Saran pointed out in [5] , the seven graphs in Figure 1 are related as follows:

$$\begin{array}{ccccccc} K_6 & \xrightarrow{\Delta Y} & G_7 & \xrightarrow{\Delta Y} & K_{4,4} - e & & \\ & & & \searrow^{\Delta Y} & & & \\ & & K_{3,3,1} & \xrightarrow{\Delta Y} & G_8 & \xrightarrow{\Delta Y} G_9 \xrightarrow{\Delta Y} & PG, \end{array}$$

where $K_{4,4} - e$ and PG are triangle free.

Proposition 3.2 implies that the application of the ΔY transformation to K_7 would yield other self-knotted graphs. By repeatedly applying this transformation until there are no triangles left, we obtain thirteen more graphs which are self-knotted. Spatial representatives of these thirteen graphs are shown in Figure 5. As Motwani-Raghunathan-Saran also pointed out in [5] , we have the following:

3.4 Theorem. *The thirteen graphs that are obtained from K_7 by repeatedly applying ΔY transformation are forbidden for $\mathcal{F}_{SK}^c$.* □

Sketch of proof of this theorem will be given in §4.

3.5 Remark. It is simple to check that if $K_{3,3,1,1} \in \mathcal{F}_{SK}$ then $K_{3,3,1,1} \in \Omega(\mathcal{F}_{SK}^c)$. By repeatedly applying ΔY transformation to $K_{3,3,1,1}$ we obtain twenty-five more graphs. By the same way as that of Theorem 3.4, we can check that these twenty-five graphs are all forbidden for $\mathcal{F}^c_{SK}$, provided that $K_{3,3,1,1} \in \Omega(\mathcal{F}^c_{SK})$.

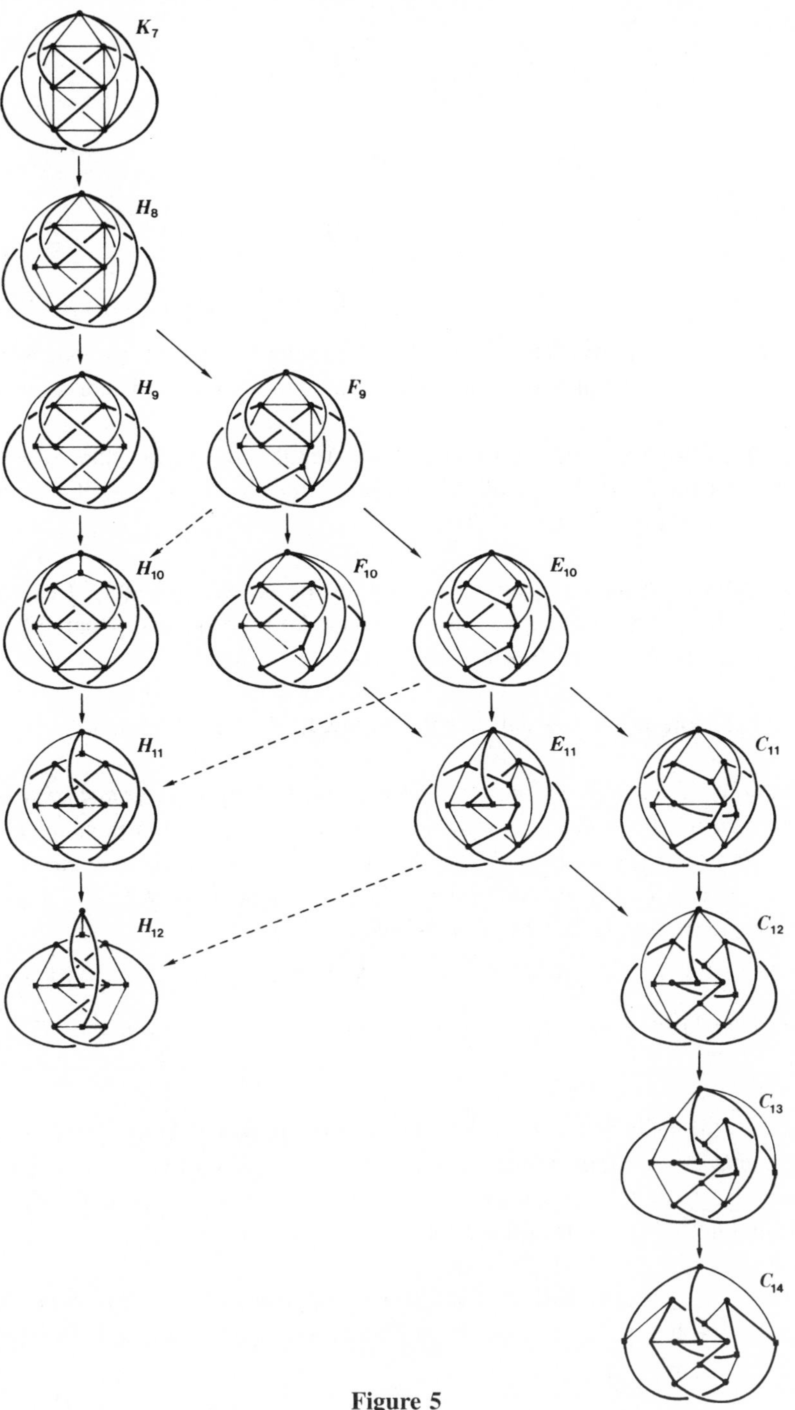

Figure 5

4. Sketch of proof of Theorem 3.4

It will be noticed that K_7 and the thirteen graphs in Theorem 3.4 are related as follows:

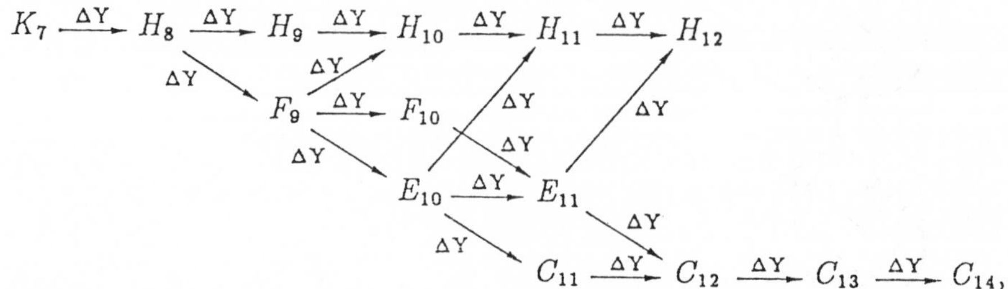

where H_{12} and C_{14} are triangle free. By comparing the degree sequences of these fourteen graphs, we can easily check that these graphs are not isomorphic to each other.

By Propositon 3.2, these thirteen graphs are all self-knotted, and so it suffices to show that these graphs are all minor minimal with respect to self-knottedness.

We will use the following definition.

4.1 Definition. A spatial graph G_f is said to be *locally unknotted* (resp. *locally unlinked*), iff every cycle in G_f is the trivial knot (resp. every pair of disjoint cycles in G_f is the trivial link).

To prove Theorem 3.4, we will use the following technical lemma.

Lemma. *Let G' be a graph obtained from a graph G by a ΔY transformation at a triangle $\triangle uvw$. We suppose that for every edge $e \in E(G) - \{uv, vw, wu\}$,*

(1) *there exist spatial embeddings f of $G - e$ and g of G/e such that $(G-e)_f$ and $(G/e)_g$ are locally unknotted (resp. locally unlinked), and*

(2) *there exist disks Δ and Δ' in $\mathbf{R}^3$ with*

$$\Delta \cap (G-e)_f = \partial\Delta = (\triangle uvw)_f, \quad \Delta' \cap (G/e)_g = \partial\Delta' = (\triangle uvw)_g.$$

If G is forbidden for $\mathcal{F}^c_{SK}$ (resp. $\mathcal{F}^c_{SL}$), then G' is also forbidden for $\mathcal{F}^c_{SK}$ (resp. $\mathcal{F}^c_{SL}$). □

Proof. By Proposition 3.2, G' is self-knotted (resp. self-linked). Thus, it suffices to show that G' is minor minimal, that is, for any edge e of G', there exist spatial embeddings f' of $G' - e$ and g' of G'/e such that $(G'-e)_{f'}$ and $(G'/e)_{g'}$ are locally unknotted (resp. locally unlinked).

Case 1. $e \in \{xu, xv, xw\} \subset E(G')$: We suppose that $e = xu$. Since G is a forbidden graph for $\mathcal{F}^c_{SK}$ (resp. $\mathcal{F}^c_{SL}$), there exist spatial embeddings f of $G -$

$\{uv, wu\}$ and g of $G - vw$ such that $(G - \{uv, wu\})_f$ and $(G - vw)_g$ are locally unknotted (resp. locally unlinked). Since $G' - e$ is homeomorphic to $G - \{uv, wu\}$ and G'/e is isomorphic to $G - vw$, required spatial embeddings f' of $G' - e$ and g' of G'/e are easily obtained from f and g, respectively.

Case 2. $e \in E(G') - \{xu, xv, xw\}$: From the conditions in Lemma, we have a spatial embedding f' of G' such that $f'|G' - \{xu, xv, xw\} = f|G - \{uv, vw, wu\}$ and $f'(xu \cup xv \cup xw)$ is contained in Δ. It follows from the natural correspondence of $\Gamma(G')$ to $\Gamma(G)$ that $(G - e)_{f'}$ is locally unknotted (resp. locally unlinked).

By the same way as that of f', we have a spatial embedding g' of G'/e such that $g'|G' - \{xu, xv, xw\} = g|G - \{uv, vw, wu\}$ and $g'(xu \cup xv \cup xw)$ is contained in Δ'. It follows also from the natural correspondence of $\Gamma(G'/e)$ to $\Gamma(G/e)$, that $(G'/e)_{g'}$ is locally unknotted (resp. locally unlinked).

This completes the proof of Lemma 4.3. □

Using Lemma 4.3, we prove Theorem 3.4 inductively according as the sequences $K_7 \xrightarrow{\Delta Y} H_8 \xrightarrow{\Delta Y} H_9 \xrightarrow{\Delta Y} H_{10} \xrightarrow{\Delta Y} H_{11} \xrightarrow{\Delta Y} H_{12}$, $H_8 \xrightarrow{\Delta Y} F_9 \xrightarrow{\Delta Y} F_{10}$, $F_9 \xrightarrow{\Delta Y} E_{10} \xrightarrow{\Delta Y} E_{11}$ and $E_{10} \xrightarrow{\Delta Y} C_{11} \xrightarrow{\Delta Y} C_{12} \xrightarrow{\Delta Y} C_{13} \xrightarrow{\Delta Y} C_{14}$.

Let G and G' be graphs in above sequences with $G \xrightarrow{\Delta Y} G'$. Let $Aut(G)$ be the automorphism group of G. Two edges e and e' of G are said to be equivalent, iff there is an element $\sigma \in Aut(G)$ with $\sigma(e) = e'$. Then, there is a partition of $E(G)$ into non-empty subsets $E_1, E_2, \cdots, E_m$ such that two edges e and e' are equivalent if and only if both e and e' belong to the same set E_i.

To prove $G' \in \Omega(\mathcal{F}^c_{SK})$, for some e in each E_i we construct spatial embeddings f of $G - e$ and g of G/e satisfying the conditions (1) and (2) in Lemma 4.3. □

4.4 Remark. Every spatial representative shown in Figure 5 has all cycles unknotted except one, which is a trefoil knot. It should be noted that this trefoil is not a Hamilton cycle except K_7 and C_{14}.

4.5 Remark. $\Omega(\mathcal{F}^c_{SL}) \subset C(\mathcal{F}_{SL})$, $\Omega(\mathcal{F}^c_{SK}) \subset C(\mathcal{F}_{SK})$. An example in $C(\mathcal{F}_{SL}) - \Omega(\mathcal{F}^c_{SL})$ is shown in Figure 6, and an example in $C(\mathcal{F}_{SK}) - \Omega(\mathcal{F}^c_{SK})$ is in Figure 7.

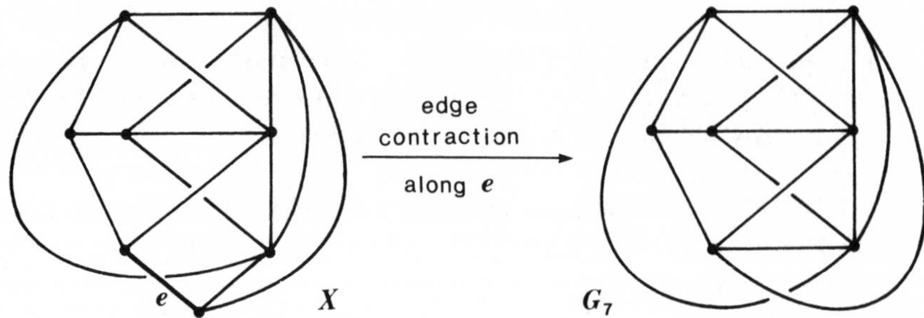

Figure 6

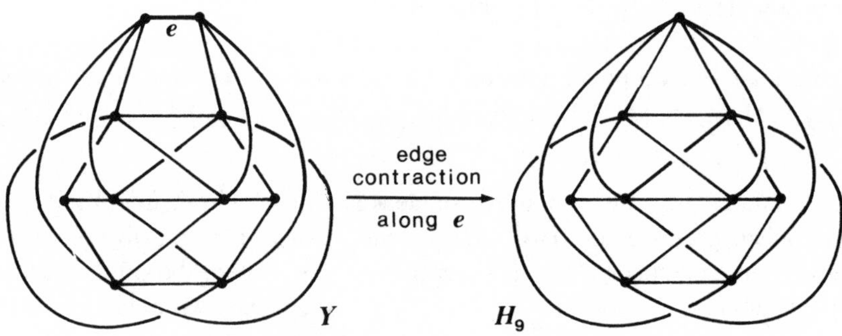

Figure 7

References

[1] J.A.Bondy and U.S.R.Murty : Graph Theory with Applications, North Holland, New York, 1976.

[2] J.H.Conway and C.McA.Gordon : Knots and links in spatial graphs, J.Graph Theory, 7(1985), 445-453.

[3] A.Kawauchi : Almost identical imitations of $(3,1)$-dimensional manifold pairs, Osaka J.Math., 26(1989), 743-758.

[4] S.Kinoshita : On elementary ideals of θ-curves in the 3-sphere, Pacific J.Math., 42(1972), 89-98.

[5] R.Motwani, A.Raghunathan and H.Saran : Constructive results from graph minors ; Linkless embeddings, in 29th annual Symposium on Foundations of Computer Science, 1988, IEEE, 398-409.

[6] N.Robertson and P.D.Seymour : Graph minors XVI. Wagner's conjecture, to appear.

[7] D.Rolfsen : Knots and Links, Math.Lecture Series 7, Publish or Perish Inc., Berkeley, 1976.

[8] H.Sachs : On a spatial analogue of Kuratowski's theorem on planar graphs – an open problem, in Graph Theory, Lagow 1981, Lecture Notes in Math., #1018, Springer-Verlag, Berlin-Heidelberg, 1983, 230-241.

[9] M.Shimabara : Knots in certain spatial graphs, Tokyo J.Math., 11(1988), 405-413.

[10] S.Suzuki : Almost unknotted θ_n-curves in 3-sphere, Kobe J.Math., 1(1984), 19-22.

A topological invariant of spatial regular graphs

Shûji Yamada

Introduction

In this paper we will introduce a polynomial invariant for 3-regular spatial graph with some good weight. This invariant is a combination of Kauffman's bracket polynomials and it is an extension of the polynomial invariant of spatial graph introduced in [4]. In the conference in Osaka, I also introduced some complex valued topological invariants for 4-regular and 5-regular spatial graphs. But, it becomes clear by a global consideration that those complex valued invariant vanish into zero. So, I am sorry my mistake.

In this paper, a graph is the total space of a finite 1-dimensional cell complex. For each point p of a graph G, we say that the degree of p is k and denote $\deg(p) = k$ if p has a simply connected neighbourhood U such that $U \setminus p$ consists of just k connected components. If $\deg(p) \neq 2$ then we say p is a vertex of G. If every vertex of G has the same degree k then G is said to be k-regular graph. $V(G)$ denotes the set of verticies of G. Each connected component of $G \setminus V(G)$ is said a edge of G and $E(G)$ denotes the set of edges of G. A spatial graph is a PL-embedded graph in $\mathbb{R}^3$.

A diagram of a spatial graph G is a regular projection image of G on a plane $\mathbb{R}^2$, whose multiple points set consists of finite transverse double points of edges. The next is a well-known fact. Let D_1 and D_2 be some diagrams of spatial graphs G_1 and G_2 respectively. Then G_1 is ambient isotopic to G_2 in $\mathbb{R}^3$ if and only if D_1 can be translated to D_2 by a finite sequence of extended Reidemeister moves which are illustrated in Figure 1.

Knots 90

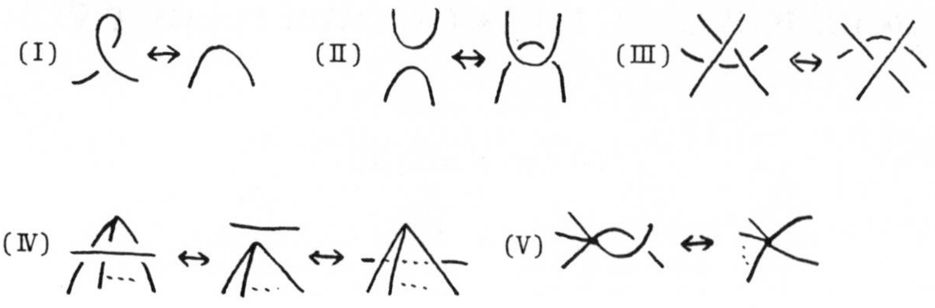

Extended Reidemeister Moves

Figure 1

The Temperley-Lieb algebra

The m-th Temperley-Lieb algebra V_m is an algebra over $\mathbb{C}$ generated by the elements 1_m, e_1, e_2, $\ldots$ e_{m-1} and has the following relations.

(1) $$e_i e_j = e_j e_i \ (|i-j| \geq 2),$$

(2) $$e_i^2 = \delta e_i,$$

(3) $$e_i e_{i\pm 1} e_i = e_i$$

here $\delta = -q - q^{-1}$ and $q \in \mathbb{C}$.

The geometrical meaning of e_i is shown in Figure 2.

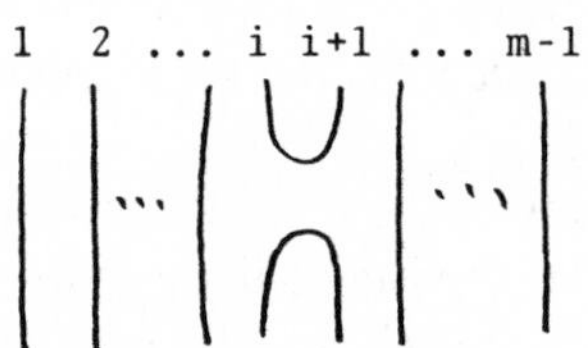

Figure 2

Let $p_1, \ldots, p_{2m}$ be $2m$ points on the boundary of a disk D^2 and D_m be the set of configurations of m disjoint arcs in D^2 joining these $2m$ points. We can regard an element of V_m as a linear combination of elements of D_m. So that, $\dim_{\mathbb{C}}(V_m)$ is the m-th Catalan number $C_m = \frac{1}{m+1}\binom{2m}{m}$.

We shall define the Gaussian polynomials, which are symmetric Laurent polynomial with indeterminate q. For any non-negative integer m, set

$$[m] = \frac{q^m - q^{-m}}{q - q-1} = q^{m-1} + q^{m-3} + \cdots + q^{3-m} + q^{1-m}.$$

Note that $[0] = 0$, $[1] = 1$ and $[2] = -\delta$. We shall use the next notation, $[m]! = \prod_{j=1}^{m}[j]$. Specially, let $[0]! = 1$. For any integers m and n, the Gaussian polynomial is defined by

$$\begin{bmatrix} m \\ n \end{bmatrix} = \begin{cases} \frac{[m]!}{[n]![m-n]!} & \text{if } 0 \le n \le m, \\ 0 & \text{otherwise.} \end{cases}$$

The above polynomials have the following properties.

Proposition 1.

(1) $$[m+n] = q^{-n}[m] + q^{m}[n] = q^{n}[m] + q^{-m}[n],$$

(2) $$[m-1] + [m+1] = [2][m] = -\delta[m],$$

(3) $$\begin{bmatrix} m \\ n \end{bmatrix} = q^{-n}\begin{bmatrix} m-1 \\ n \end{bmatrix} + q^{m-n}\begin{bmatrix} m-1 \\ n-1 \end{bmatrix} = q^{n}\begin{bmatrix} m-1 \\ n \end{bmatrix} + q^{n-m}\begin{bmatrix} m-1 \\ n-1 \end{bmatrix}.$$

The reader shall note that $[m] = (-1)^{m-1}\Delta_{m-1}$, here Δ_{m-1} is the polynomial defined in [2] with $\delta = -q - q^{-1}$.

For a positive integer $n < m$, define elements g_n and f_n of V_m by

$$g_n = \frac{1}{[n]}\sum_{j=1}^{n}[j]e_j \cdots e_{n-1}, = \frac{[1]}{[n]}e_1 \cdots e_{n-1} + \frac{[2]}{[n]}e_2 \cdots e_{n-1} + \cdots + \frac{[n-1]}{[n]}e_{n-1} + 1,$$

$$f_n = g_n g_{n-1} \cdots g_1.$$

g_n and f_n hold the following recursive formulas,

$$g_{n+1} = \frac{[n]}{[n+1]}g_n e_n + 1,$$

$$f_{n+1} = g_{n+1} f_n.$$

It follows immediately from the definitions that $g_n - 1$ and $f_n - 1$ are elements in the proper subalgebra $\mathcal{V}(e_1, \cdots, e_{n-1})$ generated by $e_1, \cdots, e_{n-1}$. So that, g_n and f_n are commute with $e_{n+1}, \cdots, e_{m-1}$. An elegant proof of the next theorem due to Wenzl [3] is given by Lickorish [2]. But we give an alternate proof.

Theorem. *f_n is the unique element in V_m which is generated by $1, e_1, \cdots, e_{n-1}$ and holds the following,*

(1) $$e_i f_n = f_n e_i = 0, \qquad \textit{for all} \quad i \le n-1,$$

(2) $$f_n^2 = f_n.$$

We say f_n is the magic knitting of degree n.

Proof. We shall prove (1) by an induction on n. If $n = 1$ then it is trivial. Let $n \geq 2$. From the hypothesis of the induction, we get

$$e_i e_j \cdots e_{n-1} f_{n-1} = \begin{cases} e_j \cdots e_{i-3} e_i \cdots e_{n-1} e_{i-2} f_{n-1} = 0, & \text{if} \quad j \leq i-2, \\ e_i \cdots e_{n-1} f_{n-1}, & \text{if} \quad j = i-1, \\ \delta e_i \cdots e_{n-1} f_{n-1}, & \text{if} \quad j = i, \\ e_i \cdots e_{n-1} f_{n-1}, & \text{if} \quad j = i+1, \\ e_j \cdots e_{n-1} e_i f_{n-1} = 0, & \text{if} \quad j \geq i+2. \end{cases}$$

So that, for any $i \leq n-1$,

$$\begin{aligned} e_i f_n &= e_i g_n f_{n-1} \\ &= e_i\Big(\frac{1}{[n]} \sum_{j=1}^{n} [j] e_j \cdots e_{n-1}\Big) f_{n-1} \\ &= \frac{[i-1] + \delta[i] + [i+1]}{[n]} e_i \cdots e_{n-1} f_{n-1} = 0, \qquad \text{by Prop.1(2).} \end{aligned}$$

By the hypothesis of the induction, if $i \leq n-2$ then $f_n e_i = g_n f_{n-1} e_i = 0$. So, we shall show that $f_n e_{n-1} = 0$. Note that $g_{n-2} f_{n-2} = f_{n-2}$, then

$$\begin{aligned} f_n e_{n-1} &= g_n g_{n-1} f_{n-2} e_{n-1} \\ &= g_n g_{n-1} e_{n-1} f_{n-2} \\ &= \frac{1}{[n]}([n-1] g_{n-1} e_{n-1} + [n]) g_{n-1} e_{n-1} f_{n-2} \\ &= \frac{1}{[n]} g_{n-1}([n-1] e_{n-1} g_{n-1} + [n]) e_{n-1} f_{n-2} \\ &= \frac{1}{[n]} g_{n-1}([n-1] e_{n-1}(\frac{[n-2]}{[n-1]} g_{n-2} e_{n-2} + 1) + [n]) e_{n-1} f_{n-2} \\ &= \frac{1}{[n]} g_{n-1}([n-2] g_{n-2} e_{n-1} e_{n-2} e_{n-1} + [n-1] e_{n-1}^2 + [n] e_{n-1}) f_{n-2} \\ &= \frac{1}{[n]} g_{n-1}([n-2] g_{n-2} e_{n-1} + \delta[n-1] e_{n-1} + [n] e_{n-1}) f_{n-2} \\ &= \frac{1}{[n]} g_{n-1} e_{n-1}([n-2] + \delta[n-1] + [n]) f_{n-2} \\ &= 0. \end{aligned}$$

Since $f_n - 1 \in \mathcal{V}(e_1 \cdots e_{n-1})$, $f_n(f_n - 1) = 0$. So that, $f_n^2 = f_n$.

The uniqueness is shown as follows. Let h_n be an element which satisfy the conditions of the theorem. Since $h_n^2 = h_n$, $h_n - 1$ is in $\mathcal{V}(e_1 \cdots e_{n-1})$. So, $h_n = h_n f_n = f_n$. Now we complete the proof.

Good weight and the invariant

Let G be a 3-regular graph and $\mu : E(G) \to \mathbb{Z}_+$ be a positive integer valued map. We say μ is a good weight on G if the following conditions are satisfied: For each vertex of G, if e_1, e_2 and e_3 are the edges incident with the vertex then

(1) $$\mu(e_1) + \mu(e_2) + \mu(e_3) \in 2\mathbb{Z},$$

(2) $$|\mu(e_1) - \mu(e_2)| \leq \mu(e_3) \leq \mu(e_1) + \mu(e_2).$$

Let μ be a good weight on a spatial 3-regular graph G and D be a diagram of G. We define D^μ as the linear combination of link diagrams derived from D by the following.

(1) Parallelize each edge e by the weight $\mu(e)$ and immerse the magic knitting of degree $\mu(e)$ as in Figure 3.

(2) Connect those parallelized edges at each vertex as in Figure 4.

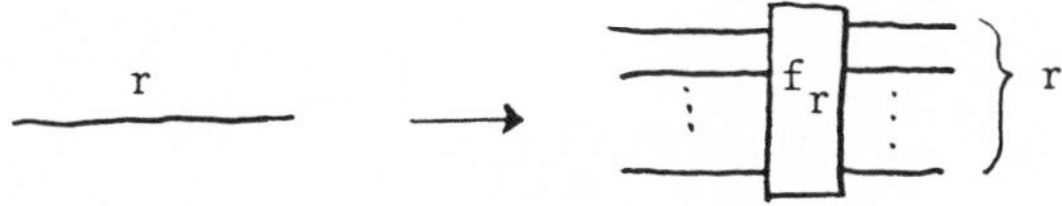

Figure 3

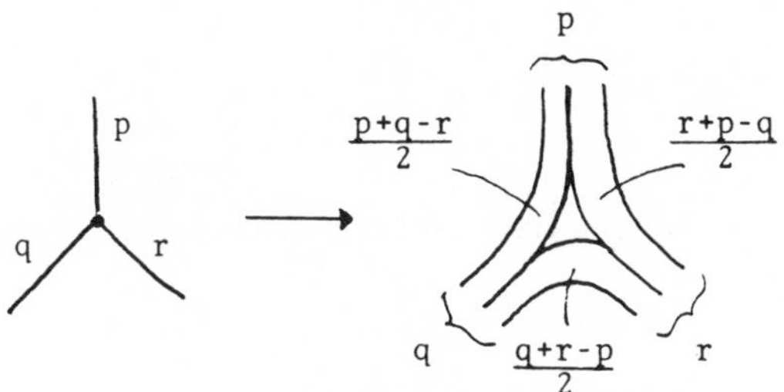

Figure 4

The bracket polynomial $< L >$ of a link diagram L was introduced in [1]. This invariant is defined by,

(1) $< \times >= A <)(> + A^{-1} < \asymp >,$

(2) $< \underbrace{\bigcirc \cdots \bigcirc}_{n} >= (-A^2 - A^{-2})^n.$

We define $< D^\mu >$ by the linear extension of the bracket polynomial.

Theorem. *$< D^{\mu} >$ is an invariant of the weighted spatial graph G up to multiplying $\pm A^n$.*

Proof. We shall show the invariance under the Reidemeister moves on D. (II), (III) and (IV) are clear from the invariance of the bracket polynomial under the regular isotopy. (I) and (V) is shown as follows. Note that the half twist of a weighted edge becomes as follows.

$$= A^{\frac{r(r-1)}{2}}$$

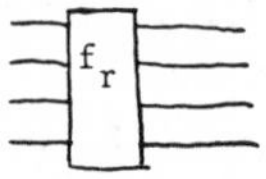

So we get,

(1)  $= (-1)^r A^{r(r+2)}$.

(2)  $= (-1)^s A^{s(s+1)+t}$,

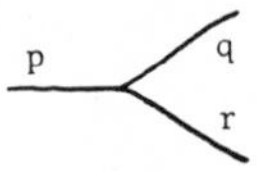

where $s = \frac{q+r-p}{2}$, $t = \frac{q^2+r^2-p^2}{2}$.

The next proposition describes the state sum of a crossing of two edges with some weights.

Proposition. *Let $p \geq q$, then*

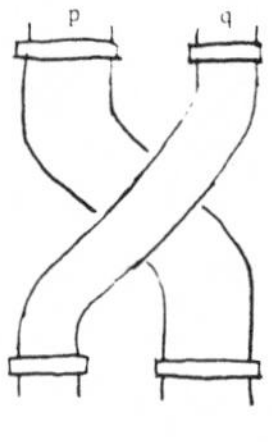

$$= \sum_{j=0}^{q} a_j^{p,q}$$.

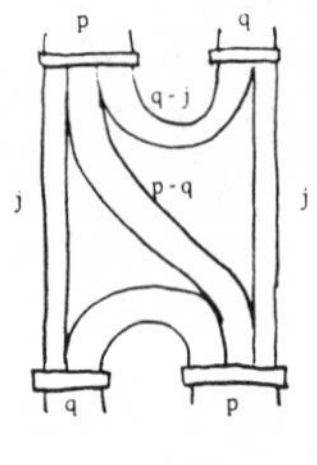

Here, $a_j^{p,q} = q^{p(j-q/2)} \begin{bmatrix} q \\ j \end{bmatrix}$.

Proof. Proof is an induction on n. If $n = 0$, there is only one term in the summention and the coefficent is $\begin{bmatrix} 0 \\ 0 \end{bmatrix} = 1$. If $n > 0$ then the left side is equal to

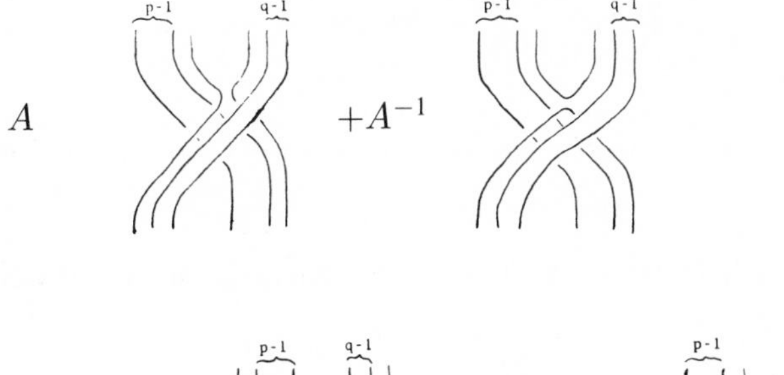

$$= A^{p+q-1} \qquad\qquad + A^{-p-q+1}$$

$$= \sum_{j=0}^{q} \left(q^{\frac{p+q-1}{2}} a_{j-1}^{p-1,q-1} + q^{\frac{-p-q+1}{2}} a_{j}^{p-1,q-1} \right) \quad .$$

We shall show that the coefficient in the summention of the terminal side is equal to $a_j^{p,q}$.

$$\begin{aligned}
&q^{\frac{p+q-1}{2}} a_{j-1}^{p-1,q-1} + q^{\frac{-p-q+1}{2}} a_{j}^{p-1,q-1} \\
&= q^{\frac{p+q-1}{2}+(p-1)(j-1-\frac{q-1}{2})} \begin{bmatrix} q-1 \\ j-1 \end{bmatrix} + q^{\frac{-p-q+1}{2}+(p-1)(j-\frac{q-1}{2})} \begin{bmatrix} q-1 \\ j \end{bmatrix} \\
&= q^{p(j-\frac{q}{2})} \left(q^{q-j} \begin{bmatrix} q-1 \\ j-1 \end{bmatrix} + q^{-j} \begin{bmatrix} q-1 \\ j \end{bmatrix} \right) \\
&= q^{p(j-\frac{q}{2})} \begin{bmatrix} q \\ j \end{bmatrix} = a_j^{p,q}.
\end{aligned}$$

We complete the proof.

Now, we shall see the relation between this invariant and the polynomial invariant R of spatial graph introduced in [4]. The invariant R is defined by the following recursive formulas.

(1) $R(\ ) = xR(\ )(\ ) + x^{-1}R(\ ) + R(\ X \ )$,

(2) $R(\ ) = R(\ ) + R(\ )(\ )$, where e is a nonloop edge,

(3) $R(B_n) = -(-\sigma)^n$, where B_n is the n-leafed bouquet (the one point wedge of n circles) and $\sigma = x + 1 + x^{-1}$.

Let D be a diagram of a spatial 3-regular graph with the good weight defined $\mu(e) = 2$ for each edge e. Then $< D^\mu >$ holds the following similar formulas.

(1) $< \ > = A^4 < \ )(\ > + A^{-4} < \ > + (A^2 + A^{-2}) < \ X \ >$,

(2) $< \ > = < \ > - \delta^{-1} < \ )(\ >$,

(3) $< B_n^\mu > = -(-A^4 - 1 - A^{-4})^n$.

Now we get the next theorem.

Theorem. *Let D be a diagram of a spatial 3-regular graph G with the good weight defined $\mu(e) = 2$ for each edge e. Then*

$$< D^\mu > (A) = (-\delta)^{\chi(G)} R_D(A^4),$$

where $\chi(G)$ the Euler characteristic of G.

References

1. L.H.Kauffman, *State models and the Jones polynomial*, Topology **26** (1987), 395–407.
2. W.B.R.Lickorish, *Three-manifolds and the Temperley-Lieb Algebra.*
3. H.Wenzl, *On sequences of projections*, C.R. Math. Rep. Acad. Sci. Canada IX (1987), 5–9.
4. S.Yamada, *An invariant of spatial graphs*, J. Graph Theory **13** (1989), 537–551.

On local relations to determine the multi-variable Alexander polynomial of colored links

Jun Murakami

§1. Introduction

A set of local axioms to determine the multi-variable (non–reduced) Alexander polynomials of colored links is introduced in [**3.**]. The aim of this paper is to give a complete proof to show that the relations in [**3.**] determine the multi–variable Alexander polynomial uniquely.

Let Δ_L be the (non–reduced) Alexander polynomial of a link L. For an n–component link L, Δ_L is an n–variable Laurent polynomial and the variables correspond to the connected components of L. At first, Δ_L is defined by using the Alexander module and so Δ_L is determined up to the multiple of monomials. Conway [**1.**] showed some local relations for the Alexander polynomial. A local relation means that a relation of polynomials of links which are equal except within a ball. For the reduced Alexander polynomial, Conway's relations are given as follows.

$$\Delta_{L_+} - \Delta_{L_-} = (t - t^{-1})\,\Delta_{L_0}, \qquad \Delta_O = 1,$$

where the links L_+, L_- and L_0 are identical except within a ball where they are as in Figure 1 and O denote the trivial knot. In this case, Δ_L is determined from Conway's relations without ambiguity. Sometimes this polynomial is called the Conway polynomial.

Knots 90

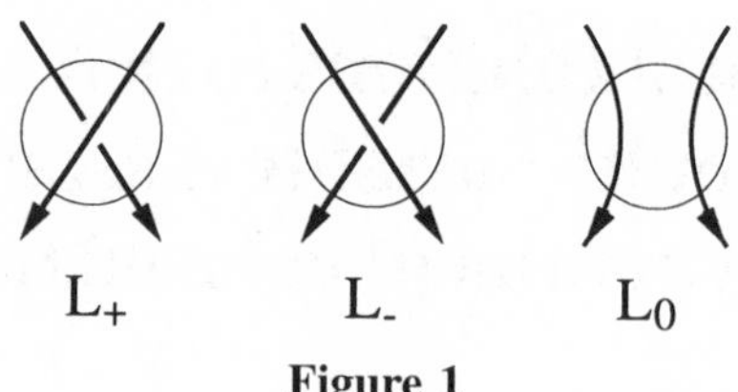

Figure 1

Now let us consider the case of colored links. A colored link is a link with mapping ϕ from the set of connected components of L to the set of positive integers $\mathbf{N}$. The value of ϕ is called the *color* of the component. For every $c \in \mathbf{N}$, we introduce a parameter t_c. Δ can be generalized to an invariant of a colored link L by substituting t_{c_i} to t_i in Δ_L where t_{c_i} is the parameter corresponding to the color c_i of the i–th component of L. The Conway's relations for the Alexander polynomial of colored links are given as follows.

(RI) $$\Delta_{L_+} - \Delta_{L_-} = (t_c - t_c^{-1})\,\Delta_{L_0},$$

where L_+, L_-, L_0 are identical except within a ball where they are as in Figure 2 and c denotes the color of the component of the string.

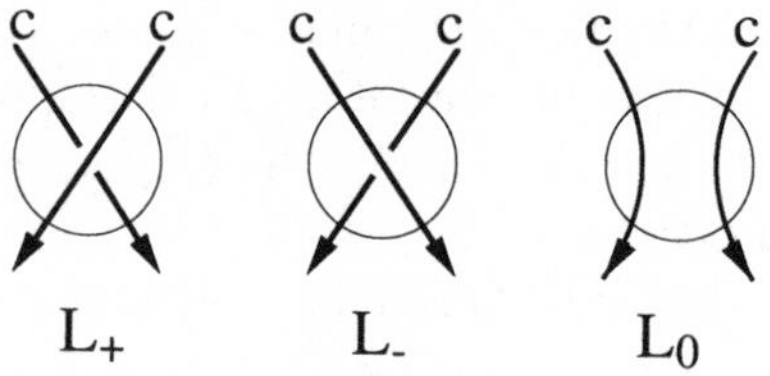

c c c c c c

L+ L- L0

Figure 2

(RII) $$\Delta_{L_{++}} + \Delta_{L_{--}} = (t_c\,t_d + t_c^{-1}\,t_d^{-1})\,\Delta_{L_0},$$

where L_{++}, L_{--}, L_{00} are identical except within a ball where they are as in Figure 3 and c, d denote the colors of the components of the strings.

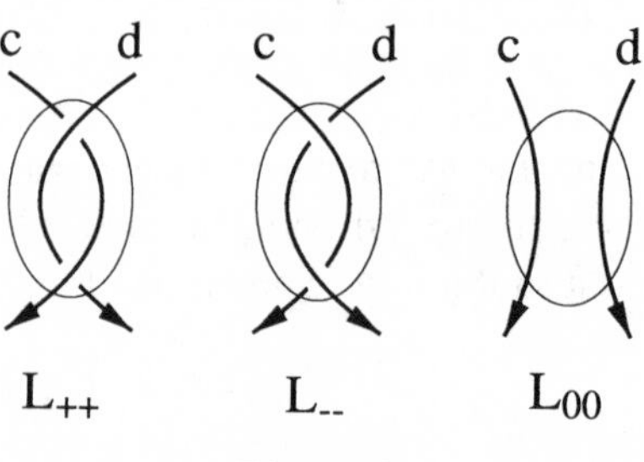

Figure 3

(RIII) $$\Delta_{L_{+++}} + \Delta_{L_{---}} = (t_c t_d^{-1} + t_c^{-1} t_d)\, \Delta_{L_{000}},$$

where L_{+++}, L_{---}, L_{000} are identical except within a ball where they are as in Figure 4 and c, d denote the colors of the components of the strings.

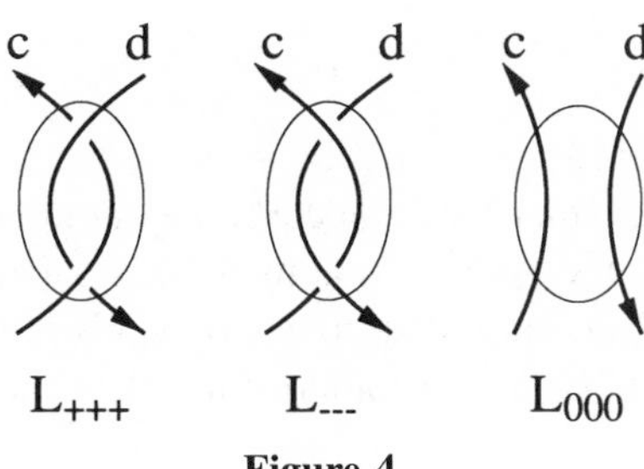

Figure 4

(RIV) $$\Delta_{L_{1-21}} - \Delta_{L_{-12-1}} = \Delta_{L_{2-12}} - \Delta_{L_{-21-2}},$$

where L_{1-21}, L_{-12-1}, L_{2-12}, L_{-21-2} are identical except within a ball where they are as in Figure 5.

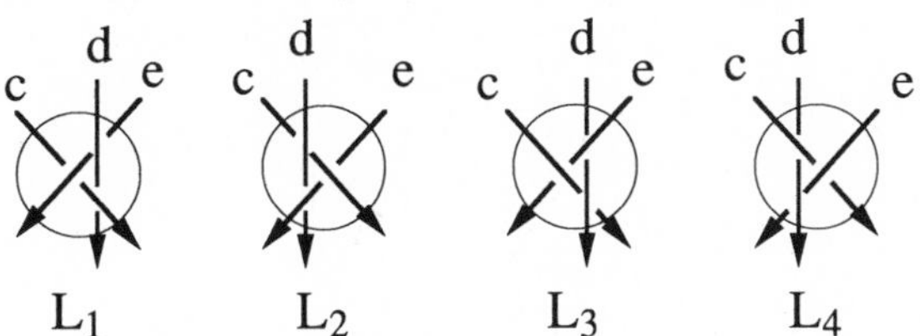

Figure 5

(RV) $$\Delta_{L \sqcup O} = 0,$$

where $L \sqcup O$ is a disjoint union of L and a trivial knot.

(RVI) $$\Delta_{L'} = (t_c - t_c^{-1})^{-1}\, \Delta_L,$$

where L and L' are identical except within a ball where they are as in Figure 6 and c is the color of the component of the string.

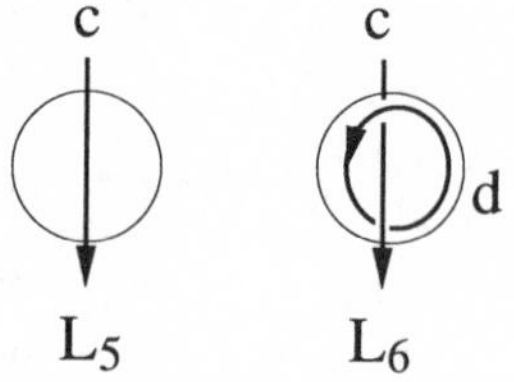

Figure 6

(RVII)
$$\Delta_O = t_c - t_c^{-1},$$

where O denote the trivial knot with color c.

But it is not known whether we can determine uniquely the Alexander polynomial of a colored link by the above relations. In this paper, a new local relation is introduced. Adding the following relation to Conway's relations, we can determine the Alexander polynomial of multi–component links. The relation is

(RVIII)
$$\begin{aligned}
&(t_c + t_c^{-1})\,(t_d - t_d^{-1})\,\Delta(L_{2112}) - (t_d - t_d^{-1})\,(t_e + t_e^{-1})\,\Delta(L_{1221}) \\
&- (t_c^{-1}t_e - t_c t_e^{-1})\,(\Delta(L_{1122}) + \Delta(L_{2211})) \\
&+ (t_c^{-1}t_d t_e - t_c t_d^{-1} t_e^{-1})\,(t_e + t_e^{-1})\,\Delta(L_{11}) \\
&- (t_c + t_c^{-1})\,(t_c t_d t_e^{-1} - t_c^{-1} t_d^{-1} t_e)\,\Delta(L_{22}) \\
&- (t_c^{-2}t_e^2 - t_c^2 t_e^{-2})\,\Delta(L_{0000}) = 0.
\end{aligned}$$

where L_{2112}, L_{1221}, L_{2211}, L_{1122}, L_{11}, L_{0000} and L_{0000} are identical except within a ball where they are as in Figure 7 and c, d, e denote the colors of the components of the strings.

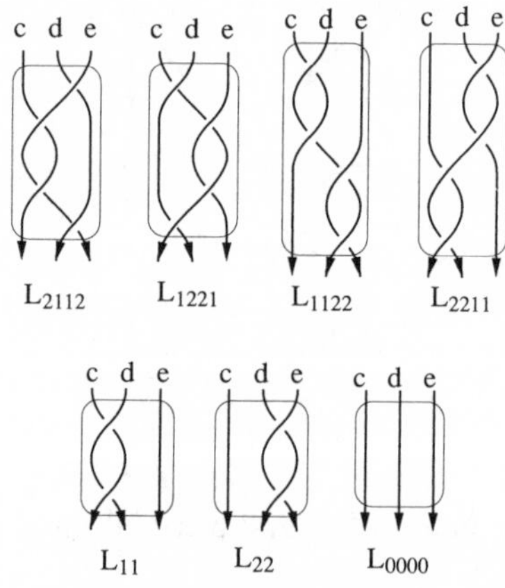

Figure 7

Precisely speaking, we get the following theorem.

Theorem. *The invariant Δ_L for a colored link L is determined uniquely by the relations* (RI), (RII), (RV), (RVI), (RVII) *and* (RVIII).

Hartley [**2.**] showed a way to determine Δ_L explicitly. The polynomial determined by our local relations is equal to the multi–variable Alexander polynomial of Hartley's sense. Turaev gives axioms in [**4.**] for the Alexander polynomial. But one of them is not a local axiom. It is still an open question whether Conway's relations are enough to determine the multi–variable Alexander polynomial or not.

I thank Professor M. Ue who pointed out a serious mistake of my first proof.

§2. The proof of the main theorem

In [**3.**], it is shown that Δ satisfies the relations (RI)–(RIII). So it remains to show that every link can be reduced to a trivial knot by the relations (RI), (RII), (RV), (RVI) and (RIII). To prove this, we use braids instead of link diagrams.

Let B_n be the braid group on n–strings, which is defined by the following generators and relations.

$$B_n = < \sigma_1, \cdots, \sigma_{n-1} \quad | \quad \sigma_i \sigma_{i+1} \sigma_i = \sigma_{i+1} \sigma_i \sigma_{i+1} \quad (1 \leq i \leq n-2), \\ \sigma_i \sigma_j = \sigma_j \sigma_i \quad (|i-j| \geq 2) > .$$

The generator σ_i corresponds to an n–string braid given in Figure 8. For b_1, $b_2 \in B_n$, the product $b_2 b_1$ corresponds to the composition of braids corresponding to b_1 and b_2 as in Figure 9.

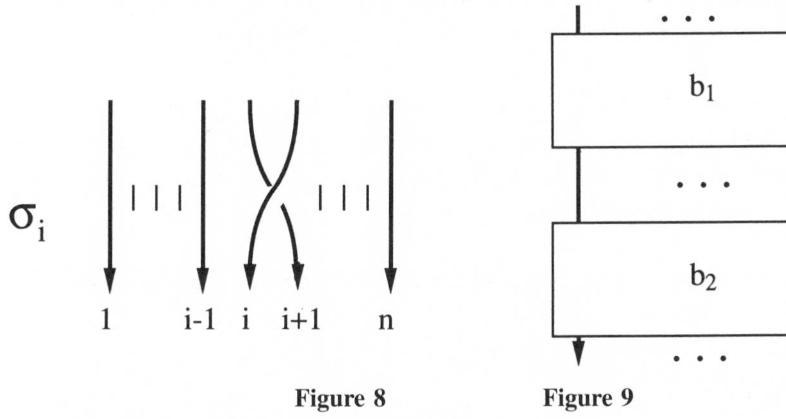

Figure 8 Figure 9

Let S_n denote the permutation group of n–letters $\{1, 2, \cdots, n\}$ and s_i denote the transposition $(i\ i+1)$ of S_n for $i = 1, \cdots, n-1$. Let θ be the group homomorphism from B_n to S_n sending σ_i to s_i.

We introduce a groupoid B'_n associated with B_n. The elements of B'_n is (b, μ) where $b \in B_n$ and μ is a permutation of n letters $\{1, 2, \cdots, n\}$. For two elemetnts (b_1, μ_1) and (b_2, μ_2) in B'_n, the product of these elements are defined as $(b_2\, b_1, \mu_1)$ if $\mu_2 = \theta(b)\mu_1$. Let $(B'_n)_{\mu_1\mu_2} = \{(b, \mu_1) | \theta(\mu_1) = \mu_2\}$. Then B'_n is a disjoint union of $(B'_n)_{\mu_1\mu_2}$ where μ_1 and μ_2 run over all permutations of n–letters. Let $(\mathbf{B}'_n)_{\mu_1\mu_2}$ be the $\mathbf{C}$–linear space whose basis are elements of $(B'_n)_{\mu_1\mu_2}$ and let $\mathbf{C}B'_n$ be the disjoint union of $(\mathbf{B}'_n)_{\mu_1\mu_2}$. A product of elements $(x, \mu_1) \in (\mathbf{B}'_n)_{\mu_1\mu_2}$ and $(y, \mu_2) \in (\mathbf{B}'_n)_{\mu_2\mu_3}$ are defined as $(y\, x, \mu_1) \in (\mathbf{B}'_n)_{\mu_1\mu_3}$ where $y\, x$ is defined as a product of elements of the group ring $\mathbf{C}B_n$. We fix non–zero complex parameters t_1, t_2, $\cdots$, t_n and introduce an equivalence relation $\sim$ on $\mathbf{C}B'_n$. On $(\mathbf{C}B'_n)_{\mu\mu}$, $\sim$ is an equivalence relation generated by the following.

$$\text{(SI)} \qquad \sigma_i^4 - (t_{\mu_i} t_{\mu_{i+1}} + t_{\mu_i}^{-1} t_{\mu_{i+1}}^{-1})\sigma_i^2 + 1 \sim 0, \qquad (1 \le i \le n-1)$$

$$\begin{aligned}
\text{(SII)} \qquad & (t_{\mu_i} + t_{\mu_i}^{-1})\,(t_{\mu_{i+1}} - t_{\mu_{i+1}}^{-1})\,\sigma_{i+1}\,\sigma_i^2\,\sigma_{i+1} \\
& - (t_{\mu_{i+1}} - t_{\mu_{i+1}}^{-1})\,(t_{\mu_{i+2}} + t_{\mu_{i+2}}^{-1})\,\sigma_i\,\sigma_{i+1}^2\,\sigma_i \\
& - (t_{\mu_i}^{-1} t_{\mu_{i+2}} - t_{\mu_i} t_{\mu_{i+2}}^{-1})\,(\sigma_i^2\,\sigma_{i+1}^2 + \sigma_{i+1}^2\,\sigma_i^2) \\
& + (t_{\mu_i}^{-1} t_{\mu_{i+1}} t_{\mu_{i+2}} - t_{\mu_i} t_{\mu_{i+1}}^{-1} t_{\mu_{i+2}}^{-1})\,(t_{\mu_{i+2}} + t_{\mu_{i+2}}^{-1})\,\sigma_i^2 \\
& - (t_{\mu_i} + t_{\mu_i}^{-1})\,(t_{\mu_i} t_{\mu_{i+1}} t_{\mu_{i+2}}^{-1} - t_{\mu_i}^{-1} t_{\mu_{i+1}}^{-1} t_{\mu_{i+2}})\,\sigma_{i+1}^2 \\
& - (t_{\mu_i}^{-2} t_{\mu_{i+2}}^2 - t_{\mu_i}^2 t_{\mu_{i+2}}^{-2}) \sim 0, \qquad (1 \le i \le n-2)
\end{aligned}$$

where $\mu = (\mu_1, \cdots, \mu_n)$. These relations (SI) and (SII) are braid versions of (RII) and (RVIII) respectively. We extend $\sim$ to $(\mathbf{C}B'_n)_{\mu_1\mu_2}$ so that $\sim$ is compatible with the multiplication. Then $\sim$ is an equivalence relation of $\mathbf{C}B'_n$. Before proving the theorem, we show the following lemma. A proof is given in the next section.

Lemma. *With respect to $\sim$, every element $(x, \mu_1) \in (\mathbf{C}B'_n)_{\mu_1\mu_2}$ is equivalent to a $\mathbf{C}$–linear combination of the elements of the form $(y\, \sigma_{n-1}^i\, z, \mu_1)$ $(y, z \in B_{n-1},\ i = 0, 1, 2, 3)$ for generic t_1, $\cdots$, t_n. Moreover, the coefficients are rational expressions of t_1, $\cdots$, t_n and they do not diverge if some of t_1, $\cdots$, t_n are equal.*

We need the second statement because, for colored braid, we specialize the parameter t_j to t_i if the colors of the i–th and j–th strings are equal.

Let c_1, c_2, $\cdots$, c_n be positive integers. Then

$$B_n^{(c_1, \cdots, c_n)} = \{b \in B_n \quad | \quad c_{\theta(b)(i)} = c_i \quad (1 \le i \le n)\}.$$

is a subgroup of B_n and is called a *colored braid group* with colors c_1, c_2, $\cdots$, c_n. The definition of the colored braid group implies that the closure of $b \in B_n^{(c_1, \cdots, c_n)}$ is a link with natural coloring coming from c_1, $\cdots$, c_n. For non–colored links,

Alexander's theorem and Markov's theorem convert the classification problem of links to that of certain equivalence classes of braids. By tracing the proof of these theorems with coloring, we get similar theorems for colored links and colored braids.

Proposition (Alexander's Theorem for colored links). *Every colored link is equivalent to the closure of a colored braid.*

Let $\sim_M$ be the equivalence relation of the set of colored braids generated by the following.

(MI) $b_1\, b_2 \sim_M b_2\, b_1$ for $b_1\, b_2 \in B_n^{(c_1,\cdots,c_n)}$ and $b_2\, b_1 \in B_n^{(c_{\theta(b_1)(1)},\cdots,c_{\theta(b_1)(n)})}$,

(MII) $b \sim_M b\,\sigma_n^{\pm 1}$ for $b \in B_n^{(c_1,\cdots,c_n)}$ and $b\,\sigma_n^{\pm 1} \in B_{n+1}^{(c_1,\cdots,c_n,c_n)}$.

Proposition (Markov's theorem for colored braids). *The closures of two colored braids are equivalent as colored links if and only if they belong to the same class with respect to* $\sim_M$.

Hence the classification problem of colored links are converted to that of equivalence classes of $\sim_M$.

Proof of the main theorem. Since every colored link is equivalent to the closure of a colored braid, it is enough to show the following.

(*) Every colored braid can be reduced to a rational expression by using the relations (RI), (RII), (RV)–(RVIII) and (MII).

This is proved by an induction on the braid index n. If $n = 1$ then the theorem is true by the relation (RVII). Now assume that $n \geq 2$. Let $\mu_n = (1\ 2\ \cdots\ n)$ be the trivial permutation. Then $B_n^{(c_1,\cdots,c_n)}$ is equal to the disjoint union $\cup_{\substack{g\in S_n \\ c_{g(i)}=c_i}} (B'_n)_{\mu_n, g(\mu_n)}$ as a set where $g(\mu_n) = (g(1)\ g(2)\ \cdots\ g(n))$. In the following, we identify both sets. The above lemma says that, by the equivalence relation generated by (RII) and (RVIII), every element of $(B'_n)_{\mu_n, g(\mu_n)}$ is equivalent to a linear combination of elements of the form $(y\,\sigma_{n-1}^i\, z, \mu_n)$ ($y, z \in B_{n-1}$, $i = 0, 1, 2, 3$). Hence it is enough to show (*) for $(y\,\sigma_{n-1}^i\, z, \mu_n)$. If $i = 0$ then $(y\,\sigma_{n-1}^i\, z, \mu_n)$ is reduced to 0 by the relation (RV). If $i = 1$ then, by (MII), $y\,\sigma_{n-1}^i\, z$ is reduced to $yz \in B_{n-1}^{(c_1,\cdots,c_{n-1})}$. If $i = 2$ then, by (RVI), $y\,\sigma_{n-1}^i\, z$ is reduced to a multiple of $yz \in B_{n-1}^{(c_1,\cdots,c_{n-1})}$. If $i = 3$ then c_{n-1} and c_n are equal. Hence we can apply (RI) and $(y\,\sigma_{n-1}^3\, z, \mu_n)$ is reduced to a linear combination of $(y\,\sigma_{n-1}^2\, z, \mu_n)$ and $(y\,\sigma_{n-1}\, z, \mu_n)$. These are reduced to a multiple of elements in $B_{n-1}^{(c_1,\cdots,c_{n-1})}$ as above. These results prove the theorem.

§3. Proof of the lemma

We prove that, for any $b \in B_n$, the element (b, μ_1) in the groupoid B'_n satisfies the statement of the lemma. An element $b \in B_n$ is a word of $\sigma_1, \cdots, \sigma_{n-1}, \sigma_1^{-1}, \cdots, \sigma_{n-1}^{-1}$. By the relation (SI), b is equivalent to a linear combination of the element presented by words consist of $\sigma_1, \cdots, \sigma_{n-1}$, which do not contain $\sigma_1^{-1}, \cdots, \sigma_{n-1}^{-1}$. Hence it is enough to prove the statement for $b \in B_n$ of the following form:

$$b = b_0 \, \sigma_{n-1}^{k_1} \, b_1 \, \sigma_{n-1}^{k_2} \cdots \sigma_{n-1}^{k_r} \, b_r$$

where $k_i \neq 0$, b_i is a word of $\sigma_1, \cdots, \sigma_{n-2}$ for $0 \leq i \leq r$ and $b_i \neq 1$ for $1 \leq i \leq r-1$. From the relation (SI), we may assume that $k_i = 1$, 2 or 3. The number r depends on b and denoted by $r(b)$. This is proved by an induction on the word length $\ell(b)$ of b, $r(b)$ and the braid index n. If $\ell(b) \leq 2$, the lemma is true for b. If $b \in B_2$, the lemma is true by the relation (SI). Now assume that $\ell(b) \geq 3$. Then $b = \sigma_i \, b_1$. Because $\ell(b_1) < \ell(b)$, the lemma is true for b_1 by the induction hypothesis. Hence the statement is also true for b if $i < n-1$. From now on, we assume $i = n-1$. In this case, $b = \sigma_{n-1}^k \, b_2$ where the first element of b_2 is not equal to $\sigma_{n-1}^{\pm 1}$ and $k = 1$, 2 or 3. If $b_2 \in B_{n-1}$, then the lemma is true for b. From now on, we assume b_2 is not contained in B_{n-1}. Then $b_2 = b_3 \, \sigma_{n-1} \, b_4$ where $b_3 \neq 1$ and $b_3 \in B_{n-1}$. From the induction hypothesis, b_2 is a linear combination of the elements of the form $b_5 \, \sigma_{n-2}^m \, b_6$ $(b_5, b_6 \in B_{n-2},\ m = 0, 1, 2, 3)$. Hence it remains to show the statement for the element of the form $b = \sigma_{n-1}^k \, \sigma_{n-2}^m \, \sigma_{n-1} \, b_7$ $(b_7 \in B_{n-1},\ k, m = 1, 2, 3)$.

Case 1. $(m = 1)$ In this case, b is equal to $\sigma_{n-2} \, \sigma_{n-1} \, \sigma_{n-2}^k \, b_7$. By the induction hypothesis, the lemma is true for $\sigma_{n-1} \, \sigma_{n-2}^k \, b_7$. Hence b also satisfies the lemma.

Case 2. $(k = 1, m = 2)$ First, we prove for the case that the first element of b_7 is not equal to σ_{n-1}. By applying the relation (SII) to $\sigma_{n-1} \, \sigma_{n-2}^2 \, \sigma_{n-1}$, we know that b is equivalent to a linear combination of six words $w_1, \cdots, w_6$ of $\sigma_1, \cdots, \sigma_{n-1}$ with $r(w_j) < r(b)$. Hence the induction hypothesis implies that b satisfies the lemma.

Now, assume that the first element of b_7 is equal to σ_{n-1}. In this case, by applying the relation (SII) to $\sigma_{n-1} \, \sigma_{n-2}^2 \, \sigma_{n-1}$, we know that b is equivalent to a linear combination of elements $w_1 = \sigma_{n-1}^2 \, \sigma_{n-2}^2 \, b_7$, $w_2 = \sigma_{n-2}^2 \, \sigma_{n-1}^2 \, b_7$, $w_3 = \sigma_{n-2} \, \sigma_{n-1}^2 \, \sigma_{n-2} \, b_7$, $w_4 = \sigma_{n-1}^2 \, b_7$, $w_5 = \sigma_{n-2}^2 \, b_7$, and $w_6 = b_7$. The induction hypothesis implies that $w_2, \cdots, w_6$ satisfies the lemma. By applying the relation (SII) to $\sigma_{n-1} \, \sigma_{n-2}^2 \, \sigma_{n-1}$, we know that w_1 is equivalent to a linear combination of elements $u_1 = b$, $u_2 = \sigma_{n-1}^3 \, \sigma_{n-2}^2 \, b_8$, $u_3 = \sigma_{n-1} \, \sigma_{n-2} \, \sigma_{n-1}^2 \, \sigma_{n-2} \, b_8 = \sigma_{n-2}^2 \, \sigma_{n-1} \, \sigma_{n-2}^2 \, b_8$, $u_4 = \sigma_{n-1}^3 \, b_8$, $u_5 = \sigma_{n-1} \, \sigma_{n-2}^2 \, b_8$ and $u_6 = \sigma_{n-1} \, b_8$ where $b_7 = \sigma_{n-1} \, b_8$. The induction hypothesis implies that $u_3, \cdots, u_6$ satisfies the lemma. If the first element of b_8 is not equal to σ_{n-1}, then $r(u_2) < r(b)$ and u_2 satisfies the lemma by the induction hypothesis. If the first element of b_8 is equal to σ_{n-1}, then $b_8 = \sigma_{n-1}\, b_9$ and $b = \sigma_{n-1}\,\sigma_{n-2}^2\,\sigma_{n-1}^3\, b_9$. Hence we may assume that the first element of b_9 is not equal to σ_{n-1} by the relation (SI). We have $u_2 = \sigma_{n-1}^3\,\sigma_{n-2}^2\,\sigma_{n-1}\, b_9$. By applying the relation (SII) to $\sigma_{n-1}\,\sigma_{n-2}^2\,\sigma_{n-1}$, we know that u_2 is a linear combination of elements $v_1 = w_1$, $v_2 = \sigma_{n-1}^4\,\sigma_{n-2}^2\, b_9$, $v_3 = \sigma_{n-1}^2\,\sigma_{n-2}\,\sigma_{n-1}^2\,\sigma_{n-2}\, b_9$, $v_4 = \sigma_{n-1}^4\, b_9$, $v_5 = \sigma_{n-1}^2\,\sigma_{n-2}^2\, b_9$ and $v_6 = \sigma_{n-1}^2\, b_9$. By the induction hypothesis, v_4, v_5, v_6 satisfy the lemma. Since $v_3 = \sigma_{n-2}\,\sigma_{n-1}^2\,\sigma_{n-2}\,\sigma_{n-1}^2\, b_9$ and $\sigma_{n-1}^2\,\sigma_{n-2}\,\sigma_{n-1}^2\, b_9$ satisfy the lemma, v_3 satisfies the lemma. Since $r(v_2) < r(b)$, the induction hypothesis implies that v_2 satisfies the lemma. Combining the above results, w_1 is equivalent to a linear combination of b, w_1, u_3, $\cdots$, u_6, v_2, $\cdots$, v_6 if the first element of b_8 is not equal to σ_{n-1}. Let $a(t_1,t_2,t_3) = (t_1^{-1}t_3 - t_1t_3^{-1})(t_1 + t_1^{-1})^{-1}(t_2 - t_2^{-1})^{-1}$. Then the coefficient of w_1 is $a(t_{\mu_{n-2}}, t_{\mu_{n-1}}, t_{\mu_n})^2$ and is not equal to 1 if the parameters $t_{\mu_{n-2}}$, $t_{\mu_{n-1}}$ and t_{μ_n} are generic. Hence w_1 is written as a linear combination of b, u_3, $\cdots$, u_6, v_2, $\cdots$, v_6. This implies that b is equivalent to a linear combination of b, w_1, u_3, $\cdots$, u_6, v_2, $\cdots$, v_6, w_2, $\cdots$, w_6. The coefficient of b is $a(t_{\mu_{n-2}}, t_{\mu_{n-1}}, t_{\mu_n})/(1 - a(t_{\mu_{n-2}}, t_{\mu_{n-1}}, t_{\mu_n})^2)$ and is not equal to 1 if the parameters $t_{\mu_{n-2}}$, $t_{\mu_{n-1}}$ and t_{μ_n} are generic. Hence b is equivalent to a linear combination of u_3, $\cdots$, u_6, v_2, $\cdots$, v_6, w_2, $\cdots$, w_6. If the first element of b_8 is equal to σ_{n-1}, b is equivalent to a linear combination of u_3, $\cdots$, u_6, w_2, $\cdots$, w_6. Hence, in the both cases, b is equivalent to a linear combination of elements satisfying the lemma and so b satisfies the lemma.

Case 3. $(k = 1, m = 3)$ First, we prove for the case that the first element of b_7 is not equal to σ_{n-1}. By applying the relation (SII) to $\sigma_{n-2}^2\,\sigma_{n-1}^2$ in $b = \sigma_{n-1}\,\sigma_{n-2}^3\,\sigma_{n-1}^2\,\sigma_{n-1}^{-1}\, b_7$, we know that b is equivalent to a linear combination of elements $w_1 = \sigma_{n-1}\,\sigma_{n-2}^2\,\sigma_{n-1}^2\,\sigma_{n-2}\,\sigma_{n-1}^{-1}\, b_7 = \sigma_{n-2}^{-1}\,\sigma_{n-1}\,\sigma_{n-2}^2\,\sigma_{n-1}^2\,\sigma_{n-2}\, b_7$, $w_2 = \sigma_{n-1}\,\sigma_{n-2}^3\,\sigma_{n-1}^{-1}\, b_7$, $w_3 = \sigma_{n-1}\,\sigma_{n-2}\,\sigma_{n-1}\,\sigma_{n-2}^2\,\sigma_{n-1}\,\sigma_{n-1}^{-1}\, b_7 = \sigma_{n-2}\,\sigma_{n-1}\,\sigma_{n-2}^3\, b_7$, $w_4 = \sigma_{n-1}\,\sigma_{n-2}\,\sigma_{n-1}^2\,\sigma_{n-2}^2\,\sigma_{n-1}^{-1}\, b_7 = \sigma_{n-2}\,\sigma_{n-1}^3\,\sigma_{n-2}\, b_7$, $w_5 = \sigma_{n-1}\,\sigma_{n-2}\,\sigma_{n-1}^2\,\sigma_{n-1}^{-1}\, b_7 = \sigma_{n-1}\,\sigma_{n-2}\,\sigma_{n-1}\, b_7$ and $w_6 = \sigma_{n-1}\,\sigma_{n-2}\, b_7$. By the induction hypothesis, w_3, $\cdots$, w_6 satisfy the lemma. By using the proof of Case 2, we know that $\sigma_{n-1}\,\sigma_{n-2}^2\,\sigma_{n-1}^2$ satisfies the lemma. Hence, the induction on r implies that w_1 satisfies the lemma. From the relation (SI), we know that w_2 is equivalent to a linear combination of $\sigma_{n-1}\,\sigma_{n-2}\,\sigma_{n-1}^{-1}\, b_7$ and $\sigma_{n-1}\,\sigma_{n-2}^{-1}\,\sigma_{n-1}^{-1}\, b_7$. But these elements are equal to $\sigma_{n-2}^{-1}\,\sigma_{n-1}\,\sigma_{n-2}\, b_7$, $\sigma_{n-2}\,\sigma_{n-1}^{-1}\,\sigma_{n-2}\, b_7$ respectively. Hence w_2 is equivalent to a linear combination of $\sigma_{n-2}\,\sigma_{n-1}\,\sigma_{n-2}\, b_7$, $\sigma_{n-2}^3\,\sigma_{n-1}\,\sigma_{n-2}\, b_7$, $\sigma_{n-2}\,\sigma_{n-1}^3\,\sigma_{n-2}\, b_7$ and $\sigma_{n-2}^3\,\sigma_{n-1}^3\,\sigma_{n-2}\, b_7$. So the induction on r implies that w_2 satisfies the lemma. Combining the above results, b is equivalent to a linear combination of elements satisfying the lemma, b also satisfies the lemma.

Case 4. ($k = 2$ and $m = 2, 3$) In this case, $b = \sigma_{n-1}^2 \sigma_{n-2}^m \sigma_{n-1} b_7$. By applying the relation (SII) to $\sigma_{n-1}^2 \sigma_{n-2}^2$, we know that b is equivalent to a linear combination of elements $w_1 = \sigma_{n-1} \sigma_{n-2}^2 \sigma_{n-1} \sigma_{n-2}^{m-2} b_7$, $w_2 = \sigma_{n-2} \sigma_{n-1}^2 \sigma_{n-2}^{m-1} b_7$, $w_3 = \sigma_{n-2}^2 \sigma_{n-1}^2 \sigma_{n-2}^{m-2} b_7$, $w_4 = \sigma_{n-1}^2 \sigma_{n-2}^{m-2} b_7$, $w_5 = \sigma_{n-2}^m b_7$ and $w_6 = \sigma_{n-2}^{m-2} b_7$. The induction hypothesis implies that $w_2, \cdots, w_6$ satisfy the lemma. From the proof of Case 2, w_1 satisfies the lemma. Hence b satisfies the lemma.

Case 5. ($k = 3$ and $m = 2, 3$) In this case, $b = \sigma_{n-1}^3 \sigma_{n-2}^m \sigma_{n-1} b_7$. By applying the relation (SII) to $\sigma_{n-1}^2 \sigma_{n-2}^2$, we know that b is equivalent to a linear combination of elements $w_1 = \sigma_{n-1}^2 \sigma_{n-2}^2 \sigma_{n-1} \sigma_{n-2}^{m-2} b_7$, $w_2 = \sigma_{n-1} \sigma_{n-2} \sigma_{n-1}^2 \sigma_{n-2}^{m-1} b_7$ $= \sigma_{n-2}^2 \sigma_{n-1} \sigma_{n-2} \sigma_{n-2}^{m-1} b_7$, $w_3 = \sigma_{n-1} \sigma_{n-2}^2 \sigma_{n-1}^2 \sigma_{n-2}^{m-2} b_7$, $w_4 = \sigma_{n-1}^3 \sigma_{n-2}^{m-2}$ b_7, $w_5 = \sigma_{n-1} \sigma_{n-2}^m b_7$ and $w_6 = \sigma_{n-1} \sigma_{n-2}^{m-2} b_7$. The induction hypothesis implies that $w_2, \cdots, w_6$ satisfy the lemma. From the proof of Case 3, we know that w_1 satisfies the lemma. Since b is equivalent to a linear combination of elements satisfying the lemma, b also satisfies the lemma.

The coefficients of (SI) and (SII) are not identically equal to 0 if some of t_1, $\cdots$, t_n are equal. Hence, except in Case 2, the coefficients of the linear combination in the proof do not diverge when some of $t_1, \cdots, t_n$ are equal. In Case 2, $a(t_{\mu_{n-2}}, t_{\mu_{n-1}}, t_{\mu_n})$ is not identically equal to a constant when two or three of $t_{\mu_{n-2}}$, $t_{\mu_{n-1}}$ and t_{μ_n} are equal. This implies that the coefficients in linear combinations in the proof of Case 2 do not diverge and are expressed as rational expressions when some of $t_1, \cdots, t_n$ are equal. Hence the lemma is proved for all the cases.

References

1. Conway, J. H., *An enumeration of knots and links, and some of their algebraic properties*, Computational Problems in Abstract Algebra, Pergamon Press, Oxford and New York, 1969, pp. 329-358..
2. Hartley, R., *The Conway potential function for links*, Comment. Math. Helvetici **58** (1983), 365–378.
3. Murakami, J., *A state model for the multi–variable Alexander polynomial*, preprint, 1990 Osaka University.
4. Turaev, T.G., *Reidemeister torsion in knot theory*, Russian Math. Surveys **41** (1986), 119–182.

Almost identical link imitations and the skein polynomial

Akio Kawauchi

Dedicated to Professor Kunio Murasugi on his 60th birthday

The imitation theory gives a method constructing from a given good (3,1)-manifold pair (M, L) a new good (3,1)-manifold pair (M, L^*) with a map $q : (M, L^*) \to (M, L)$ close to a diffeomorphism in several senses (cf.[**K0**],[**K1**]). In [**K2**], for any given good (3,1)-manifold pair (M, L), an infinite family of almost identical imitations (M, L^*) of (M, L) with exteriors $E(L^*, M)$ hyperbolic is constructed. In [**K3**] it is shown that we can take as this family an infinite family of (3,1)-manifold pairs (M, L^*) which have the hyperbolic covering property. In this paper, this result is applied to some graph in the 3-sphere S^3 to construct from any link L in S^3 an infinite family $\Im$ of almost identical imitations (S^3, L^*) of (S^3, L) with several properties, one of which is related to the skein (= two-variable Jones, HOMFLY, FLYPMOTH) polynomials (cf.[**F/Y/H/L/M/O**]) of the links L^*, L. A link L^* will be simply called an *almost identical link imitation* of a link L if (S^3, L^*) is an almost identical imitation of (S^3, L).

For an link L and any positive number C, we shall show that this family $\Im$ of almost identical link imitations L^* of L can be taken so as to have all of the following properties:

(1) Each $L^* \in \Im$ has the hyperbolic covering property, and there is a number $C^+ > C$ such that the hyperbolic volume $\mathrm{Vol}E(L^*, S^3) < C^+$ and $\sup_{L^* \in \Im} \mathrm{Vol}E(L^*, S^3) = C^+$,

(2) Each $L^* \in \Im$ is obtained as a band sum of L and a trivial knot O (In other words, L^* is a fusion of the split union $L + O$) and has the unlinking number $u(L^*) \leq \max\{u(L), 1\}$,

(3) The skein polynomial is constant on all $L^* \in \Im$, and the skein polynomial of each $L^* \in \Im$ is 'close' to the skein polynomial of L.

A certain generalization of (2) will be shown in Theorem 3.1 which is our main theorem. The precise meaning of the term 'close' in (3) will be clear in Theorem 3.1 using *coefficient polynomials* , essentially the coefficients of the skein polynomial, regarded as a polynomial in m in the convention of Lickorish/Millett

Knots 90

in [**L/M**]. Infinitely many knots with the same skein polynomial constructed by Kanenobu [**K**] have mutually non-isomorphic Alexander modules. By a property of the imitation in [**K1**], any imitation map $q:(S^3,L^*)\rightarrow(S^3,L)$ induces an isomorphism between the Alexander modules of L^* and L. Thus, for each link L, we have infinitely many links with the same skein polynomial and the same Alexander module isomorphic to the Alexander module of L.

Throughout this paper, some terminologies of [**K0**],[**K1**],[**K2**],[**K3**] will be used without mentions. However, the following terms are reconfirmed here. Namely, a compact connected oriented 3-manifold M is said to be *hyperbolic* if $\mathrm{int}_0 M = M-\partial_0 M$ (when $\partial M=\partial_0 M$) or its double $D(\mathrm{int}_0 M)$ (when $\partial M\neq\partial_0 M$) has a complete Riemannian structure of constant curvature -1, where $\partial_0 M$ denotes the union of all tori in the boundary ∂M of M. Then the volume Vol(int$_0M$) or Vol(D(int$_0M$))/2 (known to be finite) is a topological invariant of M, called the *hyperbolic volume* of M and denoted by Vol$M$. By a *good* (3,1)-manifold pair (M,L) (or a *good* 1-manifold L in M), we mean that M is a compact connected oriented 3-manifold and L is a compact proper smooth 1-submanifold of M such that any 2-sphere component of ∂M meets L with at least three points. It is said to have the *hyperbolic covering property* if for every pair of component unions L_0, L_1 (possibly, $\emptyset$) of L with $L-L_0=L_1$, any finite regular covering space of the exterior $E(L_0,M)$ branched along L_1 is hyperbolic after spherical completion.

§1. Construction

Let L be a link in S^3 and b an oriented band spanning L with orientation coherent with the orientation of L (cf. Fig.1(1)).

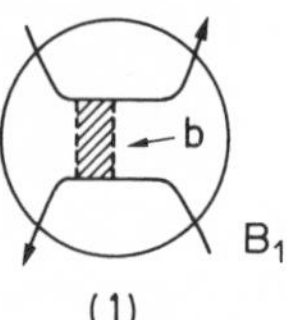

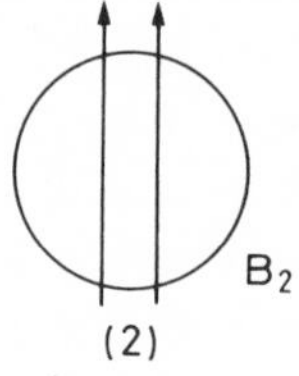

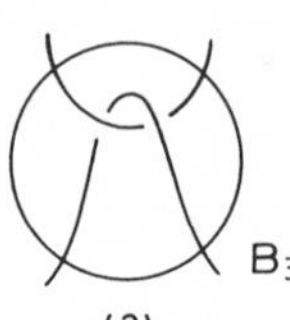

Figure 1

Let L' be a link obtained from L by surgery along b. For our application to the first half of the statement (2) in the introduction, we take b so that L' is a split union of L and a trivial knot O. We choose mutually disjoint 3-balls $B_i, i=1,2,3$, in S^3 so that $L\cap B_i$ is a 2-string tangle in B_i, as it is illustrated in Fig.1(i), i=1,2,3, where $L\cap B_2$ should belong to one component of L and we do not specify the string orientation of $L\cap B_3$. Let L'' be a link obtained from L by a crossing change at a crossing point in Fig.1(3). We call this crossing point a *clasp crossing point*. For our application to the latter half of the statement (2) in the introduction, we take L'' so that the unlinking number $u(L'') = \max\{u(L) - 1, 0\}$. Next, we replace the 2-string tangle $L \cap B_i$ by the graph G_i in Fig.2(i), i= 1,2,3.

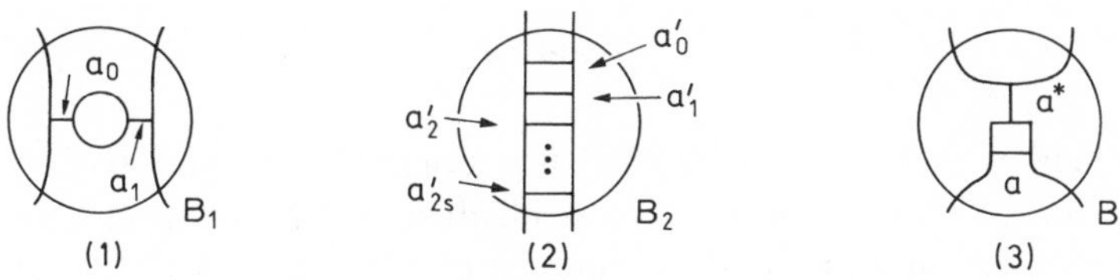

Figure 2

By Γ we denote a spatial graph occurring from L by this replacement. Let (S^3_o, Γ_o) be a good (3,1)-manifold pair obtained from (S^3, Γ) by removing a small open ball neighborhood of each vertex of Γ of degree 3. Given an almost identical imitation

$$q_o : (S^3_o, \Gamma^*_o) \to (S^3_o, \Gamma_o),$$

we have an almost identical graph imitation

$$q : (S^3, \Gamma^*) \to (S^3, \Gamma)$$

by taking a spherical completion (cf.[**K2**]). We consider an H-graph H_i in a 3-ball neighborhood V_i around the arc $a_i, i = 0, 1$ and an H-graph H'_j in a 3-ball neighborhood V'_j around the arc $a'_j, j = 0, 1, ..., 2s$, and H-graphs H, H^* in 3-ball neighborhoods V, V^* around the arcs a, a^* which are illustrated in Fig.3(1), (2) and (3), respectively.

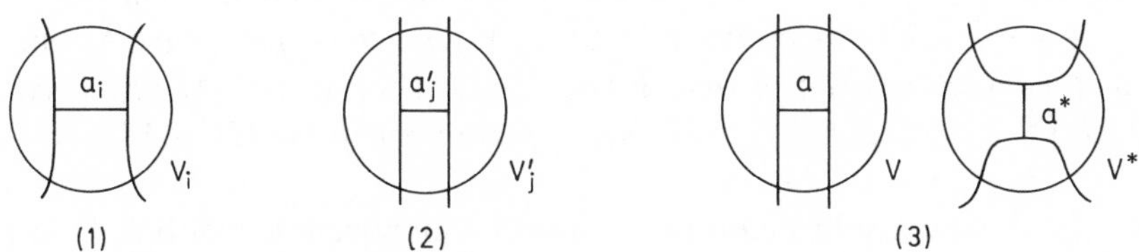

Figure 3

We replace these H-graphs by certain 2-tangles with m_i, m'_j, m and m^* full twists, respectively, as they are illustrated in Fig.4(1),(2),(3) (In the figure, the case of positive twists is illustrated and a negative twist is the mirror image of a positive twist).

We impose the following condition on m_i, m'_j, m and m^*:

(#) $$m_0 + m_1 = m'_0 + m'_1 + ... + m'_{2s} = m + m^* = 0.$$

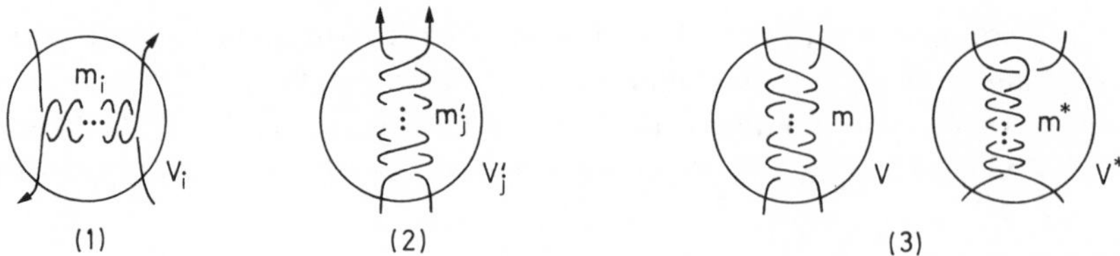

Figure 4

Then Γ changes into a link with the same oriented link type as L. Identifying this link with L, we see that any almost identical graph imitation $q : (S^3, \Gamma^*) \to (S^3, \Gamma)$ induces an almost identical imitation $q^* : (S^3, L^*) \to (S^3, L)$. Let b^* be a band spanning L^* obtained by thickening the arc a_0 in Fig.3(1) so that q^* maps b^* diffeomorphically onto b. The following lemma is obvious from our construction:

Lemma 1.1. *The map $(S^3, L^{*\prime}) \to (S^3, L')$ induced from q^* by the surgeries along b^*, b is homotopic to a diffeomorphism, and the map $(S^3, L^{*\prime\prime}) \to (S^3, L'')$ induced from q^* by the crossing changes at the clasp crossing points corresponding by q^* is homotopic to a diffeomorphism.*

Let M be a 3-manifold obtained from S^3 by removing intV_i, intV'_j for all i, j, int$V$ and intV^*. Let U, U' be 3-balls obtained from S^3 by splitting along a 2-sphere such that

(1) $L \cap U$ is a trivial $r(\geq 3)$-string tangle in U , and
(2) V_i, V'_j, V and V^* for all i, j are contained in int$U$.

Lemma 1.2. *For any positive number C, there is an almost identical graph imitation $q : (S^3, \Gamma^*) \to (S^3, \Gamma)$ extending an almost identical graph imitation $q_U : (U, (\Gamma \cap U)^*) \to (U, \Gamma \cap U)$ and an almost identical imitation $q_{U'} : (U', (L \cap U')^*) \to (U', L \cap U')$ such that $(U', (L \cap U')^*)$ has the hyperbolic covering property and the double covering spaces $M_2, (U \cap M)_2$ of $M, U \cap M$ branched along $\Gamma^* \cap M, \Gamma^* \cap U \cap M$, respectively, are hyperbolic with* Vol$M_2 > 2C$.

Proof. Let Γ^+ be a split union in S^3 of an n-component trivial link O in U and Γ. By [**K3**], we have an almost identical graph imitation $q_U^+ : (U, (\Gamma^+ \cap U)^*) \to (U, \Gamma^+ \cap U)$ such that the good (3,1)-manifold pair $(U, (\Gamma^+ \cap U)^*) \cap S_0^3$ has the hyperbolic covering property. By [**K3**] and [**K2,** Corollary 3.3], we have an almost identical imitation $q_{U'} : (U', (L \cap U')^*) \to (U', L \cap U')$ such that $(U', (L \cap U')^*)$ has the hyperbolic covering property and the extension $(S^3, (L \cap U')^* \cup (L \cap U)) \to (S^3, L)$ of $q_{U'}$ by the identity on $(U, L \cap U)$ is homotopic to a diffeomorphism. Using q_U^+ and $q_{U'}$, we have an almost identical graph imitation $q^+ : (S^3, \Gamma^{+*}) \to (S^3, \Gamma^+)$. Let O^* be the preimage of O by q^+ and $E_{U \cap M} = E(O^*, M \cap U)$ and $E_M = E(O^*, M)$. The double covering space $(E_{U \cap M})_2$ of $E_{U \cap M}$ branched along

$\Gamma^{+*}\cap U\cap M-O^*$ and lifting the tori around O^* in $\partial E_{U\cap M}$ trivially is hyperbolic by the hyperbolic covering property of $(U,\Gamma^{+*}\cap U)\cap S^3_0$. By Myers gluing lemma (cf.[**K2,** Lemma 5.3]), the double covering space $(E_M)_2$ of E_M branched along $\Gamma^{+*}\cap M-O^*$ which extends the covering $(E_{U\cap M})_2\to E_{U\cap M}$ is also hyperbolic, for $(U',\Gamma^{+*}\cap U')=(U',(L\cap U')^*)$ has the hyperbolic covering property. By Jϕrgensen's theorem [**T1**],[**T2**], we have $\mathrm{Vol}(E_M)_2>2C$ by taking n so large. By Thurston's hyperbolic Dehn surgery [**T1**],[**T2**], if $q:(S^3,\Gamma^*)\to(S^3,\Gamma)$ is obtained from q^+ by Dehn surgery along the components of O^* and O with the same coefficient $1/m$ for a large positive integer m, then q is an almost identical graph imitation and we have that M_2 and $(U\cap M)_2$ are hyperbolic with $\mathrm{Vol}M_2>2C$. This completes the proof.

From now on, we consider an almost identical imitation

$$q^*:(S^3,L^*)\to(S^3,L)$$

induced from an almost identical graph imitation

$$q:(S^3,\Gamma^*)\to(S^3,\Gamma)$$

stated in Lemma 1.2.

Lemma 1.3. *For any positive number C, there is a positive constant c such that for all m_i,m'_j,m and m^* with (#) and $|m_i|,|m'_j|,|m|,|m^*|>c$, (S^3,L^*) has the hyperbolic covering property and*

$$C<\mathrm{Vol}E(L^*,S^3)<\sup_{\{m_0,m_1\}}\mathrm{Vol}E(L^*,S^3)<+\infty$$

if we fix m'_j,m and m^* for all j.

Proof. The branched covering spaces $S^3(L^*)_2,U(L^*\cap U)_2$ of S^3,U branched along $L^*,L^*\cap U$ are obtained from $M_2,(U\cap M)_2$ by attaching solid tori whose attaching meridians vary by the values m_i,m'_j,m and m^*. Hence by Thurston's hyperbolic Dehn surgery, there is a number $c>0$ such that for all m_i,m'_j,m,m^* with $|m_i|,|m'_j|,|m|,|m^*|>c$, $S^3(L^*)_2$ and $U(L^*\cap U)_2$ are hyperbolic with $\mathrm{Vol}S^3(L^*)_2>2C$. Since the surface $\partial U(L^*\cap U)_2$ is incompressible in $S^3(L^*)_2$ which is hyperbolic, we see from [**K3,** Lemma 1.7] that (S^3,L^*) has the hyperbolic covering property. Let L^*_2 be the lift of L^* to $S^3(L^*)_2$. Then

$$\mathrm{Vol}E(L^*,S^3)=\mathrm{Vol}E(L^*_2,S^3(L^*)_2)/2>\mathrm{Vol}S^3(L^*)_2/2>C.$$

Let L^*_0 be the link L^* with $m_i=m'_j=m=m^*=0$ for all i,j. Let $L^\#$ be the link obtained from L^*_0 by adding the components k_i,k'_j,k,k^* for all i,j indicated in Fig. 5.

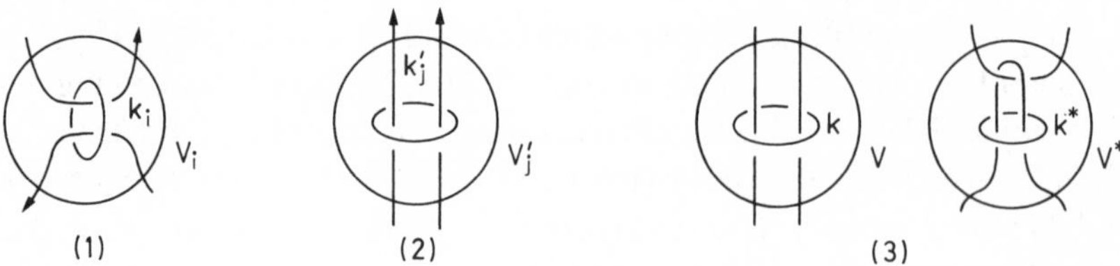

Figure 5

Note that L^* is obtained from $L^\#$ by the $1/m_i, 1/m'_j, 1/m$, and $1/m^*$-surgeries along the components k_i, k'_j, k, k^*, respectively, for all i, j. Using that $S^3(L^*)_2$ is hyperbolic, we see that $L^\#$ is a prime link. Since $E(L^*, S^3)$ is hyperbolic for all m_i, m'_j, m and m^* with $|m_i|, |m'_j|, |m|$ and $|m^*|$ greater than c, we can conclude from the torus decomposition of $E(L^\#, S^3)$ that $L^\#$ is a hyperbolic link. Hence by Thurston's hyperbolic Dehn surgery,

$$\mathrm{Vol}E(L^*, S^3) < \sup_{\{m_0, m_1\}} \mathrm{Vol}E(L^*, S^3) < +\infty$$

if we take and fix m'_j and m, m^* for all j with $|m'_j|, |m|, |m^*| > c$. This complete the proof.

§2. The coefficient polynomials

The skein polynomial $P_L(\ell, m)$ of a link L is calculable in principle by the *initial condition*

(1) $$P_O(\ell, m) = 1,$$

where O is a trivial knot, and the *skein relation*

(2) $$\ell P_{L_+}(\ell, m) + \ell^{-1} P_{L_-}(\ell, m) + m P_{L_0}(\ell, m) = 0,$$

where L_+, L_- and L_0 are links, identical except the part in a 3-ball B in which the 2-strand braids with positive half twist, negative half twist and 0 twist are occurring, respectively. We call an operation producing from one of the links L_+, L_-, L_0 the other two links a *skein move within* B. For the component number r of L, we let

$$P_\#(L; \ell, m) = (\ell m)^{r-1} P_L(\ell, m).$$

Then the initial condition and the skein relation are written as

($1_\#$) $$P_\#(O; \ell, m) = 1,$$

($2_\#$) $$-\ell^2 P_\#(L_+; \ell, m) - P_\#(L_-; \ell, m) = (\ell^2 m^2)^\delta P_\#(L_0; \ell, m),$$

where we let $\delta = (r_+ - r_0 + 1)/2$ (=1 or 0) for the component numbers r_+, r_0 of L_+, L_0, respectively.

Then we see easily that $P_\#(L; \ell, m)$ is written as

$$\Sigma_{n=0}^{+\infty} p_{2n}(L;\ell) m^{2n},$$

where $p_{2n}(L;\ell)$ is a Laurent polynomial in ℓ^2, being 0 except a finite number of n. We denote $-\ell^2$ and $-m^2$ by x and y, respectively, and then $P_\#(L;\ell,m)$ and $p_{2n}(L;\ell)$ by $C_\#(L;x,y)$ and $c_n(L;x)(-x)^n$, respectively. Clearly, $c_n(L;x)$ is a Laurent polynomial in x and we have

$$C_\#(L;x,y) = \Sigma_{n=0}^{+\infty} c_n(L;x)(xy)^n.$$

Taking $c_n(L;x) = 0$ for $n < 0$, we obtain the following, giving an alternative description of the initial condition and the skein relation of the skein polynomial:

Theorem 2.1. *There is one and only one link invariant family of Laurent polynomials in x of a link L which is denoted by $c_n(L;x), n \in Z$, and has the following identities:*

(1) *For a trivial knot O,*

$$c_n(O;x) = \begin{cases} 0 & (n \neq 0) \\ 1 & (n = 0), \end{cases}$$

(2) $xc_n(L_+;x) - c_n(L_-;x) = c_{n-\delta}(L_0;x)$ *for all n, with* $\delta = (r_+ - r_0 + 1)/2$ *(= 0 or 1) for the component numbers r_+, r_0 of L_+, L_0, respectively.*

We call the Laurent polynomial $c_n(L;x)$ the nth *coefficient polynomial* of the skein polynomial $P_L(\ell, m)$ (or simply, of the link L). Some calculations of the coefficient polynomials has been made in [**K4**]. For example, for any link L with the components $k_i, i = 1, 2, ..., r$, and the total linking number λ, we have

$$c_0(L;x) = (x-1)^{r-1} x^{-\lambda} c_0(k_1;x) c_0(k_2;x) \ldots c_0(k_r;x),$$

$$c_0(k_i;1) = 1, \frac{d}{dx} c_0(k_i;1) = 0, i = 1, 2, ..., r,$$

and this characterizes the zeroth coefficient polynomial. Since $c_0(L;x)$ determines the component number r of L, we see that for any link L, the family of the coefficient polynomials $c_n(L;x)$ for all n determines not only $P_\#(L;\ell,m)$ but also the skein polynomial $P_L(\ell,m)$ itself. From now we shall calculate the coefficient polynomial $c_n(L^*;x)$ of an almost identical link imitation L^* of a link L constructed in §1 for all $m_i (i = 0, 1), m'_j (0 \leq j \leq 2s), m, m^*$ full twists with condition (#). We show the following three lemmas:

Lemma 2.2. $c_n(L^*;x)$ *is constant on all m_0, m_1 with $m_0 + m_1 = 0$.*

Lemma 2.3. *$c_n(L^*;x) = c_n(L;x)$ for all n with $n \le s$.*

Lemma 2.4. *$c_n(L^*;x) - c_n(L;x)$ is divided by $x^m - 1$ for all n.*

Proof of Lemma 2.2. By fixing the m'_j's and m, m^*, we write L^* as $L^*_{(m_0,m_1)}$. Let $m_0 > 0, m_1 < 0$ without loss of generality. By the skein relation corresponding to a skein move in $V_i, i = 0,1$, we have

$$c_n(L^*_{(m_0,m_1)};x) = x^{-1}c_n(L^*_{(m_0-1,m_1)};x) + x^{-1}c_{n-\delta}(L';x)$$

and

$$c_n(L^*_{(m_0-1,m_1)};x) = xc_n(L^*_{(m_0-1,m_1+1)};x) - c_{n-\delta}(L';x),$$

for some δ =0 or 1. Hence

$$c_n(L^*_{(m_0,m_1)};x) = c_n(L^*_{(m_0-1,m_1+1)};x) = ... = c_n(L^*_{(0,0)};x),$$

for $m_0 + m_1 = 0$. This completes the proof of Lemma 2.2.

Next, when we fix $m_i, i = 0,1$, and m, m^*, we denote by $L^*_{(u_0,u_1,...,u_{2s})}$ a link obtained from L^* by replacing m'_j full twists in Fig.4(2) with u_j *half* twists. Thus, $L^*_{(2m'_0,2m'_1,...,2m'_{2s})} = L^*$. Take a similar presentation of L using m_i, m'_j, m, m^*. Then we also use the notation $L_{(u_0,u_1,...,u_{2s})}$ if we replace m'_j full twists with u_j half twists and fix the other $m_i, i = 0,1$, and m, m^*, so that $L_{(2m'_0,2m'_1,...,2m'_{2s})} = L$.

If a Laurent polynomial $f(x)$ in x is written as a sum

$$\Sigma_{i=1}^n \epsilon_i x^{r_i} f_i(x)$$

with some $\epsilon_i = \pm 1$, integers r_i and Laurent polynomials $f_i(x)$ in x, we say that $f(x)$ is a *unit multiple sum of* $f_i(x), i = 1,2,...,n$.

Proof of Lemma 2.3. Let

$$c_{n(u_0,u_1,...,u_{2s})} = c_n(L_{(u_0,u_1,...,u_{2s})};x)$$

and

$$c^*_{n(u_0,u_1,...,u_{2s})} = c_n(L^*_{(u_0,u_1,...,u_{2s})};x).$$

Note by our assumption in §1 that $L \cap B_2$ in Fig.1(2) belongs to one component of L. Then there is a skein move in the 3-ball V'_1 changing $c^*_{n(2m'_0,2m'_1,...,2m'_{2s})}$ into a unit multiple sum of $c^*_{n(2m'_0,u_1,2m'_2,...,2m'_{2s})}$ and $c^*_{n(2m'_0,u'_1,2m'_2,...,2m'_{2s})}$ with $|u_1| = |2m'_1| - 1$ and $|u'_1| = |2m'_1| - 2$. There is also a skein move in V'_2 changing $c^*_{n(2m'_0,u_1,2m'_2,...,2m'_{2s})}$ into a unit multiple sum of $c^*_{n-1(2m'_0,u_1,u_2,...,2m'_{2s})}$ and $c^*_{n(2m'_0,u_1,u'_2,...,2m'_{2s})}$ with $|u_2| = |2m'_2| - 1$ and $|u'_2| = |2m'_2| - 2$. By induction on $|2m'_2|$, there are skein moves in V'_2 changing $c^*_{n(2m'_0,u_1,2m'_2,...,2m'_{2s})}$

into a unit multiple sum of $c^*_{n(2m'_0,u_1,0,\ldots,2m'_{2s})}$ and $c^*_{n-1(2m'_0,u_1,u_2^{(k)},\ldots,2m'_{2s})}$ for some odd integers $u_2^{(k)}, k = 1, 2, \ldots, |m'_2|$. Hence there are skein moves in V'_1 and V'_2 changing $c^*_{n(2m'_0,2m'_1,\ldots,2m'_{2s})}$ into a unit multiple sum of $c^*_{n(2m'_0,u'_1,2m'_2,\ldots,2m'_{2s})}$ and $c^*_{n(2m'_0,u_1,0,\ldots 2m'_{2s})}$ and the $c_{n-1(2m'_0,u_1,u_2^{(k)},\ldots,2m'_{2s})}$'s. Since $|u'_1| = |2m'_1| - 2$, we have by induction on $|2m'_1|$ that there are skein moves in V'_1, V'_2 changing $c^*_{n(2m'_0,2m'_1,\ldots,2m'_{2s})}$ into a unit multiple sum of a finite number of Laurent polynomials of the following types:

$$c^*_{n(2m'_0,v_1,0,\ldots,2m'_{2s})},$$

$$c^*_{n(2m'_0,0,v_2,\ldots,2m'_{2s})}$$

and

$$c^*_{n-1(2m'_0,u_1,u_2,\ldots,2m'_{2s})}$$

with u_1, u_2 odd.

Applying this process to $2m'_{2j-1}, 2m'_{2j}, j = 2, 3, \ldots, s$, we see that there are skein moves in $V'_j, j = 1, 2, \ldots, 2s$, changing $c^*_{n(2m'_0,2m'_1,\ldots,2m'_{2s})}$ into a unit multiple sum of a finite number of Laurent polynomials of the following types:

$$c^*_{n^*(2m'_0,v_1,\ldots,v_{2s})}$$

where $n - s < n^* \leq n$ and some $v_j = 0$, and

$$c^*_{n-s(2m'_0,u_1,\ldots,u_{2s})}$$

with u_j odd. By a property of the almost identical graph imitation, we see that

$$L^*_{(2m'_0,v_1,\ldots,v_{2s})} = L_{(2m'_0,v_1,\ldots,v_{2s})}.$$

Hence

$$c^*_{n^*(2m'_0,v_1,\ldots,v_{2s})} = c_{n^*(2m'_0,v_1,\ldots,v_{2s})}.$$

Let $n \leq s$. Then if

$$c^*_{0(2m'_0,u_1,\ldots,u_{2s})} = c_{0(2m'_0,u_1,\ldots,u_{2s})}$$

is proved, we can see from the skein relations corresponding to the skein moves in the V'_j's that

$$c_n(L^*;x) = c^*_{n(2m'_0,2m'_1,\ldots,2m'_{2s})} = c_{n(2m'_0,2m'_1,\ldots,2m'_{2s})} = c_n(L;x).$$

To show that

$$c^*_{0(2m'_0,u_1,\ldots,u_{2s})} = c_{0(2m'_0,u_1,\ldots,u_{2s})},$$

we first note that for any odd $u_0, L_{(u_0,u_1,\ldots,u_{2s})}$ is a link with two or more (in fact, $r+1$) components and $L^*_{(u_0,u_1,\ldots,u_{2s})}$ is an almost identical link imitation of

$L_{(u_0,u_1,\dots,u_{2s})}$. By the characterization of the zeroth coefficient polynomial([**K4**]), we have

$$c^*_{0(u_0,u_1,\dots,u_{2s})} = c_{0(u_0,u_1,\dots,u_{2s})}.$$

There are skein moves in V_0' changing $c^*_{0(2m_0',u_1,\dots,u_{2s})}$ into a unit multiple sum of $c^*_{0(0,u_1,\dots,u_{2s})}$ and a finite number of Laurent polynomials of the type $c^*_{0(u_0,u_1,\dots,u_{2s})}$ with u_0 odd. Since

$$L^*_{(0,u_1,\dots,u_{2s})} = L_{(0,u_1,\dots,u_{2s})},$$

we see from the skein relations corresponding to the skein moves in V_0' that

$$c^*_{0(2m_0',u_1,\dots,u_{2s})} = c_{0(2m_0',u_1,\dots,u_{2s})}.$$

This completes the proof of Lemma 2.3.

Proof of Lemma 2.4. Let $L^* = L^*_{(m,m^*)}$ by fixing the other m_i and m_j' for all i,j. Let $m > 0$ and $m^* < 0$. Since $m + m^* = 0$, by skein relations corresponding to skein moves in V^*, we have

$$c_n(L^*_{(m,m^*)};x) = x^m c_n(L^*_{(m,0)};x) - (1 + x + \dots + x^{m-1})c_n(L_H;x),$$

where L_H denotes a connected sum of the link L'' appearing in our construction of §1 and the Hopf link. By skein relations corresponding to skein moves in V, we have

$$c_n(L^*_{(m,0)};x) = x^{-m}c_n(L^*_{(0,0)};x) + x^{-m}(1 + x + \dots + x^{m-1})c_n(L^*_H;x),$$

where L^*_H is a link which is an almost identical link imitation of L_H. Noting $L^*_{0,0} = L$, we have

$$c_n(L^*_{(m,m^*)};x) = c_n(L;x) + \frac{x^m - 1}{x - 1}\{x^{-m}c_n(L^*_H;x) - c_n(L_H;x)\}.$$

Note that $c_n(L^*_H;1)$ and $c_n(L_H;1)$ are the z^{2n-r+1}-coefficients of the Conway polynomials of L^*_H and L_H, respectively. Since the Conway polynomial is known to be invariant under any link imitation(cf.[**K4**]), we have $c_n(L^*_H;1) = c_n(L_H;1)$ and $x^{-m}c_n(L^*_H;x) - c_n(L_H;x)$ is divided by $x - 1$, showing that $c_n(L^*;x) - c_n(L;x)$ is divided by $x^m - 1$. The same assertion for the case that $m < 0$ and $m^* > 0$ is proved similarly. This completes the proof of Lemma 2.4.

§3. Main theorem

Theorem 3.1. *For any link L in S^3, we consider links L', L'' such that L' is obtained from L by a surgery along a band b with coherent orientation and L'' is* *obtained from L by a crossing change within a 3-ball B where $L \cap B$ is a trivial 2-string tangle. Then for any positive integers N, N' and positive number C, there is an infinite family $\Im$ of almost identical link imitations L^* of L with all of the following properties:*

(1) *Each $L^* \in \Im$ has the hyperbolic covering property and there is a number $C^+ > C$ such that the hyperbolic volume $\mathrm{Vol}E(L^*, S^3) < C^+$ and $\sup_{L^* \in \Im} \mathrm{Vol}E(L^*, S^3) = C^+$,*
(2) *For the imitation map $q : (S^3, L^*) \to (S^3, L)$, there are a band b^* spanning L^* with diffeomorphism $q|b^* : b^* \to b$ and a 3-ball B^* in S^3 with a diffeomorphism $q|B^* : B^* \to B$ such that the maps*

$$q' : (S^3, L^{*\prime}) \to (S^3, L'),$$

and

$$q'' : (S^3, L^{*\prime\prime}) \to (S^3, L'')$$

induced from q by surgeries along b^*, b and the crossing changes within B^*, B corresponding by q, respectively, are homotopic to diffeomorphisms,
(3) *For any n, the nth coefficient polynomial $c_n(L^*; x)$ is constant on all $L^* \in \Im$ and $c_n(L^*; x) = c_n(L; x)$ for all $n < N$ and the difference $c_n(L^*; x) - c_n(L; x)$ is divided by $x^{N'} - 1$ for all $n \geq N$.*

Proof. In Lemma 1.3, we take $\Im$ to be the family of L^* for all m_0, m_1 with $m_0 + m_1 = 0$ and $|m_i| > c$, by choosing and fixing m'_j, m and m^* for all $j, j = 0, 1, ..., 2s$, so that they satisfy (#) and have $s = N - 1, |m'_j| > c$ and $|m| > \max\{c, N' - 1\}$. Let $C^+ = \sup_{L^* \in \Im} \mathrm{Vol}E(L^*, S^3)$. Then for any $L^* \in \Im$, L^* is an almost identical link imitation of L with hyperbolic covering property and $C <\mathrm{Vol}E(L^*, S^3) < C^+$, showing (1). We obtain (2) from Lemma 1.1. We obtain (3) from Lemmas 2.2,2.3,2.4. This completes the proof.

Remark 3.2. For the Jones polynomial $V_L(t)$(cf.[**J**]) of a link L, we let

$$V_\#(L; t) = (\sqrt{t} - t\sqrt{t})^{r-1} V_L(t^{-1}),$$

where r is the component number of L. Then we have

$$V_\#(L; t) = \Sigma_{n=0}^{+\infty} c_n(L; t^2) t^n (t-1)^{2n}$$

(cf.[**K4**]). Thus, the statement (3) of Theorem 3.1 implies that $V_\#(L^*; t)$ and hence $V_{L^*}(t)$ are constant on all $L^* \in \Im$ and $V_\#(L^*; t) - V_\#(L; t)$ is divided by $(t - 1)^{2N}(t^{2N'} - 1)$.

The following corollary is obtained directly from Theorem 3.1 by taking L, L' and L'' to be a trivial knot, a two-component trivial link and a trivial knot, respectively.

Corollary 3.3. *For any positive integers N, N' and any positive number C, there is an infinite family $\Im$ of knots O^* with trivial Alexander polynomial and with the same skein polynomial such that*

(1) *Each $O^* \in \Im$ has the hyperbolic covering property and there is a number $C^+ > C$ such that the hyperbolic volume $\mathrm{Vol}E(O^*, S^3) < C^+$ and $\sup_{O^* \in \Im} \mathrm{Vol}E(O^*, S^3) = C^+$,*

(2) *Each $O^* \in \Im$ is a ribbon knot of 1-fusion with unknotting number $u(O^*) = 1$,*

(3)
$$c_n(O^*; x) = \begin{cases} 0 & (0 < n < N) \\ 1 & (n = 0), \end{cases}$$

and $c_n(O^*; x)$ is divided by $x^{N'} - 1$ for all $n \geq N$.

Finally, we note that we can obtain similar results, taking an infinite family of hyperbolic links (but with essential Conway spheres) as $\Im$, when we use the results of **[K2]** alone instead of **[K3]**.

References

[F/Y/H/L/M/O] P. Freyed/D. Yetter/ J. Hoste/ W.B.R. Lickorish/ K. Millett/ A. Ocneanu, *A new polynomial invariant of knots and links*, Bull. Amer. Math. Soc. **12** (1985), 239-249.

[J] V. F. R. Jones, *Hecke algebra representations of braid groups and link polynomials*, Ann. of Math. **126** (1987), 335-388.

[K] T. Kanenobu, *Infinitely many knots with the same polynomial invariant*, Proc. Amer. Math. Soc. **97** (1986), 158-161.

[K0] A. Kawauchi, *Imitations of (3,1)-dimensional manifold pairs*, Sugaku (in Japanese) **40** (1988), 193-204; Sugaku Expositions (in English) **2** (1989), 141-156.

[K1] A. Kawauchi, *An imitation theory of manifolds*, Osaka J. Math. **26** (1989), 447-464.

[K2] A. Kawauchi, *Almost identical imitations of (3,1)-dimensional manifold pairs*, Osaka J. Math. **26** (1989), 743-758.

[K3] A. Kawauchi, *Almost identical imitations of (3,1)-dimensional manifold pairs and the branched coverings*, Osaka J. Math., to appear.

[K4] A. Kawauchi, *On the coefficient polynomials of the skein polynomial of an oriented link*, preprint.

[L/M] W. B. R. Lickorish/ K. Millett, *A polynomial invariant of oriented links*, Topology **26** (1987), 107-141.

[T1] W. P. Thurston, *Three dimensional manifolds, Kleinian groups and hyperbolic geometry*, Bull. Amer. Math. Soc. **6** (1982), 357-381.

[T2] W.P. Thurston, *The geometry and topology of 3-manifolds*, preprint.

3-manifolds that are unions of three handlebodies of genus at most 1

José Carlos Gómez-Larrañaga and Wolfgang Heil

Every closed connected 3-manifold M (orientable or non-orientable) can be decomposed into 4 balls; if M is covered by 3 open balls, then M is S^3 or a connected sum of an arbitrary number of S^2-bundles over S^1 [**7**]. By [**2**] every non-orientable closed 3-manifold is a union of three orientable handlebodies along their boundaries. This kind of decomposition is a special case of Singhof's "3-fillings" [**6**]. The next simplest building blocks after balls are solid tori. In [**1**] the manifolds M that are a union of 3 orientable cubes with handles of genus at most 1 are classified. In this paper we describe all 3-manifolds that are unions of 3 pieces, each a ball, a solid torus, a solid Kleinbottle, or $P^2 \times I$.

Notations and the structure of M

B^3: 3-ball.

P^2: (real) projective plane.

$S^2 \tilde{\times} S^1$: non-orientable S^2-bundle over S^1.

B denotes $S^2 \times S^1$ or $S^2 \tilde{\times} S^1$.

$\mathcal{B}$ denotes either S^3 or a connected sum of an arbitrary number of **B**'s.

$D^2 \tilde{\times} S^1$: solid Kleinbottle.

L or $\mathbf{L}_i$: Lens space with finite non-trivial fundamental group.

$S(3)$ denotes any Seifert fibre space with at most three exceptional fibers (orientable or not, any orbit space).

S_1 denotes a (generalized) Seifert fiber space over any (compact) surface with exactly one boundary curve, which is a reflector curve, and at most one exceptional point.

S_0 denotes a (generalized) Seifert fiber space over any (compact) surface with exactly one boundary curve, which is a reflector curve, and no exceptional points.

T_i is B^3, $D^2 \times S^1$, $D^2 \tilde{\times} S^1$, or $P^2 \times I$.

Knots 90

The standard assumption throughout this paper is that M is a closed 3-manifold with a decomposition

$$(*)\qquad M = T_1 \cup T_2 \cup T_3 \quad \text{such that} \quad T_i \cap T_j = \partial T_i \cap \partial T_j \quad (\text{for} \quad i \neq j)$$

is a compact (not necessarily connected) 2-manifold.

We let $F_1 = T_2 \cap T_3$, $F_2 = T_1 \cap T_3$, $F_3 = T_1 \cap T_2$.

Lemma 1 [1]. $\chi(F_i) = \frac{1}{2}[\chi(\partial T_j) + \chi(\partial T_k) - \chi(\partial T_i)]$ *for* (i, j, k) *any permutation of* $(1, 2, 3)$.

Lemma 2. *Let* $N = T_1 \cup T_2$, *and* G_1 *and* G_2 *be unions of (non-empty) boundary components of* N *with* $\partial N = G_1 \cup G_2$. *Then either a component of* F_3 *meets both* G_1 *and* G_2, *or* $T_i = P^2 \times I$ *(for* $i = 1$ *and* 2*) and* $P^2 \times 0$ *meets* G_1, $P^2 \times 1$ *meets* G_2.

Proof. Let $\partial T_i = S_{i1} \cup S_{i2}$ (where $S_{i1} = S_{i2}$ if $T_i \neq P^2 \times I$) be the components of ∂T_i. Suppose no component of F_3 meets both G_1 and G_2. Then if a component of F_3 is contained in $S_{11} \cap S_{21}$, all components of $\overline{S_{11} - F_3}$ and $\overline{S_{21} - F_3}$ meet only one G_j. Therefore the components of $\overline{S_{i2} - F_3}$ (for $i = 1$ and 2) must meet G_k for $k \neq j$. Then $T_i = P^2 \times I$ and S_{i1} meets G_j, S_{i2} meets G_k.

Lemma 3. *Let* M *be as in* $(*)$. *If* F_3 *is not connected and one component of* F_3 *is a disk* D *with* $\partial D \simeq 0$ *on* ∂T_3, *then* $M \approx M' \sharp B$, *where* $M' = T_1' \cup T_2' \cup T_3'$ *is as in* $(*)$ *and either*

(a) $T_i' \approx T_i$ *for* $i = 1, 2, 3$; *or*
(b) $T_1 \approx T_2 \approx P^2 \times I$, $T_i' \approx D^2 \tilde{\times} S^1$ *for* $i = 1$ *or* 2, *and* $T_j' \approx T_j$ *for* $j \neq i$.

Proof. There is a disk D' in T_3 with $\partial D' = \partial D$, and since F_3 is not connected it follows that $S = D \cup D'$ is a non separating sphere. Hence $M = M' \sharp \mathbf{B}$, where $M' = \overline{M - U(S)} \cup B_1 \cup B_2 = T_1'' \cup T_2'' \cup T_3'' \cup B_1 \cup B_2$ (see Figure 1). Here $U(S) = S \times I$ is a regular neighborhood of S, $T_i'' = \overline{T_i - U(S)} \approx T_i$, B_i is a ball with $\partial B_1 = S \times 0$ consisting of a disk in $\partial T_1''$ and a disk in $\partial T_3''$, B_2 is a ball with $\partial B_2 = S \times 1$ attached to $\partial(T_1'' \cup T_2'')$ and disjoint to T_3''. Let $T_1' = T_1'' \cup B_1 \approx T_1''$, hence $M' = T_1' \cup T_2'' \cup T_3'' \cup B_2$. If a component of $F_3' = T_1' \cap T_2''$ meets both $\partial T_3''$ and ∂B_2 (see Figure 2), let C be a tubular neighborhood in T_2'' of an arc in F_3' from ∂B_2 to $\partial T_3''$, let $T_3' = T_3'' \cup C \cup B_2$, and $T_2' = \overline{T_2'' - C} \approx T_2''$. This gives $M = T_1' \cup T_2' \cup T_3'$ as in case (a).

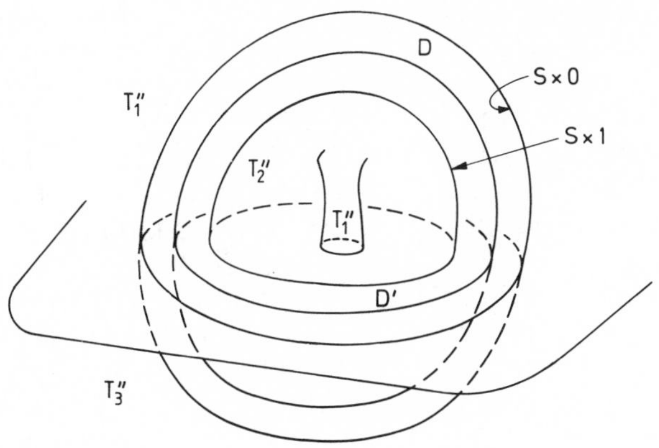

Figure 1

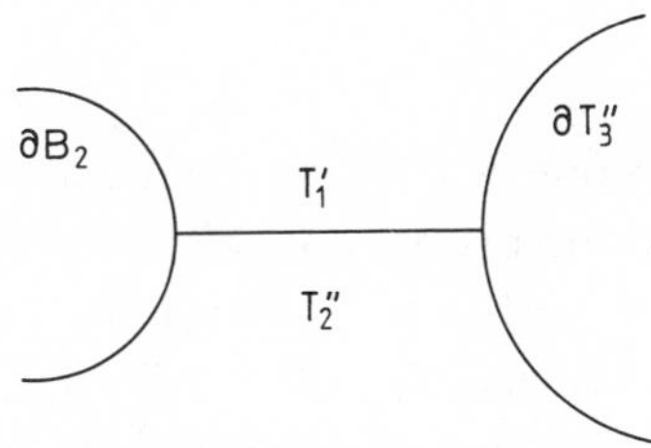

Figure 2

If no component of F_3' meets both ∂B_2 and $\partial T_3''$ then by Lemma 2 (with $N = T_1' \cup T_2''$, $\partial N = \partial B_2 \cup \partial T_3''$) we have $T_2'' \approx P^2 \times I$ and $P^2 \times 0$ meets ∂B_2, $P^2 \times 1$ meets $\partial T_3''$. Let C be a regular neighborhood of a vertical arc in T_2'', let $T_2' = \overline{T_2'' - C} \approx D^2 \tilde{\times} S^1$, and $T_3' = T_3'' \cup C \cup B_2 \approx T_3''$. Then $M = T_1' \cup T_2' \cup T_3'$ as in case (b).

Theorem 1. *If M admits a decomposition as in $(*)$ then M is as in the table below.*

	Decomposition $(*)$ of M			Homeomorphism type of M
	T_1	T_2	T_3	
$(1)(a)$	B^3	B^3	B^3	$\mathcal{B}$
(b)	B^3	B^3	$D^2 \times S^1$	$\mathcal{B}, \quad \mathbf{L}\sharp\mathcal{B}$
(c)	B^3	$D^2 \times S^1$	$D^2 \times S^1$	$\mathcal{B}, \quad \mathbf{L}\sharp\mathcal{B}, \quad \mathbf{L}_1\sharp\mathbf{L}_2\sharp\mathcal{B}$
(d)	$D^2 \times S^1$	$D^2 \times S^1$	$D^2 \times S^1$	$\mathcal{B}, \mathbf{L}\sharp\mathcal{B}, \mathbf{L}_1\sharp\mathbf{L}_2\sharp\mathcal{B}, \mathbf{L}_1\sharp\mathbf{L}_2\sharp\mathbf{L}_3\sharp\mathcal{B}, S(3)\sharp\mathcal{B}$
(2)	B^3	B^3	$P^2 \times I$	$\emptyset$
(3)	B^3	$P^2 \times I$	$P^2 \times I$	$(P^2 \times S^1)\sharp\mathcal{B}$
(4)	$P^2 \times I$	$P^2 \times I$	$P^2 \times I$	$(P^2 \times S^1)\sharp\mathcal{B}$
(5)	$P^2 \times I$	$D^2\tilde{\times}S^1$	$P^2 \times I$	$(P^2 \times S^1)\sharp\mathcal{B}$
(6)	$P^2 \times I$	$D^2\tilde{\times}S^1$	$D^2\tilde{\times}S^1$	$(P^2 \times S^1)\sharp\mathcal{B}$
(7)	$P^2 \times I$	$D^2\tilde{\times}S^1$	B^3	$(P^2 \times S^1)\sharp\mathcal{B}$
(8)	$P^2 \times I$	$D^2 \times S^1$	B^3	$\emptyset$
(9)	$P^2 \times I$	$D^2 \times S^1$	$P^2 \times I$	$(P^2 \times S^1)\sharp\mathcal{B}, \quad (P^2 \times S^1)\sharp\mathbf{L}\sharp\mathcal{B}$
(10)	$P^2 \times I$	$D^2 \times S^1$	$D^2\tilde{\times}S^1$	$(P^2 \times S^1)\sharp\mathcal{B}, \quad (P^2 \times S^1)\sharp\mathbf{L}\sharp\mathcal{B}$
(11)	$P^2 \times I$	$D^2 \times S^1$	$D^2 \times S^1$	$\emptyset$
(12)	B^3	B^3	$D^2\tilde{\times}S^1$	$\mathcal{B}$
(13)	B^3	$D^2 \times S^1$	$D^2\tilde{\times}S^1$	$\mathcal{B}, \quad \mathbf{L}\sharp\mathcal{B}$
(14)	B^3	$D^2\tilde{\times}S^1$	$D^2\tilde{\times}S^1$	$\mathcal{B}, \quad (P^2 \times S^1)\sharp\mathcal{B}$
(15)	$D^2 \times S^1$	$D^2 \times S^1$	$D^2\tilde{\times}S^1$	$\mathcal{B}, \quad \mathbf{L}\sharp\mathcal{B}, \quad \mathbf{L}_1\sharp\mathbf{L}_2\sharp\mathcal{B}$
(16)	$D^2 \times S^1$	$D^2\tilde{\times}S^1$	$D^2\tilde{\times}S^1$	$\mathcal{B}, \mathbf{L}\sharp\mathcal{B}, (P^2 \times S^1)\sharp\mathcal{B}, (P^2 \times S^1)\sharp\mathbf{L}\sharp\mathcal{B}, S_1\sharp$
(17)	$D^2\tilde{\times}S^1$	$D^2\tilde{\times}S^1$	$D^2\tilde{\times}S^1$	$\mathcal{B}, \quad (P^2 \times S^1)\sharp\mathcal{B}, \quad S_0\sharp\mathcal{B}$

Proof.

case (1) : This is Proposition 1, 2, 3 and Theorem 1 of [**1**].

case (2) : This decomposition is not possible since $\partial(B^3 \cup B^3) \neq P^2 \cup P^2$.

case (3) : By Lemma 1, $\chi(F_i) = 1$. Hence F_3 has a disk component D. If $F_3 = D$ then F_2 is a disk and $T_1 \cup T_2 \approx P^2 \times I$, $T_1 \cup T_2$ attached to T_3 along F_2 has a boundary the union of a Kleinbottle and two projective planes which is the double of F_1. Hence F_1 is the union of a Moebiusband and P^2 and $M = P^2 \times I \cup P^2 \times I$ where the four P^2's are identified in pairs. Thus $M = P^2 \times S^1$. If F_3 is not connected then since $\partial D \simeq 0$ on $\partial(P^2 \times I)$, it follows from Lemma 3 that $M = M'\sharp\mathbf{B}$ with M' admitting the same decomposition as M. Thus by the finiteness of the prime decomposition [**4**] it follows that $M \approx P^2 \times S^1\sharp\mathcal{B}$.

(4),(5),(6),(7) : In case (4); $\chi(F_i) = 1$. If no F_i has a disk component then each $F_i \approx P^2$ and $M \approx P^2 \times S^1$. Thus assume F_3 has a disk component D. Then F_3 is not connected since $\partial(T_1 \cup T_2) \neq \partial T_3$. Since $\partial D \simeq 0$ on ∂T_3 it follows from Lemma 3 that $M \approx M'\sharp\mathbf{B}$, where M' is as M or where M' has a decomposition as in (5). This step gives the three arrows in the diagram below that start at case (4). In case (5) we have $\chi(F_2) = 2$. Clearly $F_2 \neq P^2 \cup P^2$ so F_2 must contain a disk component D and F_2 is not connected. Case (a): If

$\partial D \not\simeq 0$ on ∂T_2 then $\partial D = c_1$ or $\partial D = c_2$, where c_1 bounds a disk in T_2 and c_2 separates ∂T_2 into two Moebiusbands ([**3**]). Let H be a regular neighborhood of D in $T_1 \cup T_3$ and let $T_2' = T_2 \cup H$ (which is obtained from T_2 by attaching the 2-handle H along c_i) and $T_i' = \overline{T_i - H} \approx T_i$ for $i = 1, 3$. If $\partial D = c_1$ then $T_2' = \overline{S^2 \tilde{\times} S^1 - B^3}$ and $M = \mathbf{B}\sharp(T_1' \cup T_3' \cup B^3)$ where the second term is as in case (3). Thus $M = P^2 \times S^1 \sharp \mathcal{B}$. If $\partial D = c_2$ then $T_2' \approx P^2 \times I$ and M is as in case (4). Case (b): If $\partial D \simeq 0$ on ∂T_2 then by Lemma 3, $M \approx M' \sharp \mathbf{B}$, with M' of the same type as M or M' as in (6).

In case (6), $\chi(F_i) = 1$ for $i = 2, 3$, hence F_3 has a disk component D and is not connected. Case (c): If $\partial D \not\simeq 0$ on ∂T_3 then as in (a) above by attaching the thickened D to T_3 we obtain $M = T_1' \cup T_2' \cup T_3'$ with $T_i' \approx T_i$ for $i = 1, 2$ and $T_3 = P^2 \times I$ or a punctured $S^2 \tilde{\times} S^1$. Thus M is as in case (5) or $M \approx \mathbf{B}\sharp M'$ where M' is as in (7). Case (d): If $\partial D \simeq 0$ on ∂T_3 then by Lemma 3, $M \approx \mathbf{B}\sharp M'$ where M' admits the same decomposition as M.

In case (7), $\chi(F_2) = 2$, hence F_2 is not connected and contains a disk component D. Case (e): If $\partial D \not\simeq 0$ on T_2 then as in cases (a) and (c) we obtain $M \approx \mathbf{B}\sharp M'$ where $M' = T_1' \cup T_3' \cup B^3$ as in (2) or M is as in (3). Case(f): If $\partial D \simeq 0$ on T_2 then by Lemma 3, $M \approx \mathbf{B}\sharp M'$, where M' is of the same type as M.

Note that as we chase M around the diagram it picks up factors of $\mathbf{B}$. Thus after finitely many steps there in no disc D as in cases (4)-(7) and the final M must be $P^2 \times S^1$ (as in the beginning of case (4)) or $P^2 \times S^1 \sharp \mathcal{B}$ (as in case 3).

$P^2 \times S^1$ $\longleftarrow$ case 4 $\circlearrowleft \sharp\mathbf{B}$

$\sharp\mathbf{B} \downarrow\uparrow \sharp\mathbf{B}$

case 5 $\circlearrowleft \sharp\mathbf{B}$

$\sharp\mathbf{B}$ $\swarrow$ $\quad$ $\sharp\mathbf{B} \downarrow\uparrow \sharp\mathbf{B}$

case 3 $\quad$ case 6 $\circlearrowleft \sharp\mathbf{B}$

$\sharp\mathbf{B}$ $\nwarrow$ $\quad$ $\sharp\mathbf{B} \downarrow$

case 2 $\overset{\sharp\mathbf{B}}{\longleftarrow}$ case 7 $\circlearrowleft \sharp\mathbf{B}$

Diagram for cases (4)-(7)

case (8) : $\partial(P^2 \times I) = F_2 \cup F_3$ where F_2 consists of planar surfaces and F_3 consists of planar surfaces and possibly a punctured torus. Since P^2 cannot be covered by two charts [**2**], this is impossible.

case (9),(10) : In case (9), $\chi(F_2) = 2$ and since $F_2 \neq P^2 \cup P^2$ it follows that F_2 is not connected and contains a disk component D. If $\partial D \not\simeq 0$ on ∂T_2 then attaching a thickened D to T_2 as a 2-handle gives a decomposition $M = T_1' \cup T_2' \cup T_3'$ with $T_i' \approx T_i$ for $i = 1, 3$ and T_2' a punctured lensspace (including S^3 or $S^1 \times S^2$). Thus $M \approx M'$ or $\mathbf{B}\sharp M'$ or $\mathbf{L}\sharp M'$, where M' is as in (3).

If $\partial D \simeq 0$ on ∂T_2 we apply Lemma 3 to get $M \approx M' \sharp \mathbf{B}$, where M' has a decomposition of the same type as M or is as in (10).

In case (10), $\chi(F_i) = 1$ for $i = 2, 3$ and since $F_3 \neq P^2$, F_3 contains a disc component D. If $F_3 = D$ then $\partial(T_1 \cup T_2) \neq \partial T_3$, so F_3 is not connected. If $\partial D \not\simeq 0$ on ∂T_3, then adding the thickened D as a 2-handle to T_3 yields $M = T_1' \cup T_2' \cup T_3'$ with $T_i' \approx T_i$ for $i = 1, 2$ and $T_3' \approx P^2 \times I$ or a punctured $S^2 \tilde{\times} S^1$. Hence M is as in (9) or $M \approx M' \sharp \mathbf{B}$ with M' as in (8). If $\partial D \simeq 0$ on ∂T_3, Lemma 3 implies that $M \approx M' \sharp \mathbf{B}$ with M' of the same type as M.

case (11) : $\partial T_1 = P^2 \cup P^2$ whereas $\partial(T_2 \cup T_3) = F_3 \cup F_2$ is a union of planar surfaces and punctured tori. As in case (8) this cannot happen.

case (12) : $\chi(F_3) = 2$, so F_3 is not connected and contains a disk component D. If $\partial D \not\simeq 0$ on ∂T_3 then attaching the thickened D to T_3 gives $M \approx T_1' \cup T_2' \cup T_3'$ with $T_i' \approx T_i$ for $i = 1, 2$ and $T_3' = P^2 \times I$ or a punctured $S^2 \tilde{\times} S^1$. Hence M is as in (2) or $M \approx \mathbf{B} \sharp M'$ with M' as in (1).

If $\partial D \simeq 0$ on ∂T_3 then by Lemma 3, $M \approx \mathbf{B} \sharp M'$ with M' of the same type as M.

case (13) : $\chi(F_2) = 1$, hence F_2 contains a disk component D. If $F_2 = D$ then $\partial(T_1 \cup T_3) \neq \partial T_2$ which is impossible. So F_2 is not connected. If $\partial D \not\simeq 0$ on ∂T_2 then attaching the thickened D to T_2 we obtain $M = T_1' \cup T_2' \cup T_3'$ with $T_i' \approx T_i$ for $i = 1, 3$ and T_2' a punctured lensspace (including S^3 or $S^2 \times S^1$). Hence $M \approx M'$, $\mathbf{B} \sharp M'$, or $\mathbf{L} \sharp M'$ with M' as in (12).

If $\partial D \simeq 0$ on ∂T_2 then $M \approx \mathbf{B} \sharp M'$ with M' of the same type as M.

case (14) : This proceeds exactly as in (13), except that T_2' is now either $P^2 \times I$ or a punctured $S^2 \tilde{\times} S^1$. This gives M as in (7) or $M \approx \mathbf{B} \sharp M'$ with M' as in (12).

case (15) : $\chi(F_i) = 0$ for $i = 1, 2, 3$. First suppose no F_i has a disk component. Then all components of F_i are annuli. The decomposition of ∂T_3 into annuli shows that they are compressible in T_3. Let $A = \partial D \times I$ be an annulus component of F_1, where $D \times I$ is a thickened meridian disk in T_3. Let T_2' be T_2 with the 2-handle $D \times I$ attached along A and let $T_3' = \overline{T_3 - D^2 \times I}$. Then $M = T_2' \cup (T_1' \cup T_3')$, where T_3' is a 3-ball and T_2' is a punctured lensspace. Hence $M = M'$, $\mathbf{B} \sharp M'$, or $\mathbf{L} \sharp M'$ where $M' = T_1 \cup B^3 \cup B^3$ is by case 1(b) either $\mathcal{B}$ or $\mathbf{L} \sharp \mathcal{B}$.

If F_2 has a disk component D and $\partial D \not\simeq 0$ in ∂T_2 then attaching the thickened D to T_2 gives $M = T_1' \cup T_2' \cup T_3'$ with $T_i' \approx T_i$ for $i = 1, 3$ and T_2' a punctured lensspace. Hence $M \approx M'$, $\mathbf{B} \sharp M'$, or $\mathbf{L} \sharp M'$ where M' is as in (13). The same argument applies for F_1.

If F_3 has a disk component D with $\partial D \not\simeq 0$ on ∂T_3 then we obtain $M = T_1' \cup T_2' \cup T_3'$ with $T_i' \approx T_i$ for $i = 1, 2$ and $T_3' = P^2 \times I$ or a punctured $S^2 \tilde{\times} S^1$. Hence M is as in (11) or $M = M' \sharp \mathbf{B}$ where $M' = (D^2 \times S^1) \cup (D^2 \times S^1) \cup B^3$. By case 1(c) M' is $\mathcal{B}$, $\mathbf{L} \sharp \mathcal{B}$, or $\mathbf{L}_1 \sharp \mathbf{L}_2 \sharp \mathcal{B}$.

case (16) : $\chi(F_i) = 0$ for $i = 1, 2, 3$. First assume no F_i has disk components. Then all components of F_2 and F_3 are annuli. If F_1 has no Moebiusband components then ∂T_2 and ∂T_3 decompose into annuli that are compressible in

T_2 and T_3, respectively. Let $T_3' = T_3 \cup (D^2 \times I)$, where $D^2 \times I$ is a regular neighborhood of a meridian disk D^2 in T_2 with $\partial D^2 \times I$ an annulus of F_1, and $T_2' = \overline{T_2 - D^2 \times I} \approx B^3$. Then $M = T_3' \cup (T_1 \cup T_2') = \mathbf{B}\sharp(B^3 \cup T_1 \cup T_2')$ where the second term is as in case 1(b). If F_1 has a Moebiusband then it contains exactly two Moebiusbands and its other components are annuli incompressible in T_2 and T_3. First assume that every annulus of F_2 and F_3 is incompressible in T_1. Then T_1, T_2, T_3 are as in Figure 3.

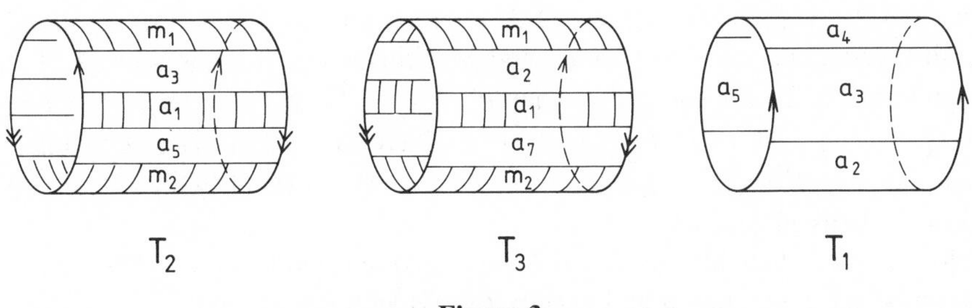

Figure 3

There is a (unique) Seifert fibration of T_i $(i = 2, 3)$ obtained from the trivial fibration $\{x\} \times I$ of $D^2 \times I$ by identifying $D^2 \times \{0\}$ and $D^2 \times \{1\}$ by a reflexion in a diameter ℓ of D. The annulus $\ell \times S^1$ consists of critical fibers and the orbit surface of T_i can be identified with the subdisk of the disk D^2 bounded by the reflectorline ℓ and an arc on ∂D^2 (see e.g. [**5**]). This fibering of T_2 and T_3 can be isotoped so that the Moebiusbands and annuli of F_i $(i = 1, 2, 3)$ are fibered and then the fibering of $F_2 \cup F_3 = \partial T_1$ can be extended to a Seifert fibration of T_1 with orbit surface a disk with at most one conepoint. This results in a Seifert fibering of $M = S_1$ as claimed (see Figure 4).

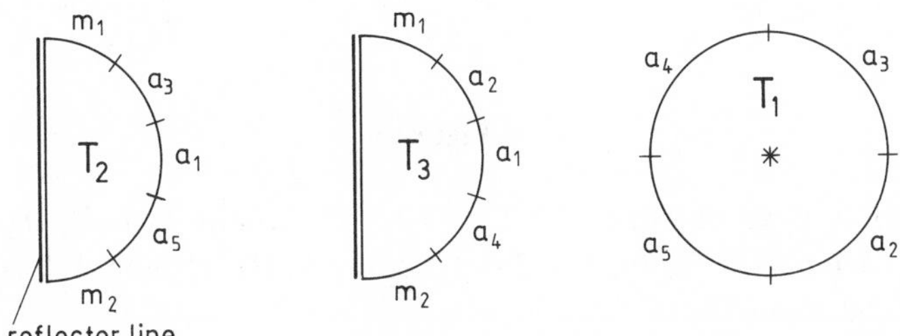

Figure 4

Now suppose an annulus A of F_2 is compressible in T_1. Let $T_3' = T_3 \cup (D^2 \times I)$ where $D^2 \times I$ is a regular neighborhood in T_1 of a compressing disk D^2 of $A = \partial D^2 \times I$, and $T_1' = \overline{T_1 - D^2 \times I} \approx B^3$. Then $M = T_3' \cup T_2 \cup T_1'$ is as in (7). The same argument applies if an annulus of F_3 is compressible in T_1.

Finally suppose that F_i has disk components. If F_1 has a disk component D with $\partial D \not\simeq 0$ on ∂T_1 then $M = T_1' \cup T_2' \cup T_3'$ where T_1' is obtained from T_1 by attaching a thickened D and $T_i' \approx T_i$ (for $i = 2, 3$). Thus $M \approx M'$, $\mathbf{B}\sharp M'$, or $\mathbf{L}\sharp M'$, where M' is as in (14).

If F_2 has a disk component D with $\partial D \not\simeq 0$ on ∂T_2 then $M = T_2' \cup T_1' \cup T_3'$ where T_2' is a punctured $S^2 \tilde{\times} S^1$ or $P^2 \times I$ and $T_i' \approx T_i$ for $i = 1, 3$. Thus $M \approx \mathbf{B}\sharp M'$ with M' as in (13) or M is as in (10). The same argument applies if F_3 has such a disk component.

If any of the disk components D of F_i has $\partial D \simeq 0$ on ∂T_i then by Lemma 3, $M \approx \mathbf{B}\sharp$(one of the above cases).

case (17) : $\chi(F_i) = 0$ for $i = 1, 2, 3$. First assume no F_i has disk components, i.e., all components of F_i are annuli and Moebiusbands. If an annulus A of F_i is compressible in T_j for some $j \neq i$, let $T_j' = \overline{T_j - D^2 \times I}$, where $\partial D^2 \times I = A$, and $T_k' = T_k \cup (D^2 \times I)$ for $i \neq k \neq j$. Then $M = T_k' \cup T_i \cup T_j'$ with T_k' a punctured $S^1 \tilde{\times} S^2$ or $P^2 \times I$ and $T_j' \approx B^3$. Thus $M \approx \mathbf{B}\sharp M'$ with M' as in case (12) or M is as in case (7).

Now assume that all annuli of F_i are incompressible in T_j. Then $F_i \cup F_j$ decompose ∂T_k into annuli and exactly two Moebiusbands and T_1, T_2, T_3 can be Seifert fibered with an induced fibration of the annuli and Moebiusbands such that the orbit surface of T_i is a disk with one reflector arc on its boundary. It follows that $M = S_0$ (see Figure 5). Finally assume that F_1 has a disk component D. If

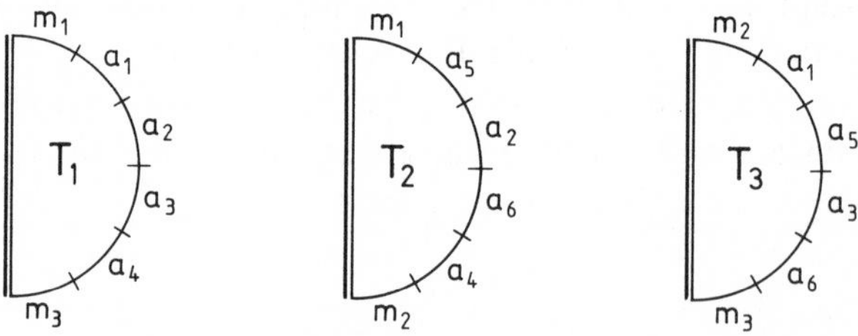

Figure 5

$\partial D \not\simeq 0$ on ∂T_1 we obtain $M = T_1' \cup T_2' \cup T_3'$, where T_1' is T_1 together with a thickened D and $T_i' \approx T_i$ for $i = 2, 3$. Then $M \approx \mathbf{B}\sharp M'$ with M' as in case (14) or M is as in case (6).

If $\partial D \simeq 0$ on ∂T_1 then by Lemma 3 and induction, $M \approx \mathcal{B}\sharp$(one of the above cases

Theorem 2. *If M is a manifold listed in the last column of the table in Theorem 1 (for any choices of $\mathcal{B}$ or $\mathbf{L}_i$) and where in cases (2)-(17) M is non-orientable, then M admits a decomposition as indicated in the same row.*

Proof. Suppose $M = T_1' \cup T_2' \cup T_3'$ and $N = T_1'' \cup T_2'' \cup T_3''$ are as in $(*)$. Let B' and B'' be 3-balls in M and N respectively as in Figure 6.

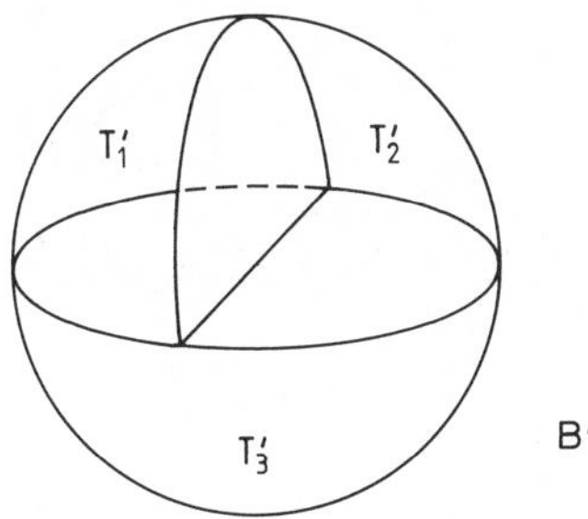

Figure 6

Then $M\sharp N = \overline{M - B'} \cup \overline{N - B'}$ where $\partial B'$ is identified with $\partial B''$ so that the disk $T_i' \cap \partial B'$ is identified with the disk $T_i'' \cap \partial B''$ for $i = 1, 2, 3$. Hence $M\sharp N = T_1 \cup T_2 \cup T_3$ where T_i is a disk sum of T_i' and T_i''. In particular if T_i'' is a 3-ball, then $T_i \approx T_i'$. Since $S^2 \tilde{\times} S^1 = T_1'' \cup T_2'' \cup T_3''$ with each $T_i'' \approx B^3$ it follows that whenever $M = T_1 \cup T_2 \cup T_3$ as in $(*)$ then $M\sharp\mathcal{B} = T_1' \cup T_2' \cup T_3'$ as in $(*)$ with $T_i' \approx T_i$ for $i = 1, 2, 3$.

So it suffices to show the existence of the decomposition in all cases without the $\mathcal{B}$-factor. We do this in cases (5), (6), (10), (14), (16) and (17). The other cases are similar or follow from the construction above or the proof of Theorem 1.

To represent $P^2 \times S^1$ as $(P^2 \times I) \cup (D^2 \tilde{\times} S^1) \cup (P^2 \times I)$ first decompose $P^2 \times S^1$ into 3 copies of $P^2 \times I$ where the boundary $P^2 \times I$ of copy k is identified with the boundary $P^2 \times 0$ of copy $k + 1 \quad (k \mod 3)$. Then remove a regular neighborhood of a vertical arc from copy 2 (to get $D^2 \tilde{\times} S^1$) and attach it to copy 1.

To represent $P^2 \times S^1$ as $(P^2 \times I) \cup (D^2 \tilde{\times} S^1) \cup (D^2 \tilde{\times} S^1)$ we start again with $P^2 \times S^1 = (P^2 \times I) \cup (P^2 \times I) \cup (P^2 \times I)$, remove a regular neighborhood of a vertical arc from copy 2 and a disjoint neighborhood of a vertical arc from copy 3 and attach both to copy 1.

To represent $P^2 \times S^1$ as $(P^2 \times I) \cup (D^2 \times S^1) \cup (D^2 \tilde{\times} S^1)$ we write $P^2 \times S^1 = (P^2 \times S^1)\sharp S^3$ and decompose $P^2 \times S^1 = (P^2 \times I) \cup B^3 \cup (D^2 \tilde{\times} S^1)$ and $S^3 = B^3 \cup (D^2 \times S^1) \cup B^3$.

To represent $P^2 \times S^1$ as $B^3 \cup (D^2 \tilde{\times} S^1) \cup (D^2 \tilde{\times} S^1)$ we decompose $P^2 \times S^1$ as $N_1 \cup N_2$ where $N_i \approx P^2 \times I$ and $N_1 \cap N_2 = (P^2 \times 0) \cup (P^2 \times 1)$. Drill out a vertical arc α_i in N_i with α_1 and α_2 disjoint, and an arc α in $P^2 \times 0$ joining α_1 and α_2. A regular neighborhood T of $\alpha_1 \cup \alpha \cup \alpha_2$ is B^3 and $\overline{N_i - T} \approx D^2 \tilde{\times} S^1$.

To represent S_1 as $(D^2 \times S^1) \cup (D^2 \tilde{\times} S^1) \cup (D^2 \tilde{\times} S^1)$ we decompose the orbit surface into three disks: first we remove from the orbit surface a regular neighborhood of the exceptional point. The remaining surface is represented as a punctured disk with bands (where the boundary of the puncture is the reflectorcurve) and is decomposed into two disks as in Figure 7.

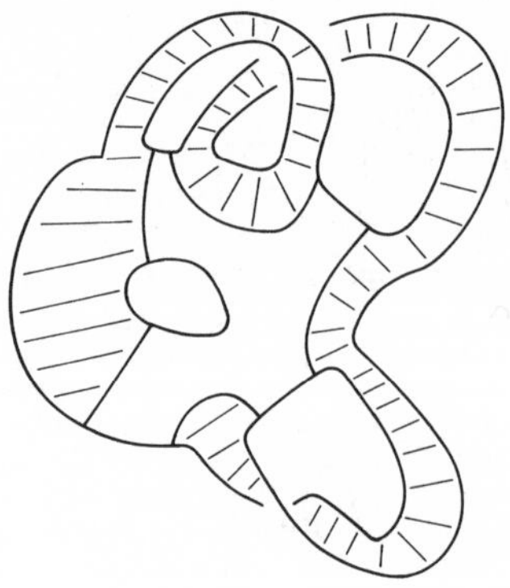

Figure 7

Finally, in order to represent any S_0 as $(D^2 \tilde{\times} S^1) \cup (D^2 \tilde{\times} S^1) \cup (D^2 \tilde{\times} S^1)$, we represent the orbit surface as a disk with bands (with boundary the reflectorcurve) and decompose the orbit surface as in Figure 8.

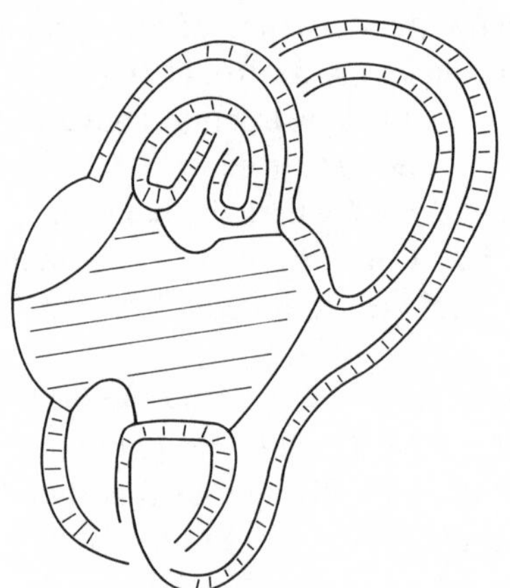

Figure 8

References

1. J.C. Gómez Larrañaga, *3-manifolds which are unions of three solid tori*, Manuscripta Math. **59** (1987), 325–330.
2. J.C. Gómez Larrañaga, F. González Acuña, J. Hoste, *Minimal atlases on 3-manifolds*, Math. Proc. Camb. Phil. Soc. **109** (1991), 105–115.
3. W.B.R. Lickorish, *Homeomorphisms of non-orientable 2-manifolds*, Proc. Cambridge Philos. Soc. **59** (1963), 307–317.
4. J. Milnor, *A unique factorization theorem for 3-manifolds*, Amer. J. Math. **84** (1962), 1–7.
5. P. Scott, *The geometries of 3-manifolds*, Bull. London Math. Soc. **15** (1983), 401–487.
6. W. Singhof, *Minimal coverings of manifolds with balls*, Manuscripta Math. **29** (1979), 385–415.
7. J. Hempel and D.R. McMillan, *Covering 3-manifolds with open cells*, Fund. Math. **64** (1969), 99–104.

On a hierarchal Jones invariant

Bohdan I. Kurpita and Kunio Murasugi

§0. Introduction

Recently Reshetikhin [**Re**], Rosso [**Ro**] and Wenzl [**We**] independently proved that we may associate with every finite dimensional irreducible representation of any complex simple Lie algebra a link invariant. In so doing they confirmed the conjecture proposed by Jones [**Jo2**].

In support of the then conjecture Jones in [**Jo2**] gave as an example a set of one-variable link invariants corresponding to the N-dimensional irreducible representation of $s\ell_2$. In Section 1 following this example of Jones we define a states model based on the vertex model in statistical mechanics. However, the Boltzmann weights obtained by Jones [**Jo2**] for each irreducible representation are the solutions of the *quantum* Yang-Baxter equation [**J1**], [**Dr**]. The partition function, after suitable normalisation, based on each set of Boltzmann weights gives rise to a hierarchal system of link invariants. The $N = 1$ case is not of much interest since it assigns the trivial polynomial to each link, but for completeness we include this case. The $N = 2$ case corresponds to the original one-variable Jones polynomial [**Jo1**], which has already been widely studied.

A hierarchal system of one-variable link invariants has also been independently discovered by M. Wadati, Y. Akutsu and T. Deguchi [**WAD**]. This hierarchal system can also be considered on a states model of the vertex type, and in fact the same model as that of Section 1 may be used to evaluate the link polynomials. The Boltzmann weights in this case correspond to the entries of a factorised S-matrix of an exactly solvable N-state model at criticality [**SAA**]. In Section 2 we confirm Jones' assertion in [**Jo2**] that these link invariants due to Wadati, Akutsu and Deguchi are the same as the link invariants obtained by Jones via the N-dimensional irreducible representation of $s\ell_2$. Hence we may define unequivocally a *hierarchal N-order Jones polynomial,* $V_L^{(N)}(t)$, using the states model and Boltzman weights of Section 1. The rest of the paper is concerned with developing some link theoretical properties of this hierarchal polynomial.

Knots 90

In Section 3 we prove a couple of results for the hierarchal N-order Jones polynomial of a periodic knot and in Section 4 evaluate the hierarchal polynomial at 1 and -1. In fact the evaluation at 1 is just a generalisation of the result for the $(N = 2)$ Jones polynomial; the result depends on the hierarchal number, N, and the number of components.

Sections 5 and 6 contain the main result of this paper the determination of the maximum and minimum degrees of the hierarchal polynomial of a (reduced) alternating link:

Proposition 5.3. *If $\widetilde{L}$ is a reduced (non-split) alternating diagram of an alternating link L, and $V_L^{(N)}(t)$ is the hierarchal N-order Jones polynomial then:*

$$d_{\max} V_L^{(N)}(t) = \frac{N(N-1)}{2} c_+(\widetilde{L}) - \frac{(N-1)}{2} \sigma(L)$$

and

$$d_{\min} V_L^{(N)}(t) = -\frac{N(N-1)}{2} c_-(\widetilde{L}) - \frac{(N-1)}{2} \sigma(L)$$

where $c_+(\widetilde{L})$ (or $c_-(\widetilde{L})$) is the number of positive (or negative) crossings of $\widetilde{L}$, and $\sigma(L)$ is the signature of L.

To prove the above proposition we introduce so called B-moves on a state, which cause the state to be changed to a new state with greater maximum degree. This process will terminate to give a state with maximal maximum degree.

The above proposition is a generalisation of the result for the $N = 2$ case (i.e. Jones polynomial, $V_L(t)$) proven in [**Mu2**]. It follows easily from the above proposition that in fact:

$$d_{\max} V_L^{(N)}(t) = (N-1) d_{\max} V_L(t) + \frac{(N-1)(N-2)}{2} c_+(\widetilde{L})$$

similarly for $d_{\min} V_L^{(N)}(t)$.

The final section discusses a couple of conjectures.

We would like to express our thanks to Dr. T. Deguchi for numerous conversations, and explanations of his work and the work he has done in collaboration with Professors M. Wadati and Y. Akutsu.

§1. A vertex model due to Jones, and Turaev

It is known that there exists a one-variable link invariant for each irreducible finite dimensional representation of a classical complex simple Lie algebra [**We**], [**Ro**], [**Re**]. Jones in [**Jo2**] gives examples of Boltzmann weights for various irreducible finite dimensional representations, and for each example an associated "states" model for a link diagram, from which a one-variable link invariant can be obtained.

The example which is of particular interest is "Example 1.20" in that paper, and the corresponding states model which is based on a vertex model derived from statistical mechanics. A similar model, also a vertex model, has been obtained independently by Wadati, Akutsu, and Deguchi [**WAD**] which we will discuss in Section 2.

In this section we will briefly review a modified version of the above mentioned vertex model due to Jones, see also [**Tu2**].

We know from statistical mechanics, see for example [**Ba**], that for a vertex model the state variables are placed on the edges.

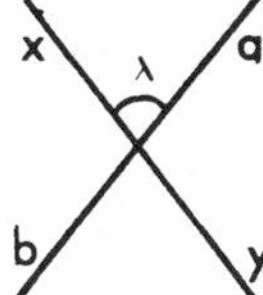

Figure 1.1

The state variables, in the above diagram $\{a, b, x, y\}$, are restricted to a finite set Θ.

To each such vertex with ascribed state variables we may assign a *Boltzmann weight* $w(ab \mid xy)(\lambda, t)$. With assigned state variables on each edge and a corresponding Boltzmann weight at each vertex of the model we obtain a *state* of the system.

In statistical mechanics we know, for example see [**Ba**], [**WAD**], that if the Boltzmann weights satisfy the Yang-Baxter equation, namely diagrammatically

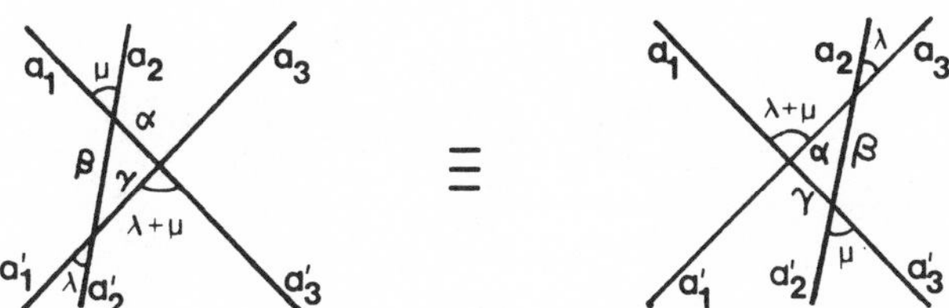

Figure 1.2

or in terms of the corresponding Boltzmann weights

$$\sum_{\alpha\beta\gamma} w(a_2\beta \mid a_1\alpha)(\mu, t) w(a_3\gamma \mid \alpha a_3')(\lambda+\mu, t) w(\gamma a_1' \mid \beta a_2')(\lambda, t)$$

$$= \sum_{\alpha\beta\gamma} w(a_3\alpha \mid a_2\beta)(\lambda, t) w(\alpha a_1' \mid a_1\gamma)(\lambda+\mu, t) w(\beta a_2' \mid \gamma a_3')(\mu, t)$$

then the system is exactly solvable.

The similarity between the above diagrammatic version of the Yang-Baxter equation and the third Reidemeister move, and the solutions of exactly solvable models have been fundamental in the construction of link invariants from exactly solvable models [**WAD**], [**Jo2**], [**Tu2**].

Given a set of Boltzmann weights, $w_{\pm}(ab \mid xy)(\lambda, t)$ which satisfy the Yang-Baxter equation, we may translate a vertex model in an obvious way onto an orientated link diagram [**Jo2**]. As above ascribe state variables on the edges of the link diagram, and let $\lambda \in (0, \pi)$ be the ingoing angle at each crossing, and t be an indeterminate. Then at each crossing we may assign a Boltzmann weight, $w_{\pm}(ab \mid xy)(\lambda, t)$, in the following manner, depending on whether it is an under-crossing or over-crossing.

$w_{+}(ab \mid xy)(\lambda, t)$ $w_{-}(ab \mid xy)(\lambda, t)$

Figure 1.3

Definition 1.1. Having assigned the states variables (from a finite set Θ) and Boltzmann weights in the above manner we may define the *partition function* for an orientated link diagram $\widetilde{L}$:

$$Z_{\widetilde{L}} = \sum \left(\prod w_{\pm}(ab \mid xy)(\lambda, t) \right) t^{\frac{1}{2\pi} \int_{\widetilde{L}} f_\sigma \, d\theta}$$

where the sum is taken over all possible states; the product is taken over all the crossings of $\widetilde{L}$; f_σ takes the constant value a along the edge with state variable a; and $d\theta$ is the change of angle or curvature 1-form on $\widetilde{L}$, usually referred to in the sequel as the *"angle contribution"*.

For the present due to Alexander's theorem we will restrict ourselves to a vertex model on a braid diagram and allow $\lambda \longrightarrow 0$.

Example 1.2.

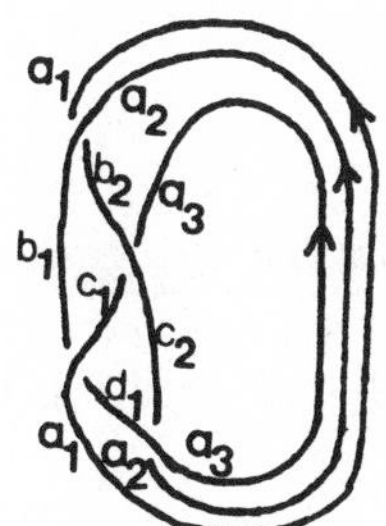

A braid diagram, $\sigma_1\sigma_2^{-1}\sigma_1\sigma_2^{-1}$, with state variables assigned and $\lambda \to 0$.

Figure 1.4

$$w_+(a_2b_1 \mid a_1b_2)(0,t)w_-(a_3c_1 \mid b_2c_2)(0,t)w_+(c_1a_1 \mid b_1d_1)(0,t)\times \\ w_-(c_2a_2 \mid d_1a_3)(0,t)t^{\frac{1}{2\pi}[2\pi(a_1+a_2+a_3)]}$$

In general for any given oriented braid diagram with state variables assigned and $\lambda \to 0$, the only angle contribution, i.e. $t^{\frac{1}{2\pi}\int_{\tilde{L}} f_\sigma d\theta}$, will come from the closure edges of the braid and the state variables that these edges carry. We will take the anti-clockwise direction to give us a positive contribution, and the clockwise direction to give us a negative contribution. This can easily be seen to be compatible with Definition 1.1.

Example 1.3. A braid $\beta \in B_3$ with angle contribution $t^{\frac{1}{2\pi}2\pi(a_2+a_3-a_1)}$.

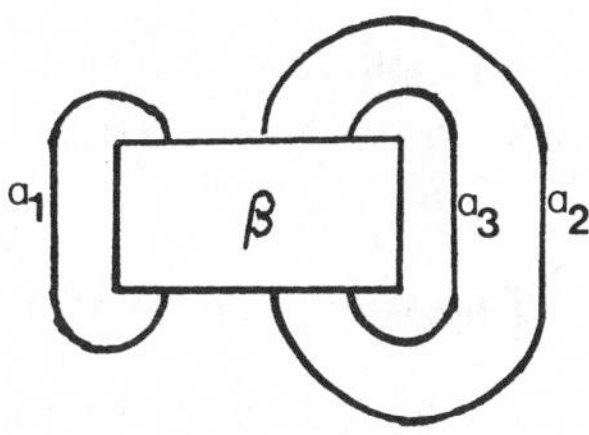

Figure 1.5

As noted above since for the present moment we will only be considering $w_\pm(ab \mid xy)(0,t)$, we may simply write this as $w_\pm(ab \mid xy)(t)$.

Definition 1.4. For each $N = 1, 2, 3, \ldots$ we may define a Θ-set to be $\{1, 2, \ldots, N\}$ and Boltzmann weights.

$$w_+^{(J;N)}(ab \mid xy)(t)$$
$$= \begin{cases} t^{-\frac{1}{2}[ax+by+(k-c+1)N-(k+c-1)]} \dfrac{(t;x-1)(t;N-y)}{(t;k)(t;y-1)(t;N-x)} \\ \qquad if \qquad a+x=c=b+y,\ k=x-y \ and\ x \geq y \\ 0 \qquad otherwise \end{cases}$$
$$w_-^{(J;N)}(ab \mid xy)(t) = w_+^{(J;N)}(xy \mid ab)(t^{-1}) ,$$

where $(t;n) = (1-t)(1-t^2)\ldots(1-t^n)$ and $(t;0) = 1$. Further, for the braid diagram the angle contribution of a given state is $t^{\sum_{i=1}^{n} \pm(a_i - \frac{(N+1)}{2})}$, where $(a_1, a_2, \ldots a_n)$ are the state variables carried by the closure edges. The sign being positive if closure is in the anti-clockwise direction, with regard to orientation, and negative if in the clockwise direction (see Example 1.3).

A few remarks need to be made concerning this definition. Firstly, these Boltzmann weights are the ones obtained by Jones in **[Jo2]** from the N-dimensional irreducible representations of $s\ell_2$, deduced from the work on quantum groups carried out by Jimbo **[Ji]**, and Drinfeld **[Dr]**. However, we have modified not only the formulation into a more calculable form, but also incorporated into the definition of the above Boltzmann weights the type I property mentioned in **[Jo2]**, i.e. this implies that the subsequent link invariant (Theorem 1.5 below) is also invariant under the first Reidemeister move.

Secondly, it should be noticed that for $N \geq 4$ the Boltzmann weights are not symmetric, for example in the case $N = 4$,

$$w_+^{(J;N)}(12 \mid 32)(t) \neq w_+^{(J;N)}(23 \mid 21)(t).$$

Finally, it is easy to see from Definition 1.4 that we may restrict the states on a link diagram to so called *contibuting states*, **[Tu2]**, and denoted by *c-states*, i.e. the states for which

$$\prod w_\pm(ab \mid xy)(t) \neq 0$$

where the product is taken over all crossings of the link diagram. In terms of Definition 1.4, this implies that we will only obtain a contributing state if $a + x = b + y$; and $x \geq y$, which implies that $b \geq a$ in the positive case, while in the negative case, $x \leq y$ and $b \leq a$.

Theorem 1.5 [Jo2] [Tu2]. *Suppose L is a closed n-braid then the following formula denoted by $V_L^{(J;N)}(t)$, and in the sequel called the hierarchal N-order Jones* *polynomial, is a link invariant:*

$$t^{\frac{N-1}{2}} \frac{(1-t)}{(1-t^N)} \sum \left(\prod w_{\pm}^{(J;N)}(ab \mid xy)(t) \right) t^{\sum_{i=1}^{n} \pm (a_i - \frac{(N+1)}{2})}$$

where the sum is taken over all contributing states on the braid diagram of L; the product is taken over all crossings on the braid diagram, and $(a_1, a_2, \ldots, a_n)$ *are the state variables carried by the closure edges of the braid diagram.* □

By Definition 1.1, it is possible to define the same link invariant using *any* link diagram. However in this general case the angle contribution is usually quite hard to explicate. So, for the present moment we will restrict ourselves to Theorem 1.5.

Note also that in the case $N = 2$, via the substitution $\sqrt{t} \mapsto -\sqrt{t}$, we obtain the "original" one-variable Jones polynomial $V_L(t)$, [**Jo1**].

§2. A set of Boltzmann weights due to Wadati, Akutsu and Deguchi

In the previous section we defined a set of Boltzmann weights obtained by V. Jones via quantum mechanics. On the other hand, Sogo, Akutsu and Abe [**SAA**] showed the existence of a set of Boltzmann weights which are the solutions of certain exactly solvable N-state models in statistical mechanics, the case $N = 2$ being the symmetric 6-vertex model. In [**WAD**] Wadati, Akutsu and Deguchi explicitly calculate these Boltzmann weights, $\sigma_{b'y',x'a'}$, and prove that from these weights one may also obtain a hierarchal N-order link invariant.

For convenience, we redefine their Boltzmann weights as follows:

Definition 2.1. For each $N = 1, 2, 3, \ldots$ define the Θ-set to be $\{1, 2, \ldots, N\}$ and Boltzmann weights:

$$w_{+}^{(W;N)}(ab \mid xy)(t) = \begin{cases} (-1)^{b+x} t^{-\frac{1}{2}[ax+by+N(k-c+1)-(k+c-1)]} \times \\ \qquad \left[\dfrac{(t;x-1)(t;N-y)}{(t;k)(t;y-1)(t;N-x)} \cdot \dfrac{(t;b-1)(t;N-a)}{(t;k)(t;a-1)(t;N-b)} \right]^{\frac{1}{2}} \\ \qquad \text{if} \quad a+x=c=b+y;\ k=x-y \ \text{and} \ x \geq y. \\ 0 \qquad \text{otherwise} \end{cases}$$

$$w_{-}^{(W;N)}(ab \mid xy)(t) = w_{+}^{(W;N)}(xy \mid ab)(t^{-1}).$$

Further, as in Definition 1.4 we may define the angle contribution to the braid diagram to be given by $t^{\sum_{i=1}^{n} \pm (a_i - \frac{(N+1)}{2})}$, again as before $(a_1, a_2, \ldots, a_n)$ are the state variables carried by the closure strands. The direction and the determination of the sign is as in Definition 1.4. This completes Definition 2.1.

The above formula for the Boltzmann weights is just a reformulation of formula (6.16) as defined in [**WAD**], so that the graphical interpretation agrees with that of §1. i.e. Figure 1.3. Further, the above Boltzmann weights satisfy the type I property in [**Jo2**], namely invariance of the link invariant in the sequel under the first Reidemeister move.

Although, at first sight the "angle contribution" in [**WAD**] which consists of assigning a crossing multiplier (defined in [**WAD**]) at the maxima of their vertex model on the braid diagram, called annihilation diagrams by them, and also at the minima, the so called creation diagrams, may seem different. However, by noticing that on a braid diagram the maxima and minima occur only on the closure strands by a suitable ambient isotopy, then their definition of values for the crossing multiplier for given N-state vertex models implies that their "angle contribution" coincides with Jones' angle contribution of Definition 1.4.

Finally note that the Boltzmann weights are symmetric, i.e. $w_{\pm}(ab \mid xy)(t) = w_{\pm}(yx \mid ba)(t)$.

Theorem 2.2 [WAD]. *Suppose that the oriented link L is a closed n-braid. Then $V_L^{(W;N)}(t)$ given by the formula:*

$$\frac{t^{\frac{N-1}{2}}(1-t)}{(1-t^N)} \sum_{c-states} \Big(\prod_{crossings} w_{\pm}^{(W;N)}(ab \mid xy)(t) \Big) t^{\sum_{i=1}^{n} \pm (a_i - \frac{(N+1)}{2})}$$

where a_i are as in Theorem 1.5, is a link invariant. □

Similarly to that noted after Theorem 1.5, we may also extend this theorem to *any* link diagram, using the creation and annihilation diagrams to calculate the angle contribution.

Definition 2.3.

(i) Let $\widetilde{V}_{\widetilde{L},S}^{(J;N)}(t)$ be defined from Theorem 1.5 as

$$\prod w_{\pm}^{(J;N)}(ab \mid xy)(t)$$

where the product runs over all crossings of a link diagram $\widetilde{L}$ with a given c-state S. Similarly we may define $\widetilde{V}_{\widetilde{L},S}^{(W;N)}(t)$, however in this case using the Boltzmann weights of Definition 2.1.

(ii) Consider an axis which is parallel to the top of the unclosed braid (see Example 2.4 below). Let $\widetilde{L}^+$ be the closed braid obtained from the closed braid $\widetilde{L}$ by rotation of S^3 by an angle π around this axis and then by a change in orientation.

Example 2.4.

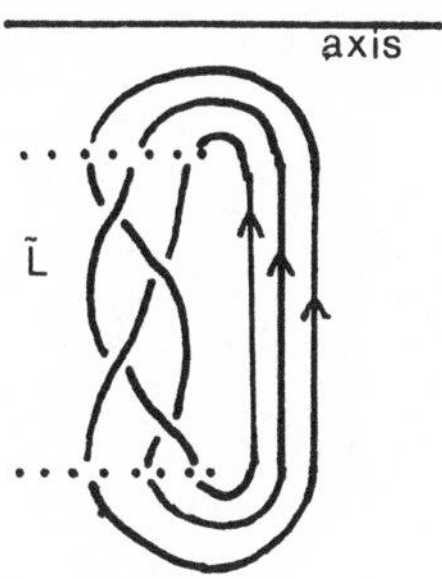

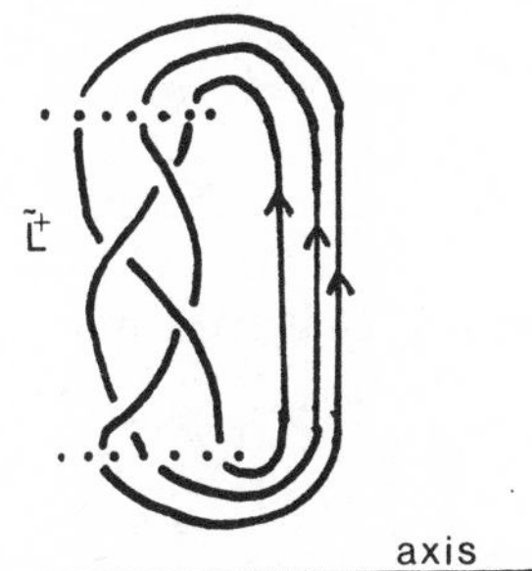

Figure 2.1

The definition of $w_+^{(W;N)}(ab \mid xy)(t)$ contains a factor of $(-1)^{b+x}$. However $a+x = c = b+y$ implies that $b+x \equiv a+y \bmod 2$. Hence we may always take the state variables assigned on the left-hand side of the crossing (see Figure 1.3) as contributing to the $(-1)^{b+x}$ factor. Note that $w_-(ab \mid xy)(t) = w_+(xy \mid ab)(t^{-1})$ and by the above $(-1)^{a+y} = (-1)^{b+x}$. Therefore the contribution to $\widetilde{V}_{\widetilde{L},S}^{(W;N)}(t)$ by the factors $(-1)^{b+x}$ may be taken along the odd-numbered strings. By definition of the vertex model, i.e. state variables appearing on the edges, each state variable on a particular odd-numbered string will contribute twice to the (-1)-factor. Hence the contribution of the (-1)-factor is negated. In other words, we have

Lemma 2.5. *The product of the (-1)-factors in $\widetilde{V}_{\widetilde{L},S}^{(W;N)}(t)$ is always equal to 1.*

□

The rotation in Definition 2.3 (ii) gives us a 1-1 correspondence between the c-states, $\{S\}$, on $\widetilde{L}$ and the c-states, $\{S^+\}$, on $\widetilde{L}^+$. This correspondence can be seen diagrammatically as

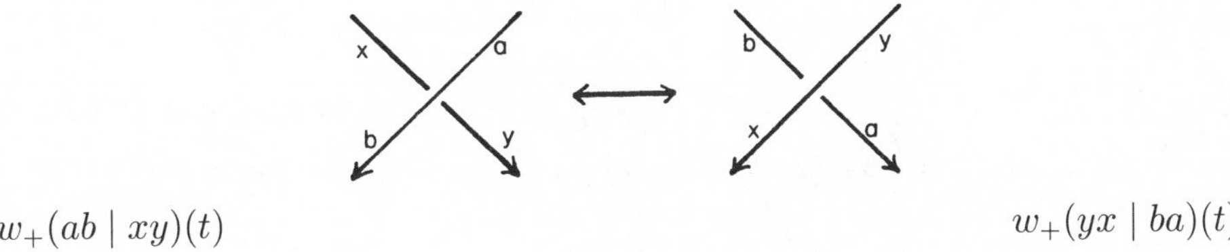

$w_+(ab \mid xy)(t)$ $\qquad$ $w_+(yx \mid ba)(t)$

Figure 2.2

and similarly $w_-(ab \mid xy)(t) \leftrightarrow w_-(yx \mid ba)(t)$. Since by definition $x \geq y$ and $a + x = c = b + y$ we also have that $b \geq a$.

By the symmetry of the Boltzmann weights in Definition 2.1, and also by the fact that k and c are invariant under the above $1-1$ correspondence we have that

$$\widetilde{V}^{(W;N)}_{\widetilde{L},S}(t) = \widetilde{V}^{(W;N)}_{\widetilde{L}^+,S^+}(t) \tag{2.1}$$

and also

Lemma 2.6. *Under the rotation by π defined in Definition 2.3 (ii) we have that*

$$\widetilde{V}^{(W;N)}_{\widetilde{L},S}(t) = \widetilde{V}^{(W;N)}_{\widetilde{L}^+,S^+}(t) \quad and \quad \widetilde{V}^{(J;N)}_{\widetilde{L},S}(t) = \widetilde{V}^{(J;N)}_{\widetilde{L}^+,S^+}(t)\,.$$

Proof. To prove that $\widetilde{V}^{(J;N)}_{\widetilde{L},S}(t) = \widetilde{V}^{(J;N)}_{\widetilde{L}^+,S^+}(t)$ it is sufficient to show $\frac{\widetilde{V}^{(J;N)}_{\widetilde{L},S}(t)}{\widetilde{V}^{(J;N)}_{\widetilde{L}^+,S^+}(t)} = 1$.

Without loss of generality we need only consider a positive crossing with given state variables, i.e.

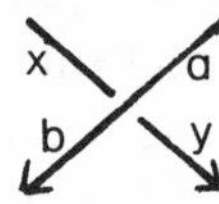

and this corresponds to $t^{-\frac{1}{2}[ax+by+N(k-c+1)-(k+c-1)]}\ \dfrac{(t;x-1)(t;N-y)}{(t;k)(t;y-1)(t;N-x)}$.

Clearly under the rotation, see Figure 2.2, the $t^{-\frac{1}{2}[ax+by+N(k-c+1)-(k+c-1)]}$ factor and $(t;k)$ terms remain the same, and so cancel in the quotient. So we need only consider $\frac{(t;x-1)(t;N-y)}{(t;y-1)(t;N-x)}$ factors.

Now consider the "y" state variable, then by continuity of the braid diagram, there exists another crossing (possibly in the extreme case σ_i itself) which also has "y" as a state variable. There are four possibilities.

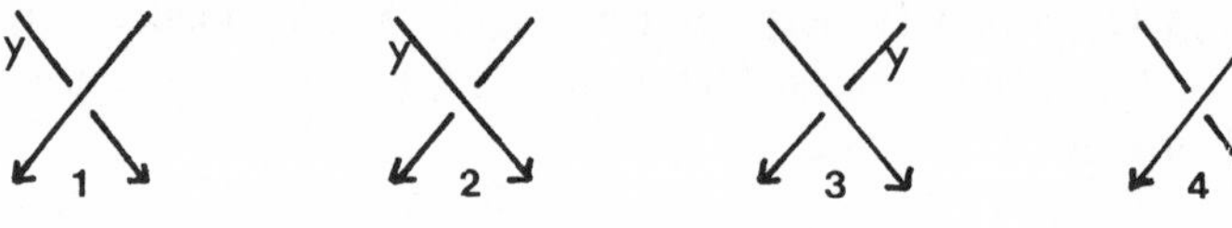

Figure 2.3

Case 1: $w_+^{(J;N)}(t)(-- \mid y-) \doteq \frac{(t;y-1)(-)}{(-)(t;N-y)}$. In this case we have cancellation in $\widetilde{V}^{(J;N)}_{\widetilde{L},S}(t)$ where $f \doteq g$ means that f and g are equal up to a factor $\pm t^m$.

Case 2: $w_-^{(J;N)}(-- \mid y-)(t) = w_+^{(J;N)}(y- \mid --)(t^{-1}) \doteq w_+^{(J;N)}(y- \mid --)(t)$, and when we rotate by π around the axis we obtain the Boltzmann weights

$$w_-^{(J;N)}(-y \mid --)(t) \doteq w_+^{(J;N)}(-- \mid -y)(t) \doteq \frac{(-)(t;N-y)}{(t;y-1)(-)}$$

hence we have cancellation in the quotient.

Case 3 is similar to Case 1, and Case 4 is similar to Case 2. The meaning of "$\doteq$" seems a bit vaque, but it is easy to see that certain factors in the Boltzmann weights due to the invariance of k and c under the rotation can be ignored since they can easily be seen to cancel in the quotient.

We may apply a similar argument to each contributing state variable in $\widetilde{V}_{\widetilde{L},S}^{(J;N)}(t)$ and $\widetilde{V}_{\widetilde{L}^+,S^+}^{(J;N)}(t)$, and hence we have cancellation either in $\widetilde{V}_{\widetilde{L},S}^{(J;N)}(t)$ (or $\widetilde{V}_{\widetilde{L}^+,S^+}^{(J;N)}(t)$ respectively) or in the quotient, and so we obtain the result.

$$\frac{\widetilde{V}_{\widetilde{L},S}^{(J;N)}(t)}{\widetilde{V}_{\widetilde{L}^+,S^+}^{(J;N)}(t)} = 1 . \qquad \square$$

From Definitions 1.4 and 2.1, Lemma 2.5 and Figure 2.2, we have that

$$\widetilde{V}_{\widetilde{L},S}^{(J;N)}(t)\widetilde{V}_{\widetilde{L}^+,S^+}^{(J;N)}(t) = \widetilde{V}_{\widetilde{L},S}^{(W;N)}(t)\widetilde{V}_{\widetilde{L}^+,S^+}^{(W;N)}(t) .$$

While Lemma 2.6 says that $[\widetilde{V}_{\widetilde{L},S}^{(J;N)}(t)]^2 = [\widetilde{V}_{\widetilde{L},S}^{(W;N)}(t)]^2$, and hence $\widetilde{V}_{\widetilde{L},S}^{(J;N)}(t) = \pm\widetilde{V}_{\widetilde{L},S}^{(W;N)}(t)$. However, on both sides the $t^{\frac{1}{2}(-)}$ factors agree and by Lemma 2.5 no (-1)-factor appears on the right-hand side. Finally, it is easy to see that the constant term coming from the rest of $\widetilde{V}_{\widetilde{L},S}^{(J;N)}(t)$ and $\widetilde{V}_{\widetilde{L},S}^{(W;N)}(t)$ is equal to 1 on both sides and hence

$$\widetilde{V}_{\widetilde{L},S}^{(J;N)}(t) = \widetilde{V}_{\widetilde{L},S}^{(W;N)}(t)$$

and we have

Proposition 2.7. *The N-order link invariant obtained by Wadati, Akutsu and Deguchi using their symmetric Boltzmann weights is equal to the hierarchal N-order Jones polynomial, i.e.*

$$V_L^{(J;N)}(t) = V_L^{(W;N)}(t) \qquad \square$$

We shall now adopt the notation $V_L^{(N)}(t)$ for the hierarchal N-order Jones polynomial whether it be in reference to $V_L^{(J;N)}(t)$ or $V_L^{(W;N)}(t)$. Further, for the rest of the paper we will restrict ourselves to the Boltzman weights of Definition 1.4, which we shall denote by $w_{\pm}^{(N)}(ab \mid xy)(t)$.

Wadati, Akutsu and Deguchi have shown that $V_L^{(N)}(t)$ satisfies an N-order skein relation i.e.

$$\underbrace{V_L^{(N)}(t)}_{N \text{ twists}} + a_{N-1}(t)\underbrace{V_L^{(N)}(t)}_{(N-1) \text{ twists}} + \cdots$$

$$+ \ a_1(t)V^{(N)}_{L\times}(t) \ + \ a_0(t)V^{(N)}_{L\,)(}(t) \ = \ 0$$

Yamagishi, Ge, and Wu [**YGW**] have obtained the same skein relation for a link invariant via Witten's [**Wi**] construction of link invariants using the Chern-Simon action in topological quantum field theory. Wenzl has also obtained this N-order hierarchal Jones polynomial but his approach is via representation theory, see Theorem 8(b) in [**We**].

§3. Periodic knots and the hierarchal Jones polynomial

In the case $N = 2$, i.e. the original Jones polynomial, Murasugi [**Mu3**], Traczyk [**Tr1**], and Yokota [**Yo**] have established criteria for periodic knots using graph theory, skein theory and representation theory respectively. Using the states model of §1, and a modification on the angle contribution, certain critera may also be obtained for the hierarchal N-order Jones polynomial.

Let K in this section denote an oriented *knot* that has prime period $r \geq 2$. In other words there exists an orientation preserving auto-homeomorphism $\phi : S^3 \to S^3$ of order r, with a set of fixed points $F \cong S^1$ disjoint from K, and which maps K onto itself. F, by the positive solution of the Smith Conjecture, is unknotted. Let $\Sigma^3 = S^3/\phi$ be the quotient space under ϕ. By the fact that F is unknotted Σ^3 is also a 3-sphere, and S^3 is the r-fold cyclic covering space of Σ^3 branched along F.

If $\psi : S^3 \to \Sigma^3$ is the covering projection then by $K_*(= \psi(K))$ denote the *factor knot.*

As noted after Theorem 1.5 the state model for the hierarchal Jones polynomial, obtained by Jones and Turaev, may be extended from the braid diagram of Theorem 1.5 to any link diagram by using the partition function of Definition 1.1. In this case the only change to Theorem 1.5 is the angle contribution, so define

$$t^{f_{\widetilde{L}}(\theta,S)} \underset{\lambda \to 0}{=} t^{\frac{1}{2\pi}\int_{\widetilde{L}} f_\sigma \, d\theta}$$

where the right-hand side is as in Definition 1.1, S being a particular c-state.

Theorem 3.1 [**Jo2**], [**Tu2**]. *Let L be an oriented link and $\widetilde{L}$ a link diagram of L, then the hierarchal N-order Jones polynomial of L is given by:*

$$\left(t^{-\frac{(N-1)}{2}} + t^{-\frac{(N-3)}{2}} + \cdots + t^{\frac{N-3}{2}} + t^{\frac{N-1}{2}}\right) V^{(N)}_L(t)$$
$$= \sum_{c-states} \Big(\prod_{crossings} w^{(N)}_{\pm}(ab \mid xy)(t)\Big) t^{f_{\widetilde{L}}(\theta,S)} . \square$$

A useful way of calculating the angle contribution in general is to use the annihilation and creation diagrams mentioned in the remarks following Definition 2.1. In our interpretation of the Boltzmann weights this amounts to assigning $t^{\pm\frac{1}{2}(x-\frac{(N+1)}{2})}$ to the maxima and minima, the sign convention being consistent with that of Definition 1.4.

Lemma 3.2.
(i) $\frac{(t;x-1)(t;N-y)}{(t;k)(t;y-1)(t;N-x)}$, *where* $k = x - y \geq 1$, $1 \leq x$, $y \leq N$ *and* $(t;n) = (1-t)(1-t^2)\dots(1-t^n)$, *can be written in either of the following two equivalent forms:* $(1-t^y)\dots(1-t^{x-1})g_1(t)$ *or* $(1-t^{N-x+1})\dots(1-t^{N-y})g_2(t)$ *where* $g_1(t)$ *is a polynomial of degree* $(N-x)k$, *and* $g_2(t)$ *is a polynomial of degree* $(y-1)k$. *By definition if* $k = 0$ *then the expression is equal to 1.*
(ii) Let S *be a contributing state on a knot diagram* $\widetilde{K}$ *such that the state variables at each crossing satisfy* $x = y$ *(and* $a = b$*), see Figure 1.3. Then if we fix a state variable on an edge,* x *say, the knot diagram will carry this fixed state variable on every edge.*
(iii) $w_{\pm}^{(N)}(ab \mid xy)(t) = w_{\pm}^{(N)}(N+1-y,\ N+1-x \mid N+1-b,\ N+1-a)(t)$.

Proof.
(i) Let $G(R, M, t) = \sum_{n\geq 0} \rho(R, M, n)t^n$ $R, M \geq 0$ where $G(R, M, t)$ is a polynomial of degree RM, and $\rho(R, M, n)$ denotes the number of partitions of n into at most M parts, with each part less than or equal to R. Then Theorem 3.1 of [**An**] states that if $R \geq 1$

$$G(R, M, t) = \frac{(1-t^{R+M})(1-t^{R+M-1})\dots(1-t^{M+1})}{(1-t^R)(1-t^{R-1})\dots(1-t)}.$$

In our case set $R = k$ and $M = N - x$ to obtain $g_1(t) = G(k, N-x, t)$, and let $M = y - 1$ to obtain $g_2(t) = G(k, y-1, t)$.

(ii) Starting with the edge with the fixed state variable x follow the knot diagram around until we return to this edge. Since $x = y$ and $a = b$ at each crossing, as we follow the knot diagram around, the state variables on the edges passed will all be equal to x. Since K is a knot the result follows.

(iii) By straightforward calculation. □

We may think of K_*, the factor knot, as the closure of a (n, n)-tangle, k_*, and K, the periodic knot, as r copies of this (n, n)-tangle joined end to end as in Figure 3.1.

If we fix N, we may define a Θ-set, and hence the values of the state variables, as in Definition 1.4. For a given c-state S on K we write $\{a_1^{(i)}, a_2^{(i)}, \dots, a_n^{(i)}\}$

for the state variables which lie on the edges joining $k_*^{(i)}$ to $k_*^{(i+1)}$. By definition $k_*^{(r+1)}$ is $k_*^{(1)}$.

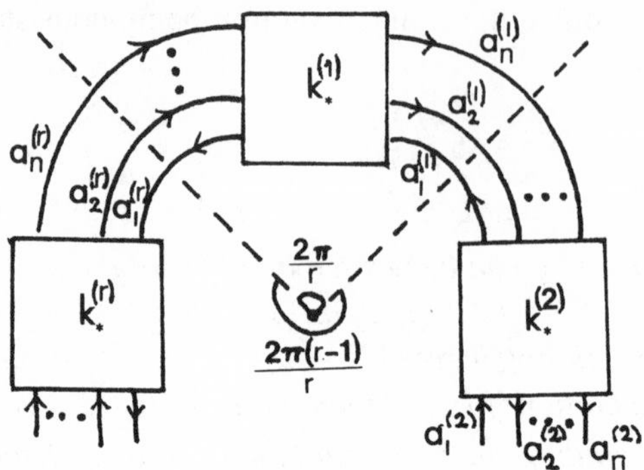

Figure 3.1

Suppose we now fix a c-state $S^{(1)}$ on the knot diagram $\widetilde{K}$, this automatically fixes c-states $\{S^{(1,1)}, \ldots, S^{(1,i)}, \ldots, S^{(1,r)}\}$ on $k_*^{(1)}, \ldots, k_*^{(i)}, \ldots, k_*^{(r)}$ respectively. Further suppose $S^{(1)}$ is not preserved under the auto-homeomorphism, see Example 3.3.

Example 3.3.

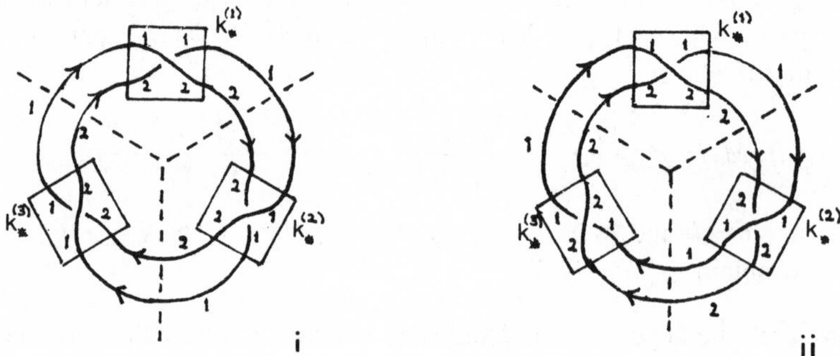

Figure 3.2

In the Figure 3.2 (i) the c-state is preserved under ϕ. While in the Figure 3.2 (ii) since the c-states on $k_*^{(1)}$, $k_*^{(2)}$, $k_*^{(3)}$ are all distinct, the c-state $S^{(1)}$ is not preserved under ϕ. N.B. *all* the c-states on $k_*^{(1)}, \ldots, k_*^{(r)}$ need *not* be distinct for $S^{(1)}$ not to be preserved under ϕ.

If we now rotate $\widetilde{K}$ by $\frac{2\pi}{r}$ keeping this state $S^{(1)}$ fixed, then we obtain a new c-state $S^{(2)}$ with c-states $\{S^{(1,r)}, S^{(1,1)}, \ldots, S^{(1,i)}, \ldots, S^{(1,r-1)}\}$ on the original $k_*^{(1)}, k_*^{(2)}, \ldots, k_*^{(i+1)}, \ldots, k_*^{(r)}$ respectively. For example if we rotate Figure 3.2 (ii) by $\frac{2\pi}{3}$ we obtain

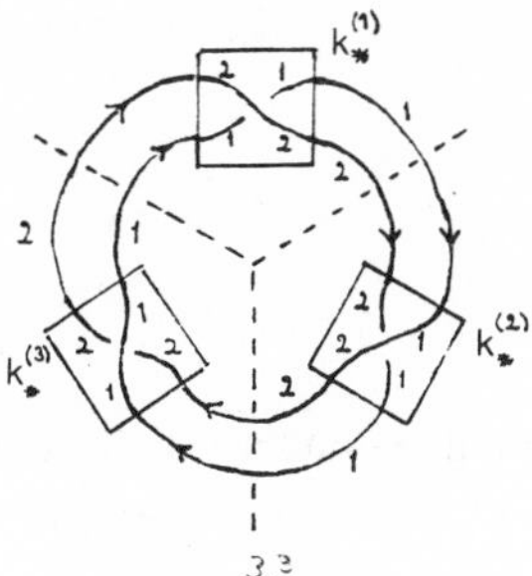

Figure 3.3

Notice under this rotation neither the Boltzmann weights nor the angle contribution is altered.

We may rotate $\widetilde{K}$ $(r-1)$ times (obviously each time in the same direction) to obtain $S^{(1)}, S^{(2)}, \ldots, S^{(r)}$. Each of these c-states of K may be distinct *but* the contribution to the partition function (and hence to the link invariant) in each case is the same.

As noticed in Example 3.3 the only occasions when we may not rotate $\widetilde{K}$ $(r-1)$ times to obtain r distinct states are when the configuration of $S^{(1,1)}$ is the same as that of $S^{(1,i)}$ for $i = 2, 3, , \ldots, r$.

Definition 3.4. Let S be a contributing state on the knot diagram $\widetilde{K}_*$ of the factor knot. The contributing state S on $\widetilde{K}_*$ induces the same contributing state on the (n,n)-tangle k_*. (A caveat to the reader: the reverse is not true, i.e. a contributing state on k_* does *not necessarily* induce a *contributing* state on $\widetilde{K}_*$.) Let $\widetilde{V}^{(N)}_{k_*,S}(t)$ denote the product of Boltzmann weights over all crossings of k_* with the contributing state S. Clearly

$$\widetilde{V}^{(N)}_{k_*,S}(t) = \widetilde{V}^{(N)}_{\widetilde{K}_*,S}(t) .$$

Lemma 3.5. *If K is a periodic knot with prime period r and k_* is the (n,n)-tangle whose closure is the factor knot K_* then*

$$\left(t^{-\frac{(N-1)}{2}} + \cdots + t^{\frac{N-1}{2}}\right) V^{(N)}_K(t) \equiv \sum \left(\widetilde{V}^{(N)}_{k_*,S}(t)\right)^r t^{rg(\theta,S)} (\text{mod } r)$$

where the summation is over all the c-states of $\widetilde{K}_*$, while $t^{g(\theta,S)}$ is the angle contribution for the c-state S on the (n,n)-tangle k_*. Nota bene the tangle is not closed.

Proof. Since K is a periodic knot using the above notation we may rewrite $V_K^{(N)}(t)$ as

$$\left(t^{-\frac{(N-1)}{2}} + \cdots + t^{\frac{N-1}{2}}\right) V_K^{(N)}(t) = \sum \left(\widetilde{V}^{(N)}_{k_*^{(1)},S^{(1,1)}}(t) \cdots \widetilde{V}^{(N)}_{k_*^{(r)},S^{(1,r)}}(t)\right) t^{g(\theta,S^{(1,1)})} \cdots t^{g(\theta,S^{(1,r)})}$$

where the sum is taken over all c-states on the knot diagram of the factor knot K_*. The result now follows from the observations noted above. □

Proposition 3.6. *Suppose K has prime period r.*
(i) If N is even and $N \not\equiv 0 \pmod{r^2}$ then

$$V_K^{(N)}(t) \equiv V_K^{(N)}(t^{-1}) \qquad \mathrm{mod}\,(r, t^r - 1)$$

(ii) If N is odd then

$$V_K^{(N)}(t) \equiv V_K^{(N)}(t^{-1}) \qquad \mathrm{mod}\,(r, t^{2r} - 1)\,.$$

Proof. By Lemma 3.5

$$\left(t^{-\frac{(N-1)}{2}} + \cdots + t^{\frac{N-1}{2}}\right) V_K^{(N)}(t) \equiv \sum \left(\widetilde{V}^{(N)}_{k_*,S}(t)\right)^r t^{rg(\theta,S)} \qquad (\mathrm{mod}\, r) \quad (3.1)$$

where the sum is taken over all c-states of $\widetilde{K}_*$. Hence

$$\left(t^{-\frac{(N-1)}{2}} + t^{-\frac{(N-3)}{2}} + \cdots + t^{\frac{N-3}{2}} + t^{\frac{N-1}{2}}\right) V_K^{(N)}(t^{-1}) \equiv \sum \left(\widetilde{V}^{(N)}_{k_*,S}(t^{-1})\right)^r t^{-rg(\theta,S)} \quad (\mathrm{mod}\, r) \qquad (3.2)$$

with the same summation as in (3.1).

Consider the states where at each crossing $x = y$. We know by Lemma 3.2 (ii), since K is a knot, each edge carries the same state variable x. Since for fixed N the Θ-set is $\{1, 2, \ldots, N\}$, there are N such states. By the remark after Theorem 3.1 the angle contribution for each of these states from the (n,n)-tangle k_* is of the form: $t^{\frac{\alpha}{2}(x - \frac{(N+1)}{2})}$ for some integer α. Note that

$$t^{-\frac{\alpha}{2}(x - \frac{(N+1)}{2})} = t^{\frac{\alpha}{2}(((N+1)-x) - \frac{(N+1)}{2})} \qquad (3.3)$$

and by Lemma 3.2 (iii)

$$w_\pm(xx \mid xx)(t) = w_\pm(N+1-x, N+1-x \mid N+1-x, N+1-x)(t)\,. \quad (3.4)$$

Hence if at each crossing $x = y$ the contribution to the right-hand side of (3.1) is

$$[w_+(xx \mid xx)(t)]^{rw(k_*)} t^{\frac{r\alpha}{2}(x - \frac{(N+1)}{2})} = t^{-\frac{rw(k_*)}{2}(2x - (2x-1)(N+1))} \cdot t^{\frac{r\alpha}{2}(x - \frac{(N+1)}{2})} \qquad (3.5)$$

where $w(k_*)$ is the writhe of the (n,n)-tangle k_*.

While the contribution to the right-hand side of (3.2) is, using the $1-1$ correspondence $x \leftrightarrow (N+1)-x$,

$$\begin{aligned}&[w_+(N+1-x, N+1-x \mid N+1-x, N+1-x)(t^{-1})]^{rw(k_*)}t^{-\frac{r\alpha}{2}(N+1-x-\frac{(N+1)}{2})}\\ =&[w_+(xx \mid xx)(t^{-1})]^{rw(k_*)}t^{\frac{r\alpha}{2}(x-\frac{(N+1)}{2})} = t^{+\frac{rw(k_*)}{2}(2x-(2x-1)(N+1))}t^{\frac{r\alpha}{2}(x-\frac{(N+1)}{2})} \quad (3.6)\end{aligned}$$

where the first equality follows from (3.3) and (3.4). Therefore we have

$$\begin{aligned}&t^{-\frac{rw(k_*)}{2}(2x-(2x-1)(N+1))}t^{\frac{r\alpha}{2}(x-\frac{(N+1)}{2})} \quad (3.7)\\ \equiv\ &t^{\frac{rw(k_*)}{2}(2x-(2x-1)(N+1))}t^{\frac{r\alpha}{2}(x-\frac{(N+1)}{2})} \begin{cases}\bmod (t^r-1) & \text{if } N \text{ is even}\\ \bmod (t^{2r}-1) & \text{if } N \text{ is odd}\end{cases}\end{aligned}$$

By the $1-1$ correspondence $x \leftrightarrow N+1-x$, (3.3) and (3.4)

$$\begin{aligned}&\sum_{x=1}^{N}[w_+(N+1-x, N+1-x \mid N+1-x, N+1-x)(t^{-1})]^{rw(k_*)}t^{\frac{r\alpha}{2}(x-\frac{(N+1)}{2})}\\ &\qquad = \sum_{x=1}^{N}[w_+(xx \mid xx)(t^{-1})]^{rw(k_*)}t^{-\frac{r\alpha}{2}(x-\frac{(N+1)}{2})}\end{aligned}$$

Hence by this and (3.7)

$$\begin{aligned}&\sum_{x=1}^{N}[w_+(xx \mid xx)(t)]^{rw(k_*)}t^{\frac{r\alpha}{2}(x-\frac{(N+1)}{2})} \quad (3.8)\\ \equiv &\sum_{x=1}^{N}[w_+(xx \mid xx)(t^{-1})]^{rw(k_*)}t^{-\frac{r\alpha}{2}(x-\frac{(N+1)}{2})} \begin{cases}\bmod (t^r-1) & \text{if } N \text{ is even}\\ \bmod (t^{2r}-1) & \text{if } N \text{ is odd}\end{cases}\end{aligned}$$

In the case when a c-state on $\widetilde{K}_*$ and hence on k_* contains a crossing at which the state variable x does not equal y, by Lemma 3.2 (i) we know that if $k = x-y \geq 2$ the positive Boltzmann weights (and analogously the negative weights) contain the factors: $(1-t^y)\dots(1-t^{x-2})(1-t^{x-1})g_1(t)$, where $g_1(t)$ is as in Lemma 3.2 (i). Then

$$\begin{aligned}&\left[(1-t^y)\dots(1-t^{x-2})(1-t^{x-1})g_1(t)\right]^r\\ &\quad\equiv (1-t^{ry})\dots(1-t^{r(x-2)})(1-t^{r(x-1)})[g_1(t)]^r \quad \bmod r\\ &\quad\equiv 0 \quad \bmod (r, t^{2r}-1) \quad (3.9)\end{aligned}$$

If $k = x-y = 1$, then again by Lemma 3.2 (i) the positive Boltzmann weight (and analogously the negative) has factors which may be written in either of the two equivalent forms:

$$(1-t^{x-1})\ g_1(t) \qquad \text{or} \qquad (1-t^{N+1-x})\ g_2(t)\ .$$

So

$$\left[\left(1-t^{x-1}\right) g_1(t)\right]^r \equiv (1-t^{r(x-1)})[g_1(t)]^r \qquad \bmod r \tag{3.10}$$

and

$$\left[\left(1-t^{N+1-x}\right) g_2(t)\right]^r \equiv (1-t^{r(N+1-x)})[g_2(t)]^r \qquad \bmod r\ . \tag{3.11}$$

Hence if x is odd, it is sufficient to consider (3.10) and then:

$$\left[\left(1-t^{x-1}\right) g_1(t)\right]^r \equiv 0 \qquad \bmod (r, t^{2r}-1) \tag{3.12}$$

While if x and N are even

$$\left[\left(1-t^{x-1}\right) g_1(t)\right]^r \equiv 0 \equiv \left[\left(1-t^{N+1-x}\right) g_2(t)\right]^r \qquad \bmod (r, t^r-1) \tag{3.13}$$

Finally in the case x is even and N is odd

$$\left[\left(1-t^{N+1-x}\right) g_2(t)\right]^r \equiv 0 \qquad \bmod (r, t^{2r}-1) \tag{3.14}$$

Combining (3.9), (3.12) and (3.13) if N is even and the c-state, S, contains two edges with distinct state variables then:

$$\widetilde{V}^{(N)}_{k_*,S}(t) \equiv 0 \equiv \widetilde{V}^{(N)}_{k_*,S}(t^{-1}) \qquad \bmod (r, t^r-1)\ . \tag{3.15}$$

If N is odd by combining (3.9), (3.12) and (3.14)

$$\widetilde{V}^{(N)}_{k_*,S}(t) \equiv 0 \equiv \widetilde{V}^{(N)}_{k_*,S}(t^{-1}) \qquad \bmod (r, t^{2r}-1)\ . \tag{3.16}$$

Since

$$t^{-\frac{(N-1)}{2}} + \cdots + t^{\frac{N-1}{2}} \not\equiv 0 \quad \begin{cases} \bmod (r, t^r-1) & \text{if} \quad N \not\equiv 0 \mod r^2 \\ \bmod (r, t^{2r}-1) & \text{if} \quad N \not\equiv 0 \mod 2r^2 \end{cases}$$

we have collecting together (3.8) and (3.15) if N is even and $N \not\equiv 0 \bmod r^2$

$$V^{(N)}_K(t) \equiv V^{(N)}_K(t^{-1}) \qquad \bmod (r, t^r-1)$$

and by (3.8) and (3.16) if N is odd

$$V^{(N)}_K(t) \equiv V^{(N)}_K(t^{-1}) \qquad \bmod (r, t^{2r}-1) \qquad \square$$

Intrinsically Lemma 3.5 shows that there exists a relationship between the periodic knot and its factor knot working modulus r. To make this relationship more precise first of all we need to close the (n, n)-tangle k_* in the natural braid way to form the factor knot K_*. The additional angle contribution due to this closure can be calculated in an analogous way to that of the closure of a braid. The calculation may be facilitated by Figure 3.1. Note that the closure causes a unit vector tangent to the closure strand to rotate through an angle of $\pm\frac{2\pi(r-1)}{r}$, rather than $\pm 2\pi$ in the braid case.

If we write $\{a_1, a_2, \ldots, a_n\}$ to be the state variables on the closure strings, then the angle contribution of the closure may be written as: $t^{\sum_{i=1}^{n}(\pm)(a_i - \frac{(N+1)}{2})\frac{(r-1)}{r}}$ where the sign depends on the direction of the string, as in Definition 1.4. Hence

$$\left(t^{-\frac{(N-1)}{2}}+\cdots+t^{\frac{(N-1)}{2}}\right)^r\left[V^{(N)}_{K_*}(t)\right]^r=\left[\sum \widetilde{V}^{(N)}_{\widetilde{K}_*,S}(t)t^{g(\theta,S)}t^{\Sigma_{i=1}^{n}(\pm)(a_i-\frac{(N+1)}{2})\frac{(r-1)}{r}}\right]^r$$

where the sum is taken over all c-states on $\widetilde{K}_*$. Therefore

$$\begin{aligned} &t^{-\frac{(N-1)r}{2}}(1+t+\cdots+t^{N-1})^r\left[V^{(N)}_{\widetilde{K}_*}(t)\right]^r \\ &\equiv \sum\left[\widetilde{V}^{(N)}_{\widetilde{K}_*,S}(t)\right]^r t^{rg(\theta,S)}t^{\Sigma_{i=1}^{n}(\pm)(a_i-\frac{(N+1)}{2})(r-1)} \qquad \bmod r\,. \end{aligned} \tag{3.17}$$

We now define $\lambda \equiv \ell k(K, F)$, where F is the set of fixed points of the auto-homeomorphism, which is the unknotted knot. Note that $\lambda \equiv n \pmod 2$. Also if $N \not\equiv 0 \bmod (r(r-1))$

$$t^{-\frac{(N-1)}{2}}+t^{-\frac{(N-3)}{2}}+\cdots+t^{\frac{N-3}{2}}+t^{\frac{N-1}{2}}\not\equiv 0 \qquad \bmod (r, t^{r-1}-1)\,.$$

Clearly if N is odd the above inequivalence always holds.

If N is odd, since $a_i - \frac{(N-1)}{2}$ is an integer, equation (3.17) becomes

$$\begin{aligned} &t^{-\frac{(N-1)}{2}}(1+t+\cdots+t^{N-1})\left[V^{(N)}_{K_*}(t)\right]^r \\ &\equiv \sum_{\substack{c-states\\ on\ \widetilde{K}_*}}\left[\widetilde{V}^{(N)}_{\widetilde{K}_*,S}(t)\right]^r t^{rg(\theta,S)} \qquad \bmod (r, t^{r-1}-1)\,. \end{aligned} \tag{3.18}$$

Next, consider the case where N is even. Note

$$\sum_{i=1}^{n}(\pm)\,(2a_i-(N+1)) \equiv n \equiv \lambda \qquad \bmod 2\,.$$

Suppose r is an *odd* prime and $N \not\equiv 0 \bmod (r(r-1))$.

(i) If $\lambda \equiv 0 \bmod 2$ equation (3.17) becomes

$$\begin{aligned} &(t^{-\frac{(N-1)}{2}}+\cdots+t^{\frac{N-1}{2}})^r\left[V^{(N)}_{K_*}(t)\right]^r \\ &\equiv \sum_{\substack{c-states\\ on\ \widetilde{K}_*}}\left[\widetilde{V}^{(N)}_{\widetilde{K}_*,S}(t)\right]^r t^{rg(\theta,S)} \qquad \bmod (r, t^{r-1}-1)\,. \end{aligned} \tag{3.19}$$

(ii) If $\lambda \equiv 1 \bmod 2$ then

$$t^{-\frac{(N-1)r}{2}} = t^{-\frac{(N-1)(r-1)}{2}}t^{-\frac{(N-1)}{2}} \equiv t^{-\frac{r-1}{2}}\,t^{-\frac{(N-1)}{2}} \qquad \bmod (t^{r-1}-1)$$

and

$$t^{\sum_{i=1}^{n}(\pm)(a_i-\frac{(N+1)}{2})(r-1)} \equiv t^{\frac{r-1}{2}} \qquad \bmod (t^{r-1}-1)\,.$$

Combining these facts equation (3.17) becomes

$$\begin{aligned}&\left(t^{-\frac{(N-1)}{2}}+\cdots+t^{\frac{N-1}{2}}\right)\left[V_{K_*}^{(N)}(t)\right]^r\\ \equiv &\sum_{\substack{c-states\\ on\ \widetilde{K}_*}}\left[\widetilde{V}_{\widetilde{K}_*,S}^{(N)}(t)\right]^r t^{rg(\theta,S)} \qquad \bmod (r,t^{r-1}-1)\end{aligned} \tag{3.20}$$

So collecting together equations (3.18), (3.19) and (3,20) and comparing with the right-hand side of Lemma 3.5 we have:

Proposition 3.7. *If K is a periodic knot with prime period r, and K_* is its factor knot, F the fixed points of the auto-homeomorphism, and $\lambda = \ell k(K,F)$ then:*
(i) If N is odd then

$$V_K^{(N)}(t) \equiv \left[V_{K_*}^{(N)}(t)\right]^r \qquad \bmod (r,t^{r-1}-1)\,.$$

(ii) Suppose N is even, and r is an odd prime such that $N \not\equiv 0 \bmod (r(r-1))$.
(a) If $\lambda \equiv 1 \bmod 2$ then

$$V_K^{(N)}(t) \equiv \left[V_{K_*}^{(N)}(t)\right]^r \qquad \bmod (r,t^{r-1}-1)\,.$$

(b) If $\lambda \equiv 0 \bmod 2$ then

$$V_K^{(N)}(t) \equiv \left(t^{-\frac{(N-1)}{2}}+\cdots+t^{\frac{N-1}{2}}\right)^{r-1}\left[V_{K_*}^{(N)}(t)\right]^r \qquad \bmod (r,t^{r-1}-1)\,.$$

□

Note, using the same arguments as those of the proof of Proposition 3.7, we may extend this proposition to periodic links, provided that under the auto-homeomorphism the orientation of each component of the link diagram is preserved.

In theory to check a given knot does not have a given period we keep on checking ad infinitum until the conditions of Proposition 3.7 are no longer satisfied. However it is possible to show that only a finite number of hierarchal polynomials need to be considered, but we do not pursue this problem in this paper.

§4. Some evaluations of the hierarchal polynomial

It is well-known, see for example [**Jo1**], that

$$V_L^{(2)}(1) = (-2)^{\mu-1} \qquad \text{and} \qquad V_L^{(2)}(-1) = \Delta_L(-1)$$

where L is a μ-component link, and $\Delta_L(t)$ is the (reduced) Alexander polynomial of L.

Using the methods of the previous section we will obtain compact formulae for the hierarchal polynomial evaluated at $t = 1$ and $t = -1$. However, first of all we need the obligatory technical lemma.

Lemma 4.1.
(i) Suppose L is a closed n-braid with $c(L)$ crossings and μ components then $c(L) + n \equiv \mu \mod 2$.
(ii) Let $c(N, \mu, m)$ denote the number of compositions of m with exactly μ parts, and each part less than or equal to N. Then $\sum_{m=\mu}^{N\mu} c(N, \mu, m) q^m = \left[\frac{q(1-q^N)}{1-q}\right]^{\mu}$.

Proof.
(i) This part may easily be proven by induction on the number of crossings $c(L)$. Clearly if $c(L) = 0$ then this implies that $n = \mu$. Now suppose Lemma 4.1(i) is true for $c(L) = k \geq 0$, and consider L, a closed n-braid, with $k + 1$ crossings. Delete a crossing by splicing, i.e.

Hence n is unchanged but μ gains or loses 1. The result now follows by a straightforward calculation and the inductive hypothesis.
(ii) See the proof of Theorem 4.2 in [**An**]. □

Corollary 4.2. *If L is a knot diagram then:* $w(L) + n \equiv 1 \pmod 2$, *where* $w(L)$ *is the writhe of L.* □

By Lemma 3.2 (i) we know that if $k = x - y \geq 1$ then the Boltzmann weight $w_{\pm}^{(N)}(ab \mid xy)(t)$ has a factor $1 - t^{x-1}$ (or in the negative case $1 - t^{a-1}$). Therefore if $k \geq 1$ and $t = 1$ then $w_{\pm}^{(N)}(ab \mid xy)(1) = 0$.

Proposition 4.3. *If L is a link with μ components then:*

$$V_L^{(N)}(1) = \left[(-1)^{N-1} N\right]^{\mu-1}.$$

Proof. We may assume without loss of generality that L is a closed n-braid. By the previous paragraph we need only consider the c-states which have at each crossing $x = y$ and $a = b$, see Figure 1.3. So by Lemma 3.2 (ii) we may fix a state variable on an edge of each component, and all the other edges on *that* component will carry the same state variable. Since for fixed N each state variable is a member of the set $\{1, 2, \ldots, N\}$, there are N^μ such c-states. Now

$$w_{\pm}^{(N)}(aa \mid xx)(t) \;=\; t^{\mp\frac{1}{2}(2ax-(c-1)(N+1))} \tag{4.1}$$

where $c = a + x$.
(i) If N is odd, then

$$w_{\pm}^{(N)}(aa \mid xx)(1) \;=\; 1\,.$$

The angle contribution for a particular c-state is of the form:

$$\left[t^{\sum\limits_{i=1}^{n}(a_i - \frac{(N+1)}{2})}\right]_{t=1} \;=\; 1\,.$$

Now using $V_L^{(N)}(t)$ of Theorem 1.5 we obtain

$$V_L^{(N)}(1) \;=\; \frac{1}{N}\,N^\mu \;=\; N^{\mu-1} \;=\; \left[(-1)^{N-1}N\right]^{\mu-1}\,.$$

(ii) If N is even we need to be slightly more careful with (4.1). Note $w_{\pm}^{(N)}(aa \mid xx)(t) = t^{\mp\frac{1}{2}(2ax-(2a-1)(N+1))}$ and $w_{\pm}^{(N)}(aa \mid aa)(1) = -1$, because $t^{\frac{1}{2}} = -1$.

Of course there may be some $w_{\pm}^{(N)}(aa \mid xx)(1) = -1$ terms to consider for a particular c-state, but these will only occur at the crossings where two distinct components meet, and all such crossings will arise in pairs.

In this the even case the angle contribution for a particular c-state is

$$\left[t^{\sum\limits_{i=1}^{n}(a_i - \frac{(N+1)}{2})}\right]_{t=1} \;=\; (-1)^n\,.$$

Therefore collecting everything together we obtain

$$V_L^{(N)}(1) \;=\; \frac{(-1)}{N}\left[(-1)^{c(L)+n}N^\mu\right]$$

but by Lemma 4.1 (i) $c(L) + n \equiv \mu \bmod 2$, hence the result. □

Let us now restrict ourselves to the case N *is odd*. We consider $w_{\pm}^{(N)}(ab \mid xy)(t)$. In the positive case, clearly if x is odd then $x - 1$ is even. While if x is even then $N + 1 - x$ is even. Therefore if $k = x - y \geq 1$ by Lemma 3.2 (i)

$$w_{\pm}^{(N)}(ab \mid xy)(-1) \;=\; 0 \tag{4.2}$$

Proposition 4.4. *Let L be a link with μ components $L_1, \ldots, L_\mu$.*
(i) If $N = 4p - 1$, $p \geq 1$ then $V_L^{(N)}(-1) \;=\; (-1)^{\mu-1}$.
(ii) If $N = 4p - 3$, $p \geq 1$ then $V_L^{(N)}(-1) \;=\; 1\,.$

Proof. As in Proposition 4.3 we may assume that L is a closed n-braid. By (4.2) we may restrict ourselves to the c-states which at each crossing have $x = y$ (and $a = b$).

Evaluating (4.1) at $t = -1$ gives us

$$w_{\pm}^{(N)}(aa \mid xx)(-1) = (-1)^{ax}(-1)^{\frac{(c-1)(N+1)}{2}} \tag{4.3}$$

(i) The case $N = 4p - 1$, $p = 1, 2, \ldots$. We see immediately that (4.3) becomes

$$w_{\pm}^{(N)}(aa \mid xx)(-1) = (-1)^{ax} .$$

Since L is a closed n-braid the angle contribution from a particular c-state is

$$\left[t^{\sum_{i=1}^{\mu} n_i(a_i - \frac{(N+1)}{2})}\right]_{t=-1} = (-1)^{\sum_{i=1}^{\mu} n_i a_i}$$

where a_i is the state variable carried by the component L_i, and n_i is the number of closure strands of L_i. Therefore the contribution to the parition function from a particular c-state is

$$(-1)^{\sum_{i=1}^{\mu} a_i(n_i + w(L_i))} \tag{4.4}$$

where $w(L_i)$ is the writhe of the component L_i. Note that $(-1)^{a_i^2} = (-1)^{a_i}$, and terms of the form $(-1)^{ax}$ $(a \neq x)$ may be ignored for the same reason as in the previous proposition. Hence by Corollary 4.2

$$(-1)^{\sum_{i=1}^{\mu} a_i(n_i + w(L_i))} = (-1)^{\sum_{i=1}^{\mu} a_i} \tag{4.5}$$

But $\sum_{i=1}^{\mu} a_i$ is just the sum of the composition $m = \sum_{i=1}^{\mu} a_i$, i.e. a composition of m with exactly μ parts each part $\leq N$.

So to calculate the total contribution to the parition function over all possible compositions, i.e. $\mu \leq \sum_{i=1}^{\mu} a_i \leq N\mu$, is the same as allowing $q = -1$ in Lemma 4.1 (ii) i.e.

$$\left[\left[q\frac{(1-q^N)}{1-q}\right]^{\mu}\right]_{q=-1} = (-1)^{\mu} .$$

Therefore

$$V_L^{(4p-1)}(-1) = (-1)^{\frac{(4p-1)-1}{2}}(-1)^{\mu} = (-1)^{\mu-1} .$$

(ii) If $N = 4p - 3$ $p = 2, 3, \ldots$. We need to modify (4.2) so that

$$w_{\pm}^{(N)}(aa \mid xx)(-1) = (-1)^{ax}(-1)^{c-1} .$$

Clearly as before we may ignore the contributions of the Boltzmann weights in the cases $a \neq x$. Further if we note $(-1)^{a^2+2a-1} = (-1)^{a-1}$ and the angle contribution, as in (i), is

$$\left[t^{\sum_{i=1}^{\mu} n_i(a_i - \frac{(N+1)}{2})}\right]_{t=-1} = (-1)^{\sum_{i=1}^{\mu}(a_i-1)} .$$

Hence as before but with a slight modification

$$(-1)^{\sum_{i=1}^{\mu}(a_i-1)(n_i+w(L_i))} = (-1)^{\sum_{i=1}^{\mu}(a_i-1)} = (-1)^{\sum_{i=1}^{\mu} a_i}(-1)^{\mu} .$$

So using the same arguments as in (i) we obtain

$$V_L^{(4p-3)}(-1) = (-1)^{\frac{(4p-3)-1}{2}}(-1)^{\mu}(-1)^{\mu} = 1$$

and since $V_L^{(1)}(t) = 1$, the result follows. □

Using similar methods and a restriction on the partition function, and hence the hierarchal polynomial it is possible to show that if ξ is a mth root of unity then $J_K^{(N)}(\xi) = 1$ where K is a knot and $N \equiv 1 \bmod m$. The case $N = 3$ corresponds to Proposition 4.4 (i).

These evaluations tend to suggest that trying to use the hierarchal polynomials to show there does or does not exist a knot (not ambient isotopic to the unknot) which has trivial original (i.e. 2-order) Jones polynomial, might not be more productive than using the original Jones polynomial.

§5. The maximum and minimum degrees of the hierarchal polynomial for an alternating link

In [**Mu2**] the following result was proved:

Theorem 5.1 [**Mu2**]. *If $\widetilde{L}$ is a reduced (non-split) alternating diagram of an (alternating) link L then*

$$d_{\max} V_L^{(2)}(t) + d_{\min} V_L^{(2)}(t) = w(\widetilde{L}) - \sigma(L) .$$

*Where $d_{\max} V_L^{(2)}(t)$, $d_{\min} V_L^{(2)}(t)$ denote the maximal and minimal degrees of $V_L^{(2)}(t)$ respectively, $w(\widetilde{L})$ is the writhe of the link diagram, and $\sigma(L)$ is the signature of L as defined in [**Mu1**].* □

Corollary 5.2 [**Mu2**]. *With the same definitions as in Theorem 5.1, we have*

$$d_{\max} V_L^{(2)}(t) = c_+(\widetilde{L}) - \frac{\sigma(L)}{2} ,$$

and

$$d_{\min} V_L^{(2)}(t) = -c_-(\widetilde{L}) - \frac{\sigma(L)}{2},$$

where $c_+(\widetilde{L})$ *(or* $c_-(\widetilde{L})$*) denotes the number of positive (or negative) crossings of* $\widetilde{L}$. □

An alternative proof of Theorem 5.1, and hence Corollary 5.2 has been given by Turaev [**Tu1**]. To prove these results Turaev uses Kauffman's states model interpretation of the 2-order Jones polynomial, while Murasugi's proof is based on a graph theoretical interpretation of the same polynomial. Also see [**Mu4**].

In the next two sections a third proof of the above results will be given, but in our case the above results will just be a special case (i.e. $N = 2$) of a general result for the N-order hierarchal Jones polynomial. The proof will be in a similar vein to most of our previous results, i.e. based on the states model of section 1.

The general results are:

Theorem 5.3. *If* $\widetilde{L}$ *is a reduced (non-split) alternating diagram of an alternating link* L, *and* $V_L^{(N)}(t)$ *is the* N*-order hierarchal Jones polynomial of* L *then:*

$$d_{\max} V_L^{(N)}(t) = \frac{N(N-1)}{2}\, c_+(\widetilde{L}) - \frac{(N-1)}{2}\, \sigma(L) \tag{5.1}$$

and

$$d_{\min} V_L^{(N)}(t) = -\frac{N(N-1)}{2}\, c_-(\widetilde{L}) - \frac{(N-1)}{2}\, \sigma(L)\,. \quad \square \tag{5.2}$$

Corollary 5.4. *With the same definitions as in Theorem 5.3 we have for* $N = 1, 2, \ldots$

$$d_{\max} V_L^{(N)}(t) + d_{\min} V_L^{(N)}(t) = \frac{N(N-1)}{2}\, w(\widetilde{L}) - (N-1)\sigma(L)\,. \quad \square$$

In order to prove Theorem 5.3 and Corollary 5.4 we need *in the following two sections* to make the following assumptions. Firstly, that all the links are *alternating*, and all alternating link diagrams are reduced (or as sometimes is written there are no nugatory crossings). By "reduced" we mean that $\widetilde{L}$ does not have a crossing of the type shown in Figure 5.1.

Figure 5.1

Also these reduced alternating diagrams will be *non-split* diagrams, in the sense that there does not exist a circle $S^1 \subset \mathbf{R}^2$ which does not intersect the diagram, but both components of $\mathbf{R}^2 - S^1$ intersect the diagram.

The set of Boltzmann weights, $w_+^{(N)}(ab \mid xy)(t)$ (of Definition 1.4) can be thought of as entries of a $N^2 \times N^2$ matrix. The matrix consists of block matrices corresponding to the *charge (or spin)* c $(= a + x = b + y)$, $2 \le c \le 2N$. Each block matrix is indexed by all compositions of c into exactly two parts, i.e. if $N = 3$ and $c = 4$ then the rows and columns will be labelled in the following manner:

$$\begin{array}{c} \\ (13) \\ (22) \\ (31) \end{array} \begin{array}{c} \begin{array}{ccc} (13) & (22) & (31) \end{array} \\ \begin{pmatrix} w_+^{(N)}(31 \mid 13)(t) & w_+^{(N)}(21 \mid 23)(t) & w_+^{(N)}(11 \mid 33)(t) \\ w_+^{(N)}(32 \mid 12)(t) & w_+^{(N)}(22 \mid 22)(t) & w_+^{(N)}(12 \mid 32)(t) \\ w_+^{(N)}(33 \mid 11)(t) & w_+^{(N)}(23 \mid 21)(t) & w_+^{(N)}(13 \mid 31)(t) \end{pmatrix} \end{array}$$

So for a given Boltzmann weight $w_+^{(N)}(ab \mid xy)(t)$ its position in the matrix is determined by its charge $c = a + x$, which puts it into a suitable block, and its row index (*by*) and column index (xa). Hence by definition of the Boltzmann weights these block matrices are triangular with the top left-hand triangle being 0. For example, if $N = 2$ we obtain the $2^2 \times 2^2$ matrix

$$\begin{array}{c} \\ (11) \\ (12) \\ (21) \\ (22) \end{array} \begin{array}{c} \begin{array}{cccc} (11) & (12) & (21) & (22) \end{array} \\ \begin{pmatrix} t^{1/2} & 0 & 0 & 0 \\ 0 & 0 & t & 0 \\ 0 & t & t^{1/2}(1-t) & 0 \\ 0 & 0 & 0 & t^{1/2} \end{pmatrix} \end{array}$$

A similar matrix may be defined for the Boltzmann weights $w_-^{(N)}(ab \mid xy)(t)$, but in this case in each block matrix the lower right-hand triangle is zero. It should be noted that these matrices are just the *S-matrices* defined in [**WAD**] and the *R-matrices* as defined in [**Tu2**] with slightly different indexing.

Lemma 5.5. *The positive Boltzmann weights have the following relations between maximum degrees:*

$$\begin{aligned} &\max\deg w_+^{(N)}(a{+}1\ b{-}1 \mid x{-}1\ y{+}1)(t) \\ &\qquad = \max\deg w_+^{(N)}(ab \mid xy)(t) - (N + 1 - k) \end{aligned} \tag{5.3}$$

$$\begin{aligned} &\max\deg w_+^{(N)}(a{+}1\ b \mid x{-}1\ y)(t) \\ &\qquad = \max\deg w_+^{(N)}(ab \mid xy)(t) - \frac{1}{2}(N + y - b) \end{aligned} \tag{5.4}$$

$$\max\deg w_+^{(N)}(a\ b{-}1 \mid x\ y{+}1)(t)$$

$$= \max\deg w_+^{(N)}(ab \mid xy)(t) - \frac{1}{2}(N + a - x) \tag{5.5}$$

$$\max\deg w_+^{(N)}(a{-}1\ b \mid x\ y{-}1)(t)$$
$$= \max\deg w_+^{(N)}(ab \mid xy)(t) + \frac{1}{2}(a + y - 2) \tag{5.6}$$

$$\max\deg w_+^{(N)}(a\ b{+}1 \mid x{+}1\ y)(t)$$
$$= \max\deg w_+^{(N)}(ab \mid xy)(t) + \frac{1}{2}(2N - b - x) \tag{5.7}$$

where $k = b - a$ as usual, and clearly these equations only make sense if the additions and subtractions do not push the state variables outside the Θ-set of Definition 1.4, i.e. we never obtain as state variables 0 *or* $N + 1$.

Proof. By Lemma 3.2 (i) we know that the maximum degree of

$$\frac{(t;\ x{-}1)(t;\ N{-}y)}{(t;\ k)(t;\ y{-}1)(t;\ N{-}x)}$$

is

$$\frac{(x + y - 1)k}{2} + (N - x)k = \frac{k}{2}(2N - 1 - k) \tag{5.8}$$

Hence the max deg $w_+(ab \mid xy)(t)$ is equal to

$$-\frac{1}{2}[ax + by + (k - c + 1)N - (k + c - 1) - 2Nk + k + k^2]$$
$$= \frac{1}{2}[Nk + (c - 1)(N + 1) - bx - ay] \tag{5.9}$$

(Note that $k^2 = (b - a)(x - y)$ and $b + y = c = a + x$.)

So if we fix $b - a = k = x - y$, and c, then by using (5.9) the above results follow by direct calculations. □

Similarly we will need comparisons of the maximum degrees of the negative Boltzmann weights.

Lemma 5.6. *The negative Boltzmann weights have the following relations:*

$$\max\deg w_-^{(N)}(a{-}1\ b{+}1 \mid x{+}1\ y{-}1)(t)$$
$$= \max\deg w_-^{(N)}(ab \mid xy)(t) - (N - k) \tag{5.10}$$

$$\max\deg w_-^{(N)}(a{-}1\ b \mid x{+}1\ y)(t)$$
$$= \max\deg w_-^{(N)}(ab \mid xy)(t) - \frac{(N + x - a)}{2} \tag{5.11}$$

$$\max\deg w_-^{(N)}(a\ b{+}1 \mid x\ y{-}1)(t)$$

$$= \max \deg w_-^{(N)}(ab \mid xy)(t) - \frac{(N+b-y)}{2} \tag{5.12}$$

$$\max \deg w_-^{(N)}(a\; b{-}1 \mid x{-}1\; y)(t)$$
$$= \max \deg w_-^{(N)}(ab \mid xy)(t) + \frac{(2N-a-y)}{2} \tag{5.13}$$

$$\max \deg w_-^{(N)}(a{+}1\; b \mid x\; y{+}1)(t)$$
$$= \max \deg w_-^{(N)}(ab \mid xy)(t) + \frac{(b+x-2)}{2} \tag{5.14}$$

The same stipulations hold as those at the end of Lemma 5.5.

Proof. The proof of this lemma follows along similar lines to those of the previous lemma. However in this case since $w_-^{(N)}(ab \mid xy)(t) = w_+^{(N)}(xy \mid ab)(t^{-1})$, it is sufficient to consider the *minimum* degree of $w_+^{(N)}(xy \mid ab)(t)$. By Lemma 3.2 (i) the minimum degree of

$$\frac{(t;\ a-1)(t;\ N-b)}{(t;\ k)(t;\ b-1)(t;\ N-1)}$$

is zero. Hence the maximum degree of $w_-^{(N)}(ab|xy)(t)$ is equal to

$$\frac{1}{2}[ax + by + (k-c+1)N - (k-c+1)] \tag{5.15}$$

As before the results follow by direct but tedious calculations. □

There is actually a purpose to these calculations.

Lemma 5.7.
(i) The Boltzmann weights $w_+^{(N)}(ab \mid xy)(t)$ attain their maximum degree

$$\frac{N-1}{2} + \frac{N(N-1)}{2}$$

uniquely when $a = y = 1$, and $b = x = N$. Also at these values of the state variable, although not exclusively their minimum degree

$$\frac{N-1}{2}$$

is obtained.
(ii) The Boltzmann weights $w_-^{(N)}(ab \mid xy)(t)$ attain their minimum degree

$$-\frac{N-1}{2} - \frac{N(N-1)}{2}$$

uniquely when $b = x = 1$ and $y = a = N$. Again as above these values of the state variables, although not exclusively, give their maximum degree

$$-\frac{(N-1)}{2}.$$

Proof.
(i) It is not too hard to see using (5.3), (5,4) and (5.5) and the block matrices indexed in the described fashion, that the maximum degree of the Boltzmann weights in a given block matrix is attained exclusively by the Boltzmann weight in the bottom right-hand corner, namely the weights: $w_+^{(N)}(11 \mid 11)(t)$, $w_+^{(N)}(12 \mid 12)(t)$, …, $w_+^{(N)}(1N \mid N1)(t)$. By (5.9) the maximum degree of $w_+^{(N)}(1x \mid x1)(t)$ is equal to

$$\frac{1}{2}[(N+1)x + N(x-1) - (x^2+1)] = \frac{1}{2}[(2N+1)x - x^2 - N - 1] \quad (5.16)$$

So by considering $x \mapsto x+1$ it is easily seen that the maximum occurs uniquely when $x = N$, and in this case

$$\max\deg w_+^{(N)}(1N \mid N1)(t) = \frac{N(N-1)}{2} + \frac{N-1}{2}.$$

Now by comparing (5.10), (5.11), and (5.12) we may similarly show that the minimum degree of max deg $w_+^{(N)}(ab \mid xy)(t)$ occurs in the bottom right-hand corner of each block matrix. In this case

$$\min\deg w_+^{(N)}(1x \mid x1)(t) = \frac{N-1}{2}$$

which clearly includes the case $w_+^{(N)}(1N \mid N1)(t)$, so (i) is proved.
(ii) If we note that $w_-^{(N)}(ab \mid xy)(t) = w_+^{(N)}(xy \mid ab)(t^{-1})$ by definition, the results follow from (i). □

Suppose we now dissolve the (non-split) reduced alternating diagram $\widetilde{L}$ by splicing into a number of disjoint oriented simple closed curves $S_1, \ldots, S_q$, called *Seifert circles* [**BZ**]. On the other hand $\widetilde{L}$ divides S^2 into $n+2$ regions $r_1, r_2, \ldots, r_{n+2}$ that are open and connected. Following [Mu 1] we classify these Seifert circles in the following manner.

Definition 5.8. S_i is of the *first type* if it bounds a region r_j. Otherwise S_i is of the *second type.*

In [**Mu1**], see also [**BZ**], Murasugi showed that it is possible to decompose $\widetilde{L}$ into $\widetilde{L}_1 * \widetilde{L}_2 * \cdots * \widetilde{L}_m * \widetilde{L}'_1 * \cdots * \widetilde{L}'_n$. Where $\widetilde{L}_i$ is a link diagram with only positive crossings called a *positive special alternating diagram.*

While $\widetilde{L}'_j$ has only negative crossings and is called a *negative special alternating diagram.* Further each $\widetilde{L}_i, \widetilde{L}'_j$ does not contain any Seifert circles of the second type.

Theorem 5.9 [**Mu1**]. *Let L be an alternating link and decompose $\widetilde{L}$ into $\widetilde{L}_1 * \widetilde{L}_2 * \cdots * \widetilde{L}_m * \widetilde{L}'_1 * \cdots * \widetilde{L}'_n$. Then*

$$\sigma(L) = \sigma(L_1) + \sigma(L_2) + \cdots + \sigma(L_m) + \sigma(L'_1) + \cdots + \sigma(L'_n) \qquad (5.17)$$

where $\sigma(L_i)$ (or $\sigma(L'_j)$) denotes the signature of L_i (or L'_j). □

Definition 5.10. Suppose $\widetilde{L}$ (or $\widetilde{L}'$) is an alternating link diagram with only positive (or negative) crossings. Then denote by $c_+(\widetilde{L})$ (or $c_-(\widetilde{L}')$) the number of crossings of $\widetilde{L}$ (or $\widetilde{L}'$), and by $s(\widetilde{L})$ (or $s(\widetilde{L}')$) the number of Seifert circles obtained by splicing $\widetilde{L}$ (or $\widetilde{L}'$).

In [**Tr2**] Traczyk shows that if L is an alternating link with only positive crossings (i.e. L has a positive special alternating diagram $\widetilde{L}$) then

$$\sigma(L) = -(c_+(\widetilde{L}) + 1 - s(\widetilde{L})) \qquad (5.18)$$

Similarly if L' is an alternating link consisting of only negative crossings then

$$\sigma(L') = c_-(\widetilde{L}') + 1 - s(\widetilde{L}') \qquad (5.19)$$

So collecting together (5.18), (5.19) and Theorem 5.9 we obtain the following proposition which may be gleaned from [**Tr2**].

Proposition 5.11. *Let L be an alternating link, and $\widetilde{L}$ its alternating (non-split) reduced link diagram which has the decomposition $\widetilde{L}_1 * \widetilde{L}_2 * \cdots * \widetilde{L}_m * \widetilde{L}'_1 * \cdots * \widetilde{L}'_n$ then*

$$\sigma(L) = -\sum_{i=1}^{m}(c_+(\widetilde{L}_i) + 1 - s(\widetilde{L}_i)) + \sum_{j=1}^{n}(c_-(\widetilde{L}'_j) + 1 - s(\widetilde{L}'_j)) \qquad \square \ (5.20)$$

Example 5.12.

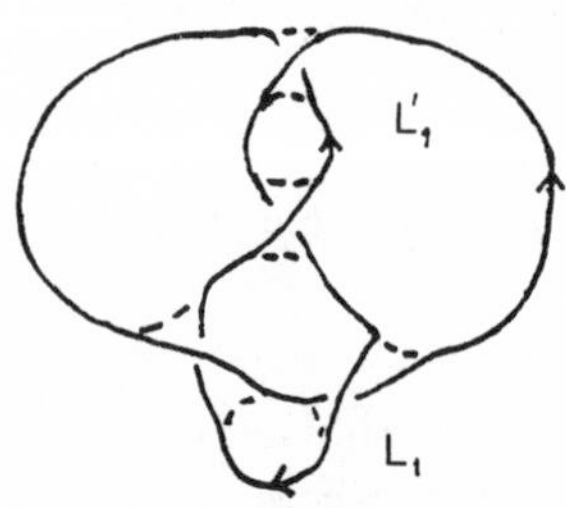

Figure 5.2

Clearly L_1 consists of two Seifert circles and two (positive) crossings, while L_1' also consists of two Seifert circles but in this case two negative crossings.

Hence $\sigma(L) = -(2+1-2)+(2+1-2) = 0$. A result which has been well known for a long time.

Let us momentarily return to Lemma 5.7 (i). There we noted the maximum degree of $w_+^{(N)}(ab \mid xy)(t)$ over all possible Boltzmann weights occurs uniquely when $a = y = 1$, $b = x = N$. While the maximum degree of $w_-^{(N)}(ab \mid xy)(t)$ occurs but not uniquely when $a = y = N$, $b = x = 1$.

Consider now that the state where at every positive crossing the state variables are assigned in the following manner:

Figure 5.3

i.e. corresponding to the maximal Boltzmann weights $w_+^{(N)}(1N \mid N1)(t)$.

While at every negative crossing assign the state variables in the following way:

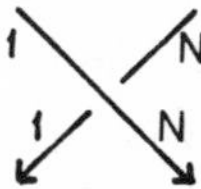

Figure 5.4

To show given any reduced alternating diagram that such a c-state exists, it is sufficient to consider the following two cases

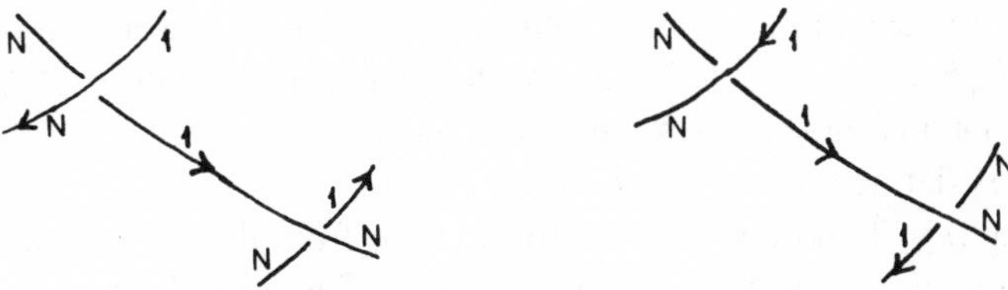

Figure 5.5

All other possibilities are variations on this theme, and all these possibilities are consistent in the sense that they allow the labelling described above. So given any reduced alternating diagram we may assign state variables in the manner described above.

Note that if the diagram is non-alternating it may not be possible to assign the state variables in such a manner. Consider for example a part of a non-alternating link which looks like

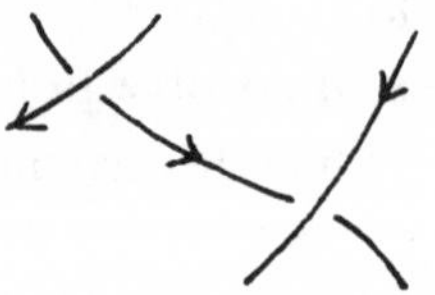

Figure 5.6

Clearly the above assignment is not possible.

Definition 5.13. Let S_N denote the c-state of a reduced alternating diagram where the state variables at the positive crossings as those of Figure 5.3, and the state variables at the negative crossings as those of Figure 5.4.

Dissolve $\widetilde{L}_1 * \widetilde{L}_2 * \cdots * \widetilde{L}_m * \widetilde{L}'_1 * \cdots * \widetilde{L}'_n$, in $\mathbf{R}^2$, by splicing into its constituent Seifert circles. These Seifert circles divide $\mathbf{R}^2$ into open and connected regions. We now think of S^2 as the one-point compactification of $\mathbf{R}^2$. Let r be the region which contains the point at infinity. Let $S_1, \ldots, S_p$ be the Seifert circles which form the frontier of r, i.e. are the limit points of r.

If $p \geq 2$ then $S_1, \ldots, S_p$ are the Seifert circles which correspond to the dissolution of some $\widetilde{L}_i$ or $\widetilde{L}'_j$ into its constituent Seifert circles. Denote this special alternating diagram by $\widetilde{L}_u$. If $p = 1$, excluding the unknot, then S_1 is one of the Seifert circles of a unique special alternating diagram, which we shall also denote by $\widetilde{L}_u$. We will call $\widetilde{L}_u$ the *underlying special alternating diagram.*

The dissolution of $\widetilde{L}_1 * \widetilde{L}_2 * \cdots * \widetilde{L}_m * \widetilde{L}'_1 * \cdots * \widetilde{L}'_n$ into its constituent Seifert circles allows us to separate $S_1, \ldots, S_p$ into two sets: the first set contains those Seifert circles of $\{S_1, \ldots, S_p\}$ which are of the first type (Definitions 5.8); while the second set contains those which are of the second type.

If $p = 1$ S_1 is defined as the *bounding circle* of $\widetilde{L}_u$.

Suppose now $p \geq 2$ and consider a Seifert circle from the second set, say S_k. This belongs to two and only two special alternating diagrams, $\widetilde{L}_u$ and $\widetilde{L}_k$ (or $\widetilde{L}'_k$) say. We say that S_k is the *bounding circle* of $\widetilde{L}_k$ (or $\widetilde{L}'_k$).

Similarly, if we consider each of these special alternating diagrams with a bounding circle in turn, we may separate their Seifert circles into the above two sets. Proceeding in this manner we will classify all the Seifert circles of $\widetilde{L}_1 * \cdots * \widetilde{L}'_n$ as being either a Seifert circle of the first type, or the unique (by construction) *bounding circle* of a special alternating diagram.

Clearly if $p = 1$ every special alternating diagram has a bounding circle, and by construction it is unique. While if $p \geq 2$ every $\widetilde{L}_i$, $\widetilde{L}'_j$ except $\widetilde{L}_u$ has a bounding circle.

Example 5.14.

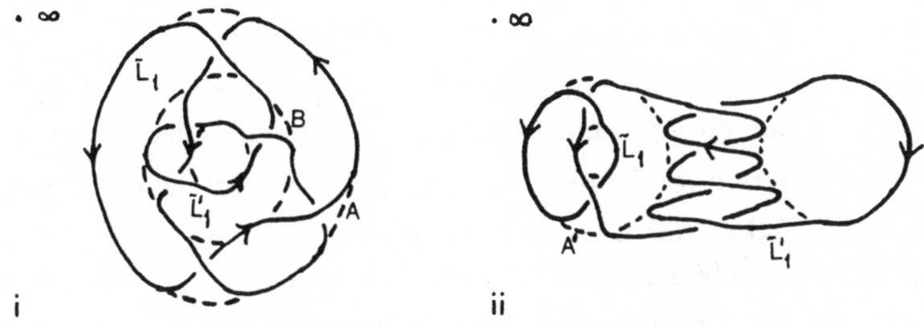

Figure 5.7

In Figure 5.7 (i) the underlying special alternating diagram, $\widetilde{L}_u$, is $\widetilde{L}_1$, while A is the bounding circle of $\widetilde{L}_u$, and B is the bounding circle of $\widetilde{L}'_1$. In Figure 5.7 (ii) $\widetilde{L}'_1$ is $\widetilde{L}_u$, while A' is the bounding circle of $\widetilde{L}_1$.

For the present we will assume that $p \geq 2$. If $p = 1$, i.e. S_1 is the bounding circle of $\widetilde{L}_u$, we may remove the bounding circle by performing a *global sphere isotopy*. This is realised in the following manner: take that part of the bounding circle which does not contain any crossings. Pull this region around the back of the sphere S^2 as in Figure 5.8 (ii) to obtain as in Figure 5.8 (iii) a new $\widetilde{L}_u$ with $p \geq 2$.

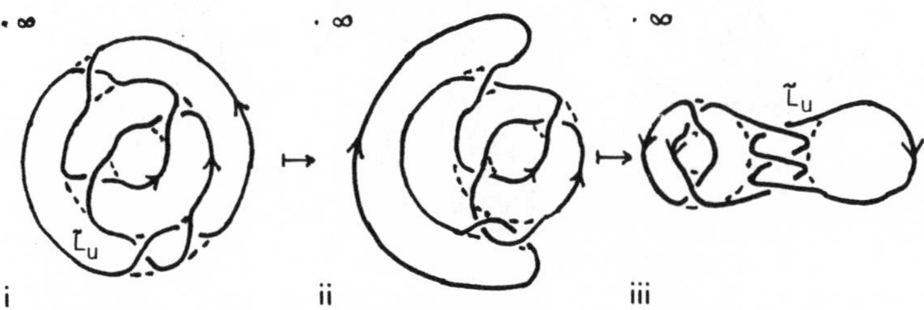

Figure 5.8

If we also for the time being restrict ourselves to the c-state S_N, and if we splice the reduced alternating diagram to form a set of Seifert circles, it is clear from Figures 5.3 and 5.4 that each Seifert circle will carry a fixed state variable, either 1 or N. So if we consider a positive special alternating diagram, $\widetilde{L}_i$, which is not $\widetilde{L}_u$, then two Seifert circles which are connected by a crossing in $\widetilde{L}_i$ will carry the state variable 1 on one of these Seifert circles and N on the other. We can make this more precise: if the bounding circle carries the state variable N, the orientation has to be in our positive anti-clockwise direction, so the angle contribution in this case will be:

$$t^{N-\frac{(N+1)}{2}} = t^{\frac{N-1}{2}}$$

The Seifert circles inside the bounding circle carrying the state variable 1 will have orientation in our positive direction, and so the angle contribution will be $t^{1-\frac{(N+1)}{2}} = t^{-\frac{N-1}{2}}$. While the Seifert circles carrying N will have orientation in our negative sense, and so the angle contribution is $t^{-\frac{N-1}{2}}$.

Similarly if the bounding circle carries the state variable 1, its orientation will be in our negative sense so the angle contribution will be $t^{\frac{N-1}{2}}$, while all the Seifert circles inside the bounding circle will each have angle contribution $t^{-\frac{N-1}{2}}$, see Figure 5.9 (i).

Similarly if we look at a negative special alternating diagram $\widetilde{L}'_j$ which is not $\widetilde{L}_u$, then if the bounding circle carries either the state variable 1 or N, the angle contribution will be $t^{-\frac{N-1}{2}}$. While all the Seifert circles inside the bounding circle each will have angle contribution $t^{\frac{N-1}{2}}$, see Figure 5.9 (ii).

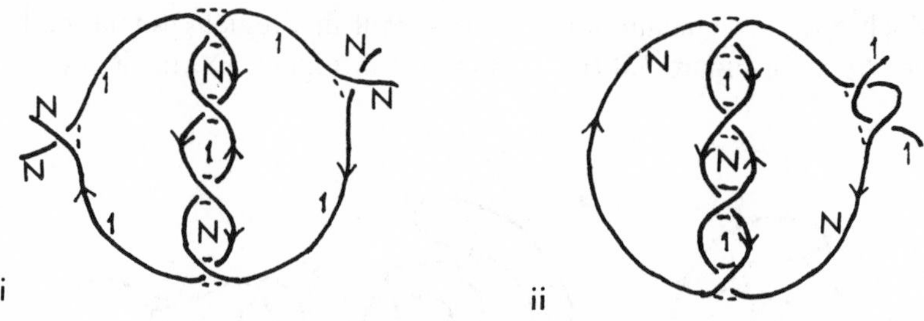

Figure 5.9

Suppose $\widetilde{L}_u$ is a *negative* special alternating diagram, clearly from the above the angle contribution from each Seifert circle is $t^{\frac{N-1}{2}}$.

Now if $\widetilde{L}_u$ is a *positive* special alternating diagram, from the above the angle contribution from each Seifert circle is $t^{\frac{-(N-1)}{2}}$. However we may change the angle contribution from one of the Seifert circles to $t^{\frac{N-1}{2}}$ by performing an isotopic deformation of $\widetilde{L}_u$ on S^2; take that part of *any* Seifert circle which does not contain any crossings and further this section consists of limit points of r. Pull this region around the back of the sphere S^2 so that this section now becomes a bounding circle, see Figure 5.10 (ii).

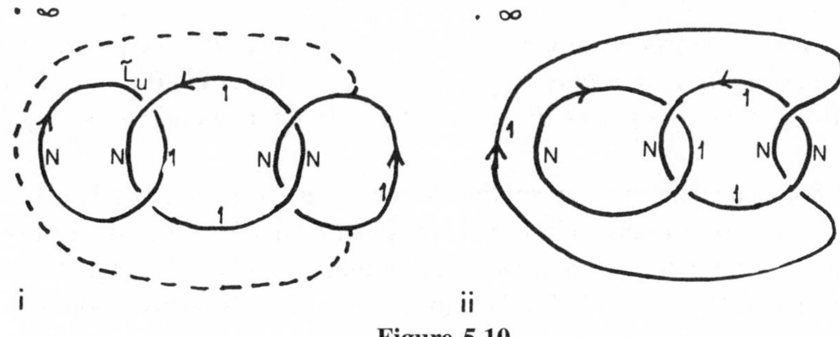

Figure 5.10

The effect on $\widetilde{L}_u$ by the above isotopy is either as in Figure 5.10 (ii) $\widetilde{L}_u$ now becomes a negative special alternating diagram, in which case from the above each Seifert circle contributes $t^{\frac{N-1}{2}}$ to the angle contribution; or $\widetilde{L}_u$ is an underlying special alternating diagram with bounding circle, i.e. $p = 1$, see Figure 5.11, where a global sphere isotopy is performed on the other Seifert circle of $\widetilde{L}_u$ in Figure 5.10 (i):

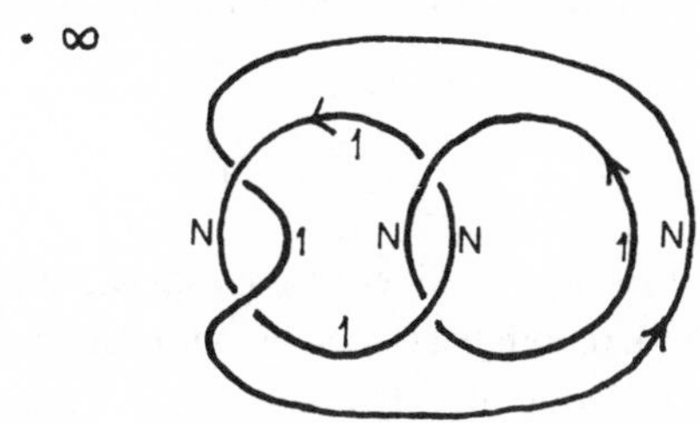

Figure 5.11

Clearly this global sphere isotopy applied to $\widetilde{L}_u$ in the positive case gives the maximal angle contribution.

So let us now fix our link diagram so that either $\widetilde{L}_u$ is a negative special alternating diagram such that $\widetilde{L}_u$ has no bounding circle, or positive with a bounding circle.

We may now argue inductively that to change the angle contribution on a Seifert circle from $t^{\frac{-(N-1)}{2}}$ to $t^{\frac{N-1}{2}}$, we need to perform a series of global sphere isotopies. However since $\widetilde{L}_u$ is now fixed to change the contribution from $t^{\frac{-(N-1)}{2}}$ to $t^{\frac{N-1}{2}}$ we must first change $t^{\frac{N-1}{2}}$ to $t^{\frac{-(N-1)}{2}}$. So no net increase occurs in the angle contribution. Hence inductively if we wish to change any angle contribution we must first change $t^{\frac{N-1}{2}}$ to $t^{\frac{-(N-1)}{2}}$. Therefore once $\widetilde{L}_u$ is fixed in the above manner there can never be any net gain in the angle contribution. So what we have shown is:

Lemma 5.15. *The maximum angle contribution on S_N occurs when either $\widetilde{L}_u$ is a negative special alternating diagram without a bounding circle (i.e. $p \geq 2$) or $\widetilde{L}_u$ is a positive special alternating diagram with a bounding circle (i.e. $p = 1$).* □

We shall now calculate the contribution to the partition function of Theorem 3.1 of S_N, with $\widetilde{L}_u$ fixed in one of the two ways of Lemma 5.15. For the rest of this section and the following section we shall assume that our link diagram is fixed in the manner of Lemma 5.15. Further, by Lemma 5.7 this contribution will give the maximum degree attainable *on* S_N.

To calculate the maximum degree of this contribution it is sufficient to consider each special alternating diagram in turn. There are in fact only four cases:
(i) $\widetilde{L}_i$ is a positive special alternating diagram, but is not $\widetilde{L}_u$. Since $\widetilde{L}_i$ is positive by Lemma 5.7 (i) the contribution to the maximum degree by the Boltzmann weights is

$$\frac{N-1}{2}\, c_+(\widetilde{L}_i) + \frac{N(N-1)}{2}\, c_+(\widetilde{L}_i)$$

where $c_+(\widetilde{L}_i)$ is as in Definition 5.10. The angle contribution from $\widetilde{L}_i$ to the maximum degree, since by the above inside the bounding circle each Seifert circle contributes $t^{\frac{-(N-1)}{2}}$, is

$$-\frac{(N-1)}{2}\,(s(\widetilde{L}_i)-1)$$

where $s(\widetilde{L}_i)$ is as in Definition 5.10. Since $\widetilde{L}_i$ is not $\widetilde{L}_u$ the contribution from the bounding circle is excluded, otherwise we would end up counting it twice. Therefore the maximum degree obtained from $\widetilde{L}_i$ is

$$\frac{N(N-1)}{2}\, c_+(\widetilde{L}_i) + \frac{(N-1)}{2}\,(c_+(\widetilde{L}_i) + 1 - s(\widetilde{L}_i)) \tag{5.21}$$

(ii) $\widetilde{L}'_i$ is a negative special alternating diagram but not $\widetilde{L}_u$. By Lemma 5.7 (ii) the contribution to the maximum degree by the Boltzmann weight is

$$-\frac{(N-1)}{2}\, c_-(\widetilde{L}'_i)$$

Again we do not consider the contribution from the bounding circle, and each Seifert circle inside contributes $\frac{N-1}{2}$ to the angle contribution. Hence the angle contribution in this case is

$$\frac{(N-1)}{2}\,(s(\widetilde{L}'_i)-1)$$

Therefore the maximum degree obtained from $\widetilde{L}'_i$ is

$$-\frac{(N-1)}{2}\,(c_-(\widetilde{L}'_i) + 1 - s(\widetilde{L}'_i)) \tag{5.22}$$

(iii) $\widetilde{L}_u$ is positive. The only change from (i) is that the angle contribution from the bounding circle, $\frac{(N-1)}{2}$, is also counted. Hence the maximum degree obtained from $\widetilde{L}_u$ is

$$\begin{aligned} &\frac{N(N-1)}{2}\, c_+(\widetilde{L}_u) + \frac{(N-1)}{2}\, (c_+(\widetilde{L}_u) + 2 - s(\widetilde{L}_u)) \\ = &\frac{N(N-1)}{2}\, c_+(\widetilde{L}_u) + \frac{(N-1)}{2}\, (c_+(\widetilde{L}_u) + 1 - s(\widetilde{L}_u)) + \frac{(N-1)}{2} \end{aligned} \tag{5.23}$$

(iv) $\widetilde{L}_u$ is negative. In this case the only change from (ii) is that there is no bounding circle, so the angle contribution from each Seifert circle to the maximum degree is $\frac{(N-1)}{2}$. So then the maximum degree obtained from $\widetilde{L}_u$ is

$$-\frac{(N-1)}{2}\, (c_-(\widetilde{L}_u) - s(\widetilde{L}_u)) = -\frac{(N-1)}{2}\, (c_-(\widetilde{L}_u) + 1 - s(\widetilde{L}_u)) + \frac{N-1}{2} \tag{5.24}$$

So collecting (5.21), (5.22), (5.23) and (5.24), and setting $\widetilde{L}_u$ to be $\widetilde{L}_1$ or $\widetilde{L}'_1$ depending if it is positive or negative, we see that the maximum degree of the contribution from S_N to the partition function is

$$\begin{aligned} &\frac{N(N-1)}{2}\, c_+(\widetilde{L}) + \frac{(N-1)}{2} \sum_{i=1}^{m} (c_+(\widetilde{L}_i) + 1 - s(\widetilde{L}_i)) \\ &- \frac{(N-1)}{2} \sum_{j=1}^{n} (c_-(\widetilde{L}'_j) + 1 - s(\widetilde{L}'_j)) + \frac{(N-1)}{2} \end{aligned}$$

which by Proposition 5.11 is equal to

$$\frac{N(N-1)}{2}\, c_+(\widetilde{L}) - \frac{(N-1)}{2}\, \sigma(L) + \frac{(N-1)}{2} \tag{5.25}$$

Proposition 5.16. *The maximum degree of the contribution from S_N to the partition function of Theorem 3.1, and hence to the non-normalised hierarchal N-order Jones polynomial is*

$$\frac{N(N-1)}{2}\, c_+(\widetilde{L}) - \frac{(N-1)}{2}\, \sigma(L) + \frac{(N-1)}{2}$$

□

§6. B-moves and the proof of Theorem 5.3

In this section we will show that in fact S_N gives a unique maximum degree over all possible contributing states. To attain this goal we will first define a sequence of moves which will allow us to transform one state into some other state, and after a finite number of these transformations we will obtain S_N.

Definition 6.1.

(i) A *B-move of the first type*, B-I, involves changing the state variables at a crossing in one of the following two ways, assuming for a positive crossing that in the first case $x \neq N$ and $a \neq 1$, and in the second $b \neq N$ and $y \neq 1$

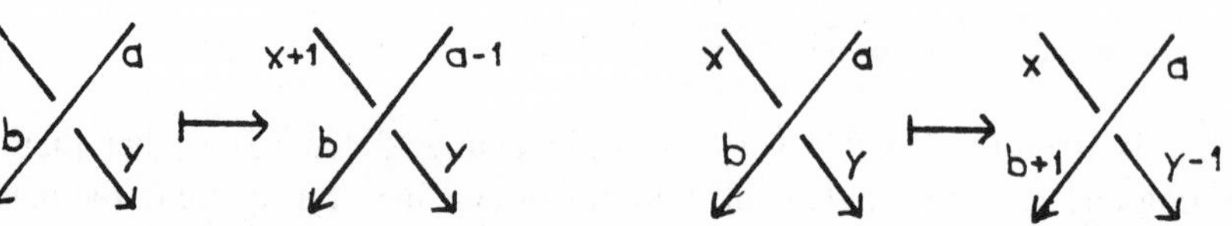

and assuming for a negative crossing that in the first case $x \neq 1$ and $a \neq N$, and in the second $b \neq 1$ and $y \neq N$

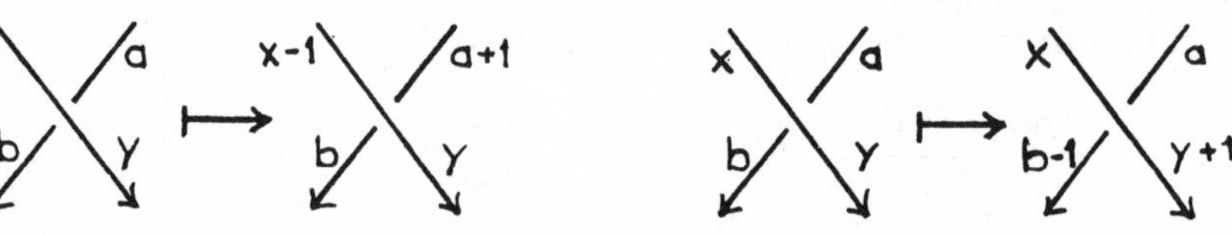

(ii) A *B-move of the second type*, B-II, involves changing the state variables at a positive crossing in one of the following two ways, assuming that $a \neq 1 \neq y$

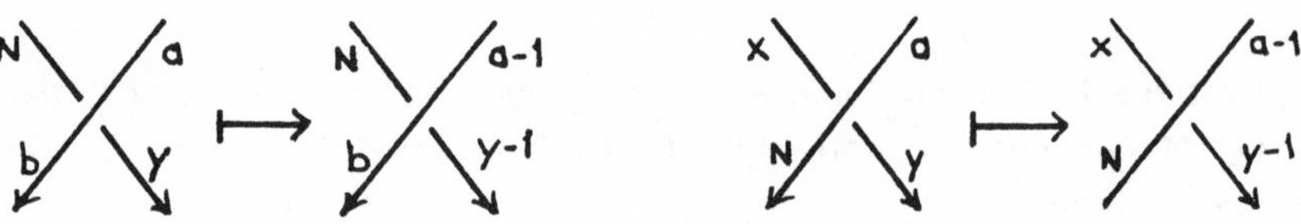

While for a negative crossing in one of the following two ways, assuming that $x \neq 1 \neq b$

(iii) A *B-move of the third type*, B-III, involves changing the state variables at a positive crossing in one of the following two ways, assuming that $x \neq N \neq b$

While for a negative crossing in one of the following two ways, assuming that $a \neq N \neq y$

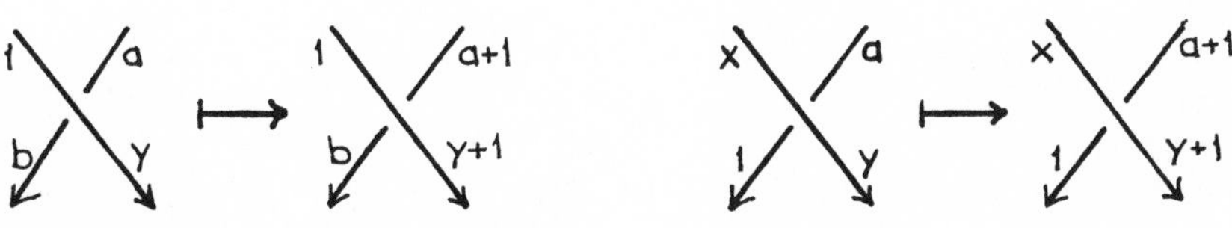

Remarks 6.2.
(i) If there exists a positive crossing with the state variables ascribed in the following manner

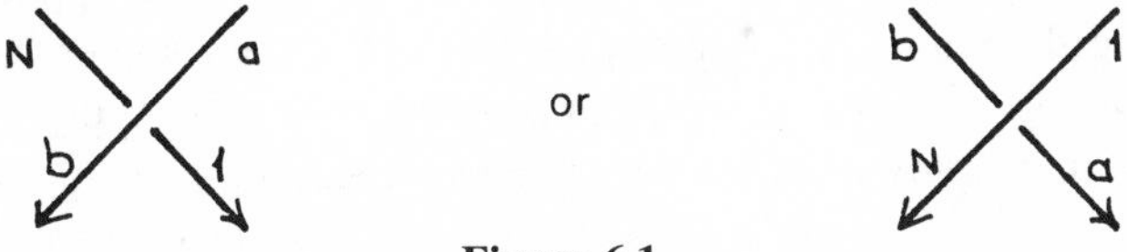

Figure 6.1

then charge conservation $N + a = b + 1$ yields $N - 1 = b - a$, but $N \geq a \geq 1$ so this implies that $b = N$ and $a = 1$. Similarly for a negative crossing.
(ii) Note that at a crossing the B-moves defined above cause the charge $(x + a = c = b + y)$ to be conserved.

Lemma 6.3. *On any state expect S_N we can always find a crossing on which we may apply one of the B-moves.*

Proof. Suppose there are no crossings on which we may apply a B-move. Then each crossing must have state variables assigned in one of the following forms

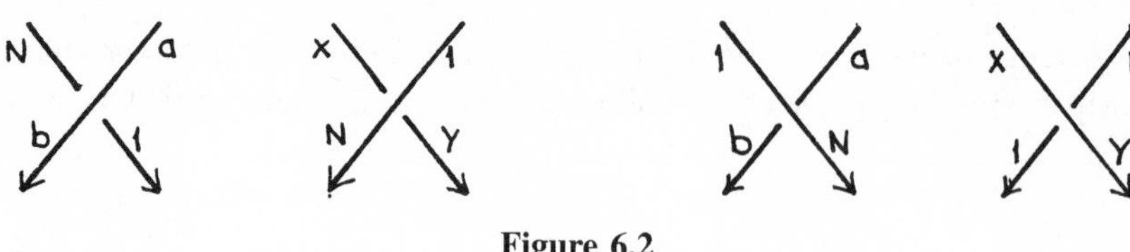

Figure 6.2

But by Remark 6.2 (i) the two positive crossings correspond to

While the two negative crossings correspond to

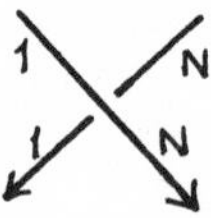

So this state is just S_N, a contradiction. □

So given any c-state, not equal to S_N, on an alternating link diagram we can always apply at least one B-move. However once we have applied our initial B-move, since the alternating diagram is reduced, we no longer have a c-state. So to conserve charge at the subsequent crossings we must apply a sequence of B-moves. We may apply a subsequent B-move because of the alternating nature of the link

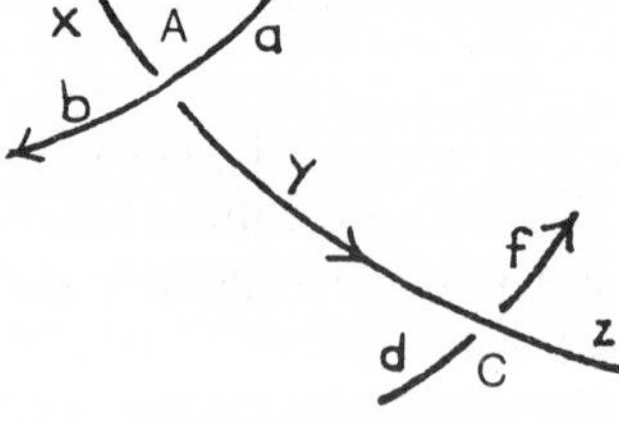

Figure 6.3

In fact, if we apply B-I to the state variables leaving the crossing A in Figure 6.3, then to conserve charge we must apply a B-move at the crossing C (otherwise we obtain one of the crossings of Figure 6.2 and hence a contradiction). We have two possibilities, see Figure 6.4.

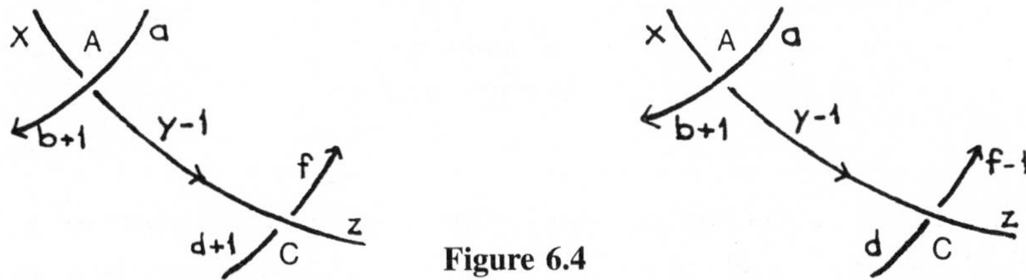

Figure 6.4

The B-moves in Figure 6.4 are well-defined because by definition of our Boltzmann weights $z \geq y$ and $d \geq f$. So increasing d or decreasing y or f will not contradict our definition of the Boltzmann weights. Since all other combinations of crossings and B-moves are variations on Figures 6.3 and 6.4, a B-move will never cause $y > z$ and $f > d$. *Note* this is not true for a non-alternating link diagram, see Example 6.7. So the B-moves cause the state variables either to increase towards N or decrease towards 1 in the correct way, i.e. towards S_N.

Definition 6.4. Given a c-state S, we define $\delta(S)$ as

$$\delta(S) \;=\; \sum(N-k-1)+\sum(N-k'-1)$$

where $k = x - y$ and the first sum is taken over all positive Boltzmann weights, $w_+^{(N)}(ab \mid xy)(t)$ in S. While $k' = y' - x'$ and the second summation is taken over all negative Boltzmann weights, $w_-^{(N)}(a'b' \mid x'y')(t)$, in S.

Clearly $\delta(S) \geq 0$ and $\delta(S) = 0$ if and only if S is equal to S_N. In a sense $\delta(S)$ measures how close S is to S_N.

Lemma 6.5. *There exists a sequence of B-moves on a c-state S, not equal to S_N, which creates a new c-state S' such that $\delta(S') < \delta(S)$.*

Proof. By Lemma 6.3 we know there exists a crossing on which we may apply a B-move. So apply this initial B-move and label one edge α, the head, and the other β, the tail. For B-I, α and β are depicted in Figure 6.5

Figure 6.5

So let α follow the knock-on effect caused by the change to the state variables at subsequent crossings by the subsequent B-moves. By the remarks preceding Definition 6.4 we may do this in a consistent fashion in terms of our definition of our Boltzmann weights.

As α now runs along the link, possibly in a way which is reverse to the orientation on some edges, we have two possibilities.

Case (i): α returns to its initial crossing via the crossing where charge conservation had not been conserved due to the change of the state variable on the edge β, i.e.

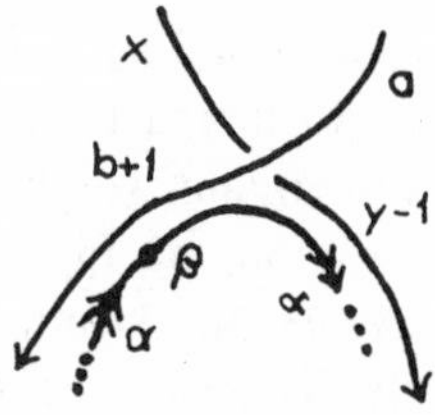

Figure 6.6

This causes the charge to be conserved at this edge also, and since by Remark 6.2 (ii) charge is conserved on all the other crossings we have obtained a new state, S'. Obviously, $\delta(S') < \delta(S)$.

Case (ii): As we keep on applying these B-moves, and assuming (i) has not occurred, α may eventually return to the same edge but in such a way that if we originally applied B-I at this crossing, we now apply a B-move of the second or third type, and vice versa, see Figure 6.7.

This may occur in only one of the following two ways.

Figure 6.7

We may assume without loss of generality that in both cases this is the first time α returns to a specific edge. However, we may discount the possibility of path due to α which causes Figure 6.7 (i).

In fact, let us consider an edge which has been effected by a B-move, i.e. the state variable on that edge has increased or decreased by 1. Label this edge with a u or o depending on if the edge, following the path α rather than the orientation of the link, has emerged from *under-crossing*, or *over-crossing*, for example

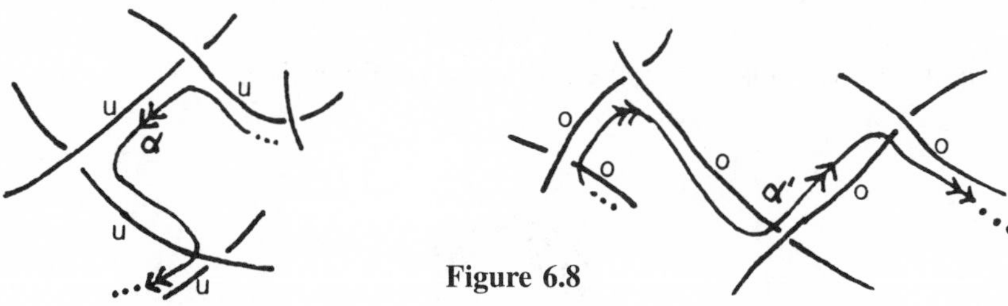

Figure 6.8

Since our link diagram is a reduced *alternating* link diagram, it is clear as we apply consecutive B-moves we will obtain either a trail of u, or trail of o, never a mixture. So consider Figure 6.7 (i), and label it with u or o where appropriate

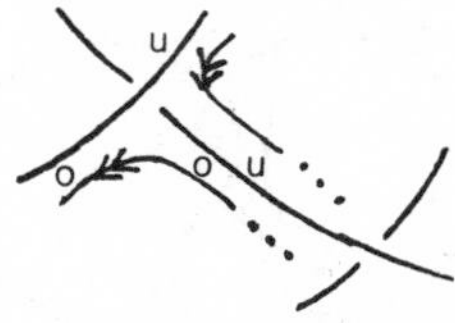

Figure 6.9

Since we obtain a mixture, and by the above comments on an *alternating* link diagram this is not possible. As stated above such a diagram as Figure 6.7 (i) cannot occur.

However Figure 6.7 (ii) may occur, for example

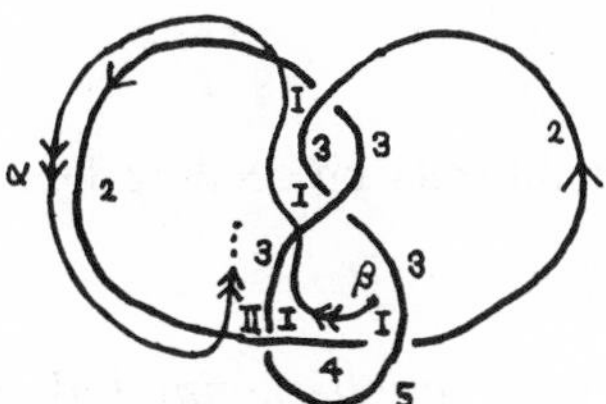

Figure 6.10

Where I indicates that at this crossing we have applied B-move of the first type, and similarly II a B-move of the second type. The crossing with a type I and type II move corresponds to Figure 6.7 (ii).

Since we are at liberty to choose our initial point, we may trace α back to our original initial point, and so returning to our original state. Now shift the initial point to the crossing of Figure 6.7 (ii); in Figure 6.10 to the crossing where we originally performed B-I and B-II. The way to choose on which edges α and β should initially lie is as follows: let α, the commencing point (i.e. we will soon allow α to run around the link diagram), be the edge that was originally covered twice. While the end point β of our initial point will correspond to the edge from which our original α entered the subsequent edge for a second time.

In Figure 6.7 (ii) this means the following configuration.

Figure 6.11

While in Figure 6.10 we obtain the following path for α

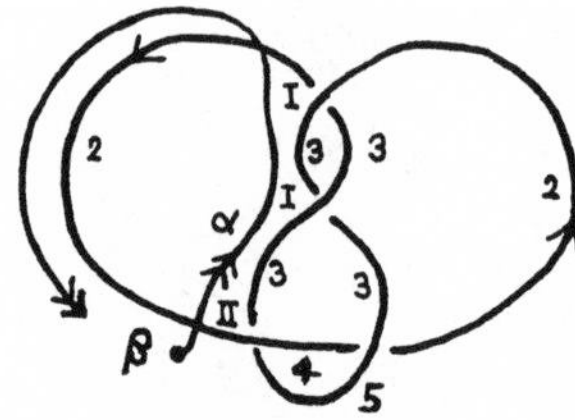

Clearly by construction we will obtain case (i), and a new well-defined states S', with $\delta(S') < \delta(S)$.

Note that because we only have a finite number of crossings in our link diagram, the path α must terminate somewhere. The only other possibility, after cases (i) and (ii), is that α reaches a crossing of the type in Figure 6.2. This now leads to a contradiction. □

Hence by Lemma 6.3, the comments proceeding Definition 6.4 and Lemma 6.5 we obtain the following:

Proposition 6.6. *Every c-state on an alternating link diagram can be changed to the state S_N by a finite number of applications of B-moves of Definition 6.1.* □

Example 6.7. It was noted before Definition 6.4 that the B-moves will not necessarily give a new c-state when applied to a non-alternating diagram. An example is the following, let $N = 5$ (for simplicity)

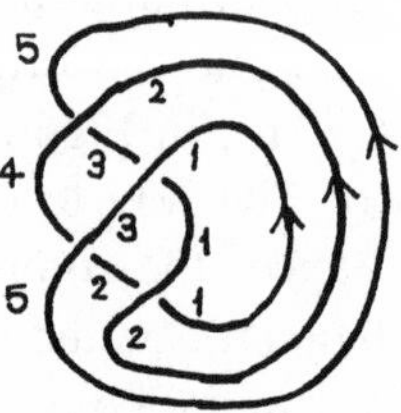

Figure 6.12

Although we have a choice of initial B-moves, after our initial B-move no further B-moves may be performed. So at some crossing charge is not conserved, hence we no longer have a c-state.

Lemma 6.8.

(a) A B-move of the first type causes the maximum degree of a positive Boltzmann weight to increase by at least $\frac{1}{2}$, and the maximum degree of a negative Boltzmann weight to also increase by at least $\frac{1}{2}$.

(b) A B-move of the second type causes the maximum degree of a positive Boltzmann weight to increase by at least 1. While in the case of a negative Boltzmann weight if $a = y = N$ then there exists no difference between the maximum degrees. While if $a \neq y$ there is an increase of at least $\frac{1}{2}$.

(c) A B-move of the third type causes the maximum degree of a positive Boltzmann weight to increase by at least 1. While if $x = 1 = b$ then there is no difference between maximum degrees of negative Boltzmann weights. While if $x \neq b$ there is an increase of at least $\frac{1}{2}$ in the maximum degree of a negative Boltzmann weight.

Proof.

(a) The result for the positive Boltzmann weight follows from Lemma 5.5 and equations (5.4) and (5.5) derived in that lemma. Similarly in the negative case the result follows from Lemma 5.6 and equations (5.11) and (5.12).

(b) For the positive case just consider equation (5.6) of Lemma 5.5, and in the negative case equation (5.13) of Lemma 5.6.

(c) Similarly in the positive case the result follows from equation (5.7) of Lemma 5.5, and in the negative case equation (5.14) of Lemma 5.6. □

The above lemma shows us how the B-moves effect the maximum degrees. Since we wish to show that S_N gives the unique maximum degree over all c-states, we need also to consider the effect of the B-moves on the angle contribution.

Suppose an edge carries the state variable x, and the angle contribution to the partition function from this edge is $t^{\pm\gamma(x-\frac{(N+1)}{2})}$ where $0 \leq \gamma \leq 1$, and $\pm$ is the dependency on the orientation. Let x be sent to $x \pm 1$, then the change in angle contribution is $t^{\pm\gamma(\pm 1)}$. To calculate γ we need to consider the ingoing angle at each crossing and allow this to go to zero, see Figure 6.13. In the limit γ is just the angle of rotation between the crossings divided by 2π.

x λ a
b+1 y-1

x a
b+1 y-1

Figure 6.13

With an ambient isotopy if necessary we may take γ to be $0, \frac{1}{2}$ or 1. For example

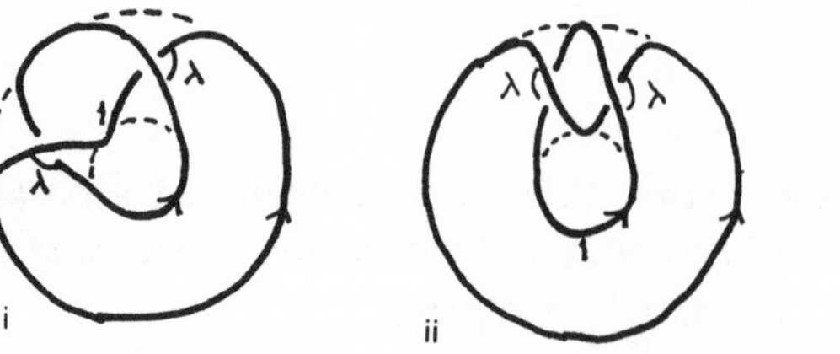

Figure 6.14

If we take the limit and consider the angle contribution from the edge carrying the state variable 1, then in Figure 6.14 (i) $\gamma = 0$; in (ii) $\gamma = 1$; and in (iii) $\gamma = \frac{1}{2}$. In fact what we have is a decomposition into annihilation and creation diagrams of [**WAD**], see also [**Tu2**].

Recall that when we were calculating the angle contribution to S_N, see remarks after Example 5.14, for a negative special alternating diagram, the edge carrying N was always in the positive (anti-clockwise) direction. While that carrying 1 was in our negative direction. Since the B-moves cause the state variables to gravitate towards N and 1 without oscillation; a state variable x, on the part of the Seifert circle of the negative special alternating diagram, that will eventually carry N is sent by any B-move to $x + 1$. So in this case we have $0, \frac{1}{2}$, or 1 as the increase in the maximum degree from the angle contribution. While if under a B-move x is sent to $x - 1$, since this edge is in our negative direction, there is an increase of $0, \frac{1}{2}$ or 1 to the maximum degree from the angle contribution.

Proposition 6.9. *Applying the B-moves to a state which does not equal S_N to form a new state will cause the maximum degree to increase.*

Proof. The proof of Lemma 6.5 shows that this sequence of B-moves may be associated with a simple (in the topological sense) loop. There are in fact three cases to consider.

(i) Suppose a loop encircles just one Seifert circle, i.e. the only B-moves we have used are all of the second type or all of the third type, see Figure 6.15.

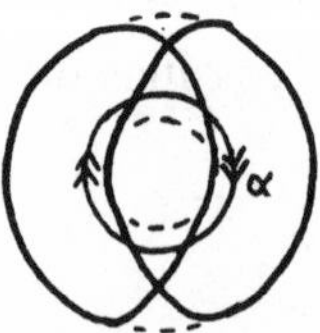

Figure 6.15

If the loop is not a bounding circle of a negative special alternating diagram, although there is no increase to the maximum degree from the Boltzmann weights, Lemma 6.8 (b) and (c), by the comments preceding this proposition the increase in angle contribution is $+1$. Now suppose the loop is inside a positive special alternating diagram. Since our alternating diagram is reduced, we must have applied at least two B-II or B-III. By Lemma 6.8 (b) and (c) this causes the maximum degree to increase by $+2$ from the change in Boltzmann weights. Recall that inside a positive special alternating diagram the edge carrying N in S_N is in the negative direction, while that carrying 1 is in the positive direction. So with similar arguments to those before this proposition the loop causes a decrease of -1 (from the angle contribution) to the maximum degree so an overall increase of at least $+1$.

(ii) Suppose that a loop does not encircle a Seifert circle, and further this loop does not leave a given positive special alternating diagram. This leads us to consider three cases, the others are variations on these themes.

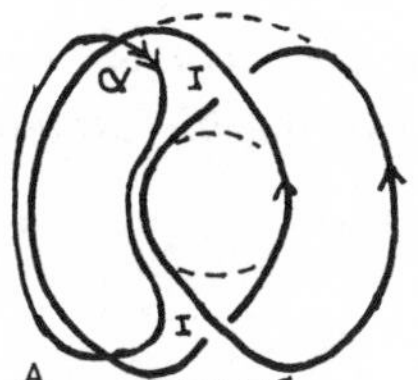

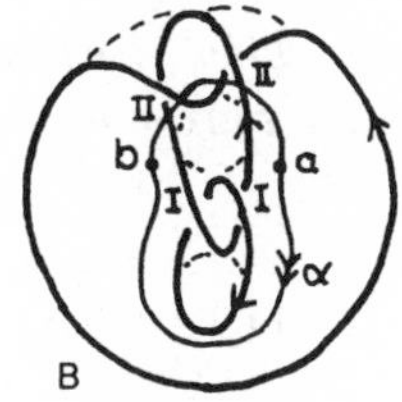

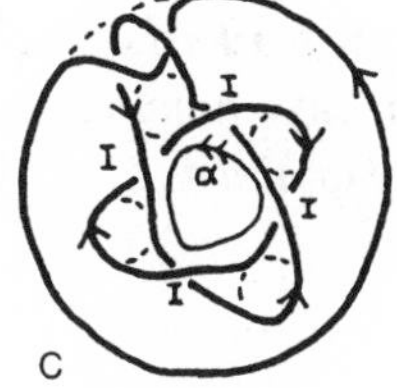

Figure 6.16

In case A the increase in the maximum degree of the Boltzmann weight by $\frac{1}{2}$, Lemma 6.8 (a), at a crossing negates the decrease of $-\frac{1}{2}$ by the subsequent angle contribution. However this leaves us a crossing which by Lemma 6.8 (a) increases the maximum degree by $\frac{1}{2}$. So we obtain a net increase. In case B, if we start at a and continue to b, the maximum degree will have decreased by -1 due to the change in angle contribution. However due to the two B-I moves required to go from a to b the maximum degree increases by $+1$, Lemma 6.8 (a). Since the link diagram is reduced, to go from b to a entails at least the use of two B-II (or B-III). Therefore by Lemma 6.8 (b) or (c) there is an increase of at least $+2$, while to go from b to a the angle contribution causes a decrease of -1. So again we have a net increase. In case C, the change in angle contribution causes the maximum degree to decrease by -1. While since we need to apply four B-I this causes the maximum degree to increase by $+2$. Hence we obtain a net increase.

(iii) Suppose now that the loop enters a positive special alternating diagram, and also exits from this positive special alternating diagram, we need to consider three cases:

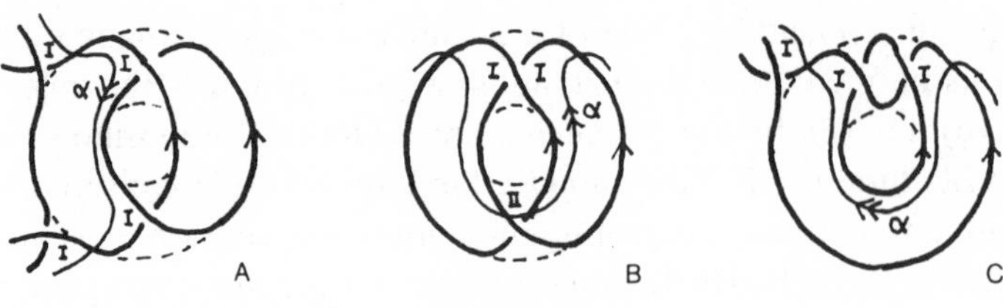

Figure 6.17

Clearly case A is just a variation of A in (ii), so we obtain a net increase in maximum degree.

In case B, two B-I and one B-II (or B-III) will increase the maximum degree by $+2$, Lemma 6.8. While the change in angle contribution will only cause a decrease of -1. So we have a net increase.

The final case due to three B-I the maximum degree will increase $\frac{3}{2}$, while the change in angle contribution will only decrease it by -1. Hence we have a net increase to the maximum degree. □

Proposition 6.10. *S_N is the state which gives the unique maximum contribution to the partition function, and so the maximum degree of the hierarchal N-order Jones polynomial is*

$$\frac{N(N-1)}{2} c_+(\widetilde{L}) - \frac{(N-1)}{2} \sigma(L) .$$

where $\widetilde{L}$ is a reduced alternating diagram of L.

Proof. By Lemma 6.9 S_N gives the unique maximum degree. Now by Theorem 3.1 the partition function must be factored by $(t^{-\frac{(N-1)}{2}} + \cdots + t^{\frac{N-1}{2}})$. By Proposition 5.16 the maximum degree of the contribution from S_N to the partition function is

$$\frac{N(N-1)}{2} c_+(\widetilde{L}) - \frac{(N-1)}{2} \sigma(L) + \frac{N-1}{2} .$$

Hence the result. □

The correspondence $x \mapsto N+1-x$ and

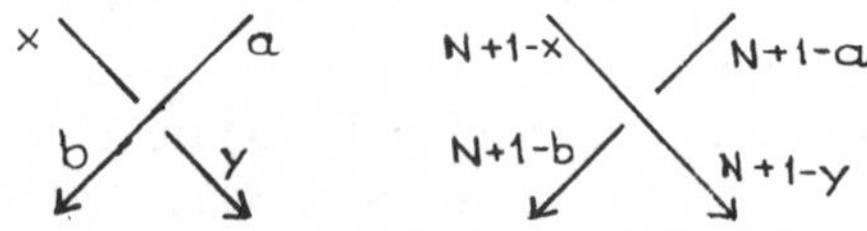

give us a $1-1$ correspondence between the states on $\widetilde{L}$, and on $r\widetilde{L}$ its mirror image. If we now apply the "flip" of Definition 2.3 (ii), to obtain $(r\widetilde{L})^+$, then Lemma 2.5 says the contribution to the partition function of the Boltzmann weights from the state on $r\widetilde{L}$, and that on $(r\widetilde{L})^{+}$ is the same, and clearly there exists a $1-1$ correspondence between the states on $\widetilde{L}$ and those on $(r\widetilde{L})^{+}$.

Now

$$\begin{aligned} & w_{-}^{(N)}(N+1-y, N+1-x \mid N+1-b, N+1-a)(t) \\ = & w_{+}^{(N)}(N+1-b, N+1-a \mid N+1-y, N+1-x)(t^{-1}) \\ = & w_{+}^{(N)}(ab \mid xy)(t^{-1}) \end{aligned}$$

by Lemma 3.2(iii).

Further since under the transformation $x \mapsto N+1-x$, $x - \frac{(N+1)}{2}$ is sent to $\frac{N+1}{2} - x$, and the normalisation factor of Theorem 3.1 is left invariant under the transformation $t \mapsto t^{-1}$. Collecting the information of these paragraphs we have the following:

Lemma 6.11. *If L is a link and rL is its mirror image then*

$$V_{rL}^{(N)}(t) = V_{L}^{(N)}(t^{-1}) \qquad \square$$

So to determine $d_{\min} V_{L}^{(N)}(t)$ it is sufficient to determine the maximum degree of $V_{rL}^{(N)}(t)$ which by Proposition 6.10 is

$$\frac{N(N-1)}{2} c_{+}(r\widetilde{L}) - \frac{(N-1)}{2} \sigma(rL) = \frac{N(N-1)}{2} c_{-}(\widetilde{L}) + \frac{(N-1)}{2} \sigma(L)$$

and so by Lemma 6.11

$$d_{\min} V_{L}^{(N)}(t) = \frac{-N(N-1)}{2} c_{-}(\widetilde{L}) - \frac{(N-1)}{2} \sigma(L) .$$

So we have finally proved our aim i.e. Theorem 5.3.

Throughout Sections 5 and 6 we restricted ourselves to alternating links, and although some of the results needed to prove Theorem 5.3 required that the link be alternating, we conjecture that:

Conjecture. *If $\widetilde{L}$ is any reduced (non-split) diagram of any link L, and $V_{L}^{(N)}(t)$ is the N-order hierarchal Jones polynomial of L then*

$$d_{\max} V_{L}^{(N)}(t) \leq \frac{N(N-1)}{2} c_{+}(\widetilde{L}) - \frac{(N-1)}{2} \sigma(L)$$

and

$$d_{\min} V_{L}^{(N)}(t) \geq -\frac{N(N-1)}{2} c_{-}(\widetilde{L}) - \frac{(N-1)}{2} \sigma(L) .$$

Remark 6.12. Using the correspondence $x \mapsto N+1-x$ on the state variables of a *particular* link component, it is possible to show, although in this paper we do not prove this result, that the reversing result, obtained in the $N=2$ case by

Lickorish and Millett, may be extended, using the techniques of this paper, to the hierarchal case.

§7. Hierarchal skein polynomials

In this last section, we briefly discuss a possible application of the hierarchal N-order Jones polynomial of a link L to the problem of determining the braid index of L.

Let $P_L(v, z)$ denote the skein polynomial, i.e. the two-variable Jones polynomial, of an oriented link L, see [**HOMFLY**] or [**Mo**]. It is known that the v-degree of $P_L(v, z)$ has some connection to the braid index of L. To be more precise, let $\mathbf{b}(L)$ denote the braid index of L, i.e. the minimum number of strings needed for L to be represented as a closed braid.

Let max v-deg P and min v-deg P denote the highest and lowest degree, respectively, of v in P. Finally if we let

$$v\text{-span } P \;=\; \max v\text{-deg } P - \min v\text{-deg } P \ ,$$

then we have the following theorem:

Theorem 7.1 [**Mo**], [**FW**]. *For any oriented link L,*

$$v\text{-span } P_L(v, z) \;\leq\; 2(\mathbf{b}(L) - 1) \ . \qquad \square$$

Yamada in [**Ya**] shows that $\mathbf{b}(L)$ is closely related to the number of *Seifert circles* of a diagram, for a definition see remarks proceeding Definition 5.8 and also [**BZ**]. Let $s(\widetilde{L})$ denote the number of Seifert circles of a link diagram $\widetilde{L}$ of L.

Theorem 7.2 [**Ya**]. *For any diagram $\widetilde{L}$ of a link L*

$$\mathbf{b}(L) \;\leq s(\widetilde{L}) \ . \qquad \square \tag{7.1}$$

Certain types of alternating links have alternating diagrams $\widetilde{L}$, for which equality holds in (7.1). For example, it is shown in [**Mu5**] that for any reduced alternating diagram of an alternating fibred link equality holds in (7.1). Hence for these alternating fibred links $\mathbf{b}(L)$ is completely determined by $s(\widetilde{L})$.

In general the inequality in (7.1) is sharp. To evaluate $s(\widetilde{L}) - \mathbf{b}(L)$ Murasugi and Przytycki [**MP**] introduced the notion of *index* of a diagram $\widetilde{L}$. However to define the index we must first introduce some concepts from graph theory.

Let G be a (finite) *graph* [**Bi**]. A *path* T in G is an alternate sequence of vertices and edges, i.e. $v_1, e_1, v_2, e_2, \ldots, v_\ell, e_\ell, v_{\ell+1}$. Where ℓ, the number of edges in T, is called the *length* of T. T is said to be *closed* if $v_{\ell+1} = v_1$, and T is called *simple* if no edge in T occurs more than once in T.

An edge e of G is called *singular* if (i) e is not a loop and (ii) none of the edges beside e join the two ends of e.

Let $F = \{e_1, e_2, \ldots, e_n\}$ be a set of (distinct) singular edges of G. F is called *independent* if any non-empty subset $F' = \{e_{i_1}, \ldots, e_{i_k}\}$ of F, $1 \leq k \leq n$, does not occur in a simple closed path of length at most $2k$.

Definition 7.3. The *index* of G, ind G, is the maximal number of independent singular edges in G.

Obviously, ind G is an invariant of a graph G. However, it is also invariant under 2-isomorphism [**MP**],

It should be noted that the index of G defined above is called the *circle* index in [**MP**], and the original definition of the index of a graph G given in [**MP**] is different from the present definition. However it is proved in [**Tr3**] that both concepts are equal for a bipartite graph. (A graph is said to be *bipartite* if every simple closed path has an even length.)

To our definition of $s(\widetilde{L})$, the number of Seifert circles of a diagram $\widetilde{L}$, let us add the definition $c(\widetilde{L})$, the number of crossings. The *Seifert graph* $\Gamma(\widetilde{L})$ (associated with $\widetilde{L}$) is the graph with $s(\widetilde{L})$ vertices $v_1, v_2, \ldots, v_{s(\widetilde{L})}$ and $c(\widetilde{L})$ edges $e_1, e_2, \ldots, e_{c(\widetilde{L})}$. Each vertex corresponds to a Seifert circle and each edge corresponds to a crossing. Two distinct vertices v_i and v_j are connected by e_k if and only if two Seifert circles S_i and S_j (corresponding to v_i and v_j respectively) are joined by a crossing c_k (corresponding to e_k). The Seifert graph $\Gamma(\widetilde{L})$ is a plane bipartite graph.

Definition 7.4. The *index of link diagram* $\widetilde{L}$, ind $\widetilde{L}$ is defined as the index of $\Gamma(\widetilde{L})$.

Theorem 7.5 [**MP**]. *For any diagram $\widetilde{L}$ of a link L,*

$$\mathbf{b}(L) \leq s(\widetilde{L}) - \operatorname{ind} \widetilde{L}\,. \qquad \square \tag{7.2}$$

Combining Theorems 7.1 and 7.5 we obtain

Theorem 7.6. *For any diagram $\widetilde{L}$ of a link L*

$$v\text{-span}\, P_L(v, z) \leq 2(s(\widetilde{L}) - \operatorname{ind} \widetilde{L} - 1)\,. \qquad \square \tag{7.3}$$

For many alternating links including 2-bridge links and a certain type of alternating algebraic links equality holds in (7.3), and hence equality also holds for these links in (7.2). However, in general the inequality in (7.3) is sharp, even for an alternating link. One of the simplest examples for which the equality of (7.3) does not hold for an alternating knot K is depicted in Figure 7.1.

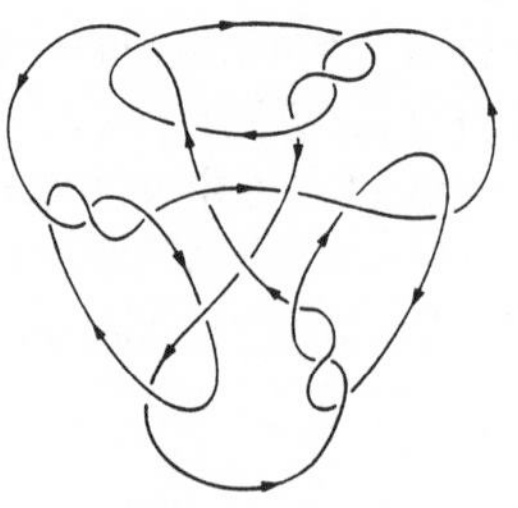

Figure 7.1

In this case $s(\widetilde{L}) = 7$ and $\operatorname{ind}\widetilde{L} = 1$, hence $s(\widetilde{L}) - \operatorname{ind}\widetilde{L} = 6$. However v-span $P_K(v,z) = 8$. Therefore by Theorem 7.6 $\mathbf{b}(K)$ is either 5 or 6. In fact $\mathbf{b}(K) = 6$. This can be shown by considering the 2-parallel K' of K. A computer calculation shows that v-span $P_{K'}(v,z) \geq 20$. If $\mathbf{b}(K) = 5$ then $\mathbf{b}(K') \leq 10$ and hence v-span $P_{K'}(v,z) \leq 18$, a contradiction. This method, of considering the 2-parallel link, works for many knots and links whose braid indices are not determined by v-span $P_L(v,z)$.

Although the inequality of (7.3) is sharp, it is plausible that the equality holds in (7.2). For example if $\operatorname{ind}\widetilde{L} = 0$ for a reduced alternating diagram $\widetilde{L}$ of an alternating link L, then $\mathbf{b}(L) = s(\widetilde{L})$.

Therefore we propose the following conjecture:

Conjecture 1. *For a (reduced) alternating diagram $\widetilde{L}$ of an alternating link L,*

$$\mathbf{b}(L) \;=\; s(\widetilde{L}) - \operatorname{ind}(\widetilde{L}) \;.$$

It may be possible to prove this conjecture by using the hierarchal N-order skein polynomial $P_L^{(N)}(v,z)$ [**WAD**]. In sections 5 and 6 it is shown that span $V_L^{(N)}(t)$ is a quadratic formula in N for alternating links L. If v-span $P_L^{(N)}(v,z)$ is a *linear* function in N for alternating links L, Conjecture 1 could be proven. To be more precise we propose the following:

Conjecture 2. *For an alternating link L, v-span $P_L^{(N)}(v,z)$ is approximately linear for sufficiently large N. In other words, there exist an integers N_0 and k_0 such that for $N > N_0$*

$$v\text{-span } P_L^{(N)}(v,z) \;=\; 2(N - N_0)(s(\widetilde{L}) - \operatorname{ind}\widetilde{L}) + k_0 \;.$$

It is easy to see that Conjecture 2 implies Conjecture 1.

References

[An] G. ANDREWS: *Theory of Partitions.* Addison-Wesley (1976).

[Ba] R. BAXTER: *Exactly Solvable Models in Statistical Mechanics.* Academic Press, London (1982).

[Bi] N. BIGGS: *Algebraic Graph Theory.* Cambridge University Press, Cambridge (1974).

[BZ] G. BURDE and H. ZIESCHANG: *Knots.* W. de Gruyter, Berlin (1985).

[Dr] V. DRINFELD: Quantum groups. Proceeding of the I.C.M., Berkeley (1986), 798-820.

[FW] J. FRANKS and R.F. WILLIAMS: Braids and the Jones polynomial. *Trans. Amer. Math. Soc.* **303** (1987), 97-108.

[HOMFLY] J. HOSTE, A. OCNEANU, K. MILLETT, P. FREYD, W. LICKORISH, and D. YETTER: A new polynomial invariant of knots and links. *Bull. A.M.S.* **12** (1985) 183-212.

[Ji] M. JIMBO: A q-analogue of $U\big(gl(N+1)\big)$, Hecke algebras and the Yang-Baxter equation. *Lett. Math. Phys.* **10** (1985) 63-69.

[Jo1] V.F.R. JONES: Hecke algebra representations of braid groups and link polynomials. *Ann. Math.* **126** (1987) 335-388.

[Jo2] V.F.R. JONES: On the knot invariants related to some statistical mechanical models. *Pacific J. Math.* **137** (1989) 311-334.

[LM] W.B.R. LICKORISH and K.C. MILLETT: The reversing result for the Jones polynomial, *Pacific J. Math.* **124** (1986) 173-176.

[Mo] H.R. MORTON: Seifert circles and knot polynomials. *Math. Proc. Camb. Phil. Soc.* **99** (1986) 107-109.

[Mu1] K. MURASUGI: On a certain numerical invariant of link types. *Trans. Amer. Math. Soc.* **117** (1965) 387-422.

[Mu2] K. MURASUGI: Jones polynomial and classical conjecture in knot theory II. *Math. Proc. Camb. Phil. Soc.* **102** (1987) 317-318.

[Mu3] K. MURASUGI: Jones polynomial of periodic links. *Pacific J. Math.* **131** (1988) 319-329.

[Mu4] K. MURASUGI: On invariants of graphs with applications to knot theory, *Trans. Amer. Math. Soc.* **314** (1989) 1-49.

[Mu5] K. MURASUGI: On the braid index of alternating links. *Trans. Amer. Math. Soc.* **326** (1991) 237-260.

[MP] K. MURASUGI and J.H. PRZYTYCKI: An index of a graph with applications to knot theory. Preprint.

[Re] N. Yu. RESHETIKHIN: Quantised universal enveloping algebras, the Yang-Baxter equation and invariants of links. LOMI preprint.

[Ro] M. ROSSO: Groupes quantiques et modeles à vertex de V. Jones en théorie des noeuds. *C.R. Acad. Sci. Paris t.* **307** (1988) 207-210.

[SAA] K. SOGO, Y. AKUTSU and T. ABE: New factorised S-matrix and its application to exactly solvable q-state model I. *Progress of Theoretical Phys.* **70** (1983) 730-746.

[Tr1] P. TRACZYK: 10_{101} has no period 7, a criterion for periodicity of links. *Proc. A.M.S.* **108** (1990) 845-846.

[Tr2] P. TRACZYK: A combinatorial formula for the signature of alternating links. Preprint.

[Tr3] P. TRACZYK: On the index of graphs: Index versus cycle index. Preprint.

[Tu1] V.G. TURAEV: A simple proof of the Murasugi and Kauffman theorems on alternating links. *L'Enseignement Mathématique t.* **33** (1987) 203-225.

[Tu2] V.G. TURAEV: The Yang-Baxter equation and invariants of links. *Invent. Math.* **92** (1988) 203-225.

[WAD] M. WADATI, Y. AKUTSU and T. DEGUCHI: Knot theory based on solvable model at criticality. *Advanced Studies in Pure Math.* **19** (1989) 193-285.

[We] H. WENZL: Representations of braid groups and the quantum Yang-Baxter equation. *Pacific J. Math.* **145** (1990) 153-180.

[Wi] E. WITTEN: Quantum field theory and the Jones polynomial. *Commun. Math. Phys.* **121** (1989) 351-399.

[Ya] S. YAMADA: The minimal number of Seifert circles equals the braid index of a link. *Invent. Math.* **89** (1987) 347-356.

[YGW] K. YAMAGISHI, M. GE, and Y. WU: New hierarchies of knot polynomials from topological Chern-Simons gauge theory. *Lett. Math. Phys.* **19** (1990) 15-24.

[Yo] Y. YOKOTA: The Jones polynomial of periodic knots. *Proc. Amer. Math. Soc.* (to appear).

Canonical surgery on alternating link diagrams

I.R. Aitchison and J.H. Rubinstein

Abstract. We associate canonically a closed orientable 3-manifold M_G^3 to every connected planar graph G. A decomposition of M_G^3 into two identical polyhedra is described, in terms of which face identifications and a polyhedral geometry occur naturally.

Introduction

Every 3-manifold can be constructed by identifying some collection of (abstract) polyhedra along pairs of faces. A single polyhedron suffices for a closed 3-manifold. However, a given abstract 3-dimensional polyhedron will not in general have faces occuring in pairs - there would appear to be no canonical way of associating a closed 3-manifold to a polyhedron, and certainly not a unique association.

In [**ALR**], it was shown that for a polyhedron defined by a connected planar 4-valent graph, there *is* a canonical way to construct an orientable 3-manifold, which is naturally identifiable as the complement of an alternating link in S^3. This general form of the construction is presumably known to Thurston [**Th**]. (See the final remark of [**AR3**].) A polyhedral structure of any unsplittable alternating link complement is determined in this way by some 4-valent planar graph obtained as an irreducible link projection. Such examples have been explored in [**La**], [**Me**], [**Ta**] and [**We**], and most recently in [**ALR**].

Consider a reduced alternating projection of an unsplittable alternating link L. The canonical 2-colouring of the regions of the plane, with unbounded region lightly shaded, determines two dual connected planar graphs G (respectively G^*), with vertices corresponding respectively to the dark (respectively light) regions, and with edges passing through the crossings of the link projection. The original link L can be reconstructed from either G or G^*, and accordingly we write $L = L_G$ given that G corresponds to the darker regions.

Knots 90

The projection determines two surfaces, not necessarily orientable, bounded by the link. We describe two 3-manifolds M^3_G and $M^3_{G^*}$, obtained by surgery on L using the framings determined by these 'checkerboard' surfaces.

A beautiful structure involving two families of disjointly embedded circles on surfaces, underlying Escher's 'Heaven and Hell', is described in [**AR1**]. Such families arise naturally when 3-valent polyhedra are identified along faces. We exploit the two canonical families of circles on a surface S_M associated with a Heegard splitting of S^3 arising from the alternating link projection. This enables us to identify closed 3-manifolds M^3_G and $M^3_{G^*}$, as obtained by surgery on the link complement $S^3 - L$, and simultaneously to recognize their decomposition as the union of two polyhedra.

In the case of an alternating *knot*, the two 3-manifolds so obtained are well-defined unoriented *knot* invariants, being independent of the alternating *projection* due to the recent resolution of the Tait conjectures ([**HKW**], [**Ka**], [**Mu**], [**Th**] and [**Tu**]).

After Witten's recent generalization of the Jones' polynomials to arbitrary 3-manifolds, it is interesting to speculate that such 'alternating' 3-manifolds may shed light on further geometric or topological interpretations of these new invariants. This is particularly so since their clearest meaning for classical links can be seen in the alternating case.

Alternating links from 4-valent graphs

Let G denote an arbitrary connected finite planar graph (Figure 1), with specified embedding. We consider G embedded in the plane $R^2 \subset S^2 \subset S^3 = R^3 \cup \{\infty\}$. G consists of a finite set V of vertices, and a finite set E of edges, each with interior embedded in the plane. A deletion of an edge corresponds to the deletion of the interior of the corresponding arc in the plane.

The medial graph $M = M(G)$ is obtained by taking one vertex v_i for each edge e_i of G, chosen at an interior point of the edge, and connecting two vertices v_1, v_2 by edges, one in each component of the complement of G for each vertex of the component at which e_1 and e_2 are incident. Alternatively, $M(G)$ is obtained by inscribing the dual polygon for each polygonal region of the 2-sphere complementary to G, as in Figure 1.

M is a 4-valent planar graph, and its complementary regions in the plane can be 2-coloured, checker-board fashion (Figure 2). The regions of $R^2 - M$ darkly shaded are those containing a vertex of G. We associate a '+'-sign to light regions, and a '$-$'-sign to each of the dark regions.

Each vertex of M can be converted to a crossing as in Figure 2, so that M is a planar representation of an alternating link $L = L_G$. Crossings viewed from

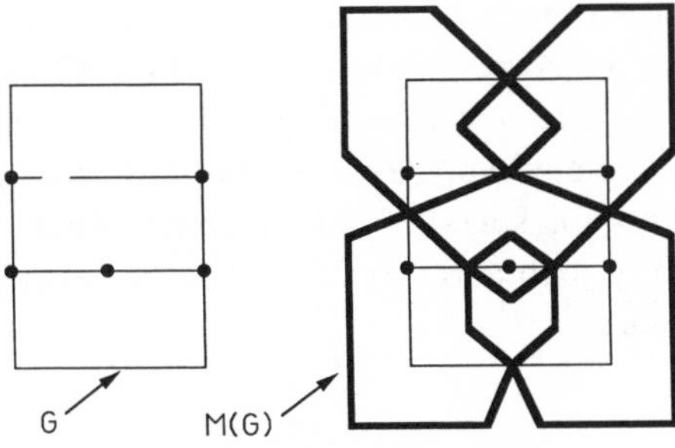

Figure 1

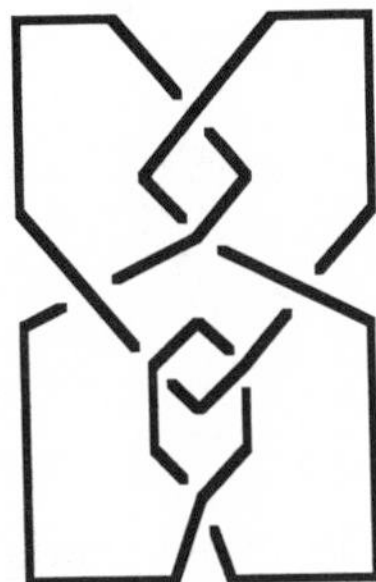

Figure 2

the interior of any region have sign that of the region. We will assume from now on that no edge-deletion disconnects the graph G, and thus that the corresponding planar projection of L_G is *reduced.* Equivalently, regions meeting at any vertex of L_G are all distinct. Menasco has proved that in these circumstances the link is unsplittable. Conversely, any reduced diagram for an unsplittable link is of this type ([**Me**]).

Face identification rules to obtain link complements

The compactified projection plane separates S^3 into two balls, the upper and lower hemispheres B^u and B^l. Readers should imagine themselves at the centre of B^u, viewing the following diagrams on $S^2 = \partial B^l$ from the outside of B^l. Each ball inherits the structure of an (abstract) 4-valent polyhedron Π_M^u, Π_M^l from the diagram M on its boundary sphere. S^3 is obtained by identifying corresponding faces of these identical polyhedra (combinatorially equivalent by reflection in the

2-sphere). Let M^* denote the planar dual of M, and $\tau(M^*)$ the truncation of M^*. Denote by $\Pi_{\tau(M^*)}$ the corresponding polyhedron. Colour red the edges dual to those of M and colour edges at the truncated vertices black. Conveniently the truncated dual can be depicted appearing as the outline of the projection of the link with all components thickened, as in Figure 3, where each 'red' edge can be associated consistently with a vertex of M. In the Figure, these edges have a white region on both sides.

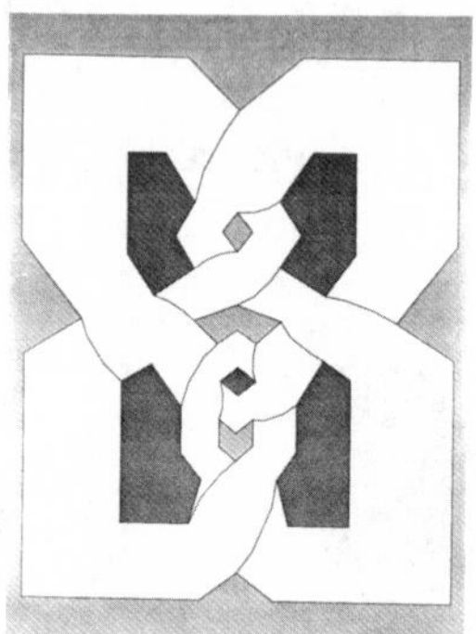

Figure 3

Following [**AR1**], to combinatorially describe any 3-manifold obtained by glueing together Π^u_M and Π^l_M along faces, we start with copies of $\Pi^u_{\tau(M^*)}$ and $\Pi^l_{\tau(M^*)}$. Faces of $\Pi^u_{\tau(M^*)}$ and $\Pi^l_{\tau(M^*)}$ are of two kinds, those with black edges corresponding to the (now truncated) vertices of $\Pi^u_{M^*}$, $\Pi^l_{M^*}$, and '2-coloured' faces. The latter, each an octagon since M is 4-valent, are in $1:1$ correspondence with vertices of Π^u_M, Π^l_M, and the former with faces of Π^u_M, Π^l_M. All of the octagons are depicted in white in Figure 3.

We embed copies of $\Pi^u_{\tau(M^*)}$ and $\Pi^l_{\tau(M^*)}$ in S^3, each in the appropriate hemisphere, with corresponding black-edged dark faces identified in the interiors of the faces of $S^2 - M$.

The resulting object is a standardly embedded handlebody H^*_M in S^3 (Figure 4), the closure of the complement of which being a handlebody H_M with 1-spine the graph M. Each edge of M is linked by a circle consisting of 2 red arcs (one in bold, the other depicted with a finer line on the back of the handlebody H_M from the viewers perspective). The boundary S_M of H_M is marked by two families of circles, one black, the other red. The black family is the intersection of S_M with the 2-sphere on which M is embedded, and thus each such circle bounds a positively- or negatively-signed region of the 2-sphere. Each region of S_M complementary to the red and black circles is an octagon with edges consecutively coloured red and black. By adding red discs in $S^3 - H^*_M$ to H^*_M along the red circles, we obtain S^3 with a deleted ball neighbourhood for each vertex of M.

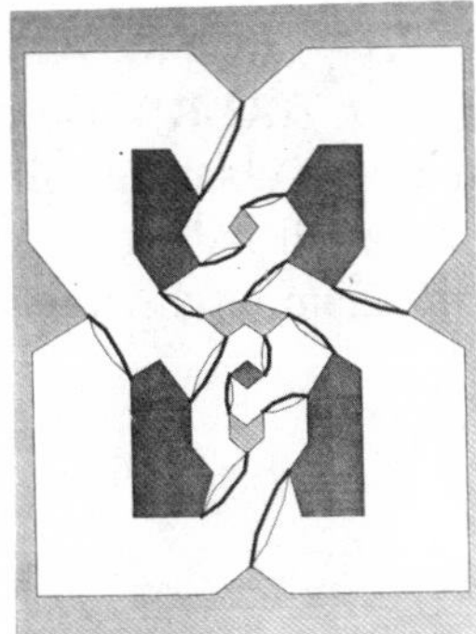

Figure 4

The handlebody H_M^* is cut into two balls by the equatorial 2-sphere S^2. Cut each of the 1-handles of H_M^* along discs of intersection with S^2, bounded by the black circles, and reglue after a rotation. Handles passing through a negatively-signed n-gon face of Π_M^u, Π_M^l are rotated by $2\pi/n$ *anticlockwise* as they pass through from Π_M^l to Π_M^u, before reglueing, while those passing through a *positively*-signed region are rotated clockwise. This does not require an orientation of S^2. Observe that the 'bold' red arcs remain fixed, whereas those on the back of the handlebody have end-points moved towards the *same* adjacent vertex of M, as in Figure 5.

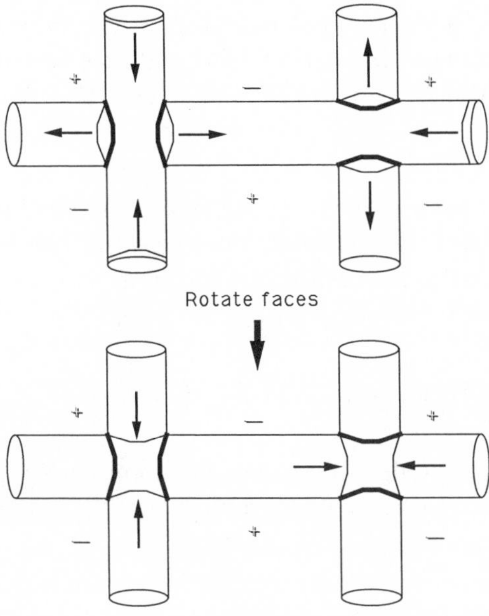

Figure 5

Adding red 2-handles in $S^3 - H_M^*$ along the resulting red circles produces the alternating link complement $S^3 - L$ ([**ALR**]). Observe in Figure 6 that the red and black families of circles on S_M so obtained still have complementary regions being 2-coloured octagons, and that all red circles consist now of *four* red arcs. (In the diagram the finer arcs are again on the far-side of the handlebody.)

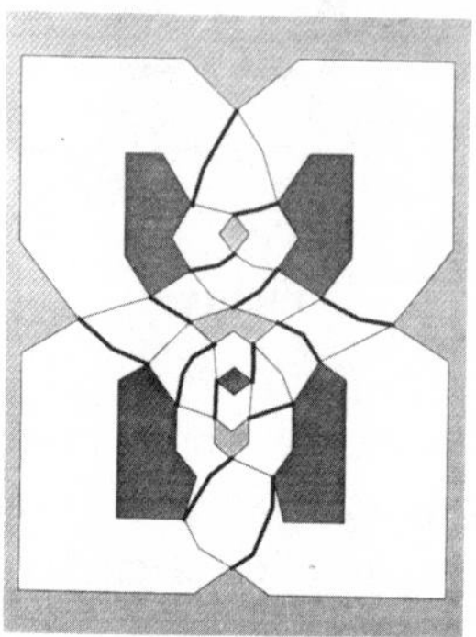

Figure 6

Thus the closure of the link complement can be obtained by identifying corresponding faces of Π_M^u, Π_M^l by rotations of $\pm 2\pi/n$ as above, and deleting open neighbourhoods of vertices. Versions of this natural way to obtain the complement of an alternating link, as the union of two ideal polyhedra with faces identified, were introduced, by example, in Thurston [**Th**], and have been developed by Lawson [**La**], Menasco [**Me**], Takahashi [**Ta**] and Weeks [**We**]. We summarize using the notation above:

Theorem. *The non-compact orientable 3-manifold obtained by identifying corresponding faces of two combinatorially identical ideal checkerboard-shaded 4-valent polyhedra Π_M^u, Π_M^l, by rotations of $\pm 2\pi/n$, according to the sign of each n-gon face, is canonically homeomorphic to $S^3 - L_G$.*

Observation [**ALR**]. *Every equivalence class of edges contains exactly four edges.*

Each red arc is dual to an edge of M, and arcs form circles linking the edge in the identification space. Corresponding vertices of Π_M^u and Π_M^l determine two elements each of an equivalence class of edges.

Remark. The same construction applied to the planar dual G^* of the graph G gives rise to the *mirror-image* L^* of the link L.

Canonical surgery: Dehn surgery curves

An octagon with edges alternatingly coloured red and black can be split into two hexagons, by adding a red arc joining the midpoints of two opposite black edges. Such an arc continues uniquely across an adjoining such octagon, splitting it in the same fashion. (Figure 7.)

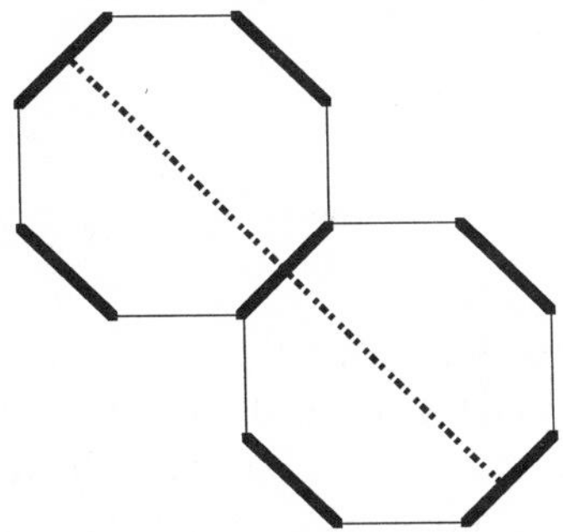

Figure 7

We can consistently and simultaneously split all octagons of S_M in this way, by adding a single red circle for each component of the link L_G. Each such circle lies on the boundary torus of a closed neighbourhood of a link component. Such a circle determines a Dehn surgery on each cusp of $S^3 - L_G$.

There are two choices for each component, corresponding to whether the added circle links the given component positively or negatively, with respect to the parallel copy in the plane, when given the same orientation. We choose the positively-linking framing circle for each component, thereby uniquely determining a *closed* 3-manifold M^3_G. (Figure 8.) In this case, the framing circle C^+ for each component is isotopic to the boundary circle of the surface spanning the link determined by the *positive* regions of the link diagram M, and so in turn by the regions in which the vertices of G do *not* lie.

Definition. The closed 3-manifold M^3_G is said to be obtained by *canonical surgery* on the link L_G determined by the alternating projection diagram $M(G)$.

Canonical surgery: identification of polyhedral faces

Adding 2-handles to each torus component surgers the tori to 2-spheres, to which we add 3-balls. Colouring these 2-handle attaching circles red, the surface S_M is decomposed into 2-coloured hexagons by the two families, red and black, each

Figure 8

of disjoint embedded circles. Reversing the argument given above, we see that this picture corresponds to glueing polyhedra together along faces. There are two polyhedra, identifiable by cutting H^*_M along discs bounded by the black circles. In Figure 9 we present the resulting structure on the boundary of B^l.

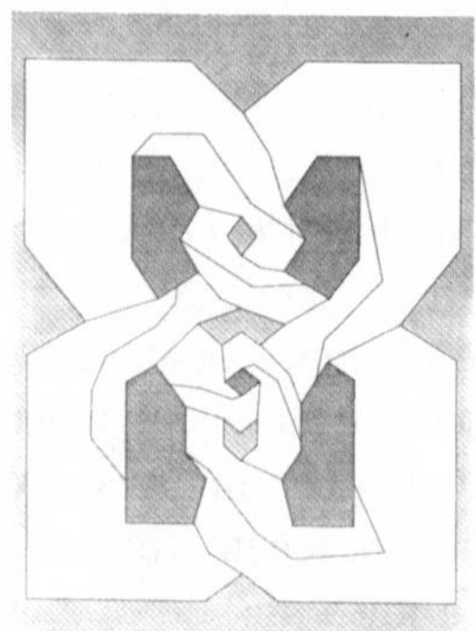

Figure 9

Polyhedral structure is now obtained by constructing the graph with one vertex in the interior of each 2-coloured hexagon, and with edges passing through red edges separating such hexagons. Figure 10 is obtained from Figure 9 in this fashion, and Figure 11 gives the result for the other polyhedron. Face identifications are determined by the circle patterns on S_M.

Denote the two polyhedra so obtained by Π^{u+}_M and Π^{l+}_M. Recalling that each of the surgery framing circles links the corresponding compatibly oriented component in a positive sense (with respect to the planar projection), we find that these new polyhedra are obtained from the originals by vertex splittings according to the diagrams of the *first* column of Figure 12. We shall discuss the significance of the other columns shortly.

Figure 10

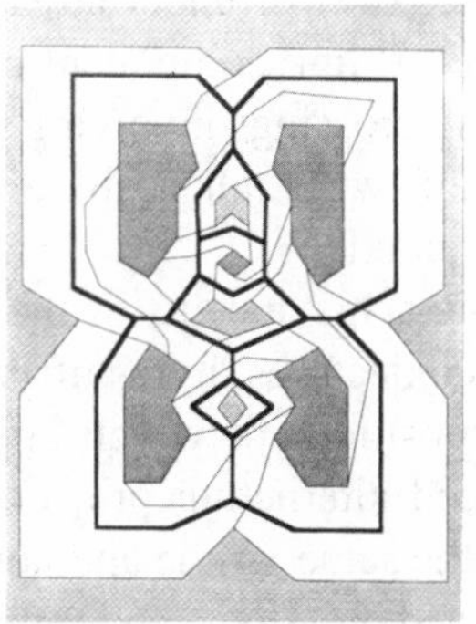

Figure 11

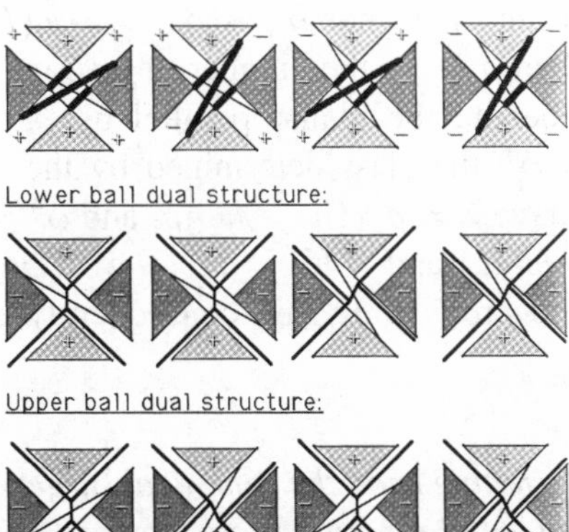

Figure 12

Proposition. *The two polyhedra* Π_M^{u+} *and* Π_M^{l+} *are determined combinatorially by the graph* $\tau(G^*)$. *Each is obtained by vertex-splitting at all vertices, so that all positive regions become isolated from each other.*

Since edges identified when the original ideal polyhedra were glued together *remain* identified, we find that face identifications are again rotations 'alternating like gears' ([**Th**]). Since each negatively-signed n-gon has become a $2n$-gon, the rotation for identification of such corresponding faces remains $2\pi/n = 2.2\pi/(2n)$. Positive faces are identified as before.

Knots and the Tait conjectures

As we follow around any component of the link diagram, we can keep track of the surgery circle relative to a planar parallel of the component. One half-twist occurs each time the component passes over or under at each crossing. Let s_i denote the number of crossings where the ith link component crosses itself, and let $d_i = 2a_i$ denote the number of remaining crossings involving the ith component (necessarily even). The total number of half-twists, all positive, is $2t_i \equiv 2s_i + 2a_i$, giving t_i as the framing relative to the parallel.

One of the most remarkable achievements of the Jones polynomial is the confirmation of three of the Tait conjectures, by Kauffman [**Ka**], Murasugi [**Mu**] and Thistlethwaite [**Th**], other expositions of which appearing in [**HKW**] and [**Tu**]. As a consequence, two reduced alternating projections of the same unsplittable oriented alternating link have the same writhe and number of crossings. The writhe is the algebraic sum of the signed crossings, and remains invariant under reversing orientations of all link components simultaneously.

In the case of an alternating *knot* K, with writhe $w(K)$, the framing of the canonical surgery circle is thus well-defined relative to the longitude of the knot. The longitude twists around the knot $n_- - n_+ = -w(K)$ times relative to the parallel in the plane, where $n_\pm$ are the number of positive and negative crossings. ([**Ro**].) In $H_1(T^2; Z)$, denote the planar parallel by ρ, the meridian by μ, the longitude by λ, and by ϕ^+ the class determined by the canonical Dehn surgery framing circle C^+. We have $\lambda = \rho + (n_- - n_+)\mu$, and $\phi^+ = \rho + (n_+ + n_-)\mu$. Thus $\phi^+ = \lambda + 2n_+\mu$. Both $w(K)$ and $c(K) = n_+ + n_-$ (the total crossing number) are invariants of reduced projections, and consequently each of n_+ and n_- are also invariant.

Theorem. *For K an alternating knot, the canonical surgery framing is well-defined independently of the reduced alternating projection of K. The closed orientable 3-manifold M^3_G is an unoriented knot invariant M^3_K of K.*

Remarks. The framing circle is the boundary of the positively-signed checkerboard surface bounded by the knot.

An analogous result for links will possibly not emerge until the final Tait conjecture is resolved.

Variations of canonical surgery

We remarked earlier that there were two choices of canonical framing circle we could make for each link component. Uniformly taking the other choice, we obtain Dehn surgery on the link specified by the boundary of the surface corresponding to the negatively signed regions of the planar link projection. Denote the resulting closed 3-manifold by M_G^-, and recall that G^* denotes the planar dual of G.

Proposition. *There is a canonical homeomorphism $M_G^- \cong M_{G^*}^3$.*

Proof. Reflecting the link in the plane of projection reverses the signs of all regions and simultaneously carries one family of Dehn surgery circles to the other, as in Figure 13.

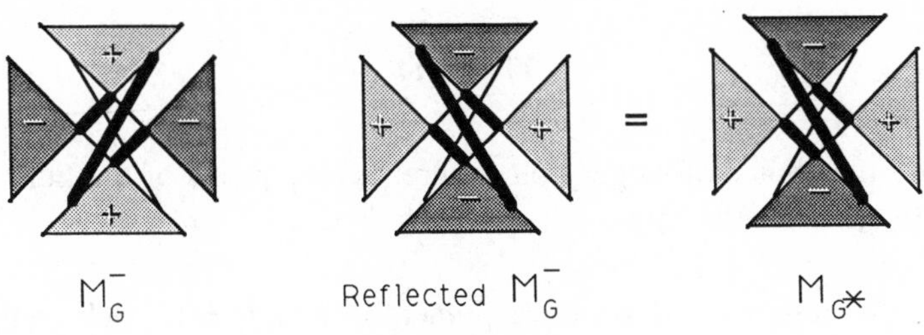

Figure 13

Remark. The polyhedra giving rise to M_G^3 are each defined by τG^*, whereas those for $M_{G^*}^3 = M_G^-$ correspond to τG. All vertices of M are split so as to isolate the negatively-signed regions.

If the original polyhedra Π_M^u and Π_M^l are truncated, the resulting new faces are all squares. All such squares can be simultaneously squeezed to become disjoint arcs, in either of two ways – each separating two negatively-signed faces, or two positively-signed regions. The effect of canonical surgery is to glue in solid tori made up of wedges, the thin ends of which being identified along the core of the solid tori, the square ends of which based on these squares. The two choices of Dehn surgery circles, producing M_G^3 and $M_{G^*}^3$, correspond to the two choices of collapsing these squares to the core of the respective solid tori.

Each torus in a cusp of the link complement inherits a combinatorial structure of squares, four meeting at each vertex ([**ALR**]). Figure 14 depicts the universal cover of such a combinatorial torus. The meridianal translation is along the diagonal of one square, whereas the surgery circles each pass through every square of the torus once, parallel to one of the sides.

In the case of a knot, M^3_G and $M^3_{G^*}$ correspond respectively to Dehn surgeries $\phi^+ = \lambda + 2n_+\mu$ and $\phi^- = \lambda - 2n_-\mu$, represented by circles C_+ and C_- arising from the signed checkerboard surfaces.

Translating either vertically or horizontally via $C_\pm$ gives another primitive generator for the group of covering transformations. (See [**ALR**] for further comments on combinatorial cusps of alternating links.)

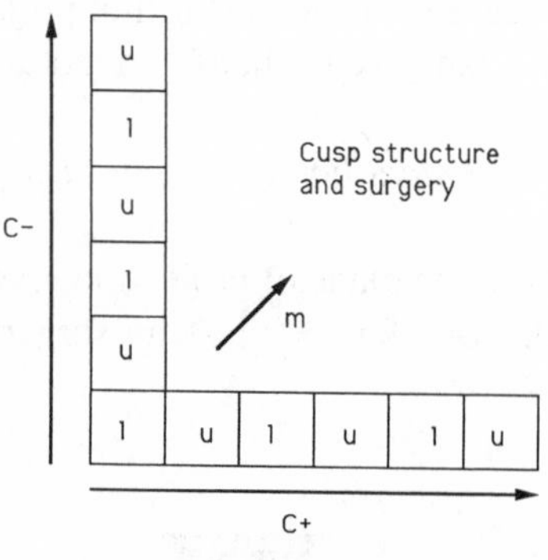

Figure 14

We can vary the choice of surgery on different components of the link, and illustrate with the Borromean rings.

Definition. A **state** σ of M is an assignment of a sign $\pm$ to each component of L_G.

A state σ defines a Dehn surgery on the link L_G enabling us to associate a closed orientable 3-manifold M^σ_G. The canonical choices described above gave rise to polyhedra derived from M by vertex-splitting in two distinct, but uniform, ways. A state σ distinguishes each vertex of M as one of 4 types, illustrated in Figure 12. Proceeding as above, we determine two polyhedra, face identifications of which describing M^σ_G. Observe that when the two components at a crossing are assigned different signs, the vertex of M is split in different ways in each of the two polyhedra.

In Figure 15, we illustrate for the $(+2, +2, -2)$-surgery on the Borromean rings.

We conclude this section with the amusing observation that many different reduced alternating link diagrams give rise to the same closed 3-manifold. Given M as above, add an extra component by drawing a round circle in the plane

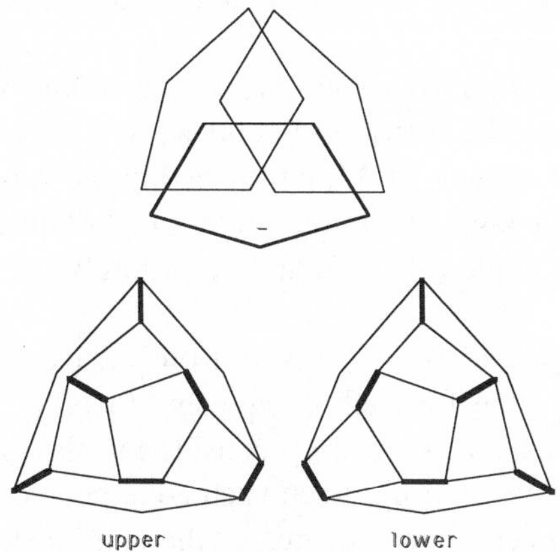

Figure 15

bounding a disc whose intersection with the graph M is an arc interior to an edge. The result is a medial graph $M(G')$ for a new graph G'.

Proposition. *The 3-manifolds M_G^3 and $M_{G'}^3$ are diffeomorphic.*

Proof. The new component is assigned the framing $+1$. Carrying out canonical surgery, we find that after handle sliding, the new component corresponds to blowing up the original link diagram by $+CP^2$. The surgered manifolds are thus diffeomorphic.

Polyhedral metrics and cubical decompositions

A canonical procedure for prescribing a canonical (singular) polyhedral metric on an arbitrary closed orientable 3-manifold, built from trivalent polyhedra by identifying faces, is given in [**AR2**]. This requires decomposing each such polyhedron into cubes, one for each 3-valent vertex, and declaring that each such cube is isometric to the standard unit cube, and that all glueings are isometries. The procedure is to introduce new vertices, one for each edge, face and polyhedron. Each n-gon face becomes n squares by joining the new face vertex to those on the edges, and the three squares around each vertex cone to the centre of the polyhedron to produce a cube. We carry out such a decomposition of M_G^3 into cubes, and describe the resulting polyhedral metric.

Each of the two polyhedra is described by the graph τG^*, or equivalently by vertex-splitting of M. The resulting polyhedral metric on M_G^3 is Euclidean in neighbourhoods of point in the interior of cubes and squares. Edges are of several kinds.

The original edges from Π_M^u and Π_M^l are identified in equivalence classes of four, and accordingly the metric is Euclidean at points in the interior of such edges. New edges on the faces of M, introduced in the cubing, are also of degree 4, and so Euclidean. An edge joining the centre of a polyhedron to the centre of an n-gon face has a cone angle of $n\pi/2$, and accordingly the metric is non-positive provided $n \geq 4$.

The edges of $\Pi^u_{\tau(G^*)}$ and $\Pi^l_{\tau(G^*)}$ arising from vertex-splitting fall into several equivalence classes e_i, one for each component L_i of L_G. The endpoints of e_i are identified in M_G^3, as can be seen by considering the glueing of faces around an edge arising by vertex-splitting. Each such edge becomes two edges when the polyhedra are cubed. Each of the squares of the combinatorial decomposition of the torus T_i corresponding to L_i contributes one edge to the equivalence class of edges in M_G^3. The number of such squares is $2t_i$. The corresponding edge e_i in M_G^3 has cone-angle $t_i\pi$.

Vertices similarly fall into several types. We describe the link of each such, as a triangulation of the 2-sphere. Vertices at the centres of edges from M are octahedra, being the double suspension of a 4-gon. A centre of an n-gon face has link the double suspension of an n-gon. The link at the centre of each of the polyhedra is $(\tau G^*)^*$. After cubing, each of the two vertices on the loop e_i has link the double suspension of a $2t_i$-gon.

With respect to this polyhedral metric on M_G^3, the cores e_i of the solid Dehn surgery tori are geodesic loops. The non-Euclidean geometry of the metric is concentrated along a graph, a subgraph of the graph consisting of the cores of all Dehn surgery tori, and the edges joining the centres of the two polyhedra to the centres of faces.

Proposition. *If no positive region in the complement of M is an n-gon, for $n \leq 3$, and L_G is a prime, unsplittable link, then M_G^3 admits a polyhedral metric of non-positive curvature arising from a cubing, and contains an immersed π_1-injective surface F_G, totally geodesic with respect to this metric. Moreover F_G satisfies the 4-plane, 1-line condition of Hass-Scott [**HS**], and accordingly M_G^3 is determined up to homeomorphy among irreducible closed 3-manifolds by its fundamental group.*

Proof. Under these conditions, all edges have degree at least 4, and so have cone angle at least 2π. Similarly, every edge-cycle of length 3 in the link of any vertex in the cubing of M_G^3 bounds a triangle, and the singular metric is non-positive at each vertex ([**AR2**]). Subdividing each cube into 8 smaller cubes by slicing with planes parallel to faces, the 12 new squares in the interior of each of the original cubes fit together to produce a surface, possibly non-orientable and disconnected. This is the surface F_G, which is totally geodesic and thus π_1-injective. The remaining properties and conclusions follow from [**AR2**].

Example. The Borromean rings do not satisfy the requirements of the Proposition, but the 3-manifold defined by $(+2, +2, -2)$-surgery on this link admits a geometric cubing with properties of the conclusion. In this example, the surgery is *not* determined by the boundary of one of the checkerboard surfaces.

References

[AR1] I.R. Aitchison and J.H. Rubinstein, *Heaven and Hell,* Proceedings VIth Int. Coll. on Differential Geometry, Santiago de Compostela, Cursos y Congresos **5** (1989), 5-24.

[AR2] I.R. Aitchison and J.H. Rubinstein, *An introduction to polyhedral metrics on 3-manifolds,* to appear in Durham Low Dimensional Topology, Cambridge Univ. Press, (1990).

[AR3] I.R. Aitchison and J.H. Rubinstein, *Combinatorial cubings, cusps, and the dodecahedral knots,* to appear in Proceedings of the Research Semester in Low Dimensional Topology, Ohio 1990.

[ALR] I.R. Aitchison, E. Lumsden and J.H. Rubinstein, *Polyhedral metrics on alternating link complements,* in preparation.

[BZ] G. Burde and H. Zieschang, *Knots,* de Gruyter Studies in Mathematics **5** 1985.

[HKW] P. de la Harpe, M. Kervaire & C. Weber, *On the Jones polynomial,* L'Enseignement Mathématique **32** (1986), 271-335.

[HS] J. Hass and P. Scott, *Homotopy equivalence and homeomorphism of 3-manifolds,* preprint MSRI July 1989.

[Jo] V. F. R. Jones, *A polynomial invariant for links via Von Neumann Algebras,* Bull. AMS **12** (1985), 103-111.

[Ka1] L.H. Kauffman, *State models and the Jones polynomial,* Topology **26** (1987), 395-407.

[Ka2] L.H. Kauffman, *New invariants in the theory of knots,* Amer. Math. Monthly **95** (1988), 195-242.

[La] T. C. Lawson, *Representing link complements by identified polyhedra,* preprint.

[Me] W. W. Menasco, *Polyhedra representation of link complements,* Amer. Math. Soc. Contemporary Math. **20** (1983), 305-325.

[Mu1] K. Murasugi, *Jones polynomials and classical conjectures in knot theory,* Topology **26** (1987), 187-194.

[Mu2] K. Murasugi, *Jones polynomials and classical conjectures in knot theory II,* Math. Proc. Cambridge Philos. Soc. **102** (1987), 317-318.

[Ro] D. Rolfsen, *Knots and Links,* Publish or Perish, 1976.

[Ta] M. Takahashi, *On the concrete construction of hyperbolic structures of 3-manifolds,* preprint.

[Th1] M.B. Thistlethwaite, *A spanning tree expansion of the Jones polynomial,* Topology **26** (1988), 297-309.

[Th2] M.B. Thistlethwaite, *Kauffman's polynomial and alternating links,* Topology **27** (1988), 311-318.

[Th] W.P. Thurston, *The geometry and topology of 3-manifolds,* Princeton University Lecture Notes 1978.

[We] J.R. Weeks, *Hyperbolic structures on three-manifolds,* PhD dissertation, Princeton 1985.

[Wi] E. Witten, *Quantum field theory and the Jones polynomial,* Commun. Math. Phys. **121** (1989), 351-399.

State models for link invariants from the classical Lie groups

David A. Meyer

Abstract. A variety of state models for link invariants have been introduced since Kauffman's original bracket polynomial for the Jones invariant. In this paper we consider a class of Yang-Baxter state models which are defined by tensors lying in the subspaces of order four tensors invariant under the action of the classical Lie groups. We compute these models for the groups $\mathbf{GL}(n, K)$, $\mathbf{SL}(n, K)$, $\mathbf{O}(n, \mathbf{C})$, $\mathbf{O}(n, \mathbf{R})$ and $\mathbf{Sp}(n, K)$, and show that the bracket polynomial arises from $\mathbf{O}(n, \mathbf{C})$ invariance. Moreover, we observe that two of the surprising features of the bracket, namely that it can be normalized to an ambient isotopy invariant and that imposing the second Reidemeister move guarantees that the third is also satisfied, are characteristic of all these Lie group invariant models.

1. Introduction

The new link invariants: the Jones [1], HOMFLY [2] and Kauffman [3,4,5] polynomials, all have skein relation definitions as does, in fact, the Alexander polynomial [6]. Because such definitions are recursive, proving that the polynomials are unambiguously defined requires somewhat complicated inductive arguments. For the Jones polynomial, however, Kauffman found an alternative definition—a state model—which is manifestly unambiguous and for which invariance is easily checked [7]. Subsequent work by a variety of authors has produced a multiplicity of state models in attempts to find a generalization to the full HOMFLY or Kauffman polynomials [8], other new invariants [9,10] or connections to statistical mechanics [11,12,13] and quantum groups [14], to give only a partial list.

Although the latter results, connecting link invariants with solutions to the Yang-Baxter equation [15], have provided a wealth of new invariants as well as some insight into the reasons for their existence, at the same time they lack the topological elegance of the earlier formulations. The goal of this paper is

Knots 90

to provide a coherent topological/diagrammatic framework within which to place some of the extant state models for link invariants. The work was motivated by a desire to organize the seemingly *ad hoc* collection of state models and informed by the tantalizing connections to Lie groups (rather than Lie algebras) suggested by Witten's topological quantum field theory for the Jones polynomial [16] and Kauffman's observation of a relation to spin-networks [17]. Nevertheless, the resulting picture is not dependent on these connections with physics nor on the accumulated knowledge about solutions to the Yang-Baxter equation.

In order to be entirely clear about the connection between the models we are considering and state models in other forms, the first several sections of this paper develop the background necessary for non-specialists (Those readers familiar with state and Yang-Baxter models will probably want to skip to Section 4 or 5.): In the next section we begin with a brief review of link invariants, recalling in particular the Reidemeister moves to which we will refer repeatedly. In Section 3 we consider Kauffman's bracket polynomial, the paradigmatic state model, systematically defining new state models, at each step proving equivalence to the preceeding model, until we arrive at a Yang-Baxter model which will motivate the eventual introduction of Lie groups. To eliminate the proliferation of indices which is introduced in the Yang-Baxter model, in Section 4 we describe Penrose's diagrammatic notation for tensors. Use of this formalism suggests that solving the tensor equations derived from the Reidemeister moves may be effectively accomplished in a Lie group invariant subspace, so in Section 5 we explain this idea and provide the necessary background in Lie group invariants. Section 6 contains the first set of diagrammatic calculations, for state models from the groups $\mathbf{GL}(n, K)$, $\mathbf{SL}(n, K)$ and $\mathbf{O}(n, \mathbf{C})$, the last of which defines an invariant we prove equivalent to the bracket polynomial. In Section 7 we discuss the formation of ambient isotopy invariants from the heretofore regular isotopy invariants, proving in particular that within this framework this can always be done. This result suggests that the apparently miraculous redundancy of the constraints implied by the third Reidemeister move might be explained similarly, so in Section 8 we prove perhaps the most interesting result in this paper: that for these group invariant state sum polynomials, invariance under the third Reidemeister move is (almost) implied by invariance under the second. Using the ideas from the proof we can perform the diagrammatic calculations in the more complicated cases of $\mathbf{O}(n, \mathbf{R})$ and $\mathbf{Sp}(n, K)$ without verifying invariance under the third move explicitly; these are done in Section 9 and the equivalence of the resulting invariants is demonstrated. The conclusions comprise Section 10.

2. Link invariants

We recall briefly the minimal background in knot theory necessary for our subsequent discussion; general surveys of the subject (from quite different points of view) may be found in [18] and [19], for example.

A *link* is an embedding (smooth, say, or piecewise linear) of a finite collection of S^1s into S^3 (or $\mathbf{R}^3$). Two links are *equivalent* or *ambient isotopic* if there is an orientation preserving diffeomorphism of S^3 (or $\mathbf{R}^3$) which maps one to the other. Composition with a general projection to S^2 (or $\mathbf{R}^2$) gives a *link diagram* L: an immersion (smooth or piecewise linear, respectively) of the collection of S^1s with singular set consisting of only finitely many transverse double points with associated crossing information [20]. The basic theorem is due to Reidemeister [21]:

Theorem (Reidemeister). *Any two link diagrams generated from a link in this way, ambient isotopy followed by projection, can be transformed into each other by a finite sequence of the three moves—RI, RII and RIII—shown in Figure 1, together with planar isotopy.*

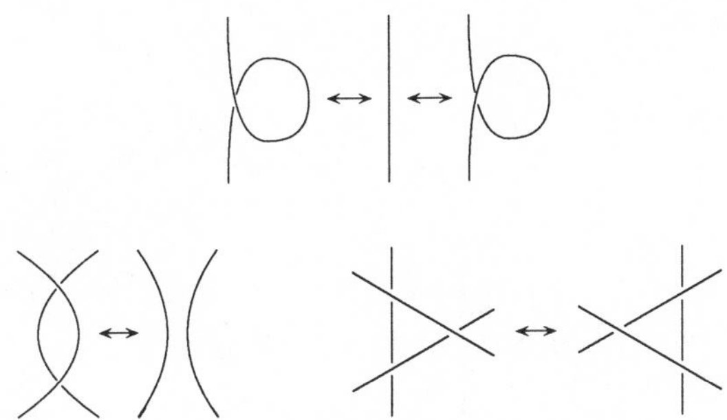

The three Reidemeister moves.

Figure 1

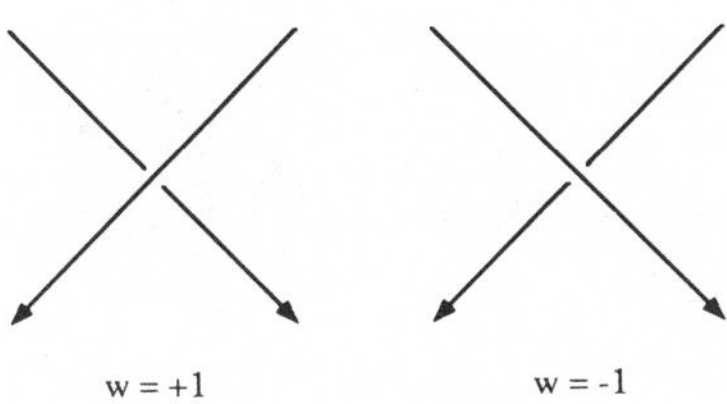

Signed crossings.

Figure 2

Thus, to study invariants of links we need only study properties of link diagrams which are invariant under the Reidemeister moves. The most trivial example is simply the number of components of the link, a property left unchanged by each of the moves.

To define a less trivial invariant, the linking number, we must first orient the diagram by assigning an orientation to each component. There are then two types of crossings, positive and negative, as shown in Figure 2, as well as four possible orientations of RII and eight of RIII. In fact, not all of these are independent moves; Reidemeister's Theorem for oriented links states that a minimal set is given by those shown in Figure 3. Figure 4 shows that there are also four distinct oriented RI moves: two each with positive and negative crossings, distinguished by another topological property of the link diagram, the *rotation number* $r(L_i)$ of a component; this is defined to be the index of the immersion map of the component (*i.e.*, the winding number of the image S^1). The *linking number* $\mathrm{lk}(L_i, L_j)$ of two components is now defined to be half the sum of the signs of the crossings between the components. It is clear that the linking numbers are left invariant by each of the Reidemeister moves, being unaffected by the first and third and changing by $(\pm 1 - \pm 1)/2 = 0$ under the second.

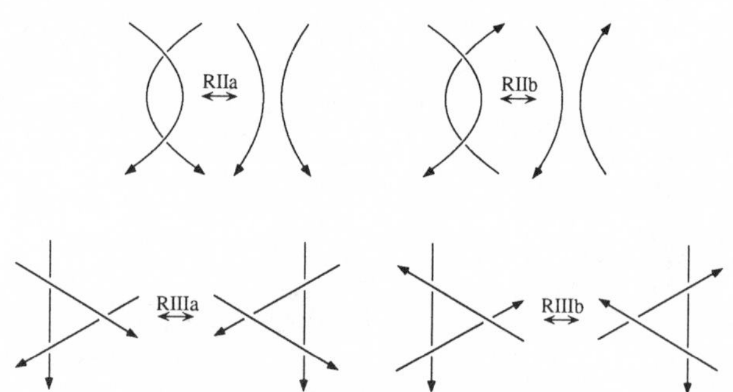

A minimal set of oriented RII and RIII moves.

Figure 3

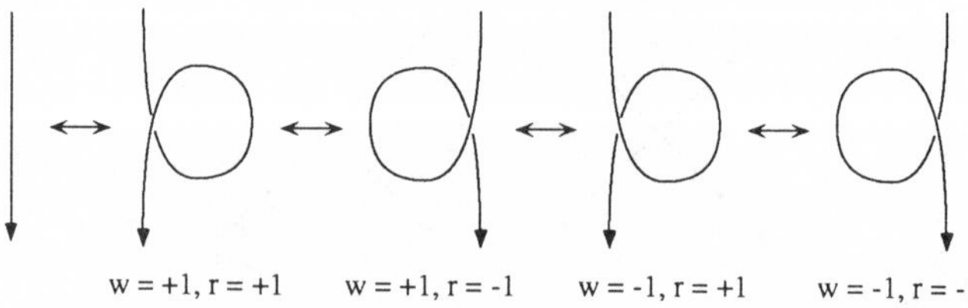

w = +1, r = +1 w = +1, r = -1 w = -1, r = +1 w = -1, r = -1

The four distinct oriented RI moves.

Figure 4

The sum of the signs of the crossings can also be computed for the whole diagram. This quantity is called the *writhe* of the link diagram, $w(L)$, and is clearly not a link invariant, being changed by ± 1 under RI moves, as is the rotation number $r(L)$. Since both the writhe and the rotation numbers are invariant under the second and third Reidemeister moves it turns out to be useful to define a new equivalence relation on link diagrams, analogous to regular homotopy: *regular isotopy*, namely equivalence by a finite sequence of RII and RIII moves [5]. For the next several sections we will consider only regular isotopy invariants of link diagrams; in Section 7 we will explain how to use the writhe and rotation numbers to convert regular isotopy invariants of the type we consider in this paper into ambient isotopy invariants.

Linking number, rotation and writhe have been introduced here primarily to motivate the definition of regular isotopy. In the next section we consider a more interesting regular isotopy invariant: Kauffman's bracket polynomial [5].

3. State models

The Kauffman bracket is a map $\langle \cdot \rangle$ from link diagrams to the commutative polynomial ring $\mathbf{Z}[A, B; n]$, defined by the three axioms diagrammed in Figure 5. Here the left hand side of the second equation denotes the bracket evaluated on a link diagram consisting of an unknotted component disjoint from the rest of the diagram L and the third equation means that the brackets of three link diagrams identical but for the region shown obey the stated relation. (The normalization is slightly different from the usual, which can be recovered by dividing by $\langle \bigcirc \rangle = n$.)

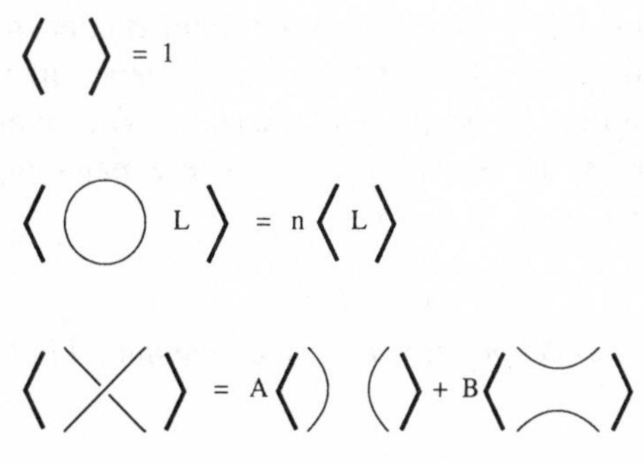

The axioms for Kauffman's bracket polynomial.

Figure 5

The bracket is well defined since these axioms give an algorithm for evaluating $\langle L\rangle$: from a diagram L with k crossings, generate the 2^k diagrams obtained by replacing each of the crossings by one of the two splicings on the right hand side of the third axiom, labelling each such diagram by the monomial of degree k in A and B corresponding to that particular choice of splicings used; since each diagram will now consist only of a set of simple closed curves, the first two axioms tell us to multiply the monomials by $n^{\#\text{ components}}$; the final expression is simply the sum of these terms.

The result of this algorithm can be expressed more compactly by a *state formula*. Define a *state* of a diagram to be a map $s : \{\text{crossings}\} \longrightarrow \{-1, 1\}$, to be interpreted as a choice of splicing at each crossing. The number of components in the resulting diagram sL is defined to be the *norm* of the state, $|s|$. Then the bracket polynomial is defined by the formula:

$$\langle L\rangle := \sum_{s} A^{|s^{-1}(1)|}B^{|s^{-1}(-1)|}n^{|s|},$$

where $|s^{-1}(x)|$ denotes the cardinality of the inverse image of x under s. We will refer to this state model for the bracket polynomial as the *splice* state model.

The usefulness of the bracket polynomial derives from the following result.

Theorem (Kauffman). *The bracket polynomial is a regular isotopy invariant provided $B \equiv A^{-1}$ and $n \equiv -A^2 - A^{-2}$.*

To prove this one simply checks that the state sum is invariant under the last two Reidemeister moves; we will do this calculation in some detail in Section 6 when we prove Proposition 7. Our interest here, however, is to give a sequence of equivalent state models which lead to the Yang-Baxter model we will associate with the orthogonal group. The first step is to generalize the definition of the states of a link diagram to be pairs of maps (s, s'), where s is as before and $s' : \{\text{components of } sL\} \longrightarrow \{1, \ldots, n\}$. These pairs of maps will be called *labelled* splice states.

Proposition 1. *The bracket polynomial can be computed in terms of labelled splice states as*

$$\langle L\rangle = \sum_{(s,s')} A^{|s^{-1}(1)|}B^{|s^{-1}(-1)|}.$$

Edge state labelling of a crossing.

Figure 6

Proof. This is just a simple computation.

$$\sum_{(s,s')} A^{|s^{-1}(1)|}B^{|s^{-1}(-1)|} = \sum_{s} A^{|s^{-1}(1)|}B^{|s^{-1}(-1)|}\sum_{s'} 1$$
$$= \sum_{s} A^{|s^{-1}(1)|}B^{|s^{-1}(-1)|}n^{|s|},$$

since the sum over s' is simply the number of maps from $|s|$ components into n labels. But the last expression is exactly the splice state model definition of the bracket. □

Our second step towards a Yang-Baxter model is to generalize the definition of a state even further: Call the segments of $L^* := L\backslash\{\text{crossings of } L\}$ the *edges* of L; then define an *edge* state of a link diagram to be a map $f : \{\text{edges}\} \longrightarrow \{1, \ldots, n\}$. Near each crossing the values of f may be indicated as shown in Figure 6. Let

$$f_1 := \#\text{ crossings where } a = c \neq b = d;$$
$$f_2 := \#\text{ crossings where } a = b \neq c = d;$$
$$f_3 := \#\text{ crossings where } a = b = c = d.$$

We will say that a crossing is of *type* 1, 2, or 3, respectively. Note that the rotation by π ambiguity in the labelling does not affect the f_i.

Proposition 2. *The bracket polynomial can be computed in terms of these edge states as*

$$\langle L\rangle = \sum_{f} A^{f_1}B^{f_2}(A+B)^{f_3}\delta(\#\text{ crossings} - f_1 - f_2 - f_3),$$

where $\delta(x) = 1$ *if* $x = 0$ *and vanishes otherwise.*

Proof. This is a little more subtle than the equivalence between the splice state model and the labelled splice state model. We will show that the terms in this sum are the same as the ones in the labelled splice state sum; then Proposition 1 will imply equivalence to the splice state model for the bracket polynomial.

Because of the presence of the delta function in the summand, the only edge states f which contribute are those in which each crossing is of type 1, 2, or 3. If we were to splice the link diagram at each type 1 crossing according to the splicing which was assigned an A in the original splice state model and each type 2 crossing according to the splicing assigned a B we would obtain a diagram in which each component (not yet simple closed curves—still having transverse double points at the type 3 crossings) would have a single label, *i.e.*, each edge in a single component would be assigned the same value by f. Finally, if we were to splice each of the type 3 crossings *both* ways, we would generate exactly the labelled splice states of the previous model, and each would contribute the same monomial in A and B to the state sum. Thus the edge state model and the labelled splice state model are equivalent and by Proposition 1, produce the bracket polynomial. □

The information contained in the classification of crossings by type and the associated contribution in the state sum can be encoded more compactly by defining

$$\begin{aligned} R_{ijkl} &:= \begin{cases} A & \text{if } i = k \neq j = l; \\ B & \text{if } i = j \neq k = l; \\ A + B & \text{if } i = j = k = l; \\ 0 & \text{otherwise} \end{cases} \\ &= A\delta_{ik}\delta_{jl} + B\delta_{ij}\delta_{kl}, \end{aligned}$$

where δ_{ik} is the Kronecker symbol: $\delta_{ik} := 1$ if $i = k$ and vanishes otherwise. Now we can define the *Yang-Baxter* model for the bracket polynomial in terms of R_{ijkl} [8,13]:

Proposition 3. *The bracket polynomial can be computed by replacing each crossing by an appropriately oriented R_{ijkl}, taking their product, and summing over each pair of indices connected by an edge of L; call this operation the* diagram trace *of R_{ijkl}.*

Proof. From the definition, it is clear that the terms in the Yang-Baxter model diagram trace are exactly those in the state sum for the edge state model. Thus Proposition 2 proves that the bracket polynomial can be computed this way. □

Now, the bracket polynomial is only a single example; the general Yang-Baxter model for an unoriented regular isotopy invariant is defined in the same way by any four index object R_{ijkl} which satisfies the conditions implied by the Reidemeister moves:

$$\text{RII} \Longrightarrow R_{ijkl}R_{lnkm} = \delta_{im}\delta_{jn}$$

$$\text{RIII} \Longrightarrow R_{aibj}R_{deik}R_{jkcf} = R_{adjk}R_{jibc}R_{keif},$$

where repeated indices are to be summed over. See Figure 7. The equation implied by RIII is the Yang-Baxter equation of statistical mechanics for vanishing or infinite spectral parameter [11,15], hence the designation "Yang-Baxter model." From our discussion the following result is clear [8,13]:

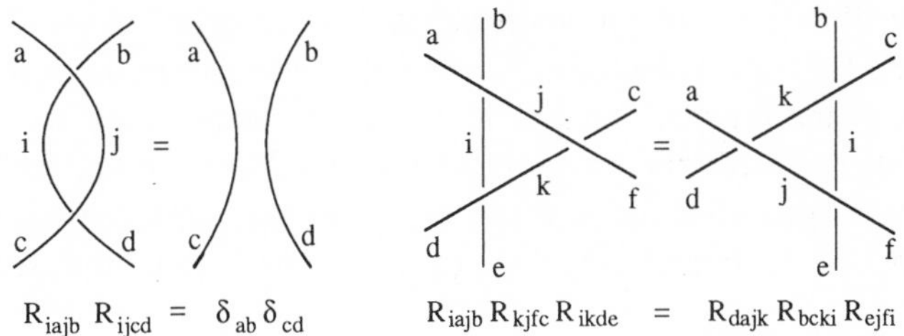

The second and third Reidemeister moves and the implied condition on R.

Figure 7

Theorem (Turaev, Kauffman). *Any R solving these equations generates a regular isotopy invariant of links.*

It is not so clear, however, how to find solutions to these conditions; in fact, since there are only n^4 elements R_{ijkl} and the Yang-Baxter equation comprises n^6 constraints, it is not even clear that there should be non-trivial solutions. Having reviewed the framework necessary to pose this problem, the rest of the paper is devoted to showing how link invariants can be associated to the classical Lie groups and how the mode of this association is related to the existence of non-trivial solutions.

4. Tensor diagrams

The indices which have appeared in our discussion of the Yang-Baxter model are not only aesthetically unappealing but also somewhat misleading. The only operations performed on R_{ijkl} in the diagram trace are recognizable as the multiplication and contraction operations of a Cartesian tensor algebra in classical index notation. That is, following Turaev [13], R_{ijkl} can be identified as the components of a K-linear map $R : V \otimes V \longrightarrow V \otimes V$, where K is the commutative ring $\mathbf{Z}[A, B]$ and V is a free K-module of rank n. As we have not distinguished between covariant and contravariant indices there must be a metric—giving a canonical isomorphism between V and V^*—which is positive definite, so that we may work in an orthonormal basis in which components of isomorphic contra- and covariant vectors are equal.

Thus the equations implied by RII and RIII are tensor equations and have basis independent meaning. Rather than writing them as conditions on the components of R, it is more suggestive to use the diagrammatic notation introduced by Penrose [22], developed also by Cvitanović [23], described in detail in the books by Penrose and Rindler [24] and Cvitanović [25] and used in this context by Reshetikhin [14] and Kauffman [8,17]. In this notation the identity map $\mathrm{id} : V \longrightarrow V$, which is a tensor of type $(1, 1)$ and whose components are δ^a_b in any basis, is represented by an immersion of the oriented interval $[0, 1]$ into the plane. If we wish to represent the components explicitly the image is labelled by a at its initial point and b at its terminal point. If the initial and terminal points are the same, the trace operation is understood; hence the immersed image of an oriented circle represents the scalar n.

Similarly, a permutation map such as $P : V \otimes V \longrightarrow V \otimes V$, with components $P^{ab}_{cd} := \delta^a_d \delta^b_c$, would be represented by a pair of immersed intervals with a single mutual intersection point (or any odd number of intersection points). Such diagrams were first used by Brauer [26] and Aitken [27]; see Figure 8.

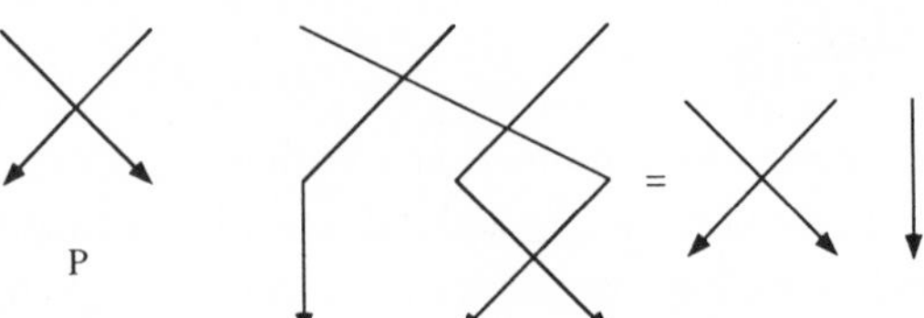

Aitken diagrams for permutation operators are constructed from products of Kronecker deltas.

Figure 8

To represent an arbitrary tensor $A \in T^p_q(V)$ we draw a vertex (or disk) which is the terminal point for p immersed intervals and the initial point for q immersed intervals. Which line corresponds to action on which V or V^* is determined either by explicit labels or by some arbitrary convention. Just as was implicit in the permutation operators shown in Figure 8, multiplication is represented by juxtaposition, while contraction is represented by the appropriate connection of tensor symbols by immersed intervals. See Figure 9.

Note that the diagrams of the Reidemeister moves shown in Figures 1, 3 and 4 can now be interpreted directly as diagrammatic tensor equations. In the unoriented case, the tensor R is represented by the crossing shown in Figure 6, while in the oriented case there are two tensors, R and $\bar{R}$ representing the crossings of Figure 3. Moreover, the particular tensor R which produced the Yang-Baxter

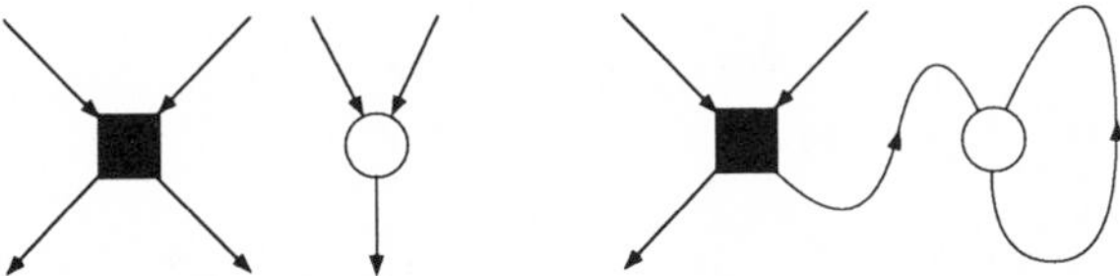

Diagrammatic represention of tensor multiplication and contraction.

Figure 9

model for the bracket polynomial is expressed diagrammatically by removing the brackets from the equation for the third axiom used in the definition. This suggests that we try to interpret the topological formulation of other state models directly as tensor diagrams and conversely, that we investigate diagrammatically which tensors determine topological link invariants.

5. Lie group invariants

The main point of this talk is that one place to look for these tensors which generate link invariants is in the subspace of $T_2^2(V)$ consisting of Lie group invariants, $T_2^2(V)^G$. Recall that $\mathbf{GL}(V)$ acts on $T_q^p(V)$ according to the tensor representation $T \longmapsto T^{\otimes p} \otimes (T^*)^{\otimes q}$. In particular, the type $(1,1)$ tensor id, the identity map considered in the previous section, is invariant under the action of $\mathbf{GL}(V)$, namely $\mathrm{id}(T \cdot (x \otimes \xi)) = \mathrm{id}(x \otimes \xi)$ for all $T \in \mathbf{GL}(V)$, $x \in V$ and $\xi \in V^*$; equivalently, id has components δ_b^a in any basis. In fact, every $\mathbf{GL}(V)$ invariant tensor is a linear combination of products of this *basic* invariant.

More generally, as described in detail in Weyl's book [28], it is a consequence of Hilbert's First and Second Main Theorems in invariant theory [29,30] that for the classical Lie groups the ring of tensor invariants is generated by a finite set of basic invariants subject to a finite number of relations. The identity tensor already discussed is invariant in each case.

In addition, $\mathbf{SL}(n,K)$ has a totally antisymmetric type $(0,n)$ tensor (an n-form) ϵ and its dual as basic invariants. These are clearly invariants, for under the $\mathbf{GL}(V)$ action they are multiplied by the Jacobian of the transformation—which is 1 for $\mathbf{SL}(n,K)$ elements. The components of these tensors are usually denoted by, respectively, $\epsilon_{i_1 \ldots i_n}$ and $\epsilon^{i_1 \cdots i_n}$, which are equal to ± 1 according to whether $(i_1 \ldots i_n)$ is an even or odd permutation of $(1 \ldots n)$.

Besides the identity tensor id, the orthogonal group $\mathbf{O}(n,K)$ leaves invariant the metric g and its dual: symmetric tensors of type $(0,2)$ and $(2,0)$ whose components are usually written as g_{ij} and g^{ij}. Similarly, the symplectic group $\mathbf{Sp}(n,K)$ has the symplectic form f and its dual as additional basic invariants:

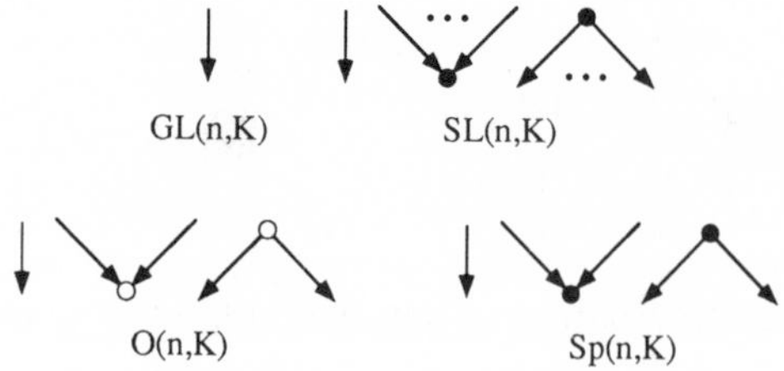

Diagrams for the basic invariants of the classical Lie groups.

Figure 10

The metric and its dual in an unoriented representation.

Figure 11

antisymmetric tensors of type $(0, 2)$ and $(2, 0)$ with components f_{ij} and f^{ij}, respectively.

The usual tensor diagrams for these basic invariants are shown in Figure 10. In the case when the metric of the orthogonal group is positive definite, there is a canonical isomorphism between V and V^*, so if we choose a basis for V in which the metric has the form $\mathrm{diag}(1, \ldots, 1)$, the diagrams for invariants are unoriented and the metric and its dual are represented as in Figure 11. We can already see that this case should be related to the bracket polynomial. In the next section we begin to determine the group invariant Yang-Baxter state models for the classical Lie groups, keeping close to the topology by making extensive use of diagrammatic tensor notation.

6. $\mathbf{GL}(n, K)$, $\mathbf{SL}(n, K)$ and $\mathbf{O}(n, \mathbf{C})$

Denote by $\mathrm{Reg}(V)$ the collection of all K-linear maps $R : V \otimes V \longrightarrow V \otimes V$ which generate regular isotopy invariants of links through a Yang-Baxter model. Note that the mapping $R \longleftrightarrow \bar{R}$ is an automorphism of $\mathrm{Reg}(V)$. In this section we compute the intersection of $\mathrm{Reg}(V)$ with the G-invariant subspaces of $T_2^2(V)$ for the groups $\mathbf{GL}(n, K)$, $\mathbf{SL}(n, K)$ and $\mathbf{O}(n, \mathbf{C})$. Note that the result of any such computation must be independent of the basis used for the G-invariant subspace since, although our calculations are diagrammatic, we are simply solving the tensor equations associated to the Reidemeister moves.

The general linear group

The only basic invariant for tensor representations of $\mathbf{GL}(n, K)$ is the identity tensor id. Thus a natural basis for the invariant subspace of $T_2^2(V)$ under the action of $\mathbf{GL}(n, K)$ is $\{P, \mathrm{id} \otimes \mathrm{id}\}$.

Proposition 4. *The intersection of* $\mathrm{Reg}(V)$ *with the* $\mathbf{GL}(n, K)$ *invariant subspace of* $T_2^2(V)$ *is the subspace spanned by* $P : V \otimes V \longrightarrow V \otimes V$.

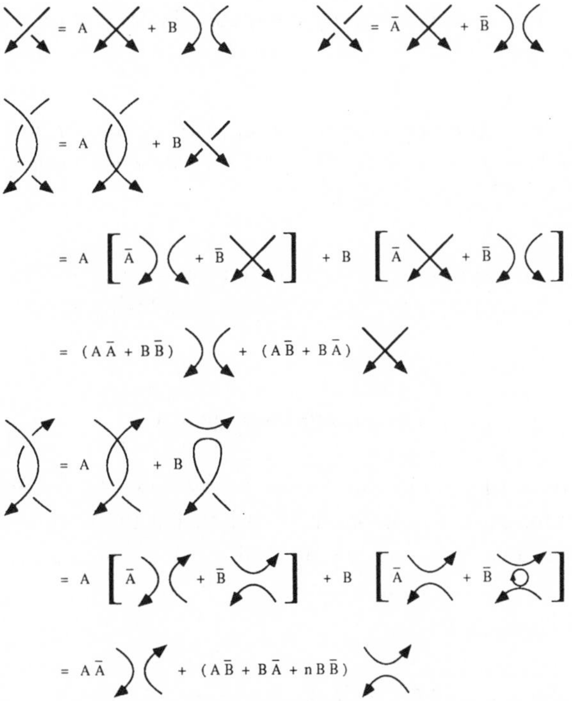

$\mathbf{GL}(n, K)$ invariant basis expansion for the oriented crossings and the expansion of the second Reidemeister moves.

Figure 12

Proof. We proceed diagrammatically. Let

$$R := AP + B\mathrm{id} \otimes \mathrm{id}$$
$$\bar{R} := \bar{A}P + \bar{B}\mathrm{id} \otimes \mathrm{id},$$

as shown in Figure 12. Expanding the oriented Reidemeister moves we find:

$$\text{RIIa} \Longrightarrow \begin{cases} 1 = A\bar{A} + B\bar{B} \\ 0 = A\bar{B} + B\bar{A} \end{cases}$$

$$\text{RIIb} \Longrightarrow \begin{cases} 1 = A\bar{A} \\ 0 = A\bar{B} + B\bar{A} + nB\bar{B} \end{cases}$$

The only solution to these equations (for $n \neq 0$) is $\bar{A} = A^{-1}$, $B = \bar{B} = 0$. This solution satisfies the oriented third Reidemeister moves for all values of A; hence it generates a regular isotopy invariant. □

It is clear that this invariant simply computes the writhe of a link diagram. In fact we have:

Proposition 5. *The evaluation of this regular isotopy invariant on an untwisted diagram of a link* $L = L_i \cup \cdots \cup L_k$ *is* $n^k A^{\mathrm{lk}(L)}$, *where the total linking number is defined by* $\mathrm{lk}(L) := \sum_{i<j} \mathrm{lk}(L_i, L_j)$.

Proof. An *untwisted* diagram of a link is defined to be one in which each component of the link has writhe zero [31]. Then the conclusion is obvious. □

The unimodular group

The additional basic invariants for tensor representations of $\mathbf{SL}(n, K)$ are the two totally antisymmetric tensors defined earlier. Up to permutations (and thus a sign), there is one, but only one, element of $T_2^2(V)$ which can be formed from these invariants: the tensor with components $\epsilon^{i_1 \ldots i_{n-2} ba} \epsilon_{i_1 \ldots i_{n-2} cd}$. Nevertheless, we have the following result:

Lemma 6. *The subspace of* $T_2^2(V)$ *invariant under the action of* $\mathbf{SL}(n, K)$ *is the same as the subspace left invariant by* $\mathbf{GL}(n, K)$.

Proof. Recall that in general there are algebraic relations among the basic invariants. According to the Second Main Theorem for the unimodular group (see [28], p. 70),

$$\epsilon^{i_1 \ldots i_n} \epsilon_{j_1 \ldots j_n} = \begin{vmatrix} \delta^{i_1}_{j_1} & \delta^{i_2}_{j_1} & \cdots & \delta^{i_n}_{j_1} \\ \delta^{i_1}_{j_2} & \delta^{i_2}_{j_2} & \cdots & \delta^{i_n}_{j_2} \\ \vdots & \vdots & \ddots & \vdots \\ \delta^{i_1}_{j_n} & \delta^{i_2}_{j_n} & \cdots & \delta^{i_n}_{j_n} \end{vmatrix}.$$

Thus $\epsilon^{i_1\ldots i_{n-2}ba}\epsilon_{i_1\ldots i_{n-2}cd} = (n-2)!(\delta^a_d\delta^b_c - \delta^a_c\delta^b_d)$; the tensor with these components can be written as a linear combination of the basis tensors we used for $\mathbf{GL}(n, K)$. See Figure 13. □

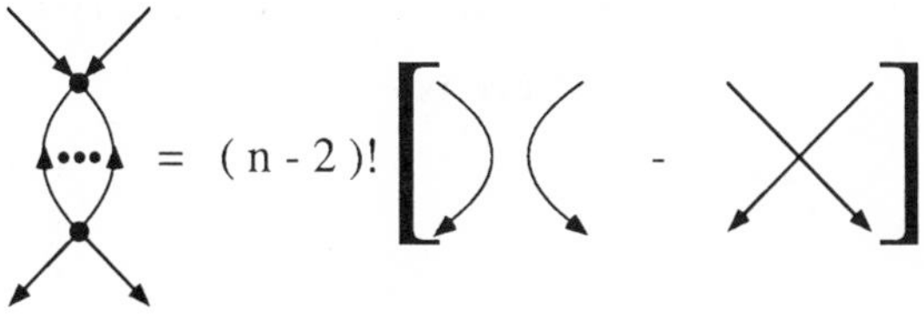

= (n - 2)! [ ) (- X]

$\mathbf{SL}(n, K)$ tensor identity in $T^2_2(V)$.

Figure 13

In conjunction with the observation at the beginning of this section that the intersection of an invariant subspace with $\mathrm{Reg}(V)$ is basis independent, this result implies that there are no additional link invariants to be derived in this way from $\mathbf{SL}(n, K)$.

The orthogonal group

The determinant form ϵ which is preserved by $\mathbf{SL}(n, K)$ is not an invariant of the full orthogonal group. Rather, in addition to id, the orthogonal group preserves a nonsingular symmetric tensor g and its dual g^*. Depending on the field K, by making a suitable choice of basis, the components g_{ij} can be put into one of a variety of canonical forms.

Let us first consider the algebraically complete case $K = \mathbf{C}$, the complex numbers, for example. In this case there is a basis in which $g_{ij} = \delta_{ij}$. This form then induces an isomorphism between V and V^* by $v^i := \delta^{ij}v_j$; thus we need make no distinction between covariant and contravariant tensor indices. Diagrammatically, this is represented by ignoring orientation; the only invariants in $T^2_2(V)$ which can be formed from these basic invariants are: P, $\mathrm{id} \otimes \mathrm{id}$ and $g \otimes g^*$ (see Figure 14). The only relations among the basic invariants are algebraic consequences of the fact that tensors totally antisymmetric in $n+1$ indices must vanish (see [28], p. 75):

$$\epsilon_{i_0\ldots i_n}\epsilon_{j_0\ldots j_n} = \begin{vmatrix} \delta_{i_0j_0} & \delta_{i_2j_0} & \cdots & \delta_{i_nj_0} \\ \delta_{i_0j_2} & \delta_{i_2j_2} & \cdots & \delta_{i_0j_2} \\ \vdots & \vdots & \ddots & \vdots \\ \delta_{i_0j_n} & \delta_{i_2j_n} & \cdots & \delta_{i_nj_n} \end{vmatrix} = 0$$

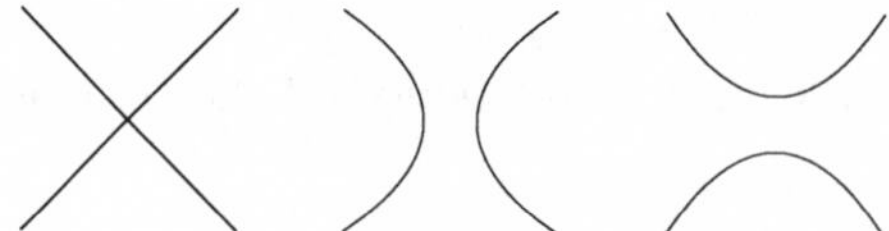

A basis for the $\mathbf{O}(n, \mathbf{C})$ invariant subspace of $T_2^2(V)$.

Figure 14

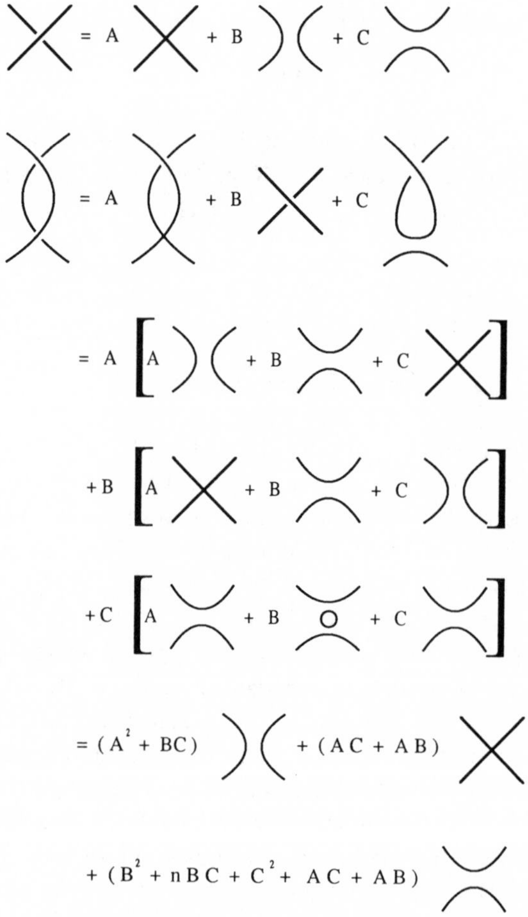

$\mathbf{O}(n, \mathbf{C})$ invariant basis expansion for the unoriented crossing and the expansion of RII.

Figure 15

so these three tensors form a basis for the $\mathbf{O}(n, \mathbf{C})$ invariant subspace.

Proposition 7. *For $n > 2$ the intersection of* $\mathrm{Reg}(\mathbf{C}^n)$ *with the* $\mathbf{O}(n, \mathbf{C})$ *invariant subspace of* $T_2^2(V)$ *consists of the two points* $\pm P$ *and the four points* $B(\mathrm{id} \otimes \mathrm{id}) + B^{-1}(g \otimes g^*)$ *where* $-B^2 - B^{-2} \equiv n$*. For* $n = 2$*, the latter four points collapse to two, but the intersection also contains the hyperbola* $AP + B\,[\mathrm{id} \otimes \mathrm{id} - g \otimes g^*]$ *where* $A^2 - B^2 \equiv 1$.

Proof. Again we proceed diagrammatically. Let $R := AP + B\mathrm{id} \otimes \mathrm{id} + Cg \otimes g^*$ as shown in Figure 15. Then the second Reidemeister move implies

$$\begin{aligned} 1 &= A^2 + BC \\ 0 &= AC + AB \\ 0 &= B^2 + nBC + C^2 + AC + AB \end{aligned}$$

There are two sets of solutions to this system of equations. The second is satisfied if $A = 0$ or if $C = -B$. When $A = 0$ we find $C = B^{-1}$ and $n = -B^2 - B^{-2}$. When $C = -B$, the only nontrivial solution satisfies $n = 2$ and $A^2 - B^2 = 1$. The first solution satisfies the third Reidemeister move; this is demonstrated diagrammatically in Figure 16. The second solution can be checked in the same way. The part of the computation which is new is shown in Figure 17; we find that the conditions $C = -B$ and $n = 2$ are necessary and sufficient to satisfy the the third Reidemeister move. □

= A + B

= A + B

=

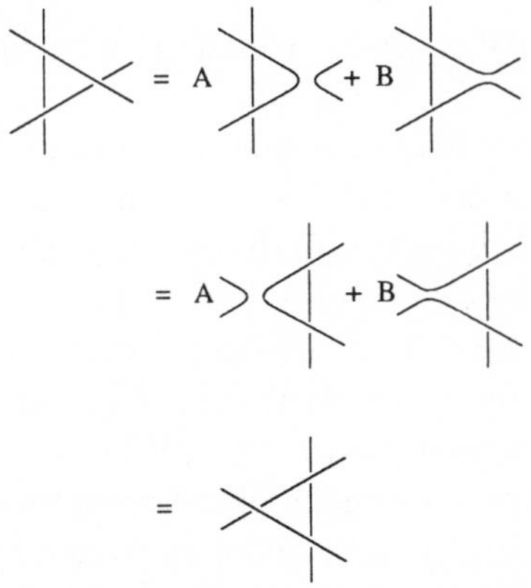

The bracket polynomial is invariant under the third Reidemeister move.

Figure 16

These regular isotopy invariants are immediately recognizable as Kauffman's bracket polynomial and its (uninteresting) generalization [5]. Thus we see that

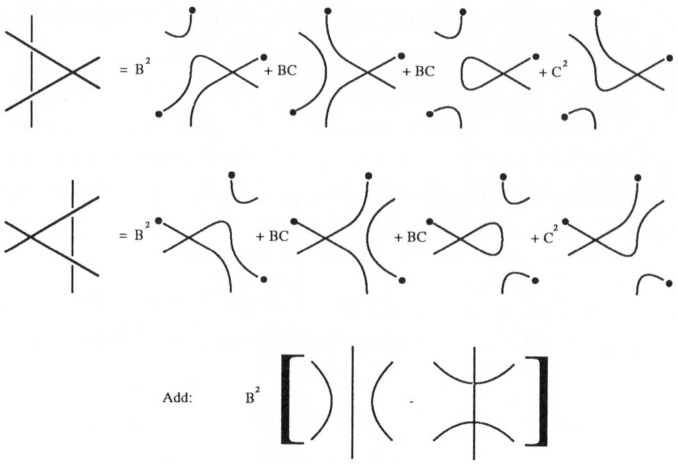

A segment of a link diagram passes under the tensor P, provided $C = -B$ and $n = 2$, for then we can add terms which make each side of the equation totally antisymmetric on three indices and hence vanishing.

Figure 17

these invariants can be thought of as arising from the imposition of orthogonal group invariance.

7. Group invariance and ambient isotopy

The two computations in the previous section demonstrate that by restricting R to lie in a G-invariant subspace of $T^2_2(V)$ we can indeed determine regular isotopy invariants of link diagrams, admittedly fairly trivial in the case of $\mathbf{GL}(n, K)$, but distinctly not so in the case of $\mathbf{O}(n, \mathbf{C})$. In both cases we start by trying to satisfy the second Reidemeister move. In the unoriented case this leads to a system of $\dim(T^2_2(V)^G)$ equations in the same number of variables, while in the oriented case there are twice as many variables (for R and $\bar{R}$) and twice as many equations. Generically then, it is no surprise to find a family of solutions parameterized by n, the rank of V, or diagrammatically, the symbol associated to a simple closed curve in the state expansion. That there is an extra degree of freedom in the $\mathbf{GL}(n, K)$ case is due to the scaling freedom in the oriented equations: if R and $\bar{R}$ satisfy the RIIa and RIIb moves, so do αR and $\alpha^{-1}\bar{R}$.

That the third Reidemeister move is also satisfied by these solutions is still surprising. Before considering why this is so, we investigate the easier question of the behaviour of the first Reidemeister move in these models. We begin with a lemma:

Lemma 8. *For each of the classical Lie group invariant Yang-Baxter models, every 1-tangle diagram evaluates to be proportional (by some element in the ring) to the identity tensor* id.

Proof. A *1-tangle* diagram is simply a link diagram with one of the immersed S^1s replaced by an immersed interval; it is a link diagram with two free ends. Evaluating such a diagram in any group invariant model produces an element in the group invariant subspace of $T_1^1(V)$, just as evaluating a link diagram trace produces a G-scalar in the ring K. But for the classical Lie groups, this subspace is one dimensional, being generated by the identity map id : $V \longrightarrow V$. □

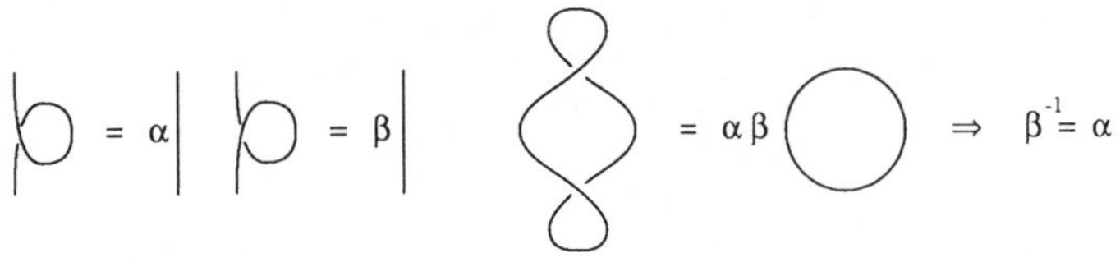

The unoriented curls appearing in RI are proportional to the identity. RII forces the proportionality constants to be inverses.

Figure 18

Thus, for example, the logo of this conference, the 1-tangle diagram constructed from the symbol for Osaka (shown on the cover), would evaluate to be proportional to the identity in any of these models. This is true for both oriented and unoriented tangles. In particular, for the curls which appear in the first Reidemeister move, as shown diagrammatically in Figure 18, we must have (in components)

$$R_{abcb} = \alpha\delta_{ac} \quad \text{and} \quad R_{abcc} = \gamma\delta_{ab},$$

where α and γ lie in the coefficient ring.

Proposition 9. *Regular isotopy invariants computed from classical Lie group invariant Yang-Baxter models for unoriented link diagrams can be normalized to obtain ambient isotopy invariants.*

Proof. First observe that the proportionality constants in which Lemma 8 tells appear in the first Reidemeister move satisfy $\gamma = \alpha^{-1}$. This follows from the second Reidemeister move as shown in Figure 18, or algebraically as follows:

$$\delta_{aa} = R_{abcb}R_{acdd} = \alpha\delta_{ac}\gamma\delta_{ac} = \alpha\gamma\delta_{aa}.$$

Second, note that the exponent of α is exactly the writhe of the curl in each case. Thus the combination $\alpha^{-w(L)}$(diagram trace) is invariant under all three Reidemeister moves and is hence an ambient isotopy invariant. □

In the oriented case the situation is slightly more complicated since there are the four curls shown in Figure 4. Lemma 8 still applies, however, so call the proportionality constants μ, ν, ρ and σ, respectively. Then we have a similar result:

Proposition 10. *Regular isotopy invariants computed from classical Lie group invariant Yang-Baxter models for oriented link diagrams can be normalized to obtain ambient isotopy invariants.*

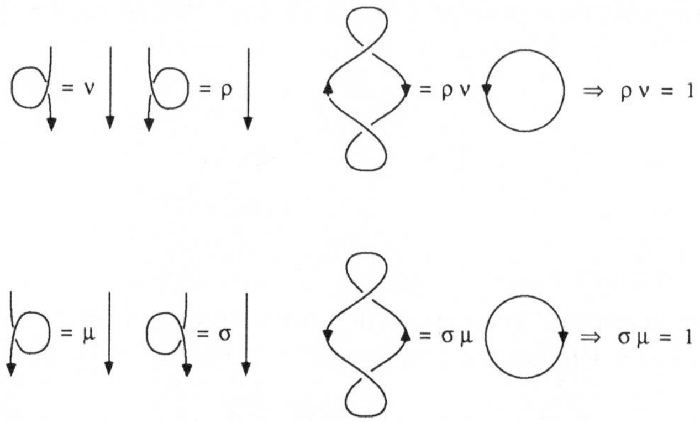

= ν = ρ = ρ ν ⇒ ρ ν = 1

= μ = σ = σ μ ⇒ σ μ = 1

RIIb determines the proportionality constants for oriented curls.

Figure 19

Proof. In this case the proportionality constants can be parameterized as $\mu = \alpha\beta$, $\nu = \alpha\beta^{-1}$, $\rho = \alpha^{-1}\beta$ and $\sigma = \alpha^{-1}\beta^{-1}$. This follows from RIIb applied to unknots of rotation numbers ± 1 as shown in Figure 19, and algebraically as follows: Beginning with an unknot diagram of rotation number $+1$ we have

$$\delta^a_a = \bar{R}^{ab}_{cb} R^{cd}_{ad} = \rho \delta^a_c R^{cd}_{ad} = \rho\nu\delta^a_a,$$

where we end with one of rotation number $+1$ as well. Thus $\rho\nu = 1$. Similarly, using an unknot diagram of rotation number -1, we find $\mu\sigma = 1$. The parameterization in terms of α and β follows immediately. Now note that the exponent of α is exactly the writhe of the curl in each case, while that of β is the rotation number. Thus the combination $\alpha^{-w(L)}\beta^{-r(L)}$(diagram trace) is invariant under all three Reidemeister moves and is hence an ambient isotopy invariant. □

These results show that for classical Lie group invariant models, once a regular isotopy invariant is found, it can be normalized to obtain an ambient isotopy invariant. The crucial ingredients in the proofs are invariance of the model under the second Reidemeister moves and the one dimensionality of the G-invariant subspace of $T_1^1(V)$. In the next section we shall use these same ingredients in our investigation of the redundancy of the third Reidemeister move.

8. On the redundancy of the third Reidemeister move

As we noted at the beginning of the previous section, satisfying the second Reidemeister move (in the unoriented case) for the models we are considering means solving a system of $\dim(T_2^2(V)^G)$ equations in the same number of variables. Similarly, the 3-tangle diagram on each side of the third Reidemeister move evaluates to an element of the G-invariant subspace of $T_3^3(V)$, a vector space of some combinatorially computable number of dimensions. If we call these two vectors u and v, the third Reidemeister move becomes the condition $u = v$, a set of $\dim(T_3^3(V)^G)$ conditions. For there to be nontrivial solutions to this system of equations, they must be degenerate since there are only $\dim(T_2^2(V)^G)$ coefficients in R all of which, generically, have already been determined by RII. Ideally, then, we would like to show that $u = v$ is simply a consequence of the second Reidemeister move. The next two results are steps in this direction, showing that the collection of equations arising from the two moves is highly degenerate in these models.

Theorem 11. *The tensor equations arising from the second and third Reidemeister moves in any unoriented Yang-Baxter model comprise no more than $2n^4$ constraints on n^4 variables. In a classical Lie group invariant model there are no more than $2\dim(T_2^2(V)^G)$ constraints on $\dim(T_2^2(V)^G)$ variables.*

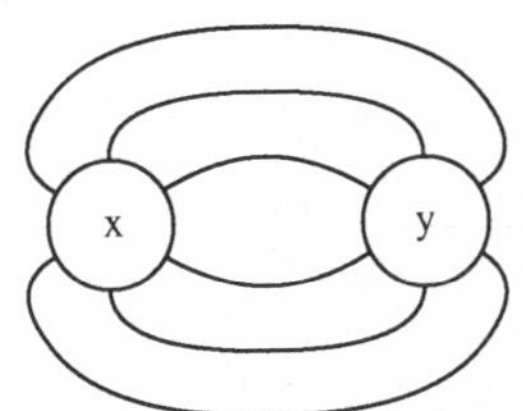

An inner product on $T_3^3(V)$.

Figure 20

Proof. The idea is to construct a positive definite inner product with the property that $\langle u - v, u - v\rangle = 0$ so we can conclude that $u = v$. Define a map $T_3^3(V) \times T_3^3(V) \longrightarrow \mathbf{R}$ as shown in Figure 20. Observe that this map is symmetric, provided we are working on S^2, for the vector x can be isotoped around the sphere to the other side of the vector y. In this unoriented case there is also another choice, which requires only planar isotopy for symmetry, namely the same diagram with y rotated through π, but the first choice is the one which will generalize to the oriented case most naturally. Both maps act as $\mathrm{id} : T_3^3(V) \longrightarrow T_3^3(V)$ and hence are positive definite. Thus we have a positive definite inner product.

Now compute $\langle u, u\rangle$, $\langle v, v\rangle$ and $\langle u, v\rangle$. Planar isotopy and the use of RII twice on $\langle u, v\rangle$ gives the results shown in Figure 21. If all three inner products were equal we would have the desired result that $\langle u - v, u - v\rangle = 0$ and hence $u = v$. Thus if we impose the tensor equation (in components)

$$R_{gaie}R_{ebjf}R_{fdlh}R_{hckg}R_{ijkl} = nR_{abcd}, \tag{CC}$$

which we shall call the *circle-crossing* condition for reasons which are more apparent in Figure 22, RIII is a consequence of RII. In fact, since CC clearly follows from RII and RIII, the two sets of constraints: $\{\mathrm{RII}, \mathrm{RIII}\}$ and $\{\mathrm{RII}, \mathrm{CC}\}$, are equivalent. Since both sides of CC lie in $T_2^2(V)$, it comprises only n^4 equations for a general Yang-Baxter model and $\dim(T_2^2(V)^G)$ for one which is Lie group invariant; hence the result is proved. □

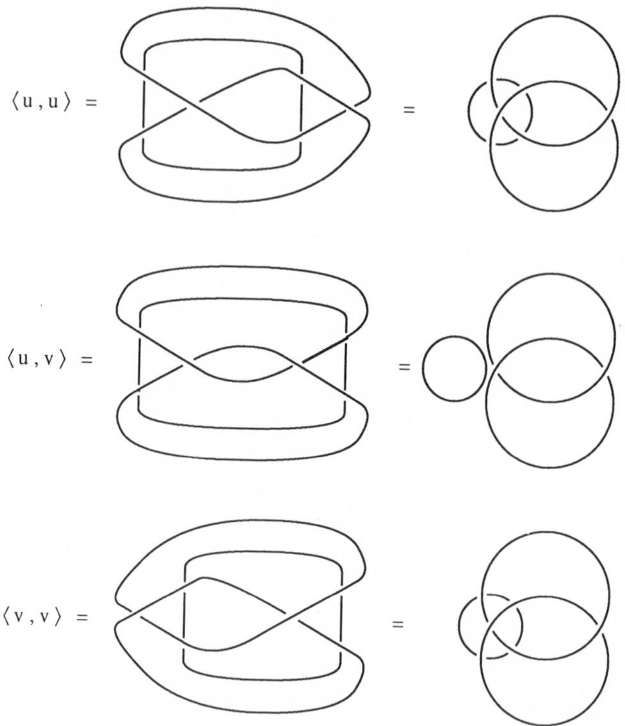

Evaluating the inner products of the vectors in RIII.

Figure 21

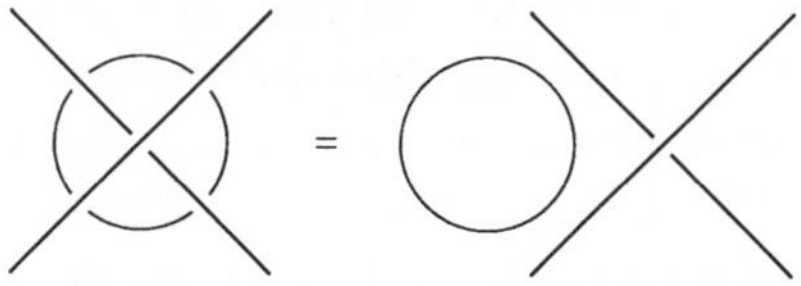

The circle-crossing condition is equivaltent to RIII (in the presence of RII).

Figure 22

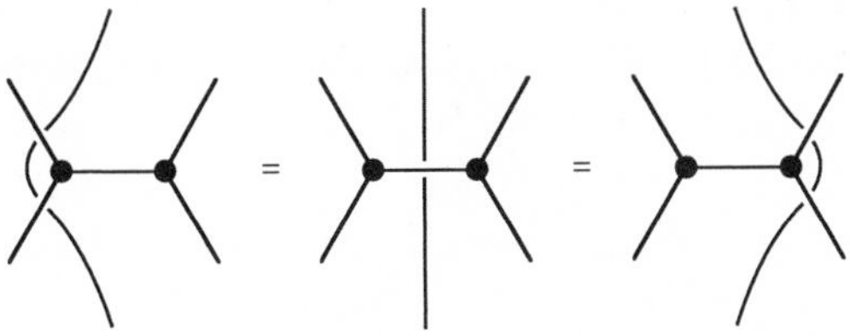

A fourth order invariant formed from a third order basic invariant and passing an undercrossing line past it.

Figure 23

Although this result reduces the degeneracy of the system of constraints considerably, generically one would still not expect a solution when there are twice as many equations as variables. We have, however, used only one of the ingredients which proved helpful in the analysis of the first Reidemeister move, namely invariance under the second. The other ingredient is the reduced dimensionality of G-invariant subspaces; this can be applied to reduce the degeneracy further.

Theorem 12. *The system of tensor equations arising from the conditions RII and RIII on a Lie group invariant oriented Yang-Baxter model is a consequence of the system comprised by RII and a single additional constraint for each basic invariant beyond* id.

Proof. Any term which appears in the G-invariant basis expansion for R is a fourth order tensor formed from basic invariants. To push the undercrossing line in the third Reidemeister move past the other crossing, one must push it past all of the basic invariants which appear in a given term in the R associated to that crossing. If it pushes past each basic invariant separately, it will push past the whole crossing.

For example, in the seven dimensional real representation of the exceptional Lie group $\mathbf{G}_2$ there is an antisymmetric third order tensor invariant [32]. Figure 23 shows one fourth order invariant tensor which would appear in the basis expansion for R. It is formed from the contraction of two of the basic third order invariants. Clearly if an undercrossing line will pass under the third order invariant it will pass under this term in R.

But now we make the same argument we made in the proof of Theorem 11: construct a positive definite inner product on $T_5(V)$ as before and observe the equivalence shown in Figure 24. Since the $\mathbf{G}_2$-invariant subspace of $T_3(V)$ is only one dimensional, this is exactly one constraint on the $\dim(T_2^2(V)^G)$ coefficients in R.

In general, there will be one such constraint for each basic invariant beyond id. With RII (which can be thought of as the necessary constraints for id), this system implies the system of RII and RIII. □

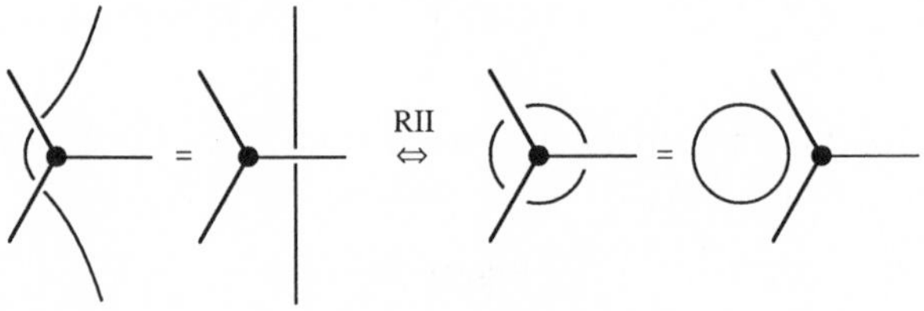

RII
$\Leftrightarrow$

Passing a line under the third order basic invariant has an equivalent condition analogous to the circle-crossing condition for R.

Figure 24

The analogous theorem holds in the oriented case as well:

Theorem 13. *The tensor equations arising from the minimal set of second and third Reidemeister moves in any oriented Yang-Baxter model comprise no more than* $4n^4 - 1$ *constraints on* $2n^4$ *variables. In a classical Lie group invariant model there are no more than* $4\dim(T_2^2(V)^G) - 1$ *constraints on* $2\dim(T_2^2(V)^G)$ *variables.*

Proof. The proof is essentially the same as in the unoriented case. The difference here is that we construct a Hermitian inner product by taking the dual of one of the $(3,3)$ tensors. This corresponds to reversing the arrows on the diagram, so the Hermitian inner product on $T_3^3(V)$ is as in Figure 25. Then RIIa and RIIIa are equivalent to RIIa and the tensor equation shown diagrammatically on the left of Figure 26, while RIIb and RIIIb are equivalent to RIIb and the equation on the right of that figure. The -1 in the statement is due to the scale freedom mentioned at the beginning of Section 7. □

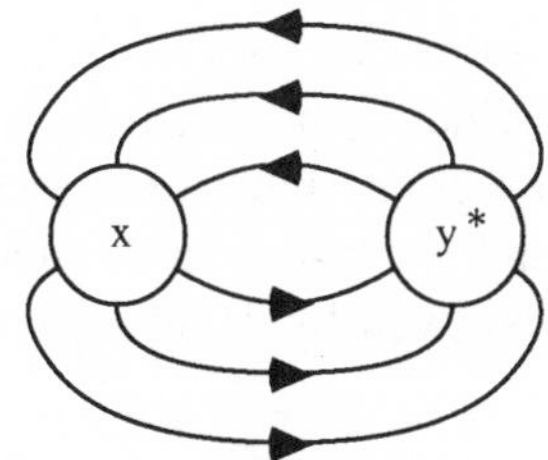

A Hermitian inner product on $T_3^3(V)$.

Figure 25

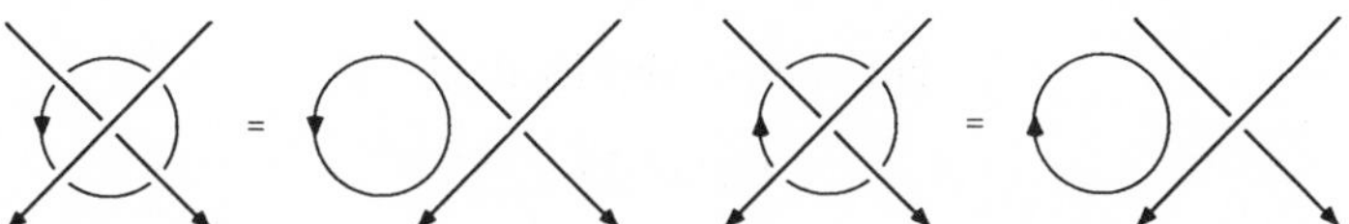

The conditions analogous to CC for the oriented case.

Figure 26

The result analogous to Theorem 12 is more complicated in the oriented case; we omit it here. Also, although Theorem 12 is a stronger result, insofar as it reduces the number of constraints further than does Theorem 11, it still does not provide a final explanation, for not only are there still more constraints than degrees of freedom, but the constraints are actually stronger than those implied by RIII. Nevertheless, these results illuminate much of the degeneracy in the system of constraints deriving from the second and third Reidemeister moves, in the classical Lie group invariant models we are considering in this paper.

9. $\mathbf{O}(n, \mathbf{R})$ and $\mathbf{Sp}(n, K)$

With the results of the previous two sections in mind we now consider the more complicated cases of regular isotopy invariants lying in the G-invariant subspaces of $T_2^2(V)$ for the classical Lie groups $\mathbf{O}(n, \mathbf{R})$ and $\mathbf{Sp}(n, K)$.

The orthogonal group

In Section 6 we considered the orthogonal group over an algebraically complete field. If the field is not algebraically complete, say $K = \mathbf{R}$, there are a variety of canonical forms for the metric g: $\mathrm{diag}(1, \ldots, 1, -1, \ldots, -1)$, differing by their signatures; if there are p 1s and q -1s, denote the group by $\mathbf{O}(p, q)$ and let

$k := p - q$. Whenever the metric is not positive definite we cannot choose a basis in which we can ignore the distinction between covariant and contravariant tensor actions. For these cases our diagrams must be oriented. The metric and its dual will be represented as in Figure 27. Just as in the unoriented case, a basis for the invariant subspace of $T_2^2(V)$ is given by P, $\mathrm{id} \otimes \mathrm{id}$ and $g \otimes g^*$ (see Figure 28).

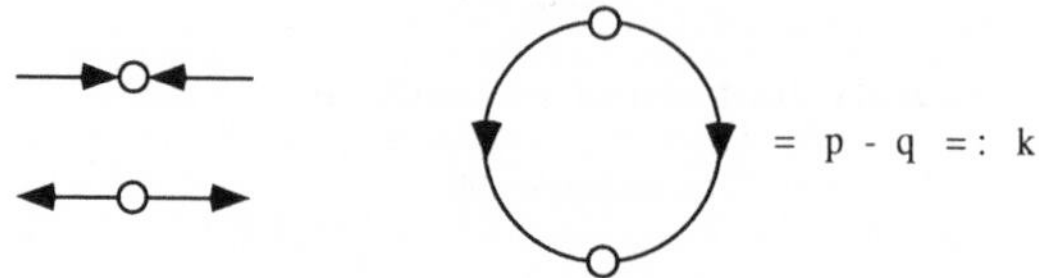

The metric and its dual.

Figure 27

Proposition 14. *For $k \neq \pm 2$ the intersection of* $\mathrm{Reg}(\mathbf{R}^n)$ *with the* $\mathbf{O}(p,q)$ *invariant subspace of $T_2^2(V)$ consists of the subspace spanned by P and the two subspaces spanned by* $\mathrm{id} \otimes \mathrm{id} + \frac{1}{2}(-k \pm \sqrt{k^2 - 4}) g \otimes g^*$. *For $k = \pm 2$ the intersection is the subspace spanned by P and* $\mathrm{id} \otimes \mathrm{id} \mp g \otimes g^*$.

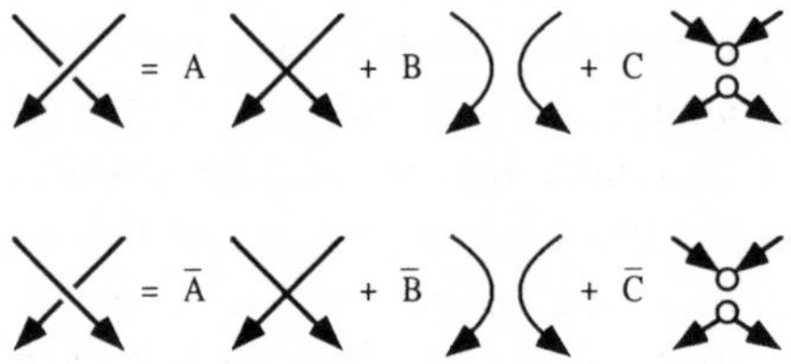

= A + B + C

= $\bar{A}$ + $\bar{B}$ + $\bar{C}$

The expansion of R and $\bar{R}$ in a basis for the $\mathbf{O}(n, \mathbf{R})$ invariant subspace.

Figure 28

Proof. As in Section 6 we proceed diagrammatically. Let

$$R = AP + B\mathrm{id} \otimes \mathrm{id} + Cg \otimes g^*$$
$$\bar{R} = \bar{A}P + \bar{B}\mathrm{id} \otimes \mathrm{id} + \bar{C}g \otimes g^*$$

as shown in Figure 28. Then Figure 29 shows the expansion of the second Reidemeister moves. We find:

$$\mathrm{RIIa} \Longrightarrow \begin{cases} 1 = A\bar{A} + B\bar{B} \\ 0 = A\bar{B} + B\bar{A} \\ 0 = A\bar{C} + C\bar{A} + B\bar{C} + C\bar{B} + kC\bar{C}, \end{cases}$$

$$\mathrm{RIIb} \Longrightarrow \begin{cases} 1 = A\bar{A} + C\bar{C} \\ 0 = A\bar{C} + C\bar{A} \\ 0 = A\bar{B} + B\bar{A} + B\bar{C} + C\bar{B} + kB\bar{B}. \end{cases}$$

These equations are degenerate and reduce to the system of five equations:

$$\begin{aligned}1 &= A\bar{A} + B\bar{B}\\0 &= A\bar{B} + B\bar{A}\\0 &= B\bar{B} - C\bar{C}\\0 &= B\bar{C} + C\bar{B} + kB\bar{B}\\0 &= A\bar{C} + C\bar{A}.\end{aligned}$$

There are three cases:

Case $A = 0$. The solution is $A = \bar{A} = 0$, $\bar{B} = B^{-1}$ and $\bar{C} = C^{-1}$, subject to the constraint $0 = k + BC^{-1} + CB^{-1}$ or, equivalently, $C = \frac{1}{2}(-k \pm \sqrt{k^2 - 4})B$.

Case $B = 0$. The solution is $\bar{A} = A^{-1}$ and $B = \bar{B} = C = \bar{C} = 0$.

Case A, B, $C \neq 0$. There is only a solution when $k = \pm 2$ in which case $C = \mp B$, $\bar{A} = A/(A^2 - B^2)$, $\bar{B} = -B/(A^2 - B^2)$ and $\bar{C} = \mp \bar{B}$.

Each of these solutions also satisfies the third Reidemeister moves. In Figure 30, for example, we verify this for the case $A = 0$ solution using the ideas of the previous section: We find that for a simple closed curve undercircling $g \otimes g^*$ to factor out the only two constraints which must be satisfied are:

$$\begin{aligned}k &= BC^{-1} + CB^{-1} + 2k\\0 &= 2 + k^2 + 2k(BC^{-1} + CB^{-1}) + B^2C^{-2} + B^{-2}C^2.\end{aligned}$$

That both of these are satisfied by the solution to the second Reidemeister moves, namely $0 = k + BC^{-1} + CB^{-1}$, reiterates that we have not completely identified the degeneracy in the third Reidemeister move. □

The symplectic group

Rather than the symmetric second order metric tensor preserved by the orthogonal groups, the symplectic group is defined by the invariance of the symplectic form f, an antisymmetric $(0, 2)$ tensor, and its dual f^*. Because f is antisymmetric there are only even dimensional representations. In components this antisymmetry is reflected in the fact that

$$f^{ab} f_{cb} = \delta^a_b = -f^{ab} f_{bc}.$$

These relations are indicated in Figure 31. Still referring to [28] (Chapter VI), a basis for the $\mathbf{Sp}(n, K)$ invariant subspace of $T^2_2(V)$ is provided by P, $\mathrm{id} \otimes \mathrm{id}$ and $f \otimes f^*$.

Proposition 15. *For* $n \neq 2$, *the intersection of* $\mathrm{Reg}(K^n)$ *with the* $\mathbf{Sp}(n, K)$ *invariant subspace of* $T^2_2(V)$ *consists of the subspace spanned by* P *and the two*

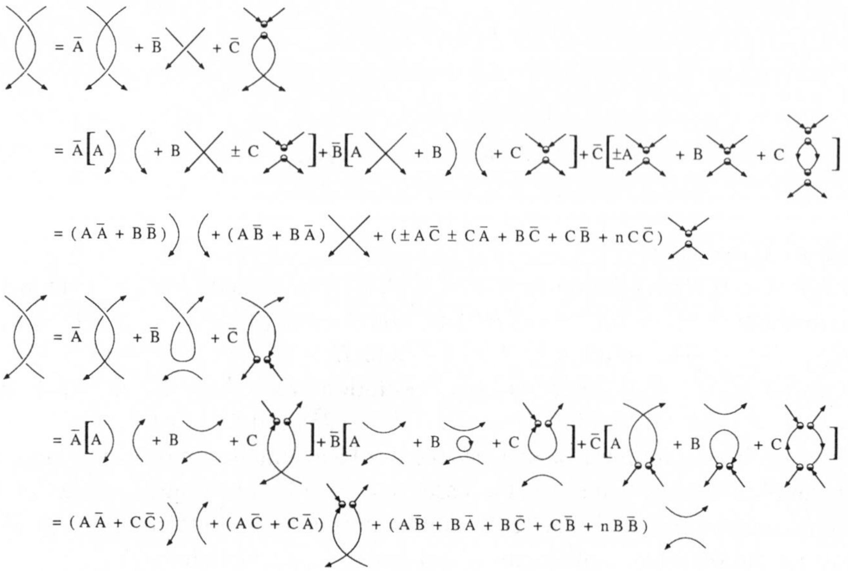

The diagrammatic computation for $\mathbf{O}(p,q)$ and $\mathbf{Sp}(n,K)$. The negative signs are for the latter case.

Figure 29

subspaces spanned by $\mathrm{id}\otimes\mathrm{id}+\frac{1}{2}(-n\pm\sqrt{n^2-4})f\otimes f^*$. *For* $n=2$ *the intersection is the subspace spanned by* P *and* $\mathrm{id}\otimes\mathrm{id}\mp f\otimes f^*$.

Proof. We proceed diagrammatically, replacing each occurrence of the symmetric tensor g in the orthogonal group calculation with the antisymmetric tensor f. Thus

$$R = AP + B\mathrm{id}\otimes\mathrm{id} + Cf\otimes f^*$$
$$\bar{R} = \bar{A}P + \bar{B}\mathrm{id}\otimes\mathrm{id} + \bar{C}f\otimes f^*$$

as shown in Figure 32. Then, using the lower signs in Figure 29, we find:

$$\mathrm{RIIa} \Longrightarrow \begin{cases} 1 = A\bar{A} + B\bar{B} \\ 0 = A\bar{B} + B\bar{A} \\ 0 = -A\bar{C} - C\bar{A} + B\bar{C} + C\bar{B} + nC\bar{C}, \end{cases}$$

$$\mathrm{RIIb} \Longrightarrow \begin{cases} 1 = A\bar{A} + C\bar{C} \\ 0 = A\bar{C} + C\bar{A} \\ 0 = A\bar{B} + B\bar{A} + B\bar{C} + C\bar{B} + nB\bar{B}. \end{cases}$$

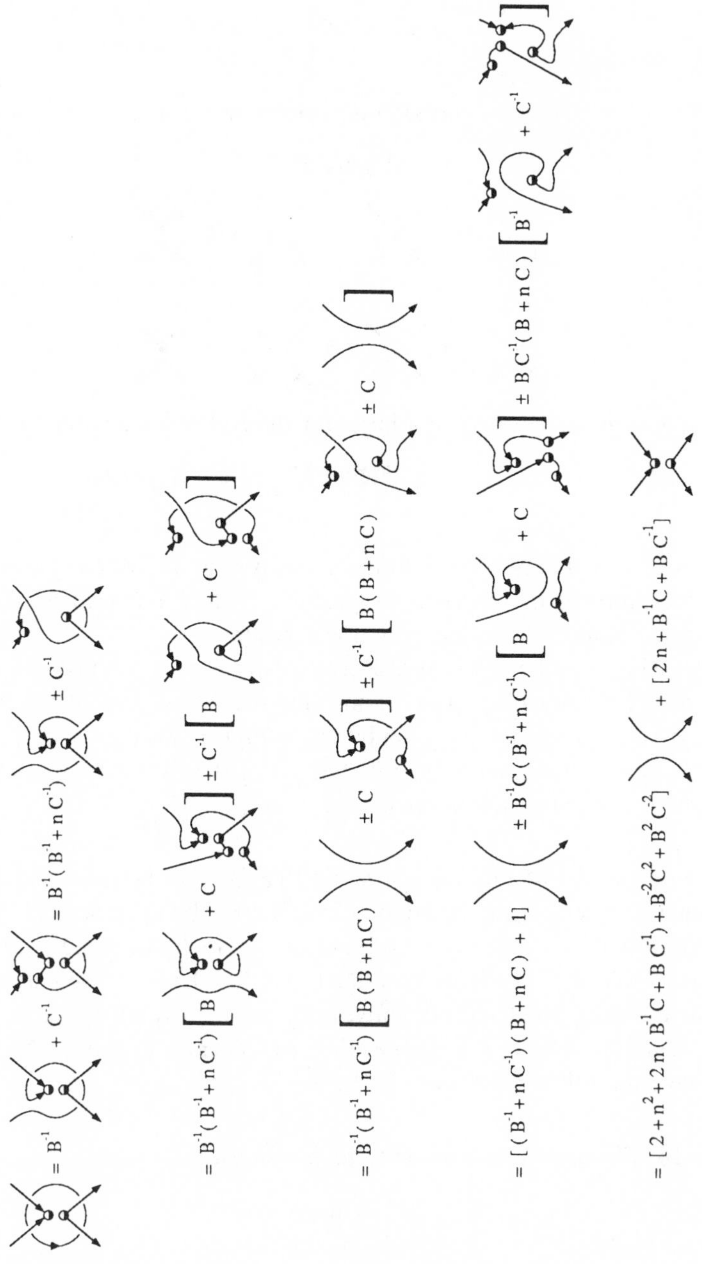

Verifying the third Reidemeister move for the $\mathbf{O}(p,q)$ and $\mathbf{Sp}(n,K)$ invariants as in Theorem 13.

Figure 30

The two partial contractions of f and f^*.

Figure 31

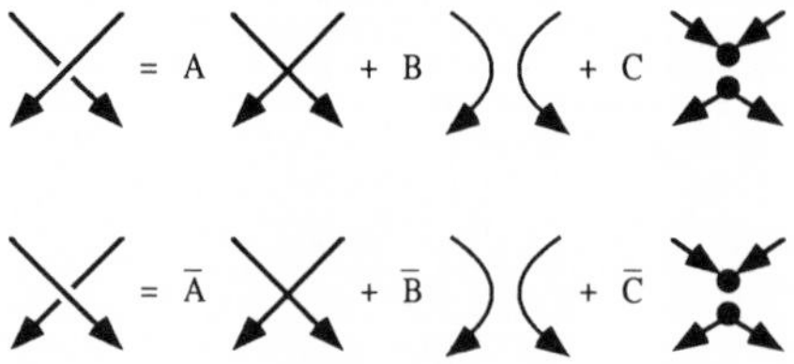

The expansion of R and $\bar{R}$ in a basis for the $\mathbf{Sp}(n, K)$ invariant subspace.

Figure 32

But this set of constraints reduces to the same system of five equations which we found in the proof of the previous proposition (with k replaced by n). Thus we find the same solutions as in the orthogonal group case.

Further, when we check the third Reidemeister move, although negative signs from the antisymmetry of f arise during intermediate stages of the calculation, as shown by the same example calculation of Figure 30, they cancel at the end, leaving the same conditions to be satisfied as in the orthogonal group case. So again we have a regular isotopy invariant. □

The computations of this section show that both orthogonal and symplectic groups generate oriented regular isotopy invariants. The solutions $(R, \bar{R})$ are very similar, differing only by the symmetry and antisymmetry of g and f, respectively. This suggests that the invariants generated may be identical; the calculation verifying that the third Reidemeister moves are satisfied reinforces this conjecture since the negative signs in the symplectic case conspire to cancel themselves. In fact, we have the following result:

Theorem 16. *The regular isotopy invariants defined by*

$$R = B\mathrm{id} \otimes \mathrm{id} + Cg \otimes g^*$$
$$\bar{R} = \bar{B}\mathrm{id} \otimes \mathrm{id} + \bar{C}g \otimes g^*$$

and

$$R = B\mathrm{id} \otimes \mathrm{id} + Cf \otimes f^*$$
$$\bar{R} = \bar{B}\mathrm{id} \otimes \mathrm{id} + \bar{C}f \otimes f^*,$$

where $0 = n + BC^{-1} + CB^{-1}$ *and* $k = n$, *are identical.*

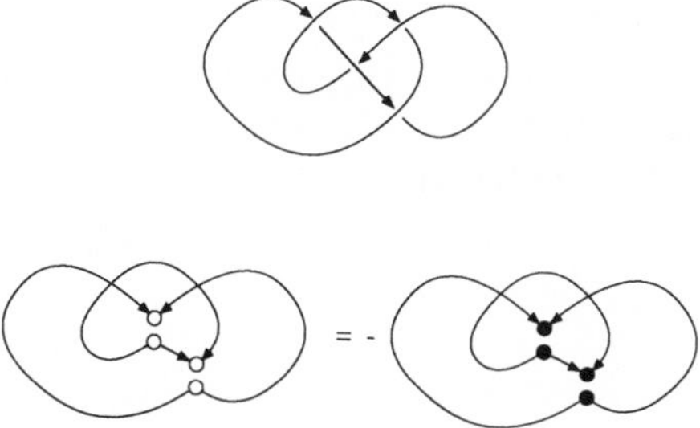

Simple closed curves evaluate with opposite sign in the orthogonal and symplectic cases if there are an odd number of $f^{ab}f_{bc}$ pairs.

Figure 33

Proof. Consider a single term in the complete state expansion of these invariants evaluated on some link diagram. It will consist of a collection of non-intersecting simple closed curves, each of which is the trace of a product of some number of gs (fs) and the same number of g^*s (f^*s). The curves are non-intersecting because there is no P in the state expansion and there are equal numbers of gs (fs) and its dual since the orientation of a curve changes at each. The term will evaluate differently (with opposite sign) in the two cases if and only if an odd number of $f^{ab}f_{bc}$ pairs occur; Figure 33 shows how this could occur if R and $\bar{R}$ had P components. But for the invariants we are comparing, this can never be the case: Consider an innermost simple closed curve in the state diagram; it can contain no such pair since each f or f^* "corner" must "point outward" as there is nothing inside the curve for them to have been created with. There are an even number of these outward pointing corners, some of which may pair off with outward pointing corners of adjacent innermost simple closed curves, leaving an even number of outward pointing corners to pair with inward pointing corners on the next most inner simple closed curve. Cancelling these inward pointing corners in pairs if they are adjacent to one another and with outward pointing corners (and an *even* number of negative signs) otherwise produces a simple closed curve with an even number of outward pointing corners. Iterating this procedure demonstrates that the two invariants evaluate this term identically, and thus are equal. □

This invariant is, moreover, related to the HOMFLY polynomial. If we multiply the defining equation for R by C^{-1}, that for $\bar{R}$ by C and subtract, we obtain the following skein relation (characteristic equation) for R:

$$C^{-1}R - C\bar{R} = \left(\frac{B}{C} - \frac{C}{B}\right) \mathrm{id} \otimes \mathrm{id},$$

which is shown diagrammatically in Figure 34. Now $0 = n + BC^{-1} + CB^{-1}$, but if we think of n as determined by B and C, we can replace the coefficient of id $\otimes$ id by a new independent variable. This is then exactly the skein relation which defines the HOMFLY polynomial [2].

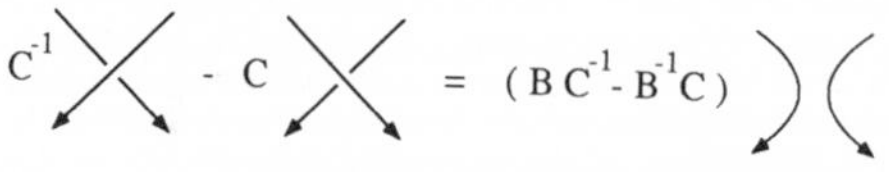

$$C^{-1} \times - C \times = (BC^{-1} - B^{-1}C) \,)(\,$$

The skein relation obtained from the $A = 0$ orthogonal/symplectic group invariant.

Figure 34

10. Conclusion

This paper began with an extensive motivation designed to show that it is very natural to define Lie group invariant Yang-Baxter models in the joint contexts of diagrammatic state models for link invariants, Penrose's diagrammatic tensor notation and some connection with Lie groups. Having defined these models, we computed the (un)oriented regular isotopy invariants of link diagrams for the classical Lie groups $\mathbf{GL}(n, K)$, $\mathbf{SL}(n, K)$, $\mathbf{O}(n, \mathbf{C})$, $\mathbf{O}(n, \mathbf{R})$ and $\mathbf{Sp}(n, K)$. The first two groups give the same invariant, one which depends only on the number of components and their linking numbers. The orthogonal group over $\mathbf{C}$ produces Kauffman's bracket polynomial for unoriented link diagrams and its (uninteresting) generalization. The last two groups generate a similar (uninteresting) invariant, but also an oriented invariant with the same skein relation as the HOMFLY polynomial.

More significant than the actual computations, however, is the result that because of their group invariance properties and invariance under the second Reidemeister move(s), these models can be normalized to give ambient isotopy invariants. Moreover, these same properties can be used to show that much of the degeneracy necessary for the constraints arising from the third Reidemeister move(s) (the Yang-Baxter equation) to have nontrivial solutions is actually present.

It is natural to try to extend these methods to produce link invariants associated with the exceptional Lie groups as well. When one tries to do so, however, there are difficulties which indicate the necessity for a broader perspective. In the case of $\mathbf{G}_2$, for example, the invariant subspace of $T^2_2(V)$ is four dimensional and one can solve the equations arising from the second Reidemeister move. But there are no interesting solutions: two are trivial and the remaining solution is simply the Kauffman bracket when $n = 7$, the dimension of the real representation of $\mathbf{G}_2$.

Only by deforming the generating relation among the basic invariants does one obtain an interesting link invariant [14].

This suggests that the group invariant approach of this paper be generalized to invariants of quantum groups. In fact, this is essentially what Kauffman has begun with his association of the quantum group $\mathbf{SL}_q(2)$ with the bracket polynomial 8. Such a generalization would identify the invariants coincident between the quantum group and classical group producing the same link invariant. Assuming such a generalization, our results illuminating the degeneracy of the third Reidemeister constraint provide a diagrammatic (partial) explanation of the existence of solutions to the Yang-Baxter equation based on deformations of Lie groups.

Acknowledgements. Although it was not until almost a year later that the details were worked out, the ideas for this approach stem from a visit to Duke University for the spring semester of 1988; I would like to thank Gerard 't Hooft and the Physics Department there for their hospitality, and especially Alfons Stahlhofen who pointed me towards Cvitanović's book. Most of this work was first presented in Michael Freedman's topology seminar at UCSD and owes much to his probing questions. Finally, it is a pleasure to thank the organizers of KNOTS 90 for producing a splendid conference.

References

[1] V.F.R. Jones, "A polynomial invariant for links via von Neumann algebras," *Bull. Amer. Math. Soc.* **12** (1985) 103–111.

[2] P. Freyd, D. Yetter, J. Hoste, W.B.R. Lickorish, K.C. Millet and A. Ocneanu, "A new polynomial invariant of knots and links," *Bull. Amer. Math. Soc.* **12** (1985) 239–246.

[3] R.D. Brandt, W.B. R. Lickorish and K. C. Millet, "A polynomial invariant for unoriented knots and links," *Invent. Math.* **84** (1986) 563–573.

[4] C.F. Ho, "A new polynomial for knots and links—preliminary report," *Abstracts Amer. Math. Soc.* **12** (1985) 300.

[5] L.H. Kauffman, "An invariant of regular isotopy," *Trans. Amer. Math. Soc.* **318** (1990) 417–471.

[6] J.H. Conway, "An enumeration of knots and links," in J. Leech, ed., *Computational Problems in Abstract Algebra* (New York: Pergamon 1969) 329–358.

[7] L.H. Kauffman, "State models and the Jones polynomial," *Topology* **26** (1987) 395–407.

[8] L.H. Kauffman, "Knots, abstract tensors and the Yang-Baxter equation," in L. Lussana, ed., *Knots, Topology and Quantum Field Theories, Proceedings of the Johns Hopkins Workshop on Current Problems in Particle Theory* **13** (Singapore: World Scientific 1990) 179–334.

[9] Ge M.-L., Wang L.-Y., Xue K. and Wu Y.-S., "Akutsu-Wadati link polynomials from Feynman-Kauffman diagrams," *Int. J. Mod. Phys.* **4** (1989) 3351–3373.

[10] Ge M.-L., Li Y.-Q. and Xue K., "Extended state model and group approach to new polynomials," *J. Phys. A.: Math. Gen.* **23** (1990) 619–639.
[11] Wadati M., Deguchi T. and Akutsu Y., "Exactly solvable models and knot theory," *Phys. Rep.* **180** (1989) 247–332.
[12] V.F.R. Jones, "On knot invariants related to some statistical mechanical models," *Pacific J. Math.* **137** (1989) 311–334.
[13] V.G. Turaev, "The Yang-Baxter equation and invariants of links," *Invent. math.* **92** (1988) 527–553.
[14] N.Yu. Reshetikhin, "Quantized universal enveloping algebras, the Yang-Baxter equation and invariants of links. I and II," LOMI preprints E-4-87 and E-17-87 (1988).
[15] R.J. Baxter, *Exactly Solvable Models in Statistical Mechanics* (New York: Academic Press 1982).
[16] E. Witten, "Quantum field theory and the Jones polynomial," *Comm. Math. Phys.* **121** (1989) 351–399.
[17] L.H. Kauffman, "Spin networks and knot polynomials," *Int. J. Mod. Phys. A* **5** (1990) 93–115.
[18] G. Burde and H. Zieschang, *Knots* (Berlin: W. de Gruyter 1985).
[19] L.H. Kauffman, *On Knots*, Annals of Mathematics Studies **115** (Princeton: Princeton University Press 1987).
[20] W.B.R. Lickorish, "Polynomials for links," *Bull. Lond. Math. Soc.* **20** (1988) 558–588.
[21] K. Reidemeister, *Knotentheorie* (New York: Chelsea 1948).
[22] R. Penrose, "Applications of negative dimensional tensors," in P. J. A. Welsh, ed., *Combinatorial Mathematics and its Applications* (New York: Academic Press 1971) 221–244.
[23] P. Cvitanović, "Group theory for Feynman diagrams in non-abelian gauge theories," *Phys. Rev. D* **14** (1976) 1536–1553.
[24] R. Penrose and W. Rindler, *Spinors and Space-Time, Vol. 1: Two-spinor calculus and relativistic fields* (Cambridge: Cambridge University Press 1984).
[25] P. Cvitanović, *Group Theory* (Copenhagen: NORDITA 1984).
[26] R. Brauer, "On algebras which are connected with the semisimple continuous groups," *Ann. Math.* **38** (1937) 857–872.
[27] A. C. Aitken, *Determinants and Matrices*, sixth edition (Edinburgh: Oliver & Boyd 1949).
[28] H. Weyl, *The Classical Groups: Their invariants and representations* (Princeton: Princeton University Press 1939).
[29] D. Hilbert, "Über die Theorie der algebraischen Formen," *Math. Ann.* **36** (1890) 473–534.
[30] D. Hilbert, "Über die vollen Invariantensysteme," *Math. Ann.* **42** (1893) 313–373.
[31] Yamada S., "An operator on regular isotopy invariants of link diagrams," *Topology* **28** (1989) 369–377.
[32] É. Cartan, *Oeuvres Completes* (Paris: Gauthier-Villars 1952).

Geometric structures and monodromy representations

*Yoav Moriah**

§0. Introduction

Every closed orientable 3-manifold can be obtained as an n-fold, $n \in \mathbb{N}$, branched cover of S^3 branched over any universal knot. In this paper we study the connections between the monodromy map of the cover $p : M \to S^3$ and geometric structures on M. The main result of this paper is:

Theorem 1.8. *Let K be a universal knot which is not a satellite knot. Let $p : M \to S^3$ be a branched cover of S^3 branched over K where M is some closed orientable 3-manifold. Then:*

(a) *If all the branching indices are equal to $n > 3$, M is a hyperbolic manifold.*
(b) *If all branching indices are larger than 3, M admits a smooth metric of strictly negative curvature.*
(c) *If K is also a 2-bridge knot and the branching indicies are smaller than 3 the manifold M is a lens space.*

We assume, throughout the paper, that the reader is familiar with chapters 3 and 4 of [**Th**]. For a reference on geometric structures see [**Sc**].

Denote by $K_1, \ldots, K_m$ the components of $p^{-1}(K)$ and set $M^* = M - \cup K_i$. The projection map $p|_{M^*} : M^* \to S^3 - K$ is a true covering map. With each covering map one can associate a monodromy map defined as follows:

Definition 0.1. Let $p : X \to Y$ be a cover. Let x_0 be a base point in Y for $\pi_1(Y)$, then $\pi_1(Y)$ acts on the fiber, $p^{-1}(x_0) = \{\widetilde{x_0}, \ldots, \widetilde{x_{d-1}}\}$, by permutations: For

* Partly supported by grant 100 - 0765 of the Technion V.R.P fund, Loewengart research fund.

Knots 90

each $\alpha \in \pi_1(Y)$ we send $\widetilde{x}_i$ to the end point of a lift of α with initial point $\widetilde{x}_i$. This process determines a homomorphism $\varphi : \pi_1(Y) \to S_d$, where S_d denotes the group of permutations on d elements $\{1, \ldots, d\} \in \mathbb{N}$. This homomorphism is called the monodromy map of the cover.

Definition 0.2. Let $p : M \to S^3$ be a branched cover of S^3 branched over K. Denote by $K_1, \ldots, K_m$ the components of $p^{-1}(K)$. The branching indices of the cover are the integers $n_1, \ldots, n_m$ which describe the degree of the covering map when restricted to a small disk transverse to K_i, $i = 1, \ldots, m$.

The branching indices are precisely the lengths of the cycles of $\varphi(\mu)$ where $\mu \in \pi_1(S^3 - K)$ is the class of a meridian of K. As the longitude of K can be covered non-trivially these cycles might be repeated. In any case the sum of the branching indices is the degree of the cover, see [**Ro**].

The manifold M^* is determined by the monodromy and will be connected exactly when the image of the monodromy is transitive on the fiber $p^{-1}(x_0)$. This gives a one to one correspondence between degree d connected branched coverings of S^3 and conjugacy classes of transitive representations of $\pi_1(S^3 - K)$ in S_d. Note that $\pi_1(M - \cup K_i) \approx \varphi^{-1}(\text{Stab}(1))$.

The monodromy map determines the covering manifold M completely as there is a unique way of gluing in the solid tori to the boundary components of $M - N(\cup K_i)$ so as to obtain M, where $N(\,)$ denotes a regular neighbourhood of $(\,)$. In particular the geometric structure of M, if it admits one, is encoded in the monodromy map.

If K is a hyperbolic universal knot then every closed orientable 3-manifold M can be obtained by surgery on some complete hyperbolic manifold M^*. The manifolds M^*, being d-fold true covers of $S^3 - K$ admit a hyperbolic structure which lifts from that of $S^3 - K$. They can be triangulated by hyperbolic tetrahedra which lift from the triangulation by q hyperbolic tetrahedra of $S^3 - K$. This triangulation determines a triangulation of the boundary components of $M - N(\cup K_i)$ and from this triangulation of $\partial(M - N(\cup K_i))$ one can obtain a system of $qd - m$ equations as in [**Th**]. In order to obtain the closed manifold M one needs to add solid tori to the boundary components. This Dehn surgery operation determines additional m equations. One gets the manifolds M by doing a particular Dehn surgery. If the equations have solutions z_i with im $z_i > 0$ then M has a complete hyperbolic structure. Otherwise M might still have a hyperbolic structure coming from a different triangulation, or not be a hyperbolic manifold at all. We call these equations the " hyperbolic Dehn surgery equations " for M. To be precise: If $M = M(r_1, \ldots, r_m)$ is a closed 3-manifold obtained by Dehn surgery on a manifolfd M^* which is the interior of a compact manifold with m tori boundary components T_i, then M will admit a complete hyperbolic structure if there is a solution, as above, to the following equations:

(1) The cone angle around each edge in the triangulation of M is 2π.
(2) If α_i, β_i are a system of generators for $H_1(T_i)$, $a_i, b_i \in \mathbb{Z}$, g.c.d.$(a_i, b_i) = 1$, $r_i = a_i/b_i$, $i = 1, \ldots, m$, and H$'(\,)$ is the lift of the derivative of the holonomy map, then:

$$a_i\mathrm{H}'(\alpha_i) + b_i\mathrm{H}'(\beta_i) = 2\pi \text{ rotation}\ , \quad \text{see[\textbf{Th}]}.$$

Remark 0.3. When M is a branched cover of S^3, then α_i is an n_i-fold cover of the meridian of $S^3 - N(K)$, $a_i = 1$ and $b_i = 0$.

In §2, Theorem 2.3, we present an algorithm for computing hyperbolic Dehn surgery equations for M from the monodromy map of the cover in the case where K is the figure eight knot. The algorithm can be modified to work for any universal 2-bridge knot. This algorithm gives access to the problem of understanding the possible geometric structures on 3-manifolds, in particular, of irregular branched covers, via homomorphisms from a fixed group, i.e. $\pi_1(S^3 - K)$ into permutation groups. The algorithm is implemented on a computer program. Using the computer program we give, in §3, the surgery equations for some manifolds contained in a list of monodromy representations obtained by John Hempel in [**He**]. This list contains monodromy representations of $\pi_1(S^3 - K) \to S_d$ for $1 \leq d \leq 10$.

Acknowledgements. I would like to thank John Hempel for making his preprint available to me, Joel Hass for his help with the proof of Proposition 1.4, and Galit Mizrahi for programming the algorithm.

§1. Branching indices

In this section we will show that there is a strong connection between the branching indices of the cover and the geometric nature of the manifold M. We first need some definitions.

Definition 1.1. A hyperbolic cone n-manifold is the metric space obtained by identifying isometrically a disjoint union of geodesic n-simplices in $\mathbb{H}^n$ along faces of dimension $n - 1$ in such a way that the space obtained is a topological manifold.

A hyperbolic cone manifold has a smooth metric of constant curvature $K = -1$ on the complement of the codimension two skeleton. An orthogonal cross section of a neighbourhood of each codimension two face is a 2-dimensional cone.

Definition 1.2. The cone angle in a hyperbolic cone manifold is the angle around the codimension two face. The cone locus of the cone manifold is the closure of the codimension two faces for which the cone angle is not 2π.

Remark. A neighbourhood of the face is obtained by gluing together simplices around the face in some combinatorial manner. The sum of their dihedral angles along the face is the cone angle. The hyperbolic metric can be extended across the codimension two faces whenever the cone angle is 2π.

Proposition 1.3. *Let K be a universal knot which is not a satellite knot. If $p : M \to S^3$ is a branched cover of S^3 branched over the knot K with branching indices all equal to $n \geq 4$, then M is a hyperbolic manifold.*

Proof. Any knot K satisfying the conditions is hyperbolic. By Theorem 4.7 of Thurston, $(n, 0)$ surgery, $n \geq 5$, $n = \text{g.c.d.}(n, 0)$, on the figure eight knot K gives a space N with a singular hyperbolic structure with cone angles equal to $2\pi/n$ along K, see [**Th**]. Hodgson [**Ho**] has improved this result to any hyperbolic knot K and to $n = 4$. This structure lifts to any branched cover of N with equal branching index $n \geq 4$ along each component K_i of $p^{-1}(K)$ by lifting the triangulation. Note that the argument is independent of whether the cover is regular or not. The cone angle around each K_i is $n \cdot 2\pi/n = 2\pi$. Thus we have an empty cone locus, and the manifolds M admit a non-singular hyperbolic structure.

If we give up the condition that the indices are all equal we still have:

Proposition 1.4. *Let K be a universal knot which is not a satellite knot. If $p : M \to S^3$ is a branched cover of S^3 branched over K with branching indices all greater or equal to 4, then M has a smooth metric of strictly negative curvature.*

Proof. Let N denote the 4-fold cyclic branched cover of S^3 branched over K. By [**Ho**] N is a hyperbolic manifold. Hence, along the preimage $p^{-1}(K)$, N has an angle of 2π. This implies that $(S^3, K) = p(N)$ has a hyperbolic cone manifold structure with cone angle $\pi/2$ along the cone locus K. Let M be any branched cover of S^3 branched along K with branching indices greater or equal to 4. We can lift the structure on (S^3, K) to M by lifting the triangulation. If the angle along K is θ then the angle along a preimage K_i with branching index n_i is $n_i\theta$. Hence we obtain a hyperbolic cone manifold structure on M with cone angles all greater than 2π. Consider a metrically regular neighbourhood of a component of the singular set. It is a solid torus admitting a hyperbolic metric singular on the geodesic core of the torus. In cylindrical coordinates this metric can be written as:

$$ds^2 = dr^2 + (n_i\theta/_{2\pi}\sinh(r))^2 d\theta^2 + c(r)^2 dt^2,$$

where r is the radius t is a parameter of the S^1 factor and $c(r) = (e^r + e^{-r})/2$. By Gromov and Thurston's Lemma 2.1, see [**GT**], we can replace this metric with a metric

$$ds'^2 = dr^2 + \beta(r)^2 d\theta^2 + c(r)^2 dt^2$$

which agrees with ds^2 outside a regular neighbourhood of the core and hence has constant curvature $K = -1$ there. Furthermore, by Lemma 2.4 of [**GT**], the metric ds'^2 has strictly negative curvature on that regular neighbourhood of the core. If we do this process for each component of the singular set the result follows.

The above two propositions suggest that when all branching indices are large the manifold possesses a hyperbolic structure. The next proposition will deal with the case when all indices are very small. Before we go on we need to describe a method for constructing Heegaard splittings for the covering manifolds M. We will describe it in detail as we will need it later on. Throughout the paper we will consider the following presentation of the figure eight knot group which we use as an example: $\pi_1(S^3 - K) =< x_1, x_3 | x_1^{-1} x_3^{-1} x_1 x_3 x_1^{-1} x_3 x_1 x_3^{-1} x_1^{-1} x_3 >$, see Fig. 1.

Given a d-fold branched cover $p : M \to S^3$ branched over a 2-bridge knot K with a monodromy map $\varphi : \pi_1(S^3 - K) \to S_d$ let $c(\tau)$ denote the number of cycles in a $\tau \in S_d$. There is an equatorial S^2 embedded in S^3 separating the pair (S^3, K) into two rational tangles . The 3-balls V^+, V^- meet K in the unknotted arcs B^+, D^+ and A^-, C^-, resp. There are arcs A, B and C, D in S^2 so that $B^+ \cup B, D^+ \cup D$, and $A^- \cup A, C^- \cup C$ bound disks D_1, D_2 and D_3, D_4 in V^+, V^-, resp. Let $H^+ = p^{-1}(V^+)$ and $H^- = p^{-1}(V^-)$, see Fig. 1.

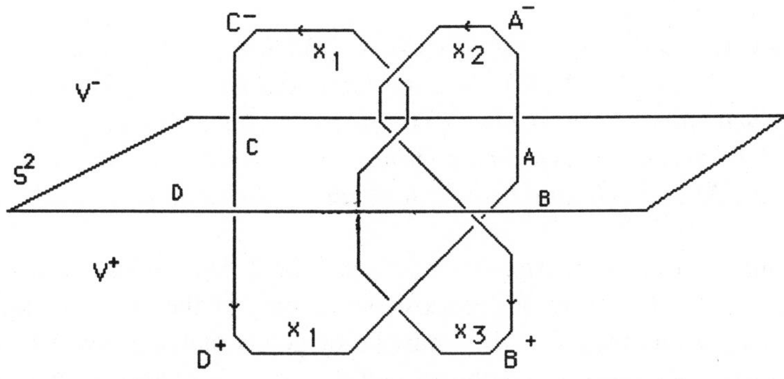

Figure 1

Now split the 3-ball V^- along the disks D_3, D_4 to obtain a 3-ball B^- as indicated in Fig. 2 below. The 3-ball B^- lifts homeomorphically to M. Hence H^- is the union of d copies of B^- denoted by $B_1^-, \ldots, B_d^-$. The copies of B_i^- will be glued along the lift of the oriented A^- with a D_i^+ glued to a D_i^- according to the cycles of $\varphi(x_2)$, similarly they will be glued along the lift of an oriented

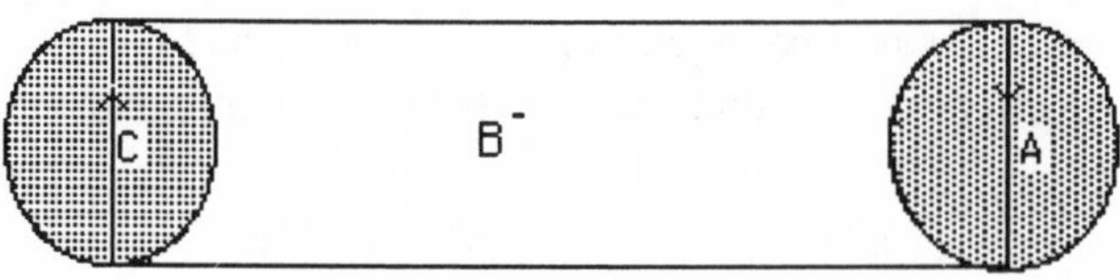

Figure 2

C^- according to the cycles of $\varphi(x_1)$. To be precise, a small loop around a lift of A^- with linking number $+1$ will encounter B_i^- according to a cycle $(i_1, \ldots, i_r)$ of $\varphi(x_2)$ and there will be a lift of A^- for each cycle. Similarly for C^- and $\varphi(x_1)$. It is clear that H^- is a handlebody. We repeat the process to obtain a handlebody H^+ from $B_1^+, \ldots, B_d^+$, the lifts of B^+, and $\varphi(x_1), \varphi(x_3)$. As the generators are all conjugate their images $\varphi(x_i)$ have the same number of cycles. The handlebodies H^-, H^+ retract to graphs. The number of vertices in each graph is $2c(\varphi(x_1))$ and the number of edges is d. It follows that the Euler characteristic of the graph satisfies

$$\chi(G) = 2c(\varphi(x_1)) - d = 1 - \text{genus}(H^-) = 1 - \text{genus}(H^+).$$

We sum up:

Theorem 1.5. *The pair $\{H^-, H^+\}$ is a Heegaard splitting of genus $d + 1 - 2c(\varphi(x_1))$ for M.*

Proposition 1.6. *If $\varphi(x_i)$ has $d/2$ cycles then M is a lens space.*

Proposition 1.7. *Let $p : M \to S^3$ be a d-fold branched cover of S^3 over a 2-bridge knot K, let $\mu \in \pi_1(S^3 - N(K))$ be a meridian and let $\varphi : \pi_1(S^3 - K) \to S_d$ be the monodromy map of the cover. If $\varphi(\mu)$ has no cycles of length ≥ 3 then M is a lens space. Furthermore such a monodromy can have at most one cycle of length one and then M is S^3. In any case M has spherical structure.*

Proof. If the indices of $p : M \to S^3$ are just 1 or 2 then around each lift of the arcs A^-, B^+, C^-, D^+ there are at most two copies of the B_i's. If there are no cycles of length one then H^-, H^+ must be of genus one and hence M is a lens space. If there are cycles of length one and the cover is connected then H^-, H^+ must be 3-balls and M is S^3.

Summing up we have:

Theorem 1.8. *Let K be a universal knot which is not a satellite knot. Let $p : M \to S^3$ be a branched cover of S^3 branched over K where M is some closed orientable 3-manifold. Then:*

(a) *If all the branching indices are equal to $n > 3$, M is a hyperbolic manifold.*
(b) *If all branching indices are larger than 3, M admits a smooth metric of strictly negative curvature.*
(c) *If K is also a 2-bridge knot and the branching indicies are smaller than 3 the manifold M is a lens space.*

Remark. All 2-bridge knots which are not torus knots are universal [**HLM1**]. The condition that the knot is not a satellite is necessary as there are universal satellite knots, see [**HLM2**].

§2. Heegaard splittings and hyperbolic tetrahedra

Given a branched d-fold covering $p : M \to S^3$ consider the Heegaard splitting $\{H^+, H^-\}$ for M as described in §1. Remove a regular neighbourhood of the branch set to obtain $M - N(\cup K_i)$. The space $M - N(\cup K_i)$ has a triangulation, induced by the lift of the complete hyperbolic structure of $S^3 - K$, by ideal hyperbolic tetrahedra Δ_i, $i = 1, \ldots, qd$, with the link of the vertices removed. This triangulation induces a triangulation of $\partial(M - N(\cup K_i))$ determined by the intersection of the 2-dimensional faces of Δ_i, $i = 1, \ldots, qd$, with $\partial(N - N(K))$. It has the property that the dihedral angles of Δ_i are the angles of the triangles on $\partial(N - N(K))$ as the tori boundary components lift to horospheres, see [**Th**]. Recall that each ideal hyperbolic tetrahedron $\Delta^3 \subset H^3$ is determined by a complex parameter $z \in \mathbb{C}$, with im $z > 0$. Each edge in Δ^3 can be labelled by a complex number determining the dihedral angle along the edge. Opposite edges will be labelled the same. The labels z^1, z^2, z^3 of Δ^3, when chosen in clockwise order, satisfy:

$$z^1 = z, \quad z^2 = (z-1)/z, \quad z^3 = 1/1 - z$$

Hence to determine the hyperbolic Dehn surgery equations for M we need to determine the labels $z_u^t, t \in \{1, 2, 3\}, u \in \{1, \ldots, qd\}$, of the qd tetrahedra around each vertex of $\partial(M - N(\cup K_i))$. The problem is how to decide which triangle on the boundary $\partial(M - N(\cup K_i))$ belongs to which tetrahedron. The idea is to use the monodromy map to determine the triangulation from the Heegaard splitting $\{H^+, H^-\}$.

Let K be the figure eight knot. The procedure of cutting $S^3 - N(K)$ along an equatorial S^2, as in § 1, cuts the boundary $\partial(S^3 - N(K))$ into four annuli $A_1^+, A_3^+, A_1^-, A_2^-$ (see Fig. 4). These annuli are "marked" by the triangulation of $S^3 - N(K)$. They are covered by "marked" annuli in the boundary of $M - N(\cup K_i)$ and induce a triangulation there as follows: Each of the two tetrahedra, in the triangulation of $S^3 - N(K)$ described in Fig. 3 below, contributes four boundary triangles denoted by $\mathcal{T}_{i,b}$, $i \in \{1,2\}, b \in \{1,\ldots,4\}$, in $\partial(S^3 - N(K))$. Cutting $S^3 - N(K)$ by an equatorial S^2 cuts each of the triangles $\mathcal{T}_{i,b}$ on $\partial(S^3 - N(K))$ into two parts (see Fig. 3). It also cuts $S^3 - N(K)$ into two genus two handlebodies with two boundary annuli in each handlebody (V^+, A_1^+, A_3^+) and (V^-, A_1^-, A_2^-) (see Fig. 4). When we cut the handlebodies along the disks D_i, $i = 1,\ldots,4$ the triangles $\mathcal{T}_{i,b}$ are cut again. Altogether each triangle is cut into three parts called **regions**, one "square" region and two "triangle" regions. Each annulus $A_i^{\pm}$, $i \in \{1,2,3\}$ inherits a pattern of six regions from the triangulation as described in Fig. 5 below.

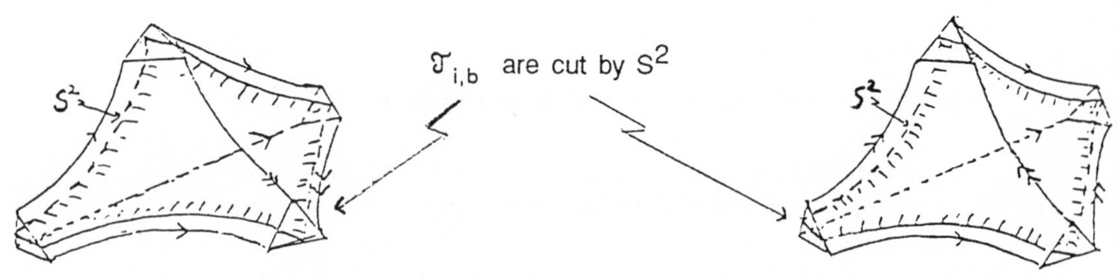

The intersection of Δ_1, Δ_2 with the equatorial S^2.

Figure 3

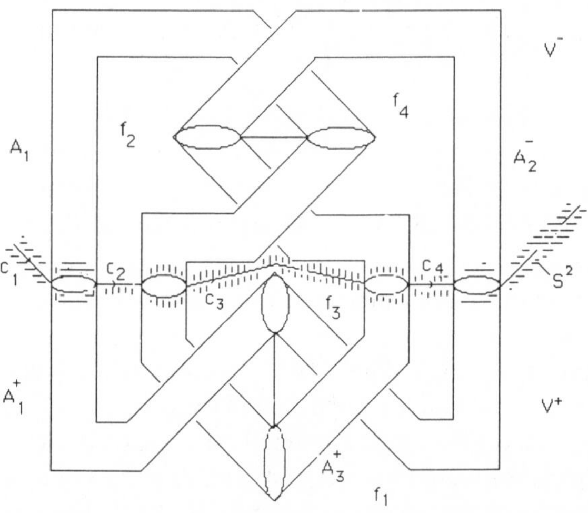

Figure 4

The triangulated space $S^3 - N(K)$ can be reconstructed from (V^+, A_1^+, A_3^+) and (V^-, A_1^-, A_2^-) by gluing the handlebodies along their boundaries matching the oriented intersection curves c_1, c_2, c_3, c_4 of the paired faces of Δ_1, Δ_2 and S^2, thus creating Δ_1, Δ_2 again. The faces of Δ_1, Δ_2, which are labelled in Fig. 4 by f_i, $i = 1, \ldots, 4$, (see [**We**]), are straightened out to look as the surfaces f_i, $i = 1, \ldots, 4$, in Fig. 6. Hence lifting (V^+, A_1^+, A_3^+) and (V^-, A_1^-, A_2^-) with the region structure to the handlebodies $\{H^+, H^-\}$ induces a region structure there. I.e.,the intersection of the 2-dimensional faces of Δ_u, $u \in \{1, \ldots, 2d\}$, and $\{H^+, H^-\}$. Gluing the handlebodies together according to the monodromy creates the triangulation of $M - N(\cup K_i)$ and in particular of $\partial(M - N(K))$. Hence the task is to reconstruct the triangles $\mathcal{T}_{u,b}$, $u \in \{1, \ldots, 2d\}, b \in \{1, \ldots, 4\}$, from the regions.

Figure 5

Consider the handlebody H^+. The combinatorial structure of H^+ is determined by $\varphi(x_1)$ and $\varphi(x_3)$. Let $c_{i,j}$, $i = 1, \ldots, c(\varphi(x_1))$ denote the ith cycle of the $\varphi(x_j)$, $j \in \{1, 3\}$, permutation. As $\varphi(x_1)$ and $\varphi(x_3)$ are conjugate we can assume that the length $n_{i,1}$ of $c_{i,1}$ is equal to the length $n_{i,3}$ of $c_{i,3}$. For each cycle $c_{i,j}$ of $\varphi(x_1)$ and $\varphi(x_3)$ there is an annulus component of $p^{-1}(A_1^+)$ and $p^{-1}(A_3^+)$, resp. in H^+ which is an n_i cover of A_1^+, A_3^+, resp. Denote these annuli by $A^+(c_{i,j})$, $j \in \{1, 3\}$. Such an annulus is represented by a strip of n_i squares, of six regions each, as in Fig. 5, with the top and bottom of the strip identified. Similarly for H^-, $\varphi(x_1), \varphi(x_2)$ and the cycles $c_{i,\ell}$, $\ell \in \{1, 2\}$.

Label each center point of each square in each annulus by an integer $k \in \{1, \ldots, d\}$ in such a way that as we follow the annulus in the direction determined by α_i, a loop covering the the meridian, we will encounter the squares in the order of the corresponding cycle of $\varphi(x_1), \varphi(x_3)$, resp. Each region of each square belongs to some triangle $\mathcal{T}_{u,b}$ in some tetrahedron z_u in the induced triangulation of $M - N(\cup K_i)$. As can be seen from Figures 5 and 6 the intersection of each region with the edges of the tetrahedron it belongs to contains exactly one point, therefore it can be labelled by a complex number z_u^t, $t \in \{1, 2, 3\}, u \in \{1, \ldots, 2d\}$, corresponding to the edge parameter of the uth tetrahedron. Hence it is sufficient to determine the correct labelling of the regions in order to label the triangles

$\mathcal{T}_{u,b}$. A careful consideration of the triangulation of $S^3 - N(K)$ by hyperbolic tetrahedra, as illustrated in Figures 3 and 6, shows the following two facts:

(*) Each center point of $A_1^+, A_3^+, A_1^-, A_2^-$ is an end point of one of the two edges of the triangulation.

(**) When we compare the labels of the regions around opposite ends of an edge we see that they are switched as if by a reflection of A_1^+, A_3^+ through a diagonal of slope -1 and A_1^-, A_2^- through a diagonal of slope $+1$.

It follows that in the covering space $M - N(\cup K_i)$:

(a) The center point of each of the n_i squares of $A^+(c_{i,j})$, $j \in \{1,3\}$, $A^-(c_{i,\ell})$, $\ell \in \{1,2\}$, $i \in \{1,\dots,c(\varphi(x_1))\}$ corresponds to an endpoint of an edge in the triangulation of M.

(b) The tetrahedra along each edge are fixed and will appear at two different squares corresponding to the two end points of the edge.

(c) The squares in $A^+(c_{i,j}), j \in \{1,3\}$, can be paired off so that the first of each pair will appear in the strip corresponding to $\varphi(x_1)$ and the second to $\varphi(x_3)$. Both squares will be labelled by the same integer k.

(d) The labels of the regions in two squares corresponding to the same integer $k \in \{1,\dots,d\}$ (i.e. the same edge) in two different annuli $A^+(c_{i,1})$ and $A^+(c_{i,3})$ will be reflected as in (**), see Fig. 7. Similarly with $H^-, A^-(c_{i,\ell})$, $\ell \in \{1,2\}$, and the permutations $\varphi(x_1)$ and $\varphi(x_2)$.

The next two technical lemmas tell us that in fact we can label the regions to create a labelling of the triangles. They are illustrated in Example 2.4 below. The reader should consult the example while reading the proofs.

Lemma 2.1. *The labelling on the annuli $A^+(c_{i,1}), A^+(c_{i,3})$, $i \in \{1,\dots,c(\varphi(x_1))\}$ will determine the labelling on all annuli $A^-(c_{i,1})$, $A^-(c_{i,2})$ and vice versa.*

Proof. When H^+ is glued to H^- to construct $M - N(\cup K_i)$, each annulus $A^+(c_{i,j})$ in H^+, $j \in \{1,3\}, i \in \{1,\dots,c(\varphi(x_1))\}$ is identified along each boundary component to some annulus $A^-(c_{i,\ell})$, $\ell \in \{1,2\}$ in H^-. The labelling of the regions in $A^+(c_{i,j})$ must coincide with the labelling of the regions on the two neighbouring annuli of H^- in such a way that any three regions which are identified along their boundary to form a triangle $\mathcal{T}_{u,b}$, $u \in \{1,\dots,2d\}$, $b \in \{1,\dots,4\}$ belonging to the u-th tetrahedron, must have parameters z_u^1, z_u^2 and z_u^3, up to a cyclic permutation. Each triangle $\mathcal{T}_{u,b}$, is composed of regions coming from exactly two annuli, $A^+(c_{i,j})$, $A^-(c_{i,\ell})$. Hence, determining the labelling on either $A^+(c_{i,j})$, $j \in \{1,3\}$, or $A^-(c_{i,\ell})$, $\ell \in \{1,2\}$, and $i \in 1,\dots,c(\varphi(x_1))$ will determine the labelling completely.

Note. Each square is divided into a right and left side by a vertical line coming from the intersection of $\partial(M - N(\cup K_i))$ with the 2-dimensional faces of the

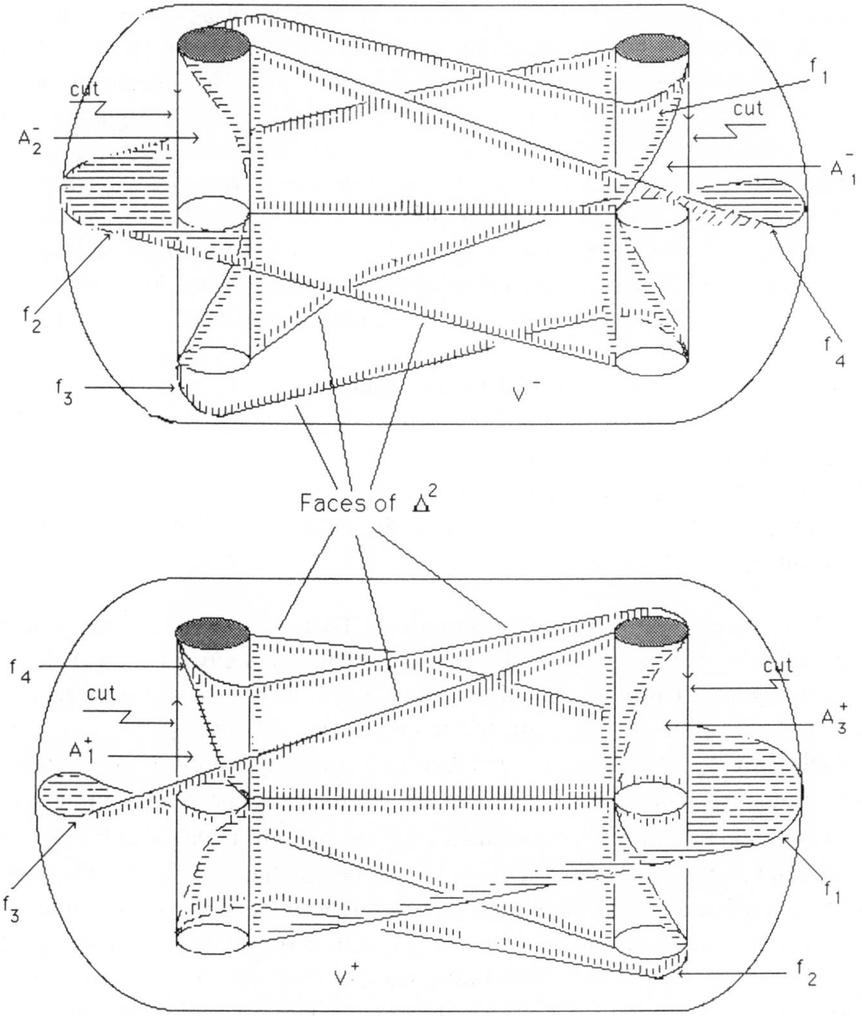

Figure 6

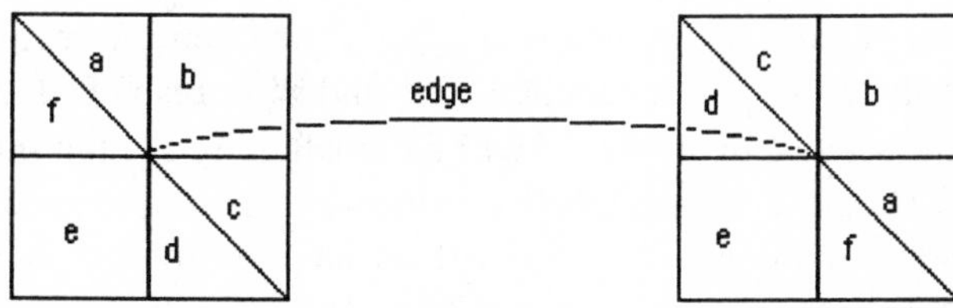

Figure 7

tetrahedra, (see Fig. 7). Hence, regions can be thought of as either right or left sided. (Regions on opposite sides of the vertical line in a square represent different triangles.) In order to refer to the different regions in each square they will be numbered counterclockwise 1 to 6 starting from the bottom on the right hand side.

Lemma 2.2. *Given a fixed $j \in \{1,3\}$, suppose that two regions, on the same side of the vertical line, in any of the annuli $A^+(c_{i,j}) \subset H^+$, are labelled by $z^t_{u_0}, t \in \{1,2,3\}$, for some fixed $u_0 \in \{1,\dots,2d\}$. Then they must be numbered either 1 and 3 or 4 and 6. Furthermore they will belong to the same triangle $\mathcal{T}_{u_0,b} \subset \Delta_{u_0}$, $b \in \{1,\dots,4\}$. Similarly for regions in annuli $A^-(c_{i,\ell}) \subset H^-$.*

Proof. Let R be a region labelled by $z^t_{u_0}$ contained in $\mathcal{T}_{u_0,b}$. Any other region labelled $z^t_{u_0}$ belongs to a triangle $\mathcal{T}_{u_0,b'}$ which shares an edge of Δ_{u_0} with $\mathcal{T}_{u_0,b}$. We shall check all such possibilities. Without loss of generality we can assume that R is located in $A^+(c_{1,3})$. Then R is one of six regions in some square labelled by an integer $k, 1 \leq k \leq d$. Throughout the proof we will write the lables as $z^{t\pm j}_{u_0}$ meaning $t \pm j$ mod 3.

Case 1. Assume that R is region number 1. Then the region number 3 in the square labelled $\varphi^{-1}(x_3)$ will be labelled $z^{t-1}_{u_0}$. However the two regions belong to the same triangle. This is in fact all that can happen. If the integer k appears in the i_1 cycle of $\varphi(x_1)$ then region number 5 in the square with label k in $A^+(c_{i_1,1})$ will be labelled by $z^t_{u_0}$. Hence region number 1 in some square of $A^-(c_{i_1,1})$ will be labelled by $z^{t-1}_{u_0}$. This implies that if that same integer appears in the i_2 cycle of $\varphi(x_2)$ then region number 1 of a square in $A^-(c_{i_2,2})$ will be labelled by $z^{t-1}_{u_0}$ and region number 5 of a square of $A^+(c_{i_3,3})$ will be labelled by $z^t_{u_0}$, but this region is on the left hand side. Continuing, this will determine some region numbered 1 in the corresponding square of some annulus, say, $A^+(c_{i_4,1})$. However this process cannot continue more than three steps, as each time we pass from annuli in $A^+(c_{i,3})$ to annuli in $A^+(c_{i,1})$, $i = 1,\dots,c(\varphi(x_1))$ we follow an edge in the z_{u_0} tetrahedron and one can follow at most three different edges in a tetrahedron without coming back to the starting point vertex.

Case 2. Assume R is region number 2 in $A^+(c_{1,3})$ then a region number 4 in some square in some annulus $A^-(c_{1,1})$ is labelled by $z^{t+1}_{u_0}$. This in turn implies that a region numbered 4 in an annulus $A^-(c_{i_1,2})$ and hence a region numbered 2 in some square in $A^+(c_{i_2,1})$ are labelled $z^{t+1}_{u_0}$ and $z^t_{u_0}$, resp. This will imply that some region numbered 4 in $A^+(c_{i_2,3})$ will be labelled by $z^t_{u_0}$ but this region is on the left side. If R is in a square labelled k then a region numbered 4 in a square labelled k in some annulus $A^+(c_{i_3,1})$ will be labelled by $z^t_{u_0}$. It follows that a region numbered 6 in an annulus $A^+(c_{i_4,1})$ will be labelled by $z^{t+1}_{u_0}$ and hence a region numbered 6 in some square in an annulus $A^+(c_{i_4,3})$ will be labelled by

$z_{u_0}^{t-1}$ which as before implies that a region numbered 4 in an annulus $A^+(c_{i_5,3})$ will be labelled by $z_{u_0}^t$, but again this region is on the left side. As before this process must end after at most three steps.

Case 3. Assume R is region number 3 in square labelled k of $A^+(c_{1,3})$. A region numbered 1 in the square $\varphi(x_3)(k)$ will be labelled $z_{u_0}^{t+1}$ but as before the two regions belong to the same triangle. Also region number 3 of square labelled k in $A^+(c_{1,1})$ will be labelled $z_{u_0}^t$ and thus region number 1 of square labelled $\varphi(x_1)(k)$ will be labelled $z_{u_0}^{t+1}$. It follows that a region numbered 5 in some square of $A^+(c_{1,3})$ will have label $z_{u_0}^{t+1}$ but it is on the other side. As before this process can go on only three steps.

A similar argument takes care of the case of a region numbered 4, 5 and 6. Making the obvious required changes and applying the same argument will finish the proof for regions in $A^-(c_{i,1})$ and $A^-(c_{i,2})$.

Theorem 2.3. *Let $K \subset S^3$ denote the figure eight knot and let M be a closed orientable 3-manifold. Let $p : M \to S^3$ be a branched cover of S^3 branched over K. There is an effective algorithm to determine hyperbolic Dehn surgery equations for M from the monodromy map of the cover.*

Proof. Given the monodromy map of a branched cover $p : M \to S^3$, represent each of the annuli $A^+(c_{i,1}), A^+(c_{i,3}), A^-(c_{i,1})$ and $A^-(c_{i,2})$ by strips of n_i squares. Arrange them so that for each $i = 1, \dots, c(\varphi(x_1))$ they will appear in the order $A^+(c_{i,1}), A^-(c_{i,2}), A^+(c_{i,3})$ and $A^-(c_{i,1})$. Label the squares in $A^+(c_{i,1})$ and $A^+(c_{i,3})$ according to the cycles from bottom to top. By Lemma 2.2 we can label regions numbered 1, 2 and 3 on the left hand side of each strip of $A^+(c_{i,1})$, $i = 1, \dots, c(\varphi(x_1))$ by ordered pairs (z_u, t), $u = 1, \dots, 2d, t = 1, 2, 3$ with no restrictions other than:

(1) If a region numbered 3, (1), in a square labelled k is labelled (z_{u_0}, t_0) then region number 1, (3), in square labelled $\varphi(x_\ell)(k)$, $\varphi^{-1}(x_\ell)(k)$, $\ell \in \{1, 3\}$, is labelled $(z_{u_0}, t_0 + 1 \bmod 3)$, $((z_{u_0}, t_0 - 1 \bmod 3))$.
(2) A number u_0 can appear in at most two regions and if it does they have to be as in case (1).

By (d) this will induce a labelling on regions 3,4 and 5 of $A^+(c_{i,3})$. Now apply rule (1) to the annuli in $A^+(c_{i,3})$ to determine regions 1 and 6 there. Apply (d) again to determine regions 5 and 6 of $A^+(c_{i,1})$ and this will in turn determine the remaining regions 4 of $A^+(c_{i,1})$ and 2 of $A^+(c_{i,3})$. By Lemma 2.1 this will induce a unique labelling on all the annuli $A^-(c_{i,1})$ and $A^-(c_{i,2})$, $i = 1, \dots, c(\varphi(x_1))$, so that all the labels match and thus will determine the gluing of the annuli to each other to create the boundary components of $M - N(\cup K_i)$ with the induced triangulation (see Example 2.4.). It is possible that a boundary component will be constructed by $4n, n > 1$, annuli. Now we can write the cone angle equation for each edge $e_s, s = 1, \ldots, 2d$ in the triangulation. For each boundary component $T_t, t = 1, \ldots, m$, delete one equation so that no two deleted equations correspond to the same edge. Hence if $z_{u_1}(e_{s'}), \ldots, z_{u_6}(e_{s'})$ are the labels around the edge s' we write:

$$(1) \qquad \begin{aligned} &z_{u_1}(e_{s'}) \cdots z_{u_6}(e_{s'}) = 1 \quad \text{and} \\ &\arg z_{u_1}(e_{s'}) + \cdots + \arg z_{u_6}(e_{s'}) = 2\pi, \quad s' = 1, \ldots, 2d - m \end{aligned}$$

For each boundary component $T_t, t = 1, \ldots, m$, denote the s.c.c. covering the meridian of K by α_t and the derivative of the holonomy map by $\mathrm{H}'(\alpha_t)$. Choose adjacent annuli $A^+(c_{i,1})$ and $A^-(c_{i,2})$ corresponding to T_t. Let z_{u_j} j = $1, \ldots, n_i$ be the labels of all regions numbered 2 in $A^+(c_{i,1})$ and $w_{u_j}, j = 1, \ldots, n_i$ be the labels of all regions numbered 5 in $A^-(c_{i,2})$ then set:

$$(2) \qquad \mathrm{H}'(\alpha_t) = \prod_{j=1}^{n_j} z_{u_j} \cdot w_{u_j}^{-1} = \text{rotation by } 2\pi, \quad t = 1, \ldots, m.$$

The collection of equations (1) and (2) are the hyperbolic Dehn surgery equations for M.

Example 2.4. Let M be the manifold in example ♯21 of [**He**], it is an 8-fold branched cover of S^3. Let φ be the monodromy map of the cover $p : M \to S^3$. M is determined by the following data: $\varphi(x_1) = (1420)(7563)$, $\varphi(x_2) = (0721)(3465)$, $\varphi(x_3) = (7146)(2503)$. As $c(\varphi(x_1)) = 2$, M admits a Heegaard splitting of genus 7 by Theorem 1.5. The graphs Γ^-, Γ^+ which are the retracts of the handlebodies are illustrated in Fig. 8 below. Both cycles in $\varphi(x_1)$ have length 4 hence by Proposition 1.3 M is a hyperbolic manifold. The fundamental group is presented by $\pi_1(M) =< x_1, \ldots, x_5 | R_1, \ldots, R_5 >$, where

$$R_1 = x_4 x_3^{-1} x_1 x_4 x_3^{-1} x_2 x_1^{-1}, \quad R_2 = x_3 x_5^{-1} x_4^{-1} x_1 x_2^{-2}, \quad R_3 = x_2 x_1 x_2^{-1} x_3 x_4^{-1} x_5,$$
$$R_4 = x_4 x_3^{-1} x_2^2 x_1^{-1} x_3 x_5^{-2}, \quad R_5 = x_5 x_1^{-1} x_2^{-1} x_4.$$

Hence $H_1(M) = Z_{24}$, see [**He**].

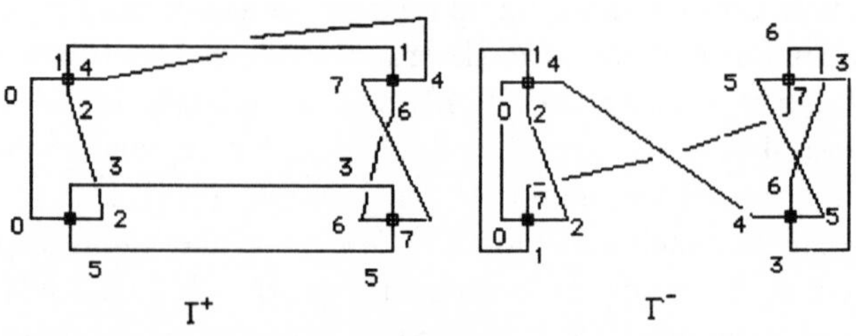

Figure 8

Now proceed to find the hyperbolic Dehn surgery equations. We shall work with H^+. As $c(\varphi(x_1)) = 2$ we have 8 annuli $A^+(c_{1,1})$, $A^+(c_{1,3})$, $A^-(c_{1,2})$, $A^-(c_{1,1})$, $A^+(c_{2,1})$, $A^+(c_{2,3})$, $A^-(c_{2,2})$, $A^-(c_{2,1})$. We draw them as in Fig. 9 and label the squares in $A^+(c_{1,1})$, $A^+(c_{1,3})$,$A^+(c_{2,1})$, $A^+(c_{2,3})$, according to the cycles. Then, using Lemma 2.2, we label the regions in the squares by ordered pairs (u, t), $t \in \{1, 2, 3\}, u \in \{1, \dots, 2d\}$, where u is short for z_u. This determines the labelling on the rest of the boundary of $M - N(K_i)$ the 8-fold irregular cover of $S^3 - N(K)$. The triangles $\mathcal{T}_{u,b}$ are reconstructed and from them we produce the equations as in [**Th**]. Note that in this case we conclude from the way the annuli are matched that there are two boundary components composed of the annuli $A^+(c_{1,1})$, $A^-(c_{1,2})$, $A^+(c_{2,3})$, $A^-(c_{2,1})$, and $A^+(c_{2,1})$, $A^-(c_{2,2})$, $A^+(c_{1,3})$, $A^-(c_{1,1})$. The labelling is described in Fig. 9 below and the equations are given in example $\sharp 21$ of §3.

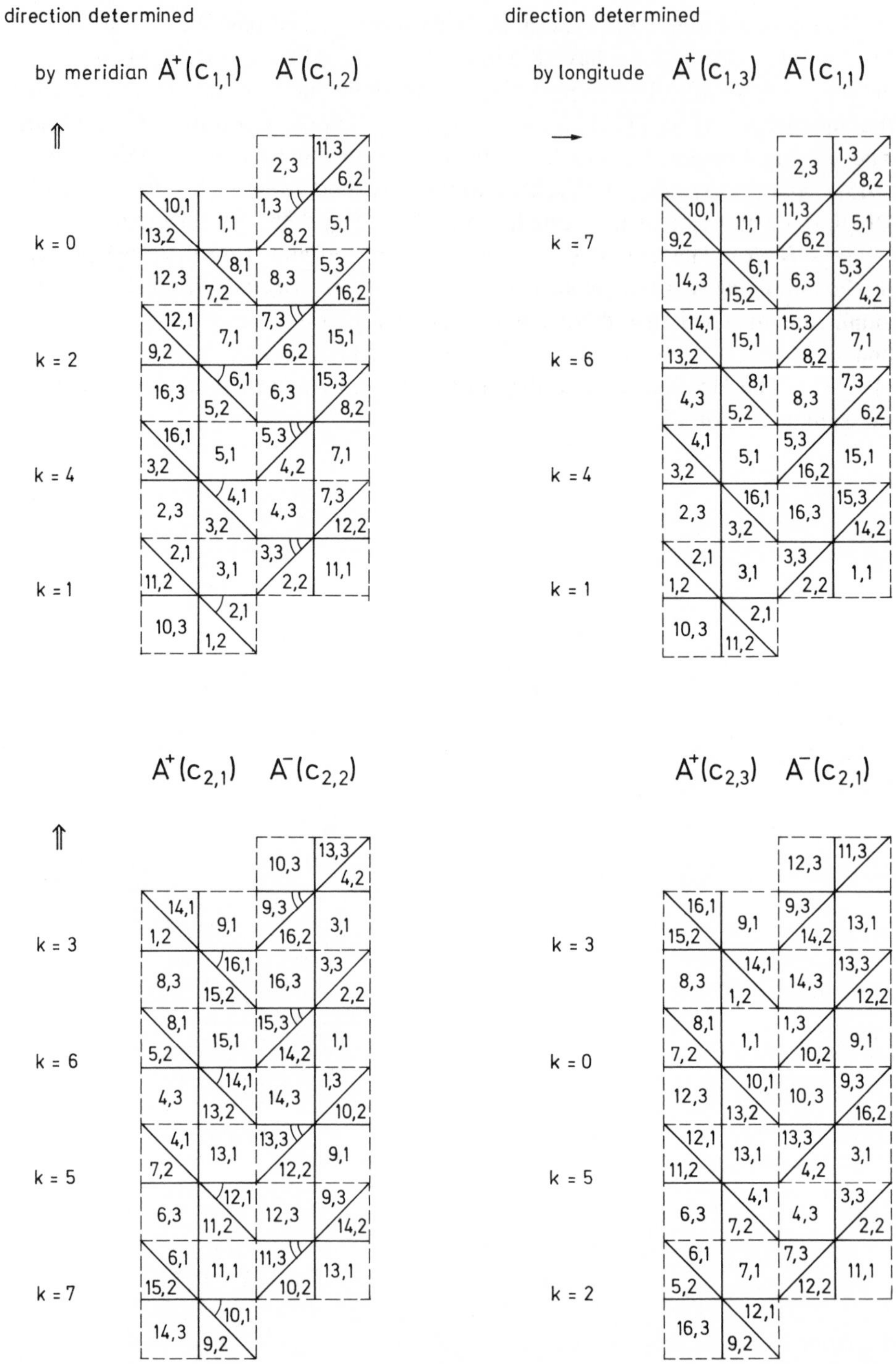
direction determined
by meridian
$A^+(c_{1,1})$ $A^-(c_{1,2})$
k = 0
k = 2
k = 4
k = 1
direction determined
by longitude
$A^+(c_{1,3})$ $A^-(c_{1,1})$
k = 7
k = 6
k = 4
k = 1
$A^+(c_{2,1})$ $A^-(c_{2,2})$
k = 3
k = 6
k = 5
k = 7
$A^+(c_{2,3})$ $A^-(c_{2,1})$
k = 3
k = 0
k = 5
k = 2

Figure 9

§3. Examples

In the table below we list the hyperbolic Dehn surgery equations for some manifolds M obtained as branched covers of S^3 of degree up to 10. We give the number of the manifold as it appears in the table at the end of [**He**], the monodromy map of the cover and the list of equations. The deleted equations appear with a " * " next to them.

(1) Example $\sharp 1$: $M = L(5,2)$

$$\varphi(x_1) = (10), \quad \varphi(x_2) = (01), \quad \varphi(x_3) = (01)$$

Equations of boundary component no. 1:

$$z_4\ (1 - z_1)\ z_2\ (1 - z_3) = 2\pi \text{ rotation}$$

(1*) $(z_3 - 1)/z_3\ z_4\ z_1\ z_4\ (z_3 - 1)/z_3\ 1/1 - z_2 = 1$

(2) $(z_1 - 1)/z_1\ z_2\ z_3\ z_2\ (z_1 - 1)/z_1\ 1/1 - z_4 = 1$

(3) $z_3\ (z_4 - 1)/z_4\ 1/1 - z_1\ 1/1 - z_2\ 1/1 - z_1\ (z_4 - 1)/z_4 = 1$

(4) $z_1\ (z_2 - 1)/z_2\ 1/1 - z_3\ 1/1 - z_4\ 1/1 - z_3\ (z_2 - 1)/z_2 = 1$

(2) Example $\sharp 3$: M = 4-fold cyclic branched cover over K, it is hyperbolic.

$$\varphi(x_1) = (3210), \quad \varphi(x_2) = (0321), \quad \varphi(x_3) = (0321)$$

Equations of boundary component no. 1:

$$z_8\ (1 - z_1)\ z_6(1 - z_7)\ z_4\ (1 - z_5)\ z_2\ (1 - z_3) = 2\pi \text{ rotation}$$

(1*) $(z_7 - 1)/z_7\ z_8\ z_1\ z_8\ (z_7 - 1)/z_7\ 1/1 - z_6 = 1$

(2) $(z_5 - 1)/z_5\ z_6\ z_7\ z_6\ (z_5 - 1)/z_5\ 1/1 - z_4 = 1$

(3) $(z_3 - 1)/z_3\ z_4\ z_5\ z_4\ (z_3 - 1)/z_3\ 1/1 - z_2 = 1$

(4) $(z_1 - 1)/z_1\ z_2\ z_3\ z_2\ (z_1 - 1)/z_1\ 1/1 - z_8 = 1$

(5) $z_7\ (z_8 - 1)/z_8\ 1/1 - z_1\ 1/1 - z_2\ 1/1 - z_1\ (z_8 - 1)/z_8 = 1$

(6) $z_5\ (z_6 - 1)/z_6\ 1/1 - z_7\ 1/1 - z_6\ 1/1 - z_7\ (z_6 - 1)/z_6 = 1$

(7) $z_3\ (z_4 - 1)/z_4\ 1/1 - z_5\ 1/1 - z_6\ 1/1 - z_5\ (z_4 - 1)/z_4 = 1$

(8) $z_1\ (z_2 - 1)/z_2\ 1/1 - z_3\ 1/1 - z_4\ 1/1 - z_3\ (z_2 - 1)/z_2 = 1$

(3) Example $\sharp 6$: $M = S^3$

$$\varphi(x_1) = (0)(21)(43), \quad \varphi(x_2) = (2)(04)(13), \quad \varphi(x_3) = (3)(02)(14)$$

Equations of boundary component no. 1:

$$z_2(1 - z_1) = 2\pi \text{ rotation}$$

(1*) $(z_1-1)/z_1\ z_2\ z_1\ z_4\ (z_5-1)/z_5\ 1/1-z_4=1$
(2) $z_9\ (z_{10}-1)/z_{10}\ 1/1-z_9\ 1/1-z_2\ 1/1-z_1\ (z_2-1)/z_2=1$

Equations of boundary component no. 2:

$$z_6\ (1-z_3)\ z_4\ (1-z_5)=2\pi \text{ rotation}$$

(1*) $(z_5-1)/z_5\ z_6\ z_3\ z_2\ (z_9-1)/z_9\ 1/1-z_8=1$
(2) $(z_3-1)/z_3\ z_4\ z_5\ z_8\ (z_1-1)/z_1\ 1/1-z_2=1$
(3) $z_7\ (z_8-1)/z_8\ 1/1-z_5\ 1/1-z_4\ 1/1-z_3\ (z_6-1)/z_6=1$
(4) $z_5\ (z_6-1)/z_6\ 1/1-z_7\ 1/1-z_6\ 1/1-z_5\ (z_4-1)/z_4=1$

Equations of boundary component no. 3:

$$z_{10}(1-z_7)z_8(1-z_9)=2\pi \text{ rotation}$$

(1*) $(z_9-1)/z_9\ z_{10}\ z_7\ z_6\ (z_7-1)/z_7\ 1/1-z_{10}=1$
(2) $(z_7-1)/z_7\ z_8\ z_9\ z_{10}\ (z_3-1)/z_3\ 1/1-z_6=1$
(3) $z_3\ (z_4-1)/z_4\ 1/1-z_1\ 1/1-z_8\ 1/1-z_7\ (z_{10}-1)/z_{10}=1$
(4) $z_1\ (z_2-1)/z_2\ 1/1-z_3\ 1/1-z_{10}\ 1/1-z_9\ (z_8-1)/z_8=1$

(4) Example ♯9: M = Seifert fibered space over S^2 with fibers of index 2,3,3

$$\varphi(x_1)=(10)(4532),\quad \varphi(x_2)=(02)(4135),\quad \varphi(x_3)=(35)(0241)$$

Equations of boundary component no. 1:

$$z_4\ (1-z_1)\ z_2\ (1-z_3)=2\pi \text{ rotation}$$

(1*) $(z_3-1)/z_3\ z_4\ z_1\ z_6\ (z_3-1)/z_3\ 1/1-z_2=1$
(2) $(z_1-1)/z_1\ z_2\ z_3\ z_2\ (z_7-1)/z_7\ 1/1-z_6=1$
(3) $z_9\ (z_{10}-1)/z_{10}\ 1/1-z_7\ 1/1-z_2\ 1/1-z_1\ (z_4-1)/z_4=1$
(4) $z_7\ (z_8-1)/z_8\ 1/1-z_9\ 1/1-z_4\ 1/1-z_3\ (z_2-1)/z_2=1$

Equations of boundary component no. 2:

$$z_{12}\ (1-z_5)\ z_{10}\ (1-z_{11})\ z_8\ (1-z_9)\ z_6\ (1-z_7)=2\pi \text{ rotation}$$

(1*)	$(z_{11}-1)/z_{11}\ z_{12}\ z_5\ z_{12}\ (z_1-1)/z_1\ 1/1-z_4=1$
(2)	$(z_9-1)/z_9\ z_{10}\ z_{11}\ z_4\ (z_9-1)/z_9\ 1/1-z_8=1$
(3)	$(z_7-1)/z_7\ z_8\ z_9\ z_8\ (z_{11}-1)/z_{11}\ 1/1-z_{10}=1$
(4)	$(z_5-1)/z_5\ z_6\ z_7\ z_{10}\ (z_5-1)/z_5\ 1/1-z_{12}=1$
(5)	$z_1\ (z_2-1)/z_2\ 1/1-z_3\ 1/1-z_6\ 1/1-z_5\ (z_{12}-1)/z_{12}=1$
(6)	$z_5\ (z_6-1)/z_6\ 1/1-z_1\ 1/1-z_{12}\ 1/1-z_{11}\ (z_{10}-1)/z_{10}=1$
(7)	$z_{11}\ (z_{12}-1)/z_{12}\ 1/1-z_5\ 1/1-z_{10}\ 1/1-z_9\ (z_8-1)/z_8=1$
(8)	$z_3\ (z_4-1)/z_4\ 1/1-z_{11}\ 1/1-z_8\ 1/1-z_7\ (z_6-1)/z_6=1$

(5) Example ♯13: M is hyperbolic

$$\varphi(x_1)=(145320), \varphi(x_2)=(243051), \varphi(x_3)=(021435)$$

Equations of boundary component no. 1:

$z_{12}\ (1-z_1)\ z_{10}\ (1-z_{11})\ z_8\ (1-z_9)\ z_6\ (1-z_7)\ z_4\ (1-z_5)\ z_2(\ 1-z_3)=2\pi$ rotation

(1*)	$(z_{11}-1)/z_{11}\ z_{12}\ z_1\ z_{10}\ (z_7-1)/z_7\ 1/1-z_6=1$
(2)	$(z_9-1)/z_9\ z_{10}\ z_{11}\ z_6\ (z_1-1)/z_1\ 1/1-z_{12}=1$
(3)	$(z_7-1)/z_7\ z_8\ z_9\ z_{12}\ (z_5-1)/z_5\ 1/1-z_4=1$
(4)	$(z_5-1)/z_5\ z_6\ z_7\ z_4\ (z_9-1)/z_9\ 1/1-z_8=1$
(5)	$(z_3-1)/z_3\ z_4\ z_5\ z_8\ (z_3-1)/z_3\ 1/1-z_2=1$
(6)	$(z_1-1)/z_1\ z_2\ z_3\ z_2\ (z_{11}-1)/z_{11}\ 1/1-z_{10}=1$
(7)	$z_5\ (z_6-1)/z_6\ 1/1-z_{11}\ 1/1-z_2\ 1/1-z_1\ (z_{12}-1)/z_{12}=1$
(8)	$z_7\ (z_8-1)/z_8\ 1/1-z_5\ 1/1-z_{12}\ 1/1-z_{11}\ (z_{10}-1)/z_{10}=1$
(9)	$z_3\ (z_4-1)/z_4\ 1/1-z_7\ 1/1-z_{10}\ 1/1-z_9\ (z_8-1)/z_8=1$
(10)	$z_1\ (z_2-1)/z_2\ 1/1-z_3\ 1/1-z_8\ 1/1-z_7\ (z_6-1)/z_6=1$
(11)	$z_9\ (z_{10}-1)/z_{10}\ 1/1-z_1\ 1/1-z_6\ 1/1-z_5\ (z_4-1)/z_4=1$
(12)	$z_{11}\ (z_{12}-1)/z_{12}\ 1/1-z_9\ 1/1-z_4\ 1/1-z_3\ (z_2-1)/z_2=1$

(6) Example ♯18: $M=L(3,1)\sharp L(3,1)\sharp L(2,1)$

$$\varphi(x_1)=(0)(2357641),\quad \varphi(x_2)=(2)(3504167),\quad \varphi(x_3)=(6)(0254713)$$

Equations of boundary component no. 1:

$$z_2(1-z_1)=2\pi \text{ rotation}$$

(1*)	$(z_1-1)/z_1\ z_2\ z_1\ z_6\ (z_7-1)/z_7\ 1/1-z_6=1$
(2)	$z_{11}\ (z_{12}-1)/z_{12}\ 1/1-z_{11}\ 1/1-z_2\ 1/1-z_1\ (z_2-1)/z_2=1$

Equations of boundary component no. 2:

$$z_{16}\,(1-z_3)\,z_{14}\,(1-z_{15})\,z_{12}\,(1-z_{13})\,z_{10}\,(1-z_{11})\,z_8\,(1-z_9)\,z_6\,(1-z_7)\,z_4\,(1-z_5) = 2\pi \text{ rotation}$$

(1*) $(z_{15}-1)/z_{15}\; z_{16}\; z_3\; z_2\; (z_{11}-1)/z_{11}\; 1/1-z_{10} = 1$

(2) $(z_{13}-1)/z_{13}\; z_{14}\; z_{15}\; z_{10}\; (z_9-1)/z_9\; 1/1-z_8 = 1$

(3) $(z_{11}-1)/z_{11}\; z_{12}\; z_{13}\; z_8\; (z_{13}-1)/z_{13}\; 1/1-z_{12} = 1$

(4) $(z_9-1)/z_9\; z_{10}\; z_{11}\; z_{12}\; (z_{15}-1)/z_{15}\; 1/1-z_{14} = 1$

(5) $(z_7-1)/z_7\; z_8\; z_9\; z_{14}\; (z_5-1)/z_5\; 1/1-z_4 = 1$

(6) $(z_5-1)/z_5\; z_6\; z_7\; z_4\; (z_3-1)/z_3\; 1/1-z_{16} = 1$

(7) $(z_3-1)/z_3\; z_4\; z_5\; z_{16}\; (z_1-1)/z_1\; 1/1-z_2 = 1$

(8) $z_1\; (z_2-1)/z_2\; 1/1-z_3\; 1/1-z_4\; 1/1-z_3\; (z_{16}-1)/z_{16} = 1$

(9) $z_5\; (z_6-1)/z_6\; 1/1-z_1\; 1/1-z_{16}\; 1/1-z_{15}\; (z_{14}-1)/z_{14} = 1$

(10) $z_{15}\; (z_{16}-1)/z_{16}\; 1/1-z_5\; 1/1-z_{14}\; 1/1-z_{13}\; (z_{12}-1)/z_{12} = 1$

(11) $z_9\; (z_{10}-1)/z_{10}\; 1/1-z_{15}\; 1/1-z_{12}\; 1/1-z_{11}\; (z_{10}-1)/z_{10} = 1$

(12) $z_{13}\; (z_{14}-1)/z_{14}\; 1/1-z_9\; 1/1-z_{10}\; 1/1-z_9\; (z_8-1)/z_8 = 1$

(13) $z_7\; (z_8-1)/z_8\; 1/1-z_{13}\; 1/1-z_8\; 1/1-z_7\; (z_6-1)/z_6 = 1$

(14) $z_3\; (z_4-1)/z_4\; 1/1-z_7\; 1/1-z_{16}\; 1/1-z_5\; (z_4-1)/z_4 = 1$

(7) Example ♯21: M is a hyperbolic manifold.

$$\varphi(x_1) = (1420)(7563), \quad \varphi(x_2) = (7210)(3465), \quad \varphi(x_3) = (0325)(1467)$$

Equations of boundary component no. 1:

$$z_8\,(1-z_1)\,z_6\,(1-z_7)\,z_4\,(1-z_5)\,z_2\,(1-z_3) = 2\pi \text{ rotation}$$

(1*) $(z_7-1)/z_7\; z_8\; z_1\; z_{10}\; (z_{13}-1)/z_{13}\; 1/1-z_{12} = 1$

(2) $(z_5-1)/z_5\; z_6\; z_7\; z_{12}\; (z_9-1)/z_9\; 1/1-z_{16} = 1$

(3) $(z_3-1)/z_3\; z_4\; z_5\; z_{16}\; (z_3-1)/z_3\; 1/1-z_2 = 1$

(4) $(z_1-1)/z_1\; z_2\; z_3\; z_2\; (z_{11}-1)/z_{11}\; 1/1-z_{10} = 1$

(5) $z_5\; (z_6-1)/z_6\; 1/1-z_{11}\; 1/1-z_2\; 1/1-z_1\; (z_8-1)/z_8 = 1$

(6) $z_{15}\; (z_{16}-1)/z_{16}\; 1/1-z_5\; 1/1-z_8\; 1/1-z_7\; (z_6-1)/z_6 = 1$

(7) $z_7\; (z_8-1)/z_8\; 1/1-z_{15}\; 1/1-z_6\; 1/1-z_5\; (z_4-1)/z_4 = 1$

(8) $z_{11}\; (z_{12}-1)/z_{12}\; 1/1-z_7\; 1/1-z_4\; 1/1-z_3\; (z_2-1)/z_2 = 1$

Equations of boundary component no. 2:

$$z_{16}\,(1-z_9)\,z_{14}\,(1-z_{15})\,z_{12}\,(1-z_{13})\,z_{10}\,(1-z_{11}) = 2\pi \text{ rotation}$$

(1*) $(z_{15}-1)/z_{15}\ z_{16}\ z_9\ z_{14}\ (z_1-1)/z_1\ 1/1-z_8=1$

(2) $(z_{13}-1)/z_{13}\ z_{14}\ z_{15}\ z_8\ (z_5-1)/z_5\ 1/1-z_4=1$

(3) $(z_{11}-1)/z_{11}\ z_{12}\ z_{13}\ z_4\ (z_7-1)/z_7\ 1/1-z_6=1$

(4) $(z_9-1)/z_9\ z_{10}\ z_{11}\ z_6\ (z_{15}-1)/z_{15}\ 1/1-z_{14}=1$

(5) $z_3\ (z_4-1)/z_4\ 1/1-z_{13}\ 1/1-z_{10}\ 1/1-z_9\ (z_{16}-1)/z_{16}=1$

(6) $z_1\ (z_2-1)/z_2\ 1/1-z_3\ 1/1-z_{16}\ 1/1-z_{15}\ (z_{14}-1)/z_{14}=1$

(7) $z_9\ (z_{10}-1)/z_{10}\ 1/1-z_1\ 1/1-z_{14}\ 1/1-z_{13}\ (z_{12}-1)/z_{12}=1$

(8) $z_{13}\ (z_{14}-1)/z_{14}\ 1/1-z_9\ 1/1-z_{12}\ 1/1-z_{11}\ (z_{10}-1)/z_{10}=1$

(8) Example ♯23: $M=?$

$$\varphi(x_1)=(10)(634752),\quad \varphi(x_2)=(67)(430251),\quad \varphi(x_3)=(35)(074162)$$

Equations of component no. 1:

$$z_4\ (1-z_1)\ z_2\ (1-z_3)=2\pi \text{ rotation}$$

(1*) $(z_3-1)/z_3\ z_4\ z_1\ z_{10}\ (z_5-1)/z_5\ 1/1-z_{16}=1$

(2) $(z_1-1)/z_1\ z_2\ z_3\ z_{16}\ (z_{11}-1)/z_{11}\ 1/1-z_{10}=1$

(3) $z_{13}\ (z_{14}-1)/z_{14}\ 1/1-z_7\ 1/1-z_2\ 1/1-z_1\ (z_4-1)/z_4=1$

(4) $z_7\ (z_8-1)/z_8\ 1/1-z_{13}\ 1/1-z_4\ 1/1-z_3\ (z_2-1)/z_2=1$

Equations of component no. 2:

$$z_{16}\ (1-z_5)\ z_{14}\ (1-z_{15})\ z_{12}\ (1-z_{13})\ z_{10}\ (1-z_{11})\ z_8\ (1-z_9)\ z_6\ (1-z_7)=2\pi \text{ rotation}$$

(1*) $(z_{15}-1)/z_{15}\ z_{16}\ z_5\ z_2\ (z_7-1)/z_7\ 1/1-z_6=1$

(2) $(z_{13}-1)/z_{13}\ z_{14}\ z_{15}\ z_6\ (z_9-1)/z_9\ 1/1-z_8=1$

(3) $(z_{11}-1)/z_{11}\ z_{12}\ z_{13}\ z_8\ (z_1-1)/z_1\ 1/1-z_4=1$

(4) $(z_9-1)z_9\ z_{10}\ z_{11}\ z_4\ (z_{13}-1)/z_{13}\ 1/1-z_{12}=1$

(5) $(z_7-1)/z_7\ z_8\ z_9\ z_{12}\ (z_{15}-1)/z_{15}\ 1/1-z_{14}=1$

(6) $(z_5-1)/z_5\ z_6\ z_7\ z_{14}\ (z_3-1)/z_3\ 1/1-z_2=1$

(7) $z_{11}\ (z_{12}-1)/z_{12}\ 1/1-z_9\ 1/1-z_6\ 1/1-z_5\ (z_{16}-1)/z_{16}=1$

(8) $z_3\ (z_4-1)/z_4\ 1/1-z_{11}\ 1/1-z_{16}\ 1/1-z_{15}\ (z_{14}-1)/z_{14}=1$

(9) $z_{15}\ (z_{16}-1)/z_{16}\ 1/1-z_3\ 1/1-z_{14}\ 1/1-z_{12}\ (z_{12}-1)/z_{12}=1$

(10) $z_5\ (z_6-1)/z_6\ 1/1-z_{15}\ 1/1-z_{12}\ 1/1-z_{11}\ (z_{10}-1)/z_{10}=1$

(11) $z_1\ (z_2-1)/z_2\ 1/1-z_5\ 1/1-z_{10}\ 1/1-z_9\ (z_8-1)/z_8=1$

(12) $z_9\ (z_{10}-1)/z_{10}\ 1/1-z_1\ 1/1-z_8\ 1/1-z_7\ (z_6-1)/z_6=1$

(9) Example ♯29: $M = L(3,1)\sharp L(3,1)$

$\varphi(x_1) = (0)(2541)(6873)$, $\varphi(x_2) = (2)(5178)(0346)$ $\varphi(x_3) = (7)(0265)(1438)$

Equations of component no. 1:

$$z_2\ (1 - z_1) = 2\pi \text{ rotation}$$

(1*) $(z_1 - 1)/z_1\ z_2\ z_1\ z_6\ (z_7 - 1)/z_7\ 1/1 - z_6 = 1$

(2) $z_{15}\ (z_{16} - 1)/z_{16}\ 1/1 - z_{15}\ 1/1 - z_2\ 1/1 - z_1\ (z_2 - 1)/z_2 = 1$

Equations of component no. 2:

$$z_{10}\ (1 - z_3)\ z_8\ (1 - z_9)\ z_6\ (1 - z_7)\ z_4\ (1 - z_5) = 2\pi \text{ rotation}$$

(1*) $(z_9 - 1)/z_9\ z_{10}\ z_3\ z_2\ (z_{15} - 1)/z_{15}\ 1/1 - z_{14} = 1$

(2) $(z_7 - 1)/z_7\ z_8\ z_9\ z_{14}\ (z_3 - 1)/z_3\ 1/1 - z_{10} = 1$

(3) $(z_5 - 1)/z_5\ z_6\ z_7\ z_{10}\ (z_{13} - 1)/z_{13}\ 1/1 - z_{12} = 1$

(4) $(z_3 - 1)/z_3\ z_4\ z_5\ z_{12}\ (z_1 - 1)/z_1\ 1/1 - z_2 = 1$

(5) $z_{13}\ (z_{14} - 1)/z_{14}\ 1/1 - z_9\ 1/1 - z_4\ 1/1 - z_3\ (z_{10} - 1)/z_{10} = 1$

(6) $z_{17}\ (z_{18} - 1)/z_{18}\ 1/1 - z_{13}\ 1/1 - z_{10}\ 1/1 - z_9\ (z_8 - 1)/z_8 = 1$

(7) $z_7\ (z_8 - 1)/z_8\ 1/1 - z_{17}\ 1/1 - z_8\ 1/1 - z_7\ (z_6 - 1)/z_6 = 1$

(8) $z_9\ (z_{10} - 1)/z_{10}\ 1/1 - z_7\ 1/1 - z_6\ 1/1 - z_5\ (z_4 - 1)/z_4 = 1$

Equations of component no. 3:

$$z_{18}\ (1 - z_{11})\ z_{16}\ (1 - z_{17})\ z_{14}\ (1 - z_{15})\ z_{12}\ (1 - z_{13}) = 2\pi \text{ rotation}$$

(1*) $(z_{17} - 1)/z_{17}\ z_{18}\ z_{11}\ z_4\ (z_9 - 1)/z_9\ 1/1 - z_8 = 1$

(2) $(z_{15} - 1)/z_{15}\ z_{16}\ z_{17}\ z_8\ (z_{17} - 1)/z_{17}\ 1/1 - z_{16} = 1$

(3) $(z_{13} - 1)/z_{13}\ z_{14}\ z_{15}\ z_{16}\ (z_{11} - 1)/z_{11}\ 1/1 - z_{18} = 1$

(4) $(z_{11} - 1)/z_{11}\ z_{12}\ z_{13}\ z_{18}\ (z_5 - 1)/z_5\ 1/1 - z_4 = 1$

(5) $z_5\ (z_6 - 1)/z_6\ 1/1 - z_1\ 1/1 - z_{12}\ 1/1 - z_{11}\ (z_{18} - 1)/z_{18} = 1$

(6) $z_{11}\ (z_{12} - 1)/z_{12}\ 1/1 - z_5\ 1/1 - z_{18}\ 1/1 - z_{17}\ (z_{16} - 1)/z_{16} = 1$

(7) $z_3\ (z_4 - 1)/z_4\ 1/1 - z_{11}\ 1/1 - z_{16}\ 1/1 - z_{15}\ (z_{14} - 1)/z_{14} = 1$

(8) $z_1\ (z_2 - 1)/z_2\ 1/1 - z_3\ 1/1 - z_{14}\ 1/1 - z_{13}\ (z_{12} - 1)/z_{12} = 1$

(10) Example ♯37: $M = ?$

$\varphi(x_1) = (120)(6948573)$, $\varphi(x_2) = (628)(0714539)$ $\varphi(x_3) = (459)(0612837)$

Equations of component no. 1:

$$z_6\ (1 - z_1)\ z_4\ (1 - z_5)\ z_2\ (1 - z_3) = 2\pi \text{ rotation}$$

(1*) $(z_5-1)/z_5\ z_6\ z_1\ z_8\ (z_{19}-1)/z_{19}\ 1/1-z_{18}=1$

(2) $(z_3-1)/z_3\ z_4\ z_5\ z_{18}\ (z_3-1)/z_3\ 1/1-z_2=1$

(3) $(z_1-1)/z_1\ z_2\ z_3\ z_2\ (z_9-1)/z_9\ 1/1-z_8=1$

(4) $z_{15}\ (z_{16}-1)/z_{16}\ 1/1-z_9\ 1/1-z_2\ 1/1-z_1\ (z_6-1)/z_6=1$

(5) $z_{11}\ (z_{12}-1)/z_{12}\ 1/1-z_{15}\ 1/1-z_6\ 1/1-z_5\ (z_4-1)/z_4=1$

(6) $z_9\ (z_{10}-1)/z_{10}\ 1/1-z_{11}\ 1/1-z_4\ 1/1-z_3\ (z_2-1)/z_2=1$

Equations of component no. 2:

$$z_{20}\ (1-z_7)\ z_{18}\ (1-z_{19})\ z_{16}\ (1-z_{17})\ z_{14}\ (1-z_{15})\ z_{12}$$
$$(1-z_{13})\ z_{10}\ (1-z_{11})\ z_8\ (1-z_9)=2\pi \text{ rotation}$$

(1*) $(z_{19}-1)/z_{19}\ z_{20}\ z_7\ z_6\ (z_{15}-1)/z_{15}\ 1/1-z_{14}=1$

(2) $(z_{17}-1)/z_{17}\ z_{18}\ z_{19}\ z_{14}\ (z_7-1)/z_7\ 1/1-z_{20}=1$

(3) $(z_{15}-1)/z_{15}\ z_{16}\ z_{17}\ z_{20}\ (z_{13}-1)/z_{13}\ 1/1-z_{12}=1$

(4) $(z_{13}-1)/z_{13}\ z_{14}\ z_{15}\ z_{12}\ (z_5-1)/z_5\ 1/1-z_4=1$

(5) $(z_{11}-1)/z_{11}\ z_{12}\ z_{13}\ z_4\ (z_{11}-1)/z_{11}\ 1/1-z_{10}=1$

(6) $(z_9-1)/z_9\ z_{10}\ z_{11}\ z_{10}\ (z_{17}-1)/z_{17}\ 1/1-z_{16}=1$

(7) $(z_7-1)/z_7\ z_{18}\ z_9\ z_{16}\ (z_1-1)/z_1\ 1/1-z_6=1$

(8) $z_{13}\ (z_{14}-1)/z_{14}\ 1/1-z_{19}\ 1/1-z_8\ 1/1-z_7\ (z_{20}-1)/z_{20}=1$

(9) $z_3\ (z_4-1)/z_4\ 1/1-z_{13}\ 1/1-z_{20}\ 1/1-z_{19}\ (z_{18}-1)/z_{18}=1$

(10) $z_1\ (z_2-1)/z_2\ 1/1-z_3\ 1/1-z_{18}\ 1/1-z_{17}\ (z_{16}-1)/z_{16}=1$

(11) $z_7\ (z_8-1)/z_8\ 1/1-z_1\ 1/1-z_{16}\ 1/1-z_{15}\ (z_{14}-1)/z_{14}=1$

(12) $z_5\ (z_6-1)/z_6\ 1/1-z_7\ 1/1-z_{14}\ 1/1-z_{13}\ (z_{12}-1)/z_{12}=1$

(13) $z_{17}\ (z_{18}-1)/z_{18}\ 1/1-z_5\ 1/1-z_{12}\ 1/1-z_{11}\ (z_{10}-1)/z_{10}=1$

(14) $z_{19}\ (z_{20}-1)/z_{20}\ 1/1-z_{17}\ 1/1-z_{10}\ 1/1-z_9\ (z_8-1)/z_8=1$

(11) Example ♯41: $M=?$

$$\varphi(x_1)=(10)(4352)(8796),\quad \varphi(x_2)=(54)(0819)(2637),$$
$$\varphi(x_3)=(67)(0514)(2938)$$

Equations of component no. 1:

$$z_4\ (1-z_1)\ z_2\ (1-z_3)=2\pi \text{ rotation}$$

(1*) $(z_3-1)/z_3\ z_4\ z_1\ z_{10}\ (z_7-1)/z_7\ 1/1-z_6=1$

(2) $(z_1-1)/z_1\ z_2\ z_3\ z_6\ (z_{11}-1)/z_{11}\ 1/1-z_{10}=1$

(3) $z_{19}\ (z_{20}-1)/z_{20}\ 1/1-z_{15}\ 1/1-z_2\ 1/1-z_1\ (z_4-1)/z_4=1$

(4) $z_{15}\ (z_{16}-1)/z_{16}\ 1/1-z_{19}\ 1/1-z_4\ 1/1-z_3\ (z_2-1)/z_2=1$

Equations of component no. 2:

$$z_{12}\ (1-z_5)\ z_{10}\ (1-z_{11})\ z_8\ (1-z_9)\ z_6\ (1-z_7)=2\pi \text{ rotation}$$

(1*) $(z_{11}-1)/z_{11}\ z_{12}\ z_5\ z_2\ (z_{15}-1)/z_{15}\ 1/1-z_{14}=1$

(2) $(z_9-1)/z_9\ z_{10}\ z_{11}\ z_{14}\ (z_1-1)/z_1\ 1/1-z_4=1$

(3) $(z_7-1)/z_7\ z_8\ z_9\ z_4\ (z_{19}-1)/z_{19}\ 1/1-z_{18}=1$

(4) $(z_5-1)/z_5\ z_6\ z_7\ z_{18}\ (z_3-1)/z_3\ 1/1-z_2=1$

(5) $z_{13}\ (z_{14}-1)/z_{14}\ 1/1-z_{11}\ 1/1-z_6\ 1/1-z_5\ (z_{12}-1)/z_{12}=1$

(6) $z_7\ (z_8-1)/z_8\ 1/1-z_{13}\ 1/1-z_{12}\ 1/1-z_{11}\ (z_{10}-1)/z_{10}=1$

(7) $z_{17}\ (z_{18}-1)/z_{18}\ 1/1-z_7\ 1/1-z_{10}\ 1/1-z_9\ (z_8-1)/z_8=1$

(8) $z_{11}\ (z_{12}-1)/z_{12}\ 1/1-z_{17}\ 1/1-z_8\ 1/1-z_7\ (z_6-1/)z_6=1$

Equations of component no. 3:

$$z_{20}\ (1-z_{13})\ z_{18}\ (1-z_{19})\ z_{16}\ (1-z_{17})\ z_{14}\ (1-z_{15})=2\pi \text{ rotation}$$

(1*) $(z_{19}-1)/z_{19}\ z_{20}\ z_{13}\ z_8\ (z_{17}-1)/z_{17}\ 1/1-z_{16}=1$

(2) $(z_{17}-1)/z_{17}\ z_{18}\ z_{19}\ z_{16}\ (z_5-1)/z_5\ 1/1-z_{12}=1$

(3) $(z_{15}-1)/z_{15}\ z_{16}\ z_{17}\ z_{12}\ (z_{13}-1)/z_{13}\ 1/1-z_{20}=1$

(4) $(z_{13}-1)/z_{13}\ z_{14}\ z_{15}\ z_{20}\ (z_9-1)/z_9\ 1/1-z_8=1$

(5) $z_9\ (z_{10}-1)/z_{10}\ 1/1-z_1\ 1/1-z_{14}\ 1/1-z_{13}\ (z_{20}-1)/z_{20}=1$

(6) $z_3\ (z_4-1)/z_4\ 1/1-z_9\ 1/1-z_{20}\ 1/1-z_{19}\ (z_{18}-1)/z_{18}=1$

(7) $z_5\ (z_6-1)/z_6\ 1/1-z_3\ 1/1-z_{18}\ 1/1-z_{17}\ (z_{16}-1)/z_{16}=1$

(8) $z_1\ (z_2-1)/z_2\ 1/1-z_5\ 1/1-z_{16}\ 1/1-z_{15}\ (z_{14}-1)/z_{14}=1$

(12) Example ♯42: M is a hyperbolic manifold

$$\varphi(x_1)=(1485973620),\quad \varphi(x_2)=(6809435271),\quad \varphi(x_3)=(0849537261)$$

Equations of component no. 1:

$$z_{20}\ (1-z_1)\ z_{18}\ (1-z_{19})\ z_{16}\ (1-z_{17})\ z_{14}\ (1-z_{15})\ z_{12}\ (1-z_{13})\ z_{10}\ (1-z_{11})\ z_8\ (1-z_9)\ z_6\ (1-z_7)\ z_4\ (1-z_5)\ z_2\ (1-z_3)=2\pi \text{ rotation}$$

(1*) $(z_{19}-1)/z_{19}\; z_{20}\; z_1\; z_{16}\; (z_3-1)/z_3\; 1/1-z_2=1$

(2) $(z_{17}-1)/z_{17}\; z_{18}\; z_{19}\; z_2\; (z_{13}-1)/z_{13}\; 1/1-z_{12}=1$

(3) $(z_{15}-1)/z_{15}\; z_{16}\; z_{17}\; z_{12}\; (z_{19}-1)/z_{19}\; 1/1-z_{18}=1$

(4) $(z_{13}-1)/z_{13}\; z_{14}\; z_{15}\; z_{18}\; (z_9-1)/z_9\; 1/1-z_8=1$

(5) $(z_{11}-1)/z_{11}\; z_{12}\; z_{13}\; z_8\; (z_{15}-1)/z_{15}\; 1/1-z_{14}=1$

(6) $(z_9-1)/z_9\; z_{10}\; z_{11}\; z_{14}\; (z_5-1)/z_5\; 1/1-z_4=1$

(7) $(z_7-1)/z_7\; z_8\; z_9\; z_4\; (z_{11}-1)/z_{11}\; 1/1-z_{10}=1$

(8) $(z_5-1)/z_5\; z_6\; z_7\; z_{10}\; (z_1-1)/z_1\; 1/1-z_{20}=1$

(9) $(z_3-1)/z_3\; z_4\; z_5\; z_{20}\; (z_7-1)/z_7\; 1/1-z_6=1$

(10) $(z_1-1)/z_1\; z_2\; z_3\; z_6\; (z_{17}-1)/z_{17}\; 1/1-z_{16}=1$

(11) $z_7\; (z_8-1)/z_8\; 1/1-z_{13}\; 1/1-z_2\; 1/1-z_1\; (z_{20}-1)/z_{20}=1$

(12) $z_9\; (z_{10}-1)/z_{10}\; 1/1-z_7\; 1/1-z_{20}\; 1/1-z_{19}\; (z_{18}-1)/z_{18}=1$

(13) $z_3\; (z_4-1)/z_4\; 1/1-z_9\; 1/1-z_{18}\; 1/1-z_{17}\; (z_{16}-1)/z_{16}=1$

(14) $z_5\; (z_6-1)/z_6\; 1/1-z_3\; 1/1-z_{16}\; 1/1-z_{15}\; (z_{14}-1)/z_{14}=1$

(15) $z_{19}\; (z_{20}-1)/z_{20}\; 1/1-z_5\; 1/1-z_{14}\; 1/1-z_{13}\; (z_{12}-1)/z_{12}=1$

(16) $z_1\; (z_2-1)/z_2\; 1/1-z_{19}\; 1/1-z_{12}\; 1/1-z_{11}\; (z_{10}-1)/z_{10}=1$

(17) $z_{15}\; (z_{16}-1)/z_{16}\; 1/1-z_1\; 1/1-z_{10}\; 1/1-z_9\; (z_8-1)/z_8=1$

(18) $z_{17}\; (z_{18}-1)/z_{18}\; 1/1-z_{15}\; 1/1-z_8\; 1/1-z_7\; (z_6-1)/z_6=1$

(19) $z_{11}\; (z_{12}-1)/z_{12}\; 1/1-z_{17}\; 1/1-z_6\; 1/1-z_5\; (z_4-1)/z_4=1$

(20) $z_{13}\; (z_{14}-1)/z_{14}\; 1/1-z_{11}\; 1/1-z_4\; 1/1-z_3\; (z_2-1)/z_2=1$

References

[GT] M. Gromov, W. Thurston, *Pinching constants for hyperbolic manifolds*, Invent. math. **89** (1987), 1–12.

[He] J. Hempel, *The lattice of branched covers over the figure eight knot*, to appear in Topology and its applications.

[HLM1] H.M Hilden, M.T. Lozano, J.M. Montesinos, *On knots that are universal*, Topology **24** (1985), 499–504.

[HLM2] H.M Hilden, M.T. Lozano, J.M. Montesinos, *Non-simple universal knots*, Math. Proc. Camb. Phil. Soc. **102** (1987), 87–95.

[HLMW] H.M Hilden, M.T. Lozano, J.M. Montesinos, W.C. Whitten, *On Universal Groups and Three manifolds*, MSRI preprint.

[Ho] C. Hodgson, *Geometric structures on 3-dimensional orbifolds: notes on Thurston's proof*, preprint.

[Ro] D. Rolfsen, *Knots and Links*, Mathematics Lecture Series 7, Publish or Perish, Berkeley. Calif., 1976.

[Sc] P. Scott, *The Geometries of 3-manifolds*, The Bull. London math. Soc. **15,** part 5, no. 56 (1983), 451- 487.

[Th] W. Thurston, *The geometry and topology of Three manifolds*, Princeton Math. Dept. (1979).

[We] J. Weeks, *Hyperbolic structures on 3-manifolds*, Ph.D. dissertation, Princeton University (1985).

The problem of finiteness for Kleinian groups in 3-space

*L. Potyagailo**

By Kleinian group G we mean a discontinuous subgroup of the group $S0(1,n+1)$ of conformal automorphisms of n-dimensional sphere $\overline{R}^n = R^n \cup \{\infty\}$. We denote by $\Omega(G)$ the domain of discontinuity of G and by $\Lambda(G) = \overline{R}^n \backslash \Omega(G)$ the limit set of G [8]. Mainly we will consider the case when group G preserves some connected component Ω_G of $\Omega(G) \subset \overline{R}^n$ and will be called function group in such a case. If the action of G on Ω_G is free then the projection $p : \Omega_G \to M_G = \Omega_G/G$ is a covering map and M_G admits a flat conformal structure inducing by that action.

One of the most important results in the classical theory of Kleinian groups on a complex plane $\mathbb{C}$ is Ahlfors' finiteness theorem [1] which states that the factor-space of finitely generated Kleinian group G consists of a finite number of Riemann surfaces of a finite hyperbolic area. In particular the factor-space of $G \subset PSL_2\mathbb{C}$ is of a finite homotopy type.

Also there are other finiteness results both for Kleinian groups on $\mathbb{C}$ and in the topology of 3-manifolds. It follows from [11,13] that finitely generated Kleinian group on a complex plane is finitely presented. On the other hand every 3-manifold with a finitely generated fundamental group F has a compact core and so F is finitely presented too [11].

The aim of this paper is to show that in higher dimension all above finiteness results don't hold.

The following theorem gives a negative answer on a question of G. Martin [7, p. 254].

Theorem 1. *There exists a Kleinian group $F \subset S0_+(1,4)$ such that F is finitely generated but isn't finitely presented, and hyperbolic manifold $M(F) = \mathbb{H}^4/F$ has no compact core.*

* Supported in part by GIF

Knots 90

It was proved in [4] that there exists a finitely generated Kleinian group $\mathcal{F} \subset S0(1,4)$ acting freely on $\Omega_{\mathcal{F}} \subset \overline{R}^3$ such that $\Pi_1\left(\Omega_{\mathcal{F}}/\mathcal{F}\right)$ is not finitely generated.

The existence of finitely generated group with infinitely many conjugacy classes of elliptic and parabolic elements was established in [5].

All the above-mentioned results in higher dimensions are based on the existence of parabolic elements which are accidental (don't correspond to the ends of the main component of a factor-space). We try to describe the class of finitely generated Kleinian groups with a factor of infinite homotopy type and therefore it's natural to ask do exist such groups in the case when there are no parabolics. The following is valid.

Theorem 2. *There exists a finitely generated Kleinian group $N \subset S0_+(1,4)$ acting freely on a component $\Omega_N \subset \overline{R}^3$ such that N contains no parabolics and $\Pi_1\left(\Omega_N/N\right)$ is not finitely generated.*

The proof of Theorem 2 is based on some new ideas and we give a sketch of it in §2.

It's interesting to describe the class of finitely generated Kleinian groups without parabolics which are hyperbolic in a sense of M. Gromov [3]. In the case of Kleinian groups on $\mathbb{C}$ it easily follows from [2] and [15] that every finitely generated group $\Gamma \subset PSL_2\mathbb{C}$ is isomorphic to a free product of geometrically finite groups so if Γ contains no parabolics then Γ is Gromov hyperbolic as well.

We prove the following.

Theorem 3. *Group N in Theorem 2 is not Gromov hyperbolic. Moreover there exists a finite extension N_0 of N such that there are infinitely many conjugacy classes of elements of a finite order in N_0 and N_0 doesn't contain parabolics.*

Detailed proof of Theorem 2 and 3 will appear in the forthcoming paper [10]. The author is very thankful to organizers and participants of the conference "Knots - 90" for useful conversations. Also I'd like to express my gratitude to Tel-Aviv University especially to Pr.V.Milman for help and the possibility to finish this work.

1. Construction of a finitely generated but infinitely presented group $F \subset S0_+(1,4)$.

Let us begin by recalling our first construction. Consider in 3-sphere $S^3 = \overline{R}^3 \cup \{\infty\}$ the configuration of four mutually tangent disjoint euclidean spheres $\Sigma_i (i = 1, \ldots, 4)$ such that $\Sigma_2 = \tau_1(\Sigma_1)$, $\Sigma_{i+2} = \tau_2(\Sigma_i)$ $(i = 1, 2)$ where τ_i is a reflection in a plane $\Pi_i = \{(x_i, x_2, x_3) \in \overline{R}^3 : x_{i+1} = 0\}$, $(i = 1, 2)$.

We have that $\Sigma_1 \cap \Sigma_2 \cap \Pi_1 = \{P_1\}$ and $\Sigma_1 \cap \Sigma_4 \cap \Pi_2 = \{P_2\}$. Let's suppose that the exterior B_1^* of Σ_1 is realized as a model of hyperbolic space $\mathbb{H}^3$ and consider the action of the group $\Gamma_0 \subset PSL_2(\mathbb{C})$ such that B_1^*/Γ_0 is a Borromean link complement [15]. Without loss of generality we can assume that the points of tangency P_i are non-conjugating parabolic fixed points of Γ_0 and $H_i = Stab(P_i, \Gamma_0) = \{h \in \Gamma_0 : hP_i = P_i\}$.

We proved in [6] that there exists a subgroup $\Gamma_1 \subset \Gamma_0$ of a finite index such that both groups $G_1 = \langle \Gamma_1, \tau_1\Gamma_1\tau_1^{-1} = \Gamma_2\rangle$ and $G = \langle G_1, \tau_2 G_1 \tau_2^{-1} = G_2\rangle$ are discontinuous. It's not hard to prove that G_1 is isomorphic to the fundamental group of some surface bundle over S^1 so there exists a normal subgroup $F_1 \lhd \Gamma_1$ which is finitely generated and of infinite index. It was shown [6, Lemma 4.c] that $F = \langle F_1, F_2\rangle$ is a normal in G and index $|G : F|$ is also infinite.

Proof of Theorem 1. By Maskit combination [1] we have that $G = G_1 *_J G_2$ where $J = \langle H_2, \tau_1 H_2 \tau_1 = H_3\rangle$ and as we have seen in [6, Lemma 4] J is a Schottky-type group and the following holds

$$I = H_2 * H_3 \simeq (\mathbb{Z} \oplus \mathbb{Z}) * (\mathbb{Z} \oplus \mathbb{Z}) ,$$
$$S_2 = \Pi_2/J, \text{genus } (S_2) = 2 . \tag{1}$$

Obviously $\Omega(F) = \Omega(G)$ and the plane Π_2 is strongly invariant under the subgroup $L = J \cap F$ in F. Repeating our consideration of Lemma 4 in [6] we obtain that $F = F_1 *_L F_2$. Group L is now a normal subgroup of J and covering $p_2 : \Pi_2/L \to \Pi_2/J$ is infinite because p_2 is just a restriction of infinite regular covering $p_1 : M_{F_1} \to M_{G_1}$. By Ahlfors' finiteness theorem or by [4, Lemma 11.2] we have that L is in fact infinitely generated. Group F is infinitely presented now by theorem of W. Neumann [9]. Theorem 1 is proved.Q.E.D.

2. Outline of the proofs of Theorems 2 and 3.

We shall consider the family of euclidean spheres $\Sigma_1, \dots, \Sigma_m$ covering the square knot $K \subset S^3$ representing by piecewise linear curve. Each sphere Σ_i is a result of the reflection τ_i of the sphere Σ_{i-1} in a plane Π_i orthogonal to K. We suppose that Γ_1 acts in the exterior of Σ_1 and $\Lambda(\Gamma_1) = \Sigma_1$. Γ_1 is a subgroup of a finite index in the Jorgensen's group Γ_0 uniformizing the fiber bundle over the circle where the fiber is a torus with one branching point [14, p. 196]. It's not hard to prove that Γ_1 is commensurable to the reflection group determined by faces of right angle dodecahedron and, so Γ_1 has the property of separability under geometrically finite subgroups [12].

Let's denote by $\Gamma_i = \tau_i \Gamma_{i-1} \tau_i^{-1}$ and by construction we can suppose that $\Sigma_i \cap \Sigma_{i+1} \cap \Pi_i = \Lambda(\Gamma_i \cap \Gamma_{i+1})$.

The choice Γ_1 is to be conditioned by the following.

Lemma. *The group* $G = \langle \Gamma'_1, \dots, \Gamma'_m \rangle$ *is discontinuous and has a normal finitely generated subgroup* N *of infinite index, where* $\Gamma'_i \subset \Gamma_i$ *is a subgroup isomorphic to a fundamental group of compact 3-manifold with geodesic boundary.*

The proof of the Lemma is based on the verification of all the conditions of Maskit's combination theorem [1]. We only point out that $H_i = \Gamma'_i \cap \Gamma'_{i+1}$ is chosen as a fuchsian subgroup of the first kind in Γ'_i.

There is an incompressible torus $T \subset \mathcal{P}(G) \subset \Omega_G$, where $\mathcal{P}(G)$ is a fundamental domain of G in Ω_G and T is a boundary of the regular neighborhood of $\mathcal{P}(G)$. We have a diagram of regular coverings:

$$\Omega_G \xRightarrow{P_1} M_N \xRightarrow{P_2} M_G \tag{2}$$

By [4,Th.11.1] the group $\Pi_1 M_N$ is infinitely generated or is isomorphic to a surface group. But the second is impossible since $\Pi_1(M_N)$ is not a finite extension of $\mathbb{Z} \oplus \mathbb{Z}$.

The proof of Theorem 3 can be obtained by considering a finite extension G_0 of group G containing an element g of a finite order. There is also a normal subgroup $F_0 \lhd G_0$ and $g \in F_0$. Now we get infinitely many non-conjugating elements $t^n g t^{-n} \in F_0,\ t \in G_0 \backslash F_0, n \in \mathbb{Z}$.

References

[1] L.V. Ahlfors, Finitely generated Kleinian groups, Amer. J. Math. 86 (1964), 413-429; 87 (1965), 759.

[2] F. Bonahon, Bouts des varietes hyperboliques de dimension 3, Ann. Math. 124 (1986), 71-158.

[3] M. Gromov, Hyperbolic groups in “Essays in group theory”: ed. S.M. Gersten, M.S.R.I. Publications N8 (1988).

[4] J. Hempel, 3-manifolds, Princeton, Princeton University Press, 1976.

[5] M. Kapovich, On the absence of Sullivan cusp finiteness theorem for Kleinian groups in higher dimensions, Preprint, 1990.

[6] M. Kapovich and L. Potyagailo, On the absence of Ahlfors' finiteness theorem for Kleinian groups in dimension 3, to appear in Topology and its Applications.

[7] G. Martin, On discrete Möbius groups in all dimensions: a generalization of Jørgensen inequality, Acta Math., 163 (1989), 253-289.

[8] B. Maskit, Kleinian groups, Springer-Verlag, 1988.

[9] W. Newmann, Some remarks on infinite groups, J. London Math. Soc. 12 (1937), 120-127.

[10] L. Potyagailo, Finitely generated Kleinian groups in 3-space and manifolds of infinite homotopy type, in preparation.

[11] P. Scott, Finitely generated 3-manifold groups are finitely presented, J. London Math. Soc. (2), 6 (1973), 437-440.

[12] P. Scott, Subgroups of surface groups are almost geometric, J. London Math. Soc. 17 (1978) 555-565; Correction ibid 32 (1985), 217-220.

[13] A. Selberg, On discontinuous groups in higher dimensional symmetric spaces, Contribution to Function Theory, Bombay, 1960, p. 147-164.

[14] D. Sullivan, Travaux de Travaux sur les groupes quasi-fuchsiens et les varietes hyperboliques de dimension 3 fibres sur S^1, Lecture Notes in Math., 842, Springer, Berlin-New York, 1981, 196-214.

[15] W. Thurston, The geometry and topology of 3-manifolds, Princeton, Princeton University, Lecture Notes, 1978.

A calculus on links in the 3-sphere

Joan S. Birman and William W. Menasco***

Abstract. This paper presents an overview of results obtained by the authors in their six papers: "Studying Links Via Closed Braids" I, II, III, IV, V and VI.

Near the end of the last century the Scottish physicist Peter Guthrie Tait initiated a project which was aimed at the premier problem in knot theory: to give an effective algorithm for determining when two links in S^3 are related by an ambient isotopy. Tait and his co-workers Little and Kirkman began with a description of a knot by a diagram, i.e. a projection onto a plane in which the singularities are at most a finite set of transverse double points. To this they added information to distinguish overpasses and underpasses as one would for highways on a map. If one fixes the number of double points, the number of knot types which can be represented by such a diagram is finite. Their idea was to systematically investigate knots of crossing number 1,2,3,4, ... by experimental techniques, hoping that in the course of such an experiment they would discover the underlying principles which determine knotting. Since a knot of crossing number c also has diagrams of crossing number m for each $m > c$, an essential part of their program was to find a calculus which would systematically simplify a link diagram to a canonical diagram of minimum crossing number. Unfortunately, the project did not suceed. In Tait's own words: "The subject is very much more difficult and intricate than at first sight one is inclined to think, and I feel that I have not succeeded in catching the key note" [**Ta,1**]. Topology was brought to bear on the subject beginning in the early 1900's, but there was scant progress. See [**B-S**] and [**C**] for partial results.

The knot tables which were produced in [**Ta,2**] and [**L**] had an abundance of alternating knots, i.e. knots in which the crossings in a diagram are alternately under, over, under, over, ... as one traverses the knot. Tait conjectured the following special calculus for such diagrams (when there are no "nugatory crossings"). The associated knots are ambient isotopic if and only if their alternating diagrams are related by a sequence of flypes (Figure 1). A salient feature of Tait flyping conjec-

* Partially supported by NSF Grant No. DMS 8805672
** Partially supported by NSF Grant No. DMS 9002673

Knots 90

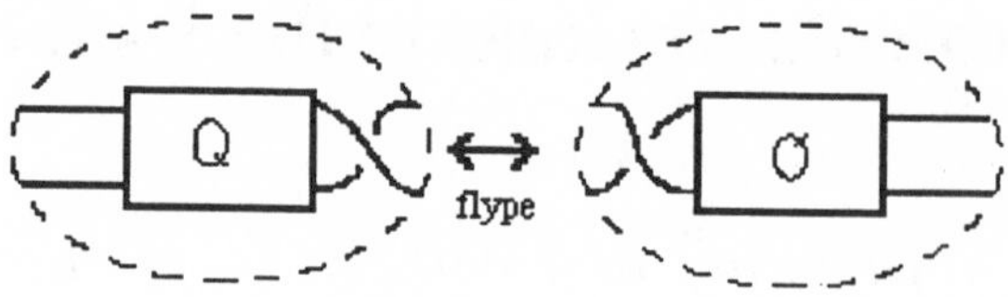

Figure 1. Flypes

ture, which has recently been proved by the second author and M. Thistlethwaite in [**M-T**], is that flypes do not change the crossing number.

In this note we give an overview of our recent partial results on Tait's original problem, i.e. the development of a calculus on oriented links in oriented S^3. Details are to be found in the manuscripts [**B-M,1**], [**B-M,2**], ..., [**B-M,6**]. Our proposed calculus is based upon the theory of braids. As we will see, braids give some extra structure to a link diagram. The extra structure enables us to find a complexity function which is strictly monotonic under the moves needed for the calculus. One of these moves is closely related to Tait's flype.

An oriented link L is a *closed braid* with braid axis A if there is an unknotted curve A in $S^3 - L$ and a fibration $D^2 \times S^1$ of $S^3 - A$ such that L intersects each fiber $D^2 \times p$ transversely. Since every link can be represented as a closed braid [**A**], a method for distinguishing link types via their closed braid representatives would be a solution to our problem. Notice that since $S^3 - A$ is fibered by the discs $D^2 \times p$ fibration will induce a (possibly singular) foliation on L and on spanning surfaces for L and on other structures in S^3. These foliations have a rich combinatorial structure. They are the chief geometric tool in our work. Since there is a one-to-one correspondence between conjugacy classes in B_n and isotopy classes of closed n-braids in the complement of the axis, an important first step toward a calculus was taken by Garside[**G**] when he solved the conjugacy problem in the braid group B_n. Another key step was the following theorem, first proved in [**Bi**]. See Figure 2.

Markov's Theorem. *Suppose L and M are braid representatives (possibly of differing index) of the same link type. Then there is a finite sequence of braids*

$$L = L_1 \to L_2 \to ... \to L_m = M$$

such that each L_{i+1} is obtained from L_i by either conjugation, or addition of a trivial loop, or deletion of a trivial loop.

The addition of trivial loops is called stabilizing. At present there is no known upper bound on the number of times it may be necessary to stabilize. Thus Garside's algorithm together with Markov's Theorem fails to establish a calculus. What is needed for a calculus on closed braids is a version of Markov's theorem without stabilization, i.e. one in which trivial loops may be deleted but not added.

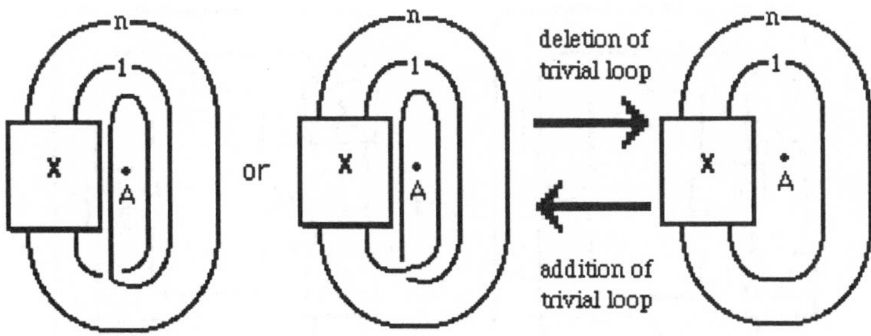

Figure 2. Deleting trivial loops

Our first result, proved in [**B-M,1**], gives form to this project. We introduce a complexity function on the collection of all closed braid representatives of a link which records features of the combinatories of the foliations. Then we prove:

Theorem 1. *A link type has only finitely many conjugacy classes of closed braid representatives of minimal complexity.*

Notice that Theorem 1 implies the existence of a complete numerical invariant for links, namely the Garside representatives of the finitely many conjugacy classes of minimum complexity.

Our next result is a lengthy example. We develop a calculus and find the invariant of Theorem 1 in the special case of links represented by closed 3-braids. We begin the project in [**B-M,2**] by studying a special type of minimum genus spanning surface for closed 3-braids. We call it a *Bennequin surface* because the idea for it came from a theorem in [**Be**]. Bennequin surfaces add geometric structure to the complement of our links, and using them, we begin to study closed 3-braids. To explain our work we first describe a new move which takes closed braids to closed braids and preserves link type. In Figure 3 the boxes denote braids on two strands. If one stares at the picture one sees a move which we call a *braid-preserving flype* (compare with the flype in Figure 1). It is an isotopy which takes the closed 3-braid on the left to the closed 3-braid on the right, crossing the axis during the isotopy. Except for a finite number of special p, q, and r, the two open 3-braids are in different conjugacy classes. If we allow three moves in our calculus, namely conjugation, deletion of trivial loops, and braid-preserving flypes, we have the following result from [**B-M,3**].

Theorem 2. *A link of braid index 3 either has a unique conjugacy class of 3-braid representatives, or admits a braid-preserving flype.*

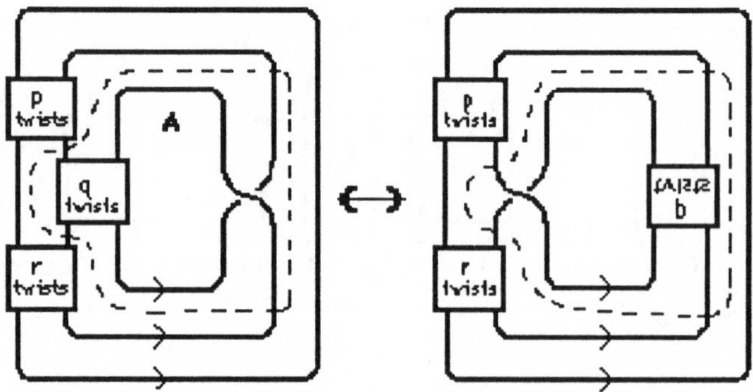

Figure 3. Braid-preserving flypes

Corollary 3. *There exists an effectively computable numerical classifying invariant for links which are represented by closed 3-braids. Using it, one can easily determine the braid index; whether the link is split or composite; whether the link has one, two or three conjugacy classes of closed 3-braid representatives; and whether the link is amphicheiral or invertible.*

One version of our 3-braid invariant uses a new solution to the conjugacy problem in the group of 3-braids, due to P.J. Xu [**X**]. It is proved in [**X**] that the spanning surface studied in [**B-M,2**] is unique up to isotopy. An algorithm is given to find it. Thus our work together with Xu's gives an algorithm to determine the genus of any link which is represented by a closed 3-braid and to find its unique incompressible spanning surface. In Figure 4 declare X, Y to be braids on more than two strands, by weighting one strand with a non-negative integer n. Thus, X and Y are arbitrary words in the braid group B_{n-1}. We call the move illustrated in the left and center pictures in Figure 4 an *exchange move*. A link L is *split* if there exists an essential S^2 in $S^3 - L$ and *composite* if there exists a twice punctured S^2 that decomposes L into two sublinks, neither of which is the trivial

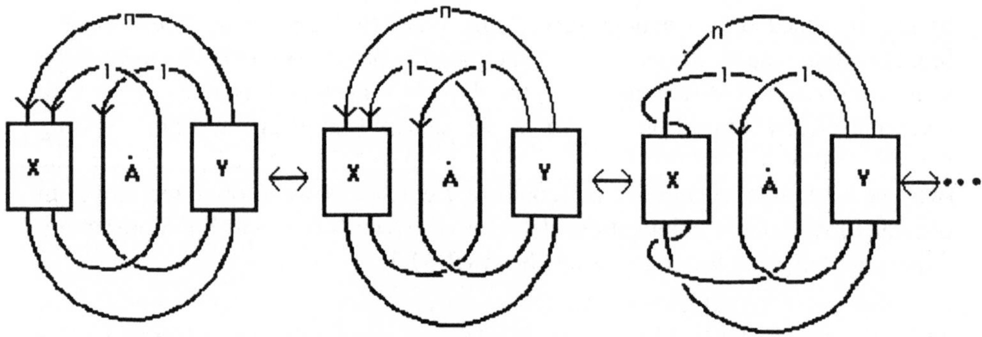

Figure 4. Exchange moves

knot. An obvious task required of any proposed calculus is the decomposition of a given link into its irreducible and prime components. To do that, we need to recognize when closed braids represent split and/or composite links. We say that a closed n-braid is a *split closed braid* if there exists a splitting S^2 for the link which is punctured exactly twice by the braid axis. In other words, up to conjugation, the braid obviously represents a split link. An closed n-braid is a *composite closed braid* if there exists a twice punctured S^2 which realizes the connected sum decomposition and which is punctured exactly twice by the braid axis. That is, up to conjugation the braid obviously represents a composite link. In general, an n-braid representative of a split or composite link will not be split or composite, since the S^2 can be extensively tangled up with the braid and essentially punctured by the braid axis arbitrarily many times. From [**B-M,4**] we have the following results which say the exchange move allows one to find nice representatives for split and composite links.

Theorem 4. *Suppose L is an n-braid representative of a split (or a composite) link. Then there is a finite sequence of n-braids:*

$$L = L_1 \to L_2 \to \dots \to L_m$$

such that each L_{i+1} is obtained from L_i by an exchange move and L_m is a split (or a composite) n-braid. Also, the complexity of L_{i+1} is less than that of L_i.

Next, any reasonable calculus must include a procedure for producing minimal braid index. We do not have a full picture of how to do this at this writing, however we do have results for the special case of the unlink. If we allow our calculus to include the deletion of trivial loops, along with the exchange move, we have the following result from [**B-M,5**] which could be called a Markov theorem without stabilization for the unlink.

Theorem 5. *Let L be an arbitrary closed n-braid representative of the unlink of r components, $r \geq 1$. Let U_r be the standard closed r-braid representative. Then there is a finite sequence of closed-braid representatives of the unlink:*

$$L = L_1 \to L_2 \to \dots \to L_m = U_r$$

such that each L_{i+1} is obtained from L_i by conjugation (or isotopy in the complement of the axis), deletion of a trivial loop, or an exchange move. Also, the complexity of L_{i+1} is less than that of L_i.

Repeated application of the exchange move can result in infinitely many distinct conjugacy classes of n-braid representatives of a link type, as indicated in the right picture in Figure 4. The strands are weighted, and so they acquire twists if one untwists the block X. Thus the closed braid in the right picture is in general not equivalent to the one in the center, if $n > 3$.

The following result from [**B-M,6**] says that this is the only way in which infinitely many conjugacy classes in the same braid group can arise. Call two conjugacy classes exchange equivalent if they have representatives which are related by a finite number set of exchange moves.

Theorem 6. *Let L be a link in S^3 which admits infinitely many conjugacy classes of n-braid representatives. Then these infinitely many conjugacy classes divide into finitely many conjugacy equivalence classes, under exchange moves.*

From Theorem 6 we are led to the following conjecture. For any integer n, there are only finitely many types of moves needed to pass between distinct conjugacy classes of braids with index n that represent the same link. Theorem 2 says that this conjecture is true for 3-braids.

Theorems 4, 5 and 6 leave open the question of how to recognize conjugacy classes which have trivial loops or admit exchange moves or flypes. In that regard, we remark that if a conjugacy class has a representative with a trivial loop, it also has a representative which admits an exchange move, and if it has a representative which admits an exchange move it also has one which admits a flype.

Knowing a complete invariant for a class of links such as links of braid index ≤ 3, and a proposed invariant for arbitrary n, it is natural to ask how our invariant compares to the Jones polynomial [**J**] and its various generalizations (e.g.[**R**]). "Jones invariants" are all constructed from traces in matrix representations of the braid group. They are all multiplicative under connected sum, and invariant under mutations. This forces them to be non-faithful. Even more, moves like our exchange move and flype (which generalizes to n-braids by weighting the strands) imply the existence of trace identities in any representation of B_n which admits a Jones invariant, and these trace identities can be expected to imply other unwanted trace identities. Jones invariants do not change when one reverses the orientation of knots or the global orientation of links, whereas our invariant does change if the knot or link in question is non-invertible. In the special case of links of braid index 3 we are able to produce, for each $N \geq 2$, examples of N links, no two of which are equivalent, which share the same 2-variable Jones polynomial, however we do not know whether other Jones invariants suffice to distinguish them.

References

[A] J.W. Alexander, *A lemma on systems of knotted curves*, Proc. Nat. Academ. Science USA, **9** , pp.93-95.

[Be] Bennequin, D., *Entrelacements et Equations de Pfaffs*, Asterisque 107-108 (1983), p. 87- 161.

[Bi] J. Birman, *Braids, links, and mapping class groups*, Annals of Math. Studies, **82**, Princeton Press, 1975.

[B-M,1] J. Birman & W. Menasco, *Studying links via closed braids I: A finiteness theorem*, Pac. J. Math., to appear.

[B-M,2] J. Birman & W. Menasco, *Studying links via closed braids II: On a theorem of Bennequin*, in Topology and Its Applications **40** (1991), 71-82.

[B-M,3] J. Birman & W. Menasco, *Studying links via closed braids III: classifying links which are closed 3-braids*, Pac. J. Math., to appear.

[B-M,4] J. Birman & W. Menasco, *Studying links via closed braids IV: split and composite links*, Inventiones Math.. 102, Fasc.1 (1990), 115-139.

[B-M,5] J. Birman & W. Menasco, *Studying links via closed braids V: Exchange moves and the unlink*, Trans. AMS, Jan. 1992, to appear.

[B-M,6] J. Birman & W. Menasco, *Studying links via closed braids VI: A non-finiteness theorem*, Pac. J. Math., to appear.

[B-S] F. Bonahon & L. Siebenmann, *New geometric splittings of classical knots (algebraic knots)*, handwritten manuscript, (1980).

[C] J.H. Conway, *An enumeration of knots and links*, Computational Problems in Abstract Algebra, ed. Leech, Pergamon press (1970), 329-358.

[G] F. A. Garside, *The braid group and other groups*, Quart. J. Math. Oxford **20** (1969), 235-264.

[J] V.F.R.Jones, *Hecke algebra representations of braid groups and link polynomials*, Annals of Math. 126 (1987), 335-388.

[L] C.N. Little, *Non-alternating ± knots*, Trans. Roy. Soc. Edinburgh, vol.39, pp.771-778.

[M-T] W. Menasco & M. Thistlethwaite, *The classification of alternating links*, preprint.

[R] N. Yu. Reshetiken, *Quantized universal enveloping algebras, the Yang-Baxter equation, and invariants of links*, LOMI Preprints, 1988, Steklov Institute Math. Leningrad.

[Ta,1] P.G. Tait, *On knots*, Trans. Royal Soc. of Edinburgh, 1876-1877.

[Ta,2] P.G. Tait, *On knots*, Collected Scientific Papers of P.G.Tait, # I. Cambridge University Press, 1898, London, pp.273-347.

[X] P. J. Xu, *The genus of closed 3-braids and the growth of the semigroup of positive braids*, PhD thesis, Columbia Univ., 1990.

List of Contributors

W.B. Raymond Lickorish, Department of Pure Mathematics, 16 Mill Lane, Cambridge, CB2 1SB, England

Tsuyoshi Kobayashi, Department of Mathematics, Osaka University, Toyonaka-city, Osaka 560, Japan

Jonathan A. Hillman, Department of Pure Mathematics, University of Sydney, NSW 2006, Australia

Hugh R. Morton, Department of Pure Mathematics, University of Liverpool, P.O. Box 147, Liverpool L69 3BX, England

P.M. Strickland, Department of Pure Mathematics, University of Liverpool, P.O. Box 147, Liverpool L69 3BX, England

Michael Farber, School of Mathematical Science, Tel Aviv University, Ramat-Aviv,Tel-Aviv 69978, Israel

Yves Mathieu, Unite d'Enseignement et de Recherche de Mathematiques, Université de Provence, 3 Place Victor Hugo-F13331 Marseille, Cedex 3, France

Masahiro Nakao, School of Science, Kwansei Gakuin University, Uegahara, Nishinomiya 662, Japan

Ulrich Koschorke, Mathematik V(FB 6), Universität-GH Siegen, Hölderlinstr. 3, D-5900, Siegen, Germany

Seiichi Kamada, Department of Mathematics, Osaka City University, Osaka 558, Japan

Toshio Harikae, School of Science, Kwansei Gakuin University, Uegahara, Nishinimiya 662, Japan

Shinji Fukuhara, Department of Mathematics, Tsuda College, Kodaira-shi, Tokyo 187, Japan

Amy Davidow, Mathematics and Computer Science Department, Macalester College, 1600 Grand Ave, Room 219A, St. Paul, MN 55105, U.S.A.

Masakazu Teragaito, Department of Mathematics, Kobe University, Rokko, Kobe 657, Japan

Yasutaka Nakanishi, Department of Mathematics, Kobe University, Rokko, Kobe 657, Japan

Gyo Taek Jin, Department of Mathematics, Korea Advanced Institute of Science and Technology, Taejon, 305-701, Korea

B.K. Kim, Department of Mathematical Education, Chongju University, Chongju, 360-764, Korea

Ki Hyoung Ko, Department of Mathematics, Korea Advanced Institute of Science and Technology, Taejon, 305-701, Korea (email: khko%kaist.kit.ac.kr@relay.cs.net)

E. Guadagnini, CERN-Geneva, Istitute Nazionale di Fisica Nucleare, Sezione di Pisa, Pisa, Italy (On leave from Dipartmento di Fisica dell'Università di Pisa, Pisa, Italy)

Maurizio U. Martellini, Dipartimento di Fisica, Universita di Milano, Via Celoria 16, 20133 Milano, Italy; Istituto Nazionale di Fisica Nucleare, Sezione di Pavia, Italy

M. Mintchev, Istituto Nazionale di Fisica Nucleare, Sezione di Pisa, Pisa, Italy; Dipartimento di Fisica dell'Università di Milano, Milano, Italy

Chichen M. Tsau, Department of Mathematics and Computer Science, Saint Louis University, St. Louis, MO 63103, U.S.A.

Katsuyuki Yoshikawa, School of Science, Kwansei Gakuin University, Uegahara, Nishinomiya 662, Japan

Yoshiaki Uchida, Department of Mathematics, Kobe University, Rokko, Kobe 657, Japan

Louis H. Kauffman, Department of Mathematics, Statistics and Computer Science, University of Illinois at Chicago, Chicago, Illinois 60680, U.S.A.

Lawrence Smolinsky, Department of Mathematics, Louisiana State University, Baton Rouge, Louisiana 70803-4918, U.S.A. (email: smolinsky@mmdehn.lsu.edu)

Masaaki Ue, Institute of Mathematics, Yoshida College, Kyoto University, Kyoto 606, Japan

Kouzi Kodama, Department of Mathematics, Faculty of Science, Kobe University, Nada, Kobe 657, Japan

Makoto Sakuma, Department of Mathematics, College of General Education, Osaka University, Toyonaka, Osaka 560, Japan

H.C. Lee, Physical Sciences, Theoretical Physics Branch, AECL Research, Chalk River Laboratories, Chalk River, Ontario K0J 1J0, Canada; Department of Applied Mathematics, University of Western Ontario, London, Ontario N6A 5B9, Canada

Jim Hoste, Department of Mathematics, Pitzer College, 1050 North Mills Avenue, Claremont, CA 91711-6110, U.S.A. (email: jhoste@pitzer.claremont.edu)

Józef H. Przytycki, Mathematics Department, University of California, Riverside, CA 92521, U.S.A.

P. Vogel, Départment de Mathématiques, Université de Nantes 2 rue de la Houssiniere, F-44072, Nantes, Cedex 03, France

N. Habegger, Départment de Mathématiques, Université de Nantes, 2 rue de la Houssiniere, F-44072, Nantes, Cedex 03, France

J.P. Levine, Department of Mathematics, Brandeis University, Waltham, Massachusetts, 02254-9110, U.S.A.

Takao Matumoto, Department of Mathematics, Hiroshima University, Hiroshima 730, Japan

Shin'ichi Kinoshita, School of Science, Kwansei Gakuin University, Uegahara, Nishinomiya 662, Japan

Steven A. Bleiler, Department of Mathematics, Portland State University, Portland, Oregon 97207-0751, U.S.A. (email: hmsb@psuorvm.bitnet)

Craig D. Hodgson, Department of Mathematics, University of Melbourne, Parkville, Victoria 3052, Australia (email : cdh@mundoe.maths.mu.oz.au)

Takashi Kohara, Department of Mathematics, School of Education, Waseda University, Sinjuku-ku, Tokyo 160, Japan

Shin'ichi Suzuki, Department of Mathematics, School of Education, Waseda University, Shinjuku-ku, Tokyo 160, Japan

Shûji Yamada, Department of Mathematics, Ehime University, Matsuyama 790, Japan

Jun Murakami, Department of Mathematics, Osaka University, Toyonaka, Osaka 560, Japan

Akio Kawauchi, Department of Mathematics, Osaka City University, Osaka 558, Japan (email: d54454@jpnkudpc)

José Carlos Gómez-Larrañaga, CICB-UAEM, Toluca, México; Instituto de Matematicas, UNAM, Ciudad Universitaria, 04510 México, D.F., México

Wolfgang Heil, Department of Mathematics, Florida State University, Tallahassee, Florida 32306, U.S.A.

Bohdan I. Kurpita, Department of Mathematics, University of Toronto, Toronto, Ontario M5S 1A1, Canada

Kunio Murasugi, Department of Mathematics, University of Toronto, Toronto, Ontario M5S 1A1, Canada

I.R. Aitchison, Department of Mathematics, University of Melbourne, Parkville, Victoria 3052, Australia (email: iain@mundoe.maths.mu.oz.au)

J.H. Rubinstein, Department of Mathematics, University of Melbourne, Parkville, Victoria 3052, Australia (email: rubin@mundoe.maths.mu.oz.au)

David A. Meyer, Institute for Pure and Applied Physical Sciences, University of California/ San Diego, La Jolla, CA 92093-0075, U.S.A.

Yoav Moriah, Department of Mathematics, Technion, Hifa 32000, Israel

L. Potyagailo, Raymond and Beverly Sackler, Faculty of Exact Science, School of Mathematical Sciences, Tel Aviv University, Ramat-Aviv, 69978, Israel

Joan S. Birman, Department of Mathematics, Columbia University, New York, New York 10027, U.S.A.

William W. Menasco, Department of Mathematics, SUNY at Buffalo, Buffalo, New York 14222, U.S.A.

List of Participants

Adachi, M. (Kyoto Univ., Japan)
Aida, H. (Tsuda College, Japan)
Aitchison, I.R. (Univ. of Melbourne, Australia)
Birman, J.S. (Columbia Univ., U.S.A.)
Bleiler, S.A. (Portland State Univ., U.S.A.)
Boileau, M. (Univ. Paul Sabatier, France)
Cappell, S. (Courant Inst. of Math. Sciences, U.S.A.)
Cochran, T.D. (Rice Univ., U.S.A.)
Cooper, D. (Univ. of California, Santa Barbara, U.S.A.)
Costes, C. (Univ. of Oxford, U.K.)
Davidow, A.L. (Univ. of Chicago, U.S.A.)
Deguchi, T. (Univ. of Tokyo, Japan)
Endo, T. (Waseda Univ., Japan)
Erbland, J.P. (Univ. of Hartford, U.S.A.)
Farber, M.S. (Tel Aviv Univ., Israel)
Floer, A. (Univ. of California, Berkeley, U.S.A.)
Fujii, M. (Tokyo Inst. of Technology, Japan)
Fukagawa, H. (Higashi-Neyagawa High School, Japan)
Fukuhara, S. (Tsuda College, Japan)
Goda, H. (Osaka Univ., Japan)
Gómez-Larrañaga, J.C. (Unam, Mexico)
González-Acuña, F. (Unam, Mexico)
Gordon, C.McA. (Univ. of Texas, Austin, U.S.A.)
Habegger, N. (Univ. de Nantes, France)
Hara, M. (Waseda Univ., Japan)
Harikae, T. (Kwansei Gakuin Univ., Japan)
Hayasaka, S. (Waseda Univ., Japan)
Hillman, J.A. (Univ. of Sydney, Australia)
Hirai, Y. (Kanazawa Univ., Japan)
Homma, T. (Aoyama Gakuin Univ., Japan)
Hosokawa, F. (Kobe Univ., Japan)
Hoste, J.E. (Pitzer College, U.S.A.)
Inagaki, A. (Tsuda College, Japan)
Ishii, I. (Keio Univ., Japan)
Jin, G.T. (Korea Inst. of Technology, Korea)
Jitsukawa, T. (Tokai Univ., Japan)

Jones, V. (Univ. of California, Berkeley, U.S.A.)
Kakimizu, O. (Hiroshima Univ., Japan)
Kamada, S. (Osaka City Univ., Japan)
Kanenobu, T. (Kyushu Univ., Japan)
Kaneto, T. (Univ. of Tsukuba, Japan)
Kanno, J. (Shirayuri Gakuen High School, Japan)
Kapovich, M. (Inst. for Applied Math., Khabarovsk, U.S.S.R.)
Kasahara, Y. (Tokyo Inst. of Technology, Japan)
Kato, M. (Kyushu Univ., Japan)
Kauffman, L.H. (Univ. of Illinois, Chicago, U.S.A.)
Kawauchi, A. (Osaka City Univ., Japan)
Kawazumi, N. (Univ. of Tokyo, Japan)
Kazama, K. (Kobe Univ., Japan)
Kearton, C. (Durham Univ., U.K.)
Kido, T. (Yao High School, Japan)
Kinoshita, S. (Kwansei Gakuin Univ., Japan)
Kirk, P.A. (California Inst. of Technology, U.S.A.)
Ko, K.H. (Korea Inst. of Technology, Korea)
Kobayashi, K. (Tokyo Women's Christian Univ., Japan)
Kobayashi, M. (Osaka City Univ., Japan)
Kobayashi, T. (Osaka Univ., Japan)
Kodama, K. (Kobe Univ., Japan)
Kodate, T. (Tokyo Women's Christian Univ., Japan)
Kohno, T. (Kyushu Univ., Japan)
Kojima, S. (Tokyo Inst. of Technology, Japan)
Komuro, H. (Kawagoe-Minami High School, Japan)
Kondo, H. (Tokai Univ., Japan)
Koschorke, U. (Univ.-GH-Siegen, Germany)
Kouno, M. (Kobe Univ., Japan)
Kurpita, B.I. (Univ. of Toronto, Canada)
Lawrence, R.J. (Harvard Univ., U.S.A.)
Lee, H.C. (Chalk River Nuclear Laboratories, Canada)
Levine, J.P. (Brandeis Univ., U.S.A.)
Lickorish, W.B.R. (Univ. of Cambridge, U.K.)
Lopez, L.M. (Tokyo Inst. of Technology, Japan)
Lozano, M.T. (Univ. de Zaragoza, Spain)
Luecke, J.E. (Univ. of Texas, Austin, U.S.A.)
Maeda, T. (Kansai Univ., Japan)
Makino, K. (Waseda Univ., Japan)
Martellini, M.U. (Univ. di Milano, Italy)
Marumoto, Y. (Osaka Sangyo Univ., Japan)
Maruyama, N. (Bunri College, Japan)

Matsumoto, S. (Nihon Univ., Japan)
Matsumoto, Y. (Univ. of Tokyo, Japan)
Matumoto, T. (Hiroshima Univ., Japan)
Mednykh, A.D. (Omsk State Univ., U.S.S.R.)
Meyer, D.A. (Univ. of California, San Diego, U.S.A.)
Miyamoto, Y. (Tokyo Inst. of Technology, Japan)
Miyazaki, K. (Univ. of Texas, Austin, U.S.A.)
Miyazawa, Y. (Osaka City Univ., Japan)
Moriah, Y. (Technion, Israel)
Morimoto, K. (Takushoku Univ., Japan)
Morita, S. (Tokyo Inst. of Technology, Japan)
Morton, H.R. (Liverpool Univ., U.K.)
Motegi, K. (Kyushu Univ., Japan)
Murakami, H. (Osaka City Univ., Japan)
Murakami, J. (Osaka Univ., Japan)
Murakami, M. (Tsuda College, Japan)
Muramoto, T. (Tokai Univ., Japan)
Murasugi, K. (Univ. of Toronto, Canada)
Nagase, T. (Tokai Univ., Japan)
Naka, H. (Kwansei Gakuin Univ., Japan)
Nakabo, S. (Kyushu Univ., Japan)
Nakagawa, Y. (Yamaguchi Women's Univ., Japan)
Nakanishi, Y. (Kobe Univ., Japan)
Nakao, M. (Kwansei Gakuin Univ., Japan)
Nishi, H. (Kyushu Univ., Japan)
Noguchi, T. (Tokai Univ., Japan)
Ochiai, M. (Nara Women's Univ., Japan)
Ohshika, K. (Tokyo Inst. of Technology, Japan)
Ohtsuki, T. (Univ. of Tokyo, Japan)
Ohyama, Y. (Waseda Univ., Japan)
Okada, M. (Osaka Univ., Japan)
Okai, T. (Hiroshima Univ., Japan)
Orr, K.E. (Univ. of Chicago, U.S.A.)
Oyaizu, H. (Japan)
Potyagailo, L. (Tel Aviv Univ., Israel)
Reid, A.W. (Ohio State Univ., U.S.A.)
Repovš, D. (Univ. of Ljubljana, Yugoslavia)
Rubinstein, J.H. (Univ. of Melbourne, Australia)
Saeki, O. (Yamagata Univ., Japan)
Saito, M. (Univ. of Texas, Austin, U.S.A.)
Sakai, S. (Osaka Univ., Japan)
Sakai, T. (Tsuda College, Japan)

Sakuma, M. (Osaka Univ., Japan)
Sanami, M. (Toba Mercantile Marine College, Japan)
Sasano, K. (Toyama Medical and Pharmaceutical Univ., Japan)
Sato, Y. (Kyushu Univ., Japan)
Scharlemann, M.G. (Univ. of California, Santa Barbara, U.S.A.)
Sekine, M. (Hiroshima Univ., Japan)
Shinohara, Y. (Kwansei Gakuin Univ., Japan)
Shiomi, T. (Osaka City Univ., Japan)
Shpilrain, V.E. (Moscow Univ., U.S.S.R.)
Smolinsky, L.J. (Louisiana State Univ., U.S.A.)
Soma, T. (Kyushu Inst. of Technology, Japan)
Suketa, M. (Kobe Univ., Japan)
Sumners, D.W. (Florida State Univ., U.S.A.)
Suzuki, S. (Waseda Univ., Japan)
Takata, T. (Kyushu Univ., Japan)
Takeuchi, Y. (Aichi Univ. of Education, Japan)
Takewaka, Y. (Kyushu Univ., Japan)
Taniyama, K. (Waseda Univ., Japan)
Tao, J. (Osaka City Univ., Japan)
Tao, Y. (Kansai Univ., Japan)
Teragaito, M. (Kobe Univ., Japan)
Thistlethwaite, M.B. (Univ. of Tennessee, U.S.A.)
Tokuda, H. (Japan)
Tsau, M.C. (Saint Louis Univ., U.S.A.)
Tsukui, Y. (Shounan Inst. of Technology, Japan)
Turaev, V.G. (Steklov Math. Inst., U.S.S.R.)
Uchida, Y. (Kobe Univ., Japan)
Ue, M. (Kyoto Univ., Japan)
Van Buskirk, J. (Univ. of Oregon, U.S.A.)
Viro, O. (Steklov Math. Inst., U.S.S.R.)
Vogel, P. (Univ. de Nantes, France)
Wada, H. (Nishi-Noda High School of Technology, Japan)
Wada, M. (Univ. of Pennsylvania, U.S.A.)
Wadati, M. (Univ. of Tokyo, Japan)
Wajnryb, B. (Technion, Israel)
Wakui, M. (Kyushu Univ., Japan)
Walker, K.M. (MSRI, U.S.A.)
Whitten, W. (Univ. of Sw. Louisiana, U.S.A.)
Yagasaki, T. (Kyoto Inst. of Technology, Japan)
Yamada, S. (Ehime Univ., Japan)
Yamamoto, M. (Osaka Women's Univ., Japan)
Yamasaki, M. (Josai Univ., Japan)

Yamashita, M. (Toyo Univ., Japan)
Yamashita, S. (Waseda Univ., Japan)
Yanagawa, T. (Kobe Univ., Japan)
Yasuda, T. (Yuge Mercantile Marine College, Japan)
Yasuhara, A. (Waseda Univ., Japan)
Yetter, D.N. (Ohio State Univ., U.S.A.)
Yokota, Y. (Waseda Univ., Japan)
Yokoyama, K. (Sophia Univ., Japan)
Yokoyama, M. (Kyushu Univ., Japan)
Yoshida, T. (Tokyo Metropolitan Univ., Japan)
Yoshikawa, K. (Kwansei Gakuin Univ., Japan)
Yoshimatsu, Y. (Tokyo Women's Christian Univ., Japan)
Yoshinaga, S. (Nagoya Univ., Japan)